GEOTECHNICAL ASPECTS OF LANDFILL DESIGN AND CONSTRUCTION

GEOTECHNICAL ASPECTS OF LANDFILL DESIGN AND CONSTRUCTION

Xuede Qian
Geotechnical Engineering Specialist
Michigan Department of Environmental Quality

Robert M. Koerner
H. L. Bowman Professor of Civil Engineering, Drexel University
Director, Geosynthetic Research Institute

Donald H. Gray
Professor of Civil and Environmental Engineering
The University of Michigan

PRENTICE HALL
Upper Saddle River, New Jersey 07458

Library of Congress Cataloging-in-Publication Data
Qian, Xuede
Geotechnical Aspects of Landfill Design and Construction
   Robert M. Koerner, Donald H. Gray.
      p. 000   cm.
   Includes bibliographical references and index.
   ISBN 0-13-012506-7
   1. Landfill design. 2. Landfill/construction I. Title.
TA337.Q63 2001
620.0042-de21

                                    2001035844
                                        CIP

Vice President and Editorial Director, ECS: *Marcia Horton*
Acquisition Editor: *Laura Fischer*
Editorial Assistant: *Erin Katchmar*
Vice President and Director of Production and Manufacturing, ESM: *David W. Riccardi*
Executive Managing Editor: *Vince O'Brien*
Managing Editor: *David A. George*
Composition: *TechBooks*
Director of Creative Services: *Paul Belfanti*
Creative Director: *Carole Anson*
Art Director: *Jayne Conte*
Art Editor: *Greg Dulles*
Cover Designer: *Bruce Kenselaar*
Manufacturing Manager: *Trudy Pisciotti*
Manufacturing Buyer: *Lisa McDowell*
Marketing Manager: *Holly Stark*
Marketing Assistant: *Karen Moon*

© 2002 by Prentice-Hall, Inc.
Upper Saddle River, New Jersey 07458

The author and publisher of this book have used their best efforts in preparing this book. These efforts include the development, research, and testing of the theories and programs to determine their effectiveness. The author and publisher make no warranty of any kind, expressed or implied, with regard to these programs or the documentation contained in this book. The author and publisher shall not be liable in any event for incidental or consequential damages in connection with, or arising out of, the furnishing, performance, or use of these programs.

Printed in the United States of America

10 9 8 7 6 5 4 3 2 1

ISBN 0-13-012506-7

Pearson Education Ltd., *London*
Pearson Education Australia Pty., Limited, *Sydney*
Pearson Education Singapore, Pte. Ltd.
Pearson Education North Asia Ltd., *Hong Kong*
Pearson Education Canada, Ltd., *Toronto*
Pearson Education de Mexico, S.A. de C.V.
Pearson Education — Japan, *Tokyo*
Pearson Education Malaysia, Pte. Ltd.
Pearson Education, *Upper Saddle River, New Jersey*

Contents

PREFACE   **XVII**

1 INTRODUCTION   **2**

    1.1   Need for Landfills 2
    1.2   Principal Landfill Requirements 4
    1.3   Landfill Components and Configuration 5
    1.4   Landfill Envelope 6
            1.4.1   *Liner System, 7*
            1.4.2   *Leachate Collection and Removal System, 8*
            1.4.3   *Gas Collection and Control System, 9*
            1.4.4   *Final Cover System, 9*
    1.5   Composite Liners 9
    1.6   Benefits of Double Composite Liners 13
    1.7   Liner Leakage Mechanisms 14
            1.7.1   *Steady Advection, 15*
            1.7.2   *Steady Diffusion, 16*
            1.7.3   *Unsteady Diffusion, 17*
            1.7.4   *Combined Advection-Diffusion, 17*
    1.8   Scope and Organization of Book 23

Problems 24
References 25

2 LANDFILL SITING AND SITE INVESTIGATION   **28**

    2.1   Siting Considerations 29
    2.2   Location Restrictions 31
            2.2.1   *Airport Safety, 31*
            2.2.2   *Floodplains, 32*
            2.2.3   *Wetlands, 32*

          2.2.4   *Fault Areas, 32*
          2.2.5   *Seismic Impact Zones, 33*
          2.2.6   *Unstable Areas, 34*
2.3   Siting Process   34
2.4   Site Investigation   36
          2.4.1   *Preinvestigation Study, 37*
          2.4.2   *Field Exploration, 37*
2.5   Borrow Source Investigation   39
          2.5.1   *Clay, 40*
          2.5.2   *Sand, 41*
          2.5.3   *Gravel, 41*
          2.5.4   *Siltysoil, 41*
          2.5.5   *Topsoil, 41*
2.6   Field Hydraulic Conductivity Tests   42
          2.6.1   *Sealed Double-Ring Infiltrometer, 42*
          2.6.2   *Two-Stage Borehole Test, 45*
          2.6.3   *Comparison of Methods, 45*
2.7   Material Laboratory Tests   46
          2.7.1   *Water Content, 46*
          2.7.2   *Particle Size Distribution, 46*
          2.7.3   *Moisture Density Relationship, 46*
          2.7.4   *Hydraulic Conductivity, 47*
          2.7.5   *Shear Strength, 48*
          2.7.6   *Compressibility, 48*

Problems   49
References   50

3   COMPACTED CLAY LINERS   52

3.1   Overview Compacted Clay Liners   52
3.2   Compaction and Permeability Considerations   55
          3.2.1   *Compaction Test, 56*
          3.2.2   *Permeability Test, 58*
3.3   Design of Compacted Clay Liners   61
          3.3.1   *Low Hydraulic Conductivity, 61*
          3.3.2   *Adequate Shear Strength, 64*
          3.3.3   *Minimal Shrinkage Potential, 65*
          3.3.4   *Acceptable Zone to Meet All Design Criteria, 68*
3.4   Influence of Clods on Hydraulic Conductivity   69
          3.4.1   *Influence of Clod Size on Compaction Curve, 69*
          3.4.2   *Influence of Clod Size on Hydraulic Conductivity, 70*
          3.4.3   *Particle Orientation versus Clod Structure, 71*
          3.4.4   *Laboratory Testing and Design Implications, 72*
3.5   Effect of Gravel Content on Hydraulic Conductivity   73
3.6   Effect of Freezing and Thawing on Hydraulic Conductivity   75
          3.6.1   *Processes Occurring during Soil Freezing, 76*
          3.6.2   *Effect of Freeze-Thaw on Hydraulic Conductivity, 77*

    3.6.3 Factors Affecting Hydraulic Conductivity during Freeze-Thaw, 78
  3.7 Summary Comments Regarding Compacted Clay Liners 80

Problems 81
References 82

4 GEOMEMBRANES   86

  4.1 Composition and Thickness of Geomembranes 87
  4.2 Current Uses of Geomembranes in Landfills 88
    4.2.1 High Density Polyethylene (HDPE) Geomembranes, 88
    4.2.2 Linear Low Density Polyethylene (LLDPE) Geomembranes, 89
    4.2.3 Coextrusion Variations of HDPE and LLDPE, 89
    4.2.4 Textured Geomembranes, 90
    4.2.5 Flexible Polypropylene (fPP) Geomembranes, 91
    4.2.6 Polyvinyl Chloride (PVC) Geomembranes, 91
    4.2.7 Chlorosulphonated Polyethylene (CSPE) Geomembranes, 92
  4.3 Tensile Behavior of Geomembranes 92
  4.4 Friction Behavior of Geomembranes 96
  4.5 Tension Stresses due to Unbalanced Friction Forces 98
  4.6 Tension Stresses due to Localized Subsidence 101
  4.7 Runout and Anchor Trenches 104
    4.7.1 Design of Runout Length, 104
    4.7.2 Design of Rectangular Anchor Trench, 106
    4.7.3 Design of V-Shaped Anchor Trench, 114
  4.8 Assessment of Leakage through Liners 119
    4.8.1 Flow Rate through Compacted Clay Liner, 120
    4.8.2 Flow Rate through Geomembrane Liner, 121
    4.8.3 Flow Rate through Composite Liner, 122
    4.8.4 Comparison of Three Types of Liners, 124
  4.9 Concluding Comments Regarding Geomembranes 127

Problems 127
References 129

5 GEOSYNTHETIC CLAY LINERS   131

  5.1 Types and Current Uses of Geosynthetic Clay Liners 132
    5.1.1 Geotextile-Encased, Adhesive-Bonded GCL, 132
    5.1.2 Geotextile-Encased, Stitch-Bonded GCL, 133
    5.1.3 Geotextile-Encased, Needle-Punched GCL, 133
    5.1.4 Geomembrane-Supported, Adhesive-Bonded GCL, 134
  5.2 Hydraulic Conductivity 134
    5.2.1 Effect of Permeating Liquids, 134

- 5.2.2 *Effect of Confining Stress*, 135
- 5.2.3 *Wet-Dry Response*, 136
- 5.2.4 *Freeze-Thaw Response*, 138
5.3 Ability to Withstand Differential Settlement 141
5.4 Shear Strength 143
- 5.4.1 *Internal Shear Strength*, 144
- 5.4.2 *Interface Shear Strength*, 148
- 5.4.3 *Design Implications*, 151
5.5 Differences between Geosynthetic Clay Liners and Compacted Clay Liners 152
5.6 Contaminant Transport through Geosynthetic Clay Liner and Compacted Clay Liner 154
- 5.6.1 *Steady Advection*, 154
- 5.6.2 *Steady Diffusion*, 155
- 5.6.3 *Advective Breakthrough Time*, 157
- 5.6.4 *Combined Advection-Diffusion*, 158
5.7 Comparison of Mass Transport through a GCL and CCL 161
5.8 Recommendations for Use of Geosynthetic Clay Liners 172
5.9 Summarizing Comments Regarding Geosynthetic Clay Liners 173

Problems 175
References 176

6 ENGINEERING PROPERTIES OF MUNICIPAL SOLID WASTE   180

6.1 Constituents of Municipal Solid Waste 181
6.2 Unit Weight of Municipal Solid Waste 182
6.3 Moisture Content of Municipal Solid Waste 185
6.4 Porosity of Municipal Solid Waste 188
6.5 Hydraulic Conductivity of Municipal Solid Waste 189
6.6 Field Capacity and Wilting Point of Municipal Solid Waste 190
6.7 Shear Strength of Municipal Solid Waste 193
6.8 Compressibility of Municipal Solid Waste 199

Problems 204
References 206

7 LEACHATE GENERATION AND EVALUATION IN MSW LANDFILLS   211

7.1 MSW Leachate Characterization 212
7.2 Factors Affecting Leachate Quantity 213

7.3 Estimation of Leachate Production Rate in an Active Condition   218
  7.3.1 Precipitation, 218
  7.3.2 Waste Squeeze Liquid, 219
  7.3.3 Evaporation, 221
  7.3.4 Waste Moisture Absorption, 221
7.4 Estimation of Leachate Production Rate in a Postclosure Condition   223
  7.4.1 Snowmelt Infiltration, 225
  7.4.2 Surface Runoff, 226
  7.4.3 Evapotranspiration, 228
  7.4.4 Soil Moisture Storage, 229
  7.4.5 Lateral Drainage, 229
  7.4.6 Moisture Extraction from Waste, 230
7.5 Hydrologic Evaluation of Landfill Performance (HELP) Model   230
  7.5.1 Versions of HELP Model, 231
  7.5.2 Data Generation and Default Values, 232
  7.5.3 Landfill Profile and Layer Descriptions, 235
  7.5.4 Modeling Procedure, 238
  7.5.5 Program Input, 239
  7.5.6 Program Output, 240
  7.5.7 Limits of Application, 241

Problems   242
References   243

8 LIQUID DRAINAGE LAYER   247

8.1 Profile of Leachate Drainage Layer   247
8.2 Soil Drainage and Filtration Layer   251
8.3 Geotextile Design for Filtration   254
  8.3.1 Geotextiles Overview, 254
  8.3.2 Allowable versus Ultimate Geotextile Properties, 256
  8.3.3 Cross-Plane Permeability, 257
  8.3.4 Soil Retention, 258
  8.3.5 Long-term Compatibility, 261
8.4 Geonet Design for Leachate Drainage   263
  8.4.1 Geonets Overview, 263
  8.4.2 Hydraulic Properties of Geonets, 264
  8.4.3 Allowable Geonet Flow Rate, 266
  8.4.4 Designing with Geonets for Drainage, 269
8.5 Estimate of Maximum Liquid Head in a Drainage Layer   274
  8.5.1 Methods for Estimating Maximum Liquid Head, 275
  8.5.2 Comparison of Various Calculation Methods, 283

Problems   289
References   291

9 LEACHATE COLLECTION AND REMOVAL SYSTEMS   294

- 9.1 Subbase Grading   294
- 9.2 Leachate Collection Trenches   295
- 9.3 Selection of Leachate Collection Pipe   297
  - 9.3.1 *Type of Pipe Material, 297*
  - 9.3.2 *Pipe Design Issues, 297*
  - 9.3.3 *Pipe Perforations, 299*
- 9.4 Deformation and Stability of Leachate Collection Pipe   304
  - 9.4.1 *Pipe Deflection, 304*
  - 9.4.2 *Pipe Wall Buckling, 311*
- 9.5 Sump and Riser Pipes   314
- 9.6 Leachate Removal Pumps   320

Problems   328
References   330

10 GAS COLLECTION AND CONTROL SYSTEMS   332

- 10.1 Gas Generation   333
- 10.2 Gas Composition   334
- 10.3 Factors Affecting Gas Generation   336
- 10.4 Gas Generation Rate   338
- 10.5 Gas Migration   339
- 10.6 Types and Components of Gas Collection Systems   341
  - 10.6.1 *Passive Gas Collection System, 341*
  - 10.6.2 *Active Gas Collection System, 342*
- 10.7 Gas Control and Treatment   249
  - 10.7.1 *Gas Flaring, 350*
  - 10.7.2 *Gas Processing and Energy Recovery, 351*
- 10.8 Design of Gas Collection System   352
  - 10.8.1 *Calculation of NMOC Emission Rate, 352*
  - 10.8.2 *Estimation of Gas Generation Rate, 354*
  - 10.8.3 *Gas Extraction Well System Layout and Spacing, 359*
  - 10.8.4 *Gas Flow Generated from Each Extraction Well or Collector, 360*
  - 10.8.5 *Collection Piping System Layout and Routing, 365*
  - 10.8.6 *Estimation of Condensate Production, 366*
  - 10.8.7 *Header Pipe Sizing and Pressure Loss Calculations, 370*
  - 10.8.8 *Valve and Fitting Pressure Loss Calculations, 376*

Problems   394
References   396

11 FINAL COVER SYSTEM   399

- 11.1 Components of Final Cover System   400
  - 11.1.1 *Erosion Control Layer, 401*
  - 11.1.2 *Protection Layer, 401*

     11.1.3   Drainage Layer, 404
     11.1.4   Hydraulic Barrier Layer, 405
     11.1.5   Gas Vent Layer, 409
     11.1.6   Foundation Layer, 409
11.2 Alternative Landfill Cover   412
     11.2.1   Water Balance of Earthen Covers, 412
     11.2.2   Capillary Barrier, 413
     11.2.3   Monolayer Barrier, 416
11.3 Field Study of Landfill Covers   417
11.4 Soil Erosion Control   417
     11.4.1   Nature of Soil Erosion, 418
     11.4.2   Soil Loss Prediction, 420
     11.4.3   Limitations of Universal Soil Loss Equation, 429
     11.4.4   Erosion Control Principles, 429
     11.4.5   Manufactured Erosion Control Materials, 430
11.5 Effects of Settlement and Subsidence   431
11.6 Differential Subsidence Case History   434

Problems   435
References   437

12 LANDFILL SETTLEMENT   440

12.1 Mechanism of Solid Waste Settlement   440
12.2 Effect of Daily Cover   442
12.3 Landfill Settlement Rate   444
12.4 Estimation of Landfill Settlement   449
     12.4.1   Settlement of New Solid Waste, 449
     12.4.2   Settlement of Existing Solid Waste, 451
12.5 Effect of Waste Settlement on Landfill Capacity   454
12.6 Other Methods for Estimating Landfill Settlement   458
     12.6.1   Empirical Functions, 459
     12.6.2   Application of Empirical Functions to Field Case Study, 461
     12.6.3   Summary for Three Empirical Functions, 463
12.7 Estimation of Landfill Foundation Settlement   469
     12.7.1   Total Settlement of Landfill Foundation, 469
     12.7.2   Differential Settlement of Landfill Foundation, 472

Problems   473
References   475

13 LANDFILL STABILITY ANALYSIS   477

13.1 Types of Landfill Failures   478
     13.1.1   Sliding Failure of Leachate Collection System, 478
     13.1.2   Sliding Failure of Final Cover System, 479
     13.1.3   Rotational Failure of Sidewall Slope or Base, 480
     13.1.4   Rotational Failure through Waste, Liner and Subsoil, 480

- 13.1.5 Rotational Failure within the Waste Mass, 480
- 13.1.6 Translational Failure by Movement along Liner System, 480
- 13.2 Factors Influencing Landfill Stability 480
- 13.3 Selection of Appropriate Properties 481
  - 13.3.1 Geosynthetic Materials Properties, 481
  - 13.3.2 Solid Waste Properties, 484
  - 13.3.3 In-Situ Soil Slope and Subsoil Properties, 485
- 13.4 Veneer Slope Stability Analysis 487
  - 13.4.1 Cover Soil (Gravitational) Forces, 487
  - 13.4.2 Tracked Construction Equipment Forces, 490
  - 13.4.3 Inclusion of Seepage Forces, 497
  - 13.4.4 Inclusion of Seismic Forces, 508
  - 13.4.5 General Remarks, 513
- 13.5 Subsoil Foundation Failures 513
  - 13.5.1 Method of Analysis, 514
  - 13.5.2 Case Histories, 514
  - 13.5.3 General Remarks, 520
- 13.6 Waste Mass Failures 520
  - 13.6.1 Translational Failure Analysis, 521
  - 13.6.2 Case Histories, 528
  - 13.6.3 General Remarks, 535
- 13.7 Concluding Remarks 537

Problems 538
References 540

14 VERTICAL LANDFILL EXPANSIONS 544

- 14.1 Considerations Involved in Vertical Expansions 545
- 14.2 Liner Systems for Vertical Expansion 546
- 14.3 Settlement of Existing Landfill 548
  - 14.4.1 Current Methods for Estimating Localized Subsidence, 551
  - 14.4.2 Elastic Solution Method Applied to a Vertical Expansion, 552
- 14.4 Estimation of Differential Settlement due to Waste Heterogeneity 551
- 14.5 Vertical Expansion over Unlined Landfills 557
- 14.6 Design Considerations for Landfill Structures 557
- 14.7 Geosynthetic Reinforcement Design for Vertical Expansions 558
- 14.8 Stability Analysis for Vertical Expansion 572

Problems 573
References 574

15 BIOREACTOR LANDFILLS 576

- 15.1 Introduction 577
- 15.2 Liquids Management Strategies 578

15.2.1 Natural Attenuation, 579
15.2.2 Remove, Treat and Discharge 579
15.2.3 Leachate Recycling, 579
15.2.4 Comparison of Strategies, 580
15.3 Concepts of Waste Degradation 581
15.3.1 Phases of Degradation, 581
15.3.2 Field Capacity Moisture Content, 583
15.3.3 Related Aspects, 584
15.4 Leachate Recycling Methods 584
15.4.1 Surface Spraying, 585
15.4.2 Surface Ponding, 585
15.4.3 Leach Fields, 585
15.4.4 Shallow Wells, 585
15.4.5 Deep Wells, 585
15.4.6 Comparison of Methods, 587
15.5 Bioreactor Landfill Issues and Concerns 587
15.5.1 Liner System Integrity, 587
15.5.2 Leachate Collection System, 588
15.5.3 Leachate Removal System, 591
15.5.4 Filter and/or Operations Layer, 591
15.5.5 Daily Cover Material, 592
15.5.6 Final Cover Issues, 592
15.5.7 Waste Stability Concerns, 595
15.6 Performance-to-Date 596
15.7 Summary Comments 598

Problems 599
References 600

16 CONSTRUCTION OF COMPACTED CLAY LINERS 603

16.1 Subgrade Preparation 605
16.2 Soil Materials for Compacted Soil Lines 606
16.3 Compaction Objectives and Choices 607
16.3.1 Destruction of Soil Clods, 607
16.3.2 Molding Water Content, 608
16.3.3 Dry Unit Weight, 610
16.3.4 Type of Compaction, 610
16.3.5 Compactive Energy, 611
16.3.6 Lift Interfaces, 612
16.4 Initial Saturation Specifications 613
16.5 Clay Liner Compaction Considerations 614
16.6 Compaction Specifications 616
16.7 Leachate Collection Trench Construction 618
16.8 Protection of Compacted Soil 620
16.8.1 Protection against Desiccation, 620
16.8.2 Protection against Freezing, 621
16.8.3 Excess Surface Water, 622

16.9	Field Measurement of Water Content and Dry Unit Weight 622	
16.10	Construction Quality Assurance and Quality Control Issues 624	
	16.10.1	Critical Quality Assurance and Quality Control Issues, 625
	16.10.2	Quality Assurance and Quality Control for Compacted Clay Liner Construction, 626
	16.10.3	Documentation Report, 629

Problems   630
References   630

17   INSTALLATION OF GEOSYNTHETIC MATERIALS 634

17.1	Material Delivery and Conformance Tests 635	
17.2	Installation of Geomembranes 637	
	17.2.1	Geomembrane Placement, 637
	17.2.2	Geomembrane Seaming, 639
	17.2.3	Geomembrane Seam Tests, 646
	17.2.4	Geomembrane Defects and Repairs, 654
	17.2.5	Geomembrane Protection and Backfilling, 656
17.3	Installation of Geonets 660	
	17.3.1	Geonet Placement, 660
	17.3.2	Geonet Joining, 660
	17.3.3	Geonet Repairs, 661
17.4	Installation of Geotextiles 662	
	17.4.1	Geotextile Placement, 662
	17.4.2	Geotextile Overlapping and Seaming, 663
	17.4.3	Geotextile Defects and Repairs, 665
	17.4.4	Geotextile Backfilling and Covering, 666
17.5	Installation of Geocompostes 667	
	17.5.1	Geocomposite Placement, 667
	17.5.2	Geocomposite Joining and Repairs, 667
	17.5.3	Geotextile Covering, 668
17.6	Installation of Geosynthetic Clay Liners 669	
	17.6.1	Geosynthetic Clay Liner Placement, 669
	17.6.2	Geosynthetic Clay Liner Joining, 671
	17.6.3	Geosynthetic Clay Liner Repairs, 672
	17.6.4	Geosynthetic Clay Liner Backfilling or Covering, 672

Problems   673
References   673

18   POSTCLOSURE USES OF MSW LANDFILLS 675

18.1	Athletic and Recreational Facilities 675	
	18.1.1	Golf Courses/Driving Ranges, 676
	18.1.2	Sport Fields, 677

		18.1.1	*Golf Courses/Driving Ranges, 676*	
		18.1.2	*Sport Fields, 677*	
		18.1.3	*Paths, Trails, and Nature Walks, 677*	
		18.1.4	*Wildlife and Conservation Areas, 679*	
		18.1.5	*Multiple Use Facilities, 679*	
	18.2	Industrial Development 680	
		18.2.1	*Parking Lots, 680*	
		18.2.2	*Equipment/Material Storage, 683*	
		18.2.3	*Light Industrial Buildings, 683*	
	18.3	Aesthetics 684	
	18.4	Concluding Remarks 686	

Problems   686
References   687

APPENDIX I   HELP MODEL INPUT AND OUTPUT—ACTIVE CONDITION   689

APPENDIX II   HELP MODEL INPUT AND OUTPUT—POSTCLOSURE CONDITION   698

INDEX   710

Preface

The United States produces about 300 million tons of solid waste per year. Up to 75 percent of the solid waste continues to be landfilled—in spite of vigorous efforts aimed at waste reduction, recycling, and re-use. A modern, well-constructed landfill can be characterized as an engineered structure that consists primarily of a composite liner, leachate collection and removal system, gas collection and control system, and final cover. A landfill also behaves as a giant in-situ bioreactor whose contents undergo complex biochemical reactions. The production of landfill gas is a major byproduct of waste decomposition processes. The adoption of suitable design and construction methods is essential not only to reduce design and construction costs, but also to minimize long term operation, maintenance, and monitoring expenses.

Geotechnical Aspects of Landfill Design and Construction addresses landfill siting, design, and construction issues in a comprehensive manner. The characteristics of landfill containment envelopes and their design/construction are treated in detail. The attributes and advantages of composite liners relative to conventional compacted clay liners are examined carefully. The book discusses both the material properties and engineering design of geosynthetic components (e.g., geomembranes, geotextiles, geocomposites, and geosynthetic clay liners) that are used in modern landfill construction. Methods of estimating landfill leachate quantities and gas generation in addition to the design of leachate and gas collection systems are also described in detail. We include other important topics as well—such as vertical expansion and bioreactor concepts—that are ways of increasing capacity at existing landfills.

Several chapters in the book are devoted to the measurement and determination of landfill performance. These performance considerations include settlement estimates, mass stability, liner leakage rates (by both hydraulic convection and chemical diffusion), envelope durability, leachate and gas collection, and drainage efficiency. Final cover design to limit rainfall infiltration, frost problems, and erosion is addressed as well.

Geotechnical Aspects of Landfill Design and Construction focuses on actual design and construction procedures, as opposed to a discussion of solid waste management issues and to general descriptions and/or conceptual designs. We present the reader with a complete, integrated package of analytical tools, design equations, and construction procedures for all elements of a landfill. The purpose of the book is to show the reader how to design and construct a real landfill step by step. To this end, we provide in the book not only design equations, but also specific guidelines and procedures, and calculation examples for constructing various elements of a modern landfill.

Since landfill design and construction in the United States uses English Computational units almost exclusively (and there is no end in sight of this practice), we have complied by using these units as primary. Worldwide, however, SI units are the norm and we have accompanied the U.S. units with SI computational units in parentheses. The conversion to SI units is "soft." The notable exception to this is hydraulic conductivity where we have used in the traditional metric unit of "cm/sec."

Geotechnical Aspects of Landfill Design and Construction is intended as (i) a reference book for practicing professionals, (ii) an agency training manual, and (iii) university textbook. A draft manuscript of the book has been used and tested by the principal author in a geoenvironmental graduate course at the University of Michigan since 1995. Carefully selected design examples, diagrams, and tables are incorporated into the book. These give the reader a better sense of the necessary site investigation, planning, analysis, and organization that go into a landfill design and construction project. In addition to worked design examples we have also included homework problems and an extensive reference list at the end of every chapter.

The authors wish to express their appreciation to the following individuals for their encouragement and support throughout the preparation of the manuscript: Professors Richard D. Woods and E. Benjamin Wylie, University of Michigan; and Jim J. Sygo, Kenneth J. Burda, Delores M. Montgomery, and Elizabeth M. Browne, Michigan Department of Environmental Quality.

The authors also would like to acknowledge and thank the following individuals for sharing their knowledge and ideas during the course of many discussions about the book: Stephen R. Blayer, V. Wesley Sherman, Jr., and Carolyn B. Parker, Michigan Department of Environmental Quality; Dr. Gary R. Schmertmann, Geosyntec Consultants; Dr. Te-Yang Soong, Earth Tech, Inc.; Dr. Jengwa Lyang, NTH Consultants, Ltd.; and Scott F. Lockhart, Hull and Associates, Inc.

XUEDE (DAN) QIAN
ROBERT M. KOERNER
DONALD H. GRAY

CHAPTER 1

Introduction

1.1 NEED FOR LANDFILLS
1.2 PRINCIPAL LANDFILL REQUIREMENTS
1.3 LANDFILL COMPONENTS AND CONFIGURATION
1.4 LANDFILL ENVELOPE
    1.4.1 LINER SYSTEM
    1.4.2 LEACHATE COLLECTION AND REMOVAL SYSTEM
    1.4.3 GAS COLLECTION AND CONTROL SYSTEM
    1.4.4 FINAL COVER SYSTEM
1.5 COMPOSITE LINERS
1.6 BENEFITS OF DOUBLE COMPOSITE LINERS
1.7 LINER LEAKAGE MECHANISMS
    1.7.1 STEADY ADVECTION
    1.7.2 STEADY DIFFUSION
    1.7.3 UNSTEADY DIFFUSION
    1.7.4 COMBINED ADVECTION-DIFFUSION
1.8 SCOPE AND ORGANIZATION OF BOOK
    PROBLEMS
    REFERENCES

Land disposal has always been and continues to be the most common form of handling and disposing of various types of waste. Land disposal can occur in the following forms:

(i) shallow burial vaults in soil
(ii) deep chambers in rock
(iii) deep well injection
(iv) surface impoundments
(v) spray irrigation and composting

This book is about the first mode of land disposal, namely, shallow burial vaults in soil, more commonly known as landfills. More specifically, this book deals primarily with municipal solid waste (MSW) landfills because this is the most common type of landfill. Furthermore, a MSW landfill must be designed and constructed to accept a highly variable waste stream. Landfill design and construction technology has advanced rapidly during the past two decades in response to more stringent regulatory requirements and demands. A solid waste landfill must be able to prevent groundwater pollution, collect leachate, permit gas venting, and provide for groundwater and gas monitoring. Most, if not all, of the design and construction principles for municipal solid waste landfills apply equally to hazardous waste landfills, so-called Subtitle C facilities.

This book also focuses on the design/construction of horizontal landfill containment systems (cover and liner) and ancillary features such as the leachate and gas collection system. It deals only indirectly with vertical barriers (e.g., clay cut-off walls and geomembrane walls), which are used primarily for remediation purposes. Readers interested specifically in the latter topic should consult other references (USEPA, 1984; Brandl, 1990; Sharma and Lewis, 1994; Manassero et al., 1995; Koerner, 1996; Burson et al., 1997; Daniel and Koerner, 2000).

A modern well-constructed landfill can be characterized as an engineered structure that consists primarily of a composite liner system, leachate collection and removal system, gas collection and control system, and final cover system. A landfill project can also be very expensive; a modern facility may cost as much as $800,000 per acre ($2 million per hectare). The adoption of suitable design and construction methods is essential not only to reduce design and construction costs, but also to minimize long-term operation, maintenance, and monitoring expenses.

1.1 NEED FOR LANDFILLS

The United States produces over 200 million U.S. tons (180 million metric tons) of solid wastes per year. The generations of municipal solid waste (MSW) in the United States since 1996 are listed in Table 1.1. We have three basic choices for handling or disposing of this waste:

(i) Bury it!
(ii) Burn it!
(iii) Recycle/Re-use it!

In the best of all possible worlds, we would attempt to minimize the amount of waste slated for burial or incineration by designing and implementing programs focused on

TABLE 1.1 Generation of Municipal Solid Waste in the United States (USEPA, 1999)

Year	Generation of Municipal Solid Waste	
	Thousands of U.S. Tons	Thousands of Metric Tons
1960	88,120	79,920
1970	121,060	109,800
1980	151,640	137,540
1990	205,210	186,130
1994	214,180	194,260
1995	211,360	191,700
1996	209,190	189,740
1997	216,970	196,790
2000	223,230*	202,470*
2005	239,540*	217,260*

*Estimated

waste reduction, recycling, and re-use. In spite of vigorous efforts in this direction during the past decade, up to 75% of the nation's solid waste is still landfilled. As an example of the density of landfills in one region, the location of municipal solid waste landfills in the Midwest and Great Lakes region of the United States is shown in Figure 1.1.

For better or worse, the need for landfilling of solid wastes will continue indefinitely for a number of reasons. Incineration is not a viable method of disposal for a wide variety of wastes, (e.g., mine and mill wastes and other inorganic noncombustibles). Furthermore, incineration may lead to air pollution problems, and it creates an ash residue that still must be landfilled. Recycling efforts eventually encounter practical limits that make further reductions in the waste stream that is slated for land disposal hard to achieve. The experience of the authors' home city, Ann Arbor, is a good case in point.

Ann Arbor, Michigan, with a population of 150,000, is typical of a mid-sized city that operates an efficient and well-managed recycling program that enjoys the active support and participation of the community. The city has invested substantial funds in recycle collection and processing facilities to service both the residential and commercial sectors. Ann Arbor also operates one of the most successful composting operations in the country. This operation largely takes care of residential yard wastes and

FIGURE 1.1 Municipal Solid Waste Landfills in the Midwest and Great Lakes Region of the United States

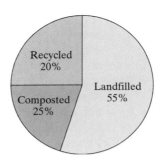

FIGURE 1.2 Fate of Municipal Solid Waste in Ann Arbor, Michigan

tree leaves from city streets. Even with this level of involvement and commitment to recycling and composting, about 55 percent of the city's wastes are still landfilled. The exact distribution between wastes that are composted, recycled, and landfilled is shown in Figure 1.2. Finally, it should be noted that the type of landfills in use today may be modified in the future depending on technological developments such as bioreactor and inward gradient landfills.

1.2 PRINCIPAL LANDFILL REQUIREMENTS

An engineered landfill is a controlled method of waste disposal. The site of the landfill must be geologically, hydrologically, and environmentally suitable. A landfill is not an open dump. Nuisance conditions associated with an open dump—such as smoke, odor, unsightliness, insect and rodent problems, seagull and other bird problems—are not present in a properly designed, operated, and maintained sanitary landfill. Professional planning and engineering supervision are required. A landfill has a carefully designed and constructed envelope that encapsulates the waste and that prevents escape of leachate into the environment. The envelope consists basically of a top cap (or cover) and bottom liner. Each of these two main components in turn comprises a system of barrier and drainage layers as shown in Figure 1.3.

The most important requirement of a landfill is that it does not pollute or degrade its environment. This requirement is achieved by both careful siting and by proper design/construction. Siting is addressed primarily in a hydrogeology investigation and report that is required of operator/owner applicants by state and Federal regulatory authorities. Although complex and sometimes costly, proper siting minimizes future impacts on public health. Good siting also yields economic dividends to the extent it reduces design and/or construction costs and long-term expenses associated with operation, maintenance, and recovery of potential leachate releases.

Design/construction/operation are addressed in an engineering investigation and report that is likewise required of landfill operator/owners. Environmental geotechnology strives to provide sound design principles and construction procedures to ensure short- and long-term stability and performance of landfill components. The strength, stability, and durability of the lining system are of paramount importance. Proper quality control of construction and construction materials that are selected according to design requirements is equally important. Landfill design based on reli-

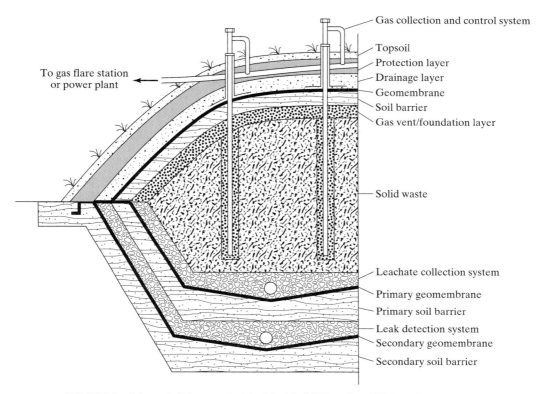

FIGURE 1.3 Schematic Diagram of a Municipal Solid Waste Landfill Containment System

able data gathered during the site investigation and using proven scientific methods significantly reduces risk to the environment.

The U.S. Environmental Protection Agency (EPA) and the various states have detailed regulations governing landfill siting, design, construction, operation, groundwater and gas monitoring, landscaping plan, closure monitoring, and maintenance for 30 years. The situation commands worldwide interest and attention in that some 28 countries have regulations of either a prescriptive or performance nature (Koerner and Koerner, 1999).

1.3 LANDFILL COMPONENTS AND CONFIGURATION

Most municipal solid waste landfills are composed of the following elements or systems:

(i) Bottom and lateral side liners system,
(ii) Leachate collection and removal system,
(iii) Gas collection and control system,
(iv) Final cover system,

(v) Stormwater management system,
(vi) Groundwater monitoring system,
(vii) Gas monitoring system.

The construction/design of a landfill requires that the following major actions be taken:

(i) Landfill footprint layout,
(ii) Subbase grading,
(iii) Cell layout and filling,
(iv) Temporary cover selection,
(v) Final cover grading,
(vi) Final cover selection.

A schematic diagram of the development and completion of a solid waste landfill is shown in Figure 1.4.

The interdependence between various landfill elements and the sequence in which they are considered plays an important role in landfill design. These elements can be combined in a variety of geometrical configurations or spatial arrays. The most common landfill types or geometrical configurations include (Repetto, 1995):

Area Fill [Figure 1.5(a)]. The landfill progresses with little or no excavation. Normally this type of landfill is used in areas with high groundwater or where the terrain is unsuitable for excavation.

Trench Fill [Figure 1.5(b)]. Solid waste is filled in a series of deep and narrow trenches for this type of landfill. It is generally used only for small waste quantities. This method is still used for hazardous waste landfills for some states.

Above and Below Ground Fill [Figure 1.5(c)]. This type of landfill is like a combination of the two previously mentioned types, Area Fill and Trench Fill. However, the excavation area is much larger than in a Trench Fill landfill. The depth of excavation normally depends on the depths of the natural clay layer and the groundwater level.

Canyon Fill [Figure 1.5(d)]. This is also called a Valley Fill. For this filling method, the solid waste is filled between hills or rolling terrain.

1.4 LANDFILL ENVELOPE

The landfill envelope is the most important consideration in the design of a modern municipal solid waste or hazardous waste landfill. The envelope encapsulates the waste and isolates it from the surrounding environment. It should prevent escape of leachate, limit rainfall infiltration, and handle gas generation. The main components of the envelope are the liner system, leachate collection and removal system, gas collection and control system, and final cover system. The function and role of each of these systems is briefly summarized in the following subsections.

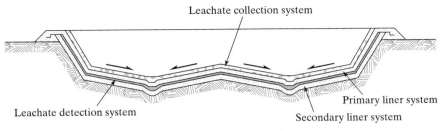

(a) Cell subgrade and leachate collection system

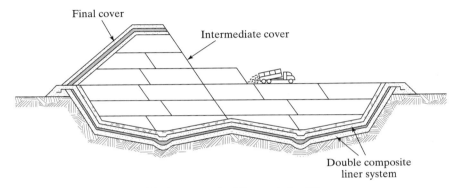

(b) Placement of solid waste in landfill

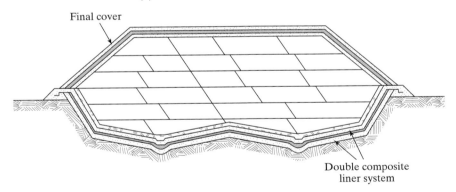

(c) Closed landfill with final cover

FIGURE 1.4 Development and Completion of a Solid Waste Landfill

1.4.1 Liner System

The liner system is placed on the bottom and lateral sides of a landfill. The liner system acts as a barrier against the advective (hydraulic) and diffusive transport of leachate solutes. Its main purpose is to isolate the solid waste and prevent contamination of the surrounding soil and groundwater. A liner consists of multiple barrier and drainage layers. The barrier may consist of a compacted clay layer, geomembrane, geosynthetic

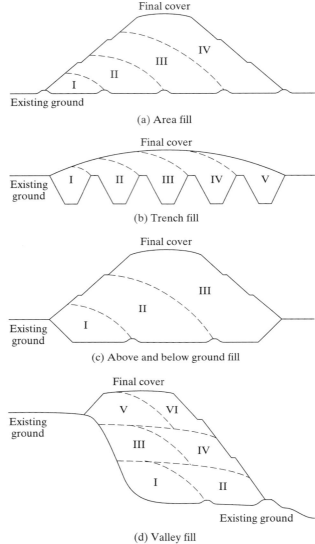

FIGURE 1.5 Four Types of Solid Waste Landfill Geometry

clay layer, and/or a combination of these. The liner system is the single most important element of a landfill.

1.4.2 Leachate Collection and Removal System

Leachate is the mobile portion of the solid waste in a landfill. Leachate is generated from liquid squeezed out of the waste itself (primary leachate) and by water that infiltrates into the landfill and that percolates through the waste (secondary leachate).

Leachate consists of a carrier liquid (solvent) and dissolved substances (solutes). A leachate collection and removal system is used to collect the leachate produced in a landfill, to prevent the buildup of leachate head on the liner, and to drain leachate to a wastewater treatment plant by a sanitary sewer line or a leachate storage tank for treatment and disposal.

1.4.3 Gas Collection and Control System

Municipal solid waste can generate large quantities of gas during decomposition. The two primary constituents in a landfill are methane (CH_4) and carbon dioxide (CO_2). The gas collection and control system is used to collect the landfill gas during decomposition of the organic components of the solid waste. Landfill gas can either be used to produce energy or flared under controlled conditions.

1.4.4 Final Cover System

The final cap or cover system consists of barrier and drainage layers. A soil layer is also included that protects the underlying layers against intrusion, damage, and the effects of frost. The main purpose of a landfill final cover is to minimize water infiltration into the landfill to reduce the amount of leachate generated after closure. The cover should be designed with the same degree of attention and care that is applied to the liner. Unlike a liner, a cover acts as a hydraulic barrier against infiltrating outside water only; it is not required to act as a barrier against leakage of leachate solutes under combined advection and diffusion.

1.5 COMPOSITE LINERS

Proper functioning of a liner system is critical to the containment effectiveness of a landfill. Considerable attention has been devoted to the development and design of different liner systems. Early liners consisted primarily of a single liner composed of a clay layer or a synthetic polymeric membrane. During the past few decades the trend has been to use composite liner systems comprising both clay and synthetic geomembranes together with interspersed drainage layers. The following is an approximate chronology showing the introduction date for each of these approaches:

- pre-1982: Single clay liner (Figure 1.6)
- 1982: Single geomembrane liner (Figure 1.7)
- 1983: Double geomembrane liner (Figure 1.8)
- 1984: Single composite liner (Figure 1.9)
- 1987: Double composite liner with primary and secondary leachate collection systems (Figure 1.10)

Double composite liners with both primary and secondary leachate collection systems (see Figure 1.10) have been widely adopted in solid waste landfills in the United States. This type of liner system is mandated by Federal and state regulations for hazardous waste, and by at least 10 state regulatory agencies for municipal and

10 Chapter 1 Introduction

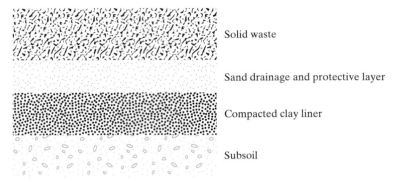

FIGURE 1.6   Single Compacted Clay Liner

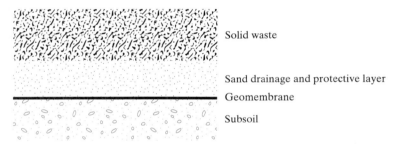

FIGURE 1.7   Single Geomembrane Liner

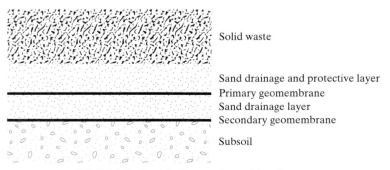

FIGURE 1.8   Double Geomembrane Liner System

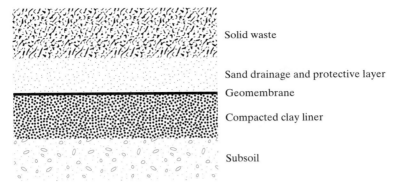

FIGURE 1.9 Single Composite Liner System

nonhazardous waste. The U.S. Environmental Protection Agency (EPA) requires composite liner systems (both clay and synthetic barriers) with leachate collection for municipal solid waste landfills [RCRA Subtitle D, 40 CFR §258.40(a)(2)] or equivalent performance [40 CFR §258.40(a)(1)], and double composite liners with leachate collection and detection systems for all hazardous waste landfills (RCRA Subtitle C, 40 CFR §264.301). Some individual states require double composite liner systems for municipal solid waste landfills as well (e.g., 6NYCRR Part 360 and Michigan NREPA Part 115).

Starting at the bottom, a double composite liner consists of a minimum 2-ft- (0.6-m)-thick compacted clay liner [or an alternative liner that can be equivalent to a 2-ft- (0.6-m)-thick compacted clay liner], followed by a secondary geomembrane,

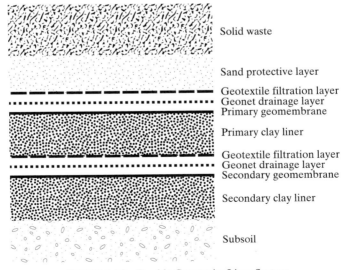

FIGURE 1.10 Double Composite Liner System

secondary leachate collection (or leak detection) layer, a minimum 2-ft- (0.6-m)-thick primary compacted clay liner [or geosynthetic clay liner that can be equivalent to a 2-ft- (0.6-m)-thick compacted clay liner], a primary geomembrane, and a primary leachate collection system. This sequence is topped off by a 2-ft- (0.6-m)-thick protective sand blanket. The leachate collection system consists of a layer of synthetic geonet and geotextile. The former provides good in-plane drainage conveyance and the latter good cross-plane drainage together with the ability to exclude (filter out) fines. The geomembrane must be at least 1.5 mm (60 mils) thick if high density polyethylene (HDPE), or 0.75 mm (30 mils) thick if made from other resins. The permeability of the subbase and compacted clay liner must not exceed 1.0×10^{-7} cm/sec.

A composite liner works in a mutually supportive way to minimize and offset the limitations, (viz., excessive seepage, holes, tears, and other void volume defects) associated with single liner systems. The manner in which a composite liner works is depicted schematically in Figure 1.11, and is contrasted with the *modus operandi* of individual geomembranes and soil (compacted clay) liners as well. If there is a hole in a geomembrane, liquid will move easily through the hole, assuming the subgrade soil does not impede seepage. With a soil liner alone, seepage takes place over the entire area of the liner. With a composite liner, on the other hand, only a limited amount of liquid will pass through any hole in the geomembrane, but it will then encounter low-permeability clay. The low-permeability clay will impede further migration of the limited amount of liquid passing through the holes.

Good composite action requires good hydraulic contact between the geomembrane and underlying clay soil. The geomembrane should not be separated from the clay with permeable material, such as a bed of sand or a thick geotextile, because this would jeopardize intimate hydraulic contact. (See Figure 1.12.) If stones of a size and shape that could puncture the geomembrane exist in the clay soil layer, the stones must be removed. To achieve intimate contact, the surface of the compacted clay soil on which the geomembrane is placed should be smooth-rolled with a steel-drum roller. Also, the geomembrane should be placed and backfilled in a way that minimizes wrinkles (Daniel, 1993).

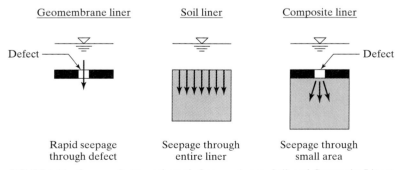

FIGURE 1.11 Seepage Patterns through Geomembrane, Soil, and Composite Liners

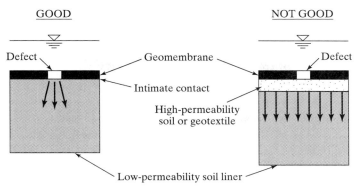

FIGURE 1.12 Proper Design of Composite Liner For Intimate Hydraulic Contact between Geomembrane and Compacted Soil (Daniel, 1993)

1.6 BENEFITS OF DOUBLE COMPOSITE LINERS

It should be stated from the outset that the authors consider double-lined composite landfill systems with leak detection capability as significantly preferable over single-lined composite systems. Many of the illustrations and much of the text will reflect this preference. It is also true that such double-lined systems are more costly initially than single-lined systems. But a leak through a primary liner is still contained within the immediate landfill area in the case of the former design. Furthermore, such leakage can be collected and properly disposed of in the case of a double-lined system.

For a single-lined system, on the other hand, the only post-construction leak detection option is downstream monitoring wells. When (even if) a leak is detected by this means, a significantly contaminated site is already at hand. This scenario begs the question, "If one goes to the cost of designing and constructing a double-lined site, why should reliance be placed on downstream monitoring wells?" In fact, "Why are downstream monitoring wells needed at all?"

The cost of design and installation of downstream monitoring wells far exceeds the additional liner/leak detection layer costs of a double-lined landfill. Further, when one considers the ongoing operational costs of down gradient monitoring wells (including sampling, testing, and analysis) the comparison of costs heavily favors the double-lined system. Even more importantly, environmental safety and security are far superior with a double-lined landfill covering the entire footprint of the site than with intermittently spaced monitoring wells.

With a leak detection layer, the quantity of liquid can be monitored over time, resulting in a calculable leak rate in units of gallons/acre/day (liters/hectare/day). Critical leak rate values can be agreed upon during the design/permitting stage with accompanying remedial actions that must be invoked. There are numerous such actions that can be implemented, depending upon site-specific conditions and the magnitude of the leakage rate.

The decision to select a double-lined composite liner system over a single-lined composite liner is an issue involving both owners and regulators. Even further, the omission of down gradient monitoring wells in lieu of a double-lined composite system needs addressing in most regulations. These regulations should be viewed as "living documents," and if change is warranted, so be it. The process should begin with (and hopefully will recognize the advantage of) double-lined landfill systems over down gradient monitoring wells—from both a cost perspective as well as from an environmental/public health concern. A site with contaminated down gradient wells is considerably more extreme and potentially dangerous than a landfill footprint by itself. Remediation of the former is a formidable task. The leakage rate at a site with a leak detection layer, on the other hand, is not only quantifiable, it can also be remediated in place with the knowledge that the secondary liner is in place to prevent leachate from leaving the site.

1.7 LINER LEAKAGE MECHANISMS

Leachate solutes, both inorganic and organic, can be transported across barriers (e.g., landfill liners) by both advection and diffusion.

> **Advection** refers to the movement of fluid (or leachate) in a porous medium under a hydraulic gradient. Leachate consists of the carrier fluid (water) and solutes or dissolved substances. Both organic and inorganic compounds (or contaminants) comprise the solutes found in landfill leachate.
>
> **Diffusion** refers to the movement of the solutes (dissolved substances) under a chemical or concentration gradient. The solutes can diffuse in the same direction as the advective movement, or they can diffuse in an opposite or counter direction. The dissolved substances in the pore water can be thought of as "swimmers." These swimmers either *go with the flow* (advection), or they *swim on their own* from a region of high to low concentration (diffusion).

In the case of barriers with very low hydraulic conductivity—characteristic of landfill liners—diffusion becomes increasingly important as a leakage pathway. An increasing body of evidence (Gillham et al., 1984; Crooks and Quigley, 1984; Shackelford, 1988; Johnson et al., 1989; Gray, 1995) has shown that in very fine-grained soils, solute flux is largely governed by diffusion. The relative importance of diffusion as a leakage pathway increases as the hydraulic conductivity of the barrier decreases. Investigative work by Goodall and Quigley (1977), Gilham et al. (1984), Rowe (1987), Johnson et al. (1989), and Shackelford (1988) has shown that at very low pore water velocities, characteristic of water movement across clay barriers (e.g., clay cut-off walls and landfill liners), solute transport is largely controlled by chemical diffusion along solute concentration gradients.

1.7.1 Steady Advection

Advection is the process by which solutes are transported simultaneously, along with the flowing fluid or solvent (typically water), in response to a gradient in total hydraulic head. Assuming that the concentration of solute of concern in the landfill leachate remains constant, the one-dimensional advective mass flux (Darcian hypothesis) of solute per unit area can be expressed as

$$J_A = C_o \cdot v = C_o \cdot k \cdot i = C_o \cdot k \cdot (H + L)/L \tag{1.1}$$

where J_A = advective mass flux, mg/cm²/sec;
  C_o = concentration of the solute in water at top of liner, mg/cm³;
  v = average seepage velocity of flow through the liner, cm/sec;
  k = hydraulic conductivity, cm/sec;
  i = hydraulic gradient;
  H = leachate head on the liner, cm; and L = thickness of the liner, cm.

Equation 1.1 is applicable only for flow through intact liners; it does not take into account flow through void volume defects, such as cracks, fissures, holes, tears, etc. (The influence of these void volume defects on advective flow and ways to calculate their contribution to total flow across a liner are addressed in Chapters 3 and 4.)

It is instructive to calculate potential advective flow rates through a liner under typical landfill operating conditions. The hydraulic gradient is usually assumed to be a 12-in. (30-cm)-high leachate head acting on top of the liner. The properties of interest necessary for this calculation for a compacted clay liner and geomembrane, respectively, are summarized as follows (Daniel and Shackelford, 1989):

For Geomembrane,

  Leachate Head, h = 1 ft (0.3 m);
  Thickness, t = 60 mil (1.5 mm);
  Porosity, n = 0.10;
  Hydraulic Conductivity, k = 1.0 × 10⁻¹² cm/sec;
  Diffusion Coefficient, D = 3.0 × 10⁻¹⁰ cm²/sec;
  Hydraulic Gradient, i = 115.

For Compacted Clay Liner,

  Leachate Head, h = 1 ft (0.3 m);
  Thickness, t = 3 ft (0.9 m);
  Porosity, n = 0.50;
  Hydraulic Conductivity, k = 1.0 × 10⁻⁷ cm/sec;
  Diffusion Coefficient, D = 2.0 × 10⁻⁶ cm²/sec;
  Hydraulic Gradient, i = 1.33.

TABLE 1.2 Advective Flow Rate across 3-foot (0.9-m) Compacted Clay Liner under 1-foot (0.3-m) Leachate Head

Hydraulic Conductivity (cm/sec)	Flow Rate	
	gallons/acre/day	liters/m^2/day
1.0×10^{-6}	1,200	1.15
1.0×10^{-7}	120	0.115
1.0×10^{-8}	12	0.0115
1.0×10^{-9}	1.2	0.00115

Table 1.2 shows advective mass flux rates through a compacted clay liner under a one-foot-high leachate head for various assumed saturated hydraulic conductivities, including the usual 1.0×10^{-7} cm/sec target value prescribed by most Federal and state regulations. Advective mass flux rates are tabulated in terms of gallons/acre/day and liter/m^2/day to provide a more realistic and understandable scale of comparison.

Several conclusions emerge from this comparison of advective flow rates. Even if regulatory requirements are met—namely, a maximum allowable hydraulic conductivity of 1.0×10^{-7} cm/sec for a clay liner—a flow rate of 120 gallons/acre/day (0.115 liter/m^2/day) is substantial. This would translate into nearly 44,000 gallons/acre/year (42 liter/m^2/year). If the regulatory target is not met, then the leakage rate becomes alarmingly large. This illustration demonstrates

1. The need for stringent quality control on liner design and construction;
2. The need to minimize leachate head buildup on a liner by means of a well-designed primary leachate collection and removal system; and
3. The need to use a composite liner system that employs a geomembrane above (on top of) the clay to greatly reduce the effective hydraulic conductivity.

1.7.2 Steady Diffusion

Solute in landfill leachate can also migrate through clay liners by molecular diffusion. Diffusion may be thought of as a transport process in which a chemical or chemical species migrates in response to a gradient in its concentration, although the actual driving force for diffusion transport is the gradient in chemical potential of the solute (Robinson and Stokes, 1959). A hydraulic gradient is not required for transport of contaminants by diffusion.

The fundamental equation for diffusion in soil is Fick's first law, which for one-dimensional transport can be written as

$$J_D = D \cdot n \cdot (\partial C / \partial z) = D \cdot n \cdot (\Delta C / L) \tag{1.2}$$

where J_D = diffusive mass flux, mg/cm²/sec;
D = effective diffusion coefficient, cm²/sec;
n = porosity;
$\partial C/\partial z$ = concentration gradient; and
ΔC = difference in concentration between top and base of the liner, mg/cm³, or

$$\Delta C = C_o - C_e \tag{1.3}$$

where C_o = solute concentration at top of the liner, mg/cm³;
C_e = solute concentration at base of the liner, mg/cm³.

The effective diffusion coefficient depends on the chemical of interest (e.g., ionic size, charge, and valence of solute) and diffusion conditions (e.g., diffusion through free solution vs. soil pore water, tortuosity of pores). Solutes diffuse more slowly through soil pores than in free solution.

1.7.3 Unsteady Diffusion

Fick's first law describes steady-state diffusive flux of solutes. For time-dependent (unsteady) transport of nonreactive solutes in soil, the following equation can be used:

$$\partial C/\partial t = D \cdot (\partial^2 C/\partial z^2) \tag{1.4}$$

This is known as Fick's second law. This equation describes the temporal and spatial variation of diffusing solute in a porous medium. The solution for this equation depends on initial and boundary conditions. If it is assumed that at time $t = 0$ the solute concentration at the top of the liner is C_o ($C_o > 0$), and the solute concentration at the base of the liner is zero, and if solute concentration C_o remains constant at the top of the liner with time, the solution for the above equation is (Crank, 1956) given by

$$C(z, t) = C_o \cdot erfc[z/(4 \cdot D \cdot t)^{0.5}] \tag{1.5}$$

where $C(z, t)$ = solute concentration at depth z and time t, mg/cm³;
z = depth, cm; and
t = time, sec.

In the above equation, *erfc* is the complimentary error function. (See Table 1.3.)

1.7.4 Combined Advection-Diffusion

The one-dimensional form of the advection-diffusion equation for nonreactive dissolved constituents in saturated, homogeneous, isotropic materials, under steady-state uniform flow is

$$\partial C/\partial t = D_1 \cdot (\partial^2 C/\partial z^2) - v_s \cdot (\partial C/\partial z) \tag{1.6}$$

TABLE 1.3 Complementary Error Function (*erfc*)

$$erf(\beta) = (2/\sqrt{\pi}) \int_0^\beta exp(-\varepsilon^2)d\varepsilon$$

$$erf(-\beta) = -erf(\beta)$$

$$erfc(\beta) = 1 - erf(\beta)$$

β	$erf(\beta)$	$erfc(\beta)$
0	0	1.0
0.05	0.056372	0.943628
0.1	0.112467	0.887537
0.15	0.167996	0.832004
0.2	0.222703	0.777297
0.25	0.276326	0.723674
0.3	0.328627	0.671373
0.35	0.379382	0.620618
0.4	0.428392	0.571608
0.45	0.475482	0.524518
0.5	0.520500	0.479500
0.55	0.563323	0.436677
0.6	0.603856	0.396144
0.65	0.642029	0.357971
0.7	0.677801	0.322199
0.75	0.711156	0.288844
0.8	0.742101	0.257899
0.85	0.770668	0.229332
0.9	0.796908	0.203092
0.95	0.820891	0.179109
1.0	0.842701	0.157299
1.1	0.880205	0.119795
1.2	0.910314	0.089686
1.3	0.934008	0.065992
1.4	0.952285	0.047715
1.5	0.966105	0.033895
1.6	0.976348	0.023652
1.7	0.983790	0.016210
1.8	0.989091	0.010909
1.9	0.992790	0.007210
2.0	0.995322	0.004678
2.1	0.997021	0.002979
2.2	0.998137	0.001863
2.3	0.998857	0.001143
2.4	0.999311	0.000689
2.5	0.999593	0.000407
2.6	0.999764	0.000236
2.7	0.999866	0.000134
2.8	0.999925	0.000075
2.9	0.999959	0.000041
3.0	0.999978	0.000022

where D_1 = dispersion coefficient, cm^2/sec; and
  v_s = seepage velocity (i.e., average velocity of flow through the voids of the liner), cm/sec.

The solution for this equation depends on initial and boundary conditions. "Worst case" conditions occur (1) when the source (inlet) concentration of contaminants on top of the liner remains constant, (2) when there is no adsorption (partitioning) of solute on the barrier solids (i.e., nonreactive solutes), and (3) when the hydrodynamic gradient acts in the same direction as the solute concentration gradient. The effect of a depleting source concentration on the spatial and temporal distribution of solute in a landfill liner has been investigated by Rowe and Booker (1985). The influence of adsorption (partitioning) of a solute has been examined by Gray (1995) and Gullick, et al. (1996).

If it is assumed that worst case conditions apply, the boundary conditions for a landfill liner are

$$C(z, 0) = 0, \quad z \geq 0;$$
$$C(0, t) = C_o, \quad t \geq 0; \text{ and}$$
$$C(\infty, 0) = 0, \quad t \geq 0;$$

For these boundary conditions, the solution to the above equation for a saturated, homogeneous porous medium is (Ogata and Banks, 1961)

$$C(z, t)/C_o = 0.5 \cdot \{erfc[(z - v_s \cdot t)/(4 \cdot D_1 \cdot t)^{0.5}] + \exp(v_s \cdot z/D_1) \cdot erfc[(z + v_s \cdot t)/(4 \cdot D_1 \cdot t)^{0.5}]\} \quad (1.7)$$

where $C(z, t)$ = solute concentration at depth z and time t, mg/cm^3.

The hydrodynamic dispersion coefficient is the sum of the mechanical dispersion coefficient, D_m, and the effective diffusion coefficient, D, written as (Freeze and Cherry, 1979)

$$D_1 = D_m + D \quad (1.8)$$

where D_m = mechanical dispersion coefficient, cm^2/sec.

The hydrodynamic dispersion coefficient is a function of the seepage velocity. As the seepage velocity approaches 0, the hydrodynamic dispersion coefficient also approaches 0. Based on a study presented by Shackelford and Redmond (1995), diffusion dominates miscible transport at the low flow rates common in fine-grained barrier materials. Under these conditions, mechanical dispersion is essentially negligible (i.e., $D_m \approx 0$ and $D_1 \approx D$). Therefore, Equation (1.7) becomes

$$C(z, t)/C_o = 0.5 \cdot \{erfc[(z - v_s \cdot t)/(4 \cdot D \cdot t)^{0.5}] + \exp(v_s \cdot z/D) \cdot erfc[(z + v_s \cdot t)/(4 \cdot D \cdot t)^{0.5}]\} \quad (1.9)$$

where $C(z, t)/C_o$ = concentration ratio at depth z and time t, dimensionless;
v_s = seepage velocity through the liner, cm/sec;
D = effective diffusion coefficient of the liner, cm²/sec;
z = depth, cm; and
t = time, sec.

For a landfill liner, the solute concentration at the base of the liner at any time t can be estimated approximately by setting $D_1 = D$ in Equation 1.7, to produce

$$C_e(t)/C_o = 0.5 \cdot \{erfc[(L - v_s \cdot t)/(4 \cdot D \cdot t)^{0.5}] + \exp(v_s \cdot L/D) \cdot erfc[L + v_s \cdot t)/(4 \cdot D \cdot t)^{0.5}]\} \quad (1.10)$$

where

$C(z, t)/C_o$ = concentration ratio at the base of the liner at time t, dimensionless;
v_s = seepage velocity through the liner, cm/sec;
D = effective diffusion coefficient of the liner, cm²/sec;
L = thickness of the liner, cm;
t = time, sec.

EXAMPLE 1.1

Given that the thickness of clay liner, $L = 0.6$ m, the seepage velocity across clay liner, $v_s = 3.0 \times 10^{-7}$ cm/sec, and the effective diffusion coefficient of clay liner, $D = 6 \times 10^{-10}$ m²/sec = 6×10^{-6} cm²/sec.,

1. Calculate the concentration ratio at $z = 0.3$ m at $t = 365$ days; and
2. Calculate the concentration ratio at the base of the liner at $t = 10$ years.

Solution:
1. From Equation 1.9,

$$z = 0.3 \text{ m} = 30 \text{ cm},$$
$$t = 365 \text{ days} = 86{,}400 \times 365 = 31{,}536{,}000 \text{ sec},$$
$$v_s = 3.0 \times 10^{-7} \text{ cm/sec, and}$$
$$D = 6 \times 10^{-6} \text{ cm}^2/\text{sec}.$$

Thus, $\exp(v_s \cdot z/D) = \exp[3.0 \times 10^{-7} \times 30/(6 \times 10^{-6})] = 4.481689$ and

$$erfc[(z - v_s \cdot t)/(4 \cdot D \cdot t)^{0.5}] = erfc[(30 - 3.0 \times 10^{-7} \times 31{,}536{,}000)/(4 \times 6 \times 10^{-6} \times 31{,}536{,}000)^{0.5}]$$
$$= erfc(0.746577)$$

From Table 1.3, $erfc(0.7) = 0.322199$, and $erfc(0.75) = 0.288844$.
So, $erfc(0.746577) = 0.322199 + (0.746577 - 0.7)(0.288844 - 0.322199)/(0.75 - 0.7)$
$$= 0.291127 \text{ and}$$

$$erfc[(z - v_s \cdot t)/(4 \cdot D \cdot t)^{0.5}] = erfc[(30 - 3.0 \times 10^{-7} \times 31{,}536{,}000)/(4 \times 6 \times 10^{-6} \times 31{,}536{,}000)^{0.5}]$$
$$= erfc(1.434356)$$

From Table 1.3, $erfc(1.4) = 0.047715$ and $erfc(1.5) = 0.033895$.
So, $erfc(1.434356) = 0.047715 + (1.434356 - 1.4)(0.033895 - 0.047715)/(1.5 - 1.4)$
$$= 0.042967$$

Hence,

$$C_e(t)/C_o = 0.5 \cdot \{erfc[(L - v_s \cdot t)/(4 \cdot D \cdot t)^{0.5}] + \exp(v_s \cdot L/D) \cdot erfc[L + v_s \cdot t)/(4 \cdot D \cdot t)^{0.5}]\} \quad (1.10)$$
$$= 0.5 \cdot (0.291127 + 4.481689 \times 0.042967)$$
$$= 0.242$$

2. From Equation 1.10,

$$L = 0.6 \text{ m} = 60 \text{ cm},$$
$$t = 10 \text{ years} = 86{,}400 \cdot 365 \cdot 10 = 315{,}360{,}000 \text{ sec},$$
$$v_s = 3.0 \times 10^{-7} \text{ cm/sec, and}$$
$$D = 6 \times 10^{-6} \text{ cm}^2/\text{sec}.$$

Thus, $\exp(v_s \cdot L/D) = \exp[3.0 \times 10^{-7} \times 60/(6 \times 10^{-6})] = 20.085537$ and

$$erfc[(L - v_s \cdot t)/(4 \cdot D \cdot t)^{0.5}] = erfc[(60 - 3.0 \times 10^{-7} \times 315{,}360{,}000)/(4 \times 6 \times 10^{-6} \times 315{,}360{,}000)^{0.5}]$$
$$= erfc(-0.397803)$$

Because $erfc(\beta) = 1 - erf(\beta)$ and $erf(-\beta) = -erf(\beta)$ (see Table 1.3),

$$erfc(-0.397803) = 1 - erf(-0.397803) = 1 + erf(0.397803)$$

From Table 1.3, $erf(0.35) = 0.379382$ and $erf(0.4) = 0.428392$.
So, $erf(0.397803) = 0.379382 + (0.397803 - 0.35)(0.428392 - 0.379382)/(0.4 - 0.35)$
$$= 0.426238 \text{ and } erfc(-0.397803) = 1 + erf(0.397803) = 1.426238$$

Also,

$$erfc[(L + v_s \cdot t)/(4 \cdot D \cdot t)^{0.5}] = erfc[(60 + 3.0 \times 10^{-7} \times 31{,}536{,}000)/(4 \times 6 \times 10^{-6} \times 31{,}536{,}000)^{0.5}]$$
$$= erfc(1.777146).$$

From Table 1.3, $erfc(1.7) = 0.016210$ and $erfc(1.8) = 0.010909$
So

$$erfc(1.777146) = 0.016210 + (1.777146 - 1.7)(0.010909 - 0.016210)/(1.8 - 1.7)$$
$$= 0.012121$$

Hence,

$$C_e(t)/C_o = 0.5(erfc[(L - v_s \cdot t/(4 \cdot D \cdot t)^{0.5}] + \exp(v_s \cdot L/D) \cdot erfc[(L + v_s \cdot t)/(4 \cdot D \cdot t)^{0.5}]\}(1.10)$$
$$= 0.5 \cdot (1.426238 + 20.085537 \times 0.012121)$$
$$= 0.835$$

The relative importance of diffusion and advection as a function of barrier hydraulic conductivity can be illustrated in a useful manner (Shackelford, 1988). Transit times (i.e., the time for the exit side concentration to reach 50% of the inlet side concentration or a relative effluent concentration, $C/C_o = 0.5$) are plotted for purely advective (i.e., Darcian), purely diffusive, and combined advective-diffusive flow in Figure 1.13. These transit times are plotted as a function of hydraulic conductivity of a hypothetical, 3-foot (0.9-m) compacted clay liner with an average porosity of 0.50. The data in Figure 1.13 were generated assuming a hydraulic gradient of 1.33 (corresponding to one foot of leachate atop a saturated compacted clay liner) and a diffusion constant of 6.0×10^{-6} cm^2/sec. The latter is representative of solutes that diffuse through fine grained porous media and that do not partition or adsorb on the barrier solids, e.g., chloride ion.

Three important conclusions can be drawn from the curves in Figure 1.13:

(i) Diffusion shortens transit time even at hydraulic conductivities of 1.0×10^{-7} cm/sec—the usual target value specified in regulations for landfill liners.

(ii) The sole use of purely advective (Darcian) flow to predict transit or breakthrough times is extremely unconservative at hydraulic conductivities of less than 5.0×10^{-8} cm/sec.

(iii) Diffusion is the dominant transport process at hydraulic conductivities of less than 2.0×10^{-8} cm/sec (i.e., the advection-dispersion curve asymptotically approaches the pure diffusion line).

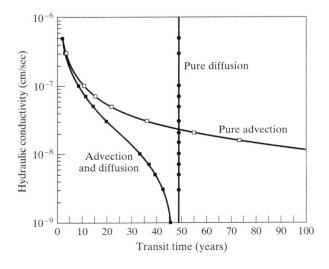

FIGURE 1.13 Transit Times (at $C/C_o = 0.5$) as A Function of Hydraulic Conductivity for Various Types of Flow (Shackelford, 1988)

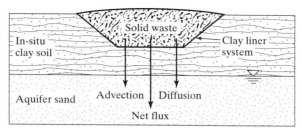

FIGURE 1.14 Schematic Diagram of Conventional Landfill Design Showing Advective and Diffusive Flow Acting in the Same Direction

Minimizing the head of leachate on the liner (and corresponding hydraulic gradient) by use of a primary leachate collection layer does not eliminate the chemical concentration difference (and resultant diffusion flux). Unfortunately, by requiring that landfills be located well above the groundwater table, we simply insure that advection and diffusion both work hand in glove with one another to enhance leakage as shown in Figure 1.14.

This dilemma imposes severe demands on the containment effectiveness of landfill liners. Several strategies can be invoked to minimize or counteract this difficulty (Gray, 1995). One possibility would be to use a controlled inward hydraulic gradient by actually setting the landfill below the groundwater table in combination with a peripheral cut-off wall and thus create a small inward head to counteract the diffusive efflux. Another possibility would be to improve the adsorption properties of clay barriers—particularly for low-molecular weight, nonpolar, non-ionic organic solutes—by adding suitable sorbents to a clay barrier, such as carbonaceous fly ash or finely ground organic shale (Gullick et al., 1997; Gray, 1995; Gray et al., 1991).

1.8 SCOPE AND ORGANIZATION OF BOOK

This book addresses landfill siting, design, and construction issues. It is broadly divided into two main parts. The first part deals primarily with the main components or elements of a landfill (viz., the liner system, leachate collection and removal system, gas collection and control system, and final cover system). The second part deals with landfill performance (viz., liner leakage rate monitoring, envelope durability, gas collection and drainage efficiency, settlement and stability analyses, and erosion control).

Landfill siting requirements are addressed in Chapter 2. Federal and state regulations that govern siting are described in addition to geoenvironmental requirements. Siting is an important consideration, but the emphasis of this book is primarily on construction and design of the landfill itself.

The characteristics of landfill containment envelopes and their design/construction are treated in detail next. The properties, advantages/limitations, and design requirements for compacted clay, geomembrane, and geosynthetic clay liners, respectively, are treated in Chapters 3 through 5. The advantages and limitations of geosynthetic clay liners (GCLs) relative to conventional compacted clay liners (CCLs) also are examined in Chapter 5. The equivalency of these two liner types is examined in terms of solute leakage across them under combined advection and diffusion.

The engineering properties of solid waste are described in Chapters 6. Landfill leachate characteristics, leachate drainage, and leachate collection and removal are treated in Chapters 7 through 9. The gas collection and control system and final cover system are described in Chapters 10 and 11. Methods of estimating landfill leachate quantities and gas generation in addition to the design of leachate and gas collection systems are described in detail in these chapters.

Methods for analyzing the settlements of solid waste and landfill foundation are addressed in Chapter 12. Several stability failures of large landfills have occurred during the past decade, such as the Kettlemen Hills Landfill failure in California, and the Rumpke Landfill failure in Ohio. These failures underscore the need for a careful understanding of stability and methods for conducting an adequate stability analysis. This topic is covered in Chapter 13 and includes a discussion of the different conditions that must be considered and evaluated as part of a landfill design.

A major trend in modern landfill design is the construction of fewer (but larger and higher) landfills. Another trend is to increase the capacity of existing landfills by filling on top, a process known as *vertical expansion*. The design of vertical expansion is far more complicated than normal or conventional landfill construction. This important topic is covered in Chapter 14, where the main technical details that must be considered by design engineers are explained there.

A landfill using the concept of greatly accelerating the degradation of municipal solid waste by adding liquid to maintain field capacity is known as a *bioreactor* landfill. This topic is covered in Chapter 15. The construction of landfill soil components—namely, liners, berms, drainage blankets, final cover, etc—are described in Chapter 16. The installation of geosynthetic components (e.g., membranes, geonets, geotextiles, etc.) are described in Chapter 17. Finally, several postclosure uses of municipal solid waste landfills are briefly described in Chapter 18.

PROBLEMS

1.1 Name and briefly describe various modes of land disposal?

1.2 What are the basic choices for handling and disposing of solid waste? What is the main choice for handling and disposing of solid waste in the United States, currently? Why?

1.3 Briefly describe the different common types of landfills.

1.4 List the main elements or systems comprising a modern municipal solid waste landfill.

1.5 What is the purpose and function of a landfill liner system?

1.6 What is the function of a landfill leachate collection and removal system?

1.7 What is the function of a landfill gas collection and control system?

1.8 What is the function of a landfill final cover system?

1.9 What is a composite liner system?

1.10 Describe the advantages of a composite liner versus a single compacted clay liner or a single geomembrane liner?

1.11 What is a double composite liner system? What are the benefits of a double composite liner system?

1.12 What is advection and diffusion? Under what conditions can diffusion become an important or even dominant leakage pathway?

1.13 Assume a clay liner with the following properties and conditions: thickness, $L = 0.6$ m, seepage velocity across clay liner, $v_s = 3.0 \times 10^{-7}$ cm/sec, and the effective diffusion coefficient of clay liner, $D = 6 \times 10^{-10}$ m²/sec $= 6 \times 10^{-6}$ cm²/sec,
  (1) determine the solute concentration ratio at $z = 0.1$ m, 0.2 m, 0.3 m, 0.4 m, 0.5 m, and 0.6 m at $t = 365$ days; and
  (2) determine the exit/source solute concentration ratio at the base of the liner at $t = 2$ years, 5 years, 10 years, 15 years, 20 years, and 50 years.

REFERENCES

Barone, F. S., Yanful, E. K., Quigley, R. M., and Rowe, R. K., (1989) "Effect of Multiple Contaminant Migration on Diffusion and Adsorption," *Canadian Geotechnical Journal,* Vol. 26, pp. 189–198.

Brandl, H., (1990) "Geomembranes for Vertical Waste Containment Sealing," *Proceedings 4th IGS Conference,* The Hague, The Netherlands, pp. 511–516.

Burson, B., Baker, A., Jones, B., and Shailer, J., (1997) "Developing and Installing a Vertical Containment System," *Geotechnical Fabrics Report,* Vol. 15, No. 3, pp. 39–45.

Crank, J., (1956) *The Mathematics of Diffusion,* Oxford University Press, Oxford England, 347 pages.

Crooks, V. E. and Quigley, R. M., (1984) "Saline Leachate Migration through Clay: A Comparative Laboratory and Field Investigation," *Canadian Geotechnical Journal,* Vol. 21, pp. 349–361.

Daniel, D. E., (1993) "Landfill and Impoundments," *Geotechnical Practice for Waste Disposal,* Chapter 5, Edited by David E. Daniel, Chapman & Hall, pp. 97–112.

Daniel, D. E. and Shackelford, C. D., (1989) "Containment of Landfill Leachate with Clay Liners," *Chapter 5.3, Sanitary Landfilling: Process, Technology and Environmental Impact,* T. H. Christensen, R. Cossu, and R. Stegmann, eds., Academic Press, London, UK, pp. 323–341.

Daniel, D. E. and Koerner, R. M. (2000) "On the Use of Geomembranes in Vertical Barriers," *Proceedings GeoDenver 2000,* Advances in Transportation and Geoenvironmental Systems Using Geosynthetics, J. G. Zornberg and B. R. Christoper Eds., GSP No. 103, GeoInstitute of ASCE, Reston, VA, pp. 81–93.

Freeze, R. A. and Cherry, J. A., (1979) *Groundwater,* Prentice-Hall Inc., Englewood Cliffs, N.J., 604 pages.

Gillham, R. W., Robin, M. J., and Dytynyshyn, D. J., (1984) "Diffusion of Non-Reactive and Reactive Solutes through Fine Grained Barrier Materials," *Canadian Geotechnical Journal*, Vol. 21, pp. 541–550.

Gullick, R. W., Weber, W. J. and Gray, D. H., (1996) "Organic Contaminant Transport through Clay Liners and Slurry Cutoff Walls," in Clay Minerals Society Lectures, Vol. 8, Organic Pollutants in the Environment, Sahwney, B., ed., The Clay Mineral Society, Boulder, CO, pp. 95–136.

Gray, D. H., (1995) "Strategies for Effective Containment of Landfilled Wastes," *Proceedings of GeoEnvironment 2000,* ASCE Geotechnical Special Publication No. 46, New Orleans, LA, U.S.A., February 24 to 26, pp. 384–498.

Gray, D. H., Mott, H. V., and Weber, W. J., (1991) "Fly Ash Utilization in Cut-Off Wall Backfill Mixes," *Proceedings, 9th International Ash Use Symposium,* EPRI GS-7162, Vol. II, Orlando, Florida, pp. 38:1–15.

Johnson, R. L., Cherry, J. A., and Pankow, J. F., (1989) "Diffusive Contaminant Transport in Natural Clay: A Field Example and Implications for Clay Lined Waste Disposal Sites," *Environ. Sci.Technology,* Vol. 23, No. 3, pp. 340–349.

Koerner, R. M., (1996) "Remediation vs. Containment of Solid/Liquid Waste Sites," Geosynthetics World, Vol. 6, No. 1, March/April, pp. 6–7.

Koerner, J. R. and Koerner, R. M., (1999) "A Survey of Solid Waste Landfill Liner and Cover Regulations: Part II–Worldwide Status," GRI Report #23, Geosynthetics Institute, Folsom, PA, 177 pages.

Manassero, M., Fratalocchi, E., Pasgualini, E., Spann, C., and Verga, F., (1995) "Containment with Vertical Cut-Off Walls," *Proceedings of Geoenvironment 2000,* Y.B. Acar and D.E. Daniel, Eds., ASCE Geotechnical Special Publication No. 46, pp. 1142–1172.

Ogata, A. and Banks, R. B., (1961) "A Solution to the Differential Equation of Longitudinal Dispersion in Porous Media," U.S. Geol. Survey Professional Paper No. 411A, D.C.

Repetto, P. C., (1995) "Geo-Environment," Section III Geotechnical Engineering, *The Civil Engineering Handbook,* Editor-of-Chief: W.F. Chen, CRC Press, Inc., 2000 Corporate Blvd., N. W., Boca Raton, FL 33431, pp. 883–902.

Robinson, R. A. and Stokes, R. H., (1959) "Electrolyte Solution," 2nd Edition, Butterworths Scientific Publications, London, England.

Rowe, R. K. and Booker, J. R., (1985) "1-D Pollutant Migration in Soils of Finite Depth," *Journal of Geotechnical Engineering,* ASCE, Vol. 111, No. 4, pp. 479–499.

Rowe, R. K., (1987) "Pollutant Transport through Barriers," Geotechnical Practice for Waste Disposal '87. Proceedings, Specialty Conference, Geotechnical Engineering Division (ASCE), Ann Arbor, MI, June 1987, Geotechnical Special Publication No. 13, ASCE, New York, NY, pp. 159–181.

Shackleford, C. D., (1988) "Diffusion as a Transport Process in Fine Grained Materials," *Geotechnical News,* Vol. 6, No. 2, pp. 24–27.

Shackelford, C. D. and Daniel, D. E., (1991) "Diffusion in Saturated Soil: I. Background," *Journal of Geotechnical Engineering,* ASCE, Vol. 117, No. 3, pp. 467–484.

Shackelford, C. D. and Redmond, P. L., (1995) "Solute Breakthrough Curves for Processed Kaolin at Low Flow Rates," *Journal of Geotechnical Engineering,* ASCE, Vol. 121, No. 1, pp. 17–32.

Sharma, H. D. and Lewis, S. P., (1994) *Waste Containment System, Waste Stabilization, and Landfills,* John Wiley & Sons, Somerset, NJ.

USEPA, (1984) "Slurry Trench Construction for Pollution Migration Control," EPA-54/2-84-001, U.S. Environmental Protection Agency, Office of Emergency and Remedial Response, Washington D.C. and Office of Research and Development, Municipal Environmental Research Laboratory, Cincinnat, OH, February.

USEPA, (1999) "Characterization of Municipal Solid Waste in the United States: 1998 Update," EPA530-R-99, U.S. Environmental Protection Agency, Municipal and Industrial Solid Waste Division, Office of Solid Waste, Prepared by Franklin Associates, Prairie Village, KS, July.

CHAPTER 2

Landfill Siting and Site Investigation

2.1 SITING CONSIDERATIONS
2.2 LOCATION RESTRICTIONS
    2.2.1 AIRPORT SAFETY
    2.2.2 FLOODPLAINS
    2.2.3 WETLANDS
    2.2.4 FAULT AREAS
    2.2.5 SEISMIC IMPACT ZONES
    2.2.6 UNSTABLE AREAS
2.3 SITING PROCESS
2.4 SITE INVESTIGATION
    2.4.1 PREINVESTIGATION STUDY
    2.4.2 FIELD EXPLORATION
2.5 BORROW SOURCE INVESTIGATION
    2.5.1 CLAY
    2.5.2 SAND
    2.5.3 GRAVEL
    2.5.4 SILTSOIL
    2.5.5 TOPSOIL
2.6 FIELD HYDRAULIC CONDUCTIVITY TESTS
    2.6.1 SEALED DOUBLE-RING INFILTROMETER
    2.6.2 TWO-STAGE BOREHOLE TEST
    2.6.3 COMPARISON OF METHODS
2.7 MATERIAL LABORATORY TESTS
    2.7.1 WATER CONTENT
    2.7.2 PARTICLE SIZE DISTRIBUTION
    2.7.3 MOISTURE DENSITY RELATIONSHIP
    2.7.4 HYDRAULIC CONDUCTIVITY
    2.7.5 SHEAR STRENGTH
    2.7.6 COMPRESSIBILITY
PROBLEMS
REFERENCES

Selection of a suitable site is crucial for a landfill project. Although the process of selecting a landfill site is complex and sometimes costly, it can greatly minimize the future impact on public health. Proper siting can also be economical to the extent it contributes to the reduction in design and/or construction costs, as well as in long-term expenses, such as those associated with operation, maintenance, and recovery of potential leachate releases.

2.1 SITING CONSIDERATIONS

Landfill siting is one of the most difficult and often controversial tasks faced by many communities when implementing an integrated solid waste disposal program. Factors that must be considered in evaluating potential sites for the long-term disposal of solid waste include haul distance, location restrictions, available land area, site access, soil conditions, topography, climatological conditions, surface water hydrology, geologic and hydrogeologic conditions, local environmental conditions, and potential end uses for the completed site, which are discussed as follows (Tchobanoglous et al., 1993):

1. Haul Distance. Haul distance is an important variable in the selection of a disposal site because it can significantly affect the overall design and operation of a waste management system. A minimum haul distance is desirable but other factors must be considered as well. Landfill siting is governed increasingly by environmental and political concerns, which of themselves may necessitate long-distance hauling. The development of intermediate transfer stations is a method whereby haul distance is made somewhat more economical by using very large trucks for the final transfer of waste.

2. Location Restrictions. Location restrictions may constrain landfill placement. These restrictions apply to siting landfills near airports, in floodplains, in wetlands, in areas with known faults, in seismic impact zones, and in unstable areas. Applicable federal requirements are contained in Subpart B—Location Restrictions of Part 258 of Subtitle D of the Resource Conservation and Recovery Act (RCRA) (USEPA, 1995). Many states have also adopted additional location restrictions. All current restrictions must be reviewed carefully during the preliminary siting process to avoid wasted time and expense evaluating a site that will not conform with regulatory requirements. Specific requirements that apply to landfill location restrictions will be discussed in the next section.

3. Available Land Area. There are no fixed rules with regard to required area for disposal; however, it is advisable to have enough area, including an adequate buffer zone, to operate a site for at least five years. Unit disposal costs become considerably more expensive for shorter periods, especially with respect to site preparation, provision of auxiliary facilities, such as platform scales, and completion of final cover. The extent of any waste diversion that is likely to occur in the future should be projected during initial assessment of a potential site. Furthermore, it is important to determine the impact of that diversion on the quantity and condition of the residual disposal.

4. Site Access. There has been a tendency for new landfill sites to increase in size (so-called "megafills") as the number of operational landfills continues to decrease. Unfortunately, it is often difficult to find large enough land areas near existing developed roadways and cities. As a result, the construction of access roadways and the use of long-haul equipment has become an important aspect of landfill siting. Rail lines often pass near remote sites that are suitable for use as landfills, and thus the use of rail haul for transporting wastes to these remote sites has attracted increased interest.

5. Soil Conditions and Topography. Solid wastes placed in a landfill must be covered each day, and a final soil cover layer or cap must be placed after the landfilling operation is terminated. If soil under the proposed landfill area itself is to be used for cover material, its geologic and hydrogeologic characteristics must be ascertained. On the other hand, if the cover material is to be obtained from a borrow pit, test borings will be needed to characterize this off-site material. The local topography must be considered as well, because it will affect the type of landfill operation that is adopted, the equipment requirements, and the extent of work necessary to make the site usable.

6. Climatologic Conditions. Local weather conditions can greatly affect the operation and management of a disposal site. In many locations, winter conditions will affect access to the site. Wet weather may necessitate the use of separate landfill areas. Where freezing is severe and excavation is impractical, landfill cover material must be stock-piled in advance. Wind strength and wind patterns also play critical roles; to avoid blowing or flying debris, adequate windbreaks must be established.

7. Surface Water Hydrology. Surface water hydrology refers to the pattern of natural drainage and runoff characteristics of a site. Potential flooding problems and the limits of the 100-year flood must be identified carefully. Mitigation measures may be required to divert surface runoff exiting the landfill away from lakes and streams. Conversely, measures may be required to limit storm water runoff flow into the landfill. Accordingly, planners must take great care in defining existing and intermittent flow channels in the area and the characteristics of the contributing watershed.

8. Geologic and Hydrogeologic Conditions. Geologic and hydrogeologic conditions are extremely important in establishing the environmental suitability of an area for a landfill site. The principal hydrogeologic siting requirements for a landfill are noted in Table 2.1. Hydrogeologic data are required to assess the pollution potential of

TABLE 2.1 Principal Hydrogeologic Landfill Siting Requirements

Considerations	Site Requirement Descriptions and Rationale
1	Significant Thickness of Vadose (Unsaturated) Zone beneath Landfill • Permits adsorption/attenuation of heavy metals and complex organic pollutants on clay solids and soil organic matter; • Allows biodegradation of organic substances by oxidation and bacteria in vadose zone; • Minimizes danger of hydraulic uplift and rupture during excavation.
2	Underlain by Strata with Low Hydraulic Conductivity • Strata with low hydraulic conductivity limit rate of downward and lateral movement of escaped leachate and possible contamination of underlying aquifer (if present).
3	Does *Not* Overlie Sole Source or Usable Aquifer • Even if landfill is separated from aquifer by tight cap layer with low hydraulic conductivity, it is prudent to avoid the site if it overlies a sole source or important fresh water aquifer.
4	Located outside Floodplain Areas • Limits storm water runoff (SWRO) to the landfill site.
5	Has Adequate Set Back from Populations, Lakes, Streams, and Wetlands • Limits impact of both SWRO and leakage from the landfill site.

a proposed site and to establish what must be done to the site to ensure that the movement of leachate or gases from the landfill will not impair the quality of local groundwater or contaminate underlying bedrock aquifers. U.S. Geological Survey maps and state or local geologic information is useful for the preliminary assessment of alternative sites. Geologic drilling logs of nearby wells can also be used for this purpose.

9. Local Environmental Conditions. Landfill sites have and can be built in close proximity to both residential and industrial developments. To do so, they must be planned and operated very carefully to be environmentally acceptable with respect to traffic, noise, odor, dust, airborne debris, visual impact, vector control, and surrounding property values. To minimize the impact of filling operations, landfills now tend to be sited in more remote locations where adequate buffer zones can be maintained.

10. Ultimate Use for Completed Landfills. A capped and decommissioned landfill represents a sizeable area of land that could be used for other purposes. More and more communities are now examining the use of old landfills for such things as, recreation sites, sport fields, and parking areas. Ideally, these potential end uses should be taken into account during the design, layout, and operation of a landfill. Choices for the ultimate use of completed landfills are becoming less limited by state and Federal regulations, providing that appropriate measures are taken dealing with landfill closure and postclosure maintenance. If a completed landfill is to be used for some municipal function (e.g., a toboggan hill), a staged landscaping and planting program should be initiated and continued as portions of the landfill are completed. (Chapter 18, the concluding chapter, describes this topic.)

2.2 LOCATION RESTRICTIONS

The RCRA Subtitle D rule (40 CFR Part 258) establishes six location restrictions applicable to municipal solid waste (MSW) landfills in order to protect human health and the environment. The six location restrictions are MSW landfills in the vicinity of airports and in 100-year floodplains, wetlands, fault areas, seismic impact zones, and unstable areas (USEPA, 1995). These criteria generally apply to other landfill wastes as well (e.g., hazardous, ash, and construction/demolition wastes).

2.2.1 Airport Safety

Owners or operators of MSW landfill units located within 10,000 feet (3,000 meters) of any airport runway end used by turbojet aircraft, or within 5,000 feet (1,500 meters) of any airport runway end used only by piston-type aircraft must demonstrate that the unit does not pose a bird hazard to aircraft. The owner or operator must notify the state director that the demonstration has been placed in the operating record.

In addition, the Subtitle D rule requires that owners or operators proposing new MSW landfill units within an 8-km- (5-mile)-radius of any airport runway end used by turbojet or piston-type aircraft must notify the affected airport and the appropriate Federal Aviation Administration (FAA) office.

2.2.2 Floodplains

Owners or operators of MSW landfill units located in 100-year floodplains must demonstrate that the unit will not restrict the flow of the 100-year flood, reduce the temporary water storage capacity of the floodplain, or result in washout of solid waste so as to pose a hazard to human health and the environment. The owner or operator must place the demonstration in the operating record and notify the state director that it has been placed in the operating record.

Floodplain means the lowland and relatively flat areas adjoining inland and coastal waters, including flood-prone areas of offshore islands that are inundated by the 100-year flood. The 100-year flood is a flood that has a 1-percent or greater chance of recurring in any given year, or a flood of a magnitude equaled or exceeded once in 100 years, on average, over a significantly long period.

2.2.3 Wetlands

New MSW landfill units must not be located in wetlands unless the owners or operators can make the following demonstrations to the director of an approved state:

1. The owner or operator must rebut the presumption that a practicable alternative to the proposed landfill is available that does not involve wetlands.
2. The owner or operator must show that the construction or operation of the landfill will not cause or contribute to violations of any applicable state water quality standard, violate any applicable toxic effluent standard or prohibition, jeopardize the continued existence of endangered or threatened species or critical habitats, or violate any requirement for the protection of a marine sanctuary.
3. The owner or operator must demonstrate that the MSW landfill unit will not cause or contribute to significant degradation of wetlands. To this end, the owner or operator must ensure the integrity of the MSW landfill unit, minimize impacts on fish, wildlife, and other aquatic resources and their habitat from release of the solid waste, and assure that the ecological resources in the wetland are sufficiently protected.
4. The owner or operator must demonstrate that steps have been taken in an attempt to achieve no net loss of wetlands by first avoiding impacts to wetlands to the maximum extent practicable, then minimizing unavoidable impacts to the maximum extent practicable, and finally offsetting remaining unavoidable wetland impacts through all appropriate and practicable compensatory mitigation actions.

2.2.4 Fault Areas

New MSW landfill units must not be located within 200 feet (60 meters) of a fault that has exhibited displacement in recent geological (Holocene) time, unless the owner or operator demonstrates to the director of an approved state that an alternative setback distance of less than 200 feet (60 meters) will prevent damage to the structural integrity of the MSW landfill unit and will be protective of human health and the environment.

Fault means a fracture or a zone of fractures in any material along which strata on one side have been displaced with respect to that on the other side. *Holocene* means the most recent epoch of the Quaternary period, extending from the end of the Pleistocene Epoch to the present.

2.2.5 Seismic Impact Zones

New MSW landfill units must not be located in a seismic impact zone, unless the owner or operator demonstrates to the director of an approved state/tribe that all containment structures—including liner systems, leachate collection and removal systems, gas collection and control systems, final cover systems, and surface water control systems—are designed to resist the maximum horizontal acceleration in lithified earth material for the site. The owner or operator must place the demonstration in the operating record and notify the state director that it has been placed in the operating record.

Seismic impact zone means an area with 10 percent or greater probability that the maximum horizontal acceleration in lithified earth material, expressed as a percentage of the earth's gravitational pull, will exceed 0.10g in 250 years. Figure 2.1 shows the seismic impact zones of the United States. Maximum horizontal acceleration in lithified earth material is the maximum expected horizontal acceleration depicted on a seismic hazard map, with a 90 percent or greater probability that the acceleration will not be exceeded in 250 years, or the maximum expected horizontal acceleration based on a site-specific seismic risk assessment. *Lithified earth material* means all rock, including all naturally occurring and naturally formed aggregates or masses of minerals or small particles of older rock that formed by crystallization of

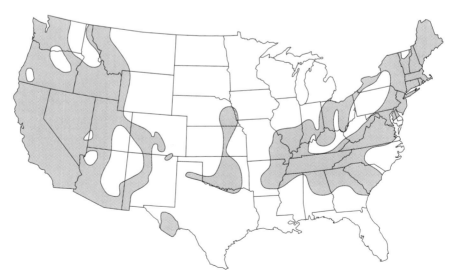

FIGURE 2.1 Seismic Impact Zones within the Continental United States (Area with a 10% or greater probability that the maximum horizontal acceleration will exceed 0.10g in 250 years) (USEPA, 1993a)

magma or by induration of loose sediments. This term does not include man-made materials, such as fill, concrete, and asphalt, or unconsolidated earth materials, soil, or regolith lying at or near the earth surface.

2.2.6 Unstable Areas

The owner or operator of MSW landfill units located in an unstable area must demonstrate to the satisfaction of the state director that engineering measures have been incorporated into the MSW landfill unit's design to ensure that the integrity of the structural components of the landfill unit will not be disrupted. The demonstration must show that the structural components of the MSW landfill can withstand the impacts of establishing events, such as landslides. The structural components include liner systems, leachate collection and removal systems, gas collection and control systems, final cover systems, run-on and runoff control systems, and any other component used in the construction and operation of the landfill unit that is necessary for protection of human health and the environment.

Unstable area means a location that is susceptible to natural or human-induced events or forces capable of impairing the integrity of some or all of the landfill structural components responsible for preventing releases from a landfill. Unstable areas can include those with poor foundation conditions, areas susceptible to mass movements, and Karst terrain. Areas susceptible to mass movement are those areas of influence (i.e., areas characterized as having an active or substantial possibility of mass movement) where the movement of earth material at, beneath, or adjacent to the landfill unit, because of natural or man-induced events, results in the downslope transport of soil and rock material by means of gravitational influence. Areas of mass movement include, but are not limited to, landslides, avalanches, debris slides and flows, solifluction, block sliding, and rock fall. *Karst terrain* refers to areas where karst topography, with its characteristic surface and subterranean features, is developed as the result of dissolution of limestone, dolomite, or other soluble rock. Characteristic physiographic features present in karst terrain include, but are not limited to, sinkholes, sinking streams, caves, large springs, and blind valleys.

The above six location restrictions constitute the basic constraints on landfill siting. Some states have proposed stricter locational criteria for landfill siting. For instance, some states require that landfills keep some specific distance away from water supply wells, lakes, and streams; a minimum clearance is also required between landfill bottom and groundwater level (MDEQ, 1999). These are site-specific situations that must be considered at the very earliest stages in the siting process.

2.3 SITING PROCESS

Three major issues tend to influence and affect landfill siting: environmental impact, economic consequences, and political considerations. Geotechnical and hydrogeological considerations can be subsumed within the environmental category. Public opinion and attitude strongly influence political considerations. To achieve public acceptance, citizen groups should participate in the identification and evaluation of prospective sites. Citizens' concerns about a particular site should not be dismissed out of hand.

The ultimate goal is to select a secure site that will provide the greatest public health and environmental protection in the event of a landfill containment failure.

Oweis and Khera (1998) have identified the following elements and stages of a landfill siting process:

  (i) Define project and its needs;
  (ii) Identify major environmental factors;
  (iii) Identify candidate sites;
  (iv) Collect and analyze environmental, economic, and socioeconomic data;
  (v) Evaluate and compare candidate sites;
  (vi) Screen candidate sites to a small number;
  (vii) Collect and assess site-specific engineering and environmental data;
  (viii) Recommend one or more sites for final selection; and
  (ix) Determine final location.

Oweis and Khera note that public participation is usually needed for some stages, especially (6), (8), and (9). The purpose of this staged inquiry is to narrow the scope of the study from a wide geographic area to one or more sites for detailed investigation and analysis.

According to Oweis and Khera (1998) the main factors to consider in assessing the suitability of a site are as follows:

Economic Factors:

  (i) Access to highways and available highway capacity,
  (ii) Compatibility with existing solid waste management systems,
  (iii) Cost of police, fire, and road maintenance,
  (iv) Development, operation, and maintenance cost,
  (v) Distance to waste generation locations,
  (vi) Economic effects on community,
  (vii) Effect on property value,
  (viii) Land development,
  (ix) Highly productive agricultural areas,
  (x) Flexibility.

Socioeconomic Factors:

  (i) Archaeological and historical sites,
  (ii) Cultural patterns,
  (iii) Dedicated land,
  (iv) Economic and community resources,
  (v) Emergency response,

- (vi) Land use and zoning,
- (vii) Noise impact,
- (viii) Proximity to school and residents,
- (ix) Public safety and health,
- (x) Sensitive receptors.

Environmental/Geotechnical Factors:

- (i) Aesthetic impact,
- (ii) Agriculture preservation areas,
- (iii) Air quality, gas compositions, and particulate matter,
- (iv) Areas with high groundwater level,
- (v) Climate and atmospheric conditions,
- (vi) Distance from water supply wells,
- (vii) Fault area,
- (viii) Flood plains and wetlands,
- (ix) Forest, wilderness, and scenic areas,
- (x) Geology.

Useful data sources are available in the United States to assist in siting landfills. U.S. Geological Survey (USGS) data include geological index maps, professional papers, water supply papers, topographic maps, and other data. U.S. Department of Agriculture (USDA) Natural Resources Conservation Service (NRCS) publishes surveys of surface soils on a county-by-county basis. Data such as pH, cation exchange capacity (CEC), seasonal water levels, etc., are provided for the counties surveyed. The NRCS survey maps also provide use-suitability information for each of the mapped soil series.

Other federal agencies, such as the Bureau of Sport, Fisheries & Wildlife, Bureau of Mines, Bureau of Indian Affairs, National Oceanic and Atmospheric Administration, Federal Aviation Administration, provide data and information that may be useful. Many state geological surveys also provide publications with excellent detailed geological maps covering local areas (Oweis and Khera, 1998).

2.4 SITE INVESTIGATION

The intent of a geotechnical and hydrogeological investigation at a proposed landfill site is to obtain data about the soil conditions at the site and to prepare a stratigraphic profile (or cross-section) and a groundwater map for the site. Surface and subsurface conditions of soils must be evaluated for design, operation and maintenance, and future utilization of a landfill facility. The techniques used to collect the required data include remote sensing, geophysical methods, test pits, test boring and penetrometer resistance, sampling of soils and rocks, and in-situ tests for measuring properties of soils, rocks and waste materials, and groundwater quality.

2.4.1 Preinvestigation Study

Available topographic and geological data, aerial photographs, site reconnaissance, and data from previous investigations should be reviewed during this phase of the investigation. Topographic and geological data, including water supply papers, are available from the United States Geologic Survey and each state's geological survey. Useful data on near-surface soils are available from the soil surveys published by the United States Department of Agriculture (USDA) and the Natural Resources Conservation Service (NRCS). Other useful data are sometimes available from highway departments, universities, and public libraries.

Conventional topographic and geological maps at scales of 1:62,500 or smaller are useful for initial screening and site selection. Their usefulness is more limited, however, for detailed site evaluation because they lack definition for geotechnical purposes. Greater detail can be gleaned from black-and-white or color aerial photographs, which are the most common type of remote sensing technique. An experienced air photo interpreter can extract valuable information about a site, such as depth to rock, drainage patterns, site morphology, and surface soil type (Way, 1973). Aerial photographs are usually available in nine-inch frames in scales of 1:12,000 to 1:80,000. Other types of imagery (NAVFAC DM 7.1, 1982) include Skylab, NASA, SLAK, and thermal infrared. These are generally used for more detailed investigations, such as fault studies and water resources planning (Oweis and Khera, 1998).

The outcome of the preinvestigation study usually reveals the geomorphology and broad geological features at the site. It also helps determine the scope of the field investigation, such as the number of borings needed, where they should be located, and their required depth.

2.4.2 Field Exploration

The aerial extent and depth of the preliminary field investigation partially depends on the nature of the geological details identified in the initial phase or map study. However, for a new land disposal facility (a so-called "greenfield" site), the minimum scope of the investigation is often dictated by regulations such as the minimum number of borings shown in Table 2.2, which are recommended by the New Jersey Department of Environmental Protection.

TABLE 2.2 Minimum Number of Borings

Acreage		Total Number of Borings	Number of Deep Borings
acres	hectares		
Less than 10	Less than 4	4	1
10 ~ 49	4 ~ 19	8	2
50 ~ 99	20 ~ 39	14	4
100 ~ 200	40 ~ 80	20	5
More than 200	More than 80	24 plus a boring each additional 10 acres (4 hectares)	6 plus 1 boring each additional 10 acres (4 hectares)

In the absence of specific requirements regarding the number of borings, Bagchi (1994) recommends the following guidelines:

(i) The borings should be distributed in such a way that they cover an area at least 25% larger than the proposed waste limits.
(ii) Five borings should be done for the first hectare (ha) (2.5 acres) or less and two additional borings for each additional hectare. The borings should be well distributed over the entire area.
(iii) The borings should extend at least 25 feet (7.5 m) below the proposed base of the landfill. Some selected borings should be drilled to bedrock to verify the depth of bedrock.

With regard to landfill design, soil borings help to identify soil types, bedrock depth, and the depth/thickness of usable groundwater aquifers. One or two borings should extend at least 6 feet (2 m) into the aquifer/bedrock for verification purposes. For sites with very deep bedrock or aquifers, regional geologic data may be used to assess the bedrock and water table depth.

Knowledge of local geology and hydrogeology is essential for proper planning of a soil boring program. The recommendations in Table 2.2 provide a guideline only. The exact number of boreholes and groundwater table wells may far exceed the suggested minimum number. The reverse may also be true in some cases—that is, fewer borings may be sufficient to define the soil stratigraphy of a site with uniform conditions. Mobilization costs for bringing a drill rig on site can be minimized by scheduling all the necessary borings while the rig is present on site.

During drilling, split-spoon soil samples are generally collected, either at 5-foot- (1.5-m)-intervals or continuously sampled, to determine lithologies encountered during drilling. Split-spoon soil samples are collected following the Standard Penetration Test procedure, ASTM D1586. A 1-3/8-inch- (35-mm)-I. D. split spoon is driven into the soil with a 140-pound- (600 N)-hammer falling freely for a distance of 30 inches (750 mm). The sampler is generally driven at three successive 6-inch (150-mm) increments, with a record made of the number of blows for each increment. The number of blows required to advance the sampler the last 12 inches (300 mm) is termed the Standard Penetration Resistance (N). In instances of split-spoon sampling for 24 inches (600 mm), the second and third increments are used to calculate N. Where very hard subsurface materials are encountered, the split spoon is driven a minimum of 50 blows and to a distance that would result in sample recovery. Soil samples are recovered directly from the split-spoon sampler.

Samples from borings can be broadly grouped as *disturbed* or *undisturbed* samples. Disturbed samples are usually adequate for classification purposes, physical property testing such as water content and plasticity characteristics (liquid limit and plastic limit), and chemical testing (chemical corrosivity and contamination). Disturbed samples are usually recovered from test borings at typically 5-ft (1.5 m) intervals using a standard split barrel (ASTM D1586) or a split-spoon sampler.

Undisturbed testing preserves the texture and structure of the in-situ soil. This texture significantly affects certain engineering properties. Undisturbed samples are required, for example, to conduct laboratory strength, permeability, and compressibility testing, Shelby tube samples can be obtained using techniques that minimize sample disturbance (NAVFAC DM 7.1, ASTM D1587).

Test pits may be used to visually examine soil strata or waste materials at shallow depths. Borings for sites where foundation instability is not a concern should extend to a minimum depth of 50 ft (15m) beneath the base of the disposal facility. Deeper borings are required to obtain data for preliminary stability assessment. The site investigation should define the underlying stratigraphy, groundwater flow regime, and general site characteristics. Sufficient data must be collected to develop a preliminary design for submission to appropriate regulatory agencies (Oweis and Khera, 1998).

At the completion of drilling activities, all borings not being used for monitoring well or piezometer installation are grouted throughout their lengths. A grout of Portland cement mixed with bentonite (2% by weight) is tremied into the borehole as the hollow-stem auger is removed. The auger acts as a surface casing, preventing the cave-in of any upper water-bearing formations and/or noncohesive soils.

At least one water observation well should be installed at the beginning of the subsurface exploration program. Groundwater levels are monitored periodically (some state regulations require at least a one-year duration). The elevation of the highest groundwater level must be considered with respect to the lowest proposed elevation of the landfill. Many state regulations call for a minimum 3-m- (10 ft)-separation distance.

2.5 BORROW SOURCE INVESTIGATION

A borrow source investigation may be required to locate suitable materials for both the liner and cap. If the specified liner material is a type of clay and the drainage blanket is clean sand, then identification of acceptable sources for these two materials is critical to the construction of the landfill. Local geologic maps can be used to identify potential soil borrow sources. The next step is a detailed investigation of the properties of these materials at candidate borrow areas.

In order to determine the properties of the borrow soil, samples are often obtained from the potential borrow area for laboratory analysis prior to actual excavation. Samples may be obtained in several ways. One method of sampling is to recover samples of soil from borings. This procedure can be very effective in identifying major strata and substrata within the borrow area. Small samples obtained from the borings are excellent for index property testing, but often do not provide a very good indication of subtle stratigraphic changes in the borrow area. Test pits excavated into the borrow soil with a backhoe, front-end loader, or other excavation equipment can expose a large cross-section of the borrow soil. One can obtain a much better idea of the variability of soil in the potential borrow area by examining exposed cuts rather than viewing small soil samples obtained from borings.

Large bulk samples of soil are required for compaction testing in the laboratory. Small samples of soil taken with soil sampling devices do not provide a sufficient volume of soil for laboratory compaction testing. Some engineers combine samples of soil taken at different depths or from different borings to produce a composite sample of adequate volume. This technique is not recommended, because a degree of mixing takes place in forming the composite laboratory test sample that would not take place in the field. Other engineers prefer to collect material from auger borings for use in performing laboratory compaction tests. This technique is likewise not recommended without careful borrow pit control, because vertical mixing of material takes place during auguring in a way that would not be expected to occur in the field unless controlled vertical cuts are made. The best method for obtaining large bulk samples of material for laboratory compaction testing is to take a large sample of material from one location in the borrow source. A large bulk sample can be taken from the wall or floor of a test pit that has been excavated into the borrow area. Alternatively, a large piece of drilling equipment such as a bucket auger can be used to obtain a large volume of soil from a discreet point in the ground (USEPA, 1993b).

Bagchi (1994) has suggested guidelines for examining each major type or classification group of geologic material used in constructing a landfill; the subsections that follow describe these guidelines.

2.5.1 Clay

Clay may be used as a primary or secondary liner material. Test pits and borings are used to find the vertical and horizontal extent of a potential clay borrow source. The locations of the test pits or borings should be well distributed on a uniform grid pattern. Logs should identify the geologic origin, testing results, soil classification, and visual description of each major soil layer. The layer or layers of soil should be identified based on test pit logs and boring logs. Avoid the use of a clay layer for liner construction that is less than 5 feet (1.5 m) in thickness. Bagchi (1994) furthermore recommends obtaining grain size distribution curves and Atterberg limits for at least two to three samples from the candidate clay layers.

When compacted in the field, clay should provide a low permeability (1.0×10^{-7} cm/sec or less) layer. Bagchi (1994) recommends the following specifications for identifying a suitable clay borrow: liquid limit between 20 and 30%; plasticity index between 10 and 20%; 0.074-mm-or-less fraction (P#200) content: 50% or more; clay fraction (0.002 mm or less size): 25% or more. Grain size distribution also significantly influences the permeability of soil. Accordingly, a soil that has a classical *inverted S-type* particle size distribution can be compacted relatively easily to achieve low permeability. On the other hand, a soil with a very high clay content, but with poor particle size distribution, may not be as readily compacted to achieve low permeability.

Representative samples should be used to develop Modified Proctor compaction curves. Each Proctor curve should be developed based on a minimum of five points. It is essential to study the relationship between compactive effort, molding moisture content, and recompacted permeability of the clay layers. Published findings (Mitchell, 1976) indicate that a soil compacted wet of optimum moisture will develop lower permeability than the same soil compacted dry of optimum. Compaction at wet of optimum moisture provides a better kneading action in the field, which in turn results in a soil fabric with a more uniform pore size distribution and lower hydraulic conductivity.

2.5.2 Sand

Sand has large enough particles to make it suitable for use as a drainage blanket and as a protective layer over a clay cap. A drainage layer should have a permeability of at least 10^{-2}-cm/sec. Sand containing no more than 5% of minus 0.074-mm-size fraction (P#200) material usually exhibits a relatively high permeability. Coarse sand can be washed free of fines to satisfy the permeability criteria.

The number of test pits and borings used to identify a clay borrow source is also applicable for sand borrow source identification. Borrowing from a sand layer less than 5 feet (1.5 m) thick should be avoided. Two or three samples from each test pit/boring should be collected and tested individually for grain size distribution. The permeability of samples compacted or re-constituted to 80 to 90% relative density should be determined as well.

2.5.3 Gravel

Gravel is used for leachate collection systems where rapid transmission is required (e.g., in bioreactor landfills). Gravel is also used wherever the hydraulic head in the liner system must be minimized. Due to relatively strict regulations against river dredging, most gravel is crushed rock available from quarries.

The type of rock can be important insofar as the generation of fines and long-term durability is concerned. Thus, the L.A. Abrasion test (ASTM C33) is generally used to assess rock soundness; durability is determined by repeated immersion in a sodium or magnesium sulfate solution. Maximum allowable losses recommended by the Corps of Engineers are as follows:

$$\text{Abrasion} \leq 50\% \text{ loss}$$
$$\text{Durability} \leq 18\% \text{ loss}$$

The latter is particularly important to avoid a long-term bonding of the gravel particles together, forming what some have termed *bio-rock*.

2.5.4 Silty Soil

A silty soil layer is required over the barrier (clay) layer in the landfill final cover to protect the barrier from freeze-thaw and desiccation. This layer is sometimes referred to as the *thermal* protection layer. Freeze-thaw cycles and desiccation cracks can increase the permeability of compacted clay by several orders of magnitude. The number of test pits and borings necessary to identify a suitable silty soil borrow source are also similar to those for clay and sand. Soils classified as silt loams under the USDA or textural classification system usually make suitable thermal protection layers.

2.5.5 Topsoil

Topsoil, sometimes referred to as the *vegetative support* layer, is used as the final cover of a landfill. The topsoil must be capable of supporting grass or herbaceous vegetation. Often, the area stripped for the construction of a landfill has suitable topsoil that can be collected and stockpiled for future use. If there is insufficient topsoil at the site, additional borrow sources should be identified. Local agricultural experts/horticulturists can provide advice about likely sources. It may be necessary to amend a prospective topsoil to take care of soil nutrient deficiencies or to adjust pH before a seed mix can be planted. Although detailed nutritional testing is not necessary at the feasibility stage, it is essential that a backup or supplementary source be identified if the need arises.

2.6 FIELD HYDRAULIC CONDUCTIVITY TESTS

The most important single parameter for compacted clay liners for waste facilities is *hydraulic conductivity*. Hydraulic conductivity should be measured on representative compacted clay samples or test pads. In typical practice, the vertical conductivity normally governs the containment and isolation effectiveness of a barrier. Regulatory agencies are increasingly requiring in-situ tests in addition to laboratory tests to verify hydraulic conductivity. The purpose of the field hydraulic conductivity tests is to measure the hydraulic conductivity of in-situ materials. The Sealed Double-Ring Infiltrometer (SDRI) and the Two-Stage Borehole (TSB) test have received the widest acceptance and use in field hydraulic conductivity testing.

2.6.1 Sealed Double-Ring Infiltrometer

The SDRI is the most common method of measuring field hydraulic conductivity on test pads. A schematic diagram of the SDRI is shown in Figure 2.2. The SDRI was developed specifically for soils with infiltration rates of less than 1.0×10^{-6} cm/sec (Daniel and Trautwein, 1986). This procedure has been formalized as ASTM Standard Test Method D5093. A double ring design, incorporating large moderately deep rings (12 ft and 5 ft or 3.7 m and 1.5 m), is used to overcome the problem of lateral flow. Water flow is measured by connecting a flexible bag filled with a known weight of water to a port on the inner ring. Water infiltrates downward, and after a known interval of time, the bag is removed and weighed. The weight loss, converted to a volume, is equal to the amount of water that has infiltrated the ground. An infiltration rate is

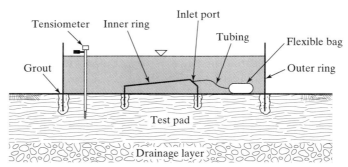

FIGURE 2.2 Schematic Diagram of Sealed Double-Ring Infiltrometer (USEPA, 1993b)

then determined from this volume of water, the area of the inner ring, and the interval of time.

Evaporation effects are eliminated with the SDRI device because the inner ring is sealed. The head at any elevation in the inner or outer ring is the same, and thus there is no gradient to cause water to flow in the lateral direction from one ring to the other. Since the head in the inner and outer rings is always the same, the pressure difference across the wall of the inner ring is constant; hence, the inner ring will not expand or contract, even though the water level in the outer ring may change (Trautwein and Boutwell, 1994).

With this method, the quantity of water that flows into the test pad over a known period of time is measured. This flow rate, which is called the infiltration rate I, is computed as:

$$I = Q/(A \cdot t) \tag{2.1}$$

where Q is the quantity of water entering the surface of the soil through a cross-sectional area A and over a period of time t.

Hydraulic conductivity k is computed from the infiltration rate and hydraulic gradient i as

$$k = I/i \tag{2.2}$$

Three procedures have been used to compute the hydraulic gradient: (1) apparent gradient method, (2) wetting front method, and (3) suction head method. The equation for computing hydraulic gradient from each method is shown in Figure 2.3.

The *apparent gradient method* (Figure 2.3a) is the most conservative of the three methods, because this method yields the lowest estimate of the hydraulic gradient i and, therefore, the highest estimate of hydraulic conductivity k. The apparent gradient method assumes that the test pad is fully soaked with water over the entire depth of the test pad. For relatively permeable test pads, the assumption of full soaking is reasonable, but for compacted clay liners with $k < 1.0 \times 10^{-7}$ cm/s, the assumption of full soaking is excessively conservative and should not be used unless verified.

The second and most widely used method is the *wetting front method*. The wetting front is assumed to partly penetrate the test pad (Figure 2.3b), and the water pressure at the wetting front is conservatively assumed to equal atmospheric pressure. Tensiometers are used to monitor the depth of wetting of the soil over time, and the variation of water content with depth is determined at the end of the test. The wetting front method is conservative, but in most cases not excessively so. The wetting front method is the method that is usually recommended.

The third method, called the *suction head method*, is the same as the wetting front method except that the water pressure at the wetting front is not assumed to be atmospheric pressure. The suction head (which is defined as the negative of the pressure head) at the wetting front is H_S and is added to the static head of water in the infiltration ring to calculate hydraulic gradient (Figure 2.3c). The suction head H_S is identical to the wetting front suction head employed in analyzing water infiltration with the Green-Ampt theory. The suction head H_S is not the ambient suction head in the unsaturated soil and is generally very difficult to determine (Brakensiek, 1977). Following are two techniques available for determining H_S:

FIGURE 2.3 Three Procedures for Computing Hydraulic Gradient from Sealed Double-Ring Infiltrometer Test (USEPA, 1993b)

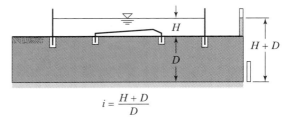

$$i = \frac{H + D}{D}$$

(a) Apparent hydraulic conductivity method

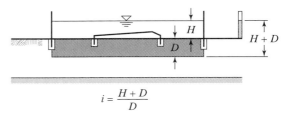

$$i = \frac{H + D}{D}$$

(b) Wetting front method

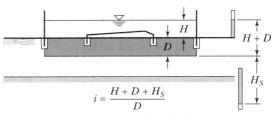

$$i = \frac{H + D + H_S}{D}$$

(c) Suction head method

(i) Integration of the hydraulic conductivity function (Neuman, 1976), viz.,

$$H_s = \int_{h_{sc}}^{0} k_r \cdot dh_s \tag{2.3}$$

where h_{sc} is the suction head at the initial (presoaked) water content of the soil, k_r is the relative hydraulic conductivity (k at particular suction divided by the value of k at full saturation), and h_s is suction.

(ii) Direct measurement with air entry permeameter (Daniel, 1989, and references therein).

Reinbold (1988) found that H_S was close to zero for two compacted soil liner materials. Because proper determination of H_S is very difficult, the suction head method cannot be recommended, unless the testing personnel take the time and make the effort to determine H_S properly and reliably.

Corrections may be made to account for various factors. For example, if the soil swells, some of the water that infiltrated into the soil was absorbed into the expanded soil. No consensus exists on various corrections and these should be evaluated case by case (USEPA, 1993b).

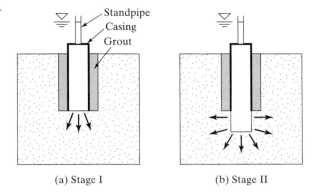

FIGURE 2.4 Schematic Diagram of Two-Stage Borehole (or Boutwell) Test (USEPA, 1993b)

(a) Stage I

(b) Stage II

2.6.2 Two-Stage Borehole Test

The two-stage borehole (TSB) hydraulic conductivity was developed by Boutwell (Boutwell and Tsai, 1992). The test is also called the *Boutwell Test*. The device is installed by drilling a hole (which is typically 100 to 150 mm in diameter), placing a casing in the hole, and sealing the annular space between the casing and borehole with grout as shown in Figure 2.4(a). A series of falling head tests is performed and the hydraulic conductivity from this first stage (k_1) is computed. Stage one is complete when k_1 ceases to change significantly. The maximum vertical hydraulic conductivity may be computed by assuming that the vertical hydraulic conductivity is equal to k_1. However, the test may be continued for a second stage by removing the top of the casing and extending the hole below the casing as shown as in the Figure 2.4(b). The casing is reassembled, the device is again filled with water, and falling head tests are performed to determine the hydraulic conductivity from stage two (k_2). Both horizontal and vertical hydraulic conductivity may be computed from the values of k_1 and k_2. Further details on methods of calculation are provided by Boutwell and Tsai (1992).

The two-stage borehole test permeates a smaller volume of soil than the sealed double-ring infiltrometer. The required number of two-stage borehole tests per pad is a subject of current research. At the present time, it is recommended that at least five two-stage borehole tests be performed on a test pad if the two-stage test is used. If five two-stage borehole tests are performed, then one would expect that all five of the measured vertical hydraulic conductivity would be less than or equal to the required maximum hydraulic conductivity for the soil liner (USEPA, 1993b).

2.6.3 Comparison of Methods

An advantage of the SDRI device is the ability to test large volumes. The TSB device, on the other hand, is able to determine both vertical and horizontal hydraulic conductivities. The SDRI is preferable for testing thin layers, but the TSB is better for detecting variations in hydraulic conductivity with depth. The SDRI test is considerably more costly and requires longer testing times. Testing times to reach equilibrium field hydraulic conductivity range from one to six months for the SDRI versus one to six weeks for the TSB. Trautwein and Boutwell (1994) recommend that at least one SDRI test or five TSB tests be performed to determine field hydraulic conductivity.

2.7 MATERIAL LABORATORY TESTS

Samples of soil must be taken for laboratory testing to ensure conformance with specifications for parameters such as percentage fines and the plasticity index. The material tests that are normally performed are water content, Atterberg limits, particle size distribution, moisture density relationship, and hydraulic conductivity. Shear strength and consolidation tests are also needed for the on-site materials. Each of these tests is discussed in the subsections that follow (USEPA, 1993b):

2.7.1 Water Content

It is important to know the water content of the borrow soils so that the need for wetting or drying the soil prior to compaction can be identified. The water content of the borrow soil is normally measured following the procedures outlined in ASTM D2216 if one can wait overnight for results. Alternatively, moisture content can be determined rapidly by a microwave drying technique per ASTM specifications if a suitable calibration is determined beforehand.

Construction specifications for compacted soil liners often require a minimum value for the liquid limit and/or plasticity index of the soil. These parameters are measured in the laboratory with the procedures outlined in ASTM D4318.

2.7.2 Particle Size Distribution

Construction specifications for soil liners often place limits on the minimum percentage of fines, the maximum percentage of gravel, and in some cases the minimum percentage of clay. Particle size analysis is performed following the procedures in ASTM D422. Normally, the requirements for the soil material are explicitly stated in the construction specifications. An experienced inspector can often judge the percentage of fine material and the percentage of sand or gravel in the soil. However, compliance with specifications is best documented by laboratory testing.

2.7.3 Moisture-Density Relationship

Moisture-density relationships (or compaction curves) are developed utilizing the method of laboratory compaction testing required in soil liner construction specifications. Standard compaction (ASTM D698) and modified compaction (ASTM D1557) are two common methods of laboratory compaction specified for soil liners. (Additional details will be provided in Chapter 16.) Other compaction methods (particularly those unique to state highway or transportation departments) are sometimes specified as well.

Great care should be taken to follow the procedures for soil preparation outlined in the relevant test method. In particular, the drying of a cohesive material can change the Atterberg limits as well as the compaction characteristics of the soil. If the test procedure recommends that the soil not be dried, the soil should not be dried. Also, care must be taken when sieving the soil not to remove clods of cohesive material. Rather, clods of soil retained on a sieve should be broken apart by hand if necessary to cause them to pass through the openings of the sieve. Sieves should only by used to remove stones or other large pieces of material following ASTM procedures.

2.7.4 Hydraulic Conductivity

The hydraulic conductivity of compacted samples of borrow material may be measured periodically to verify that the soil liner material can be compacted to achieve the required low hydraulic conductivity. Several methods of laboratory permeation are available, and others are under development. ASTM D5084 is a widely used procedure currently available. Care should be taken not to apply excessive effective confining stress to test specimens. If no value is specified in the testing plan, a maximum effective stress of 5 lb/in^2 (35 kPa) is recommended for both liner and cover systems.

Care should be taken to properly prepare specimens for hydraulic conductivity testing. In addition to water content and dry unit weight, the method of compaction and the compactive energy can have a significant influence on the hydraulic conductivity of laboratory-compacted soils. It is particularly important not to deliver too much competitive energy to attain a desired dry unit weight. The purpose of the hydraulic conductivity test is to verify that borrow soils can be compacted to the desired hydraulic conductivity using a reasonable compactive energy.

No ASTM compaction method exists for preparation of hydraulic conductivity test specimens. The following procedures are recommended (USEPA, 1993b):

1. Obtain a large, bulk sample of representative material with a mass of approximately 44 lb (20 kg).
2. Develop a laboratory compaction curve using the procedure specified in the construction specifications for compaction control (e.g., ASTM D698 or ASTM D1557).
3. Determine the target water content w_{target} and dry unit weight $(\gamma_d)_{target}$ for the hydraulic conductivity test specimen. The value of w_{target} is normally the lowest acceptable water content and $(\gamma_d)_{target}$ is normally the minimum acceptable dry unit weight (Figure 2.5).
4. Enough soil to make several test specimens is mixed to w_{target}. The compaction procedure used in Step 2 is used to prepare a compacted specimen, except that the energy of compaction is reduced (e.g., by reducing the number of drops of the ram per lift). The dry unit weight γ_d is determined. If $\gamma_d \approx (\gamma_d)_{target}$, the compacted specimen may be used for hydraulic conductivity testing. If $\gamma_d \neq (\gamma_d)_{target}$, then another test specimen is prepared with a larger or smaller (as appropriate) compactive energy. Trial-and-error preparation of test specimens is repeated until $\gamma_d \approx (\gamma_d)_{target}$. The procedure is illustrated in Figure 2.5. The actual compactive effort should be documented along with hydraulic conductivity.
5. Atterberg limits and percentage fines should be determined for each bulk sample. Water content and dry density should be reported for each compacted specimen.

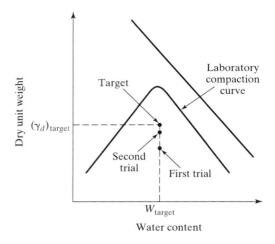

FIGURE 2.5 Recommended Procedure for Preparation of a Compaction Test Specimen Using Variable Compactive Energy for Each Trial (USEPA, 1993b)

2.7.5 Shear Strength

Strength tests are performed on Shelby tubes of cohesive soils collected from the proposed excavation area of the landfill. Strength tests are conducted to develop shear strength parameters to be used in the sideslope and foundation stability analyses. These shear strength tests consist of unconfined compression tests and consolidated-undrained triaxial tests with pore pressure measurements. The results of the unconfined compression tests are used to develop total-strength parameters for the short-term or undrained condition, which is representative of the conditions immediately after excavation. These tests are performed in accordance with ASTM D2166. The results of the consolidated-undrained triaxial tests with pore pressure measurements are used to develop effective-strength parameters for the long-term or drained condition. These tests are performed in accordance with ASTM D2850.

2.7.6 Compressibility

Consolidation tests should be performed on undisturbed samples. These tests are conducted to obtain compression parameters of the foundation soils for landfill foundation settlement analysis. These analyses are performed in accordance with ASTM D2435.

Recommended minimum values for the frequency of testing are shown in Table 2.3. The tests listed in Table 2.3 are normally performed prior to construction as part of the characterization of the borrow source. However, if time or circumstances do not permit characterization of the borrow source prior to construction, the samples for testing are obtained during excavation or delivery of the soil materials.

The common soil laboratory and field tests used for landfill site investigation, design and construction are listed as follows:

Unified Soil Classification System (ASTM D2487)
AASHTO Soil Classification System (ASTM D3282)

TABLE 2.3 Recommended Minimum Testing Frequencies for Investigation of Borrow Source (USEPA, 1993b)

Parameter	Frequency
Water Content	1 Test per 2,000 m^3 or Each Change in Material Type
Atterberg Limits	1 Test per 5,000 m^3 or Each Change in Material Type
Percentage Fines	1 Test per 5,000 m^3 or Each Change in Material Type
Percentage Gravel	1 Test per 5,000 m^3 or Each Change in Material Type
Compaction Curve	1 Test per 5,000 m^3 or Each Change in Material Type
Hydraulic Conductivity	1 Test per 10,000 m^3 or Each Change in Material Type

Note: 1 yard^3 = 0.76 m^3

Moisture Content (ASTM D2216)
Specific Gravity (ASTM D854)
Atterberg Limits (Plastic and Liquid Limits) (ASTM D4318)
Sieve Analysis (ASTM D421)
Hydrometer Analysis (ASTM D422)
Shrinkage Limit (ASTM D427)
Standard Proctor Compaction (ASTM D698)
Modified Proctor Compaction (ASTM D1557)
Permeability of Granular Soil (ASTM D2434)
Consolidation (ASTM D2435)
Unconfined Compression (ASTM D2166)
Direct Shear (Granular Soil) (ASTM D3080)
Triaxial Compression (ASTM D2850)
Flexible Wall Permeability Test (ASTM D5084)
Sealed Double-Ring Infiltrometer (ASTM D5093)
Field Density by Nuclear Method (ASTM D3017)
Field Density by Sand Cone (ASTM D1556)
Field Density by Rubber Balloon (ASTM D2167)

PROBLEMS

2.1 What factors must be considered in evaluating potential landfill sites?

2.2 Describe what locational criteria or restrictions must be considered for landfill siting according to the RCRA Subtitle D rules.

2.3 List important geologic and hydrogeologic conditions that must be considered when selecting a landfill site.

2.4 Explain what the 100-year floodplain means.

2.5 What is the definition of a seismic impact zone for landfill design in the RCRA Subtitle D rules?

2.6 What are three major issues (apart from regulatory and hydrogeologic requirements) that affect landfill siting?

2.7 The area of a proposed landfill is 372 acres (149 hectares). According to the two methods presented in Section 2.4.2, what is the minimum number of exploratory borings required?

2.8 How should all borings be treated after the drilling activities are completed? Why?

2.9 Explain why a soil with an *inverted S-type* particle size distribution can be easily compacted to achieve low hydraulic conductivity.

2.10 What comparisons can be drawn between a sealed double-ring infiltrometer (SDRI) and a two-stage borehole (TSB) test?

2.11 What common soil laboratory tests should be conducted as part of a landfill site investigation?

REFERENCES

Bagchi, A., (1994) "Design, Construction, and Monitoring of Sanitary Landfill," 2nd Edition, John Wiley & Sons, Inc., 605 Third Avenue, New York, NY 10158-0012.

Boutwell, G. P. and Tsai, C. N., (1992) "The Two-Stage Field Permeability Test for Clay Liners," *Geotechnical News,* Vol. 10, No. 2, pp. 32–34.

Brakensiek, D. L., (1977) "Estimating the Effective Capillary Pressure in the Green and Ampt Infiltration Equation," *Water Resources Research,* Vol. 12, Vol. 3, pp. 680–681.

Daniel, D. E., (1989) "In-Situ Hydraulic Conductivity Tests for Compacted Clay," *Journal of Geotechnical Engineering,* ASCE, Vol. 115, No. 9, pp. 1205–1226.

Daniel, D. E. and Trautwein, S. J., (1986) "Field Permeability Test for Earthen Liner," *Proceedings, In-Situ '86, ASCE Specialty Conference on Use of In-Situ Tests in Geotechnical Engineering,* Virginia Polytechnic Institute and State University, Blacksburg, VA, ASCE New York, pp. 146–160.

MDEQ, (1999) "Solid Waste Management Act Administrative Rules Promulgated Pursuant to Part 115 of the Natural Resources and Environmental Protection Act, 1994 PA 451, as amended (Effective April 12, 1999)," Michigan Department of Environmental Quality, Waste Management Division, Lansing, MI.

Mitchell, J. K., (1976) "Fundamentals of Soil Behavior," John Wiley & Sons, Inc., New York, NY.

Neuman, S. P., (1976) "Wetting Front Pressure Head in the Infiltration Model of Green and Ampt," *Water Resources Research,* Vol. 11, No. 3, pp. 564–565.

NAVFAC DM 7.1, (1982) "Soil Mechanics," Chapter 6, Naval Facilities Engineering Command, Alexandria, VA.

Oweis, I. S. and Khera, R. P., (1998) *Geotechnology of Waste Management,* 2nd Edition, PWS Publishing Company, 20 Park Plaza, Boston, MA 02116-4324.

Reinbold, M. W., (1988) "An Evaluation of Models for Predicting Infiltration Rates in Unsaturated Compacted Clay Soil," M. S. Thesis, The University of Texas at Austin, Austin, TX.

Tchobanoglous, T., Theisen, H., and Vigil, S., (1993) *Integrated Solid Waste Management, Engineering Principles and Management Issues,* McGraw-Hill, Inc.

Trautwein, S. J. and Boutwell, G. P., (1994) "In-Situ Hydraulic Conductivity Tests for Compacted Soil Liners and Caps," *Hydraulic Conductivity and Waste Contaminant Transport in Soil,* ASTM ATP 1142, David E. Daniel and Stephen J. Trautwein, Eds., American Society for Testing and Materials, Philadelphia, PA, pp. 184–223.

USEPA (1993a), "Technical Manual: Solid Waste Disposal Facility Criteria," United States Environmental Protection Agency, EPA 530-R93-017, Washington, DC.

USEPA, (1993b) "Quality Assurance and Quality Control for Waste Containment Facilities," Technical Guidance Document, EPA/600/R-93/182, U.S. Environmental Protection Agency, Office of Research and Development, Washington, DC, September.

USEPA, (1995) "Code of Federal Regulations, 40 CFR Parts 190 to 259," Revised as of July 1, 1995, U.S. Environmental Agency, Washington, DC.

Way, S. G., (1973) "Terrain Analysis: A Guide to Site Selection Using Aerial Photographic Interpretation," Dowden, Hutchinson, and Russ, Inc., Stroudsburg, PA.

CHAPTER 3

Compacted Clay Liners

3.1 OVERVIEW COMPACTED CLAY LINERS
3.2 COMPACTION AND PERMEABILITY CONSIDERATIONS
    3.2.1 COMPACTION TEST
    3.2.2 PERMEABILITY TEST
3.3 DESIGN OF COMPACTED CLAY LINERS
    3.3.1 LOW HYDRAULIC CONDUCTIVITY
    3.3.2 ADEQUATE SHEAR STRENGTH
    3.3.3 MINIMAL SHRINKAGE POTENTIAL
    3.3.4 ACCEPTABLE ZONE TO MEET ALL DESIGN CRITERIA
3.4 INFLUENCE OF CLODS ON HYDRAULIC CONDUCTIVITY
    3.4.1 INFLUENCE OF CLOD SIZE ON COMPACTION CURVE
    3.4.2 INFLUENCE OF CLOD SIZE ON HYDRAULIC CONDUCTIVITY
    3.4.3 PARTICLE ORIENTATION VERSUS CLOD STRUCTURE
    3.4.4 LABORATORY TESTING AND DESIGN IMPLICATIONS
3.5 EFFECT OF GRAVEL CONTENT ON HYDRAULIC CONDUCTIVITY
3.6 EFFECT OF FREEZING AND THAWING ON HYDRAULIC CONDUCTIVITY
    3.6.1 PROCESSES OCCURRING DURING SOIL FREEZING
    3.6.2 EFFECT OF FREEZE-THAW ON HYDRAULIC CONDUCTIVITY
    3.6.3 FACTORS AFFECTING HYDRAULIC CONDUCTIVITY DURING FREEZE-THAW
3.7 SUMMARY COMMENTS REGARDING COMPACTED CLAY LINERS
    PROBLEMS
    REFERENCES

Compacted clay soil is widely used to line landfills and waste impoundments, to cap new waste disposal units, and to close old waste disposal sites. Most regulatory agencies require that compacted clay liners and covers be designed to have a hydraulic conductivity of less than or equal to a specified maximum value. Typically, clay liners used to contain hazardous waste, industrial waste, and municipal solid waste must have a hydraulic conductivity of less than or equal to 1.0×10^{-7} cm/sec.

3.1 OVERVIEW OF COMPACTED CLAY LINERS

Previous practice required that soil liners be compacted within a specified range of water content and to a minimum dry unit weight. The "acceptable zone" shown in Figure 3.1 represented the zone of acceptable water content and dry unit weight based on practice prior to 1990. A designer usually required that the dry unit weight of the compacted soil be greater than or equal to a percentage P of the maximum dry unit

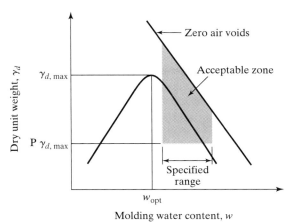

FIGURE 3.1 Traditional Method for Specification of Acceptable Water Contents and Dry Densities for Compacted Clay Liners (Daniel and Benson, 1990). Used with permission of ASCE.

weight from a laboratory compaction test. According to Herrmann and Elsbury (1987), the percentage P was usually 95% of the maximum dry unit weight based on standard Proctor compaction (ASTM D698), or 90% of the maximum dry unit weight based on modified Proctor compaction (ASTM D1557). The range of acceptable water content varies with the characteristics of the soil. For soil liners and covers, the range typically varies from zero to four percentage points wet of standard or modified Proctor optimum.

The acceptable zone in Figure 3.1 evolved empirically from construction practice applied to roadway bases, structural fills, embankments, and earth dams. The specification is based primarily upon the need to achieve a minimum dry unit weight for adequate strength and limited compressibility. Soil liners are compacted wet of optimum because wet-side compaction minimizes hydraulic conductivity (Lambe, 1958; Mitchell et al., 1965; Boynton and Daniel, 1985).

An early and extensive study illustrating how molding water content and dry unit weight influence the hydraulic conductivity of compacted clay was published by Mitchell et al. (1965). They demonstrated that the energy and method of compaction significantly influence the hydraulic conductivity of compacted clay. For a given method of compaction, increasing the compactive energy decreases the hydraulic conductivity of the soil. Figure 3.2 (after Daniel and Benson, 1990) illustrates the concept.

The compaction data from Figure 3.2 were replotted by Daniel and Benson in Figure 3.3. Open symbols are used to denote compacted specimens that had a hydraulic conductivity greater than 1.0×10^{-7} cm/sec, and solid symbols are used for compacted specimens with a hydraulic conductivity less than or equal to 1.0×10^{-7} cm/sec. The acceptable zone in Figure 3.3 encompasses the soil symbols (specimens with hydraulic conductivity less than or equal to 1.0×10^{-7} cm/sec). The shape of the acceptable zone in Figure 3.3 bears little or no resemblance to the one shown in Figure 3.1 (Daniel and Benson, 1990).

Boutwell and Hedges (1989) plotted contours of hydraulic conductivity and shear strength for a clay soil as shown in Figure 3.4. The conventional acceptable zone

54 Chapter 3 Compacted Clay Liners

FIGURE 3.2 Effect of Compactive Energy on Hydraulic Conductivity (Daniel and Benson, 1990). Used with permission of ASCE.

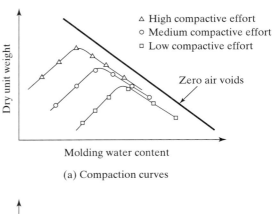

(a) Compaction curves

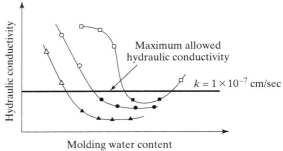

(b) Hydraulic conductivity versus molding water content

in Figure 3.4 applies to $P = 95\%$ from standard Proctor compaction test, and an acceptable water content of 0 to 4% wet of optimum. All water content-dry unit weight (w-γ_d) points contained within the acceptable zone correspond to test specimens with hydraulic conductivity less than or equal to 1.0×10^{-7} cm/sec, but the shape and boundaries of the acceptable zone in Figure 3.4 correlate with neither hydraulic conductivity nor shear strength.

The above research results show that the traditional geotechnical approach does not succeed very well in discriminating between w-γ_d points that correspond to soils

FIGURE 3.3 Replot of Compaction Curves Using Solid Symbols for Compacted Specimens with Hydraulic Conductivities ≤ Maximum Allowable Value (Daniel and Benson, 1990). Used with permission of ASCE.

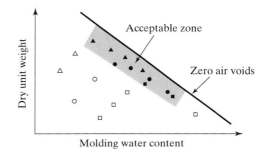

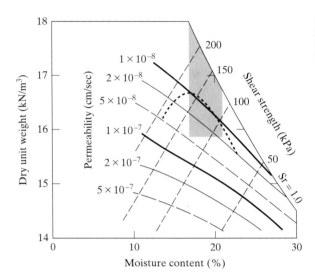

FIGURE 3.4 Contours of Constant Hydraulic Conductivity (cm/sec) and Shear Strength (kPa) (Boutwell and Hedges, 1989)

with hydraulic conductivity less than or equal to 1.0×10^{-7} cm/sec, which is a common regulatory maximum, and w-γ_d points corresponding to hydraulic conductivity greater than 1.0×10^{-7} cm/sec.

Shear strength and hydraulic conductivity are not the only parameters of concern to an engineer who designs compacted clay liners and covers for landfills and waste containment facilities. Other parameters that require consideration include (1) potential for desiccation, (2) resistance to chemical attack, (3) interfacial friction with overlying geomembranes, and (4) ability to deform without cracking during settlement. A good way to establish an appropriate or overall acceptable zone is to measure the parameters of interest on compacted specimens and then to relate those parameters to water content and dry unit weight. Hydraulic conductivity is the key parameter for most compacted clay liners and covers; accordingly, great attention is generally focused on achieving low hydraulic conductivity. After the acceptable range of water content and dry unit weight is established based on hydraulic conductivity, one must carefully consider all other relevant parameters and adjust the acceptable range of water content/dry unit weight to account for those factors as well. Table 3.1 shows a comparison between geotechnical compacted clay and landfill compacted clay liners.

3.2 COMPACTION AND PERMEABILITY CONSIDERATIONS

One of the most important aspects of constructing compacted clay liners that have low hydraulic conductivity is the proper remolding and compaction of the soil. Therefore, a key requirement for designing an acceptable landfill compacted clay liner is to establish the relationship among dry density, water content, and hydraulic conductivity for the selected clayey soil. The most important laboratory tests that are used to develop this relationship are the *compaction test* and the *permeability test*.

Chapter 3 Compacted Clay Liners

TABLE 3.1 Comparison of Geotechnical Compacted Clay and Landfill Compacted Clay Liners

Geotechnical Compacted Clay	Landfill Compacted Clay
Design Criteria	
Bearing capacity (shear strength), compressibility.	Permeability ($\leq 1.0 \times 10^{-7}$ cm/sec), shear strength, shrinkage potential, chemical resistance, and chemical compatibility.
Construction Requirements	
90 ~ 95% of maximum dry density (both sides of optimum water content).	90% ~ 95% of maximum dry density (wet side of optimum water content).
Lift thickness is generally 9 to 18 inches (230 to 300 mm).	Lift thickness is not more than 6 inches (150 mm) after compaction.

3.2.1 Compaction Test

Rational design of compacted clay liners should be based upon test data developed for each soil that is under consideration. Field test data is preferable to laboratory data, but the determination of compaction criteria in the field through a series of test sections can be very costly. For this reason, design engineers normally rely on laboratory tests using the most appropriate method of compaction (to match field compaction as closely as possible). It is important to recognize, however, that laboratory-scale compaction can never perfectly duplicate the repeated passage of heavy compaction equipment over a lift of soil in the field. Laboratory compaction methods can simulate field compaction reasonably well, but the compactive effort in the field is impossible to determine in advance and will undoubtedly vary from point to point. On the basis of these facts, it is hard to justify a single arbitrary compactive effort for use in laboratory testing.

A logical solution to this dilemma is to select several compactive efforts in the laboratory that span the range of compactive effort anticipated in the field. If done in this manner, the water content and dry unit weight criterion will apply to any intermediate compactive effort. This approach is similar to the one described by Mundell and Bailey (1985).

The Standard Proctor Test (ASTM D698) and the Modified Proctor Test (ASTM D1557) are the two most commonly used compaction tests in the laboratory. Both of these tests use the same size mold—that is, one with a 4-inch- (100-mm)-diameter and 4.5-inch- (120-mm)-height, or one that is 1/30 ft^3 (9.5×10^5 mm^3) in volume. The main difference between the standard and modified Proctor test is compaction energy. The standard Proctor test uses a 5.5-lb- (24-N)-hammer, a 12-inch- (300-mm)-drop distance, three layers of soil in the mold, and 25 blows (or drops) per layer. The modified Proctor test, on the other hand, uses a 10-lb- (45-N)-hammer, an 18-inch- (460-mm)-drop distance, five layers of soil, and 25 blow per layer. Thus, the compaction energy of the modified Proctor test is approximately four times that of the Standard test.

In order to span the range of the possible compaction energies more widely, a new compaction test method called the *reduced* Proctor test was developed by Daniel and Benson (1990). The procedure for the reduced Proctor test is almost identical to the Standard Proctor test except that 15 blows per layer are used instead of 25 blows per layer. This means the compaction energy of the reduced Proctor test is lower than that of the standard Proctor test. A comparison of the three types of Proctor tests is presented in Table 3.2.

For most geotechnical landfill earthworks, the modified Proctor effort represents an upper limit on the compactive effort in the field. The standard Proctor effort represents a medium compactive effort in the field. The reduced Proctor effort represents a minimum level of compactive energy for a typical compacted clay liner or cover. The basic concept behind the use of these three types of Proctor tests is to span the range of compactive effort expected in the field.

The three compaction curves in Figure 3.5 represent the relationship between dry density and moisture content obtained from the modified, standard and reduced Proctor tests, respectively. The main reason for including the reduced Proctor test is to achieve a better distribution of paired density-moisture points used to plot the compaction curves.

A compaction curve is developed by preparing several soil samples (usually four, five, or six samples) at different water contents and compacting each of the samples in a mold of known volume using the same compaction procedure. The dry density (γ_d) and water content (w) of each compacted sample are then measured and recorded. The water content and dry unit weight points are next plotted in a diagram with dry unit weight as a vertical axis and water content as a horizontal axis. A smooth curve is drawn between the points to define the compaction curve. Four to five points (samples) are usually needed to define a smooth compaction curve. These points should be located on both sides of the peak value.

The peak value of the dry density is called the *maximum dry density*. The water content corresponding to the maximum dry unit weight $(\gamma_d)_{max}$ is called the *optimum water content* w_{opt}. The main reason for developing a compaction curve is to determine the optimum water content and maximum dry unit weight for a selected soil under a specific compaction energy.

The zero-air-voids curve, also known as the 100% saturation curve, is a curve that relates dry unit weight to water content for a saturated soil that contains no air. The equation for the zero air voids curve is

$$\gamma_d = \gamma_w/[w + (1/G_s)] \tag{3.1}$$

TABLE 3.2 Three Types of Proctor Tests

Type of Test	Weight of Hammer		Drop Distance		Layers	Blows Per Layer	Specific Energy	
	lb	N	inch	mm			lb-ft/ft³	kN-m/m³
Modified Proctor	10	45	18	450	5	25	56,250	2,690
Standard Proctor	5.5	24	12	300	3	25	12,380	590
Reduced Proctor	5.5	24	12	300	3	15	7,430	360

Chapter 3 Compacted Clay Liners

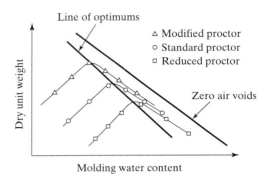

FIGURE 3.5 Compaction Curves with Modified, Standard, and Reduced Proctor Compactive Effort (Daniel and Benson, 1990). Used with permission of ASCE.

where G_s is the specific gravity of solids (typically 2.6 to 2.8) and γ_w is the unit weight of water. If the specific gravity of solids of a soil changes, the zero-air-voids curve will also change. The right-hand branch of the compaction curve lies slightly below and approximately parallel to the zero-air-voids curve. Theoretically, no points on a plot of dry unit weight versus water content should lie above the zero-air-voids curve, but in practice some points usually lie slightly above the zero-air-voids curve as a result of soil variability and inherent limitations in the accuracy of water content and unit weight measurements.

Benson and Boutwell (1992) summarized the maximum dry unit weights and optimum water content measured on soil liner materials from 26 soil liner projects and found that the degree of saturation at the point of optimum water content and maximum dry unit weight ranged from 71% to 98%, based on an assumed G_s-value of 2.75. The average degree of saturation at the optimum point was 85%. That means a typical compacted clay liner is very close to a fully saturated condition.

EXAMPLE 3.1

For a clayey soil used as liner material, the
Modified Proctor Test results were $(\gamma_d)_{max} = 116$ lb/ft^3 (18.24 kN/m^3), $w_{opt} = 14\%$; and the Index Properties were $G_s = 2.75$, $\gamma_w = 62.4$ lb/ft^3 (9.81 kN/m^3).
Thus, the Void Ratio is

$$e = G_s \cdot \gamma_w / \gamma_d - 1 = (2.75)(62.4)/(116) - 1 = 0.48$$

and the Degree of Saturation is

$$S = (w \cdot G_s / e) \times 100\% = [(0.14)(2.75)/(0.48)] \times 100\% = 80.2\%$$

3.2.2 Permeability Test

Hydraulic conductivity or permeability is the coefficient of proportionality between flow rate and hydraulic gradient in Darcy's law, which can be written as

$$q = k \cdot i \cdot A = k \cdot (\Delta H / L) \cdot A \tag{3.2}$$

where q = rate of flow, cm³/sec;
   k = hydraulic conductivity, cm/sec;
   i = hydraulic gradient = $\Delta H/L$;
   ΔH = head loss across specimen, cm;
   L = length of specimen, cm; and
   A = cross-section area of specimen perpendicular to direction of flow, cm².

Upon completion of the targeted compaction conditions, the hydraulic conductivity of each soil specimen should be measured. The hydraulic conductivity of a compacted clay sample can be determined by either a constant head or falling head permeability test in a rigid-wall permeameter in the laboratory (see ASTM D2434). The hydraulic conductivity of a compacted clay sample can also be determined in the laboratory by a constant head permeability test in a flexible-wall permeameter (see ASTM D5084). The hydraulic conductivity of soil can be determined by field tests as described previously in Section 2.6.

A schematic diagram of a rigid-wall permeameter is shown in Figure 3.6. A rigid-wall permeameter consists of a rigid tube or box that contains the specimen to be permeated. The tube is always circular and constructed of metal or plastic. A cylindrical glass tube can also be used when testing with chemicals or waste liquids. The permeating liquid flows along the axis of the cylindrical test specimen. Flow may be from top-to-bottom or bottom-to-top. Upward flow may help to dislodge entrapped gas, but care must be taken not to liquefy the specimen nor to displace an unrestrained test specimen upward.

The main advantages of a rigid-wall permeameter over a flexible-wall device are (1) it is less expensive, (2) it is easier to use, and (3) it can be fabricated from a wide

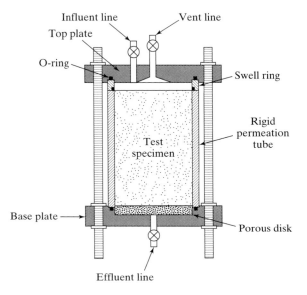

FIGURE 3.6 Rigid-Wall, Compaction Mold Permeameter (Daniel, 1994)

range of chemically resistant materials. However, *sidewall leakage* caused by sample contamination from the permeating liquid is always possible, and the applied stress cannot be controlled with most types of rigid-wall permeameters. In principle, one could eliminate gas bubbles in a rigid-wall cell with backpressure (i.e., by pressurization of both the influent and effluent liquid). Pressurizing the water in the test specimen reduces the volume of air present by compressing air bubbles and dissolving air into the pore water. However, backpressure has not been found to work well in rigid-wall cells (Edil and Erickson, 1985).

A schematic diagram of a flexible-wall permeameter is shown in Figure 3.7. In this test configuration, the specimens that are prepared for testing typically have a diameter of 2.85 inches (72 mm) and a length of 2 to 3 inches (50 to 75 mm). The clay soil specimen is confined between porous disks and end caps on the top and bottom, surrounded by a flexible rubber membrane on the sides, and placed in a pressure cell. Double drainage lines at both ends of the specimen can be used to flush air from the system, apply a back pressure to saturate the soil sample, or connect a pressure transducer to measure the pore pressure in the soil sample for checking saturation of the sample. The cell is filled with water and pressurized to press the flexible membrane against the test specimen and thereby minimize or eliminate sidewall leakage.

After the specimen is fully saturated, it should be consolidated under a desired pressure. The consolidation pressure is determined from known or expected field overburden pressures. At the end of the consolidation phase, a hydraulic gradient is imposed by applying a different pressure to the top and bottom of the sample. The flow of permeant normally proceeds from bottom to top. The flow rate is determined by the quantities of influent and effluent collected in a given time period.

The flexible-wall permeameter is generally more versatile than the rigid-wall cell and eliminates problems with sidewall leakage. Confining pressures simulating those likely to be experienced in the field can be applied with this device. All fine-grained

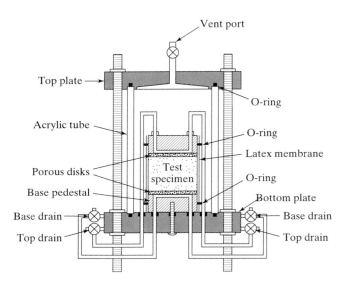

FIGURE 3.7 Flexible-Wall Permeameter (Daniel, 1994)

silts and clays can be permeated in a flexible wall cell. Because the soil can be saturated fairly rapidly with backpressure in a flexible-wall cell, testing times tend to be less with flexible-wall cells. Furthermore, saturation of the test specimen can be verified prior to permeation (Daniel, 1994). The flexible-wall permeameter is usually the preferred type of permeameter to use in situations that involve testing of undisturbed samples obtained from the field and returned to the laboratory.

Flexible-wall permeameters are superior to rigid-wall permeameters in many important regards, but they also are costly, more complex to operate, difficult to use at very low effective confining stress, and vulnerable to membrane integrity problems. The latter can occur when permeating a test specimen with certain chemicals or waste liquids. Testing at low effective confining stress is a problem, because enough confining stress must always be applied to press the membrane tightly against the soil. Reliable testing with a flexible-wall permeameter requires a minimum of 2 psi (14 kPa) effective confining stress to prevent sidewall leakage.

3.3 DESIGN OF COMPACTED CLAY LINERS

A well-designed compacted clay liner must have a low permeability to prevent or minimize leachate leakage, adequate shear strength for stability, and minimal shrinkage potential to prevent desiccation cracking. Accordingly, a design objective for a compacted soil liner is to determine the range of water content and dry unit weight within which compacted test specimens will have (1) low hydraulic conductivity ($\leq 1.0 \times 10^{-7}$ cm/sec), (2) adequate shear strength, and (3) minimal shrinkage upon drying.

3.3.1 Low Hydraulic Conductivity

The procedure for establishing water content and dry unit weight ranges at which the hydraulic conductivities of compacted specimens would be less than or equal to 1.0×10^{-7} cm/sec is presented as follows (after Daniel and Benson, 1990):

(i) Compact the soil in the laboratory following modified, standard, and reduced Proctor compaction procedures to develop compaction curves similar to those shown in Figure 3.5. Approximately five or six different specimens should be compacted using each effort.

(ii) The compacted specimens should be permeated to determine their hydraulic conductivity. Care should be taken to ensure that permeation procedures are accurate, with important details such as degree of saturation, backpressure, and effective confining pressure carefully selected. The measured hydraulic conductivities should be plotted as a function of molding water content similar to those shown in Figure 3.8.

(iii) The dry unit weight and water content points should be replotted similar to those shown in Figure 3.9 with different symbols used to represent compacted specimens that meet specified hydraulic conductivity requirements. In Figure 3.9, the

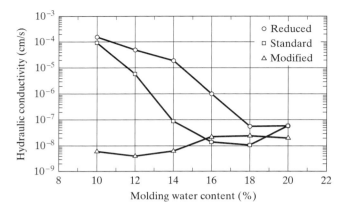

FIGURE 3.8 Hydraulic Conductivity versus Molding Water Content (Daniel and Wu, 1993). Used with permission of ASCE.

open symbols represent the specimens with the hydraulic conductivities greater than the maximum acceptable value (1.0×10^{-7} cm/sec), and the solid symbols represent the specimens with the hydraulic conductivity less than or equal to maximum acceptable value. An acceptable zone should be drawn to encompass the data points representing test results meeting or exceeding the design criteria.

To illustrate the importance of the above three-step procedure, a database consisting of 85 full-scale compacted clay liners was assembled to evaluate field performance based on field hydraulic conductivity k_F (Benson et al., 1999). Large-scale field hydraulic conductivity tests (lysimeters and sealed double-ring infiltrometers) were conducted on each liner. The database indicated that 26% of the clay liners had $k_F \geq 1.0 \times 10^{-7}$ cm/sec despite all of the clay liners being constructed for the explicit purpose of achieving this level of low hydraulic conductivity. Contrary to a belief that achieving a hydraulic conductivity of $\leq 1.0 \times 10^{-7}$ cm/sec is routine, the database indicated otherwise, with a failure rate of roughly one in four. In most cases, the primary cause for $k_F \geq 1.0 \times 10^{-7}$ cm/sec was that compaction was achieved dry of the

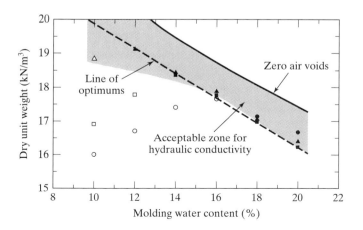

FIGURE 3.9 Acceptable Zone Based on Hydraulic Conductivity Considerations (Solid Symbols Correspond to Compacted Specimens with Hydraulic Conductivity less than or equal to 1×10^{-7} cm/sec) (Daniel and Wu, 1993). Used with permission of ASCE.

line of optimums. None of the liners compacted with w-γ_d points falling on or above the line of optimums had $k_F > 1.0 \times 10^{-7}$ cm/sec. In spite of these findings, many clay liner construction specifications still employ a percent compaction specification, wherein a range of water content (usually wet of optimum) and a minimum percent compaction are specified. Although laboratory data have demonstrated that percent-compaction specifications can result in compaction dry of the line of optimums, the use of this type of specification is pervasive in practice. Analysis of the information in the database points to the clear advantage of examining compaction results not in terms of water content relative to an optimum or percent compaction, but rather in terms of the position of the field compaction data relative to the line of optimums. When most of the field compaction data points lie on or above the line of optimums, there is a high probability that k_F will indeed be $\leq 1.0 \times 10^{-7}$ cm/sec (Benson et al., 1999). Thus, the acceptable zone for meeting the permeability criterion should be adjusted as shown in Figure 3.10.

The recommended procedure involves compaction and permeation of approximately 5 or 6 soil samples at 3 different compactive efforts, or a total of 15 to 18 compaction and permeability tests for each soil to be investigated. A typical geotechnical engineering laboratory should be able to complete such a scope of testing in a few weeks. The number of tests required, the time needed to develop the data, and the cost of developing the data should not be prohibitive for an important soil liner or cover project.

Tests to determine compatibility with chemical wastes are occasionally performed on laboratory-compacted soils as well. The purpose of the compatibility tests is to determine the effect, if any, of the chemicals on hydraulic conductivity and other engineering properties of the compacted clay. To test for chemical compatibility, one or two test specimens near the lower bound of the acceptable zone should be selected for hydraulic conductivity tests using a permeant that is representative of leachate produced from the chemical waste.

Finally, there is more to good construction than control of water content and dry unit weight. Good mixing of the soil, use of appropriate compaction equipment, effective bonding of lifts, proper protection of compacted lifts from desiccation and freez-

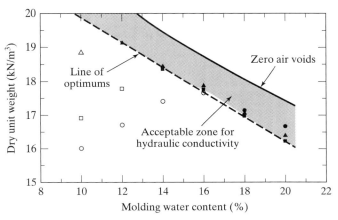

FIGURE 3.10 Adjusted Acceptable Zone Based on Hydraulic Conductivity Considerations (Solid Symbols Correspond to Compacted Specimens with Hydraulic Conductivity less than or equal to 1.0×10^{-7} cm/sec) (modified from Daniel and Wu, 1993). Used with permission of ASCE.

ing, and careful inspection by qualified personnel also are essential ingredients to good quality assurance of soil liners and covers.

3.3.2 Adequate Shear Strength

Most modern sanitary landfills are built quite high to create sufficient filling space. The heights of some landfills may exceed 200 feet (60 m). The soil liners of these landfills must have adequate strength to withstand high overburden pressures and shear stresses. If the height of a municipal solid waste landfill is 250 feet (75 m), and the average unit weight of the solid waste is 65 lb/ft^3 (10 kN/m^3), the bearing stress acting on the underlying compacted clay liner will be 16,250 lb/ft^2 (750 kN/m^2).

The following example serves to illustrate how shear strength requirements can be met. The required strength for a clay liner to support the maximum bearing stress in a landfill project was calculated to be 30 psi (200 kPa) (Daniel and Wu, 1993). The procedure for establishing the water content and dry unit weight range within which the compacted soil specimens have the required shear strength is as follows:

(i) Compact soil in the laboratory using the modified, standard, and reduced Proctor compaction procedures to develop compaction curves as shown in Figure 3.5. Approximately five or six different specimens should be compacted using each effort.

(ii) Unconfined compression or unconsolidated-undrained triaxial tests are run on each specimen to determine the respective shear strengths. The measured shear strengths are next plotted as a function of molding water content as shown in Figure 3.11.

(iii) The dry unit weight and water content points are then replotted as shown in Figure 3.12, with different symbols used to represent compacted specimens that meet specified shear strength requirements. In Figure 3.12, the open symbols represent the specimens with the shear strengths less than the maximum acceptable value, 30 psi (200 kPa), and the solid symbols represent the specimens with the shear strengths greater than or equal to the maximum acceptable value. The

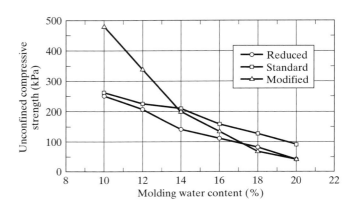

FIGURE 3.11 Unconfined Compressive Strength versus Molding Water Content (Daniel and Wu, 1993). Used with permission of ASCE.

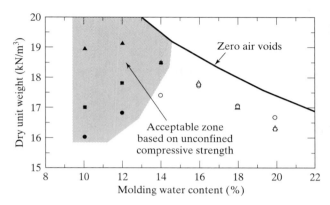

FIGURE 3.12 Acceptable Zone Based on Unconfined Compressive Strength Considerations (Solid Symbols Correspond to Compacted Specimens with Unconfined Compressive Strengths greater than 30 psi [200 kPa]) (Daniel and Wu, 1993). Used with permission of ASCE.

acceptable zone should be drawn to encompass the data points representing test results meeting or exceeding the design shear strength criterion.

This procedure involves compaction and strength testing of approximately five or six soil specimens at three different compactive efforts, or a total of 15 to 18 compaction and shear strength tests for each soil to be investigated.

3.3.3 Minimal Shrinkage Potential

Clay-rich soils used in constructing low-hydraulic-conductivity liners and covers for solid waste landfills are typically placed and compacted wet of optimum water content. This practice minimizes the hydraulic conductivity of the compacted soil at the time of construction. However, at relatively arid sites or sites where the clay could be subjected to seasonal drying, this practice could be counterproductive if the liner desiccates or dries out. Large cracks can occur when wet compacted clays are allowed to dry. Cracks and fissures in a compacted clay soil can increase the hydraulic conductivity by several orders of magnitude, compared with the same clay in an intact condition. (Other causes of cracks in clays and methods for preventing or mitigating this problem are discussed in Chapter 4.)

The crux of the shrinkage problem is that as the molding water content of a compacted soil is increased, the shrinkage potential of the soil increases as well. The greater the amount of remolding or shear strain during compaction, the greater will be the shrinkage potential. Soil liners are almost always compacted wet of optimum water content to comply with requirements that the hydraulic conductivity of the compacted soil not exceed 1.0×10^{-7} cm/sec. This practice of wet-side compaction, while effective in terms of regulatory compliance with regard to hydraulic conductivity, may also cause problems in arid regions or even in less arid locations where near-surface clays may desiccate during periods of drought. Desiccation or shrinkage cracking can also occur during lift placement if the lifts are left uncovered too long in hot weather. Therefore, it is very important to find a way to compact clay soil with both low hydraulic conductivity and low shrinkage potential.

Based on research results of several groups, the available design principles for designing low-hydraulic-conductivity, compacted soil liners and covers for sites in arid areas or for facilities where desiccation might be expected for a variety of reasons are summarized as follows (Daniel and Wu, 1993):

(i) *Use soils rich in sand.* DeJong and Warkentin (1965) and Kleppe and Olson (1984) have demonstrated that desiccation shrinkage increases with increasing clay content. Conversely, clay sand (SC) combines the attributes of low hydraulic conductivity and low compressibility to minimize the amount of shrinkage that takes place upon drying. If clayey sands are unavailable, mixtures of locally available sandy materials mixed with processed clay (e.g., a nonexpandable clay, such as kaolinite) should be given consideration. These are referred to as *amended clay liners.*

(ii) *Place soil at the lowest practical water content.* The results of Daniel and Wu's (1993) investigation demonstrate that some soils can be compacted relatively dry, with high compactive energy, and achieve both low hydraulic conductivity and low shrinkage potential. Even if some drying of the soil takes place, the soil will undergo minimal shrinkage and desiccation cracking when compacted in this manner.

(iii) *Be especially cautious with cover systems.* Boynton and Daniel (1985) demonstrated that desiccated clay swells when rewetted, but it does not regain the original low hydraulic conductivity if compressive stress from the overburden is low. Desiccation has a greater potential for permanently damaging clay barriers in cover systems (where the clay is subjected to a low stress) than in liner systems subjected to large compressive stress from the loading provided by the overlying waste.

(iv) *Protect the soil barrier.* Neither a thin layer of topsoil (Hawkins and Horton, 1965; Montgomery and Parsons, 1989) nor an uncovered geomembrane (Corser and Cranston, 1991) can be counted on to prevent an underlying compacted soil barrier from drying out and cracking. On the other hand, an adequately thick soil cover in a humid region (Hawkins and Horton, 1965), or alternatively, a protective geomembrane overlaid by topsoil in an arid region (Corser and Cranston, 1991), can effectively stop desiccation of an underlying compacted soil liner. Designers should consider a composite geomembrane/protective soil layer to minimize drying of the barrier layer in cover systems. A special problem is posed by low hydraulic conductivity soil liners placed on very dry soils with suctions that are far greater than the compacted liner. Consideration should be given in this case to separating the soil liner from the dry subsoil with a geomembrane. At arid sites, the only practical way of stopping a clay-rich, low hydraulic conductivity barrier soil from drying out may be to place geomembranes above and below the soil barrier (Daniel and Wu, 1993). Even then, side slopes may be of concern with moisture transfer from the top to the bottom of these slopes.

(v) *Avoid using highly plastic clays.* Clay soils with high plasticity indices (PIs) or liquid limits (LLs) have poor volume stability and high shrink-swell potentials. In addition, these clay soils are difficult to compact in the field. As a general rule, clay soils with a PI $>$ 50 should not be used in a landfill compacted clay cover or liner.

In principle, the best approach for designing a compacted clay liner with both low hydraulic conductivity and low shrinkage potential is to place and compact the soil at the lowest practical water content. In any event, the resulting permeability of this clay liner still must be less than 1.0×10^{-7} cm/sec. The key, therefore, is how to determine this practical water content for a selected soil.

Daniel and Wu (1993) investigated a clayey soil in west Texas in 1993. According to their findings, an acceptable limiting value of volumetric strain to prevent desiccation for this soil was less than or equal to 4%. The recommended procedure for establishing the water content and dry unit weight range within which the compacted soil specimens achieve a limiting value of volumetric strain is similar to the procedures presented previously for hydraulic conductivity and shear strength:

(i) Compact soil in the laboratory with modified, standard, and reduced Proctor compaction procedures to develop compaction curves as shown in Figure 3.5. Approximately five or six different specimens should be compacted using each effort.

(ii) Run volumetric shrinkage tests (ASTM D427) to determine the volumetric strain for each compacted specimen. The measured volumetric strains should be plotted as a function of molding water content as shown in Figure 3.13.

(iii) The dry unit weight and water content points should be replotted as shown in Figure 3.14, with different symbols used to represent compacted specimens with different volumetric strains. In Figure 3.14, the open symbols represent the specimens with the volumetric strains greater than the maximum acceptable value, 4%, and the solid symbols represent the specimens with the volumetric strains less than or equal to maximum acceptable value. The acceptable zone should be drawn to encompass the data points representing test results meeting or exceeding the design criteria.

The procedure involves conducting compaction and volumetric shrinkage tests for approximately 5 or 6 soil specimens at 3 different compactive efforts, or a total of

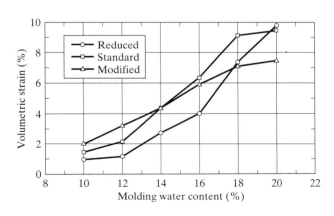

FIGURE 3.13 Volumetric Strain Caused by Drying versus Molding Water Content (Daniel and Wu, 1993). Used with permission of ASCE.

68 Chapter 3 Compacted Clay Liners

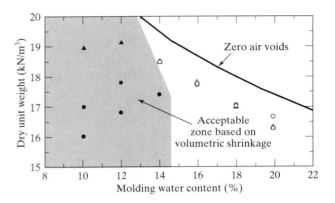

FIGURE 3.14 Acceptable Zone Based on Volumetric Shrinkage Considerations (Solid Symbols Correspond to Compacted Specimens with Desiccation-Induced Volumetric Shrinkage less than or equal to 4%) (Daniel and Wu, 1993). Used with permission of ASCE.

15 to 18 compaction and volumetric shrinkage tests for each soil. A limiting value of volumetric strain to prevent desiccation must be determined ahead of time for each soil under consideration.

3.3.4 Acceptable Zone to Meet All Design Criteria

The overall objective is to relate compaction criteria (water content and dry unit weight) to hydraulic conductivity (maximum of 1.0×10^{-7} cm/sec), unconfined compressive strength [e.g., minimum of 30 psi (200 kPa)], and volumetric shrinkage upon drying (e.g., maximum of 4%). Following Daniel and Wu (1993), an acceptable zone that meets all of these criteria can be established by superposition.

Hydraulic conductivity is plotted as a function of molding water content and method of compaction in Figure 3.8. The acceptable zone of water contents and dry unit weights that yielded the required hydraulic conductivity of 1.0×10^{-7} cm/sec or less is shown in Figure 3.10.

Unconfined compressive strength varied between 8 and 72 psi (45 and 480 kPa). (See Figure 3.11.) The range of molding water contents and dry unit weights that yields compacted specimens with unconfined compressive strengths in excess of 30 psi (200 kPa) was shown in Figure 3.12.

The relationship between desiccation-induced volumetric shrinkage strain and molding water content was presented in Figure 3.13 for three compaction energies. The range of acceptable water contents and dry unit weights based on a threshold volumetric shrinkage strain of 4% was shown in Figure 3.14.

Thus, the acceptable zones based on hydraulic conductivity, volumetric shrinkage strain, and unconfined compressive strength are all superimposed and presented in Figure 3.15. The results of this exercise illustrate that it is possible to compact a clayey soil to a low hydraulic conductivity and adequate strength, and to simultaneously produce a compacted material with minimal shrinkage potential. However, the exercise also suggests that it is not a simple task to develop this information in the laboratory or to implement the results in the field.

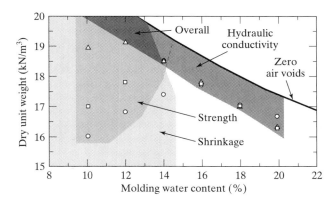

FIGURE 3.15 Acceptable Zone Based on Design Objectives for Hydraulic Conductivity, Volumetric Shrinkage and Unconfined Compressive Strength (modified from Daniel and Wu, 1993). Used with permission of ASCE.

3.4 INFLUENCE OF CLODS ON HYDRAULIC CONDUCTIVITY

Geotechnical engineers are often challenged to translate soil properties measured at a laboratory scale to field-scale applications. The in-situ hydraulic conductivity of compacted soil, for example, is frequently different from the results of laboratory tests. One factor that contributes to this difference is the presence of void-volume defects (viz., fissures, cracks, and macropores) that are present in the field, but whose influence may not be captured in relatively small laboratory test specimens. Another factor is the structure of soil compacted in the laboratory, compared with the structure of soil compacted in the field. Unless soil in the field and laboratory are in a similar state of compaction, the hydraulic conductivity may be significantly different. To create soil conditions that are similar, the variables that influence the structure of compacted soil must be carefully controlled.

Several investigations of the in-situ hydraulic conductivity of compacted soils have been performed by Daniel (1984, 1987). An important finding of his work is that laboratory measurements of hydraulic conductivity often underestimate the in-situ hydraulic conductivity by one order of magnitude or more. Daniel suggested that discrepancies between the hydraulic conductivity of laboratory and field compacted soil may be partly due to differences in clod size between the field and laboratory. Additional work by Benson and Daniel (1990) indicated that the hydraulic conductivity of a highly plastic, compacted clay soil is significantly influenced by the size of clods used in preparing the soil for compaction.

3.4.1 Influence of Clod Size on Compaction Curve

Two standard Proctor compaction curves are shown in Figure 3.16 for soils prepared with small (4.8 mm or No. 4 sieve) and large (19 mm or 3/4 inch sieve) clods (Benson and Daniel, 1990). Clod size significantly influenced the compaction curves. The shape and optimum moisture content of the two curves are different, but the maximum dry unit weights are almost identical. The soil with initially smaller clods had a compaction curve with a much flatter peak, suggesting less sensitivity to molding water content.

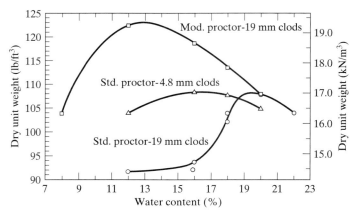

FIGURE 3.16 Standard and Modified Proctor Compaction Curves (Benson and Daniel, 1990). Used with permission of ASCE.

The soil with smaller clods also exhibited an optimum water content that was 3% lower than the soil with larger clods.

For water contents dry of optimum, the soil with initially smaller clods could be compacted more effectively than the soil with larger clods. This conclusion was supported by a significantly higher dry density on the dry side of optimum for the soil with smaller clods. The clods in both soils were hard when the water content was below optimum (Benson and Daniel, 1990).

3.4.2 Influence of Clod Size on Hydraulic Conductivity

Daniel (1984) provided the earliest documentation of the influence of clod size on the hydraulic conductivity of compacted clay by performing tests on clayey soil compacted with clods of different sizes. Results of the tests are shown in Table 3.3. In a later study, Benson and Daniel (1990) reported that clod size had a significant influence on hydraulic conductivity of soils compacted dry of optimum using Standard Proctor effort (Figure 3.17). At molding water contents between 12% and 16%, the hydraulic conductivities of soils with small clods (4.8 mm or 3/16 inches) were four to six orders of magnitude lower, compared with soil with larger (19 mm or 3/4 inches) clods.

TABLE 3.3 Influence of Clod Size on Hydraulic Conductivity of Compacted Clay (Daniel, 1984)

Average Diameter of Clods		Hydraulic Conductivity (cm/sec)
9.5 mm	3/8 inches	3.0×10^{-7}
4.8 mm	3/16 inches	2.0×10^{-8}
1.6 mm	1/16 inches	9.0×10^{-9}

Used with permission of ASCE.

Section 3.4 Influence of Clods on Hydraulic Conductivity 71

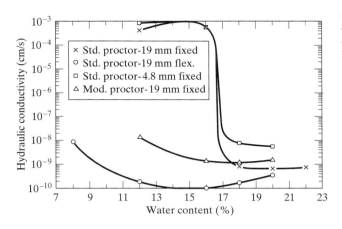

FIGURE 3.17 Hydraulic Conductivity versus Molding Water Content (Benson and Daniel, 1990). Used with permission of ASCE.

At molding water contents between 18% and 20%, the clods were moist, soft, and sticky (Benson and Daniel, 1990). The size of clods had little influence on the hydraulic conductivity of specimens at these molding water contents. Apparently, a relatively high moisture content rendered clods soft and easily compressible into a relatively homogeneous mass of low-permeability soil, regardless of clod size. Thus, the size of the clods in this water content range was unimportant with regard to hydraulic conductivity after compaction.

3.4.3 Particle Orientation versus Clod Structure

Based on the clod concepts proposed by Olsen (1962), most of the flow of water in compacted clay occurs in relatively large pore spaces located between clods of clay, rather than between the particles of clay within the clods (Figure 3.18). Soft, wet clods of soil are easier to remold than hard, dry clods (Benson and Daniel, 1990). Therefore, when a soil is compacted wet of optimum, the clods are remolded with relative ease, which results in smaller interclod voids and lower hydraulic conductivity.

Benson and Daniel (1990) concluded that the status of clods and interclod voids controlled the hydraulic conductivity after they examined photographs of specimens of soil compacted with standard Proctor and modified Proctor procedures at various water contents. They observed that the driest specimens (12% water content) com-

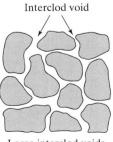
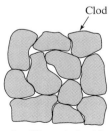

FIGURE 3.18 Flow of Water through Relatively Large Pores between Clods of Soil

pacted with standard Proctor effort looked more like granular material than clay soil; standard Proctor effort was not sufficient to reconstitute and remold the dry, hard clods together and eliminate large interclod voids. These specimens had large hydraulic conductivities. Compaction of soil at the same water content, but with modified Proctor effort, resulted in greater deformation of clods, reduction of large voids, and lower hydraulic conductivity. (Recall Figure 3.17.) The energy imparted by modified Proctor effort was sufficient to reconstitute and breakdown the dry, hard clods. Furthermore, soils compacted wet of optimum (20% water content) by standard or modified Proctor effort showed no evidence of remnant clods or interclod pores. The hydraulic conductivity of all specimens compacted wet of optimum was very low in this case.

The dry unit weight of the soil at the time of compaction provides an indication of the presence of large interclod pores. Low dry unit weight is associated with soils having large, visible voids between pores, whereas soils with no large, visible interclod pores have much higher dry unit weights.

Benson and Daniel (1990) have plotted the hydraulic conductivity of test specimens as a function of dry unit weight at the time of compaction. (See Figure 3.19.) Test specimens with initial dry unit weight greater than 101 pcf (16.0 kN/m^3) had a hydraulic conductivity of less than 1.0×10^{-7} cm/sec. A dry unit weight of 101 pcf corresponds to 93% of the maximum dry unit weight based on standard Proctor compaction, and 82% of the maximum dry unit weight based on modified Proctor compaction. In order to achieve low hydraulic conductivity in highly plastic soils that form clods, the clods must be destroyed and large interclod voids eliminated. The dry unit weight of the compacted soil provides an indirect measure of the degree to which large voids have been eliminated.

3.4.4 Laboratory Testing and Design Implications

How should soil be processed in the laboratory so as to adequately simulate field conditions? Should the soil be predried and crushed to pass the No. 4 sieve (a standard reproducible soil preparation technique), or should the soil be prepared at water con-

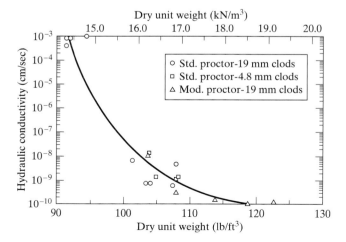

FIGURE 3.19 Hydraulic Conductivity versus Dry Unit Weight (Benson and Daniel, 1990). Used with permission of ASCE.

tents close to those expected in the field and with clods of a size more nearly comparable to the size of clods anticipated in the field? The logical answer is that if laboratory tests are intended to produce data representative of field results, the clod sizes, drying conditions, and other details of compaction in the laboratory should match field conditions as closely as possible.

Field construction practices should ensure that clods of clay are thoroughly remolded and that large interclod voids are eliminated. Daniel (1987) recommended that compaction be accomplished with heavy equipment that remolds the soil during compaction. Benson and Daniel (1990) suggest that clods may be destroyed in one of two ways:

(i) The soil can be wetted to a high molding water content to produce soft, weak clods that can be easily remolded into a mass that is free of large interclod pores (provided that the soil is reasonably workable at high water content).

(ii) The soil can be compacted at a lower water content, but with an extremely heavy roller that crushes the clods and thereby eliminates large interclod pores.

The water content of the soil should be matched carefully with the compactive energy used in field construction, or vice versa. If the soil is relatively dry, extremely high compactive effort will likely be required to achieve a low hydraulic conductivity. On the other hand, less compactive effort is required with relatively wet soil. The choice of molding water content also depends on other factors (e.g., concern over long-term desiccation or swelling and shear strength considerations).

3.5 EFFECT OF GRAVEL CONTENT ON HYDRAULIC CONDUCTIVITY

The hydraulic conductivity of a compacted clay liner must be less than 1.0×10^{-7} cm/sec for most types of regulated waste disposal units. For economic reasons, locally available, clay-rich soils are the preferred material for construction of compacted clay liners. Unfortunately, many deposits of clayey soil (e.g., glacially deposited materials) are gravelly. According to the Unified Classification System, gravel is defined as material that will not pass through a No. 4 sieve. Daniel (1990) recommends that soil liner materials be restricted to soils containing less than 10 to 20% gravel-size particles. Many state regulatory agencies place tight restrictions on the amount of gravel that a liner material may contain.

If the gravel content of a soil is too large, the soil may have to be screened to remove stones, or specialized equipment may be required that extracts stone from a loose lift of soil. With wet, sticky materials, mechanical processors tend to become plugged by the clayey material and may be ineffective. Shelly and Daniel (1993) reported that on some projects, teams of laborers have removed stones by hand from a loose lift of soil. In any case, the removal of gravel particles is difficult and expensive; the question of how much gravel is too much is an issue that has important economic and logistical ramifications.

Holtz and Lowitz (1957) performed compaction tests on sandy, silty, and clayey soils containing varying amounts of gravel. They found that when the gravel content approached one-third (by weight) of the total material, the gravel began to interfere with compaction of the finer fraction. Furthermore, when the gravel content approached two-thirds of the total material, there was often insufficient fine material to completely fill the void spaces between the large particles. Holtz and Lowitz also found that the void ratio of the finer fraction increased when gravel contents exceeded 40 to 50%, even though the gravel/soil mixture was compacted at the optimum water content. The void ratio of the fine fraction can influence the hydraulic conductivity of a mix because the hydraulic conductivity of the matrix of fines is generally quite sensitive to void ratio.

Shakoor and Cook (1990) performed permeability tests on a silty clay (CL) of glacial origin mixed with 0.5- to 0.75-inch- (13- to 19-mm)-diameter gravel particles in 10% increments from 10 to 80 percent gravel by weight. The tests were performed on specimens compacted by Standard Proctor compactive effort to a minimum of 95 percent of the maximum dry density, and at a water content no greater or less than 2 percent of optimum. Hydraulic conductivity ranged between 1.0×10^{-7} and 2.0×10^{-7} cm/sec for gravel contents between 0 and 50%. When the gravel content reached 50%, hydraulic conductivity increased sharply. At this level of gravel content, coarse particles interfere significantly with compaction, and the voids between gravel particles are not completely filled with finer material.

Shelly and Daniel (1993) performed compaction and hydraulic conductivity tests on two clayey soils mixed with varying percentages of gravel. The gravel itself had a hydraulic conductivity of 170 cm/sec. Hydraulic conductivity of the compacted gravel/soil mixtures was less than 1.0×10^{-7} cm/sec for gravel contents as high as 60% (Figure 3.20). For gravel contents less than 60%, gravel content was not important. For gravel contents greater than 60%, the clayey soil does not fill voids between gravel particles, and much higher hydraulic conductivity is observed. The results of this study are consistent with those of Shakoor and Cook (1990). The water content of the nongravel fraction provided a useful means of assessing the hydraulic conductivity of the clay-gravel mixtures. The minimum hydraulic conductivity of the two soil/gravel mixtures was obtained when the molding water content of the nongravel fraction was about 4 percentage points greater than the optimum water content determined from

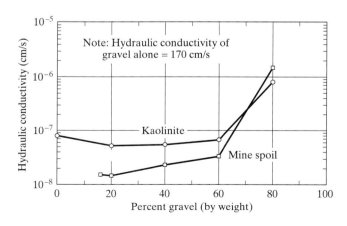

FIGURE 3.20 Hydraulic Conductivity of Kaolinite/Gravel and Mine Spoil/Gravel Mixtures as Function of Percent Gravel (Shelley and Daniel, 1993). Used with permission of ASCE.

separate compaction tests performed on the nongravel fraction alone. Therefore, for purposes of construction quality control, it is better to focus on the water content of the nongravel fraction than the water content of the overall gravel/soil mixture (Shelly and Daniel, 1993).

The results presented by Shelly and Daniel (1993) demonstrate that under carefully controlled laboratory conditions, clayey soils containing up to 60% gravel can be compacted to achieve a hydraulic conductivity of less than 1.0×10^{-7} cm/sec. Shakoor and Cook (1990) found that the presence of as much as 50% gravel did not adversely affect the hydraulic conductivity of a glacial till mixed with gravel. Thus, two independent investigations performed on three different soils demonstrate that clayey soils containing up to 50 to 60% gravel can be compacted to achieve a hydraulic conductivity of less than 1.0×10^{-7} cm/sec. Based on these results, Shelly and Daniel (1993) conclude that the amount of gravel in a soil liner material should not exceed approximately 50%.

Some caution is required in the interpretation of these studies that were developed under carefully controlled laboratory conditions (i.e., not field conditions). The laboratory experiments did not allow an opportunity for gravel particles to segregate. Isolated pockets of segregated gravel particles, whose voids are not filled with clayey material, would tend to increase the overall hydraulic conductivity of a soil liner. Thus, even though it is theoretically possible to achieve a hydraulic conductivity of less than 1.0×10^{-7} cm/sec using soil with up to 50% gravel, the possibility of gravel segregation during construction must be considered carefully. Fortunately, machinery is available that will blend materials thoroughly (e.g., pugmills and rototillers), and field procedures can be established to minimize, if not eliminate, segregation of gravel. Furthermore, some materials are probably more vulnerable to segregation of gravel particles than others. The greater the degree of uniformity of the gravel/soil mixture, the less the possibility of segregation of gravel particles and the larger the amount of gravel that can be tolerated (Shelly and Daniel, 1993). Ultimately, the decision about how much gravel to allow in a clay liner must be based on the individual circumstances of a given project.

3.6 EFFECT OF FREEZING AND THAWING ON HYDRAULIC CONDUCTIVITY

Many laboratory and field studies have shown that compacted clays used for construction of liners and covers at landfills can be severely affected when exposed to freezing and thawing (Zimmie and LaPlante, 1990; Chamberlain et al., 1990; Wong and Haug, 1991; Kim and Daniel, 1992; Zimmie et al., 1992). These studies have shown that ice lenses formed during freeze-thaw cycles result in a network of cracks. As a result, the hydraulic conductivity increases. An increase in hydraulic conductivity of one order of magnitude or more can result after a single freeze-thaw cycle and a conductivity increase of one to three orders of magnitude after three to five cycles (Benson and Othman, 1993; Benson et al., 1995; Chamberlain et al., 1995).

Environmental regulations for some states located in the northern part of the United States require that soil liners be protected from freezing because of its effect on hydraulic conductivity. A common precaution is to increase the thickness of the protective layer over the hydraulic barrier of the final cover so that it exceeds the local frost penetration depth.

3.6.1 Processes Occurring during Soil Freezing

The nature and distribution of soil water changes markedly when it freezes. When the temperature in moist soil drops below 0°C, the water supercools and ice crystals nucleate in larger pores. As the water phase changes to ice, its volume increases about 9% due to the development of a hexagonal crystalline structure. The crystals grow to form ice lenses as long as water is available and heat is being extracted. The thickness and spacing of the ice lenses depend on the relative magnitudes of the rate of freezing, temperature gradient, pressure, and availability of water (Penner, 1960). The growing ice crystals interact with each other and the surrounding soil particles (Andersland and Anderson, 1978).

In the region adjacent to the growing ice lenses, large pore water suctions are generated, pulling water from the unfrozen soil into the freezing zone (Benson and Othman, 1993; Williams, 1966; Konrad and Morgenstern, 1980; Chamberlain and Gow, 1979). Effective pore water suctions as large as 72 psi (500 kPa) have been observed (Chamberlain, 1981). This often results in extremely high effective stresses in the zone adjacent to the growing ice lenses, which in turn cause drying and consolidation of the unfrozen soil in this region. Drying and consolidation produce changes in the soil structure; in particular, the formation of shrinkage cracks that form perpendicular to the freezing zone (Benson and Othman, 1993; Chamberlain and Gow, 1979). As the freezing zone advances into the unfrozen soil mass, these cracks become filled with ice and the soil develops an aggregated structure, with the aggregates bounded by the ice lenses and ice-filled cracks (Othman, 1994).

Ice continues to grow in the frozen zone even when the flow of water from the unfrozen soil is cut off by the leading ice layer. This occurs because water progressively freezes as the temperature is lowered (Andersland and Anderson, 1978). Some film water adjacent to the soil particles remains unfrozen even at very cold temperatures. The expansion of this unfrozen water in the frozen zone exerts pressure that further deforms the soil (Andersland and Anderson, 1978; Chamberlain, 1981). As a result, the thickness of the ice lenses and the cracks increases.

The freezing of soil water produces cracks and other structural modifications in a soil. These cracks and other structural changes may not be visible to the naked eye after thawing, but they have been observed under SEM magnification (Hunsicker, 1987). At low magnification, distinct cracks spaced at 0.5 mm were observed in clay soil specimens. At higher magnifications, voids as small as 0.005 mm were observed within the aggregates formed during the freezing process. Thus, freezing causes changes in the visible macrostructure and the microstructure. The changes in the macrostructure, however, are most likely the cause of large increases in hydraulic conductivity that occur in clayey soils after freezing and thawing.

The cracks and voids that form in soil during freezing can change its effective porosity, which is defined as the volume of fluid-conducting pores divided by the total volume of soil. Kim and Daniel (1992) used tracer tests to calculate the effective porosity of compacted clay specimens before and after freeze-thaw. They found the effective porosity ranged between 0.18 and 0.22 for control specimens and between

Section 3.6   Effect of Freezing and Thawing on Hydraulic Conductivity   77

0.23 and 0.33 for specimens subjected to five cycles of freeze-thaw. They concluded that more fluid-conducting pores are present in the soil after freeze-thaw, which causes the hydraulic conductivity to increase (Othman et al., 1994).

3.6.2 Effect of Freeze-Thaw on Hydraulic Conductivity

Several investigators have conducted laboratory studies to evaluate the effect of freeze-thaw on the hydraulic conductivity of compacted clays. Figure 3.21 summarizes the change in hydraulic conductivity measured in each study in terms of a hydraulic conductivity ratio, which is defined as the hydraulic conductivity after freeze-thaw divided by the hydraulic conductivity before freeze-thaw (Chamberlain et al., 1990). The graph shows hydraulic conductivity ratio as a function of initial hydraulic conductivity (i.e., hydraulic conductivity before freeze-thaw). As shown in Figure 3.21, the hydraulic conductivity ratio ranges between 1 and 1400. The ratio is generally largest when the initial hydraulic conductivity is lowest, and it decreases as the initial hydraulic conductivity becomes larger.

Clayey soils with low hydraulic conductivity are generally compacted at water contents wet of optimum. In this condition, clods easily deform during compaction, which results in a dense, relatively homogeneous mass with a very fine pore size. The fine (perhaps microscopic) pore size limits the conduction of fluid. Hence, large pores and cracks formed during freezing significantly increase the effective pore size. Consequently, the hydraulic conductivity increases dramatically.

On the other hand, specimens with relatively high initial hydraulic conductivity are generally compacted dry of optimum or with a low compactive effort. The drier aggregates of clay have higher strength and thus are more resistant to deformation during compaction. As a result, compaction dry of optimum results in a heterogeneous network of macroscopic pores and high hydraulic conductivity (Lambe, 1962; Benson and Daniel, 1990). In this condition, freeze-thaw results in greater aggregate formation and possibly even crack formation, but the size of the pores formed during freezing is

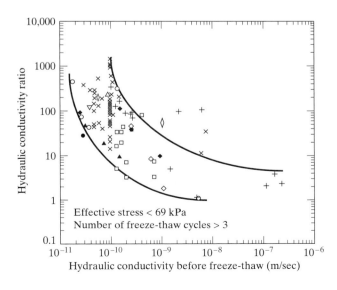

FIGURE 3.21   Hydraulic Conductivity Ratio versus Hydraulic Conductivity before Freeze-Thaw from Several Laboratory Studies (modified from Othman et al., 1994)

not dramatically different from the size of the existing pores before freezing. Thus, for specimens with high initial hydraulic conductivity, the hydraulic conductivity remains unchanged or increases only slightly after freeze-thaw (Othman et al., 1994).

3.6.3 Factors Affecting Hydraulic Conductivity during Freeze-Thaw

Laboratory and field studies indicate that the number of freeze-thaw cycles, rate of freezing, and state of stress have the largest effect on the change in hydraulic conductivity. The hydraulic conductivity increases as the number of freeze-thaw cycles and rate of freezing are increased and as the overburden pressure is decreased. Other factors such as the ultimate temperature, direction of freezing, and availability of an external supply of water do not appear to have a significant effect on the change in hydraulic conductivity.

3.6.3.1 Number of Freeze-Thaw Cycles. The number of cycles of freeze-thaw occurring in a compacted clay liner depends on the local climate. In regions where the temperature varies significantly, many cycles of freeze-thaw can be expected. Alternatively, if the temperature remains below freezing for a significant portion of the winter season, soil will be subjected to fewer freeze-thaw cycles. The number of cycles also depends on the depth of the barrier. For barriers located near the surface, such as a landfill cap, the soil temperature is more sensitive to changes in air temperature and thus more cycles of freeze-thaw occur. Deeper in the soil, fluctuations in temperature that exist at the surface are damped out and hence less cycling occurs.

Experiments have shown that an increase in hydraulic conductivity of one order of magnitude or more may occur in the first cycle of freeze-thaw (see Figure 3.22), and that most changes in hydraulic conductivity occur during the first few cycles. Increases in hydraulic conductivity after 3 to 10 freeze-thaw cycles are usually not significant (Chamberlain et al. 1990; Zimmie and LaPlante 1990; Wong and Haug 1991; Benson and Othman, 1993; Benson et al., 1995; Chamberlain et al., 1995).

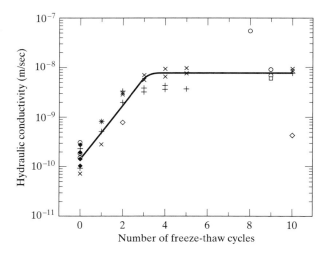

FIGURE 3.22 Hydraulic Conductivity Measured on Soils Frozen and Thawed in the Laboratory and Field (Benson et al., 1995). Used with permission of ASCE.

The most significant increase in hydraulic conductivity occurs in the first cycle because of the radical change in soil structure that occurs during the initial freeze. The network of cracks formed during the first cycle of freeze-thaw appears to cause the greatest change in a soil's fabric. During subsequent cycles, more cracks form, but these additional cracks have much less impact on hydraulic conductivity. Furthermore, at some point, the soil becomes so aggregated that no more cracking occurs, and thus the hydraulic conductivity ceases to change (Othman et al., 1994).

3.6.3.2 Rate of Freezing. Rate of freezing affects the magnitude and rate of frost heave. At greater rates of freezing, larger pore water suctions develop and a greater number of ice lenses form—provided a sufficient quantity of water is available (Dirksen and Miller, 1966). Othman and Benson (1992) studied this effect of rate of freezing on the hydraulic conductivity of compacted clays subjected to three-dimensional freeze-thaw. They used the free-standing method and controlled the rate of freezing by varying the thickness of fiberglass insulation surrounding the specimen. Specimens were subjected to rates of freezing of 2.0×10^{-6} m/s (fast), 4.0×10^{-7} m/s (moderate), and 2.6×10^{-7} m/s (slow). The rate of freezing was defined as the distance to the farthest point to which the freezing front travels divided by the time it takes to change the temperature of this point from $0°C$ ($32°F$) to the ambient temperature of the cold environment (Othman et al., 1994). The hydraulic conductivity was largest in the case of fast freezing. Examination of the specimens showed that fast freezing resulted in more ice lense formation than occurred during slow freezing. Whether the hydraulic conductivity would continue to increase or if the freezing rate was increased beyond the maximum value used by Othman and Benson (1992) is unknown.

3.6.3.3 State of Stress. *Overburden pressure after freeze-thaw.* Increases in hydraulic conductivity caused by freeze-thaw can be reversed if a large overburden pressure is applied to the soil. Overburden pressure reduces the size of voids and cracks; thus, the hydraulic conductivity decreases. Figure 3.23 shows the influence of effective stress applied after freeze-thaw on the hydraulic conductivity of 21 clays. As the effective stress increases, the hydraulic conductivity ratio decreases significantly. However, very large stresses (> 200 kPa or > 30 psi) are required to reach hydraulic conductivity ratios near unity.

The sensitivity of hydraulic conductivity to effective stress has important ramifications. For compacted clay liners used in municipal and hazardous waste landfills, exposure to freeze-thaw will result in only temporary increases in hydraulic conductivity, provided a sufficient depth of waste is placed on the liner. However, in cases where overburden pressure is low, such as a final cover or surface impoundment, changes in hydraulic conductivity are likely to be permanent (Othman et al., 1994).

Overburden pressure during freeze-thaw. The effect of overburden pressure applied during freeze-thaw is three-fold. First, it inhibits the formation and growth of ice lenses because it reduces suctions below the growing ice lenses (Konrad and Morgenstern, 1982). Second, overburden pressure reduces the hydraulic conductivity of the frozen fringe, a partially frozen zone separating frozen soil from unfrozen soil (Penner and Ueda, 1977; Konrad and Morgenstern, 1982). Both of these factors impede the migration of water to the frozen soil. Third, large pores and cracks formed

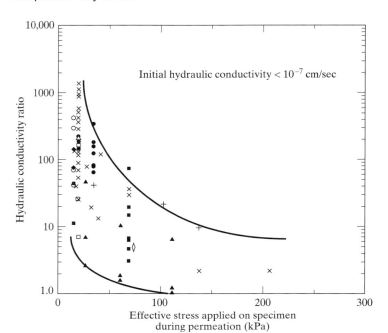

FIGURE 3.23 Influence of Effective Stress Applied during Permeation on Hydraulic Conductivity Ratio of 21 Clay Soils (modified from Othman et al., 1994)

during freeze-thaw compress during thawing when overburden pressure is applied. Hence, movement of water to the freezing front during the next freezing cycle is restricted (Othman et al., 1994).

Othman and Benson (1993) evaluated the effect of overburden pressure applied during freeze-thaw on the hydraulic conductivity of a compacted clay. They froze specimens three-dimensionally in a flexible-wall permeameter under different overburden pressures. They found that the hydraulic conductivity after freeze-thaw was largest at zero overburden pressure and decreased significantly with increasing overburden pressure. (See Figure 3.24.) Furthermore, the decrease in hydraulic conductivity caused by pressure applied during freezing is greater than the reduction in hydraulic conductivity that occurs when pressure is applied after freezing (Figure 3.24). Similar results were obtained by Chamberlain et al. (1990).

3.7 SUMMARY COMMENTS REGARDING COMPACTED CLAY LINERS

This chapter illustrates for the reader the simplicity of stipulating two or three feet of compacted clay liner with a hydraulic conductivity of no more than 1.0×10^{-7} cm/sec under laboratory conditions versus the reality of achieving the same results in the field. Beginning with the critical issue of balancing compacted density and molding water to achieve a specified hydraulic conductivity, volume stability, and shear strength (recall

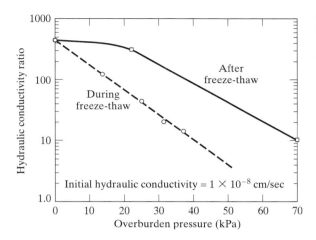

FIGURE 3.24 Effective of Overburden Pressure during and after Freeze-Thaw on Hydraulic Conductivity of a Wisconsin Clay (Othman et al., 1994)

Figure 3.15), this poses a most difficult task even under ideal (laboratory) conditions. To then take these parameters and translate them into the proper field context is a much more daunting task.

Field conditions such as the effect of clods, excessive amounts of gravel, and freeze-thaw cycling are extremely difficult to predict and control on many actual projects. The construction of a properly compacted clay liner is quite difficult and requires careful control of operating parameters, placement equipment, and compaction procedures. The difference in difficulty between placement of a low permeability geotechnical layer (e.g., a clay core in an earth dam) versus a landfill compacted clay liner is huge. The intent of this chapter is to illustrate this difference and to identify caveats and procedures associated with the proper design and construction of a landfill compacted clay liner.

PROBLEMS

3.1 The compaction requirement for clay soils in traditional applications is 90 to 95% of maximum dry density at compaction water contents on both sides of optimum, whereas landfill clay liner specifications normally require 90 to 95% of maximum dry density at compaction water contents only on the wet side of optimum water content. Why?

3.2 What is the general effect or trend of increasing compactive effort on the maximum dry unit weight and on the optimum water content, respectively, for a clayey soil?

3.3 The following test results were obtained on a clayey soil that has been proposed for a landfill liner:

- Modified Proctor Test: $(\gamma_d)_{max}$ = 122 lb/ft^3 (19.2 kN/m^3), w_{opt} = 12.0%,
- Standard Proctor Test: $(\gamma_d)_{max}$ = 117 lb/ft^3 (18.4 kN/m^3), w_{opt} = 14.0%,
- Reduced Proctor Test: $(\gamma_d)_{max}$ = 113 lb/ft^3 (17.7 kN/m^3), w_{opt} = 16.0%,
- Index Properties: G_s = 2.78, γ_w = 62.4 lb/ft^3 (9.81 kN/m^3).

Estimate the degree of saturation of the soil at the point of optimum water content and maximum dry unit weight based on these three compaction tests. What influence does increasing compactive effort have on the saturation at optimum water content?

3.4 Describe the advantages and limitations of rigid-wall permeameters.

3.5 Describe the advantages and limitations of flexible-wall permeameters.

3.6 What are the main design criteria or requirements for a compacted clay liner in a landfill?

3.7 The zone shown in Figure 3.9 seems to meet the hydraulic conductivity requirement for a compacted clay liner. Why will the final acceptable zone for hydraulic conductivity be adjusted as shown in Figure 3.10?

3.8 Why do laboratory measurements of hydraulic conductivity often underestimate the in-situ hydraulic conductivity by an order of magnitude or more?

3.9 What methods can be used to break down soil clods during construction of a compacted clay liner?

3.10 According to laboratory testing results, it is possible to achieve a hydraulic conductivity less than 1.0×10^{-7} cm/sec using a soil with up to 50 to 60% gravel. Do you think this soil can be used directly in field construction specifications for a clay liner? Why or why not?

3.11 Explain why the hydraulic conductivity of compacted clay increases after being exposed to freezing and thawing? What are the factors that affect the hydraulic conductivity of compacted clay during freezing and thawing?

3.12 Describe how the hydraulic conductivity of compacted clay is affected by the number of freeze-thaw cycles?

REFERENCES

Andersland, O. B. and Anderson, D. M., (1978) "Geotechnical Engineering for Cold Regions," McGraw-Hill Book Co., Inc., New York, NY.

Benson, C. H. and Daniel, D. E., (1990) "Influence of Clods on Hydraulic Conductivity of Compacted Clay," *Journal of Geotechnical Engineering,* ASCE, Vol. 116, No. 8, pp. 1231–1248.

Benson, C. H. and Bosscher, P. J., (1992) "Effect of Winter Exposure on the Hydraulic Conductivity of A Test Pad," *Environmental Geotechnics Report,* No. 92-8, Department of Civil and Environmental Engineering, University of Wisconsin, Madison, WI, 33 pages.

Benson, C. H. and Boutwell, G. P., (1992) "Compaction Control and Scale-Dependent Hydraulic Conductivity of Clay Liners," *Proceedings of Fifth Annual Madison Waste Conference,* University of Wisconsin, Madison, WI, pp. 62–83.

Benson, C. H. and Othman, M. A., (1993) "Hydraulic Conductivity of Compacted Clay Frozen and Thawed In Situ," *Journal of Geotechnical Engineering,* ASCE, Vol. 119, No. 2, pp. 276–294.

Benson, C. H., Abichou, T. H., and Olson, M. A., (1995) "Winter Effects on Hydraulic Conductivity of Compacted Clay," *Journal of Geotechnical Engineering,* ASCE, Vol. 121, No. 1, pp. 69–79.

Benson, C. H., Daniel, D. E., and Boutwell, G. P., (1999) "Field Performance of Compacted Clay Liners," *Journal of Geotechnical and Geoenvironmental Engineering,* ASCE, Vol. 125, No. 5, pp. 390–403.

Boutwell, G. P. and Hedges, C., (1989) "Evaluation of Waste-Retention Liners by Multivariate Statistics," *Proceedings of 12th International Conference on Soil Mechanics and Foundation Engineering,* Rio De Janerio, Brazil, Vol. 2, pp. 815–818.

Bowders, J. J. Jr. and McClelland, S., (1994) "The Effects of Freeze-Thaw Cycles on the Permeability of Three Compacted Soils," *Hydraulic Conductivity and Waste Contaminant*

Transport in Soils, ASTM STP 1142, David E. Daniel and Stephen J. Trautwein, Eds., American Society for Testing and Materials, Philadelphia, PA, pp. 461–481.

Boynton, S. S. and Daniel, D. E., (1985) "Hydraulic Conductivity Test on Compacted Clay," *Journal of Geotechnical Engineering,* ASCE, Volume 111, No. 4, pp. 465–478.

Chamberlain, E. J., (1981) "Overconsolidation Effects of Ground Freezing," *Engineering Geology,* Vol. 18, pp. 97–110.

Chamberlain, E. J., (1992) "Freeze-Thaw Effects on the Permeability of Shakopee and Rosemont II Soils," CRREL Final Report, US Army Cold Regions Research and Engineering Laboratory, Hanover, NH, 9 pages.

Chamberlain, E. J. and Gow, A. J., (1979) "Effect of Freezing and Thawing on the Permeability and Structure of Soils," *Engineering Geology,* Vol. 13, No. 1, pp. 73–92.

Chamberlain, E. J., Iskander, I., and Hunsiker, S. E., (1990) "Effect of Freeze-Thaw on the Permeability and Macrostructure of Soils," *Proceedings, International Symposium on Frozen Soil Impacts on Agriculturlal, Range, and Forest Lands,* March 21–22, Spokane, WA, pp. 145–155.

Chamberlain, E. J., Erickson, A. E., and Benson, C. H., (1995) "Effects of Frost Action on Compaction on Compacted Clay Barriers," *Proceedings of GeoEnvironment 2000,* ASCE Geotechnical Special Publication No. 46, New Orleans, LA, U.S.A., February 24–26, pp. 702–717.

Corser, P. and Cranston, M., (1991) "Observations on Long-Term Performance of Composite Clay Liners and Covers," *Geosynthetic Design and Performance,* Vancouver Geotechnical Society, Vancouver, British Columbia, Canada.

Daniel, D. E., (1984) "Predicting Hydraulic Conductivity of Clay Liners," *Journal of Geotechnical Engineering,* ASCE, Volume 110, No. 4, pp. 285–300.

Daniel, D. E., (1987) "Earth Liners for Waste Disposal Facilities," *Geotechnical Practice for Waste Disposal '87,* R. D. Woods, ed., ASCE, Ann Arbor, Michigan, pp. 21–39.

Daniel, D. E., (1990) "Summary Review of Construction Quality Control for Compacted Soil Liners," *Waste Containment System: Construction, Regulation, and Performance,* ASCE, R. Bonaparte, ed., New York, NY, pp. 175–189.

Daniel, D. E., (1994) "State-of-the-Art: Laboratory Hydraulic Conductivity Tests for Saturated Soils," *Hydraulic Conductivity and Waste Contaminant Transport in Soil,* ASTM STP 1142, Edited by D. E. Daniel and S. J. Trautwein, American Society for Testing and Materials, Philadelphia, PA, pp. 30–78.

Daniel, D. E. and Benson, C. H., (1990) "Water Content-Density Criteria for Compacted Soil Liners," *Journal of Geotechnical Engineering,* ASCE, Volume 116, No. 12, pp. 1811–1830.

Daniel, D. E. and Wu, Y. -K., (1993) "Compacted Clay Liners and Covers for Arid Sites," *Journal of Geotechnical Engineering,* ASCE, Volume 119, No. 2, pp. 223–237.

DeJong, E. and Warkentin, B. P., (1965) "Shrinkage of Soil Samples with Varying Clay Content," *Canadian Geotechnical Journal,* Vol. 2, No. 1, pp. 16–22.

Dirksen, C. and Miller, R. D., (1966) "Closed-System Freezing of Unsaturated Soil," *Proceedings, Soil Science Society of America,* Vol. 30, pp. 168–178.

Edil, T. B. and Erickson, A. E., (1985) "Procedure and Equipment Factors Affecting Permeability Testing of A Bentonite-Sand Liner Material," *Hydraulic Barriers in Soil and Rock,* ASTM STP 874, Edited by A. I. Johnson, R. K. Frobel, N. J. Cavalli, and C. B. Pettersson, American Society for Testing and Materials, Philadelphia, PA, pp. 155–170.

Hawkins, R. H. and Horton, J. H., (1965) "Bentonite as a Protective Cover for Buried Radioactive Waste," *Health Physics,* Vol. 13, No. 3, pp. 287–292.

Herrmann, J. G. and Elsbury, B. R., (1987) "Influence Factors in Soil Liners Construction for Waste Disposal Facilities," *Geotechnical Practice for Waste Disposal,* ASCE, Ann Arbor, MI, pp. 522–536.

Holtz, W. G. and Lowitz, C. W., (1957) "Compaction Characteristics of Gravelly Soils," Special Technical Publication No. 232, ASTM, Philadelphia, PA, pp. 70–83.

Hunsicker, S. E., (1987) "The Effect of Freeze-Thaw Cycles on the Permeability and Macrostructure of Fort Edwards Clay," Master of Engineering Thesis, Thayer School of Engineering, Dartmouth College, Hanover, NH, 168 pages.

Kim, W. H. and Daniel, D. E., (1992) "Effects of Freezing on the Hydraulic Conductivity of a Compacted Clay," *Journal of Geotechnical Engineering,* American Society of Civil Engineers, Vol. 118, No. 7, pp. 1083–1097.

Kleppe, J. H. and Olson, R. E., (1984) "Desiccation Cracking of Soil Barriers," Hydraulic Barriers in Soil and Rock, ASTM 874, pp. 263–275.

Konrad, J. -M. and Morgenstern, N. R., (1980) "Mechanistic Theory of Ice Lense Formation in Fine-Grained Soils," *Canadian Geotechnical Journal,* Vol. 17, No. 4 pp. 473–486.

Konrad, J. -M. and Morgenstern, N. R., (1982) "Effects of Applied Pressure on Freezing Soils," *Canadian Geotechnical Journal,* Vol. 19, No. 4, pp. 494–505.

Lambe, T. W., (1958) "The Permeability of Compacted Fine-Grained Soils," Special Technical Publication 163, ASTM, Philadelphia, PA, pp. 55–67.

Lambe, T. W., (1962) "Foundation Engineering," Leonards, G. A., Editor, Chapter 4, Soil Stabilization, pp. 351–437.

Liang, W., Bomeng, X., and Zhijin, W., (1983) "Properties of Frozen and Thawed Soil and Earth Dam Construction in Winter," *Proceedings, 4th International Conference on Permafrost,* pp. 1366–1372.

Mitchell, J. K., Hopper, D. R., and Campanella, R. G., (1965) "Permeability of Compacted Clay," *Journal of Soil Mechanics and Foundation Engineering,* ASCE, Vol. 91, No. 4, pp. 41–65.

Montgomery, R. J. and Parson, L. J., (1989) "The Omega Hills Final Cover Test Plot Study: Three-Year Data Summary," Presented at the 1989 Annual Meeting of the National Solid Waste Management Association, Washington, DC.

Mundell, J. A. and Bailey, B., (1985) "The Design and Testing of a Compacted Clay Barrier Layer to Limit Percolation through Landfill Covers," *Hydraulic Barriers in Soil and Rock,* Special Technical Publication 874, ASTM, Philadelphia, PA, pp. 246–262.

Olsen, H. W., (1962) "Hydraulic Flow through Saturated Clays," *Clays and Clay Minerals,* Vol. 9, No. 2, pp. 131–161.

Othman, M. A. and Benson, C. H., (1992) "Effect of Freeze-Thaw on the Hydraulic Conductivity of Three Compacted Clays from Wisconsin," *Transportation Research Record,* No. 1369, pp. 118–129.

Othman, M. A. and Benson, C. H., (1993) "Effect of Freeze-Thaw on the Hydraulic Conductivity and Morphology of Compacted Clay," *Canadian Geotechnical Journal,* Vol. 30(2), pp. 236–246.

Othman, M. A., Benson, C. H., and Chamberlain, E. J., (1994) "Laboratory Testing to Evaluate Changes in Hydraulic Conductivity of Compacted Clays Caused by Freeze-Thaw: State-of-the-Art," *Hydraulic Conductivity and Waste Contaminant Transport in Soil,* ASTM ATP 1142, David E. Daniel and Stephen J. Trautwein, Eds., American Society for Testing and Materials, Philadelphia, PA, pp. 227–254.

Penner, E., (1960) "The Importance of Freezing Rate in Frost Action in Soils," *Proceedings of ASTM,* Vol. 60, pp. 1151–1165.

Penner, E. and Ueda, T., (1977) *Proceedings, Symposium on Frost Action in Soils,* University of Lulea, Lulea, Sweden, Vol. I, pp. 91–100.

Shakoor, A. and Cook, B. D., (1990) "The Effect of Stone Content, Size, and Shape on the Engineering Properties of a Compacted Silty Clay," Bulletin of Association of Engineering, Geologists, XXVII(2), pp. 245–253.

Shelley, T. L. and Daniel, D. E., (1993) "Effect of Gravel on Hydraulic Conductivity of Compacted Soil Liners," *Journal of Geotechnical Engineering,* ASCE, Volume 119, No.1, pp. 54–68.

Williams, P. J., (1966) "Pore Pressures at a Penetrating Frost Line and Their Prediction," *Geotechnique,* London, England, Vol. XVI, No. 3, pp. 187–208.

Wong, L. C. and Haug, M. D., (1991) "Cyclical Closed-System Freeze-Thaw Permeability Testing of Soil Liner and Cover Materials," *Canadian Geotechnical Journal,* Vol. 28, pp. 784–793.

Zimmie, T. F., (1992) "Freeze-Thaw Effects on the Permeability of Compacted Clay Liners and Covers," *Geotechnical News,* pp. 28–30.

Zimmie, T. F. and LaPlante, C. M., (1990) "The Effect of Freeze-Thaw Cycles on the Permeability of a Fine-Grained Soil," *Proceedings, 22nd Mid-Atlantic Industrial Waste Conference,* Philadelphia, PA, pp. 580–593.

Zimmie, T. F., LaPlante, C. M., and Bronson, D., (1992) "The Effects of Freezing and Thawing on the Permeability of Compacted Clay Landfill Covers and Liners," *Environmental Geotechnology,* Proceedings of the Mediterranean Conference on Environmental Geotechnology, A. A. Balkema Publishers, pp. 213–218.

CHAPTER 4

Geomembranes

4.1	COMPOSITION AND THICKNESS OF GEOMEMBRANES	
4.2	CURRENT USES OF GEOMEMBRANES IN LANDFILLS	
	4.2.1	HIGH DENSITY POLYETHYLENE (HDPE) GEOMEMBRANES
	4.2.2	LINEAR LOW DENSITY POLYETHYLENE (LLDPE) GEOMEMBRANES
	4.2.3	COEXTRUSION VARIATIONS OF HDPE AND LLDPE
	4.2.4	TEXTURED GEOMEMBRANES
	4.2.5	FLEXIBLE POLYPROPYLENE (FPP) GEOMEMBRANES
	4.2.6	POLYVINYL CHLORIDE (PVC) GEOMEMBRANES
	4.2.7	CHLOROSULPHONATED POLYETHYLENE (CSPE) GEOMEMBRANES
4.3	TENSILE BEHAVIOR OF GEOMEMBRANES	
4.4	FRICTION BEHAVIOR OF GEOMEMBRANES	
4.5	TENSION STRESSES DUE TO UNBALANCED FRICTION FORCES	
4.6	TENSION STRESSES DUE TO LOCALIZED SUBSIDENCE	
4.7	RUNOUT AND ANCHOR TRENCHES	
	4.7.1	DESIGN OF RUNOUT LENGTH
	4.7.2	DESIGN OF RECTANGULAR ANCHOR TRENCH
	4.7.3	DESIGN OF V-SHAPED ANCHOR TRENCH
4.8	ASSESSMENT OF LEAKAGE THROUGH LINERS	
	4.8.1	FLOW RATE THROUGH COMPACTED CLAY LINER
	4.8.2	FLOW RATE THROUGH GEOMEMBRANE LINER
	4.8.3	FLOW RATE THROUGH COMPOSITE LINER
	4.8.4	COMPARISON OF THREE TYPES OF LINERS
4.9	CONCLUDING COMMENTS REGARDING GEOMEMBRANES	
	PROBLEMS	
	REFERENCES	

A geomembrane (also known as a flexible membrane liner or FML) is one of five major types of geosynthetic products used in landfill engineering (the others being geosynthetic clay liners, geonets, geotextiles, and geogrids). Geomembranes are relatively thin sheets of flexible thermoplastic or thermoset polymeric materials that are manufactured and prefabricated at a factory and transported to the site. They are placed directly on the subgrade or another geosynthetic and seamed accordingly. Because of their inherent impermeability, geomembranes have been widely used as landfill liners or covers. The primary function of a geomembrane in landfill engineering is as a liquid and/or vapor barrier.

4.1 COMPOSITION AND THICKNESS OF GEOMEMBRANES

Geomembranes are made of one or more polymers along with a variety of other ingredients such as carbon black, pigments, fillers, plasticizers, processing aids, crosslinking chemicals, antidegradants, and biocides. The polymers used to manufacture geomembranes include a wide range of plastics and rubbers differing in properties such as chemical resistance and tensile strength. The polymeric materials may be categorized as follows:

1. *Semicrystalline thermoplastics,* such as high density polyethylene (HDPE), linear low-density polyethylene (LLDPE), very low density polyethylene (VLDPE), and flexible polypropylene (fPP);
2. *Thermoplastics,* such as polyvinyl chloride (PVC);
3. *Thermoplastic elastomers,* such as chlorosulfonated polyethylene (CSPE).

The polymeric materials used most frequently in geomembranes are HDPE, LLDPE, PVC, fPP, and CSPE. The thickness of geomembranes range from 30 to 120 mil (1 mil = 0.001 inch) (0.75 to 3 mm). The recommended minimum thickness for all geomembranes is 30 mil (0.75 mm), with the exception of HDPE, which should be at least 60 mil (1.5 mm) to allow for extrusion seaming. Some geomembranes can be manufactured by a calendering process with fabric reinforcement, called scrim, to provide additional tensile strength and dimensional stability. This is the case for fPP and CSPE, which are then referred to as fPP-R and CSPE-R, respectively.

The issue of permeability is sometimes discussed for geomembranes. It should be recognized that for an intact geomembrane, the transfer of moisture across the membrane occurs by diffusion and the rates are extremely low. Testing of water vapor transmission through geomembranes is performed according to ASTM E-96. The testing results for different geomembranes are listed in Table 4.1.

A related measurement is solvent gas transmission through geomembranes [e.g., methane (CH_4)]. This lighter-than-air gas will rise up from the waste and interface with the geomembrane in the cover system. Methane gas transmission rates for different geomembranes are listed in Table 4.2.

TABLE 4.1 Water Vapor Transmission for Different Geomembranes (USEPA, 1988)

Geomembrane	Thickness		Water Vapor Transmission Rate	
	mil	mm	gallon/acre/day	$g/m^2/day$
PVC (polyvinyl chloride)	30	0.75	2.03	1.9
CPE (chlorinated polyethylene)	40	1.0	0.43	0.4
CSPE (chlorosulfonated polyethylene)	40	1.0	0.43	0.4
HDPE (high density polyethylene)	30	0.75	0.021	0.02
HDPE (high density polyethylene)	98	2.45	0.0064	0.006

TABLE 4.2 Methane Gas Transmission for Different Geomembranes (USEPA, 1988)

Geomembrane	Thickness mil	Thickness mm	Methane Gas Transmission Rate (ml/m^2-day-atm)
PVC (polyvinyl chloride)	10	0.25	4.4
PVC (polyvinyl chloride)	20	0.5	3.3
LLDPE (linear low density polyethylene)	18	0.45	2.3
CSPE (chlorosulfonated polyethylene)	32	0.8	0.27
CSPE (chlorosulfonated polyethylene)	34	0.85	1.6
HDPE (high density polyethylene)	24	0.6	1.3
HDPE (high density polyethylene)	34	0.85	1.4

4.2 CURRENT USES OF GEOMEMBRANES IN LANDFILLS

From these tables, one can see that even relatively thick HDPE is not absolutely impervious, albeit the values of vapor transmission are very low. These diffusion related values can be converted to an approximate Darcian, or advection, related permeability. [See Koerner (1998)]. The comparable values for the geomembranes in Table 4.1 range from 0.5×10^{-10} to 0.5×10^{-13} cm/sec (1.0×10^{-10} to 1.0×10^{-13} ft/min). Thus, the geomembranes listed are from 10^3 to 10^6 times lower in equivalent advective permeability than the compacted clay liners discussed in Chapter 3. In this context, they are essentially impermeable. Geomembranes act in the primary function of a barrier to aqueous liquids or vapors. The subsections that follow describe and classify the types of geomembranes used in landfill projects.

4.2.1 High Density Polyethylene (HDPE) Geomembranes

The most widely used geomembrane in the waste management industry is High Density Polyethylene (HDPE). HDPE offers excellent performance for landfill liners (primary and secondary), landfill caps, lagoon liners, wastewater treatment facilities, canal linings, floating covers, tank linings, and so on.

HDPE geomembranes are manufactured on a flat sheet extruder or by blown film process using approximately 97% high molecular weight polyethylene, from 2 to 3% carbon black, and 0.5 to 1.0% stabilizers and antioxidants. It is available in thicknesses of 30 mil (0.75 mm), 40 mil (1.0 mm), 50 mil (1.25 mm), 60 mil (1.5 mm), 80 mil (2.0 mm), 100 mil (2.5 mm), 120 mil (3.0 mm), and 140 mil (3.5 mm). Some salient features of HDPE geomembrane include the following:

Chemical Resistance: Often the chemical resistance of the liner is the most critical aspect of the design process. HDPE is the most chemically resistant of all geomembranes. Typical landfill leachates pose no threat to a liner made of HDPE.

Low Permeability: The low permeability of HDPE (recall Tables 4.1 and 4.2) strongly suggests that groundwater will not penetrate the liner; rainwater will not infiltrate a cap; and methane gas will not migrate away from the gas venting system.

Ultraviolet Resistance: HDPE has excellent resistance to ultraviolet degradation. The carbon black and antioxidants provide a synergistic effect that is very beneficial in this regard.

Generic Specification: A generic specification for HDPE geomembranes for all of the above thicknesses, in both smooth and textured styles, is available from the Geosynthetic Institute under the designation GRI-GM13.

4.2.2 Linear Low-Density Polyethylene (LLDPE) Geomembranes

Linear low-density polyethylene (LLDPE) liners are designed for waste containment applications requiring greater flexibility than HDPE. The lower molecular weight resin allows LLDPE to conform to nonuniform surfaces, making it well suited for landfill caps, pond liners, leach pads, lagoon liners, potable water containment, tunnels and tank linings.

LLDPE geomembranes are extruded as flat sheets or in a blown film manner. LLDPE geomembranes are available in thicknesses of 40 mil (1.0 mm), 50 mil (1.25 mm), 60 mil (1.5 mm), 80 mil (2.0 mm), and 100 mil (2.5 mm). Some salient features of LLDPE geomembrane include the following:

Multi-Axial Tension Properties: LLDPE exhibits high out-of-plane elongation, allowing it to conform to differential settlement and non-uniform surfaces without jeopardizing the integrity of the liner.

Puncture Resistance: LLDPE provides excellent puncture resistance due to its flexibility. It conforms readily to gravel and other irregularities in a subgrade.

Stress Crack Resistance: LLDPE is relatively unaffected by stress cracking.

Chemical Resistance: In most applications, LLDPE is resistant to leachate and other chemicals that can have harmful effects on the environment. Its chemical resistance is surpassed only by HDPE.

Low Permeability: LLDPE has a very low permeability to both liquids and gases.

Ultraviolet Resistance: LLDPE exhibits good resistance to ultraviolet degradation.

Generic Specification: A generic specification for LLDPE geomembranes for all thicknesses, in both smooth and textured styles, is available from the Geosynthetic Institute under the designation GRI-GM17.

4.2.3 Coextrusion Variations of HDPE and LLDPE

By utilizing auxiliary extruders, both HDPE and LLDPE geomembranes can be coextruded, using flat dies or blown film methods, into a number of interesting variations.

Coextruded sheet is manufactured such that a HDPE/LLDPE/HDPE geomembrane results. The HDPE on the upper and lower surfaces is approximately 10 to 20% of the total sheet thickness. The objective is to retain the excellent chemical resistance of HDPE on the geomembrane's surfaces, with the flexibility of LLDPE in the core. Coex, as with all of these coextruded variations, is not a laminated material. The bonding of the various streams of molten polyethylene occurs at the molecular level (i.e., primary bonding occurs), and thus the materials cannot be separated.

White/black coextruded HDPE and LLDPE geomembranes are produced to take advantage of the white surface facing the ambient atmosphere. In this manner, the surface temperature increase is greatly reduced (to approximately 50% of a black surface), and the sheet expansion is reduced by a similar amount. Waves, or wrinkles, are reduced accordingly. In addition, black areas showing through the white surface provides for excellent visual detection of scratches, cuts, or holes.

High/low carbon black HDPE and LLDPE geomembranes are produced to provide for field spark detection of cuts, holes, and seaming defects. The lower portion of the sheet has a 20 to 25% carbon black loading, while the remaining portion is the standard 2 to 3%. By means of an electric circuit connected to the underside of the sheet, the entire facility can be tested by the electric spark method.

4.2.4 Textured Geomembranes

Both HDPE and LLDPE can be made with textured surfaces. The textured surface greatly improves slope stability by increasing interface friction between the geomembrane and soils, geotextiles, and other geosynthetics. Cover soils placed on top of a geomembrane liner are held more securely because of greatly increased friction. This enables design engineers to improve factors of safety on slopes of varying steepness. Table 4.3 lists the peak friction angle and peak adhesion between a textured geomembrane and various materials by direct shear box testing. Textured surfaces can be made on one or both sides of a geomembrane. There are four different methods that can be used for texturing:

(i) blown film coextrusion
(ii) impingement by hot PE particles
(iii) lamination by a PE foam
(iv) structured by using embossed calendars

Textured geomembranes are generally produced with a 6-inch (150-mm) ± nontextured border on both sides of the sheet. The smooth border provides a better surface for welding than a textured surface. The smooth edges also permit quick verification of the core's thickness and strength before installation. Textured geomembranes are available in thicknesses of 40 mil (1.0 mm), 60 mil (1.5 mm), 80 mil (2.0 mm), and 100 mil (2.5 mm), for a wide range of applications.

TABLE 4.3 Peak Shear Strength Parameters between Textured Geomembranes and Various Materials (USEPA, 1991)

Textured Geomembrane Against	Friction Angle (degree)	Adhesion lb/ft^2	Adhesion kN/m^2
Drainage Sand	37	25	1.2
Clay	29	150	7.2
Nonwoven Geotextile	32	55	2.6

4.2.5 Flexible Polypropylene (fPP) Geomembranes

Still within the polyolefin family of polymers is flexible polypropylene which is often reinforced with an open weave fabric called *scrim*. Its designation thus becomes fPP-R. It is available in thicknesses of 36 mil (0.91 mm) and 45 mil (1.14 mm). By virtue of the scrim, the tensile strength is quite high. The material is used in many applications such as floating covers for surface impoundments (e.g., in conditions where tear stresses are high). Some of its characteristics are as follows:

Chemical Resistance: As a member of the polyolefin family, chemical resistance is quite good in this product.

Ultraviolet Resistance: The resistance to high temperatures and UV radiation also is good, as attested by the use of this type of geomembrane for exposed flat roofs of buildings.

Installation Survivability: With tear resistance being high, puncture, impact and other survivability characteristics are comparably high as well.

Generic Specification: A generic specification of fPP-R geomembranes in both of the above thicknesses is available from the Geosynthetic Institue under the designation of GRI-GM18.

4.2.6 Polyvinyl Chloride (PVC) Geomembranes

One of the original geomembranes used for waste containment applications was polyvinyl chloride (PVC). It has been used in almost all applications requiring a barrier material, including landfills. Thicknesses range from 20 mils (0.50 mm) to 100 mils (2.5 mm).

PVC geomembranes are formulations consisting of a number of ingredients including plasticizers to impart flexibility in what is ordinarily a relatively stiff polymer. On the basis of weight percentages, a typical PVC formation consists of the following components:

resin = 35%
plasticizer = 30%
filler = 25%
pigment = 5–10%
additions = 2–3%

Some important features of PVC geomembranes:

Flexibility: The material exhibits excellent out-of-plane flexibility and conformance to irregular and harsh subgrades.

Seameability: PVC geomembranes can be seamed by solvent methods as well as thermal techniques.

Constructability: PVC geomembranes are arguably the easiest to install of all types of geomembranes.

Generic Specification: A generic specification for PVC geomembranes in various thicknesses is available from the PVC Geomembrane Institute.

4.2.7 Chlorosulphonated Polyethylene (CSPE) Geomembranes

As with PVC geomembranes, chlorosulphonated polyethylene geomembranes have been used for almost 30 years. They are always scrim reinforced and carry the designations of CSPE-R. Similar to fPP-R they are available in thicknesses of 36 mil (0.91 mm) and 45 mil (1.14 mm). CSPE-R geomembranes are marketed under the trade name Hypalon®. They have been used for essential all-barrier (or waterproofing) applications, including landfill liners and covers.

CSPE-R geomembranes are somewhat unique in that they are initially thermoplastic polymers (and thus can be thermally seamed), but over time they crosslink and become thermoset polymers. Their approximate weight percent formulation is as follows:

>   resin = 45%
>   filler = 20–25%
>   carbon black (or pigment) = 20–25%
>   additives = 5–7%

Some important features of CSPE-R geomembranes include the following:

> *Chemical Resistance:* The chemical resistance to a wide range of landfill leachate is excellent.
>
> *Ultraviolet Resistance:* UV and high temperature resistance are excellent as well. CSPE-R geomembranes are regularly used in exposed conditions like floating covers and uncovered surface impoundment liners.
>
> *Installation Survivability:* As with other scrim-reinforced geomembranes, tear, puncture, and impact strengths are high. Seaming can be performed by both solvent and thermal methods, although after full crosslinking an adhesive bonding agent must be used.
>
> *Generic Specification:* Although it is currently depreciated, the NSF specification is still available for specifying CSPE-R geomembranes.

4.3 TENSILE BEHAVIOR OF GEOMEMBRANES

There are a number of tensile tests that have been developed to determine the strength of geomembranes. The index tension test is used routinely for quality control and quality assurance of the manufactured sheet materials. A relatively narrow sample is used in this test. The width of the specimen used in index tensile test is 0.25 inch (6.35 mm) for HDPE and LLDPE and 1.0 inch (25.4 mm) for fPP-R, PVC, and CSPE-R. The results of an index tension test of HDPE, LLDPE, PVC, and CSPE-R geomembranes are shown in Figure 4.1. Figure 4.1 shows that the scrim-reinforced geomembrane CSPE-R exhibited the highest strength, but failed abruptly when the scrim broke. The HDPE geomembrane responded in characteristic fashion by displaying a pronounced yield point, dropping slightly, and then straining at nearly constant stress to approximately 1,000% where failure or rupture occurred. The LLDPE and

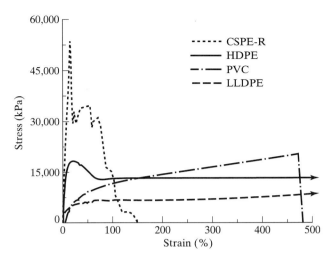

FIGURE 4.1 Index Tension Test Results Commonly Used Geomembranes (Koerner, 1998)

PVC geomembranes gave relatively smooth responses, gradually increasing in stress until failure at 700% and 450% strain, respectively. Quantitative information gained from these curves include the following (Koerner, 1994):

Stress at maximum stress (at ultimate for PVC and LLDPE, at scrim break for CSPE-R, and at yield for HDPE);

Maximum strain (at ultimate for PVC and LLDPE, at scrim break for CSPE-R, and at yield for HDPE);

Modulus (the slope of the initial portion of the stress-strain curve);

Ultimate stress (at complete failure);

Ultimate strain (at complete failure).

Values of index stress-strain parameters for four different geomembrane materials are given in the first part of Table 4.4 (Part a). While all of the listed values of strength are significant, attention is often focused on the stress at a particular allowable strain for materials like PVC and LLDPE, the scrim-breaking stress for reinforced materials like CSPE-R, and the yield stress for HDPE materials. Polymers are viscoelastic materials, and strain invariably plays an important role in the determination of stress levels that are selected for analysis and design purposes.

A major criticism of narrow index test specimens is their tendency to contract or "neck down" at the center, a behavior not experienced by geomembranes in field situations. Uniform width and wider specimens are desirable for testing purposes. Accordingly, specimens with a width of 8 inches (200 mm) have been adopted for testing geomembranes. Figure 4.2 presents tensile stress versus strain curves on the same four geomembrane materials that were shown in Figure 4.1, but instead with a uniform 8-inch (200-mm) width. This test is called a wide-width tension test. The results of this test are tabulated in the second part of Table 4.4 (Part b). The 8-inch (200-mm) wide-width type of test specimen minimizes the influence of "necking" and simulates in-situ conditions much better than do narrow-width specimens. This simulation is

TABLE 4.4 Tensile Behavior of 60-mil (1.5-mm) HDPE, 40-mil (1-mm) LLDPE, 30-mil (0.75-mm) PVC, and 36-mil (0.9-mm) CSPE-R (Koerner, 1994)

(a) Index Tension Tests (Figure 4.1)

Test Property	HDPE	LLDPE	PVC	CSPE-R
Maximum Stress (lb/in^2)	2,700	1,200	3,000	7,900
(kPa)	18,600	8,300	21,000	54,500
Corresponding Strain (%)	17	500+	480	19
Modulus (lb/in^2)	48,000	11,000	4,500	48,000
(Mpa)	330	76	31	330
Ultimate Stress (lb/in^2)	2,000	1,200	3,000	830
(kPa)	13,800	8,300	20,700	5,700
Corresponding Strain (%)	500+	500+	480	110

(b) Wide-Width Tension Tests (Figure 4.2)

Test Property	HDPE	LLDPE	PVC	CSPE-R
Maximum Stress (lb/in^2)	2,300	1,100	2,000	4,500
(kPa)	15,900	7,600	13,800	31,000
Corresponding Strain (%)	15	400+	210	23
Modulus (lb/in^2)	61,000	10,000	2,900	43,000
(Mpa)	450	69	20	300
Ultimate Stress (lb/in^2)	1,600	1,100	2,000	410
(kPa)	11,000	7,600	13,800	2,800
Corresponding Strain (%)	400+	400+	210	79

(c) Axisymmetric Tension Tests (Figure 4.4)

Test Property	HDPE	LLDPE	PVC	CSPE-R
Maximum Stress (lb/in^2)	3,400	1,500	2,100	4,500
(kPa)	23,500	10,300	14,500	31,000
Corresponding Strain (%)	12	75	100	13
Modulus (lb/in^2)	105,000−	24,000−	15,000−	50,000−
(Mpa)	720−	170−	100−	350−
Ultimate Stress (lb/in^2)	3,400	1,500	2,100	4,500
(kPa)	23,500	10,300	14,500	31,000
Corresponding Strain (%)	25	75	100	13

Note: " + " = did not fail; " − " = values felt to be high.

particularly important when plane-strain conditions are assumed in the design process (e.g., for side slope stability calculations).

In some situations geomembrane's tensile resistance is mobilized by out-of-plane stresses. Localized deformation beneath a geomembrane is such a case. This type of behavior would be applicable for a geomembrane used in a landfill cover or vertical

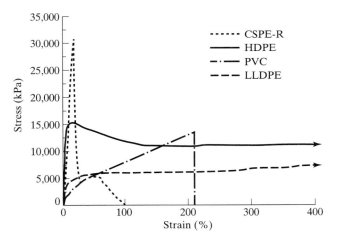

FIGURE 4.2 Tensile Test Results on 200-mm Wide-Width Specimens of Commonly Used Geomembranes Using ASTM D4885 Test Method (Koerner, 1998)

expansion placed over subsiding solid waste material. This condition can be modeled by placing the geomembrane in an empty container, as shown in Figure 4.3. An appropriate seal is made with the cover section, and water is introduced under pressure above the geomembrane.

The pressure is increased until failure of the test specimen occurs. Figure 4.4 presents stress versus strain curves on the same four geomembrane materials that were

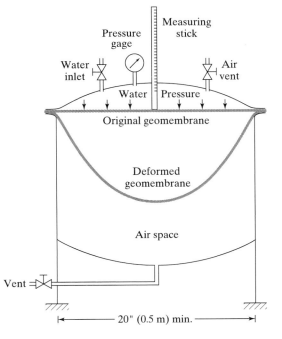

FIGURE 4.3 Schematic Diagram of Three-Dimensional Axisymmetric Geomembrane Tension Test Apparatus (Koerner et al., 1990)

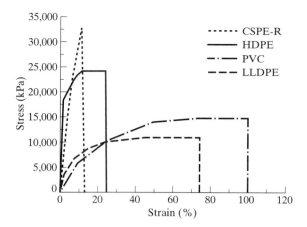

FIGURE 4.4 Three-Dimensional Axisymmetric Stress-versus-Strain Response Curves of Various Types of Geomembranes (after Koerner et al., 1990)

shown in Figures 4.1 and 4.2, but now under axisymmetric hydrostatic pressure. These test results have also been transcribed onto Table 4.4 (Part c) where the different tensile test results can be compared. Note the relatively large differences between the axisymmetric behavior and the other tests presented earlier. These results underscore the importance of modeling or simulating field conditions with appropriate laboratory tests.

4.4 FRICTION BEHAVIOR OF GEOMEMBRANES

Correct estimates of soil-geomembrane and geomembrane-geotextile interface friction are critical for the proper design of geomembrane-lined side slopes of landfills. The data of Table 4.3 in this regard was very general. Cover soils can often slide over the surface of geomembranes. Alternatively, the geomembrane itself can fail by pulling out of the anchor trench and slipping downhill on a lower contact surface beneath the geomembrane. There currently are many papers available on interface friction between geomembranes and numerous other surfaces (soils and geosynthetics). (See Mitchell et al., 1990; Seed et al., 1990; Byrne et al., 1992; Stark and Poeppel, 1994.) Results from an early effort in this regard by Martin et al. (1984) are presented in Table 4.5.

As seen in Table 4.5(a), peak friction angles of soil to geomembrane are always less than soil to soil, with the smoother, harder geomembranes being the lowest (e.g., HDPE). Conversely, the rough, softer geomembranes (EPDM-R, PVC, and CSPE-R) have relatively high friction values. It is important to note that textured HDPE was not evaluated in this early reference. Its values are significantly higher than those shown in the table. Tables 4.5(b) and (c) give the geomembrane-to-geotextile and soil-to-geotextile friction values that are necessary for design of slopes constructed with a geotextile under or over the geomembrane liner.

An important consideration, and one that has been over looked in early interface friction testing studies (like Martin et al., 1984), is the issue of peak versus large defor-

TABLE 4.5 Peak Friction Values and Efficiencies (in parentheses) (Martin et al., 1984)

(a) Soil-to-Geomembrane Friction Angle

	Soil Types		
Geomembrane	Concrete Sand ($\phi = 30°$)	Ottawa Sand ($\phi = 28°$)	Mica Schist Sand ($\phi = 26°$)
EPDM-R	24° (0.77)*	20° (0.68)	24° (0.91)
PVC			
Rough	27° (0.88)	–	25° (0.96)
Smooth	25° (0.81)	–	21° (0.79)
CSPE-R	25° (0.81)	21° (0.72)	23° (0.87)
HDPE (smooth)	18° (0.56)	18° (0.61)	17° (0.63)

(b) Geomembrane-to-Geotextile Friction Angle

	Geomembrane				
		PVC			HDPE
Geotextile	EPDM-R	Rough	Smooth	CSPE-R	(smooth)
Nonwoven, Needle-Punched	23°	23°	21°	15°	8°
Nonwoven, Heat-Bonded	18°	20°	18°	21°	11°
Woven, Monofilament	17°	11°	10°	9°	6°
Woven, Slit Film	21°	28°	24°	24°	10°

(c) Soil-to-Geotextile Friction Angle

	Soil Types		
Geotextile	Concrete Sand ($\phi = 30°$)	Ottawa Sand ($\phi = 28°$)	Mica Schist Sand ($\phi = 26°$)
Nonwoven, Needle-Punched	30° (1.00)	26° (0.92)	25° (0.96)
Nonwoven, Heat-Bonded	26° (0.84)	–	–
Woven, Monofilament	26° (0.84)	–	–
Woven, Slit Film	24° (0.77)	24° (0.84)	23° (0.87)

*Efficiency values in parentheses are based on the relationship $E = \tan\delta/\tan\phi$.

mation (or residual) interface strengths. Shear tests must be carried out to such deformations that the possible drop-off of shear strength is assessed. This type of "strain softening" behavior can occur in geosynthetic interfaces such as textured geomembranes, thick nonwoven geotextiles, and geosynthetic composite materials. (More information about geomembrane interface friction can be found in Chapter 13.)

4.5 TENSION STRESSES DUE TO UNBALANCED FRICTION FORCES

The shear stresses from the cover soil above the liner act downward on the underlying geomembrane and mobilize upward shear stresses beneath the geomembrane from the underlying soil (Koerner and Hwu, 1991). This interaction is depicted in Figure 4.5, and the stresses are shown in Figure 4.6.

The designations in Figures 4.5 and 4.6 are as follows:

Tw = total tension force in geomembrane, lb or kN;
W = total weight acting on geomembrane, lb or kN;
L = length of geomembrane, ft or m;
w = width of geomembrane, ft or m;
β = slope angle, degree;
τ_U = shear stress between geomembrane and upper soil, lb/ft² or kN/m²;
τ_L = shear stress between geomembrane and lower soil, lb/ft² or kN/m²;
c_{aU} = adhesion between geomembrane and upper soil, lb/ft² or kN/m²;
c_{aL} = adhesion between geomembrane and lower soil, lb/ft² or kN/m²;
δ_U = friction angle between geomembrane and upper soil, degree; and
δ_L = friction angle between geomembrane and lower soil, degree.

Three different scenarios can be envisioned for the situation shown in Figure 4.5:

If $\tau_U = \tau_L$, the geomembrane goes into a state of pure shear which should not be of great concern for most types of geomembranes.

If $\tau_U < \tau_L$, the geomembrane goes into a state of pure shear up to a magnitude of τ_U and the balance of $\tau_L - \tau_U$ is simply not mobilized.

If $\tau_U > \tau_L$, the geomembrane goes into a state of pure shear equal to τ_L and the balance of $\tau_U - \tau_L$ must be carried by the geomembrane in tension.

This last named case is the focus of this part of the design process. This situation generally occurs when a material with high interface friction (like sand or gravel) is placed above the geomembrane and a material with low interface friction (like high moisture content clay) is placed beneath the geomembrane. The essential equation for this design scenario is shown next. The tensile stress acting on a geomembrane due to

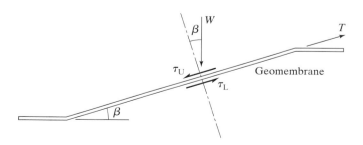

FIGURE 4.5 Shear and Tensile Stresses Acting on a Geomembrane

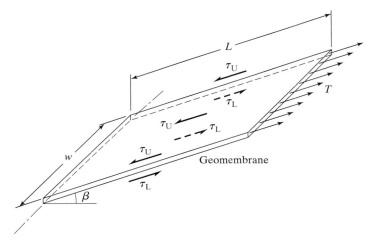

FIGURE 4.6 Geomembrane Tensile Stress due to Unbalanced Friction Forces

unbalanced friction forces is shown in Figure 4.6. The factor of safety against sliding in this case is given by

$$(FS)_S = \frac{\text{Resisting Force}}{\text{Driving Force}} = \frac{Tw + \tau_L \cdot w \cdot L}{\tau_U \cdot w \cdot L} \quad (4.1)$$

where $(FS)_S$ = factor of safety against sliding.

Based on the relative magnitudes of the upper and lower shear stresses, one of the following safety factors will be applicable:

(a) If $\tau_U = \tau_L$, $(FS)_S > 1$, pure shear @ $\tau_U = \tau_L$.
(b) If $\tau_U < \tau_L$, $(FS)_S > 1$, pure shear $= \tau_U$, rest of $(\tau_L - \tau_U)$ is not mobilized.
(c) If $\tau_U > \tau_L$, $(FS)_S$ may be $\geq$ or $<$ 1, depending on the Tw value.

For condition (c), when $(FS)_S = 1 \Rightarrow \tau_U \cdot w \cdot L = Tw + \tau_L \cdot w \cdot L$

$$Tw = \tau_U \cdot w \cdot L - \tau_L \cdot w \cdot L$$
$$Tw/w = (\tau_U - \tau_L) \cdot L$$
$$T = (\tau_U - \tau_L) \cdot L = \tau_U \cdot L - \tau_L \cdot L = S_U - S_L$$

where T = tension force per unit width in geomembrane, lb/ft or kN/m;
 S_U = shear force per unit width between geomembrane and upper soil, lb/ft or kN/m; and
 S_L = shear force per unit width between geomembrane and lower soil, lb/ft or kN/m.

In the derivation that follows,

 S = shear force per unit width between geomembrane and adjacent material, lb/ft or kN/m;

C = adhesion force per unit width between geomembrane and adjacent material, lb/ft or kN/m;
δ = friction angle between geomembrane and adjacent material, degree;
c = adhesion between geomembrane and adjacent material, lb/ft² or kN/m²;
N = normal force per unit width acting on geomembrane, lb/ft or kN/m;
γ_s = unit weight of cover soil, lb/ft³ or kN/m³; and
H = thickness of cover soil (perpendicular to slope), ft or m.

The unit shear force comprises a frictional and cohesive component, and can be written as

$$S = C + N \cdot \tan \delta \tag{4.2}$$

where $N = \gamma_s \cdot H \cdot (L \cdot \cos \beta) = \gamma_s \cdot H \cdot L \cdot \cos \beta$
and $C = c \cdot L$

Therefore, $S = c \cdot L + \gamma_s \cdot H \cdot L \cdot \cos \beta \cdot \tan \delta \tag{4.3}$

The mobilized unit shear resistance on the upper and lower surfaces of the membrane will be given by corresponding equations:

$$S_U = c_{aU} \cdot L + \gamma_s \cdot H \cdot L \cdot \cos \beta \cdot \tan \delta_U$$
$$S_L = c_{aL} \cdot L + \gamma_s \cdot H \cdot L \cdot \cos \beta \cdot \tan \delta_L$$

The mobilized or required tension force in the geomembrane is the difference between the unit shear at the top and bottom; that is,

$$T_{reqd} = (S_U - S_L)$$

or $T_{reqd} = [(c_{aU} - c_{aL}) + \gamma_s \cdot H \cdot \cos \beta \cdot (\tan \delta_U - \tan \delta_L)] \cdot L \tag{4.4}$

The allowable unit tension force, on the other hand, is given by the equation

$$T_{allow} = \sigma_{allow} \cdot t \tag{4.5}$$

The factor of safety against tensile failure in the geomembrane is expressed by the ratio of allowable-to-required unit tensile forces, written as

$$(FS)_T = T_{allow} / T_{reqd} \tag{4.6}$$

where T_{reqd} = required tension force per unit width in the geomembrane, lb/ft or kN/m;
T_{allow} = allowable tension force per unit width in the geomembrane, lb/ft or kN/m;
σ_{allow} = allowable tensile stress in the geomembrane, lb/ft² or kN/m²;
t = thickness of geomembrane, ft or m; and
$(FS)_T$ = factor of safety for geomembrane against tension.

The target values are T_{break} for scrim-reinforced geomembranes, T_{yield} for semicrystalline geomembranes, and T_{allow} (at a certain value of strain) for nonreinforced flexible geomembranes. Note that these curves should be obtained from a wide-width tensile test such as that illustrated in Figure 4.2.

EXAMPLE 4.1

The following are given: a soil slope angle $\beta = 18.4°$ [i.e. 3(H):1(V)], slope length $L = 100$ m (330 ft), cover soil thickness $H = 0.90$ m (3 ft), and cover soil unit weight $\gamma_s = 18$ kN/m^3 (115 lb/ft^3). A geomembrane has an allowable tensile strength (T_{allow}) of 30 kN/m (2,055 lb/ft). The shear strength parameters of the geomembrane with respect to the upper soil are $c_{aU} = 0$ and $\delta_U = 14°$, and with respect to the lower soil are $c_{aL} = 2.3$ kPa (48 lb/ft^2) and $\delta_L = 5°$. Calculate the tension in the geomembrane and the resulting FS against tensile geomembrane failure.

Solution:

$$T_{\text{reqd}} = [(c_{aU} - c_{aL}) + \gamma_s \cdot H \cdot \cos\beta \cdot (\tan\delta_U - \tan\delta_L)] \cdot L \quad (4.4)$$

$$= (0 - 2.3) + (18)(0.9)(\cos 18.46°)(\tan 14° - \tan 5°)(100)$$

$$= [(0 - 2.3) + (18)(0.9)(0.949)(0.249 - 0.087)](100)$$

$$= [(0 - 2.3) + (18)(0.9)(0.949)(0.162)](100)$$

$$= (-2.3 + 2.49)(100)$$

$$= 19 \text{ kN/m (1,300 lb/ft)};$$

$$(FS)_T = T_{\text{allow}}/T_{\text{reqd}} \quad (4.6)$$

$$= 30/19$$

$$= 1.58$$

If the factor of safety is too low, an alternative design for increasing $(FS)_T$ is to bench the cover soil (thereby decreasing the slope length) or to use a liner whose lower surface has a higher adhesion or a higher friction surface than the one used in the example. One-sided textured geomembrane (with the textured surface down) is a possibility in this regard.

4.6 TENSION STRESSES DUE TO LOCALIZED SUBSIDENCE

Whenever localized subsidence occurs beneath a geomembrane that is supporting a cover soil, some induced tensile stresses will occur due to out-of-plane forces from the overburden. This type of subsidence can be expected in landfill closure situations above completed or abandoned landfills where the underlying waste has been poorly compacted. The magnitude of the induced tensile stresses in the geomembrane depends upon the dimensions of the subsidence zone and on the cover soil properties (Koerner and Hwu, 1991).

The subsidence in question and accompanying deformation in the geomembrane is shown schematically in Figure 4.7. In the analysis, the deformed shape of the geomembrane is assumed to be that of a spheroid of gradually decreasing center point along the symmetric axis of the deformed geomembrane. As a worst case assumption, the geomembrane is assumed to be fixed at the circumference of the subsidence zone. The required tensile strength σ_{reqd} in the geomembrane can be derived as follows:

FIGURE 4.7 Tensile Stresses in a Geomembrane Mobilized by Cover Soil and Caused by Subsidence

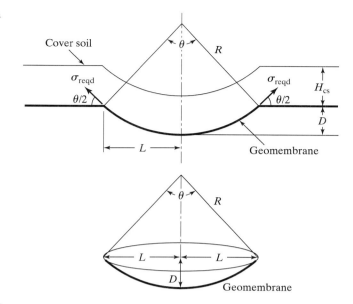

Circumference: $C = 2 \cdot \pi \cdot L$

$$R^2 = (R - D)^2 + L^2$$
$$R^2 = R^2 - 2 \cdot R \cdot L + D^2 + L^2$$
$$R = \frac{D^2 + L^2}{2 \cdot D}$$

From Figure 4.7, $\Sigma M_o = 0$. So,

$$\int_0^L (2 \cdot \pi \cdot r) \cdot dr \cdot \gamma_s \cdot H_{cs} \cdot r = \sigma_{reqd} \cdot t \cdot C \cdot R$$

or
$$(2/3) \cdot \pi \cdot L^3 \cdot \gamma_s \cdot H_s = \sigma_{reqd} \cdot t \cdot (2 \cdot \pi \cdot L) \cdot R$$

Hence,

$$\sigma_{reqd} = \frac{(2/3) \cdot \pi \cdot L^3 \cdot \gamma_s \cdot H_{cs}}{t \cdot (2 \cdot \pi \cdot L) \cdot R}$$

$$= \frac{L^2 \cdot \gamma_s \cdot H_{cs}}{3 \cdot t \cdot R}$$

or
$$\sigma_{reqd} = \frac{2 \cdot D \cdot L^2 \cdot \gamma_s \cdot H_{cs}}{3 \cdot t \cdot (D^2 + L^2)} \quad (4.7)$$

For multilayered cover soil, the preceding equation becomes

$$\sigma_{reqd} = \frac{2 \cdot D \cdot L^2 \cdot [\Sigma(\gamma_s)_i \cdot H_i]}{3 \cdot t \cdot (D^2 + L^2)} \quad (4.8)$$

where σ_{reqd} = required tensile strength of geomembrane caused by subsidence, lb/ft² or kN/m²;

γ_s = unit weight of cover soil, lb/ft³ or kN/m³;
H_{cs} = thickness of cover soil, ft or m;
t = thickness of geomembrane, ft or m;
R = radius of spheroid cause by subsidence, ft or m;
D = depth of subsidence, ft or m;
L = distance between symmetric axis and the top edge of the subsidence, ft or m;
C = circumference of the top edge of the subsidence, ft or m;
$(\gamma_s)_i$ = unit weight of any layer of cover soil, lb/ft³ or kN/m³; and
H_i = depth of any layer of cover soil, ft or m.

The calculated value of σ_{reqd} for a specific site situation is compared with an appropriate laboratory simulation test for allowable tension. The recommended test in this case is a three-dimensional axisymmetric tension test as shown in Figure 4.3. The results of the three-dimensional axisymmetric tension test for various types of geomembranes are shown in Figure 4.4. The factor of safety against tensile failure during subsidence is then given by the expression

$$(FS)_{sub} = \sigma_{allow}/\sigma_{reqd} \tag{4.9}$$

where $(FS)_{sub}$ = factor of safety of geomembrane against the mobilized tensile stresses caused by subsidence;
σ_{allow} = allowable tensile strength of geomembrane obtained from three-dimensional axisymmetric tension test, lb/in² or kN/m²; and
σ_{reqd} = required tensile strength of geomembrane caused by subsidence, lb/in² or kN/m².

EXAMPLE 4.2

Given cover soil thickness H_{cs} = 3.0 ft (0.9 m), and cover soil unit weight γ_s = 120 lb/ft³ (18.9 kN/m³). A local subsidence occurs which is estimated to be 1.0-ft (0.3-m) deep by 3.0-ft (0.9-m) radius. The geomembrane is 40-mils (1-mm) thick and has a σ_{allow} of 1,000 lb/in² (6,900 kN/m²). Determine the factor of safety of the geomembrane against the mobilized tensile stresses.

Solution:

$$\sigma_{reqd} = \frac{2 \cdot D \cdot L^2 \cdot \gamma_s \cdot H_{cs}}{3 \cdot t \cdot (D^2 + L^2)} \tag{4.7}$$

$$\sigma_{reqd} = \frac{(2)(1.0)(3.0)^2(120)(3.0)}{(3)(0.040/12)[(1.0)^2 + (3.0)^2]} = 64{,}800 \text{ lb/ft}^2 = 450 \text{ lb/in}^2 \text{ (3,100 kN/m}^2\text{)}$$

$$(FS)_{sub} = \sigma_{allow}/\sigma_{reqd} \tag{4.9}$$

$$= 1{,}000/450 = 2.2$$

4.7 RUNOUT AND ANCHOR TRENCHES

Geomembranes (and other geosynthetics if they are also involved) are usually terminated by a horizontal runout at the top of the slope followed by a short drop into an anchor trench. The anchor trench is backfilled with soil and suitably compacted to hold installed geosynthetics in place against applied loads. Concrete anchor trenches with full fixity to the liner should generally not be used, since soil backfill is invariably adequate to mobilize the full strength of the geosynthetics.

Two separate cases will now be analyzed—one with a geomembrane runout only and no anchor trench (as is often used with ditch liners), and the other with both runout and an anchor trench (as typically used with reservoirs and landfills). For the intermediate case or thick geomembrane, a V-shaped anchor trench is also possible. The holding capacity of runout length and anchor trenches is developed by the soil placed over the geomembrane, which creates frictional resistance between geomembrane and soil. The key factors affecting the anchor trench holding include runout length, cover soil depth, anchor trench shape and depth, type of soils underlying and overlying the geomembrane, and type of geomembrane used.

4.7.1 Design of Runout Length

Figure 4.8 depicts the case with a geomembrane runout and no anchor, as well as the forces and stresses involved. With regard to the cover soil, it is assumed that it applies normal stress because of its weight, but that it does not contribute frictional resistance above the liner. This is because the soil above moves along with the geomembrane as it deforms and thus provides no lateral (tangential) resistance.

From Figure 4.8, the following force summations can be observed, which in turn lead to the appropriate design equations:

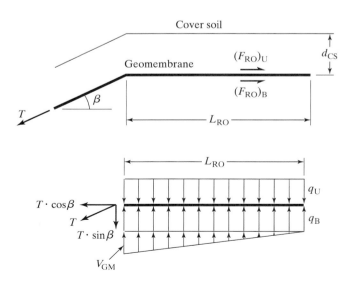

FIGURE 4.8 Cross Section of Geomembrane Runout Section and Related Stresses and Forces Involved

From $\Sigma F_V = 0$,
$$T \cdot \sin\beta = 0.5 \cdot V_{GM} \cdot L_{RO}$$

Since $q_U = q_B = \gamma_s \cdot d_{CS}$,
$$V_{GM} = \frac{2 \cdot T \cdot \sin\beta}{L_{RO}} \quad (4.10)$$

From $\Sigma F_H = 0$,
$$T \cdot (\cos\beta) = (F_{RO})_U + (F_{RO})_B$$

The friction force above runout geomembrane is given by
$$(F_{RO})_U = q_U \cdot L_{RO} \cdot \tan\delta_F$$

The friction force above runout geomembrane, $(F_{RO})_U$, is neglected in the following analysis since the cover soil probably moves along with the geomembrane as it deforms neglected in the following analysis.

The friction force beneath runout geomembrane is
$$(F_{RO})_B = (q_B + 0.5 V_{GM}) \cdot L_{RO} \cdot \tan\delta_C$$

or
$$(F_{RO})_B = [q_B + 0.5 \cdot (2 \cdot T \cdot \sin\beta / L_{RO})] \cdot L_{RO} \cdot \tan\delta_C$$

which leads to
$$T \cdot (\cos\beta) = q_B \cdot L_{RO} \cdot \tan\delta_L + T \cdot \sin\beta \cdot \tan\delta_C \quad (4.11)$$

or
$$T \cdot (\cos\beta - \sin\beta \cdot \tan\delta_C) = \gamma_s \cdot d_{CS} \cdot L_{RO} \cdot \tan\delta_C$$

so that
$$T = \frac{\gamma_s \cdot d_{CS} \cdot L_{RO} \cdot \tan\delta_C}{\cos\beta - \sin\beta \cdot \tan\delta_C} \quad (4.12)$$

where
T = geomembrane tensile force (i.e., runout resistance force) per unit width;
V_{GM} = vertical force due to geomembrane force;
L_{RO} = length of runout (unknown);
δ_C = friction angle between geomembrane and underlying soil;
γ_s = unit weight of cover soil;
d_{CS} = depth of cover soil; and
β = sideslope angle, measured from horizontal.

The following example illustrates the use of the concept and equations just developed.

EXAMPLE 4.3

Consider a 30-mil (0.75-mm) LLDPE geomembrane with an allowable stress 350 lb/in² (2,400 kN/m²) (which equals one half of its ultimate stress or strength), which is on a 3(H):1(V) side slope. Determine the required runout length. In this analysis, use 12 inches (0.3 m) of cover soil weight 100 lb/ft³ (15.7 kN/m³) and a friction angle of 20 degrees between the geomembrane and the underlying soil.

Solution:

Assume the runout resistance force is equal to the geomembrane allowable tensile force. From the design equations just presented,

$$T \cdot (\cos\beta) = 350(144)(0.030/12)\cos 18.4°$$
$$= 120 \text{ lb/ft } (1.75 \text{ kN/m})$$
$$T \cdot (\sin\beta) = 39.8 \text{ lb/ft } (0.58 \text{ kN/m})$$
$$q_B = \gamma_s \cdot d_{CS} = (100)(1.0) = 100 \text{ lb/ft } (1.46 \text{ kN/m})$$

which, when substituted into Equation 4.11, gives

$$T \cdot (\cos\beta) = q_B \cdot \tan\delta_C(L_{RO}) + T \cdot \sin\beta \cdot \tan\delta_C$$
$$120 = 100(\tan 20°)(L_{RO}) + 39.8(\tan 20°) \tag{4.11}$$

$$120 = 36.4 \cdot L_{RO} + 14.5$$

from which it follows that

$$L_{RO} = 2.9 \text{ ft } (0.88 \text{ m}); \text{ use } 3.0 \text{ ft } (\text{use } 1 \text{ m})$$

Note that the runout lengt\h is strongly dependent on the value of allowable stress used in the analysis. To mobilize the full strength of the geomembrane would require a longer runout length or an anchor trench. However, this might not be desirable. Pullout, without geomembrane failure, might be preferable to tensile rupture and separation of the geomembrane. Thus, the design runout or anchor resistance capacity should fall between the ultimate strength and allowable strength of a geosynthetic liner (Qian, 1995). That is,

Ultimate Strength > Runout and/or Anchor Resistance Capacity > Allowable Strength

Runout and/or Anchor Resistance Capacity = T/t

$$\sigma_{allow} = \sigma_{ult}/FS, \text{ and } T_{allow} = \sigma_{allow} \cdot t,$$

where T = geomembrane tensile force (i.e., runout or anchor resistance force) per unit width;
  t = geomembrane thickness;
  σ_{ult} = ultimate geomembrane stress (e.g., yield or break);
  FS = factor of safety based on geomembrane strength;
  σ_{allow} = allowable geomembrane stress; and
  T_{allow} = allowable geomembrane force per unit width.

4.7.2 Design of Rectangular Anchor Trench

The situation with a rectangular anchor trench in place at the end of the runout section is illustrated in Figure 4.9. The configuration requires some important assumptions regarding the state of stress within the anchor trench and its resistance mechanism. In order to establish static equilibrium, an imaginary and frictionless pulley is assumed at

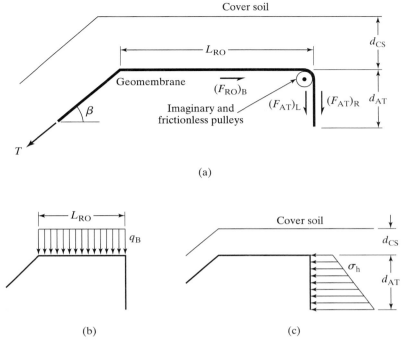

FIGURE 4.9 Cross Section of Geomembrane Runout Section with a Rectangular Anchor Trench and Related Stresses and Forces Involved

the top edge of the anchor trench, as shown in Figure 4.9 (Qian, 1995), which allows the geomembrane to be considered as a continuous member along its entire length.

From Figure 4.9, the following force summations lead to the appropriate design equations:

From $\Sigma F_V = 0$,

$$T \cdot (\sin \beta) = 0.5 \cdot V_{GM} L_{RO}$$

The cover soil pressure on the runout length is

$$q_B = \gamma_s \cdot d_{CS}$$

The lateral earth force acting on both sides of the geomembrane buried in the anchor trench is

$$P_L = P_R = K_o \cdot \gamma_s \cdot (d_{CS} + 0.5 \cdot d_{AT}) \cdot d_{AT}$$

The vertical force due to the geomembrane force is

$$V_{GM} = \frac{2 \cdot T \cdot \sin \beta}{L_{RO}}$$

The friction force above the runout geomembrane is always neglected in the anchor trench design, since the cover soil probably moves along with the geomembrane as it deforms.

From $\Sigma F_H = 0$,

$$T \cdot (\cos\beta) = (F_{RO})_B + (F_{AT})_L + (F_{AT})_R \qquad (4.13)$$

and
$$(F_{RO})_B = q_B \cdot L_{RO} \cdot \tan\delta_C + 0.5 \cdot V_{GM} \cdot L_{RO} \cdot \tan\delta_C$$
$$= q_B \cdot L_{RO} \cdot \tan\delta_C + 0.5 \cdot (2 \cdot T \cdot \sin\beta / L_{RO}) \cdot L_{RO} \cdot \tan\delta_C$$

or
$$(F_{RO})_B = q_B \cdot L_{RO} \cdot \tan\delta_C + T \cdot \sin\beta \cdot \tan\delta_C \qquad (4.14)$$

Because $q_B = \gamma_s \cdot d_{CS}$, the friction force beneath the runout geomembrane is

$$(F_{RO})_B = \gamma_s \cdot d_{CS} \cdot L_{RO} \cdot \tan\delta_C + T \cdot \sin\beta \cdot \tan\delta_C \qquad (4.15)$$

The friction force between the left side of the geomembrane and the side wall of the anchor trench is

$$(F_{AT})_L = (\sigma_h)_{ave} \cdot d_{AT} \cdot \tan\delta_C$$

The friction force between the right side of the geomembrane and the side wall of the anchor trench is

$$(F_{AT})_R = (\sigma_h)_{ave} \cdot d_{AT} \cdot \tan\delta_F$$

where $(\sigma_h)_{ave} = K_o \cdot (\sigma_v)_{ave}$

Because $K_o = 1 - \sin\phi$ and $(\sigma_v)_{ave} = \gamma_s \cdot (d_{CS} + 0.5 \cdot d_{AT})$

$$(\sigma_h)_{ave} = (1 - \sin\phi) \cdot \gamma_s \cdot (d_{CS} + 0.5 d_{AT}) \qquad (4.16)$$

So
$$(F_{AT})_L = (1 - \sin\phi) \cdot \gamma_s \cdot (d_{CS} + 0.5 \cdot d_{AT}) \cdot d_{AT} \cdot \tan\delta_C \qquad (4.17)$$

and
$$(F_{AT})_R = (1 - \sin\phi) \cdot \gamma_s \cdot (d_{CS} + 0.5 \cdot d_{AT}) \cdot d_{AT} \cdot \tan\delta_F \qquad (4.18)$$

Substituting Equations 4.15, 4.17, and 4.18 into Equation 4.13 gives

$$T \cdot (\cos\beta - \sin\beta \cdot \tan\delta_L) = \gamma_s \cdot d_{CS} \cdot L_{RO} \cdot \tan\delta_C +$$
$$(1 - \sin\phi) \cdot \gamma_s \cdot (d_{CS} + 0.5 \cdot d_{AT}) \cdot d_{AT} \cdot (\tan\delta_C + \tan\delta_F)$$

which leads to

$$T = \frac{\gamma_s \cdot d_{CS} \cdot L_{RO} \cdot \tan\delta_C + (1 - \sin\phi) \cdot \gamma_s \cdot (d_{CS} + 0.5 \cdot d_{AT} \cdot (\tan\delta_C + \tan\delta_F)}{\cos\beta - \sin\beta \cdot \tan\delta_C} \qquad (4.19)$$

or

$$T = \frac{q_B \cdot L_{RO} \cdot \tan\delta_C + K_o \cdot (\sigma_v)_{ave} \cdot d_{AT} \cdot (\tan\delta_C + \tan\delta_F)}{\cos\beta - \sin\beta \cdot \tan\delta_C} \qquad (4.20)$$

When $\delta_C = \delta_F = \delta$, Equation 4.19 becomes

$$T = \frac{\gamma_s \cdot d_{CS} \cdot L_{RO} \cdot \tan\delta + 2 \cdot (1 - \sin\phi) \cdot \gamma_s + 0.5 \cdot d_{AT} \cdot \tan\delta}{\cos\beta - \sin\beta \cdot \tan\delta} \qquad (4.21)$$

and Equation 4.20 becomes

$$T = \frac{q_B \cdot L_{RO} \cdot \tan\delta + 2 \cdot K_o \cdot (\sigma_v)_{ave} \cdot d_{AT} \cdot \tan\delta}{\cos\beta - \sin\beta \cdot \tan\delta} \quad (4.22)$$

where T = geomembrane tensile force (i.e., anchor trench resistance force) per unit width;
$(F_{RO})_B$ = friction force beneath runout geomembrane;
$(F_{AT})_L$ = friction force between the left side of the geomembrane and the side wall of the anchor trench;
$(F_{AT})_R$ = friction force between the right side of the geomembrane and the side wall of the anchor trench;
$(\sigma_h)_{ave}$ = average horizontal stress in anchor trench;
$(\sigma_v)_{ave}$ = average vertical stress in anchor trench;
H_{ave} = average depth of anchor trench;
K_o = coefficient of at-rest earth pressure;
L_{RO} = runout length;
d_{CS} = depth of cover soil;
d_{AT} = anchor trench depth;
γ_s = unit weight of cover and backfill soil;
ϕ = friction angle of backfill soil in anchor trench;
δ_C = friction angle between geomembrane and underlying soil;
δ_F = friction angle between geomembrane and backfill soil;
δ = friction angle between geomembrane and soil; and
β = sideslope angle, measured from horizontal.

Note that because this situation results in one equation with two unknowns, thus a choice of L_{RO} or d_{AT} is necessary to calculate the other.

EXAMPLE 4.4

A 60-mil (1.5-mm) HDPE geomembrane of allowable stress 840 lb/in² (5,800 kN/m²) is placed on a 3(H) to 1(V) sideslope. There is a cover soil of 12 inches (0.3 m) placed over the geomembrane. The unit weight of cover soil and backfill soil in the anchor trench is 110 lb/ft³ (17.3 kN/m³). The friction angle between the geomembrane and the underlying soil is 18 degrees, and the friction angle between the geomembrane and the backfill soil in the anchor trench is 22 degrees. The friction of the backfill soil is 30 degrees. Determine the required runout length for a 24-inch-deep (0.6-meter-deep) anchor trench.

Solution:
Assume the anchor resistance force is equal to the geomembrane allowable tensile force. Using the previously developed design equation from Figure 4.9,

$$T \cdot (\cos\beta) = (F_{RO})_B + (F_{AT})_L + (F_{AT})_R \quad (4.13)$$

where $T = T_{allow} = \sigma_{allow} \cdot t$

From Equation 4.19, we have

$$T = \frac{\gamma_s \cdot d_{CS} \cdot L_{RO} \cdot \tan\delta_C + (1 - \sin\phi) \cdot \gamma_s \cdot (d_{CS} + 0.5 \cdot d_{AT}) \cdot d_{AT} \cdot (\tan\delta_C + \tan\delta_F)}{\cos\beta - \sin\beta \cdot \tan\delta_C} \quad (4.19)$$

and

$$\sigma_{\text{allow}} \cdot t \cdot (\cos\beta - \sin\beta \cdot \tan\delta_C) = \gamma_s \cdot d_{CS} \cdot L_{RO} \cdot \tan\delta_C \\ + (1 - \sin\phi) \cdot \gamma_s \cdot (d_{CS} + 0.5 \cdot d_{AT}) \cdot d_{AT} \cdot (\tan\delta_C + \tan\delta_F)$$

so that

$$\sigma_{\text{allow}} \cdot t = (840)(144)(0.060)/12 = 605 \text{ lb/ft (8.83 kN/m) and } (605)[(\cos 18.4°) - \\ (\sin 18.4°)(\tan 18°)] = (110)(1)(\tan 18°)(L_{RO}) + (0.5)(110)(2)(2)(\tan 18° + \tan 22°)$$

or

$$(605)(0.846) = (35.74) \cdot L_{RO} + (220)(0.729) \text{ which yields } 512.83 = (35.74) \cdot L_{RO} + 160.38 \text{ or} \\ L_{RO} = 9.86 \text{ ft (2.96 m)}$$

Thus, use the runout length $L_{RO} = 10$ ft (3 m).

The geomembrane can also be extended along the trench bottom to increase resistance force, which is called an L-shaped rectangular anchor trench. A typical layout in an L-shaped rectangular anchor trench, which is widely used in landfill projects, is shown in Figure 4.10. In order to establish the static equilibrium equation, two imaginary and frictionless pulleys are assumed at the top edge and the bottom corner of the anchor trench, as shown in Figure 4.10 (Qian, 1995). This assumption again allows the geomembrane to be considered as a continuous member.

The friction force above a runout geomembrane is always neglected in the anchor trench design, since the cover soil probably moves together with the geomembrane as it deforms.

From $\Sigma F_H = 0$

$$T \cdot (\cos\beta) = (F_{RO})_B + (F_{AT})_L + (F_{AT})_R + (F_{AB})_B + (F_{AB})_U \quad (4.23)$$

The friction force between the geomembrane and the underlying soil at the bottom of the anchor trench is

$$(F_{AB})_B = \sigma_{vB} \cdot L_{AT} \cdot \tan\delta_C \quad (4.24)$$

The friction force between the geomembrane and the overlying soil at the bottom of the anchor trench is

$$(F_{AB})_U = \sigma_{vB} \cdot L_{AT} \cdot \tan\delta_F \quad (4.25)$$

Because $\sigma_{vB} = \gamma_s \cdot (d_{CS} + d_{AT})$,

$$(F_{AB})_B = \gamma_s \cdot (d_{CS} + d_{AT}) \cdot L_{AT} \cdot \tan\delta_C \quad (4.26)$$

and

$$(F_{AB})_U = \gamma_s \cdot (d_{CS} + d_{AT}) \cdot L_{AT} \cdot \tan\delta_F \quad (4.27)$$

Substituting Equations 4.15, 4.17, 4.18, 4.26, and 4.27 into Equation 4.23 gives

$$T \cdot (\cos\beta - \sin\beta \cdot \tan\delta_L) = \gamma_s \cdot d_{CS} \cdot L_{RO} \cdot \tan\delta_C + \gamma_s \cdot (\tan\delta_C + \tan\delta_F) \\ [(1 - \sin\phi) \cdot \gamma_s \cdot (d_{CS} + 0.5 \cdot d_{AT}) \cdot d_{AT} + (d_{CS} + d_{AT}) \cdot L_{AT}]$$

Section 4.7 Runout and Anchor Trenches

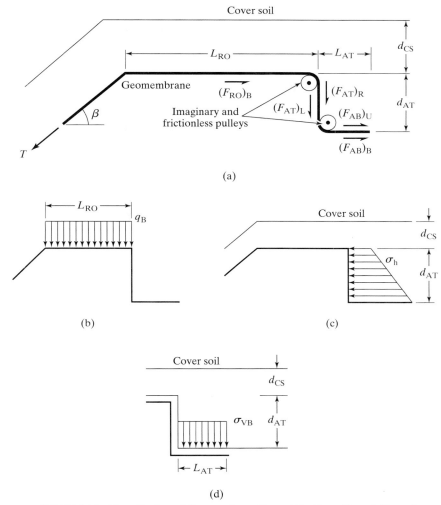

FIGURE 4.10 Cross Section of Geomembrane Runout Section with an L-Shaped Rectangular Anchor Trench and Related Stresses and Forces Involved

which leads to

$$T = \frac{\gamma_s \cdot d_{CS} \cdot L_{RO} \cdot \tan\delta_C + \gamma_s \cdot [(1 - \sin\phi) \cdot \gamma_s \cdot (d_{CS} + 0.5 \cdot d_{AT}) \cdot d_{AT} + (d_{CS} + d_{AT}) \cdot L_{AT}](\tan\delta_C + \tan\delta_F)}{\cos\beta + \sin\beta \cdot \tan\delta_C} \quad (4.28)$$

or

$$T = \frac{q_B \cdot L_{RO} \cdot \tan\delta_C + [K_o \cdot (\sigma_v)_{ave} \cdot d_{AT} + \sigma_{vB} \cdot L_{AT}](\tan\delta_C + \tan\delta_F)}{\cos\beta - \sin\beta \cdot \tan\delta_C} \quad (4.29)$$

When $\delta_C = \delta_F = \delta$, Equation 4.28 becomes

$$T = \frac{\gamma_s \cdot d_{CS} \cdot L_{RO} \cdot \tan\delta + 2\cdot\gamma_s\cdot[(1-\sin\phi)\cdot\gamma_s\cdot(d_{CS} + 0.5\cdot d_{AT})\cdot d_{AT} + (d_{CS} + d_{AT})\cdot L_{AT}]\cdot\tan\delta}{\cos\beta - \sin\beta\cdot\tan\delta}$$

(4.30)

and Equation 4.29 becomes

$$T = \frac{q_B\cdot L_{RO}\cdot\tan\delta + 2\cdot[K_o\cdot(\sigma_v)_{ave}\cdot d_{AT} + \sigma_{vB}\cdot L_{AT}]\cdot\tan\delta}{\cos\beta - \sin\beta\cdot\tan\delta}$$

(4.31)

where T = geomembrane tensile force (i.e., anchor trench resistance force) per unit width;
$(F_{RO})_B$ = friction force beneath runout geomembrane;
$(F_{AT})_L$ = friction force between the left side of the geomembrane and the side wall of the anchor trench;
$(F_{AT})_R$ = friction force between the right side of the geomembrane and the side wall of the anchor trench;
$(F_{AB})_B$ = friction force between the geomembrane and the underlying soil at the bottom of the anchor trench;
$(F_{AB})_U$ = friction force between the geomembrane and the overlying soil at the bottom of the anchor trench;
$(\sigma_v)_{ave}$ = average vertical stress in anchor trench;
K_o = coefficient of at-rest earth pressure;
L_{RO} = runout length;
d_{CS} = depth of cover soil;
d_{AT} = anchor trench depth;
γ_s = unit weight of cover and backfill soil;
ϕ = friction angle of backfill soil in anchor trench;
δ_C = friction angle between the geomembrane and the underlying soil;
δ_F = friction angle between the geomembrane and the backfill soil;
δ = friction angle between the geomembrane and the soil; and
β = sideslope angle, measured from horizontal.

The design of an anchor trench is considered to be adequate if mobilized stress lies between the yield stress and allowable stress of the geosynthetic components. It should be mentioned that many manufacturers specify 1.5-feet- (0.45-m)-deep anchor trenches and a 3.0-feet- (0.90-m)-long runout section.

EXAMPLE 4.5

Calculate the resistant capacity of a given geomembrane in a L-shaped rectangular anchor trench of known dimensions. The geomembrane is 60-mil (1.5-mm) HDPE with an ultimate strength (at yield) 2,100 lb/in^2 (14,500 kN/m^2) and an allowable strength 840 lb/in^2 (5,800 kN/m^2).

Section 4.7 Runout and Anchor Trenches

The runout length is 3 feet (0.9 m). The cover soil is 1 foot (0.3 m). The anchor trench is 2 feet (0.6 m) wide and 2 feet (0.6 m) deep. The side slope angle is 18.4 degrees [3(H):1(V)]. The unit weight of soil is 110 lb/ft^3 (17.3 kN/m^3). The soil friction angle is 30 degrees. The friction angle between the soil and the geomembrane is 20 degrees.

Solution:
The resistance capacity of the geomembrane in the anchor can be calculated from Equation 4.31 as

$$T = \frac{q_B \cdot L_{RO} \cdot \tan\delta + 2 \cdot [K_o \cdot (\sigma_v)_{ave} \cdot d_{AT} + \sigma_{vB} \cdot L_{AT}] \cdot \tan\delta}{\cos\beta - \sin\beta \cdot \tan\delta}$$

where
$q_B = \gamma_s \cdot d_{CS} = 110 \times 1 = 110$ lb/ft^2 (5.27 kN/m^2)
$K_o = 1 - \sin\phi = 1 - 0.5 = 0.5$
$(\sigma_v)_{ave} = \gamma_s \cdot (d_{cs} + 0.5 \cdot d_{AT})$
$= 110 \times (1 + 0.5 \times 2) = 110 \times 2 = 220$ lb/ft^2 (10.53 kN/m^2)
$\sigma_{vB} = \gamma_s \cdot (d_{cs} + d_{AT}) = 110 \times (1 + 2) = 330$ lb/ft^2 (15.80 kN/m^2)

Substituting these calculated values into Equation 4.31 yields

$$T = \frac{q_B \cdot L_{RO} \cdot \tan\delta + 2 \cdot [K_o \cdot (\sigma_v)_{ave} \cdot d_{AT} + \sigma_{vB} \cdot L_{AT}] \cdot \tan\delta}{\cos\beta - \sin\beta \cdot \tan\delta}$$

$$= \frac{(110)(2)(\tan 20°) + 2[(0.5)(220)(2) + (330)(2)](\tan 20°)}{\cos 18.4° - (\sin 18.4°)(\tan 20°)}$$

$$= \frac{(110)(2)(0.364) + 2(220 + 660)(0.364)}{0.949 - (0.316)(0.364)}$$

$$= \frac{80.08 + 640.64}{0.834}$$

$$= \frac{720.72}{0.834}$$

$$= 864 \text{ lb/ft } (12.61 \text{ kN/m})$$

So,
Anchor Resistance Capacity = 864 lb/ft = 72 lb/in ÷ 0.06 in = 1,200 lb/in^2 (8,270 kN/m^2), which leads to the following inequalities:

Ultimate Strength > Anchor Resistance Capacity > Allowable Strength
2,100 lb/in^2 > 1,200 lb/in^2 > 840 lb/in^2
(14,500 kN/m^2 > 8,270 kN/m^2 > 5,800 kN/m^2)

The results of the calculation indicate the design anchor resistance capacity falls between the yield stress and allowable stress of a geosynthetic membrane liner. Therefore, the anchor trench dimensions are acceptable.

By using a model as presented here, any set of conditions can be used to analyze and arrive at an acceptable design solution. Even situations in which geotextiles and geonets or geocomposites are used in conjunction with a geomembrane can be analyzed in a similar manner.

114   Chapter 4   Geomembranes

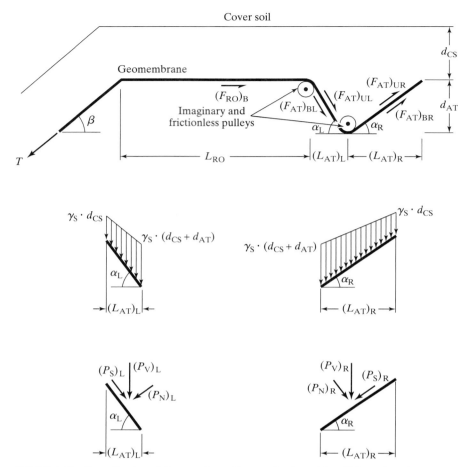

FIGURE 4.11 Cross Section of Geomembrane Runout Section with a V-Shaped Anchor Trench and Related Stresses and Forces Involved

4.7.3 Design of V-Shaped Anchor Trench

As the thickness of geomembranes increase, they become more difficult to bend and bury in narrow rectangular anchor trenches. This is particularly the case for HDPE geomembranes. A V-shaped anchor trench may be suitable for this case. (See Figure 4.11.) In order to establish the static equilibrium equation, two imaginary and frictionless pulleys are assumed to be at the top edge and at the bottom of the V-shaped anchor trench, as shown in Figure 4.11. This assumption again allows the geomembrane to be considered as a continuous member.

The vertical force acting on the left wing of the geomembrane in the trench is given by

$$(P_V)_L = 0.5 \cdot \gamma_s \cdot (2 \cdot d_{CS} + d_{AT}) \cdot (L_{AT})_L / \cos \alpha_L$$

or

$$(P_V)_L = \gamma_s \cdot (d_{CS} + 0.5 \cdot d_{AT}) \cdot (L_{AT})_L / \cos \alpha_L \quad (4.32)$$

Section 4.7   Runout and Anchor Trenches   115

The tangential component force along the geomembrane surface of $(P_V)_L$ is

$$(P_S)_L = (P_V)_L \cdot \sin\alpha_L = \gamma_s \cdot (d_{CS} + 0.5 \cdot d_{AT}) \cdot (L_{AT})_L \cdot \sin\alpha_L / \cos\alpha_L$$

or

$$(P_S)_L = \gamma_s \cdot (d_{CS} + 0.5 \cdot d_{AT}) \cdot (L_{AT})_L \cdot \tan\alpha_L$$

Because $(L_{AT})_L = d_{AT}/\tan\alpha_L$,

$$(P_S)_L = \gamma_s \cdot (d_{CS} + 0.5 \cdot d_{AT}) \cdot d_{AT} \tag{4.33}$$

The component force perpendicular to the geomembrane of $(P_V)_L$ is

$$(P_N)_L = (P_V)_L \cdot \cos\alpha_L = \gamma_s \cdot (d_{CS} + 0.5 \cdot d_{AT}) \cdot (L_{AT})_L \cdot \cos\alpha_L / \cos\alpha_L$$

or

$$(P_N)_L = \gamma_s \cdot (d_{CS} + 0.5 \cdot d_{AT}) \cdot (L_{AT})_L \tag{4.34}$$

The friction force at the bottom of the left wing of the geomembrane in the trench caused by $(P_N)_L$ is given by

$$(F_{AT})_{BL} = (P_N)_L \cdot \tan\delta_C = \gamma_s \cdot (d_{CS} + 0.5 \cdot d_{AT}) \cdot (L_{AT})_L \cdot \tan\delta_C \tag{4.35}$$

The friction force at the top of the left wing of the geomembrane in the trench caused by $(P_N)_L$ is

$$(F_{AT})_{UL} = (P_N)_L \cdot \tan\delta_F = \gamma_s \cdot (d_{CS} + 0.5 \cdot d_{AT}) \cdot (L_{AT})_L \cdot \tan\delta_F \tag{4.36}$$

The vertical force acting on the right wing of the geomembrane in the trench is

$$(P_V)_R = 0.5 \cdot \gamma_s \cdot (2 \cdot d_{CS} + d_{AT}) \cdot (L_{AT})_R / \cos\alpha_R$$

or

$$(P_V)_R = \gamma_s \cdot (d_{CS} + 0.5 \cdot d_{AT}) \cdot (L_{AT})_R / \cos\alpha_R \tag{4.37}$$

The tangential component force along the geomembrane surface of $(P_V)_R$ is

$$(P_S)_R = (P_V)_R \cdot \sin\alpha_R = \gamma_s \cdot (d_{CS} + 0.5 \cdot d_{AT}) \cdot (L_{AT})_R \cdot \sin\alpha_R / \cos\alpha_R$$

or

$$(P_S)_R = \gamma_s \cdot (d_{CS} + 0.5 \cdot d_{AT}) \cdot (L_{AT})_R \cdot \tan\alpha_R$$

Because $(L_{AT})_R = d_{AT}/\tan\alpha_R$,

$$(P_S)_R = \gamma_s \cdot (d_{CS} + 0.5 \cdot d_{AT}) \cdot d_{AT} \tag{4.38}$$

The component force perpendicular to the geomembrane of $(P_V)_R$ is given by

$$(P_N)_R = (P_V)_R \cdot \cos\alpha_R = \gamma_s \cdot (d_{CS} + 0.5 \cdot d_{AT}) \cdot (L_{AT})_R \cdot \cos\alpha_R / \cos\alpha_R$$

or

$$(P_N)_R = \gamma_s \cdot (d_{CS} + 0.5 \cdot d_{AT}) \cdot (L_{AT})_R \tag{4.39}$$

The friction force at the bottom of the right wing of the geomembrane in the trench caused by $(P_N)_R$ is

$$(F_{AT})_{BR} = (P_N)_R \cdot \tan\delta_C = \gamma_s \cdot (d_{CS} + 0.5 \cdot d_{AT}) \cdot (L_{AT})_R \cdot \tan\delta_C \tag{4.40}$$

The friction force at the top of the left wing of the geomembrane in the trench caused by $(P_N)_R$ is

$$(F_{AT})_{UR} = (P_N)_R \cdot \tan\delta_F = \gamma_s \cdot (d_{CS} + 0.5 \cdot d_{AT}) \cdot (L_{AT})_R \cdot \tan\delta_F \tag{4.41}$$

116 Chapter 4 Geomembranes

The friction force above the runout portion geomembrane is always neglected in the anchor trench design, since the cover soil probably moves along with the geomembrane as it deforms.

From $\Sigma F_H = 0$,

$$T \cdot (\cos\beta) + (P_S)_R = (F_{RO})_B + (F_{AT})_{BL} + (F_{AT})_{UL} + (F_{AT})_{BR} + (F_{AT})_{UR} + (P_S)_L$$

Because Equation 4.33 = Equation 4.38 [i.e., $(P_S)_L = (P_S)_R$],

$$T \cdot (\cos\beta) = (F_{RO})_B + (F_{AT})_{BL} + (F_{AT})_{UL} + (F_{AT})_{BR} + (F_{AT})_{UR} \quad (4.42)$$

From Equation 4.15,

$$(F_{RO})_B = \gamma_s \cdot d_{CS} \cdot L_{RO} \cdot \tan\delta_C + T \cdot \sin\beta \cdot \tan\delta_C \quad (4.15)$$

Substituting Equations 4.15, 4.35, 4.36, 4.40, and 4.41 into Equation 4.42 gives

$$T \cdot (\cos\beta - \sin\beta \cdot \tan\delta_L) = \gamma_s \cdot d_{CS} \cdot L_{RO} \cdot \tan\delta_C + \gamma_s \cdot (d_{CS} + 0.5 \cdot d_{AT}) \cdot \\ [(L_{AT})_L \cdot \tan\delta_C + (L_{AT})_L \cdot \tan\delta_F + (L_{AT})_R \cdot \tan\delta_C + (L_{AT})_R \cdot \tan\delta_F]$$

$$T \cdot (\cos\beta - \sin\beta \cdot \tan\delta_L) = \gamma_s \cdot d_{CS} \cdot L_{RO} \cdot \tan\delta_C + \gamma_s \cdot (d_{CS} + 0.5 \cdot d_{AT}) \cdot \\ [(L_{AT})_L + (L_{AT})_R] \cdot (\tan\delta_C + \tan\delta_F)$$

Because $(L_{AT})_L = d_{AT}/\tan\alpha_L$ and $(L_{AT})_R = d_{AT}/\tan\alpha_R$,

$$T \cdot (\cos\beta - \sin\beta \cdot \tan\delta_L) = \gamma_s \cdot d_{CS} \cdot L_{RO} \cdot \tan\delta_C \\ + \gamma_s \cdot (d_{CS} + 0.5 \cdot d_{AT}) \cdot (d_{AT}/\tan\alpha_L + d_{AT}/\tan\alpha_R) \cdot (\tan\delta_C + \tan\delta_F)$$

or

$$T \cdot (\cos\beta - \sin\beta \cdot \tan\delta_L) = \gamma_s \cdot d_{CS} \cdot L_{RO} \cdot \tan\delta_C \\ + \gamma_s \cdot (d_{CS} + 0.5 \cdot d_{AT}) \cdot d_{AT} \cdot (\tan\delta_C + \tan\delta_F) \cdot (\cot\alpha_L + \cot\alpha_R)$$

Thus,

$$T = \frac{\gamma_s \cdot d_{CS} \cdot L_{RO} \cdot \tan\delta_C + \gamma_s \cdot (d_{CS} + 0.5 \cdot d_{AT}) \cdot d_{AT} \cdot (\tan\delta_C + \tan\delta_F) \cdot (\cot\alpha_L + \cot\alpha_R)}{\cos\beta - \sin\beta \cdot \tan\delta_C}$$

(4.43)

When $\alpha_L = \alpha_R = \alpha$ [i.e., a symmetric V-shaped anchor trench (see Figure 4.12)], Equation 4.43 becomes

$$T = \frac{\gamma_s \cdot d_{CS} \cdot L_{RO} \cdot \tan\delta_C + 2 \cdot \gamma_s \cdot (d_{CS} + 0.5 \cdot d_{AT}) \cdot d_{AT} \cdot (\tan\delta_C + \tan\delta_F)/\tan\alpha}{\cos\beta - \sin\beta \cdot \tan\delta_C} \quad (4.44)$$

When $\delta_C = \delta_F = \delta$, Equation 4.43 becomes

$$T = \frac{\gamma_s \cdot d_{CS} \cdot L_{RO} \cdot \tan\delta + 2 \cdot \gamma_s \cdot (d_{CS} + 0.5 \cdot d_{AT}) \cdot d_{AT} \cdot \tan\delta \cdot (\cot\alpha_L + \cot\alpha_R)}{\cos\beta - \sin\beta \cdot \tan\delta} \quad (4.45)$$

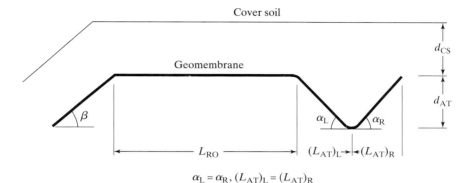

FIGURE 4.12 Cross Section of a Symmetric V-Shaped Anchor Trench

When $\alpha_L = \alpha_R = \alpha$ [i.e., a symmetric V-shaped anchor trench (see Figure 4.12)] and $\delta_C = \delta_F = \delta$, Equation 4.43 becomes

$$T = \frac{\gamma_s \cdot d_{CS} \cdot L_{RO} \cdot \tan\delta + 4 \cdot \gamma_s \cdot (d_{CS} + 0.5 \cdot d_{AT}) \cdot d_{AT} \cdot \tan\delta/\tan\alpha}{\cos\beta - \sin\beta \cdot \tan\delta} \quad (4.46)$$

When $(L_{AT})_R = 0$ [i.e., a one-sided V-shaped anchor trench (see Figure 4.13)], Equation 4.43 becomes

$$T = \frac{\gamma_s \cdot d_{CS} \cdot L_{RO} \cdot \tan\delta_C + \gamma_s \cdot (d_{CS} + 0.5 \cdot d_{AT}) \cdot d_{AT} \cdot (\tan\delta_C + \tan\delta_F) \cdot (\cot\alpha_L)}{\cos\beta - \sin\beta \cdot \tan\delta_C} \quad (4.47)$$

When $\delta_C = \delta_F = \delta$, Equation 4.47 becomes

$$T = \frac{\gamma_s \cdot d_{CS} \cdot L_{RO} \cdot \tan\delta + 2 \cdot \gamma_s \cdot (d_{CS} + 0.5 \cdot d_{AT}) \cdot d_{AT} \cdot \tan\delta \cdot (\cot\alpha_L)}{\cos\beta - \sin\beta \cdot \tan\delta} \quad (4.48)$$

where T = geomembrane tensile force (i.e., anchor trench resistance force) per unit width;
$(F_{RO})_B$ = friction force beneath runout geomembrane;

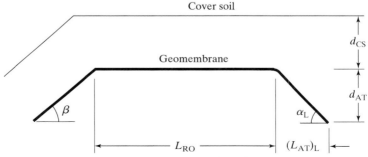

FIGURE 4.13 Cross Section of a One-Side V-Shaped Anchor Trench

$(F_{AT})_{BL}$ = friction force at the bottom of the left wing of the geomembrane in the trench;
$(F_{AT})_{UL}$ = friction force at the top of the left wing of the geomembrane in the trench;
$(F_{AT})_{BR}$ = friction force at the bottom of the right wing of the geomembrane in the trench;
$(F_{AT})_{UR}$ = friction force at the top of the right wing of the geomembrane in the trench;
L_{RO} = runout length;
d_{CS} = depth of cover soil;
d_{AT} = anchor trench depth;
γ_s = unit weight of the cover and the backfill soil;
δ_C = friction angle between the geomembrane and the underlying soil;
δ_F = friction angle between the geomembrane and the backfill soil;
δ = friction angle between the geomembrane and the soil;
α_L = left bottom angle of V-shaped anchor trench, measured from horizontal;
α_R = right bottom angle of V-shaped anchor trench, measured from horizontal;
α = bottom angle of symmetric V-shaped anchor trench, measured from horizontal; and
β = sideslope angle, measured from horizontal.

EXAMPLE 4.6

Calculate the resistant capacity of a V-shaped anchor trench of known dimensions (Figure 4.11). The side slope angle is 18.4 degrees [3(H):1(V)]. The runout length is 3 feet (0.9 m). The cover soil is 1-foot (0.3-m) deep. The anchor trench is 2.5-feet (0.75-m) deep. The left bottom angle of the V-shaped anchor trench is 60 degrees, and the right bottom angle of the V-shaped anchor trench is 30 degrees. The unit weight of the cover and the backfill soil is 110 lb/ft³ (17.3 kN/m³). The friction angle between the geomembrane and the underlying soil is 18 degrees, and the friction angle between the geomembrane and the backfill soil is 22 degrees.

Solution:

$$\gamma_s = 110 \text{ lb/ft}^3, L_{RO} = 3 \text{ ft}, d_{CS} = 1 \text{ ft}, d_{AT} = 2.5 \text{ ft};$$
$$\sin\beta = \sin 18.4° = 0.316, \cos\beta = \cos(18.4°) = 0.949;$$
$$\tan\delta_C = \tan(18°) = 0.325, \tan\delta_F = \tan(22°) = 0.404;$$
$$\cot\alpha_L = \cot(60°) = 0.577, \cot\alpha_R = \cot(30°) = 1.732.$$

The resistance capacity of the geomembrane in a V-shaped anchor can be calculated from Equation 4.43 as follows (see Figure 4.11):

$$T_{AT} = \frac{\gamma_s \cdot d_{CS} \cdot L_{RO} \cdot \tan\delta_C + \gamma_s \cdot (d_{CS} + 0.5 \cdot d_{AT}) \cdot d_{AT} \cdot (\tan\delta_C + \tan\delta_F) \cdot (\cot\alpha_L + \cot\alpha_R)}{\cos\beta - \sin\beta \cdot \tan\delta_C}$$

(4.43)

$$= \frac{(110)(1)(3)(0.325) + (110)[1 + (0.5)(2.5)](2.5)(0.325 + 0.404)(0.577 + 1.732)}{0.949 - (0.316)(0.325)}$$

$$= \frac{107.25 + (110)(2.25)(2.5)(0.729)(2.309)}{0.949 - 0.103}$$

$$= \frac{107.25 + 1{,}041.52}{0.846}$$

$$= \frac{1{,}148.77}{0.846}$$

$$= 1{,}358 \text{ lb/ft } (19.8 \text{ kN/m})$$

This value can now be compared with the ultimate tensile strength of the geomembrane, $T_{GM(ult)}$, to determine which value governs:

(i) If $T_{AT} > T_{GM(ult)}$, the geomembrane will fail if sufficient tensile stresses are mobilized.
(ii) If $T_{AT} < T_{GM(ult)}$, the geomembrane will pull out of the anchor trench if sufficient tensile stresses are mobilized.
(iii) If $T_{AT} \approx T_{GM(ult)}$, a balanced design results.

4.8 ASSESSMENT OF LEAKAGE THROUGH LINERS

A leakage assessment (the volumetric release of leachate from the proposed containment envelope design) should be based on analytical approaches supported by empirical data from other existing operational facilities of similar design, particularly double liner systems that have leak detection monitoring capability. Since such information is just becoming available, conservative analytical assumptions may be used to estimate anticipated leakage rates.

The movement of fluids like waste leachate through geomembranes differs in principle from movement through compacted clay liner materials. The dominant mode of leachate movement through liner components is flow through holes and penetrations of the geomembrane, whereas Darcian flow characterizes movement through soil components. In the absence of tears, punctures, imperfections, or seam failures, transport through geomembranes is governed by molecular diffusion. Diffusion occurs in response to a concentration gradient and is governed by Fick's first law. Diffusion rates through geomembranes are very low in comparison to hydraulic flow rates in compacted clay liners. For geomembranes, the most significant factors influencing liner performance are defects or penetrations of the liner, including imperfect seams, punctures or pinholes caused by construction defects in the geomembrane. Figure 4.14 illustrates three types of hydraulic barriers (liners):

(i) A low hydraulic conductivity compacted clay liner;
(ii) A geomembrane liner; and
(iii) A geomembrane/compacted clay (or composite) liner.

Flow rates for each of these types of liners have been calculated (USEPA, 1991) and are noted below for the purpose of comparing the effectiveness of the barriers.

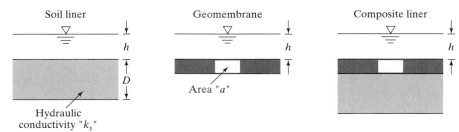

FIGURE 4.14 Soil Liner (a), Geomembrane Liner (b), and Composite Liner (c)

4.8.1 Flow Rate through Compacted Clay Liner

Flow rates through compacted clay liners are calculated using Darcy's law, which is the basic equation used to describe the flow of fluids through porous materials. Darcy's law states that

$$Q = k_s \cdot i \cdot A \tag{4.49}$$

where Q = flow rate through the liner, cm³/sec;
  k_s = hydraulic conductivity of the soil, cm/sec;
  i = hydraulic gradient; and
  A = area over which flow occurs, cm².

If the soil is saturated and there is no soil suction, the hydraulic gradient is given by

$$i = (h + D)/D \tag{4.50}$$

where i = hydraulic gradient;
  h = leachate head over the liner (see Figure 4.15); and
  D = thickness of the soil liner.

For example, if 1 ft (0.3 m) of liquid is ponded on a 3-ft (0.9-m) thick soil liner that has a hydraulic conductivity of 1.0×10^{-7} cm/sec, the flow rate is 120 gallon/acre/day (0.115 liter/m²/day). If the hydraulic conductivity is increased or decreased, the flow rate is changed proportionally. (See Table 4.6.)

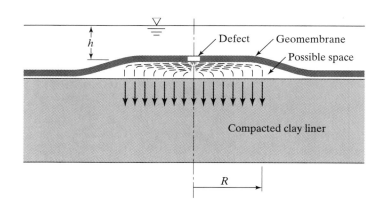

FIGURE 4.15 Liquid Flow through a Composite Liner with poor contact between geomembrane and compacted clay liner

TABLE 4.6 Calculated Flow Rates through Soil Liners with 1 foot (0.3 m) of Water Ponded on the Liner (USEPA, 1991)

Hydraulic Conductivity (cm/sec)	Rate of Flow	
	gallons/acre/day	liter/m^2/day
1.0×10^{-6}	1,200	1.15
1.0×10^{-7}	120	0.115
1.0×10^{-8}	12	0.0115
1.0×10^{-9}	1.2	0.00115

4.8.2 Flow Rate through Geomembrane Liner

The second liner depicted in Figure 4.14 is a geomembrane liner. It was assumed (USEPA, 1991) that the geomembrane has one or more circular holes (defects) in the liner, that the holes are sufficiently widely spaced that leakage through each hole occurs independently from the other holes, that the head h of liquid ponded above the liner, is constant, and that the soil that underlies the geomembrane has a relatively large hydraulic conductivity (i.e., the subsoil offers no resistance to flow through a hole in the geomembrane). In this case, flow rates through holes in geomembranes can be estimated using the Bernoulii equation assuming the size and shape of the holes are known. The Bernoulii equation is

$$Q = C_b \cdot a \cdot (2 \cdot g \cdot h)^{0.5} \qquad (4.51)$$

where Q = flow rate through geomembrane, cm^3/sec;
C_b = flow coefficient with a value approximately 0.6 for a circular hole;
a = area of a circular hole in geomembrane, cm^2;
g = acceleration due to gravity, 981 cm/sec^2; and
h = liquid head above the liner, cm.

For example, if there is a single hole with an area of 1 cm^2 and the head is 30 cm (1 ft), the calculated rate of flow is 3,300 gallon/day (12,500 liter/day). If there is one hole per acre (2.5 holes per hectare), then the flow rate is 3,300 gallon/acre/day (3 liter/m^2/day). Flow rate for other circumstances are calculated (USEPA, 1991) in Table 4.7. Giroud and Bonaparte (1989a) report that with good quality control, one hole per acre (2.5 holes per hectare) is typical. With poor quality control, 30 holes per acre (75 holes per hectare) is possible. They also note that most defects are small (< 0.1 cm^2), but that larger holes are occasionally observed. In calculating the rate of flow for "No holes" in Table 4.7, it was assumed (USEPA, 1991) that any flux of liquid was controlled by water vapor transmission; flux of 0.01 gallon/acre/day (9.35×10^{-6} liter/m^2/day) corresponds to a typical water vapor transmission rate for a 60-mil (1.5-mm) HDPE geomembrane. It is important to note that this information is based on the investigation of geomembrane installation prior to 1989. With wedge welders, certified technicians, and construction quality assurance, these estimates are generally very high.

TABLE 4.7 Calculated Flow Rates through a Geomembrane with a Water Head of 1 foot (0.3 m) above Geomembrane (USEPA, 1991)

Size of Hole (cm^2)	Number of Holes		Flow Rate	
	hole/acre	hole/hectare	gallon/acre/day	liter/m^2/day
No holes	0	0	0.01	9.4×10^{-6}
0.1	1	2.5	330	0.31
0.1	30	75	10,000	9.4
1	1	2.5	3,300	3.1
1	30	75	100,000	94
10	1	2.5	33,000	31

4.8.3 Flow Rate through Composite Liner

The third type of liner depicted in Figure 4.14 is a *composite* geomembrane/compacted clay liner. A composite liner is a liner composed of a geomembrane layer and a layer of low-permeability soil placed in intimate contact with the geomembrane. Composite liners have been widely used in both hazardous and municipal solid waste landfills.

Leakage through a composite liner can result from flow through geomembrane defects or permeation through the geomembrane. In the case of a geomembrane defect, the rate of leakage through a composite liner is significantly less than the rate of leakage through a similar defect in a geomembrane placed on a high-permeability soil like sand or gravel (Giroud and Badu-Tweneboah, 1992).

The contained liquid is on the geomembrane side of the composite liner in landfills. If there is a defect in the geomembrane, the liquid flows first through the geomembrane defect, laterally some nominal distance between the geomembrane and the low-permeability soil, and finally, into and through the low-permeability soil layer, as shown in Figure 4.15. Flow in the space between the geomembrane and the soil, if there is such a space, is called *interface flow*, and the area covered by the interface flow is called the *wetted area*. The flow in the soil is assumed to be vertical and R is the radius of the wetted area.

The quality of the contact between the two components of a composite liner (i.e. the geomembrane and the low-permeability soil) is one of the key factors governing the rate of flow through the composite liner, because it governs the radius of the wetted area (Figure 4.15). Good and poor contact conditions have been characterized by Bonaparte et al. (1989) as follows:

(i) Good contact conditions correspond to a geomembrane installed, with as few waves or wrinkles as possible, on top of a low-permeability soil layer that has been adequately compacted and has a smooth surface.

(ii) Poor contact conditions correspond to a geomembrane that has been installed with a certain number of waves or wrinkles, and/or placed on a low-permeability soil that has not been well compacted and is not smooth.

Other factors affecting the rate of flow (Giroud and Badu-Tweneboah, 1992) through a composite liner are the size of the defect, the hydraulic conductivity of the low-permeability soil underlying the geomembrane, and the head of liquid on top of

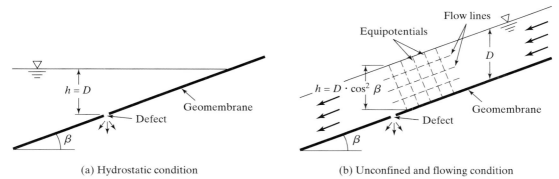

(a) Hydrostatic condition   (b) Unconfined and flowing condition

FIGURE 4.16 Liquid Head on Geomembrane

the geomembrane. If hydrostatic conditions prevail, the head of liquid is equal to the depth of liquid [Figure 4.16(a)], and if the liquid is unconfined and flowing along a slope [Figure 4.16(b)], the head of liquid is given by

$$h = D \cdot \cos^2\beta \tag{4.52}$$

where h = liquid head above the liner, cm;
D = depth of liquid on liner, cm;
β = slope angle.

Analytical studies and model tests have led to two empirical equations proposed by Giroud and Bonaparte (1989b) and Giroud et al. (1989) for the rate of leakage through a circular hole in the geomembrane component of a composite liner.

For "good" contact conditions,

$$Q = 0.21 \cdot a^{0.1} \cdot h^{0.9} \cdot k_s^{0.74} \tag{4.53}$$

For "poor" contact conditions,

$$Q = 1.15 \cdot a^{0.1} \cdot h^{0.9} \cdot k_s^{0.74} \tag{4.54}$$

where Q = leakage rate through a hole in the geomembrane component, m³/sec;
a = area of a circular hole in geomembrane, m²;
h = liquid head on top of the geomembrane, m; and
k_s = hydraulic conductivity of the low-permeability soil component of the composite liner, m/sec.

These equations assume that the hydraulic gradient through the soil is 1.0. The equations are not dimensionally homogeneous and can only be used with the following units: Q (m³/sec), a (m²), h (m), and k_s (m/sec).

If there is good contact between the geomembrane and the underlying soil, the flow rate is approximately one-fifth the value computed from the equations just presented. For example, suppose the geomembrane component of a composite liner has one hole per acre (2.5 holes per hectare), with an area of 1 cm² per hole, the hydraulic conductivity of subsoil is 1.0×10^{-7} cm/sec (1.0×10^{-9} m/sec), the head of water is

TABLE 4.8 Calculated Flow Rates for Composite Liners with A Head of Water of 1 foot (0.3 m) (USEPA, 1991)

Hydraulic Conductivity of Subsoil (cm/sec)	Size of Hole in Geomembrane (cm²)	Number of Holes		Flow Rate	
		hole/acre	hole/hectare	gallon/acre/day	liter/m²/day
1×10^{-6}	0.1	1	2.5	3	2.81×10^{-3}
	0.1	30	75	102	9.54×10^{-2}
	1	1	2.5	4	3.74×10^{-3}
	1	30	75	130	0.122
	10	1	2.5	5	4.68×10^{-3}
1×10^{-7}	0.1	1	2.5	0.6	5.61×10^{-4}
	0.1	30	75	19	1.78×10^{-2}
	1	1	2.5	0.8	7.48×10^{-4}
	1	30	75	24	2.25×10^{-2}
	10	1	2.5	1	9.35×10^{-4}
1×10^{-8}	0.1	1	2.5	0.1	9.35×10^{-5}
	0.1	30	75	3	2.81×10^{-3}
	1	1	2.5	0.1	9.35×10^{-5}
	1	30	75	4	3.74×10^{-3}
	10	1	2.5	0.2	1.87×10^{-4}
1×10^{-9}	0.1	1	2.5	0.2	1.87×10^{-4}
	0.1	30	75	0.6	5.61×10^{-4}
	1	1	2.5	0.03	2.81×10^{-5}
	1	30	75	0.6	5.61×10^{-4}
	10	1	2.5	0.03	2.81×10^{-5}

0.3 m (1 ft), and a poor contact exists between the geomembrane and soil. The calculated flow rate is 0.8 gallon/acre/day (7.5×10^{-4} liter/m²/day). Table 4.8 shows other calculated flow rates (USEPA, 1991) for composite liners with a head of water of 0.3 m (1 ft).

4.8.4 Comparison of Three Types of Liners

The performance of three types of liners under a variety of assumed conditions have been compared by USEPA (1991) as illustrated in Table 4.9. For discussion purposes, each liner type is classified as *poor, good,* or *excellent.* The U.S. Environmental Protection Agency (U.S. EPA) requires that low-permeability compacted soil liners used for both hazardous and municipal solid waste landfills have a hydraulic conductivity no greater than 1.0×10^{-7} cm/sec. Therefore, a soil liner with a hydraulic conductivity of 1.0×10^{-7} cm/sec is described in Table 4.9 as a *good* liner. A compacted soil liner with 10-fold higher hydraulic conductivity is described as a *poor* liner, and a soil liner with a 10-fold lower hydraulic conductivity is described as an *excellent* liner.

For geomembrane liners, a liner with a large number of small holes (30 holes per acre or 75 holes per hectare, with each hole having an area of 0.1 cm²) is described as a *poor* liner because Giroud and Bonaparte (1989a) suggest that such a large number of

TABLE 4.9 Calculated Flow Rates for Soil Liners, Geomembrane Liners, and Composite Liners with Poor Contact (USEPA, 1991)

Type of Liner	Overall Quality of Liner	Assumed Values of Key Parameters	Flow Rate gal/acre/day	Flow Rate liter/m²/day
Compacted Soil	Poor	$k_s = 1 \times 10^{-6}$ cm/sec	1,200	1.12
Geomembrane	Poor	30 hole/acre or 75 hole/hectare, $a = 0.1$ cm²	10,000	9.35
Composite	Poor	$k_s = 1 \times 10^{-6}$ cm/sec, 30 hole/acre or 75 hole/hectare, $a = 0.1$ cm²	100	9.35×10^{-2}
Compacted Soil	Good	$k_s = 1 \times 10^{-7}$ cm/sec	120	0.112
Geomembrane	Good	1 hole/acre or 2.5 hole/hectare, $a = 1$ cm²	3,300	3.09
Composite	Good	$k_s = 1 \times 10^{-7}$ cm/sec, 1 hole/acre or 2.5 hole/hectare, $a = 1$ cm²	0.8	7.48×10^{-4}
Compacted Soil	Excellent	$k_s = 1 \times 10^{-8}$ cm/sec	12	1.12×10^{-2}
Geomembrane	Excellent	$k_s = 1 \times 10^{-8}$ cm/sec, 1 hole/acre or 2.5 hole/hectare, $a = 0.1$ cm²	330	0.309
Composite	Excellent	$k_s = 1 \times 10^{-8}$ cm/sec, 1 hole/acre or 2.5 hole/hectare, $a = 0.1$ cm²	0.1	9.35×10^{-5}

defects would be expected only with minimal construction quality control and essentially no construction quality assurance. A *good* geomembrane liner was assumed to have been constructed with good quality control and quality assurance, and an *excellent* geomembrane liner was assumed to have one small hole per acre (or 2.5 small holes per hectare), with excellent quality control and quality assurance. For all of the flow rates computed for composite liners in Table 4.9, it was assumed that there was poor contact between the geomembrane and soil.

As Table 4.9 illustrates, a composite liner (even one built by a poor-to-mediocre standard) significantly outperforms a soil liner or a geomembrane liner alone. Even so, good quality control and quality assurance during construction is absolutely essential when using a composite liner; otherwise, leakage rates can increase several fold as either the total area of holes in the membrane or the hydraulic conductivity of the underlying soil increases. This increase in leakage rate can be observed by comparing the flow rate through an excellent composite liner (0.1 gal/acre/day or 9.35×10^{-5} liter/m²/day) versus a poor composite liner (100 gal/acre/day or 9.35×10^{-2} liter/m²/day) in Table 4.9.

To maximize the effectiveness of a composite liner, the geomembrane must be placed to achieve a good hydraulic contact (often called "intimate contact") with the underlying layer of low-permeability soil. As shown in Figure 1.12, the composite liner works by limiting the flow of fluid in the soil to a very small area. Fluid must not be allowed to spread laterally along the interface between the geomembrane and soil. To ensure good hydraulic contact, the soil liner should be smooth-rolled with a steel-drummed roller before the geomembrane is placed, and the geomembrane should have a minimum number of waves or wrinkles when it is finally covered (see Soong and Koerner, 1998). In addition, high-permeability material, such as a sand bedding layer or a thick geotextile, should not be placed between the geomembrane and low-permeability soil (Figure 1.12), because this will destroy the composite action of the two materials.

If there are concerns that rocks or stones in the soil material may punch holes in the geomembrane, the stones must be removed, or a stone-free material with a low hydraulic conductivity must be placed on the surface. Vibratory screening also can be used to sieve stones prior to placement. Alternatively, mechanical devices that sieve stones or move them to a row in a loose lift of soil may be used. A different material, or a differently processed material that has fewer and smaller stones, may be used to construct the uppermost lift of the compacted clay liner (i.e., the lift that will serve as a foundation for the geomembrane).

Figure 4.17 presents a comparison of leakage liquid collected below a compacted clay liner and a composite liner from a field test. Two landfill cells were built with the same area at the same location, but one cell used a compacted clay liner as its primary liner and the other one used a composite liner as its primary liner. These two cells were filled with the same type and the same volume of solid waste, using the same filling procedure. Figure 4.17 illustrates that the leakage volume below the composite liner is significantly less than that below a compacted clay liner.

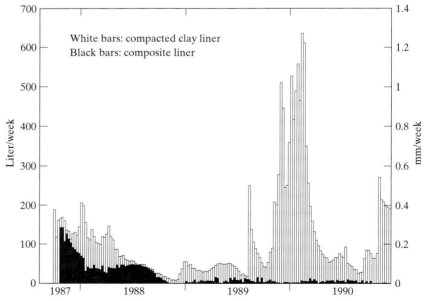

FIGURE 4.17 Comparison of Leakage Liquid Collected below Compacted Clay Liner and Composite Liner from A Field Test (Melchior and Miehlich, 1994)

4.9 CONCLUDING COMMENTS REGARDING GEOMEMBRANES

Beginning with the first U.S. EPA regulations in the early 1980's, geomembranes have been shown to drastically reduce leakage rates in landfill liner systems. (Additional data to that given in this chapter will be offered in Chapter 5 on geosynthetic clay liners). Even further, the use of a geomembrane with an underlying compacted clay liner provides the optimal liner system. This approach is referred to as a *composite liner system,* that is, a geomembrane/compacted clay liner or GCL. These benefits notwithstanding, the introduction of the geomembrane complicates the situation from both a design and a construction viewpoint.

Design Issues: A complication arises here because of the introduction of a relatively low interface shear strength material oriented exactly in the direction of the slope. When on steep side slopes, this fact is usually the controlling design feature. The introduction of textured geomembranes has significantly helped this situation, allowing for 3(H):1(V) side slopes and even steeper ones in some cases. Unbalanced shear stresses can also mobilize tensile forces in the geomembrane, and this situation was addressed in this chapter. Finally, the termination of the geomembrane in its anchor trench at the top of the slope was also described and calculated accordingly.

Construction Issues: The placement of a geomembrane on a compacted clay liner is not a trivial matter. The compacted clay liners are quite wet (see Chapter 3), and trafficking and placement of the geomembrane rolls are difficult. After seaming, the exposed geomembrane can elongate from sunlight and high ambient temperature. This has the effect of creating waves, or wrinkles. In turn, this challenges the concept of intimate contact of the two materials. Such intimate contact was seen to be the essence of extremely low leakage rates. The calculations herein clearly illustrate the situation. This issue will be further addressed in Chapter 17 on installation of geomembranes and other geosynthetics.

PROBLEMS

4.1 Which type of geomembrane is the most chemically resistant of all geomembranes?

4.2 What are the advantages of a textured geomembrane? Where should a textured geomembrane be used in a landfill?

4.3 A landfill cover slope has the following characteristics: slope angle = 4(H):1(V), slope length $L = 105$ ft (32 m). A 2-foot- (0.6-m)-thick silt ($\gamma_{silt} = 110$ lb/ft^3 or 17.3 kN/m^3) and 6 inches (0.15 m) of topsoil ($\gamma_{top} = 90$ lb/ft^3 or 14 kN/m^3) cover are placed over a 40-mil (1-mm) geomembrane. The geomembrane has an allowable tensile stress of 2,200 lb/in^2 (15,000 kN/m^2). The interface shear strength parameters of the geomembrane with the upper soil are $c_{aU} = 0$ and $\delta_U = 14°$, and with the lower soil are $c_{aL} = 0$ and $\delta_L = 10°$. Calculate the tension in the geomembrane and the resulting *FS* against geomembrane failure. If the factor of safety is less than 1.5, suggest an effective measure to increase the value of *FS* to 1.5 or more.

4.4 A composite landfill cover consisting of an 18-inch (0.45-m) compacted clay layer ($\gamma_{clay} = 120$ lb/ft^3 or 19 kN/m^3), 24 inches (0.6 m) of silt ($\gamma_{silt} = 110$ lb/ft^3 or 17.3 kN/m^3) and 6 inches (0.15 m) of topsoil ($\gamma_{top} = 90$ lb/ft^3 or 14 kN/m^3) is placed over a PVC geomembrane. A local subsidence occurs, which is estimated to be 1.2-ft (0.36-m) deep by 3.5 ft (1.05 m) in radius. The PVC geomembrane is 30-mils (0.75-mm) thick and has an allowable strength of 1,000 lb/in² (6,900 kN/m²). Determine the factor of safety of the geomembrane against the mobilized tensile stresses caused by the subsidence.

4.5 Given a 60-mil (1.5-mm) textured HDPE geomembrane with an allowable tensile stress of 1,100 lb/in² (7,600 kN/m²), which rests on a 2.5(H):1(V) sideslope, determine the required runout length. In this analysis, use 18 inches (0.45 m) of cover soil with a unit weight of 115 lb/ft³ (18 kN/m³) and a friction angle of 25 degrees with respect to the geomembrane.

4.6 A 60-mil (1.5-mm) HDPE geomembrane with an allowable stress 1,000 lb/in² (6,900 kN/m²) rests on a 3.5(H):1(V) side slope. A 24-inch (0.6-m) thick cover soil with a unit weight of 110 lb/ft³ (17.3 kN/m³) (also the unit weight of the backfill soil) is placed over the geomembrane. The friction angle between the geomembrane and soil is 18 degrees and that of the soil itself is 25 degrees. Determine the required length of runout for a 24-inch (0.6-m) deep rectangular anchor trench like the one shown in Figure 4.9.

4.7 An 80-mil (2-mm) HDPE geomembrane with an allowable stress of 1,300 lb/in² (9,000 kN/m²) rests on a 3(H):1(V) side slope. A 30-inch (0.75-m) thick cover soil with a unit weight of 110 lb/ft³ (17.3 kN/m³) (also the unit weight of the backfill soil) is placed over the geomembrane. The friction angle between the geomembrane and soil is 15 degrees and that of the soil itself is 30 degrees. Make the runout length equal to 60 inches (1.5 m), and the depth of anchor trench the same as the width of anchor trench. Calculate the depth and width of an L-shaped rectangular anchor trench similar to the one shown in Figure 4.10.

4.8 An 80-mil (2-mm) HDPE geomembrane with an allowable stress of 1,300 lb/in² (9,000 kN/m²) rests on a 3(H):1(V) side slope. A 12-inch (0.3-m) thick cover soil with a unit weight of 110 lb/ft³ (17.3 kN/m³) (also the unit weight of the backfill soil in the anchor trench) is placed over the geomembrane. The friction angle between the geomembrane and soil is 18 degrees and that of the soil itself is 30 degrees. Make the depth and the width of the L-shaped rectangular anchor trench (similar to the one shown in Figure 4.10) equal to 2 feet (0.6 m). Calculate the runout length.

4.9 An 80-mil (2-mm) HDPE geomembrane with an allowable stress of 1,300 lb/in² (9,000 kN/m²) rests on a 3(H):1(V) side slope. A 18-inch- (0.45-m)-thick cover soil with a unit weight of 110 lb/ft³ (17.3 kN/m³) (also the unit weight of the backfill soil in the anchor trench) is placed over the geomembrane. The friction angle between geomembrane and underlying soil is 18 degrees, and the friction angle between geomembrane and backfill soil in the anchor trench is 20 degrees. Make the depth of symmetric V-shaped anchor trench (similar to the one shown in Figure 4.12) equal to 1.5 feet (0.45 m) and the bottom angle of symmetric V-shaped anchor trench equal to 45 degrees. Calculate the runout length.

4.10 Compare the pullout forces for the following five types of geomembrane anchor measures:

1. *Runout anchor.* Runout length: 8 ft (2.4 m) (Figure 4.8);
2. *Rectangular anchor trench.* Runout length: 6 ft (1.8 m), depth of anchor trench: 2 ft (0.6 m) (Figure 4.9);
3. *L-shaped rectangular anchor trench.* Runout length: 4 ft (1.2 m), depth of anchor trench: 2 ft (0.6 m), width of anchor trench: 2 ft (0.6 m) (Figure 4.10).
4. *One-side V-shaped anchor trench.* Runout length: 6 ft (1.8 m), depth of anchor trench: 2 ft (0.6 m), bottom angle of one-side V-shaped anchor trench: 45 degrees (Figure 4.13).
5. *Symmetric V-shaped anchor trench.* Runout length: 4 ft (1.2 m), depth of anchor trench: 2 ft (0.6 m), bottom angle of symmetric V-shaped anchor trench: 45 degrees (Figure 4.12).

TABLE 4.10 Leakage Conditions for Three Types of Liners for Problem 4.11

Type of Liner	Size of Hole (cm^2)	Number of Holes		Contact Condition
		hole/acre	hole/hectare	
Geomembrane	0.1	40	100	–
Composite Liner	0.1	40	100	Poor
Composite Liner	0.1	40	100	Good

Use a textured 80-mil (2.0-mm) HDPE geomembrane on a 3.5(H):1(V) side slope. A 24-inch (0.6-m) thick cover soil with a unit weight of 115 lb/ft^3 (18 kN/m^3) (also the unit weight of the backfill soil) is placed over the membrane. The friction angle between the geomembrane and soil is 25 degrees and that of the soil itself is 30 degrees.

4.11 Compare the leakage rates (gallon/acre/day or liter/m^2/day) for the three types of liners listed in Table 4.10. The leachate head on the bottom liner is 12 inches (0.3 m). The hydraulic conductivity of compacted clay soil beneath the liners is 1.0×10^{-7} cm/sec.

4.12 Figure 4.17 shows a comparison of leakage volumes collected below a composite liner and a single compacted clay liner. The total leakage volume collected below the composite liner is much lower than that collected below the compacted clay liner; however, the leakage below the composite liner during the first year was relatively high. The leakage then decreased in subsequent years to a very low number. What accounts for this observed behavior, namely, the similarity in leakage volumes the first year?

REFERENCES

Bonaparte, R., Giroud, J. P. and Gross, B. A., (1989) "Rates of Leakage through Landfill Liners," *Proceedings of Geosynthetics '89*, Vol. I, IFAI, St. Paul, MN, pp. 18–29.

Byrne, R. J., Kendall, J., and Brown, S., (1992) "Cause and Mechanism of Failure of Kettleman Hills Landfill," *Proceedings of ASCE Specialty Conference on Stability and Performance of Slope and Embankments-II,* Berkeley, CA, June 28–July 1, pp. 1–23.

Geosynthetic Institute Specifications (ongoing), GSI, Folsom, PA.

Giroud, J. P. and Bonaparte, R., (1989a) "Leakage through Liners Constructed with Geomembranes—Part I. Geomembrane Liners," *Geotextiles and Geomembranes,* Vol. 8, Elsevier Science Publishers Ltd., England, pp. 27–67.

Giroud, J. P. and Bonaparte, R., (1989b) "Leakage through Liners Constructed with Geomembranes—Part II. Composite Liners," *Geotextiles and Geomembranes,* Vol. 8, Elsevier Science Publishers Ltd., England, pp. 71–111.

Giroud, J. P., Khatami, A., and Badu-Tweneboah, K., (1989) "Evaluation of Leakage through Composite Liners," *Geotextiles and Geomembranes,* Vol. 8, Elsevier Science Publishers Ltd., England, pp. 333–340.

Giroud, J. P., Badu-Tweneboah, K., and Bonaparte, R., (1992) "Rate of Leakage through a Composite Liner due to Geomembrane Defects," *Geotextiles and Geomembranes,* Vol. 11, No. 1, Elsevier Science Publishers Ltd., England, pp. 1–28.

Koerner, R. M., (1990) "Designing with Geosynthetics," 2nd Edition, Prentice Hall Inc., Englewood Cliffs, NJ.

Koerner, R. M., (1994) "Designing with Geosynthetics," 3rd Edition, Prentice Hall Inc., Englewood Cliffs, NJ.

Koerner, R. M., (1998) "Designing with Geosynthetics," 4th Edition, Prentice Hall Inc., Upper Saddle River, NJ.

Koerner, R. M., Koerner, G. R., and Hwu, B. L., (1990) "Three-Dimensional Axi Symmetric Geomembrane Tension Test," ASTM STP 1081, R. M. Koerner ed., Philadelphia, PA, pp. 170–184.

Koerner, R. M. and Hwu, B. L., (1991) "Stability and Tension Considerations Regarding Cover Soils on Geomembrane Lined Slopes," *Geotextiles and Geomembranes,* Elsevier Science Publishers Ltd., England, pp. 335–355.

Martin, J. P., Koerner, R. M., and Whitty, J. E., (1984) "Experimental Friction Evaluation of Slippage between Geomembranes, Geotextiles and Soils," *Proceedings of International Conference on Geomembranes,* IFAI, Denver, CO, pp. 191–196.

Melchior, S. and Miehlich, G., (1994) "Hydrological Studies on the Effectiveness of Different Multilayered Landfill Caps," *Landfilling of Waste: Barriers,* Edited by T. H. Christensen, R. Cossu, and R. Stegmann, E & FN Spon, London, U.K., pp. 115–137.

Mitchell, J. K., Seed, R. B., and Seed, H. B., (1990) "Kettleman Hills Waste Landfill Slope Failure I: Liner System Properties," *Journal of Geotechnical Engineering,* ASCE, Volume 116, No. 4, pp. 647–668.

National Sanitation Foundation (NSF) International, Ann Arbor, MI.

Polyvinyl Chloride Geomembrane Institute Specifications, PGI, University of Illinois, Urbana-Champagne, IL.

Qian, Xuede, (1995) "Geotechnical Aspects of Landfill Design," Course Pack, Department of Civil and Environmental Engineering, The University of Michigan, Ann Arbor, MI, 485 pages.

Seed, R. B., Mitchell, J. K., and Seed, H. B., (1990) "Kettleman Hills Waste Landfill Slope Failure II: Stability Analyses," *Journal of Geotechnical Engineering,* ASCE, Volume 116, No. 4, pp. 669–690.

Stark, T. D. and Poeppel, A. R., (1994) "Landfill Liner Interface Strengths from Torsional-Ring-Shear Tests," *Journal of Geotechnical Engineering,* ASCE, Vol. 120, No. 3, pp. 597–615.

USEPA, (1988) "Final Cover on Hazardous Waste Landfills and Surface Impoundments," U. S. EPA Guide to Technical Resources for the Design of Land Disposal Facilities, EPA/530-SW-88-047.

USEPA, (1991) "Design and Construction of RCRA/CERCLA Final Covers," EPA/625/4-91/025, Office of Research and Development, U. S. Environmental Protection Agency, Washington, DC, May.

CHAPTER 5

Geosynthetic Clay Liners

5.1 TYPES AND CURRENT USES OF GEOSYNTHETIC CLAY LINERS
    5.1.1 GEOTEXTILE-ENCASED, ADHESIVE-BONDED GCL
    5.1.2 GEOTEXTILE-ENCASED, STITCH-BONDED GCL
    5.1.3 GEOTEXTILE-ENCASED, NEEDLE-PUNCHED GCL
    5.1.4 GEOMEMBRANE-SUPPORTED, ADHESIVE-BONDED GCL
5.2 HYDRAULIC CONDUCTIVITY
    5.2.1 EFFECT OF PERMEATING LIQUIDS
    5.2.2 EFFECT OF CONFINING STRESS
    5.2.3 WET-DRY RESPONSE
    5.2.4 FREEZE-THAW RESPONSE
5.3 ABILITY TO WITHSTAND DIFFERENTIAL SETTLEMENT
5.4 SHEAR STRENGTH
    5.4.1 INTERNAL SHEAR STRENGTH
    5.4.2 INTERFACE SHEAR STRENGTH
    5.4.3 DESIGN IMPLICATIONS
5.5 DIFFERENCES BETWEEN GEOSYNTHETIC CLAY LINERS AND COMPACTED CLAY LINERS
5.6 CONTAMINANT TRANSPORT THROUGH GEOSYNTHETIC CLAY LINER AND COMPACTED CLAY LINER
    5.6.1 STEADY ADVECTION
    5.6.2 STEADY DIFFUSION
    5.6.3 ADVECTIVE BREAKTHROUGH TIME
    5.6.4 COMBINED ADVECTION-DIFFUSION
5.7 COMPARISON OF MASS TRANSPORT THROUGH A GCL AND CCL
5.8 RECOMMENDATIONS FOR USE OF GEOSYNTHETIC CLAY LINERS
5.9 SUMMARIZING COMMENTS REGARDING GEOSYNTHETIC CLAY LINERS
    PROBLEMS
    REFERENCES

Geosynthetic clay liners (GCLs) are thin hydraulic barriers containing approximately 1 lb/ft^2 (5 kg/m^2) of bentonite, sandwiched between two geotextiles or attached with an adhesive to a geomembrane. Sodium bentonite is used primarily in North America, while calcium bentonite is used more frequently worldwide. The sodium type has a lower hydraulic conductivity. Geosynthetic clay liners are manufactured in continuous sheets and are installed by unrolling and overlapping the edges and ends of the panels. Overlaps self-seal when the bentonite hydrates. Geosynthetic clay liners are increasingly used in bottom liners for landfills and impoundments, in final covers for landfills and remediation projects, and as liners for secondary containment around liquid storage tanks.

Geosynthetic clay liners can be installed by a number of methods (Tranger and Tawes, 1995), minimizing the risk of damage to underlying components, and can easily be placed on side slopes. Also, a geosynthetic clay liner, which is initially placed dry, does not yield consolidation water upon loading, unlike compacted clay liners (CCLs). Under consolidation loads, significant amounts of pore water from a saturated compacted clay can flow into the underlying drainage layer and be misinterpreted as liner leakage. Also, in a composite landfill liner system, geosynthetic clay liners reduce the thickness of compacted clay liners and caps, which increases landfill airspace in the same footprint and allows less excavation work for a given landfill volume.

5.1 TYPES AND CURRENT USES OF GEOSYNTHETIC CLAY LINERS

At present, there are four types of geosynthetic clay liner (GCL) available in the North American market. Distinguished by their structures, the four types are geotextile-encased, adhesive-bonded GCL; geotextile-encased, stitch-bonded GCL; geotextile-encased, needle-punched GCL; and geomembrane-supported, adhesive-bonded GCL (Figure 5.1). Each type of GCL has its own unique properties.

5.1.1 Geotextile-Encased, Adhesive-Bonded GCL

A geotextile-encased, adhesive-bonded GCL is a nonreinforced GCL with two light woven or nonwoven geotextiles encapsulating a layer of sodium bentonite [Figure 5.1(a)]. As seen in Table 5.1, one of these products also includes a thin geomembrane contained within the upper and lower geotextile. The table also contains information about geotextile-encased, adhesive-bonded GCL products currently available on the market.

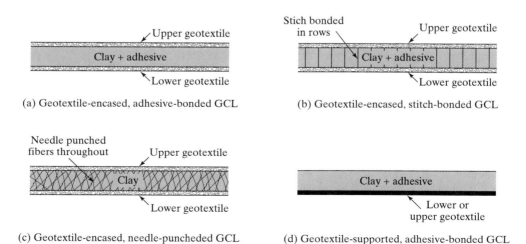

FIGURE 5.1 Four Types of Geosynthetic Clay Liners (Koerner and Daniel, 1997). Used with permission of ASCE.

TABLE 5.1 Types of Geosynthetic Clay Liner (products are subject to change)

Name	Manufacturer	Bentonite Type	Bentonite Form	Upper Geotextile	Lower Geotextile
Geotextile-Encased, Adhesive-Bonded GCL					
Claymax® 200R	Colloid Environmental Technologies Company	sodium	granules	woven	woven
Claymax® 600CL		sodium	granules	woven	woven with a film lamination
Geotextile-Encased, Stitch-Bonded GCL					
Claymax® 500SP	Colloid Environmental Technologies Company	sodium	granules	woven	woven
NaBento®	Huesker Synthetic GmbH & Co., Germany	sodium or soda activated	powder	woven	woven
Geotextile-Encased, Needle-Punched GCL					
Bentofix® NS	Naue Fasertechnik GmbH & Co., Germany	sodium	granules	woven	nonwoven
Bentofix® WP		sodium	granules	woven	nonwoven
Bentofix® NW		sodium	granules	nonwoven	nonwoven
Bentomat® ST	Colloid Environmental Technologies Company	sodium	granules	woven	nonwoven
Bentomat® DN		sodium	granules	nonwoven	nonwoven
Bentomat® CL		sodium	granules	woven	nonwoven with a film lamination
Geomembrane-Supported, Adhesive-Bonded GCL					
Gundseal®	GSE Lining Technology, Inc.	sodium	granules	none	polyethylene geomembrane

5.1.2 Geotextile-Encased, Stitch-Bonded GCL

A geotextile-encased, stitch-bonded GCL is a reinforced GCL consisting of a layer of sodium bentonite encapsulated between two geotextiles [Figure 5.1(b)]. The two woven or nonwoven geotextiles on this material are stitched together with parallel rows of stitches. The stitches are at 100 mm in one of the products given in Table 5.1 and 25 mm in the other. The increase in internal strength achieved by stitch-bonding the encapsulating geotextiles together is very substantial.

5.1.3 Geotextile-Encased, Needle-Punched GCL

A geotextile-encased, needle-punched GCL is also a reinforced GCL consisting of a layer of sodium bentonite encapsulated between two woven or nonwoven geotextiles, needle punched together [Figure 5.1(c)]. This integrated matrix of bentonite and needle-punched fibers provides high internal shear strength. The needle punching is provided by penetrating the nonwoven, needle-punched geotextile on one side of the GCL and "dragging" fibers through the bentonite and the lower geotextile.

For one product, the emergent fibers are thermally bonded to the adjacent fibers. This type of GCL can consist of one nonwoven geotextile and one woven geotextile, or two nonwoven geotextiles. When two nonwoven geotextiles are used, the manufacturer recommends that 0.4 kg/m (0.3 lb/ft) of loose, dry bentonite be placed along the centerline of the overlap when installing the GCL to ensure that the material self-seals along the overlap and forms a continuous barrier. The edge overlap is usually 150 mm (6 inches) and the end overlap is usually 600 mm (2 feet).

5.1.4 Geomembrane-Supported, Adhesive-Bonded GCL

A geomembrane-supported, adhesive-bonded GCL is a bentonite-geomembrane composite GCL. The geomembrane can be of any type (it is usually HDPE or LLDPE), of any thickness, and either smooth or textured. This type of GCL can be installed in two ways. The first configuration is with the geomembrane side facing downward against the subgrade, and with the bentonite side facing upward against an overlying field-deployed geomembrane, forming a three-layer system. The second configuration is with the bentonite side facing downward against the subgrade.

The sealing between installed rolls is made by overlapping the material. The lengthwise seams are typically overlapped a minimum 150 mm (6 inches), and the widthwise seams are overlapped 300 mm (1 foot). When this type of GCL is installed with the bentonite side facing downward, the manufacturer recommends that tape be placed along the seam to prevent overlying cover soils from separating the seams (USEPA, 1993).

5.2 HYDRAULIC CONDUCTIVITY

Hydraulic conductivity is the most important parameter to evaluate in deciding whether a GCL can be used as alternative material to substitute a conventional compacted clay liner for various environmental projects. There are several factors that affect the hydraulic conductivity of GCLs, such as permeant liquids and confining stress. The effects of wet-dry cycling and freeze-thaw cycling on the hydraulic conductivity of the GCLs also must be considered when using GCLs in the liner and final cover systems for landfill and site remediation projects.

5.2.1 Effect of Permeating Liquids

Studies dealing with the effects of chemicals on the hydraulic conductivity of bentonite have been reported by Alther et al. (1985), Anderson et al. (1985), and Keren and Singer (1988). Findings from these studies are consistent with diffuse double-layer theory: liquids with low dielectric constant, high cation valence, and high electrolyte concentration tend to increase the hydraulic conductivity of bentonite.

Schubert (1987) permeated a geotextile-encased GCL with hazardous waste leachate from a landfill and found that the hydraulic conductivity remained unchanged. Details about the chemistry of the leachate were not provided. Shan and

Daniel (1991) permeated a geotextile-encased GCL with water and then with various chemicals, including methanol (pure and diluted with water), pure heptane, sulfuric acid, 0.01 N $CaSO_4$, and 0.5 N $CaCl_2$. The maximum effective confining stress was 35 kPa, and the pore volumes of flow varied from 0.2 to 24.2. In some cases, hydraulic conductivity increased tenfold (using $CaSO_4$, and $CaCl_2$ solutions), and in other cases hydraulic conductivity remained constant or dropped slightly. All test specimens were hydrated with water prior to introduction of the chemical solution.

Daniel et al. (1993) permeated the bentonite component of a geomembrane-supported GCL with various nonaqueous-phase liquids (NAPLs), including gasoline. The bentonite was hydrated to different initial water contents (including tests with no additional water for direct permeation of the dry GCL) and then was permeated with NAPLs. The initial water content had a large effect on the hydraulic conductivity; at water contents less than or equal to 50%, the hydraulic conductivity to the NAPLs was very high ($\approx 1.0 \times 10^{-5}$ cm/sec), but when the bentonite was moistened to a water content of 100% or more prior to introducing the NAPL, the hydraulic conductivity was less than 1.0×10^{-9} cm/sec.

Ruhl and Daniel (1997) performed hydraulic conductivity tests on five GCLs using seven permeating liquids and three conditions of hydration to investigate the effects of chemical solutions and leachate on the hydraulic conductivity of geosynthetic clay liners. The permeating liquids used in the testing program were tap water, hydrochloric acid (HCl) (pH = 1), sodium hydroxide (NaOH) (pH = 13), a real municipal solid waste leachate obtained from a midwestern municipal landfill that had been in operation for several years (pH = 7), a simulated municipal solid waste leachate based on the formulation for synthetic municipal solid waste leachate by Stanforth et al. (1979) (pH = 4.4), a simulated hazardous waste leachate based on the formulation data published by Bramlett et al. (1987), McNann et al. (1987), and Sai and Anderson (1991), and a simulated fly ash leachate developed following procedure of Texas Water Commission (1985) (pH = 11.5 ~ 12). The hydraulic conductivity of GCLs to tap water was 3.0×10^{-10} to 2.0×10^{-9} cm/sec. The condition of hydration was found to be very important: much lower hydraulic conductivity generally resulted when the first wetting liquid was water rather than the chemical solutions or leachate. The GCLs exhibited relatively high hydraulic conductivities (10^{-7} to 10^{-5} cm/sec) when permeated directly with (1) simulated municipal solid waste leachate that was rich in calcium, (2) a strong acid solution, or (3) a strong base solution. On the other hand, the GCLs maintained low hydraulic conductivities (1×10^{-8} to 1×10^{-10} cm/sec) when they were permeated with (1) simulated hazardous waste leachate, (2) real municipal solid waste leachate, or (3) simulated fly ash leachate. GCLs containing contaminant-resistant bentonite maintained a lower hydraulic conductivity than GCLs that contained regular bentonite for some, but not all, permeating liquids (Ruhl and Daniel, 1997).

5.2.2 Effect of Confining Stress

Several investigators (e.g., Shan and Daniel, 1991; Estornell and Daniel, 1992; USEPA, 1993; Rad et al., 1994; Petrov et al., 1997) have reported significant decreases in the GCL hydraulic conductivity with increased effective confining stress σ'.

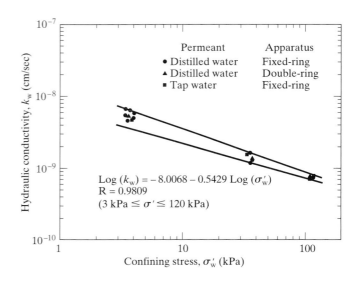

FIGURE 5.2 GCL Hydraulic Conductivity versus Confining Stresses for Confined Hydraulic Conductivity Tests (modified from Petrov et al., 1997). Used with permission of ASCE.

Presumably this is the result of lower bentonite void ratios. The impact of low ($\sigma' = 3 \sim 4$ kPa or $0.44 - 0.58$ psi), intermediate ($\sigma' = 34 \sim 37$ kPa or $4.9 \sim 5.4$ psi), and high confining stresses ($\sigma' = 109 \sim 117$ kPa or $15.8 \sim 17.0$ psi) on the hydraulic conductivity of a geotextile-encased, needle-punched GCL is shown in Figure 5.2. The hydraulic conductivity decreased from 3.7×10^{-9} cm/sec to 6.4×10^{-9} cm/sec for low stress conditions, and from 7.1×10^{-10} cm/sec to 7.9×10^{-10} cm/sec for high stress conditions. The almost one order-of-magnitude difference in GCL hydraulic conductivity from low to high stress conditions is significant compared with compacted clayey soils, and indicates the importance of the confining stress in controlling the advective flow through GCLs. For these GCLs confined prior to hydration and permeation, there is a well-defined linear relationship on a logarithmic scale between the GCL hydraulic conductivity and the confining stress with some experimental scatter, as shown in Figure 5.2. The reduction in GCL hydraulic conductivity can be attributed (Petrov et al., 1997) to lower bentonite void ratios resulting from higher confining stresses. A GCL used in the landfill liner system should have a lower hydraulic conductivity than the same GCL used in the landfill cover system because the GCL in the liner system is subjected to much higher confining stress than that in the final cover system. For instance, the overburden pressure acting on a GCL in the liner is approximately 45 psi (310 kN/m^2) for a 100-ft (30 m) municipal solid waste landfill, whereas the overburden pressure acting on a GCL in a municipal solid waste cap is only approximately 1.8 psi (12.4 kN/m^2).

5.2.3 Wet-Dry Response

Dry bentonite swells when wetted and shrinks when dried. If a GCL is subjected to several wet-dry cycles, a hydrated GCL will become severely desiccated when it dries. Alternate wetting and drying may occur in a GCL in final cover systems of landfills and site remediation projects. Boardman and Daniel (1996) performed laboratory hydraulic conductivity tests on three large-scale GCLs to investigate the effect of wetting and drying on the hydraulic conductivity of GCLs. Non-overlapped GCLs and overlapped panels were tested. Each GCL was buried under 2 ft (600 mm) of pea gravel and permeated with water for several weeks. Then, the water was removed from the gravel and the GCLs were desiccated by circulating heated air through the gravel. Severe drying and cracking occurred in the bentonite component of the GCLs. After drying, each GCL was slowly rehydrated. The hydraulic conductivity was then monitored to determine the ability of the desiccated GCL to rehydrate and self-seal.

Three types of GCL were used in this study. The testing results showed that the geotextile-encased GCLs swelled and self-sealed upon rehydration, after a cycle of wetting and drying. When the desiccated GCLs were rehydrated, water initially flowed rapidly through the cracks in most of the desiccated samples, but the bentonite quickly expanded and the hydraulic conductivity decreased as the cracked bentonite began to absorb water and swell. According to Boardman and Daniel's (1996) test results, the long-term steady value of hydraulic conductivity was essentially the same before and after the desiccation cycle (Figures 5.3 and 5.4). Shan and Daniel (1991) performed laboratory hydraulic conductivity tests on small samples of one type of geotextile-encased GCL that had been subjected to several wet-dry cycles and reported that severe desiccation cracks developed when the wet GCLs were dried. They likewise reported that the hydraulic conductivity after several wet-dry cycles was the same as the conductivity of the nondesiccated material.

Based on these test results, the wetting and drying cycle did not appear to cause any irreversible shrinkage to occur along the overlap for overlapping samples of any of the GCLs tested. However, samples were partially attached to a rigid, steel frame in these tests, and the performance of these materials in the field might be different. Although the bentonite did form open cracks upon drying, the cracks swelled and closed upon wetting. The geosynthetic component of the GCL (geotextile or geomembrane) prevented any intrusion of overlaying pea gravel into the cracks. Designers should be careful that the openings in the geotextile component of the GCL are small enough to prevent the overlying soil from migrating into cracks that develop in the bentonite. (The design method for selection of geotextiles is described in Chapter 8.)

Boardman and Daniel (1996) maintain that the initially high value of hydraulic conductivity of the desiccated GCLs shown in Figures 5.3 and 5.4 may not be representative of true field conditions because the overlying cover soils would likely adsorb

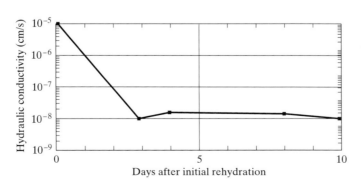

FIGURE 5.3 Long-Term Hydraulic Conductivity versus Time for Rehydration of a Desiccated Geotextile-Encased, Adhesive-Bonded GCL, (Intact Sample with No Overlap) (Boardman and Daniel, 1996). Used with permission of ASCE.

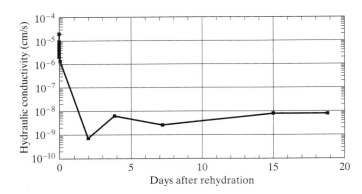

FIGURE 5.4 Long-Term Hydraulic Conductivity versus Time for Rehydration of a Desiccated Geotextile-Encased, Adhesive-Bonded GCL (Overlapped Panels) (Boardman and Daniel, 1996). Used with permission of ASCE.

some of the incoming rainfall and cause a more gradual wetting of the GCL. In addition, the rehydration rate of 40 mm/hr used in these tests would correspond to an extreme infiltration rate, and the GCL would either have to be overlain by extremely permeable material (e.g., gravel) or buried at extremely shallow depth for a flux of water of 40 mm/hr to be applied to the GCL in the field. If the GCL is slowly wetted (which would be the case in many field situations), the GCL would have time to absorb water and to swell without allowing seepage through the GCL. The significance of high initial hydraulic conductivity should be considered on a project-specific basis.

The self-sealing capability of GCLs make them a potentially viable hydraulic barrier for situations in which the barrier may undergo cyclic wetting and drying (e.g., within a landfill final cover). However, the reader is cautioned not to inappropriately extrapolate the results of these tests. The tests carried out by Boardman and Daniel (1996) were performed under carefully controlled conditions with a single severe wetting and drying cycle. Such a severe cycle of wetting and drying is not likely to occur in the field. Numerous less severe cycles of wetting and drying are more likely to occur in the field. Further research (particular field data) is needed before a final conclusion can be drawn concerning the ability of GCLs to safely withstand numerous wetting and drying cycles under the full range of possible field conditions. Nevertheless, these results are encouraging and suggest that GCLs may be an attractive material to use when some degree of cycling in water content is anticipated within the hydraulic barrier.

5.2.4 Freeze-Thaw Response

Resistance to increases in hydraulic conductivity caused by freeze-thaw cycling is a characteristic of particular importance in cold regions. Numerous studies have shown that compacted clay liners undergo large increases in hydraulic conductivity when exposed to freeze-thaw cycling. (Recall Chapter 3.) For compacted clays having an initial hydraulic conductivity less than 1.0×10^{-7} cm/sec, freezing and thawing generally increases the hydraulic conductivity one to three orders of magnitude (Zimmie and LaPlante, 1990; Kim and Daniel, 1992; Othman et al., 1994; Chamberlain et al., 1995; Benson et al., 1995; Kraus et al., 1997). Bentonite appears to be less vulnerable to freeze-thaw damage than other types of soil. Wong and Haug (1991) found that the hydraulic conductivity of compacted sand-bentonite mixture did not increase after five freeze-thaw cycles.

GeoServices, Inc. (1989) evaluated how freeze-thaw cycling affects the hydraulic conductivity of a geotextile-encased, adhesive bonded GCL using flexible-wall permeameters. An initial hydraulic conductivity of 4.0×10^{-10} cm/sec was measured at an effective confining pressure of 196 kPa (28.4 psi) and a hydraulic gradient of 1,000. The saturated specimen was then repeatedly frozen and thawed three-dimensionally. After 10 freeze-thaw cycles, the hydraulic conductivity was 1.5×10^{-10} cm/sec. Similar findings have been reported by Shan and Daniel (1991).

GeoSyntec Consultants (1991) studied the effect of freeze-thaw cycling on the hydraulic conductivity of a geotextile encased, needle-punched GCL under an effective confining pressure of 35 kPa (5 psi) and a hydraulic gradient of 30. The specimen was subjected to four freeze-thaw cycles, with the hydraulic conductivity of each specimen measured after each cycle. The hydraulic conductivity of the specimens ranged between 1.0×10^{-9} and 6.0×10^{-9} cm/sec after each cycle, with no increasing or decreasing trends.

Tests to evaluate the effect of freeze-thaw cycling on a similar GCL have also been performed by Robert L. Nelson and Associates (1993). Two sets of tests were conducted. In the first set, six specimens were evaluated after undergoing up to six freeze-thaw cycles with no initial hydration (i.e., no initial saturation or permeation). In the second set, only one specimen was tested. It was exposed to 10 freeze-thaw cycles, with its hydraulic conductivity being measured after each thaw. No significant increase or decrease in hydraulic conductivity was observed in either set of tests. The hydraulic conductivity ranged between 1.1×10^{-9} and 4.0×10^{-9} cm/sec for the specimens in the first set of tests, and 1.9×10^{-9} and 3.3×10^{-9} cm/sec for the second set.

Hewitt and Daniel (1997) performed hydraulic conductivity tests in large tanks on intact (single panel) and overlapped samples of three GCLs that had been subjected to freeze-thaw cycles. Three GCLs were selected for testing to cover the range of types of commercial GCLs available. The tested materials included a geotextile-encased, needle-punched GCL, a geotextile-encased, stitch-bonded GCL, and a geomembrane-supported GCL with a 20-mil- (0.5-mm)-thick smooth, HDPE component. The compressive stress applied to the GCLs (7.6 ~ 12.4 kPa or 1.1 ~ 1.8 psi) was selected to simulate final cover systems for landfills. The hydraulic conductivity tests were performed using a laboratory flexible-wall type apparatus. According to Hewitt and Daniel (1997) with the exception of one overlapped GCL, all three GCLs withstood three freeze-thaw cycles without a significant change in hydraulic conductivity. An overlapped geotextile-encased, stitch-bonded GCL did undergo a 1,000-fold increase in hydraulic conductivity after one freeze-thaw cycle, but the overlapped area contained stitches, which are left off the edge of the full-sized material that is deployed in the field. In general, the tests showed that GCLs can withstand at least three freeze-thaw cycles without significant changes in hydraulic conductivity.

Kraus et al. (1997) conducted hydraulic conductivity tests in both the laboratory and field on GCLs to determine if the hydraulic conductivity was affected by freezing and thawing. Three geotextile-encased GCLs were used in the laboratory portion of this study. The field tests were conducted using geotextile-encased, needle-punched GCLs and a geomembrane-supported GCL. In the laboratory, an average effective

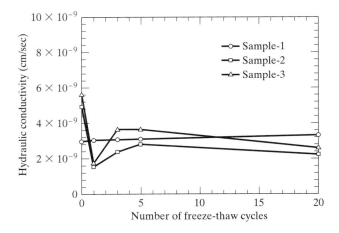

FIGURE 5.5 Hydraulic Conductivity of a Geotextile-Encased, Needle-Punched GCL-1 versus Number of Freeze-Thaw Cycles (Kraus et al., 1997). Used with permission of ASCE.

stress of 14 kPa (2 psi) and hydraulic gradient of 75 were applied. The effective stress of 14 kPa (2 psi) was assumed to simulate field conditions where a GCL would be used in a final cover. Test results showed that all of the hydraulic conductivities were very low, ranging between 2.9×10^{-9} and 4.9×10^{-9} cm/sec for the initial condition, and 1.7×10^{-9} and 3.3×10^{-9} cm/sec for the specimens exposed to 20 freeze-thaw cycles. Furthermore, for all three GCLs, a small decrease in hydraulic conductivity apparently occurred as a consequence of freeze-thaw cycling (Figures 5.5, 5.6, and 5.7). Kraus et al. (1997) suggest that this decrease in hydraulic conductivity is probably the result of thaw consolidation (Chamberlain and Gow, 1979) or particle reorientation that occurred when the stress changed slightly between permeation and freezing. In the field, two types of GCLs were exposed to one freeze-thaw cycling (one winter of exposure). The hydraulic conductivity of the GCLs ranged from 1.0×10^{-8} to 2.8×10^{-8} cm/sec prior to winter, and from 1.0×10^{-8} to 3.0×10^{-8} cm/sec after winter. The only exception was one of the seamed GCLs, which had a hydraulic conductivity of 7.0×10^{-7} cm/sec.

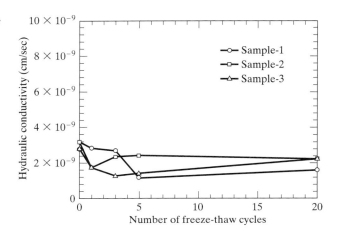

FIGURE 5.6 Hydraulic Conductivity of a Geotextile-Encased, Needle-Punched GCL-2 versus Number of Freeze-Thaw Cycles (Kraus et al., 1997). Used with permission of ASCE.

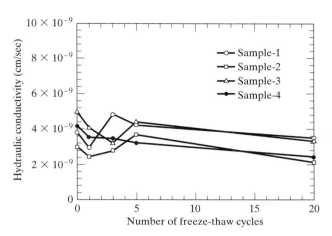

FIGURE 5.7 Hydraulic Conductivity of a Geotextile-Encased, Adhesive-Bonded GCL-3 versus Number of Freeze-Thaw Cycles (Kraus et al., 1997). Used with permission of ASCE.

Examination of the GCLs, while frozen and after thawing, revealed why these materials do not incur the increases in hydraulic conductivity typical of compacted clays. Ice segregation does occur in GCLs, but the cracks formed during ice segregation close when the bentonite thaws because the thawed bentonite is very soft and compressible (Kraus et al., 1997). Furthermore, the geotextile-encased GCL prevents soil from entering into the cracks during the freezing cycle. Consequently, GCLs do not undergo increases in hydraulic conductivity.

Although the findings from these studies are encouraging, designers are cautioned not to inappropriately extrapolate the results of these tests. Kraus et al. (1997) caution that the tests were performed under carefully controlled conditions in laboratory devices, and the conditions of freeze-thaw were not superimposed with other environmental stresses (e.g., differential settlement and desiccation). They recommended that designers carefully consider the use of GCLs in situations where freezing will occur. This is particularly important in applications where GCLs are the sole hydraulic barrier. Only long-term field tests, in which GCLs are monitored for an extended number of years, will provide the definitive information regarding the long-term performance of GCLs subjected to freezing conditions. For geomembrane-covered GCLs, the situation is quite different and, as a result, composite barriers are generally looked upon favorably.

5.3 ABILITY TO WITHSTAND DIFFERENTIAL SETTLEMENT

GCLs serve an important function as a hydraulic barrier in final covers for landfills. The ability of hydraulic barriers to withstand differential settlement is an important issue in the case of final covers. Gilbert and Murphy (1987) demonstrated that settlement over short distances in landfill covers is more threatening to the performance of the barrier than relatively uniform settlement over longer distances. Differential settlement may be characterized by the distortion Δ/L, which is defined as the settlement Δ over a horizontal distance L (Figure 5.8). The average tensile strain ε_t caused by distortion can be calculated over the deflected shape to determine the arc length of the deformed section (Gilbert and Murphy, 1987), or from simple mechanics, as will be

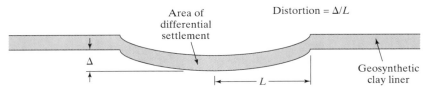

FIGURE 5.8 Definition of Distortion, Δ/L (LaGatta et al., 1997). Used with permission of ASCE.

shown later. If tensile strains are large enough, the barrier layer (e.g., CCL or GCL) may crack and lose its low hydraulic conductivity (LaGatta et al., 1997). The tensile strain at failure of compacted clay liner is typically between 0.1% and 4% according to a summary of published data (Tschebotarioff et al., 1953; Leonards and Narain, 1963; Ajaz and Parry, 1975; Scherbeck et al., 1991; Scherbeck and Jessberger, 1993).

In order to investigate the ability of GCLs to withstand differential settlement, five GCLs were tested to measure the hydraulic conductivity as the GCLs were subjected to differential settlement in either a dry state or a hydrated state (LaGatta et al., 1997). Two of the GCLs tested were geotextile-encased, needle-punched GCLs with a woven geotextile on the upper side and a nonwoven geotextile on the lower side. Another GCL tested was a geotextile-encased, adhesive-bonded GCL. The fourth GCL tested was a geotextile-encased, stitch-bonded GCL. The fifth GCL tested was a geomembrane-supported, adhesive-bonded GCL. Overlaps were 225 mm (9 inches) wide. For the two needle-punched GCLs, dry bentonite was applied in the overlap at an application rate of 0.4 kg/m (0.3 lb/ft) as recommended by the manufacturers.

Test results reported by LaGatta et al. (1997) showed that intact and overlapped geotextile-encased, needle-punched GCLs maintained a final hydraulic conductivity of 1.0×10^{-7} cm/sec or less when subjected to a distortion Δ/L as large as 0.35 ~ 0.6, which corresponds to a tensile strain ε_t of 5 ~ 16% (Figure 5.9). The tensile strength of the geotextile components in some cases enabled the intact GCL to bridge over the underlying subsidence. Needle-punched fibers limited migration of bentonite within the GCL. The overlapped area maintained its hydraulic conductivity integrity despite >25 mm slippage along the overlap.

An intact and overlapped geotextile-encased, adhesive-bonded GCL maintained a hydraulic conductivity of 1.0×10^{-7} cm/sec or less up to a distortion Δ/L of about 0.1, which corresponds to a tensile strain ε_t of 1%. Significant bentonite migration occurred, but this migration was strongly influenced by lack of confinement as would have been provided from needle-punched fibers or sewn stitches.

An intact and overlapped geotextile-encased, stitch-bonded GCL generally maintained a hydraulic conductivity of 1.0×10^{-7} cm/sec or less up to a distortion Δ/L of about 0.35, which corresponds to a tensile strain ε_t of 5%. This material performed significantly better than the unstitched version. The tensile strength of the geotextile components enabled the intact GCL to bridge over the underlying subsidence. The sewn stitches limited the movement of bentonite within the GCL. Overlapped and nonoverlapped materials performed about the same.

An intact and overlapped geomembrane-supported, adhesive-bonded GCL maintained an equivalent hydraulic conductivity of 1.0×10^{-7} cm/sec or less for a distortion Δ/L of up to 0.8, which translates to tensile strain ε_t of nearly 30%. The tensile

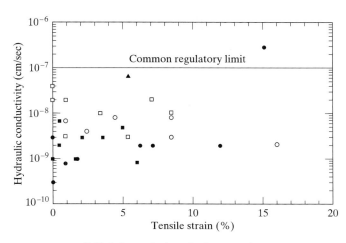

FIGURE 5.9 Final Hydraulic Conductivity versus Tensile Strain for Two Types of Geotextile-Encased, Needle-Punched GCLs (LaGatta et al., 1997). Used with permission of ASCE.

- ■ GCL-1, intact, hydrated prior to settlement
- □ GCL-1, intact, unhydrated prior to settlement
- ● GCL-1, overlapped, unhydrated prior to settlement
- ▲ GCL-1, overlapped, hydrated prior to settlement
- ○ GCL-2, intact, hydrated prior to settlement
- □ GCL-2, overlapped, hydrated prior to settlement

strength of the geomembrane component allowed the intact GCL to bridge over areas of subsidence, and the swelling and self-healing ability of the bentonite component caused overlaps to maintain their hydraulic integrity despite approximately 100 mm (4 inches) of slippage along the overlap.

According to LaGatta et al. (1997), tensile strains at failure of compacted clay range from $0.1 \sim 4\%$, whereas tensile strains at failure for geomembranes subjected to multiaxial testing range from $20 \sim 100\%$. The GCLs tested maintained hydraulic conductivity of 1.0×10^{-7} cm/sec or less while subjected to a tensile strain of $1 \sim 10\%$ or more, depending on the material and conditions of testing. Independent testing by Koerner et al. (1996) show that GCLs can withstand 10 to 20% tensile strain in axisymetric strength tests. In general, GCLs appear to fall between compacted clay liners and geomembranes in terms of ability to maintain their hydraulic integrity during distortion such as that induced by differential settlement in landfill final covers. Because the data indicates that GCLs can withstand more differential settlement than compacted clay, GCLs appear to be an attractive alternative to compacted clay liners in landfill final covers, assuming that other issues (such as slope stability) do not preclude the use of a GCL.

5.4 SHEAR STRENGTH

Stability of GCLs is an important design consideration because of the low shear strength of the bentonite after hydration. Unreinforced GCLs are held together by chemical adhesives and, once hydrated, provide relatively low resistance to shear. For higher shear-strength applications, reinforced GCLs are required in which the carrier geosynthetics are connected by stitched or needle-punched fibers (Fox et al., 1998) that transmit shear stress across the bentonite layer. Regardless of the type of GCL used, internal shear strength of the GCL and interface shear strengths between the GCL and adjacent materials need to be considered for stability analysis.

5.4.1 Internal Shear Strength

Numerous studies have been conducted to investigate the internal shear strength of various types of unreinforced and reinforced GCLs in recent years. The unreinforced GCLs include geotextile-encased, adhesive-bonded GCLs and geomembrane-supported, adhesive-bonded GCLs. The reinforced GCLs include both geotextile-encased, stitch-bonded GCLs and geotextile-encased, needle-punched GCLs.

The bentonite in the manufactured geosynthetic clay liners is initially considered "dry" although its water content can range from 15 to 30% due to its hydrophilic nature. Dry bentonite is strong, but bentonite saturated with water has a very low shear strength. The information contained in Figure 5.10, by Daniel et al (1993), is important in illustrating that GCLs will absorb moisture very quickly beyond their dry state. This is caused by the very large suction potential of bentonite. The internal shear strength of bentonite, particularly when in unreinforced GCLs decreases as the moisture content of the bentonite increases. Shan and Daniel (1991) reported that the peak internal friction angle ϕ for a dry, unreinforced GCL is about 30°, but for water-saturated bentonite, the internal friction angle drops to approximately 9°. Daniel et al. (1993) found that for a geomembrane supported GCL, the drained shear strength decreased as the moisture content increased from 17% (typical for the manufactured product) up to about 100% and then remained relatively constant for much higher moisture contents. (See Table 5.2.)

If the bentonite side of geomembrane-supported GCLs is placed facing downward against the subgrade soil, the dry bentonite has a very high suction value of

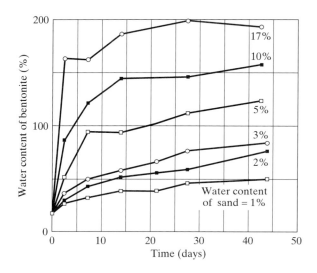

FIGURE 5.10 Water Content versus Time for the Bentonite Side of a Geomembrane-Supported, Adhesive-Bonded GCL Placed in Contact with Sand at Various Water Contents (Daniel et al., 1993)

TABLE 5.2 Summary of Peak Internal Shear Strength of a Geomembrane Supported GCL with Various Water Contents (Daniel et al., 1993)

Nominal Water Content (%)	Shearing Speed	Cohesion (kPa)	Friction Angle (degrees)
17	Fast	13	36
17	Slow	10	22
50	Fast	15	27
50	Slow	15	7
100	Fast	19	12
100	Slow	8	7
145	Slow	5	9

7,500 kPa and will draw moisture from the soil to increase the moisture content with time. That means the internal shear strength will decrease with an increase of bentonite moisture until the equilibrium water content of bentonite is reached. The equilibrium water content of bentonite ranges from 50% for the driest sand to 190% for the wettest sand. (Recall Figure 5.10.) It also seems that the equilibrium time for bentonite moisture adsorption when placed in contact with soil is 1 to 3 weeks (Daniel et al., 1993).

The shear strength of an unreinforced GCL also apparently decreases with decreasing shear rate. Daniel et al. (1993) found that measured peak strengths at the 0.003 mm/min displacement rate were approximately 50% of those at the 0.26 mm/min rate for specimens of a geomembrane supported GCL with moisture content of 100%.

Fox et al. (1998) conducted a study of the internal shear strength of adhesive-bonded, stitch-bonded, and needle-punched GCLs. Tests were performed using a large direct shear machine capable of measuring peak and large displacement (or near residual) shear strengths. All specimens were hydrated prior to testing. The peak internal shear-strength failure envelope for each GCL tested is shown in Figure 5.11. The adhesive-bonded GCL had the lowest peak internal shear strength at any normal stress.

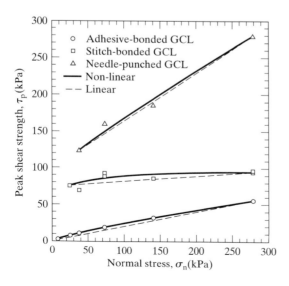

FIGURE 5.11 Failure Envelopes for Peak Internal Shear Strength of Three Geotextile-Related GCL Products (Fox et al., 1998). Used with permission of ASCE.

The measured values are in close agreement with the peak internal strength for two unreinforced GCL products presented by Shan and Daniel (1991) and Daniel et al. (1993) ($\phi \approx 9°$). The peak internal strength of a geotextile-encased, adhesive-bonded GCL increased with normal stress for $\sigma_n <$ 72 kPa (10 psi) and was nearly constant at approximately 91 kPa (13 psi) for $\sigma_n >$ 72 kPa (10 psi). For a geotextile-encased, needle-punched GCL, the peak internal strength increased sharply with normal stress.

Fox et al. (1998) reported that each peak strength envelope exhibited modest nonlinearity, as shown by the solid lines in Figure 5.11. The peak strength data can be approximated using linear relationships, shown as dashed lines in Figure 5.11. To be conservative, each line was drawn between the endpoints of the nonlinear curves using the equation

$$\tau_p = c + \sigma_n \cdot \tan\phi$$

where τ_p = peak internal shear strength (kPa); σ_n = normal stress (kPa); and c (kPa) and ϕ (degrees) are constants.

For the adhesive-bonded GCL,

$$\tau_p = 2.4 + \sigma_n \cdot \tan(10.2°)$$

For the stitch-bonded GCL,

$$\tau_p = 71.6 + \sigma_n \cdot \tan(4.3°)$$

For the needle-punched GCL,

$$\tau_p = 98.2 + \sigma_n \cdot \tan(32.6°)$$

The importance of the reinforcement (stitching or needling) can be readily assessed by noting the y-axis intercept of these response curves. For the nonreinforced product, the intercept is very low (i.e., 2.4 kPa). This is the intrinsic cohesion of the bentonite. For the two reinforced products, on the other hand, the intercept is very high (i.e., 71.6 and 98.2 kPa, respectively). This result reflects predominantly the effect of the reinforcement. Reinforced GCLs are indeed reinforced textile structures and should be considered in that context.

Residual internal shear strength for all three GCL products are shown in Figure 5.12. The data lie in a narrow band in close proximity with one another, indicating that stitch-bonded and needle-punched reinforcement does not affect the residual strength τ_r of the bentonite (i.e., assuming that the reinforcement fails). Linear envelopes can be fitted approximately to the data in the same fashion as for the peak strength τ_r giving the following equation:

$$\tau_r = 1.0 + \sigma_n \cdot \tan(4.7°) \quad [7 \text{ kPa (1 psi)} \leq \sigma_n \leq 280 \text{ kPa (40 psi)}]$$

For each product, failure occurred at the woven geotextile/bentonite interface and excess pore pressures remained zero on the failure plane during shear (Fox et al., 1998). The peak shear strength of the needle-punched GCL increases significantly with increasing normal stress because of the frictional connection of the reinforcing fibers. The peak shear strengths of the adhesive-bonded and stitch-bonded GCLs showed smaller corresponding increases. The residual shear-strength failure envelope was essentially independent of product type. Figure 5.13 shows shear strength versus

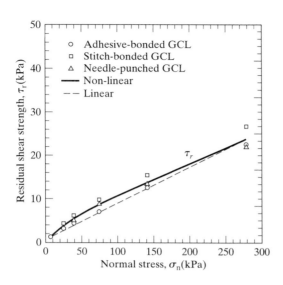

FIGURE 5.12 Failure Envelopes for Residual Internal Shear Strength of Three Geotextile-Related GCL Products (Fox et al., 1998). Used with permission of ASCE.

displacement for each GCL product at normal stress of 140 kPa (20 psi). A detail of the initial portion of these curves is shown in the upper right corner. Small decreases in peak and residual shear strengths were observed (Fox et al., 1998), with decreasing displacement rates for the reinforced products.

Internal GCL shear strengths generally are obtained from laboratory direct shear tests and are dependent on many variables, including product type, hydration fluid, hydration time, consolidation time, shear rate, shear displacement, normal stress, and drainage conditions during hydration and shear. Other equipment-specific

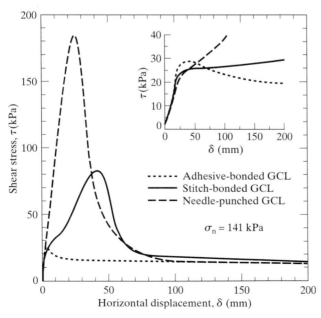

FIGURE 5.13 Stress-Displacement Curves for Three Geotextile-Related GCL Products at Normal Stress of 140 kPa (Fox et al., 1998). Used with permission of ASCE.

factors (e.g., specimen gripping system) may affect results as well. As such, design values of internal GCL shear strength should be measured on a product-specific basis under conditions closely simulating those expected in the field (Fox et al., 1998).

5.4.2 Interface Shear Strength

A shearing failure involving a GCL can occur at three possible locations: (1) the external interface between the top of the GCL and the overlying material (soil or geosynthetic); (2) internally within the GCL; and (3) the external interface between the bottom of the GCL and the underlying material (soil or geosynthetic) (Daniel et al., 1998). The discussion in Section 5.4.1 focused on internal strength.

Gilbert et al. (1996) conducted large-scale direct shear tests with 290-mm-wide by 430-mm-long shear boxes to evaluate the interface strength between reinforced GCLs and other geosynthetic materials. The materials used in the interface shear testing included a needle-punched GCL with a 110 g/m^2 upper, woven, slit film, polypropylene geotextile and a 220 g/m^2 lower, nonwoven, needle-punched polypropylene geotextile, 1.5 mm (60 mil) HDPE smooth geomembrane (GM), 1.5 mm (60 mil) HDPE textured geomembrane (GMX), and drainage geocomposite (GN) that was a 5.6-mm-thick, HDPE geonet heat bonded to a 190 g/m^2 nonwoven, polypropylene geotextile. The nonwoven geotextile of the drainage composite was placed against the nonwoven geotextile of the GCL in the interface shear tests. The GCL specimens were prepared by hydrating them with deionized water under the normal stress to be applied during shear.

Failure envelopes from the three sets of interface shear tests and an internal shear test for a reinforced GCL are shown in Figure 5.14 (peak shear stress) and Figure 5.15 (large-displacement shear stress) for normal stresses between 3.45 and

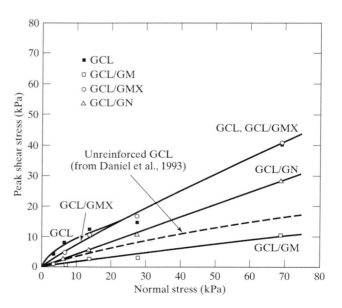

FIGURE 5.14 Failure Envelopes for Peak Interface Shear Strength (Gilbert et al., 1996). Used with permission of ASCE.
(Note: GMX refers to textured HDPE geomembrane)

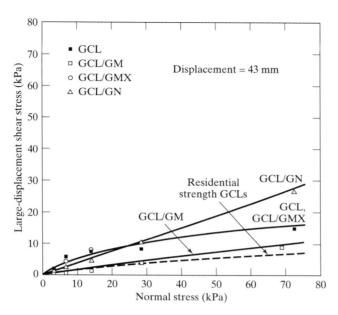

FIGURE 5.15 Failure Envelopes for Large-Displacement Interface Shear Strength (after Gilbert et al., 1996). Used with permission of ASCE.
(Note: GMX refers to textured HDPE geomembrane)

69.0 kPa (0.5 and 10 psi) (Gilbert et al., 1996). A peak failure envelope ($\phi = 9 \sim 10°$) from two unreinforced GCL products presented by Daniel and Shan (1991) and Daniel et al. (1993) and a residual failure envelope ($\phi = 4.7°$) of both unreinforced and reinforced GCL products presented by Fox et al. (1998) are also plotted in Figures 5.14 and 5.15, respectively. The relationship between shear stress and shear displacement for GCL/GM, GCL/GMX, and GCL/GN interfaces under a normal stress of 69.0 kPa (10 psi) is shown in Figure 5.16. For the GCL/GM and GCL/GN interfaces, a lin-

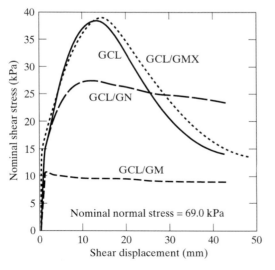

FIGURE 5.16 Shear Stress versus Displacement (Gilbert et al., 1996). Used with permission of ASCE.
(Note: GMX refers to textured HDPE geomembrane)

ear failure envelope was fit through the data, leading to

$$\tau = c + \sigma_n \cdot \tan\phi$$

where τ = shear strength; σ_n = normal stress; and c and ϕ are constants determined from linear regression analysis. For the other cases, a nonlinear model was adopted (Duncan et al., 1978), written as

$$\tau = \sigma_n \cdot \tan[\phi_0 + \Delta\phi \cdot \log(\sigma_n/P_a)]$$

where τ = shear strength; σ_n = normal stress; ϕ_0 and $\Delta\phi$ are constants determined from regression analysis; and P_a = atmospheric pressure (101 kPa). The ϕ_0 coefficient represents the secant friction angle at $\sigma_n = P_a$, while $\Delta\phi$ describes the change in the secant friction angle with increasing σ_n. Results from fitting linear and nonlinear failure envelopes to the test data are summarized in Table 5.3.

Gilbert et al. (1996) observed that the interface between a smooth geomembrane and the reinforced GCL (woven geotextile) produced the lowest peak shear strength—in fact even lower than the peak internal shear strength for the unreinforced GCLs ($\phi = 9 \sim 10°$) (Figure 5.14). This interface also gave the lowest large-displacement shear strength compared with other interfaces (e.g., GCL/GMX and GCL/GH), but higher than the residual internal shear strength for either unreinforced or reinforced GCLs ($\phi \approx 4.7°$) (Figure 5.15) (Fox et al., 1998). Gilbert et al. (1996) concluded that extrusion of bentonite through the GCL-woven geotextile probably contributed to the low shear resistance at large displacements.

At normal stresses up to 13.8 kPa (2.0 psi), failure occurred at the interface between the textured geomembrane and the reinforced GCL; however, the peak strength of the interface was not realized for normal stresses exceeding 13.8 kPa (2.0 psi) because the failure surface moved into the GCL. At these stresses, the shear strength of the interface between a textured geomembrane and the reinforced GCL was comparable to or greater than the interface strength of the GCL (Figures 5.14 and 5.15).

The interface between the GCL (nonwoven geotextile) and a drainage geocomposite (nonwoven geotextile) provided a peak strength that was less than the internal

TABLE 5.3 Summary of Failure Envelope Parameters (Gilbert et al., 1996)

Interface	Normal Stress Range (kPa)	Peak Envelope				Large-Displacement Envelope			
		Linear		Nonlinear		Linear		Nonlinear	
		c (kPa)	ϕ (deg.)	ϕ_o (deg.)	$\Delta\phi$ (deg.)	c (kPa)	ϕ (deg.)	ϕ_o (deg.)	$\Delta\phi$ (deg.)
GCL	3.45–23.0	–	–	18	–23	–	–	9.8	–16
GCL	23.0–69.0	–	–	30	–4.7	–	–	9.8	–16
GCL/GM	3.45–69.0	0.0	8.4	–	–	0.0	8.1	–	–
GCL/GMX	3.45–69.0	–	–	30	–4.7	–	–	9.8	–16
GCL/GN	3.45–69.0	0.38	23	–	–	0.0	22	–	–

GMX: textured geomembrane.
Used with permission of ASCE.

strength of the reinforced GCL (Figure 5.14). The peak internal strength of the reinforced GCL was not mobilized and the fibers were not damaged because slippage occurred at the interface within the geocomposite. Hence, substantial reductions in strength with further displacement were not realized (Figure 5.16) (Gilbert et al., 1996).

5.4.3 Design Implications

The shear strength of GCLs is an important consideration for design because these products are expected to withstand transient in-plane shear stresses during construction and, in some cases, permanent shear stresses over the life of a facility. The information presented herein has important implications for the design of waste-containment facilities and other facilities incorporating GCLs, as well as for the manufacturing and testing of GCL products.

Each GCL product or interface between a GCL and adjacent geosynthetic materials strain softens in direct shear. The reduction of internal or interface shear strength from peak to residual is dependent on the reinforcement type of the GCLs or contact materials at the interface. As a general guideline, unreinforced GCLs are not recommended for slopes steeper than 10(H):1(V) (Frobel, 1996; Richardson, 1997). Based on these findings and from an internal stability standpoint, Fox et al. (1998) conclude the following: stitch-bonded and needle-punched GCLs probably are suited equally for applications involving a low normal stress (e.g., pond and lagoon liners and cover systems), whereas needle-punched GCLs are probably the better choice for applications where a high normal stress is applied (e.g., landfill bottom liners).

The location of a potential failure surface is controlled by the internal shear strength of a GCL and the interface shear strength between a GCL and adjacent materials. The surface with the minimum shear resistance for a composite liner or cover system with a textured geomembrane against the woven geotextile of the reinforced GCL depends on the normal stress. For normal stresses up to 14 kPa (2.0 psi), such as those expected in typical cover systems, the interface has the minimum peak shear strength. However, for higher normal stresses, such as those expected in lining systems, the failure surface is expected to move into the reinforced GCL (Gilbert et al., 1996). Thus, interface failures are more likely for reinforced GCL products at low normal stress. This situation was indicated by failures of the Cincinnati GCL test plots (Daniel et al., 1998). At high normal stress, internal strength may become the limiting factor, causing the failure plane to move into the GCL. The potential for both internal and interface failures must be evaluated for designs that incorporate GCLs (Fox et al., 1998).

Waste containment systems can be designed to minimize the potential for strain-softening in the reinforced GCL. If an interface with a lower peak strength is located within the lining system, then the peak shear strength of the GCL cannot be exceeded and strain-softening may be prevented. The test results from the interface of the reinforced GCL and the drainage geocomposite demonstrated this concept (Figure 5.16). This approach could be applied to a proposed primary liner by using a geomembrane with a textured lower surface against the nonwoven geotextile of a reinforced GCL and a smooth upper surface against the cushion geotextile. The interface between the nonwoven geotextile and the smooth surface of the geomembrane will have a lower peak strength than the peak internal strength of the reinforced GCL. This approach is only valid if the peak shear strength of the isolating interface is greater than the large-displacement strength of the reinforced GCL. In addition, even though the shear stress applied to the reinforced GCL is less than its peak internal strength, the potential for creep and bentonite extrusion still requires consideration (Gilbert et al., 1996). However, bentonite extrusion can be avoided using nonwoven geotextiles on both surfaces of the reinforced GCLs.

Engineers can design stable waste containment facilities with GCLs and take advantage of the excellent hydraulic properties of these materials, but they must not design slopes that exceed the safe slope angle for the GCLs or their respective interfaces within the systems.

5.5 DIFFERENCES BETWEEN GEOSYNTHETIC CLAY LINERS AND COMPACTED CLAY LINERS

Geosynthetic clay liners are often viewed as possible, or even preferable, alternatives to compacted clay liners. Each system has its advantages and limitations. Some of the differences between compacted clay liners and geosynthetic clay liners and the relative advantages of one system over another are listed in Table 5.4. The relative advantages of compacted clay liners (CCLs) versus geosynthetic liners (GCLs) depend upon the specific application and the specific type of GCL.

TABLE 5.4 Differences between Compacted Clay Liners and Geosynthetic Clay Liners (after USEPA, 1993)

Characteristic	Compacted Clay Liner	Geosynthetic Clay Liner
Materials	Native soils or blend of soil and bentonite	Bentonite clay, adhesive, geotextile, and geomembrane
Construction	Construction in the field	Manufactured and then installed in the field
Thickness	Approximately 2 to 3 ft (600 to 900 mm)	Approximately 0.5 inches (13 mm)
Hydraulic conductivity	$\leq 1.0 \times 10^{-7}$ cm/sec	$\leq 1.0 \times 10^{-9}$ to 5.0×10^{-9} cm/sec
Availability of materials	Suitable materials not available at all sites	Materials easily shipped to any site
Speed and ease of construction	Slow, complicated construction	Rapid, simple installation
Vulnerability to damage during construction as a result of desiccation	CCLs are nearly saturated. CCLs can desiccate during construction and crack severely. CCLs can produce consolidation water	GCLs are essentially dry. GCLs cannot desiccate during construction, but there can be problems with overlap width for some GCLs. GCLs produce no consolidation water
Ease of quality Assurance	Complex QA procedures, requiring highly skilled and knowledgeable people	Relatively simple, straight-forward, common-sense procedures
Cost	Highly variable, estimated range: $0.50 to $5.00 per square foot	Typically $0.42 to $0.60 per square foot for a large site
Experience Level	Has been used for many years	Limited

Advantages of Compacted Clay Liners (USEPA, 1993):

- **(i)** Many regulatory agencies require CCLs; use of another type of liner may require demonstration of equivalency to a CCL.
- **(ii)** A CCL should be considered if large quantities of suitable clay are available locally.
- **(iii)** The large thickness of CCLs makes them virtually puncture proof.
- **(iv)** The large thickness of CCLs and the fact that they are constructed of multiple layers makes them relatively insensitive to small imperfections in any one layer.
- **(v)** There is a long history of use of CCLs.
- **(vi)** Quality assurance procedures are reasonably well established for CCLs.
- **(vii)** The large thickness of CCLs (relative to GCLs) greatly increases break-through time by diffusion that varies as the square of barrier thickness.

Advantages of Geosynthetic Clay Liners (USEPA, 1993):

- **(i)** The small thickness of GCLs tends to conserve landfill space. This volume reduction has an economical advantage because more space is available for waste disposal per unit area of landfill.
- **(ii)** Construction of GCLs is rapid and simple.
- **(iii)** GCLs can be shipped to any location. Their use is not dependent upon local availability of materials.
- **(iv)** Heavy equipment is not needed to install a GCL, which is very helpful for final covers underlain by compressible waste (where compaction with heavy equipment is difficult) and for primary liner systems underlain by geosynthetic components (which are vulnerable to damage from construction equipment).
- **(v)** Installation of a GCL requires less vehicular traffic and less energy use than placement and compaction of a CCL. This also leads to less air pollution when using a GCL (may be important in air-pollution-sensitive areas).
- **(vi)** Some i clement weather delays (e.g., freezing temperatures) that stop construction of CCLs are not a problem with GCLs.
- **(vii)** Construction water is not needed with a GCL, which can be critical in arid areas where water resources are scarce.
- **(viii)** Because a GCL is a manufactured material, a consistent and uniform material can be produced. There is no need to engage in detailed material characterization for each site (which is necessary for CCLs).
- **(ix)** Quality assurance is simpler for a GCL than for a CCL.
- **(x)** GCLs can better withstand freeze/thaw and wet/dry cycles than CCLs.
- **(xi)** GCLs are not vulnerable to desiccation damage during construction and may be less vulnerable after construction.
- **(xii)** Because GCLs are lighter than CCLs, GCLs tend to cause less settlement of underlying waste in a final cover system. In addition, the GCLs themselves can tolerate significantly more differential settlement than CCLs.

5.6 CONTAMINANT TRANSPORT THROUGH GEOSYNTHETIC CLAY LINER AND COMPACTED CLAY LINER

Both Federal and state regulations presently require the use of compacted clay liners in landfills; these regulations contain detailed standards for their use. Substitution of a geosynthetic clay liner (GCL) for a compacted clay liner (CCL) generally requires an applicant to demonstrate that the proposed GCL will provide equivalent or better performance to a CCL (Qian, 1999).

One concept or approach that has gained popularity among engineers is that equivalency is based solely on advective mass flux through liners. In other words, two liner systems are equivalent if they have the same advective mass flux rates under the same hydraulic head difference. This approach neglects, however, possible differences in diffusive mass flux (i.e., differences in solute breakthrough and leakage rates under a chemical or concentration difference). Neglect of the diffusion mass flux contribution in the equivalency calculation is inappropriate. Diffusion can provide an important and significant leakage pathway in low hydraulic conductivity liners such as CCLs and GCLs.

The main objective of this section is to provide an explanation and details of contaminant transport through a GCL versus a CCL by an overall numerical analysis. This analysis accounts for changes of distribution of solute concentrations in the liner with time, changes of both advective and diffusive mass fluxes with time, and the accumulated contaminant mass through the liner with time. The contaminant transport equivalency between a GCL and CCL liner is evaluated based on comparing not only mass fluxes through the liner but also the solute (contaminant) concentrations at the base of the liner during the lifetime of the landfill (Qian, 1999).

Leakage mechanisms for solute transport across a landfill liner were identified previously in Chapter 1. Equations were presented there describing the transport of leachate solutes across a liner by advection, diffusion, and by combined advection and diffusion.

5.6.1 Steady Advection

As noted earlier (recall Section 1.7.1), advection is the process by which solutes are transported simultaneously along with the flowing fluid or solvent in response to a gradient in total hydraulic head. The one-dimensional advective mass flux of solute per unit area can be expressed as

$$J_A = C_o \cdot v = C_o \cdot k \cdot i = C_o \cdot k \cdot (H + L)/L \tag{5.1}$$

where J_A = advective mass flux, mg/cm²/sec;
  C_o = concentration of the solute in water at top of liner, mg/cm³;
  v = average seepage velocity of flow through the liner, cm/sec;
  k = hydraulic conductivity, cm/sec;
  i = hydraulic gradient;
  H = leachate head on the liner, cm; and
  L = thickness of the liner, cm.

For a geosynthetic clay liner, Equation 5.1 is applicable only for flow through the bentonite component. If the GCL contains a geomembrane, water flux will be controlled by water vapor diffusion through the geomembrane component. Equation 5.1 is for liquid flow through porous material, and not flow through a geomembrane/clay composite liner. At the present time, verified equations for computation of flow rate through geomembrane and GCL composites are still not available. Thus, this discussion is for CCLs alone and for nonmembrane-associated GCLs.

An advective mass flux ratio R_A is usually used to evaluate whether a GCL is equivalent to the CCL in terms of advective mass flux through the liner. The advective mass flux ratio for water is defined as (USEPA, 1993)

$$R_A = \frac{(J_A)_{GCL}}{(J_A)_{CCL}} = \frac{k_{GCL} \cdot L_{CCL} \cdot (H + L_{GCL})}{k_{CCL} \cdot L_{GCL} \cdot (H + L_{CCL})} \tag{5.2}$$

If $R_A \leq 1$, the GCL is equivalent or superior to the CCL in terms of steady advective mass flux through the liner. If $R_A > 1$, the GCL is not equivalent to the CCL in terms of steady advective mass flux through the liner.

EXAMPLE 5.1

Check whether the GCL is equivalent to the CCL in terms of advective mass flux using the following information:

GCL: $k_{GCL} = 1.0 \times 10^{-9}$ cm/sec;

$L_{GCL} = 0.7$ cm.

CCL: $k_{CCL} = 1.0 \times 10^{-7}$ cm/sec;

$L_{CCL} = 90$ cm (3 ft).

Leachate head on top of the liner, $H = 30$ cm (1 ft).

Solution

$$R_A = \frac{(1.0 \times 10^{-9})(90)(30 + 0.7)}{(1.0 \times 10^{-7})(0.7)(30 + 90)} = 0.33$$

Since $R_A < 1$, the GCL is superior to the CCL in terms of steady advective mass flux through the liner. There would be approximately one-third the advective mass flux through the GCL compared with the CCL in a steady-state condition.

5.6.2 Steady Diffusion

Solute in landfill leachate can also migrate through clay soils, like bentonite, by molecular diffusion. As noted earlier, diffusion may be thought of as a transport process in which a chemical or chemical species migrates in response to a gradient in its concentration (Robinson and Stokes, 1959). The fundamental equation for diffusion in soil (see Section 1.7.2) is Fick's first law, which can be written as

$$J_D = D \cdot n \cdot (\partial C / \partial z) = D \cdot n \cdot (\Delta C / L) \tag{5.3}$$

where J_D = diffusive mass flux, mg/cm²/sec;
D = effective diffusion coefficient, cm²/sec;
n = porosity;
$\partial C/\partial z$ = concentration gradient;
$\Delta C = C_o - C_e$, difference in concentration between top and base of the liner, mg/cm³;
C_o = solute concentration at top of the liner, mg/cm³; and
C_e = solute concentration at base of the liner, mg/cm³.

The effective diffusion coefficient depends on the chemical of interest (e.g., ionic size, charge, and valence of solute) and diffusion conditions (e.g., diffusion through free solution versus soil pore water, tortuosity of pores). Solutes diffuse more slowly through soil pores than in free solution. Some typical values of effective diffusion coefficient for CCLs and GCLs are as follows (USEPA, 1993):

Compacted Kaolin:

$D_{CCL} = 6 \times 10^{-10}$ m²/sec = 6×10^{-6} cm²/sec for the non-reactive chloride solutes.

Sodium Bentonite used in GCLs:

$D_{GCL} = 2 \times 10^{-10}$ m²/sec = 2×10^{-6} cm²/sec.

A diffusive mass flux ratio R_D is usually used to evaluate whether a GCL is equivalent to the CCL in terms of steady diffusive mass flux through the liner. The diffusive mass flux ratio, is defined as (USEPA, 1993)

$$R_D = \frac{(J_D)_{GCL}}{(J_D)_{CCL}} = \frac{D_{GCL} \cdot n_{GCL} \cdot L_{CCL}}{D_{CCL} \cdot n_{CCL} \cdot L_{GCL}} \qquad (5.5)$$

If $R_D \leq 1$, the GCL is equivalent or superior to the CCL in terms of steady diffusive mass flux through the liner. If $R_D > 1$, the GCL is not equivalent to the CCL in terms of steady diffusive mass flux through the liner.

EXAMPLE 5.2

Check whether a GCL is equivalent to a CCL in terms of diffusive mass flux using following information:

GCL: $D_{GCL} = 2 \times 10^{-10}$ m²/sec = 2×10^{-6} cm²/sec

$n_{GCL} = 0.6$

$L_{GCL} = 0.7$ cm

CCL: $D_{CCL} = 6 \times 10^{-10}$ m²/sec = 6×10^{-6} cm²/sec

$n_{CCL} = 0.5$

$L_{CCL} = 90$ cm (3 ft)

Solution

$$R_D = \frac{(2 \times 10^{-6})(0.6)(90)}{(6 \times 10^{-6})(0.5)(0.7)} = 51.4$$

$R_D > 1$, therefore the GCL is not equivalent to the CCL in terms of steady diffusive mass flux through the liner. There would be much more diffusive mass flux through the GCL than CCL in a steady-state condition. Note that the diffusion coefficient for the CCL and GCL are the same order of magnitude. The increased diffusive mass flux through the GCL relative to the CCL is due completely to the greatly reduced thickness of the former (viz., 0.7 cm versus 90 cm (3 ft), i.e., a factor of more than 128). As will be seen, however, the situation is more complex.

5.6.3 Advective Breakthrough Time

A comparison of advective and diffusive mass flux under steady-state conditions between the two types of liners shows that a GCL can be considered equivalent to a CCL in terms of advective mass flux, but not in terms of diffusive mass flux. The diffusive mass flux through the GCL appears to be 51 times higher than that through the CCL. On the other hand, the advective mass flux through the CCL is about 3 times higher than that through the GCL.

Actually, both advective and diffusive mass fluxes do not reach steady-state conditions during the life of a typical landfill. This means that the advective and diffusive mass flux change with time, and these changes will not be synchronous with elapsed time. Furthermore, a GCL and CCL have different advective breakthrough times—that is, the solute flux through the liners by advection will occur at a different time in each case. Thus, using the advective and diffusive mass flux ratios alone as a criterion for equivalency may be misleading.

The average velocity of flow through a liner or approach velocity v is defined as the volumetric flux Q divided by the full cross-sectional area (voids and solids alike). In fact, the flow passes only through that portion of the cross-sectional area occupied by voids. The seepage velocity, or average velocity of flow through the voids of the liner, is

$$v_s = v/n = k \cdot i/n = k \cdot (H + L)/(L \cdot n) \tag{5.5}$$

where v_s = seepage velocity (i.e., average velocity of flow through the voids of the liner), cm/sec;
v = average seepage velocity of flow through the liner, cm/sec;
n = porosity;
k = hydraulic conductivity, cm/sec;
i = hydraulic gradient;
H = leachate head on the liner, cm; and
L = thickness of the liner, cm.

The breakthrough time by the advection can be calculated as

$$T_B = L/v_s \tag{5.6}$$

where v_s = seepage velocity, cm/sec; and
T_B = advective breakthrough time, sec.

5.6.4 Combined Advection-Diffusion

For a landfill liner, the solute concentration ratio at any depth of the liner z at any time t under combined advection-diffusion can be estimated approximately (recall Section 1.6.4) by

$$C(z, t)/C_o = 0.5 \cdot \{erfc[(z - v_s \cdot t)/(4 \cdot D \cdot t)^{0.5}] + \exp(v_s \cdot z/D) \cdot erfc\ [(z + v_s \cdot t)/(4 \cdot D \cdot t)^{0.5}]\} \quad (5.7)$$

where $C(z, t)/C_o$ = concentration ratio at depth z and time t, dimensionless;
  v_s = seepage velocity through the liner, cm/sec;
  D = effective diffusion coefficient of the liner, cm²/sec;
  z = depth, cm; and
  t = time, sec.

For a landfill liner, the solute concentration ratio at the base of the liner at any time t under combined advection-diffusion can be estimated approximately (recall Section 1.6.4) by

$$C_e(t)/C_o = 0.5 \cdot \{erfc[(L - v_s \cdot t)/(4 \cdot D \cdot t)^{0.5}] + \exp(v_s \cdot L/D) \cdot erfc\ [(L + v_s \cdot t)/(4 \cdot D \cdot t)^{0.5}]\} \quad (5.8)$$

where $C_e(t)/C_o$ = concentration ratio at the base of the liner at time t, dimensionless;
  v_s = seepage velocity through the liner, cm/sec;
  D = effective diffusion coefficient of the liner, cm²/sec;
  L = thickness of the liner, cm; and
  t = time, sec.

The advective mass flux at time t is

$$J_A = C_e(t) \cdot v \quad (5.9)$$

where J_A = advective mass flux, mg/cm²/sec;
  $C_e(t)$ = solute concentration at the base of the liner at any time t, mg/cm³; and
  v = average seepage velocity of flow through the liner, cm/sec.

The contaminant mass through the liner per unit area in time period Δt by advection is

$$dM_A = C_e(t) \cdot v \cdot \Delta t \quad (5.10)$$

The contaminant mass through the liner per unit area in time period Δt by advection is approximately equal to

$$\Delta M_A = C'_e \cdot v \cdot \Delta t \quad (5.11)$$

where ΔM_A = contaminant mass throughput per unit area in a given elapsed time period by advection, mg/cm²;
  C'_e = average solute concentration at the base of the liner in time period Δt, mg/cm³, so that $C'_e = 0.5 \cdot [C_e(t_2) + C_e(t_1)]$; and
  Δt = time period, sec, $\Delta t = t_2 - t_1$.

The accumulated contaminant mass throughput per unit area by advection is

$$M_A = \int_{T_B}^{T} C_e(t) \cdot v \cdot dt \tag{5.12}$$

where M_A = accumulated contaminant mass through the liner per unit area by advection, mg/cm²;
   T_B = breakthrough time, sec; and
   T = post breakthrough time ($T > T_B$), sec.

The accumulated contaminant mass throughput per unit area by advection can be also approximately expressed as

$$M_A = \sum_{i=1}^{n} (C'_e)_i \cdot v \cdot (\Delta t)_i \qquad (t \geq T_B) \tag{5.13}$$

where M_A = accumulated contaminant mass through the liner per unit area by advection, mg/cm²;
   $(C'_e)_i$ = average solute concentration at the base of the liner in time period $(\Delta t)_i$, mg/cm³, or $(C'_e)_i = 0.5 \cdot [C_e(t_i) + C_e(t_{i-1})]$;
   v = average seepage velocity of flow through the liner, cm/sec; and
   $(\Delta t)_i$ = time period, sec, $(\Delta t)_i = t_i - t_{i-1}$.

The diffusive mass flux at time t is

$$J_D = D \cdot n \cdot [\Delta C(t)/L] \tag{5.14}$$

where J_D = diffusive mass flux, mg/cm²/sec;
   D = effective diffusion coefficient, cm²/sec;
   n = porosity;
   $\Delta C(t)$ = difference in solute concentration between top and base of the liner at time t, mg/cm³, $\Delta C(t) = C_o - C_e(t)$; and
   L = thickness of the liner, cm.

The contaminant mass through the liner per unit area during the time period Δt by diffusion is

$$\Delta M_D = D \cdot n \cdot [\Delta C(t)/L] \cdot \Delta t \tag{5.15}$$

The contaminant mass through the liner per unit area during the time period Δt by diffusion is approximately equal to

$$\Delta M_D = D \cdot n \cdot (\Delta C'/L) \cdot \Delta t \tag{5.16}$$

where ΔM_D = contaminant mass through the liner per unit area in a certain time period by diffusion, mg/cm² and
   $\Delta C'$ = average concentration difference between the top and base of the liner in time period Δt, mg/cm³, $\Delta C' = C_o - C'_e$.

The accumulated contaminant mass throughput per unit area by diffusion is

$$M_D = \int_{0}^{T} D \cdot n \cdot [\Delta C(t)/L] \cdot \Delta t \tag{5.17}$$

where M_D = accumulated contaminant mass through the liner per unit area by diffusion, mg/cm^2.

The accumulated contaminant mass throughput by diffusion can also be approximately expressed by the equation

$$M_D = \sum_{i=1}^{n} D \cdot n \cdot [(\Delta C')_i/L] \cdot (\Delta t)_i \quad (5.18)$$

The contaminant mass through the liner per unit area in time period Δt by the combination of diffusion and advection is

$$\Delta M = \Delta M_D + \Delta M_A = D \cdot n \cdot [\Delta C(t)/L] \cdot \Delta t + C_e(t) \cdot v \cdot \Delta t \quad (5.19)$$

(*Note*: When $t < T_B$, $\Delta M_A = 0$.)
The contaminant mass throughput per unit area in time period Δt by the combination of diffusion and advection is approximately equal to

$$\Delta M = \Delta M_D + \Delta M_A = D \cdot n \cdot (\Delta C'/L) \cdot \Delta t + C'_e \cdot v \cdot \Delta t \quad (5.20)$$

where ΔM = contaminant mass throughput per unit area in a given time period by the combination of diffusion and advection, mg/cm^2.

(*Note*: When $t < T_B$, $\Delta M_A = 0$.)
The total accumulated contaminant mass through the liner per unit area by the combination of diffusion and advection is

$$M = \int_0^T D \cdot n \cdot [\Delta C(t)/L] \cdot \Delta t + \int_{T_B}^T C_e(t) \cdot v \cdot \Delta t \quad (5.21)$$

where M = total accumulated contaminant mass through the liner per unit area by the combination of diffusion and advection, mg/cm^2.

The total accumulated contaminant mass throughput per unit area by the combination of diffusion and advection can also be expressed approximately as

$$M = \sum_{i=1}^{n} D \cdot n \cdot [(\Delta C')_i/L] \cdot (\Delta t)_i + \sum_{j=1}^{n} (C'_e)_j \cdot v \cdot (\Delta t)_j \quad (5.22)$$

where M = total accumulated contaminant mass through the liner per unit area by the combination of diffusion and advection, mg/cm^2;
  D = effective diffusion coefficient, cm^2/sec;
  n = porosity;
  $(C'_e)_i$ = average solute concentration at the base of the liner in time period $(\Delta t)_i$, mg/cm^3, $(C'_e)_i = 0.5 \cdot [C_e(t_i) + C_e(t_{i-1})]$;
  L = thickness of the liner, cm;
  $(\Delta t)_i$ = time period, sec, $(\Delta t)_i = t_i - t_{i-1}$;
  $(C'_e)_j$ = average solute concentration at the base of the liner in time period $(\Delta t)_j$, mg/cm^3, $(C'_e)_j = 0.5 \cdot [C_e(t_j) + C_e(t_{j-1})]$;

v = average seepage velocity of flow through the liner, cm/sec; and
$(\Delta t)_j$ = time period, sec, $(\Delta t)_j = t_j - t_{j-1}$.

(*Note*: Advective mass transport should not be calculated until $t = T_B$.)

5.7 COMPARISON OF MASS TRANSPORT THROUGH A GCL AND CCL

A comparison of contaminant transport through GCL and CCL can be conducted on the basis of the following information:

Typical GCL properties:

$k_{GCL} = 1.0 \times 10^{-9}$ cm/sec,
$D_{GCL} = 2 \times 10^{-10}$ m²/sec = 2×10^{-6} cm²/sec,
$n_{GCL} = 0.6$,
$L_{GCL} = 0.7$ cm.

Typical CCL properties:

$k_{CCL} = 1.0 \times 10^{-7}$ cm/sec,
$D_{CCL} = 6 \times 10^{-10}$ m²/sec = 6×10^{-6} cm²/sec,
$n_{CCL} = 0.5$,
$L_{GCL} = 90$ cm (3 ft).

For leachate head on top of the liner, $H = 30$ cm (1 ft), and landfill lifetime, $T = 50$ years (20 years operation period and 30 years maintenance period after closure).

The seepage velocity for the GCL and CCL can be calculated using Equation 5.5:

GCL: $(v_s)_{GCL} = k \cdot (H + L)/(L \cdot n)$
$= (1.0 \times 10^{-9})(30 + 0.7)/(0.7 \times 0.6)$
$= 7.310 \times 10^{-8}$ cm/sec

CCL: $(v_s)_{CCL} = (1.0 \times 10^{-7})(30 + 90)/(90 \times 0.5)$
$= 2.667 \times 10^{-7}$ cm/sec

The advective breakthrough time for the GCL and CCL can be calculated using Equation 5.6:

GCL: $(T_B)_{GCL} = L/v_s = 0.7/(7.310 \times 10^{-8}) = 2,660$ hrs = 110.8 days;
CCL: $(T_B)_{CCL} = 90/(2.667 \times 10^{-7}) = 93,738$ hrs = 3906 days = 10.7 years.

The values of the concentration ratio at different depths in a liner at any time were calculated using Equations 1.9 and 5.7, and are listed in Table 5.5 for a GCL, and in Table 5.6 for a CCL.

Strictly speaking, Equations 5.7 and 5.8 are suitable only for an infinitely thick, single homogeneous barrier. When a GCL and CCL are used as the secondary liner in a double composite liner system, they are usually placed directly on the subsoil. The value of the effective diffusion coefficient of the subsoil beneath the GCL and CCL may not be the same as that of GCL and CCL. In general, the value of the effective diffusion coefficient of the subsoil is relatively close to that of a CCL and greater than

162 Chapter 5   Geosynthetic Clay Liners

TABLE 5.5   Changes of Concentration Ratio in a Geosynthetic Clay Liner

Time		Concentration Ratio (C/C_o)					
Year	Day	0	1.4 mm	2.8 mm	4.2 mm	5.6 mm	7 mm
	0.125	1.0	0.502	0.179	0.044	0.007	0.001
	0.25	1.0	0.636	0.343	0.155	0.058	0.018
	0.5	1.0	0.738	0.503	0.315	0.180	0.094
	1	1.0	0.814	0.637	0.479	0.345	0.237
	2	1.0	0.869	0.740	0.618	0.506	0.405
	4	1.0	0.908	0.816	0.727	0.640	0.559
	8	1.0	0.935	0.871	0.807	0.744	0.683
	30	1.0	0.968	0.935	0.903	0.871	0.838
	90	1.0	0.982	0.965	0.947	0.929	0.911
1	365	1.0	0.992	0.985	0.977	0.969	0.962
5	1825	1.0	0.998	0.995	0.993	0.991	0.988
10	3650	1.0	0.999	0.998	0.996	0.995	0.994
20	7300	1.0	0.999	0.999	0.998	0.998	0.997
50	18250	1.0	1.0	1.0	1.0	1.0	1.0

that of a GCL. Thus, the values of concentration ratio for a CCL shown in Table 5.6 calculated from Equations 5.7 and 5.8 will be quite close to actual field conditions. Because a GCL is very thin (viz., only 7 mm), even if the effective diffusion coefficient of the subsoil is different from that of a GCL, the concentration ratio at the base of a GCL will be relatively unaffected by this disparity.

Changes occurring with time in distribution of concentration ratio in a liner for a GCL and for a CCL are shown in Figures 5.17 and 5.18, respectively. Figure 5.19 presents the changes of the concentration ratio at the base of a geosynthetic clay liner and compacted clay liner with time. Figures 5.17, 5.18, and 5.19 show that the concentration ratio in a GCL increases much faster than in a CCL because a GCL is much thinner. The concentration ratio at the base of a GCL can increase to over 0.1 in only a half

TABLE 5.6   Changes of Concentration Ratio in a Compacted Clay Liner

Time		Concentration Ratio (C/C_o)					
Year	Day	0	15 cm	30 cm	45 cm	60 cm	75 cm	90 cm
	90	1.0	0.166	0.004	0	0	0	0
	180	1.0	0.371	0.053	0.003	0	0	0
1	365	1.0	0.590	0.226	0.053	0.007	0	0
2	730	1.0	0.768	0.485	0.245	0.097	0.029	0.007
5	1825	1.0	0.918	0.796	0.642	0.479	0.325	0.202
7	2555	1.0	0.951	0.873	0.769	0.644	0.510	0.379
10	3650	1.0	0.974	0.931	0.871	0.792	0.697	0.589
15	5475	1.0	0.989	0.972	0.933	0.909	0.859	0.799
20	7300	1.0	0.995	0.987	0.975	0.957	0.931	0.897
30	10950	1.0	0.999	0.997	0.994	0.989	0.982	0.978
50	18250	1.0	1.0	1.0	1.0	0.999	0.998	0.998

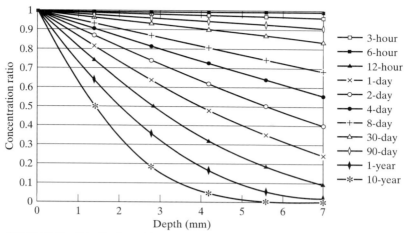

FIGURE 5.17 Distribution of Concentration Ratio in a Geosynthetic Clay Liner with Time

tion ratio in a GCL increases much faster than in a CCL because a GCL is much thinner. The concentration ratio at the base of a GCL can increase to over 0.1 in only a half day and to almost 0.95 in 250 days, whereas the concentration ratio at the base of a CCL takes 1,380 days to reach 0.1 and approximately 25 years to reach 0.95.

The values of advective and diffusive mass fluxes at any time across a GCL and CCL were calculated using Equations 5.9 and 5.14, respectively, and are listed in Table 5.7. The diffusive mass flux data in Table 5.7 were used to plot the diffusion flux-versus-time curves for a CCL and GCL shown in Figure 5.20. A comparison of diffusive mass flux across a GCL and CCL is presented in Figure 5.20. The data show that a GCL

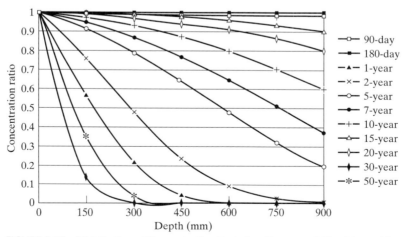

FIGURE 5.18 Distribution of Concentration Ratio in a Compacted Clay Liner with Time

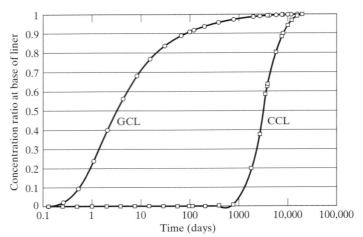

FIGURE 5.19 Changes of Concentration Ratio at Base of Liner for a GCL and a CCL

below that of a CCL after 1,050 days (2.9 years) and then increases to slightly higher than that of a CCL after 5,560 days (15.2 years) (see Figure 5.21). From Figure 5.20 it can be seen that although the diffusive mass flux of a GCL is much higher than that of a CCL during the early stages, the decrease of the diffusive mass flux of a GCL is also much faster than that of a CCL. After 5,560 days (15.2 years) the diffusive mass flux value of a GCL is very close to that of a CCL. Figure 5.22 presents a comparison of advective mass flux between a GCL and CCL. Figure 5.22 also shows that although a GCL has a shorter advective breakthrough time $[(T_B)_{GCL} = 111$ days$]$ than a CCL $[(T_B)_{CCL} = 3,906$ days $= 10.7$ years$]$, the advective mass flux of the GCL is much lower than that of the CCL. The maximum advective mass flux after 50 years of elapsed time is only $(0.438 \times 10^{-7}) \cdot C_0$ mg/cm^2/sec for a GCL, for a GCL, whereas the maximum advective mass flux after 50 years may increase to $(1.330 \times 10^{-7}) \cdot C_o$ mg/cm^2/sec for a CCL, which is 3 times as that of a GCL.

Based on the preceding results, it is incorrect to assume that both advection and diffusion reach steady state conditions simultaneously during the life of a landfill liner. Accordingly, a simple, steady state analysis cannot be used to compare and evaluate contaminant transport through a GCL and CCL during the entire life of a landfill. The mass fluxes will change with time under actual conditions. The rate of change of advective mass flux and diffusive mass flux for a GCL and CCL are not synchronous with one another; accordingly, a GCL and CCL will have different advective breakthrough times. In other words, the solute fluxes through the two types of liners by advection will attain their maximum, steady state values at different times. Thus, using a comparison of advective and diffusive mass flux ratios alone as a criterion for equivalency will yield an incorrect evaluation.

The contaminant mass through the liner per unit area in time period dt or Δt by advection and diffusion can be calculated using Equation 5.10 or 5.11, and Equation 5.15 or 5.16, respectively. The advective contaminant mass throughput across the two

Section 5.7 Comparison of Mass Transport Through a GCL and CCL

TABLE 5.7 Diffusive and Advective Mass Flux at Bases of CCL and GCL

Time		C_e/C_o		ΔC ($\times C_o$ mg/cm^3)		$J_D (\times 10^{-7} C_o$ mg/cm^2/sec)		$J_A (\times 10^{-7} C_o$ mg/cm^2/sec)	
Year	Day	GCL	CCL	GCL	CCL	GCL	CCL	GCL	CCL
	0.125	0.001	0	0.999	1.0	17.009	0.333	0	0
	0.25	0.018	0	0.982	1.0	16.841	0.333	0	0
	0.5	0.094	0	0.906	1.0	15.537	0.333	0	0
	1	0.237	0	0.763	1.0	13.082	0.333	0	0
	2	0.405	0	0.595	1.0	10.198	0.333	0	0
	4	0.559	0	0.441	1.0	7.563	0.333	0	0
	8	0.683	0	0.317	1.0	5.428	0.333	0	0
	15	0.768	0	0.232	1.0	3.974	0.333	0	0
	30	0.838	0	0.162	1.0	2.771	0.333	0	0
	60	0.889	0	0.111	1.0	1.906	0.333	0	0
	90	0.911	0	0.089	1.0	1.520	0.333	0	0
	111	0.921	0	0.079	1.0	1.357	0.333	0.404	0
	180	0.941	0	0.059	1.0	1.018	0.333	0.413	0
1	365	0.961	0	0.038	1.0	0.658	0.333	0.422	0
2	730	0.976	0.007	0.024	0.993	0.414	0.331	0.428	0
5	1825	0.988	0.202	0.011	0.798	0.204	0.267	0.433	0
7	2555	0.991	0.379	0.009	0.621	0.153	0.207	0.435	0
10	3650	0.994	0.589	0.006	0.411	0.105	0.137	0.436	0
10.7	3906	0.994	0.628	0.006	0.372	0.099	0.124	0.436	0.837
11	4015	0.994	0.643	0.006	0.357	0.098	0.119	0.436	0.857
15	5475	0.996	0.799	0.004	0.201	0.065	0.067	0.437	1.065
19	6935	0.997	0.884	0.003	0.116	0.052	0.039	0.437	1.178
20	7300	0.997	0.897	0.003	0.103	0.047	0.034	0.437	1.197
25	9125	0.998	0.948	0.002	0.052	0.034	0.017	0.438	1.264
30	10950	0.998	0.978	0.002	0.022	0.027	0.007	0.438	1.304
40	14600	0.999	0.992	0.001	0.008	0.015	0.003	0.438	1.323
50	18250	1.000	0.998	0.000	0.002	0.008	0.001	0.438	1.330

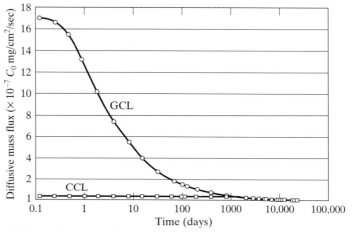

FIGURE 5.20 Comparison of Diffusive Mass Flux between a GCL and a CCL

166 Chapter 5 Geosynthetic Clay Liners

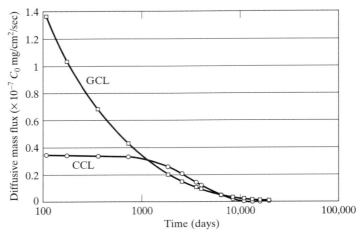

FIGURE 5.21 Comparison Details of Comparison of Diffusive Mass Flux between a GCL and a CCL

types of liners can be calculated from the point of breakthrough onward [viz., $t = 111$ days for the GCL and $t = 10.7$ years (3,906 days) for the CCL, respectively].

The accumulated contaminant mass throughput per unit area by diffusion and advection are calculated using Equation 5.12 or 5.13, and Equation 5.17 or 5.18, respectively. These individual contributions to mass throughput accumulation are listed in Table 5.8. The total accumulated contaminant mass throughputs per unit area by the combination of both diffusion and advection are calculated using Equation 5.21 or 5.22. These total accumulations are listed in the final column of Table 5.8.

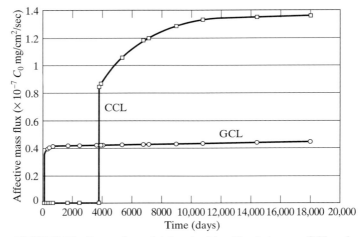

FIGURE 5.22 Comparison of Advective Mass Flux between a GCL and a CCL

Section 5.7 Comparison of Mass Transport Through a GCL and CCL

TABLE 5.8 Contaminant Transport through a GCL and a CCL

Time		C_e/C_o		$M_D(\times C_o \text{ mg/cm}^2)$		$M_A(\times C_o \text{ mg/cm}^2)$		$M(\times C_o \text{ mg/cm}^2)$	
Year	Day	GCL	CCL	GCL	CCL	GCL	CCL	GCL	CCL
	0	0	0	0	0	0	0	0	0
	1	0.237	0	0.131	0.003	0	0	0.131	0.003
	2	0.405	0	0.231	0.006	0	0	0.231	0.006
	4	0.559	0	0.385	0.012	0	0	0.385	0.012
	8	0.683	0	0.609	0.023	0	0	0.609	0.023
	15	0.768	0	0.893	0.043	0	0	0.893	0.043
	30	0.838	0	1.330	0.086	0	0	1.330	0.086
	60	0.889	0	1.936	0.173	0	0	1.936	0.173
	90	0.911	0	2.380	0.259	0	0	2.381	0.259
	111	0.921	0	2.629	0.317	0	0	2.629	0.317
	180	0.941	0	3.347	0.518	0.247	0	3.594	0.518
1	365	0.961	0	4.649	1.051	0.915	0	5.564	1.051
2	730	0.976	0.007	6.339	2.099	2.254	0	8.594	2.099
5	1825	0.988	0.202	9.056	4.973	6.334	0	15.390	4.973
7	2555	0.991	0.379	10.165	6.463	9.072	0	19.237	6.463
10	3650	0.994	0.589	11.368	8.072	13.190	0	24.558	8.072
10.7	3906	0.994	0.628	11.593	8.361	14.155	0	25.748	8.361
11	4015	0.994	0.643	11.686	8.475	14.565	0.798	26.251	9.273
15	5475	0.996	0.799	12.705	9.620	20.072	13.037	32.776	22.658
19	6935	0.997	0.884	13.439	10.270	25.585	27.259	39.024	37.529
20	7300	0.997	0.897	13.595	10.385	26.964	31.004	40.559	41.389
25	9125	0.998	0.948	14.238	10.791	33.863	50.402	48.101	61.193
30	10950	0.998	0.978	14.723	10.986	40.766	70.648	55.489	81.633
40	14600	0.999	0.992	15.387	11.141	54.580	112.073	69.967	123.214
50	18250	1.0	0.998	15.744	11.195	68.402	153.906	84.146	165.101

Figures 5.23 and 5.24 show the accumulated mass throughput with elapsed time through a GCL and CCL, respectively, by diffusion and advection separately, and by the combination of diffusion and advection. Figure 5.23 shows that during the first 8.2 years (3,000 days) the accumulated mass throughput across a GCL by diffusion is greater than that by advection. On the other hand, after 8.2 years (3,000 days) the accumulated mass throughput across the liner by advection accounts for most of the total contaminant transport in the case of a GCL. Figure 5.24 shows that during the first 14 years, the accumulated contaminant mass through a CCL by diffusion is greater than that by advection, and after 14 years the accumulated contaminant mass through the liner by advection quickly dominates. After 50 years, the accumulated contaminant mass across a GCL is $15.744 \cdot C_o$ mg/cm^2 by diffusion, and $68.402 \cdot C_o$ mg/cm^2 by advection (see last row of Table 5.8). In the case of a GCL, the accumulated contaminant mass throughput by diffusion is only 19% of the total contaminant transport for a CCL. After 50 years, the accumulated mass throughput across a CCL by diffusion is $11.195 \cdot C_o$ mg/cm^2, and $153.906 \cdot C_o$ mg/cm^2 by advection. (See the last row of Table 5.8.) Thus, in the case of a CCL, the accumulated contaminant mass throughput by diffusion is only 7% of the total.

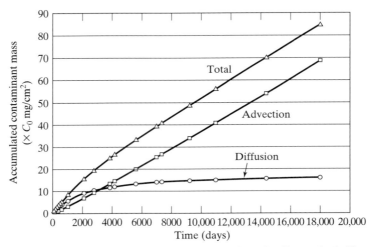

FIGURE 5.23 Accumulated Contaminant Mass through a Geosynthetic Clay Liner

Comparisons of the accumulated contaminant mass throughputs across GCL and CCL liners by diffusion, advection, and the combination of advection and diffusion at different times are shown in Figures 5.25, 5.26, and 5.27, respectively. Figure 5.25 shows that the accumulated mass throughput across a GCL by diffusion is always greater than that across a CCL during the entire 50-year life of a landfill. After 50 years, the ratio of the accumulated diffusive contaminant mass throughput across a GCL to that across a CCL is 1.41. Figure 5.26 shows that the accumulated contaminant mass throughput across a GCL by advection is greater than that across a CCL during

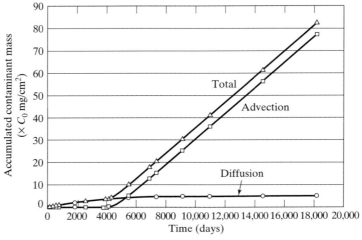

FIGURE 5.24 Accumulated Contaminant Mass through a Compacted Clay Liner

Section 5.7   Comparison of Mass Transport Through a GCL and CCL   169

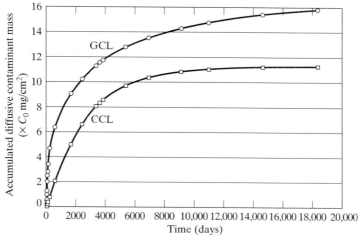

FIGURE 5.25   Comparison of Accumulated Contaminant Mass by Diffusion through a GCL and a CCL

the first 19 years. After 19 years, the accumulated mass throughput across a CCL by advection becomes greater than that across a GCL. After 50 years, the ratio of the accumulated advective contaminant mass throughput by advection across a GCL to that across a CCL is 0.44. Figure 5.27 shows that the total accumulated contaminant mass throughput (due to both advection and diffusion) across a GCL is greater than that across a CCL during the first 20 years. After 20 years, the total accumulated contaminant mass throughput across a CCL becomes greater than that through a GCL. Furthermore, the rate of increase of the total accumulated contaminant mass throughput across a CCL is much higher than that through a GCL after 20 years. After 50

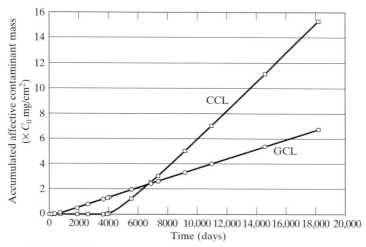

FIGURE 5.26   Comparison of Accumulated Contaminant Mass by Advection through a GCL and a CCL

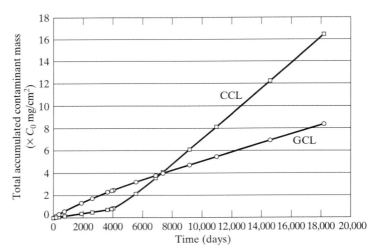

FIGURE 5.27 Comparison of Total Accumulated Contaminant Mass through a GCL and a CCL

years, the ratio of the total accumulated contaminant mass passing through a CCL is up to two times that passing through a GCL. The difference continues to widen with still greater service times.

According to the preceding analysis, if the total accumulated contaminant mass throughput is used as a criterion to evaluate equivalency between a GCL and a CCL, the GCL is not equivalent to a CCL for a landfill lifetime of less than 20 years. Conversely, when the landfill lifetime exceeds 20 years, the GCL is superior to the CCL (i.e., less contaminant mass is transported across a GCL, compared with a CCL over this longer period of time).

On the other hand, if diffusive and advective mass fluxes are used as a criterion to evaluate equivalency, when the landfill lifetime is less than 1,050 days (2.9 years), both diffusive and advective mass flux of a GCL are greater than that of a CCL. This means the GCL is not equivalent to the CCL during this initial period of time. When the time is longer than 1,050 days (2.9 years), but less than approximatly 10.7 years (3,900 days), the diffusive mass flux across a GCL is slightly less than that of a CCL (see Figure 5.21); the advective mass flux across a GCL is still greater than that across a CCL because the advective breakthrough time for the CCL has not been reached. The GCL may not be equivalent to a CCL during this period as well. Finally, when the time exceeds 10.7 years (3,900 days), the advective mass flux across a GCL is much less than that across a CCL, and the diffusive mass flux across a GCL is slightly higher than that of a CCL after 5,560 days (15.2 years) (see Figure 5.21); under these conditions the GCL is superior to the CCL.

The contaminant concentration ratio at the base of the liner can also be considered as a possible criterion. If the concentration ratio at the base is used as an equivalency criterion, a GCL will not be equivalent to the CCL, because the concentration ratio at the base of the GCL is almost always greater than that at the base of a CCL during a 50-year landfill lifetime.

With respect to the issue of contaminant transport and equivalency for geosynthetic clay liners and compacted clay liners, the following findings and tentative conclusions are offered (Qian, 1999):

(i) The concentration ratio at the base of a GCL increases much faster than that at the base of a CCL. The solute concentration at the base of a GCL can increase to over $0.1C_o$ in only a half day and to almost $0.95C_o$ in 250 days, whereas the solute concentration at the base of a CCL takes 1,380 days (3.8 years) to reach $0.1C_o$ and 25 years to reach $0.95C_o$.

(ii) The diffusive mass flux decreases and the advective mass flux increases with time for both a GCL and CCL. The changes of diffusive and advective mass fluxes for the GCL and CCL are not synchronous with one another because a GCL and CCL have different thicknesses, effective diffusion coefficients, advective breakthrough times, and hydraulic conductivities. Thus, a simple steady state analysis of advection and diffusion over the life of a landfill will yield misleading comparisons about contaminant transport through a GCL and CCL.

(iii) The accumulated contaminant mass through the liner by advection dominates solute transport through the liner for both a GCL and a CCL. The accumulated contaminant mass through liner by advection is 81% of the total for a GCL and up to 93% of the total for a CCL during a 50-year landfill lifetime.

(iv) Contaminant transport by diffusion is important during the early stages. However, the contaminant mass transported by diffusion is relatively small and the effect of diffusion can be ignored during later stages.

(v) If a comparison of concentration ratio at the base of the liner is used as a criterion to evaluate equivalency, a GCL will not be equivalent to the CCL because the concentration ratio at the base of the GCL is almost always greater than at the base of the CCL during a 50-year landfill lifetime.

(vi) If a comparison of diffusive and advective mass fluxes is used as a criterion to evaluate equivalency, a GCL is not equivalent to a CCL when the time is less than 3,900 days (10.7 years). However, the GCL will be superior to the CCL after 3,900 days (10.7 years). Furthermore, longer times will favor the GCL by a gradually improving margin.

(vii) If a comparison of total accumulated contaminant mass throughput is used as a criterion to evaluate equivalency, a GCL is not equivalent to a CCL for a landfill lifetime of less than 20 years. On the other hand, a GCL is definitely equivalent to or better than a CCL when the landfill lifetime exceeds 20 years.

It is important to note that these findings and conclusions are based on "worst case scenario" assumptions, namely, that the solute or contaminant does not react with or adsorb on the soil particles composing the liner, that the source concentration remains constant over time, and that hydraulic and concentration gradients always act in the same direction. Partitioning or adsorption of contaminants or solutes can greatly increase breakthrough times and decrease the total amount of solute transported across a barrier during a given time period. Decreasing or depleting source concentrations of solute will likewise decrease the total amount of solute transported across a liner during a given period of time. The implication of these modifications as part of possible strategies to improve liner performance and containment effectiveness are discussed elsewhere by Gray (1995).

It should also be remembered that this analysis is for a CCL versus a nongeomembrane associated GCL (i.e., neither has an overlying geomembrane as is required in composite liners). Also, for the geomembrane associated GCLs, a thick or thin geomembrane within the GCL structure complicates the analysis greatly. Clearly, this is a topic for future research. Yet another situation to consider is a growing tendency to use a GCL with an underlying, low-permeability soil layer, albeit one in which the hydraulic conductivity is low, but not as low as 1.0×10^{-7} cm/sec. If the locally available low-permeability soil has a hydraulic conductivity in the range of 1.0×10^{-5} to 1.0×10^{-6} cm/sec, the combination of this soil layer with the GCL will significantly change the results presented in this section.

5.8 RECOMMENDATIONS FOR USE OF GEOSYNTHETIC CLAY LINERS

Geosynthetic clay liners can be used in waste containment systems for the following applications: (1) as primary and secondary liners for solid waste landfills, (2) as primary and secondary liners for vertical expansion landfills, and (3) as final cover for all types of landfills. The contaminant transport equivalencies for various uses of a GCL in a landfill are summarized as follows. To be noted in all of these summary applications is that neither the GCL nor the CCL have a covering geomembrane (i.e., these statements do not refer to composite liners). Even further, none of the GCLs that are referred to have an associated thick geomembrane on the top or bottom, or a thin geomembrane contained within its manufactured structure. Thus, these comparisons are made on the basis that the GCLs are acting only as single-layer barrier materials (Qian, 1999).

GCL used as a primary liner for solid waste landfills. A GCL may not be equivalent to a CCL according to the comparison criteria presented before a time corresponding to complete breakthrough for the CCL. This elapsed time is approximately 11 years for a 3-foot compacted clay liner with a hydraulic conductivity of 1.0×10^{-7} cm/sec. After complete breakthrough across the CCL (in approximately 11 years) the solute concentration in the secondary leachate collection layer at the base of the primary liner will approach that at the top. Under these conditions the concentration gradient becomes negligible and the diffusive mass flux approaches zero—a condition reached much earlier in the GCL. After approximately 11 years, the only contribution to total mass flux in both types of liners will be by advection. However, the advective mass flux of a GCL is always much lower than that of a CCL. Consequently, a GCL will be equivalent to, or better than, a CCL according to all criteria—namely, accumulated mass throughput and advective and diffusive mass flux rates—after a time corresponding to complete breakthrough across the CCL.

GCL used as secondary liner for solid waste landfills. The initial solute concentration at the base of the GCL used as the secondary liner is approximately equal to zero, and the solute concentration at the top of the GCL is the same as that of the secondary leachate drainage layer. Under these conditions, a GCL is equivalent to a CCL based on the criterion regarding accumulated contaminant mass through the liner if the landfill lifetime is greater than 20 years, or based on the criterion of diffusive and advective mass fluxes if the landfill lifetime is greater than 11 years. But the GCL may not be equivalent to the CCL based on the criterion of concentration ratio at the base of the liner.

GCL used as primary liner for vertical expansion landfills. This situation is the same as the GCL used for primary liner for normal landfills. Thus, the GCL may not be equivalent to a CCL based on any of the criteria before a time corresponding to complete breakthrough (approximately 11 years). But after complete breakthrough for the CCL, a GCL is equivalent to, or better than, a CCL according to all comparison criteria.

GCL used as secondary liner for vertical expansion landfills. The solute concentration at the top of the GCL used as a secondary liner beneath the secondary leachate drainage layer is the same as that at the base of the GCL because the solid waste in the existing landfill underlies the secondary liner. The GCL is equivalent to the CCL for all criteria—namely, accumulated contaminant mass throughput, diffusive and advective mass fluxes, and concentration ratio at the base of the liner.

GCL used as final cover for all types of landfills. Because the solute concentration at the top of the GCL in this condition is equal to zero, both advective and diffusive mass fluxes of solute through the cover liner are equal to zero. The GCL is equivalent to the CCL according to all criteria—namely, accumulated contaminant mass throughput, diffusive and advective mass fluxes, and concentration ratio at the base of the liner.

5.9 SUMMARIZING COMMENTS REGARDING GEOSYNTHETIC CLAY LINERS

GCLs, which were introduced to the waste containment area in the mid-1980s, represent a distinct difference and challenge to compacted clay liners. As noted in Table 5.4, there are many interesting reasons for their use and for their substitution for CCLs. A dilemma is created in this regard because CCLs are embedded in most governmental regulations, and only on the basis of technical equivalency can substitutions be justified. In this regard, this chapter has presented the physical, mechanical and hydration properties of GCLs in an attempt to (1) describe the materials, (2) illustrate some of their salient features, and (3) establish a framework for assessing technical equivalency to CCLs. The last three sections of the chapter were devoted to technical equivalency issues of contaminant transport.

Procedures were presented for estimating and evaluating contaminant transport across landfill liners under combined advection and diffusion. It should be emphasized, however, that the comparison evaluation presented herein is for GCLs and CCLs acting by themselves. Essentially all governmental regulations now require a geomembrane above the GCL or CCL, and this complicates the analysis greatly. Clearly, a composite GM/GCL or GM/CCL outperforms a GCL or CCL by itself.

Environmental performance of these composite systems has been excellent in this regard. Table 5.9 presents field data from 287 single or multiple cells of double-lined landfills for up to 10 years of service performance. There are three-types of primary liners:

TABLE 5.9 Flow Rates from the Leachate Detection Systems (LDS's) of Modern Double-Lined Landfills (all flow rates in gallons/acre/day (GPAD) (after Bonaparte et al., 2000)

Liner and LDS Type	Type I (GM-Sand)			Type II (GM-GN)			Type III (GM/CCL-Sand)		
Life Cycle Stage	1	2	3	1	2	3	1	2	3
Average Flow	41	18	6.8	10	11	ND	23	15	6.8
Minimum Flow	0.81	0.00	0.02	0.51	0.15	ND	0.13	2.4	0.00
Maximum Flow	229	158	26	40	38	ND	126	71	29
# of "points"	30	32	8	7	11	ND	31	41	15
# of landfills	11	11	4	4	6	ND	11	11	4
Liner and LDS Type	Type IV (GM/CCL-GN)			Type V (GM/GCL-Sand)			Type VI (GM/GCL-GN)		
Life Cycle Stage	1	2	3	1	2	3	1	2	3
Average Flow	18	8.9	7.0	14	2.38	0.03	0.70	0.28	ND
Minimum Flow	0.00	0.00	0.00	0.00	0.00	0.00	0.00	0.00	ND
Maximum Flow	74	54	14	104	30	0.10	3.6	1.0	ND
# of "points"	21	27	12	19	19	4	6	4	ND
# of landfills	6	9	3	3	3	1	1	2	ND

NOTES: *Life Cycle Stages*
Stage 1 - Initial Life
Stage 2 - Active Life
Stage 3 - Post Closure
ND = No Data

("points" = Number of measuring points i.e., outlets of single or multiple cells)
GM = geomembrane;   GN = geonet;
GCL = geosynthetic clay liner;
CCL = compacted clay liner

(i) geomembrane alone (GM)
(ii) geomembrane/compacted clay liner (GM/CCL)
(iii) geomembrane/geosynthetic clay liner (GM/GCL)

There are two types of leak detection systems:

(i) 12 inches (300 mm) of sand
(ii) geonet or geonet composite

The results of the average flow values are plotted in Figures 5.28(a) and (b) for the sand and geonet leak detection systems, respectively. The data is striking in that the GM/GCL composite outperforms—by a wide margin and over all life cycle stages—both the geomembrane alone (as expected) and the GM/CCL composite (somewhat surprisingly). It remains for future research to analytically describe and assess such composite systems.

Furthermore, it is important to note that some GCLs now come directly from the factory with an associated geomembrane. Neither thick geomembranes on top or bottom of the bentonite component, nor a thin geomembrane encased with the stitched or needled GCLs, were the focus of the last part of this chapter. Both of these styles of GCLs provide significantly enhanced containment capability over a bentonite layer by itself.

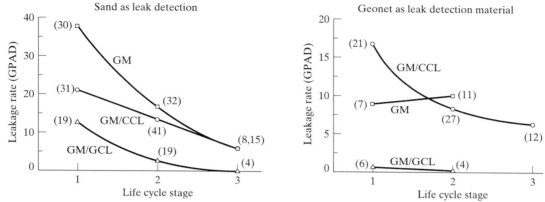

FIGURE 5.28 Average Flow from Leak Detection Layers of Double Lined Landfills (Note that each plotted point represents the average of the number of monitored cells given in parentheses) (from Bonaparte et al., 2000)

Currently, GCLs are being used on a regular basis for the following applications:

(i) GM/GCL primary liners in double-lined waste containment facilities
(ii) GM/GCL primary liners, or sole liners, for vertical expansion landfills
(iii) GM/GCL liners for final covers of all types of landfills

Current trends notwithstanding, CCLs are still the main barrier for single GM/CCL secondary liners of landfills where the clay can be economically located, placed, compacted, and graded on firm soil subgrades (as described in Chapter 3).

PROBLEMS

5.1 List the main advantages and limitations of a geotextile-encased, needle-punched GCL and a geomembrane-supported, adhesive-bonded GCL by comparing their attributes and characteristics.

5.2 A GCL has a hydraulic conductivity of 5.0×10^{-9} cm/sec when subjected to a confining pressure of 0.44 psi (3 kPa). If this GCL is used as a liner beneath a 55-ft- (16.5-m)-high landfill with a waste unit weight of 45 lb/ft^3 (7 kN/m^3), what will be the hydraulic conductivity of this GCL at the bottom of the landfill?

5.3 Describe the wet-dry and freeze-thaw response of GCLs in comparison to the response behavior of compacted clay liners (CCLs).

5.4 How does the ability of GCLs to withstand differential settlement compare with that of compacted clay liners (CCLs)?

5.5 What is the difference between peak and residual internal shear strength among different types of GCLs? Consider each of the following: geotextile-encased, adhesive-bonded GCL, geotextile-encased, stitch-bonded GCL, and geotextile-encased, needle-punched GCL?

5.6 Why are the residual internal shear strengths for unreinforced and reinforced GCLs almost the same (as shown in Figure 5.12)?

5.7 Explain why the unreinforced GCLs are not recommended for slopes steeper than 10(H):1(V).

5.8 How will the normal stress affect the location of the failure surface in a multi-layered liner comprised of a GCL and various geosynthetic materials?

5.9 List three major advantages and two major limitations of using a geosynthetic clay liner (GCL) in a landfill.

5.10 List two major advantages and three major limitations of using a compacted clay liner (CCL) in a landfill.

5.11 A municipal solid waste landfill has a double composite liner system. The thickness of the primary compacted clay liner is 3 feet (0.9 m). The total area of this landfill is 72 acres (29 hectares). Assume that a geosynthetic clay liner with 0.25-inch-thick (6.4-mm-thick) bentonite will be used in lieu of a 3-ft (0.9-m) primary clay liner, calculate the capacity increment or additional filling space. Assume the municipal solid waste gate fee is \$11.50/yard^3 (\$15/m^3) and the compaction ratio after tipping is 2:1. How much additional revenue can be attributed to the use of the geosynthetic clay liner?

REFERENCES

Ajaz, A. and Parry, R. H. G., (1975) "Strain-Strain Behavior of Two Compacted Clays in Tension and Compression," *Geotechnique,* London, England, Vol. 25, No. 3, pp. 495–512.

Alther, G. R., Evans, J. C., Fang, H. Y., and Witmer, K., (1985) "Influence of Organic Permeants upon the Permeability of Bentonite," *Hydraulic Barriers in Soil and Rock,* ASTM STP 874, A. I. Johnson, N. J. Cavalli, and C. B. Pettersson, eds., West Conshohocken, PA, pp. 64–73.

Anderson, D. C., Crawley, W., and Zabcik, J. D., (1985) "Effects of Various Liquids on Clay Soil: Bentonite Slurry Mixtures," *Hydraulic Barriers in Soil and Rock,* ASTM STP 874, A. I. Johnson, N. J. Cavalli, and C. B. Pettersson, eds., West Conshohocken, PA, pp. 93–103.

Benson, C. H., Abichou, T. H., and Olson, M. A., (1995) "Winter Effects on Hydraulic Conductivity of Compacted Clay," *Journal of Geotechnical Engineering,* ASCE, Vol. 121, No. 1, pp. 69–79.

Boardman, B. T. and Daniel, D. E., (1996) "Hydraulic Conductivity of Desiccated Geosynthetic Clay Liners," *Journal of Geotechnical Engineering,* ASCE, Vol. 122, No. 3, pp. 204–208.

Bonaparte, R., Daniel, D. E., and Koerner, R. M., (2000) "Assessment and Recommendations for Optimal Performance of Waste Containment Facilities," Final Report, USEPA Grant No. CR-821448-01-0, in press.

Bramlett, J. A., Furman, C., Johnson, A., Ellis, W. D., Nelson, H., and Vick, W. H., (1987) "Composition of Leachates from Actual Hazardous Waste Sites," Report, U. S. Environmental Protection Agency, Cincinnati, OH.

CETCO, (1995) "Bentomat®/Claymax® Technical Manual," Colloid Environmental Technologies Company, Arlinton Heights, IL.

Chamberlain, E. J., Erickson, A. E., and Benson, C. H., (1995) "Effects of Frost Action on Compaction on Compacted Clay Barriers," *Proceedings of GeoEnvironment 2000,* ASCE Geotechnical Special Publication No. 46, New Orleans, LA, February 24–26, pp. 702–717.

Chamberlain, E. J., and Gow, A. J., (1979) "Effects of Freezing and Thawing on the Permeability and Structure of Soils," *Engineering Geology,* Vol. 12, No. 2, pp. 73–92.

Daniel, D. E., Koerner, R. M., Bonaparte, R., Landreth, R. E., Carson, D. A., and Scranton, H. B., (1998) "Slope Stability of Geosynthetic Clay Liner Test Plots," *Journal of Geotechnical and Geoenvironmental Engineering,* ASCE, Vol. 124, No. 7, pp. 628–637.

Daniel, D. E., Shan, H. Y., and Anderson J. D., (1993) "Effects of Partial Wetting on the Performance of the Bentonite Component of a Geosynthetic Clay Liner," *Proceedings of Geosynthetics '93,* Vol. 3, Vancouver, March 1993, Industrial Fabrics Association International, St. Paul, MN, pp. 1483–1496.

Duncan, J. M., Byrne, P., Wong, K. S., and Mabry, P., (1978) "Strength, Stress-Strain and Bulk Modulus Parameters for Finite Element Analyses of Stresses and Movements in Soil Masses," *Report No. UCB/GT/78-02,* University of California, Berkeley, CA.

Estornell, P. M. and Daniel, D. E., (1992) "Hydraulic Conductivity of Three Geosynthetic Clay Liners," *Journal of Geotechnical Engineering,* ASCE, Vol. 118, No. 10, pp. 1592–1606.

Fox, P. J., Rowland, M. G., and Scheithe, J. R., (1998) "Internal Shear Strength of Three Geosynthetic Clay Liners," *Journal of Geotechnical and Geoenvironmental Engineering,* ASCE, Vol. 124, No. 10, pp. 933–944.

Frobel, R. K., (1996) "Geosynthetic Clay Liners, Part Four: Interface and Internal Shear Strength Determination," Geotechnical Fabrics Report, Industrial Fabrics Association International, Roseville, MN, Vol. 14, No. 8, pp. 20–23.

GeoServices, Inc., (1989) "Freeze-Thaw Effect on Claymax® Liner Systems," *Report to James Clem Corporation,* Technical Note No. 3, Norcross, GA.

GeoSynthec Consultants, (1991) "Final Report Laboratory Testing of Bentomat®," *Report to American Colloid Company,* Norcross, GA.

Gilbert, P. A. and Murphy, W. L., (1987) "Prediction Mitigation of Subsidence Damage to Hazardous Waste Landfill Covers," EPA/600/2-87/025, U. S. Environmental Protection Agency, Cincinnati, OH, March.

Gilbert R. B., Fernandez, F., and Horsfield, D. W., (1996) "Shear Strength of Reinforced Geosynthetic Clay Liner," *Journal of Geotechnical Engineering,* ASCE, Vol. 122, No. 4, pp. 259–266.

Gray, D. H., (1995) "Containment Strategies for Landfilled Wastes," *Proceedings of GeoEnvironment 2000,* Geotechnical Special Publication No. 46, ASCE, New Orleans, LA, February 24–26, pp. 484–498.

Hewitt, R. D. and Daniel, D. E., (1997) "Hydraulic Conductivity of Geosynthetic Clay Liners after Freeze-Thaw," *Journal of Geotechnical and Geoenvironmental Engineering,* ASCE, Vol. 123, No. 4, pp. 305–313.

Keren, R. and Singer, M. J., (1988) "Effect of Low Electrolyte Concentration on Hydraulic Conductivity of Sodium/Calcium-Montmorillonite-Sand System," *Soil Sci. Soc. of Am. J.,* Vol. 52, No. 2, pp. 368–373.

Kim, W. H. and Daniel, D. E., (1992) "Effects of Freezing on the Hydraulic Conductivity of a Compacted Clay," *Journal of Geotechnical Engineering,* American Society of Civil Engineers, Vol. 118, No. 7, pp. 1083–1097.

Koerner, R. M. and Daniel, D. E., (1997) "Final Covers for Solid Waste Landfills and Abandoned Dumps," ASCE Press, Reston, VA, and Thomas Telford, London, UK.

Koerner, R. M., Koerner, G. R., and Eberle, M. A., (1996) "Out-of-Plane Tensile Behavior of Geosynthetic Clay Liners," *Geosynthetics International,* Vol. 3, No. 2, pp. 277–296.

Kraus, J. F., Benson, C. H., Ericksaon, A. E., and Chamberlain, E. J., (1997) "Freeze-Thaw Cycling and Hydraulic Conductivity of Bentonite Barrier," *Journal of Geotechnical and Geoenvironmental Engineering,* ASCE, Vol. 123, No. 3, pp. 229–238.

LaGatta, M. D., Boardman, B. T., Cooley, B. H., and Daniel, D. E., (1997) "Geosynthetic Clay Liners Subjected to Differential Settlement," *Journal of Geotechnical and Geoenvironmental Engineering,* ASCE, Vol. 123, No. 5, pp. 402–410.

Leonards, G. A. and Narain, J., (1963) "Flexibility of Clay and Cracking of Earth Dams," *Journal of Soil Mechanics and Foundation Engineering,* ASCE, Vol. 89, No. 2, pp. 47–98.

McNabb, G. D., Payne, J. B., Harkins, P. C., Ellis, W. D., and Bramlett, J. A., (1987) "Composition of Leachates from Actual Hazardous Waste Sites," *Land Disposal, Remedial Action, Incineration and Treatment of Hazardous Waste,* Proc., 13th Annu. Res. Symp., EPA/600/9-87/015, U. S. Environmental Protection Agency, Cincinnati, OH, pp. 130–138.

Othman, M. A., Benson, C. H., and Chamberlain, E. J., (1994) "Laboratory Testing to Evaluate Changes in Hydraulic Conductivity of Compacted Clays Caused by Freeze-Thaw: State-of-the-Art," *Hydraulic Conductivity and Waste Contaminant Transport in Soil,* ASTM ATP 1142, David E. Daniel and Stephen J. Trautwein, Eds., American Society for Testing and Materials, Philadelphia, PA, pp. 227–254.

Petrov, R. J., Rowe, R. K., and Quigley, R. M., (1997) "Selected Factors Influencing GCL Hydraulic Conductivity," *Journal of Geotechnical and Geoenvironmental Engineering,* ASCE, Vol. 123, No. 8, pp. 683–695.

Qian, Xuede, (1999) "Equivalent Evaluation between Geosynthetic Clay Liner and Compacted Clay Liner," Michigan Department of Environmental Quality, Waste Management Division, Lansing, MI, June.

Rad, N. S., Jacobson, B. D., and Bachus, R. C., (1994) "Compatibility of Geosynthetic Clay Liners with Organic and Inorganic Permeants," *Proceedings of 5th International Conference on Geotextiles, Geomembranes and Related Products,* International Geosynthetics Society, Singapore, pp. 1165–1168.

Richardson, G. N., (1997) "GCL Internal Shear Strength Requirements," Geotechnical Fabrics Report, Industrial Fabrics Association International, Roseville, MN, Vol. 15, No. 2, pp. 20–25.

Robert L. Nelson and Associates, (1993) "Report of Bentomat® Freeze-Thaw Results," *Report to CETCO, Inc. by Robert L. Nelson and Associates, Inc.,* Schaumburg, IL.

Robinson, R. A. and Stokes, R. H., (1959) *Electrolyte Solution,* 2nd Edition, Butterworths Scientific Publications, London, England.

Ruhl, J. L. and Daniel, D. E., (1997) "Geosynthetic Clay Liners Permeated with Chemical Solutions and Leachates," *Journal of Geotechnical and Geoenvironmental Engineering,* ASCE, Vol. 123, No. 4, pp. 369–381.

Sai, J. O. and Anderson, D. C., (1991) "Long-Term Effect of an Aqueous Landfill Leachate on the Permeability of a Compacted Clay Liner," *Hazardous Waste and Hazardous Materials,* Vol. 8 No. 4, pp. 303–312.

Scherbeck, R. and Jessberger, H. L., (1993) "Assessment of Deformed Mineral Sealing Layers," *Proceedings of Green '93,* Bolton University, Manchester, England.

Scherbeck, R., Jessberger, H. L. and Stone, K. J. L., (1991) "Mineral Liner Reaction from Settlement Induced Deformation," *Centrifuge 91,* H. Y. Ko and F. G. McLean, eds., A. A. Balkema, Rotterdam, The Netherlands, pp. 121–128.

Schubert, W. R., (1987) "Bentonite Matting in Composite Lining Systems," *Geotechnical Practice for Waste Disposal '87,* R. D. Woods ed., ASCE, New York, NY, pp. 784–796.

Shan, H. Y. and Daniel. D. E., (1991) "Results of Laboratory Tests on a Geotextile/Bentonite Liner Material," *Geosynthetics 91,* Vol. 2, Industrial Fabrics Association International, St. Paul, MN, pp. 517–535.

Stanforth, R., Ham, R., Anderson, M., and Stegmann, R., (1979) "Development of a Synthetic Municipal Leachate," *J. Water Pollution Control Fed.,* Vol. 51, No. 7, pp. 1965–1975.

Texas Water Commission, (1985) "Waste Evaluation/Classification," Technical Guideline, No.1, Austin, TX.

Trauger, R. and Tawes, K., (1995) "Design and Installation of a State-of-the-Art Landfill Liner System," *in Geosynthetic Clay Liners,* R. M. Koerner, E. Gartung, and H. Zanzinger Eds., A. A. Balkema, Rotterdam, pp. 175–182.

Tschebotarioff, G. P., Ward, E. R., and DePhilippe, A. A., (1953) "The Tensile Strength of Disturbed and Recompacted Soils," *Proceedings of 3rd International Conference on Soil Mechanics and Foundation Engineering,* Switzerland, Vol. 1, pp. 207–210.

USEPA (1993) "Report of Workshop on Geosynthetic Clay Liners," EPA/600/R-93/171, August 1993, Office of Research and Development, U.S. Environmental Protection Agency, Washington, DC.

Wong, L. C. and Haug, M. D., (1991) "Cyclical Closed-System Freeze-Thaw Permeability Testing of Soil Liner and Cover Materials," *Canadian Geotechnical Journal,* Vol. 28, pp. 784–793.

Zimmie, T. F. and LaPlante, C. M., (1990) "The Effect of Freeze-Thaw Cycles on the Permeability of a Fine-Grained Soil," *Proceedings, 22nd Mid-Atlantic Industrial Waste Conference,* Philadelphia, PA, pp. 580–593.

CHAPTER 6

Engineering Properties of Municipal Solid Waste

6.1 CONSTITUENTS OF MUNICIPAL SOLID WASTE
6.2 UNIT WEIGHT OF MUNICIPAL SOLID WASTE
6.3 MOISTURE CONTENT OF MUNICIPAL SOLID WASTE
6.4 POROSITY OF MUNICIPAL SOLID WASTE
6.5 HYDRAULIC CONDUCTIVITY OF MUNICIPAL SOLID WASTE
6.6 FIELD CAPACITY AND WILTING POINT OF MUNICIPAL SOLID WASTE
6.7 STRENGTH OF MUNICIPAL SOLID WASTE
6.8 COMPRESSIBILITY OF MUNICIPAL SOLID WASTE
PROBLEMS
REFERENCES

The correct selection of engineering properties of waste materials is critical in the analysis and design of municipal solid waste (MSW) landfills to meet long-term performance requirements. The safety and cost of landfills are sensitive to variations in these properties. Unfortunately, municipal solid waste often contains hard inclusions and tends to be quite heterogeneous. Furthermore, waste properties change with time and are difficult to measure. Published data is limited, and the conditions under which the properties were measured or back-calculated are often unclear (Fassett et al., 1994).

It is important to note that this chapter, and indeed the book itself, is focused mainly on municipal solid waste (MSW). This type of waste is also called domestic waste, household waste, and (sometimes) non-hazardous waste. However, the book is fully applicable to hazardous or radioactive waste, and to unique wastes such as incinerator ash, sludges, and related viscous materials.

Waste properties have both engineering and economic significance. In addition to their impact on the assessment of landfill performance, the unit weight and compressibility of waste materials influence storage capacity and the resulting economic evaluation of landfill projects. This influence stems from the fact that the waste stream is generally measured as gate weight, whereas landfill capacity is measured by volume (in-place, compacted). Therefore, selection of these engineering properties affects issues such as financing, tipping fees, cell life, and construction scheduling.

The engineering properties of MSW materials required to perform engineering analyses for landfill design are listed in Table 6.1. From Table 6.1, it is apparent that unit weight of the solid waste is critically involved in all aspects of landfill design (Qian and Guo, 1998).

TABLE 6.1 Use of Engineering Properties of Municipal Solid Waste

Analysis	Unit Weight	Moisture Content	Porosity	Hydraulic Conductivity	Field Capacity	Shear Strength	Compressibility
Landfill Capacity	×		×				×
Liner Design	×					×	
Leachate Evaluation	×	×	×	×	×		
Leachate Collection System Design	×						
Gas Collection System Design	×	×	×	×	×		
Final Cover Design	×	×	×	×	×	×	
Foundation Settlement	×						
Landfill Settlement	×		×				×
Foundation Stability	×					×	
Landfill Stability	×				×	×	×
Leachate Recirculation	×	×	×	×	×	×	×
Vertical Expansion	×	×	×	×	×	×	×

According to Fassett, et al. (1994), the following characteristics and conditions of municipal solid waste make determination of its engineering properties difficult:

   (i) The inconsistent and heterogeneous composition of landfill material results in widely variable properties;
   (ii) Samples of sufficient size representative of field condition are difficult to obtain;
   (iii) The erratic nature of the waste particles makes sampling and testing difficult; there are no generally accepted sampling and testing procedure for waste materials; and
   (iv) Waste properties change with time, depth, and location.

A discussion of engineering properties of the waste including unit weight, moisture content, porosity, hydraulic conductivity, field capacity, wilting point, shear strength and compressibility of MSW is presented in the following sections of this chapter.

6.1 CONSTITUENTS OF MUNICIPAL SOLID WASTE

Municipal solid waste generally consists of many different constituents, and these constituents are often porous and not fully saturated. Based on an analysis of numerous types of waste and a comprehensive review of the literature, the following categories have been selected for classification purposes: (i) paper products, (ii) glass, (iii) metal, (iv) plastics, (v) rubber and leather, (vi) textiles, (vii) wood, (viii) food wastes, (ix) yard trimmings and (x) miscellaneous inorganic wastes. The exact constituent composition varies with each type of waste and also with time.

Some waste constituents are readily biodegradable, others are slowly biodegradable, and some are not degradable. Based on this observation, the following broad classes have been suggested (Landva and Clark, 1990) for use in engineering applications that require an estimate of the relative ease of waste breakdown over time:

Organic (O)

    Putrescible (monomers and low-resistance polymers, readily biodegradable) (OP):

        Food waste,

        Garden waste,

        Animal waste,

        Material contaminated by such wastes.

    Non-puterscible (highly resistant polymers, slowly biodegradable) (ON):

        Paper,

        Wood,

        Textiles,

        Leather,

        Plastic, rubber,

        Paint, oil, grease, chemicals, organic sludge.

Inorganic (IN)

    Degradable (ID):

        Metals (corrodible to varying degree)

    Non-degradable (IN):

        Glass, ceramics,

        Mineral soil, rubble,

        Tailings, slims,

        Ash,

        Concrete, masonry (construction debris).

The last three groups (ON, ID, IN) may contain numerous void-forming constituents that will affect the geotechnical behavior of the fill:

Hollow containers:	boxes, crates, cans, bottles, jars, drums, barrels, pipe, tubing, etc.
Platy or elongated items:	beams, sheets, plates, etc.
Bulky items:	furniture, appliances, automobile bodies, etc.

The compositions of municipal solid waste in the United States from 1960 to 1997 are listed in Table 6.2.

6.2 UNIT WEIGHT OF MUNICIPAL SOLID WASTE

The unit weight of municipal solid waste varies within a broad range because its components vary widely and placement procedures as well as environmental conditions can influence outcomes. Each landfill operator handles incoming waste differently, and

TABLE 6.2 Compositions of Municipal Solid Waste in the United States (USEPA, 1999)

Composition		Percent of Total Weight							
		1960	1970	1980	1990	1994	1995	1996	1997
Paper and Paperboard		34.0%	36.6%	36.4%	35.4%	37.7%	38.6%	38.1%	38.6%
Glass		7.6%	10.5%	10.0%	6.4%	6.2%	6.1%	5.9%	5.5%
Metals	Ferrous	11.7%	10.2%	8.3%	6.2%	5.5%	5.5%	5.7%	5.7%
	Aluminum	0.4%	0.7%	1.1%	1.4%	1.4%	1.4%	1.4%	1.4%
	Other Nonferrous	0.2%	0.6%	0.8%	0.5%	0.6%	0.6%	0.6%	0.6%
Plastics		0.4%	2.4%	4.5%	8.3%	9.0%	8.9%	9.4%	9.9%
Rubber and Leather		2.1%	2.5%	2.8%	2.8%	2.9%	2.9%	3.0%	3.0%
Textiles		2.0%	1.7%	1.7%	2.8%	3.4%	3.5%	3.7%	3.8%
Wood		3.4%	3.1%	4.6%	6.0%	5.3%	4.9%	5.2%	5.3%
Food Wastes		13.8%	10.6%	8.6%	10.1%	10.0%	10.3%	10.4%	10.1%
Yard Trimmings		22.7%	19.2%	18.1%	17.1%	14.7%	14.0%	13.3%	12.8%
Miscellaneous Inorganic Wastes		1.5%	1.5%	1.5%	1.4%	1.4%	1.5%	1.5%	1.5%
Other*		0.1%	0.6%	1.7%	1.6%	1.7%	1.7%	1.8%	1.7%
Total		100%	100%	100%	100%	100%	100%	100%	100%

*Includes electrolytes in batteries and fluff pulp, feces, and urine in disposable diapers.

this in turn can result in different levels of compaction. Common difficulties in assessing the unit weight of municipal solid waste include (Fassett et al., 1994) the following:

(i) Determining the contribution of daily soil cover to the unit weight of the waste;
(ii) Assessing the changes in unit weight with time and depth; and
(iii) Obtaining data on moisture content of the waste.

Some conditions that must be specified for meaningful estimates of the unit weight of municipal solid waste (Fassett et al. 1994):

(i) Composition of municipal solid waste including daily cover and moisture content;
(ii) Method and degree of compaction;
(iii) Depth at which the unit weight was measured; and
(iv) Age of the waste.

The unit weight of municipal solid waste can be measured in several ways:

(i) In the field (full-scale landfill test cells, test pits, and samples from bucket augers);
(ii) From laboratory samples;
(iii) By surveying landfill volume and weight of incoming waste and cover material;
(iv) By in-situ unit weight logging with gamma rays; and
(v) By measuring the unit weight of individual components of the waste and making an estimate of the overall unit weight by using percentages of each component.

Methods based on direct field or laboratory measurements tend to be more reliable and, in general, test methods which involve controlled conditions and large samples, such as test pits or test cells, are the most reliable. According to Fassett et al. (1994) the least reliable values are those computed by indirect methods (e.g. values based on incoming weights and in-place volume estimates).

A summary of average unit weights of municipal solid waste is shown in Table 6.3. The unit weight values presented in Table 6.3 range from 20 to 84 lb/ft^3 (3.1 to 13.2 kN/m^3). The wide range is likely caused by the diversity of material in the waste stream, the variable amount of daily cover, and varying moisture content and compaction efforts. More information about the unit weight of solid waste can be found in Table 6.5.

The initial in-place unit weight will increase immediately when compressed by the application of overburden pressure from subsequent waste placement. The in-place unit weight may also increase with further compression that occurs over time. Earth Technology (1988) reported the results of field and laboratory studies of unit weight performed at the Puente Hills landfill near Los Angeles. They developed an interpreted profile of unit weight versus depth for that landfill. The interpreted profile was derived from measurements of unit weight of drive samples recovered for laboratory testing and down-hole geophysical gamma-gamma logging. The interpreted unit weight profile shown in Figure 6.1, varied from 21 lb/ft^3 (3.3 kN/m^3) at the surface to 81.4 lb/ft^3 (12.8 kN/m^3) at depths greater than 200 ft (60 m). Based on reported initial in-place waste densities for modern landfills and the waste compressibility values reported by Fassett et al. (1994), Kavazanjian et al. (1995) developed a profile to show the relationship between the unit weight of the waste and landfill depth. In the absence of site specific information, this profile can be used to estimate the unit weight of the municipal solid waste in engineering analysis for modern municipal solid waste landfills (Kavazanjian et al., 1995).

TABLE 6.3 Average Unit Weight of Municipal Solid Waste (Sharma et al., 1990)

Source	Waste Placement Conditions	Unit Weight	
		lb/ft^3	kN/m^3
Sowers (1968)	Sanitary Refuse: Depending on compaction effort	30 to 60	4.7 to 9.4
NAVFAC (1983)	Sanitary Landfill (a) Not Shredded		
	• Poor compaction	20	3.1
	• Good compaction	40	6.3
	• Best compaction	60	9.4
	(b) Shredded	55	8.6
NSWMA (1985)	Municipal Refuse		
	• In a landfill	44 to 49	6.9 to 7.7
	• After degradation and settlement	63 to 70	9.9 to 11
Landva and Clark (1986)	Refuse Landfill (Refuse to soil cover ratio varied from about 2:1 to 10:1)	57 to 84	9 to 13.2
EMCON Associates (1989)	For 6:1 refuse to daily cover soil	46	7.2

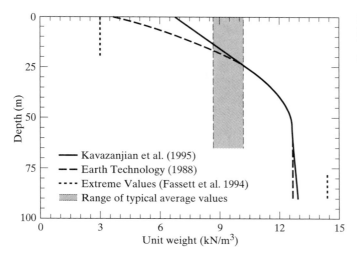

FIGURE 6.1 Unit Weight Profile for Municipal Solid Waste (Kavazanjian et al., 1995) Used with permission of ASCE.

Zornberg et al. (1999) conducted direct field measurements and spectral surface wave analysis (SASW) surveys to characterize the unit weight profile for the waste at a MSW landfill located in southern California. The waste unit weight obtained from direct field measurements ranged approximately from 64 lb/ft³ to 95 lb/ft³ (10 kN/m³ to 15 kN/m³) at a depth of between 26 ft and 164 ft (8 m and 50 m) below the landfill surface.

Based on the compaction equipment used in most landfills at present, the compaction ratio of loose waste to compacted waste usually ranges from 2:1 to 3:1. The average unit weight of compacted solid waste is usually 55 to 70 lb/ft³ (55 to 70 kN/m³) for modern solid waste landfills.

6.3 MOISTURE CONTENT OF MUNICIPAL SOLID WASTE

Two types of moisture content are used in landfill design. The first type of moisture content is defined as the percent by weight of water in the waste based on the dry weight of the waste. This dry gravimetric moisture content definition, commonly used in geotechnical engineering analyses, is written as

$$w_d = (W_w/W_s) \times 100 \qquad (6.1)$$

where w_d = dry gravimetric moisture content, %;
W_w = weight of water; and
W_s = dry weight of solid waste.

In some references, moisture content is defined on the basis of wet weight of the wastes (i.e., w_w), written as

$$w_w = [W_w/(W_s + W_w)] \times 100 \qquad (6.2)$$

where w_w = wet gravimetric moisture content, %;
W_w = weight of water; and
W_s = dry weight of solid waste.

This can be somewhat misleading in that it gives moisture content values much lower than those computed based on dry weight (i.e., in Equation 6.1). Unless stated otherwise, the dry weight definition will be used in this book, but the distinction must be kept in mind since many references use a wet weight basis for unit weight.

A different type of moisture content is defined as the percent by volume of water in the waste based on the total volume of the waste. This volumetric moisture content definition is widely used in hydrology and environmental engineering analyses. Mathematically,

$$\theta = (V_w/V) \times 100 \tag{6.3}$$

where θ = volumetric moisture content of solid waste, %;
V_w = volume of water; and
V = total volume of solid waste.

If the moisture content is known by either a weight or volume basis, the following equations can be used to convert dry gravimetric moisture content to volumetric moisture content or vice versa:

$$\theta = \frac{w_d \cdot \gamma}{(1 + w_d) \cdot \gamma_w} \tag{6.4}$$

$$w_d = \frac{\theta \cdot \gamma_w}{\gamma - \theta \cdot \gamma_w} \tag{6.5}$$

Where θ = volumetric moisture content of solid waste;
w_d = dry gravimetric moisture content of solid waste;
γ = unit weight of solid waste, lb/ft^3 or kN/m^3; and
γ_w = unit weight of water, 62.4 lb/ft^3 or 9.81 kN/m^3.

The moisture content discussed herein is the volumetric or dry gravimetric moisture content of MSW that has been placed in a landfill. The moisture content of MSW in landfills is highly dependent on several interrelated factors (Fassett, et al., 1994): composition of the waste (e.g., organic content), types of waste (e.g., liquid waste or sludge), waste properties (e.g., hydraulic conductivity and field capacity), local climatic conditions (e.g., precipitation and season), landfill operating procedure (e.g., leachate circulation), effectiveness of leachate collection and removal system (e.g., leachate head control), and amount of moisture generated by biological processes within the landfill (Mitchell and Mitchell, 1992).

The gravimetric water content on a dry-weight basis of MSW reported by Sowers (1968) and CalRecovery, Inc. (1993) usually ranged between 21% and 35%. Zornberg et al. (1999) reported an average dry gravimetric water content of 28% in a MSW landfill in Southern California. His measured water content values did not increase significantly with depth. The relationship between the values of dry gravimetric water content and organic content determined on the various refuse samples from different areas of Canada are compiled in Figure 6.2 (Landva and Clark, 1990). In general, the water content seems to increase with increasing organic content, as might be expected. The moisture content may also vary with location as well because of differences in climate, landfill operating procedures, and waste compositions at each location.

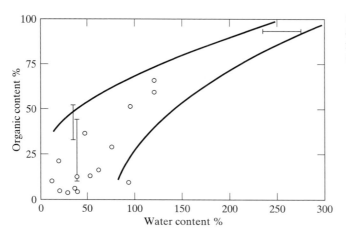

FIGURE 6.2 Organic Content and Water Content of Samples from Old Landfills in Canada (Landva and Clark, 1990)

If the dry gravimetric moisture content and weight percent of each waste component are known (see Table 6.4), the average dry gravimetric moisture content of a particular type of solid waste can be calculated as

$$w_d = \frac{\sum (w_d)_i \cdot \eta_i}{100} \qquad (6.6)$$

where w_d = dry gravimetric moisture content of solid waste, %;
$(w_d)_i$ = dry gravimetric moisture content of a given component of the waste, %; and
η_i = weight percent of a given component of the waste, %.

TABLE 6.4   Municipal Solid Waste Composition and Moisture Content

Component	Percent by Wet Weight	Dry Gravimetric Moisture Content
Organics		
Food Waste	10.4%	60.0%
Garden Waste	19.1%	50.0%
Paper Waste	34.6%	20.0%
Plastics/Rubber	6.0%	10.0%
Textile	5.0%	15.0%
Wood	9.5%	15.0%
Inorganics		
Metals	4.0%	2.0%
Glass/Ceramics	7.2%	2.0%
Ash/Dirt/Rock	2.8%	8.0%
Other	1.4%	3.0%

TABLE 6.5 Index Properties of Solid Waste

Source	Unit Weight		Volumetric Moisture Content	Porosity	Void Ratio
	lb/ft^3	kN/m^3			
Rovers and Farquhar (1973)	59	9.3	0.16	–	–
Fungaroli (1979)	63	9.9	0.05	–	–
Wigh (1979)	73	11.5	0.08	–	–
Walsh and Kinman (1979)	90	14.1	0.17	–	–
Walsh and Kinman (1981)	89	14.0	0.17	–	–
Schroeder et al. (1984a, b)	–	–	0.28	0.52	1.08
Oweis et al. (1990)	40 to 90	6.3 to 14.1	0.10 to 0.20	0.40 to 0.50	0.67 to 1.0
Schroeder et al. (1994a, b)	–	–	0.29	0.67	2.03
Zornberg et al. (1999)	64 to 95	10 to 15	0.30	0.49 to 0.62	1.02 to 1.65

Based on its constituent composition the average moisture content of the solid waste shown in Table 6.4 can be calculated as follows:

$$w_d = [(60.0)(10.4) + (50.0)(19.1) + (20.0)(34.6) + (10.0)(6.0) + (15.0)(5.0)$$
$$+ (15.0)(9.5) + (2.0)(4.0) + (2.0)(7.2) + (8.0)(2.8) + (3.0)(1.4)]/100$$
$$= (624 + 955 + 692 + 60 + 75 + 142.5 + 8 + 14.4 + 22.4 + 4.2)/100$$
$$= 2597.5/100$$
$$= \underline{26.0\%}$$

Thus, the average dry gravimetric moisture content of the solid waste shown in Table 6.4 is 26.0%.

More information about the moisture content of solid waste can be found in Table 6.5. It should be noted that the values of moisture content listed in Table 6.5 are calculated on a volume basis and differ from those calculated on a weight basis, which is more common to geotechnical analyses.

6.4 POROSITY OF MUNICIPAL SOLID WASTE

Porosity is defined as the ratio of the volume of voids to the total volume occupied by a solid waste or soil. Void ratio is defined as the ratio of the volume of voids to the volume of solids. Porosity can be related to the void ratio by using the relationships

$$n = \frac{e}{1 + e} \tag{6.7}$$

and

$$e = \frac{n}{1 - n} \tag{6.8}$$

where n = porosity of solid waste; and
e = void ratio of solid waste.

The porosity of MSW varies typically from 0.40 to 0.67 depending on the compaction and composition of the waste. For comparison, a typical compacted clay liner material will have a porosity of about 0.40. Table 6.5 shows a summary of the index properties of municipal solid waste, which includes initial volumetric moisture content, initial porosity, initial void ratio and unit weight data.

6.5 HYDRAULIC CONDUCTIVITY OF MUNICIPAL SOLID WASTE

Proper assessment of the hydraulic conductivity of municipal solid waste is important in the design of leachate collection systems and in leachate recirculation planning particularly for bioreactor landfills (see Chapter 15). The hydraulic conductivity can be measured using a field leachate pumping test and a large-scale percolation test in test pits or by using large-diameter permeameters in the laboratory.

Hydraulic conductivity measured in test pits at several landfills in Canada by Landva and Clark (1990) is plotted against unit weight in Figure 6.3. The values shown are based on an intermediate stage of water level recession, after the flow had stabilized and before any debris could clog the voids. The measured coefficients of hydraulic conductivity (1.0×10^{-3} to 4.0×10^{-2} cm/sec) correspond to those associated with clean sand and gravel. Qian (1994) used three-year field data from an active landfill in the state of Michigan to develop a relationship between precipitation and leachate volume from a primary leachate collection system with time. With this information, the hydraulic conductivity of the waste can be calculated based on the water travel time, hydraulic gradient, and waste thickness. The hydraulic conductivity calculated in this way was estimated to be about 9.2×10^{-4} to 1.1×10^{-3} cm/sec. Table 6.6 summarizes the hydraulic conductivity of different types of MSW taken from the

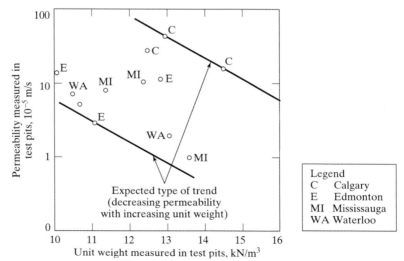

FIGURE 6.3 Unit Weight and Permeability (from Percolation) as Measured in Landfill Test Pits (Landva and Clark, 1990)

190 Chapter 6 Engineering Properties of Municipal Solid Waste

TABLE 6.6 Summary of Hydraulic Conductivity of Municipal Solid Waste

Source	Unit Weight		Hydraulic Conductivity (cm/sec)	Method
	lb/ft^3	kN/m^3		
Fungaroli et al. (1979)	7 to 26 milled waste	1.1 to 4.1 milled waste	1.0×10^{-3} to 2.0×10^{-2}	Lysimeter determination
Schroder et al. (1984a, b)	–	–	2.0×10^{-4}	Estimated based on summary
Oweis and Khera (1986)	41 (estimated)	6.4 (estimated)	Order of 10^{-3}	Estimated based on field data
Landva and Clark (1990)	64 to 92	10.1 to 14.5	1.0×10^{-3} to 4.0×10^{-2}	Test pits
Oweis et al. (1990)	41	6.4	1.0×10^{-3}	Pumping tests
Oweis et al. (1990)	60 to 90 (estimated)	9.4 to 14.1 (estimated)	1.5×10^{-4}	Falling head field test
Oweis et al. (1990)	40 to 60 (estimated)	6.3 to 9.4 (estimated)	1.1×10^{-3}	Test pit
Qian (1994)	–	–	9.2×10^{-4} to 1.1×10^{-3}	Estimated based on field data
Schroder et al. (1994a, b)	–	–	1.0×10^{-3}	Estimated based on summary

literature. The average hydraulic conductivity of municipal solid waste in landfills is approximately 1.0×10^{-3} cm/sec. The hydraulic conductivities of other types of waste are listed in Table 6.7 for comparison.

6.6 FIELD CAPACITY AND WILTING POINT OF MUNICIPAL SOLID WASTE

Field capacity (θ_{FC}) is defined as the residual volumetric water content after a prolonged period of gravity drainage. It is an important parameter in bioreactor landfill technology (Chapter 15) for it signifies the amount of moisture that can be absorbed into, and surrounding, each waste particle. MSW at field capacity represents the target moisture state for optimal waste degradation. Wilting point is defined as the lowest volumetric water content that can be achieved by moisture removal via plant transpiration. The difference between the moisture content at field capacity and the wilting point is the moisture-holding capacity or available water content of a soil or waste. The

TABLE 6.7 Hydraulic Conductivities of Different Waste Materials (Schroder et al., 1994a & 1994b)

Waste Material	Hydraulic Conductivity (cm/sec)	Reference
Stabilized Incinerator Fly Ash	8.8×10^{-5}	Poran and Ahtchi-Ali (1989)
High-Density Pulverized Fly Ash	2.5×10^{-5}	Swain (1979)
Electroplating Sludge	1.6×10^{-5}	Bartos and Palermo (1977)
Nickel/Cadmium Battery Sludge	3.5×10^{-6}	Bartos and Palermo (1977)
Inorganic Pigment Sludge	5.0×10^{-6}	Bartos and Palermo (1977)
Brine Sludge–Chlorine Production	8.2×10^{-5}	Bartos and Palermo (1977)
Calcium Fluoride Sludge	3.2×10^{-5}	Bartos and Palermo (1977)
High Ash Papermill Sludge	1.4×10^{-6}	Perry and Schultz (1977)

Section 6.6 Field Capacity and Wilting Point of Municipal Solid Waste

amount of water that a given soil will be able to hold is a function of its texture. Fine soils such as silty clays can hold more water than coarse soils such as sands. Figure 6.4 illustrates the variability in water retention for different soil types.

The field capacity of waste material is of critical importance in determining the formation of leachate in landfills. Water in excess of the field capacity will be released as leachate. It is also a very important parameter for design of a leachate recirculation program—a key feature of a bioreactor landfill. The field capacity varies with the degree of applied pressure and the state of decomposition of the waste. The volumetric field capacity of municipal solid waste ranges from approximately 22% to 55%, while the volumetric field capacity for compacted clay liners is about 35.6% (Tchobanoglous et al., 1993; Sharma and Lewis, 1994).

Sharma and Lewis (1994) suggested a typical value of volumetric field capacity for municipal solid waste of 22.4%. A volumetric field capacity of 29.2% for municipal solid waste is used in the Hydrologic Evaluation of Landfill Performance (HELP) Model, Version 3 (Schroder, et al., 1994a, 1994b). The volumetric field capacity for compacted electric plant fly ash is typically 18.7%, and for electric plant bottom ash is 26.6% (Sharma and Lewis, 1994). The volumetric field capacity of uncompacted commingled wastes from residential and commercial sources lies in the range of 50 to 60 percent. The field capacity of municipal solid waste from published materials is summarized in Table 6.8.

The wilting point for municipal solid waste ranges approximately from 8.4% to 17% (by volume), while the wilting point for compacted clay liners is about 29% (by volume). A wilting point of 7.7% (by volume) for municipal solid waste is used in the Hydrologic Evaluation of Landfill Performance (HELP) Model, Version 3 (Schroder et al., 1994a, 1994b). The value of wilting point suggested by Sharma and Lewis (1994) for municipal solid waste, compacted coal-burning electric plant fly ash and bottom

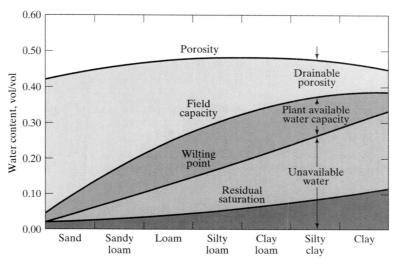

FIGURE 6.4 Relation among Moisture Retention Parameters and Soil Texture Class (Schroder, et al., 1994b)

TABLE 6.8 Summary of Field Capacity for Municipal Solid Waste

Data Source	Volumetric Field Capacity (%)
Rovers and Farquhar (1973)	30.2
Fungaroli and Steiner (1979)	34.2
Walsh and Kinman (1979)	31.8
Wigh (1979)	36.7
Walsh and Kinman (1981)	40.4
Blight et al. (1992)	36.2 (for old waste) 70.6 (for fresh waste)
Schroder, et al., (1994a & b)	29.2
Sharma and Lewis (1994)	22.4
McBean et al. (1995)	55.0
Zornberg et al. (1999)	47.4 to 53.0

ash are 8.4%, 4.7%, and 6.5%, respectively. McBean et al. (1995) reported a typical wilting point value for municipal solid waste of 17%.

The field capacities and wilting points of ash and slag wastes can be calculated using the following empirical equations reported by Brakensiek et al. (1984):

$$Field\ Capacity = 0.1535 - (0.0018)(\%\ Sand) + (0.0039)(\%\ Clay) + (0.1943)(Porosity) \quad (6.9)$$

$$Wilting\ Point = 0.0370 - (0.0004)(\%\ Sand) + (0.0044)(\%\ Clay) + (0.0482)(Porosity) \quad (6.10)$$

In these equations, the following conditions hold: $0.05\ mm < $ Sand Particles $ < 2\ mm$ and Clay Particles $ < 0.002\ mm$ (McAneny et al., 1985).

These equations were developed for natural soils having a sand content between 5 and 70% and a clay content between 5 and 60%. While the particle size distribution of some of the ash and slag wastes fell outside this range, the effects of this variation on water retention are probably minimal. The applicability of these equations to MSW materials has not been verified (Schroder et al., 1994b). Table 6.9 lists the porosity, field capacity, wilting point, and hydraulic conductivity for various waste materials used in the HELP model, Version 3.

Field capacity can be determined in the laboratory by subjecting a saturated sample to 100 cm of capillary suction head and measuring the resulting moisture content. In the field it usually is calculated using a water balance approach (Reinhart and Townsend, 1998).

Field capacity is a function of the waste composition, age, density, porosity, overburden pressure, and compaction effort. It is a difficult parameter to assess, and the literature contains somewhat conflicting reports on its trends mainly because of the highly variable nature of MSW. Generally, field capacity of the waste decreases with increasing overburden pressure and compaction effort provided the composition of the solid waste does not change. Clearly, the more organic material present in the waste, such as paper, cardboard, and textiles, the greater will be the field capacity

TABLE 6.9 Summary of Porosity, Field Capacity, Wilting Point, and Hydraulic Conductivity for Various Waste Materials (Schroeder et al., 1994b)

Waste Material	Porosity	Volumetric Field Capacity	Volumetric Wilting Point	Hydraulic Conductivity (cm/sec)
Municipal Solid Waste	0.671	0.292	0.077	1.0×10^{-3}
Municipal Solid Waste with Channeling	0.168	0.073	0.019	1.0×10^{-3}
High-Density Electric Plant Coal Fly Ash*	0.541	0.187	0.047	5.0×10^{-5}
High-Density Electric Plant Coal Bottom Ash*	0.578	0.076	0.025	4.1×10^{-3}
High-Density Municipal Solid Waste Incinerator Fly Ash**	0.450	0.116	0.049	1.0×10^{-2}
High-Density Fine Copper Slag**	0.375	0.055	0.020	4.1×10^{-2}

*All values, except hydraulic conductivity, are at maximum dry density. Hydraulic conductivity was determined in-situ.
**All values are at maximum dry density. Hydraulic conductivity was determined by laboratory methods.

(Qian and Guo, 1998). Holmes (1983) reported findings from an analysis of samples obtained from nineteen landfills. Field capacity was observed to decline with age of the landfill due to degradation of organic fractions in the waste. Field capacity also decreased with increasing density of the waste due to the collapse of void spaces available for moisture migration and retention. Other researchers have observed a significant decrease in moisture retention in baled waste (Hentrich et al., 1979). Conversely to these decreasing trends in field capacity, Fungaroli and Steiner (1979) also found that as mean particle size of the waste decreases, field capacity increases. Also Hentrich et al. (1979) reported that shredding of waste increases the field capacity of the waste. Thus, the creation of finer particle sizes increases the specific surface area and the field capacity. These irregularities in field capacity will be addressed further in Chapter 15 on Bioreactor Landfills.

6.7 SHEAR STRENGTH OF MUNICIPAL SOLID WASTE

Solid waste is a particulate material and its behavior resembles that of soils in many ways. Like soils, the strength of municipal solid waste appears to increase with increasing normal load applied on the waste. However, due to its high organic content and fibrous nature, municipal solid waste behaves more like a fibrous peat than a typical soil (Howland and Landva, 1992). Factors believed to affect the strength properties of municipal solid waste include the following (Fassett et al., 1994):

  (i) The organic and fiber content in the waste;
  (ii) The age of the waste placed in the landfills, and the extent to which it has decomposed; and
  (iii) The mode of placement (i.e. the compaction effort, lift thickness, and amount and type of daily cover).

The strength of municipal solid waste is also a function of the direction of shear stress. Direct shear tests (Landva et al., 1984) have shown that the shear strength is at a minimum parallel to the bedding plane (i.e. parallel to the lift surface).

According to Howland and Landva (1992), the strength of the solid waste is primarily frictional in nature. However, Mitchell and Mitchell (1992) point out that, while the cohesive nature of refuse has not yet been adequately characterized it clearly is not "true cohesion" as exists in clay particles. Instead it is better referred to as "adhesion," which is the result of interlocking or overlapping of the waste particles. It is reasonable to include an adhesive (or apparent cohesive) component in municipal solid waste shear strength evaluations. This interpretation is supported by the fact that extremely high and steep vertical cuts in landfills have been observed to remain stable for long periods of time. Estimates of the solid waste strength have been made using three approaches (Singh and Murphy, 1990; Howland and Landva, 1992):

(i) Direct laboratory and field testing;
(ii) Back-calculation from failures and load tests; and
(iii) Indirect in-situ testing.

The literature is replete with laboratory testing of reconstituted or totally disturbed samples, triaxial testing on samples obtained using Shelby tubes and drive samplers (Singh and Murphy, 1990), and unconfined compression and tensile testing of bailed waste (Fang et al., 1977). Large scale, direct shear field testing was conducted at a landfill in central Maine (after a failure had occurred) using a 16 ft^2 concrete shear box constructed at the site. Six direct shear tests were performed in the trash using large concrete blocks to vary the normal force on the waste (Richardson and Reynolds, 1991).

Results of shear tests obtained with a large direct shear laboratory apparatus are shown in Figures 6.5 and 6.6 (Landva and Clark, 1990). An examination of the waste material in both its natural and dry conditions showed that it possessed both frictional and adhesive attributes. As seen from Figures 6.5 and 6.6, this was indeed the case; the friction angle (ϕ) varied between 24 and 41 degrees. These materials also had a y-axis intercept (c_a) of between zero and 480 lb/ft^2 (23 kPa).

Back-calculation of failures and load tests performed at several landfill sites provide an indication of MSW shear strength. Many slope stability studies performed by consultants for landfill owners have used strength parameters obtained from a field load test conducted on a landfill in Monterey Park, California (Singh and Murphy, 1990). Additional data has been obtained from the failure of the Global Landfill in New Jersey (Divinoff and Munion, 1986; Fassett et al., 1994).

As reported by Singh and Murphy (1990), back-calculated strength data has also been obtained from observing the satisfactory performance of many landfills in southern California during earthquakes. Since the landfill slopes survived the events, back-calculated values of apparent cohesion (c_a) and friction angle (ϕ) obtained by assuming a factor of safety equal to 1.0, represent minimum available strength and therefore were assumed to be conservative.

When strength values are obtained by direct measurements (i.e. direct shear tests), the data can be presented in terms of shear strength versus normal stress.

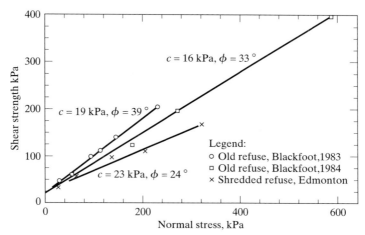

FIGURE 6.5 Large Direct Shear Tests on Samples from Old Landfill in Calgary and from Fresh Shredded Landfill in Edmonton (Landva and Clark, 1990)

However, back-calculated values of shear strength from failures and load tests are often reported as apparent cohesion (c_a) and friction angle (ϕ) pairs that satisfy equilibrium. This uncertainty results from using a known value (viz., a factor of safety = 1.0) to determine two unknowns (viz., c_a and ϕ). The data summarized by Singh and Murphy (1990) are presented separately as pairs of c and ϕ determined from laboratory tests, from back-calculations, and from in-situ testing. All of the results were

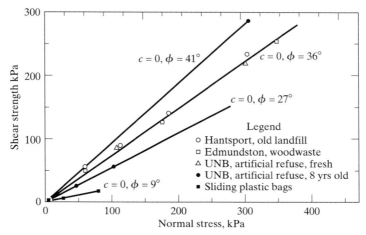

FIGURE 6.6 Large Direct-Shear Tests on Samples from Old Woodwaste/Refuse Landfill in Hantsport, N. S., Fraser Woodwaste Stockpile in Edmundston, N. B., Artificial UNB Samples, and Sliding Garbage Bags (Landva and Clark, 1990)

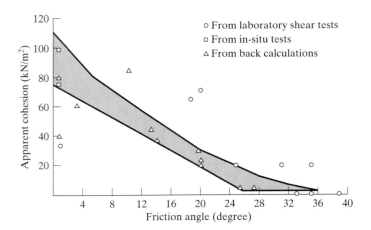

FIGURE 6.7 Summary of Municipal Solid Waste Strength Data (Singh and Murphy, 1990)

then plotted on a single figure, reproduced herein as Figure 6.7. The shaded zone in Figure 6.7 is the range of strength parameters Singh and Murphy recommend for use in stability analyses.

In an assessment of the shear strength of municipal solid waste, Kavazanjian et al. (1995) recognized that factors such as waste compressibility and strain compatibility should be considered when developing strength parameters for limit equilibrium analysis. Their assessment of the waste shear strength was based primarily on shear strengths back-calculated from case histories and the results of in-situ testing. With the exception of one set of data from large-scale tests, laboratory test data on the waste shear strength was not used. The excluded laboratory data was not considered reliable due to either the use of processed waste in the testing program or the small size of the waste samples relative to size nonhomogeneity in municipal solid waste landfills. Field and laboratory test data that they considered being reliable and used in their re-evaluation of the waste shear strength are summarized in Table 6.10.

Kavazanjian et al. (1995) supplemented the data in Table 6.10 with data from back-analyses of existing solid waste landfill slopes known to be stable. Table 6.11 presents back-calculated waste shear strength from four existing landfills. The back-calculated friction angles for municipal solid waste presented in Table 6.11 were obtained assuming a cohesion of 5 kPa (100 psf) using the modified Bishop method of slices. As the slopes at these landfills have been standing for up to 15 years without excessive deformation or other signs of impending instability, the factors of safety against slope failure within the waste are certainly larger than 1.0 and probably greater than 1.3. To be conservative, the results for a factor of safety of 1.2 were used in the waste shear strength assessment.

The shear strength from Tables 6.10 and 6.11 are plotted versus normal stress in Figure 6.8. The data plotted in Figure 6.8, combined with the observation from landfill operations with stable trenches excavated in the waste with vertical faces in excess of 6 m in height, appear to support a bilinear representation of the waste shear strength (Kavazanjian et al., 1995). A vertical slope in a soil (or waste) requires the presence of some apparent cohesion. A "sliding wedge" analysis shows that the minimum cohe-

TABLE 6.10 Data Used in MSW Shear Strength Re-Evaluation (Kavazanjian et al., 1995)

Reference	Data Type	Results	Comments
Pagotto & Rimoldi (1987)	Back-calculation from plate bearing tests	$\phi = 22°$, $c = 605$ psf (29 kPa)	No data on waste types or test procedures are provided.
Landva & Clark (1990)	Laboratory direct shear tests on MSW	$\phi = 24°$, $c = 460$ psf (22 kPa) to $\phi = 39°$, $c = 400$ psf (19 kPa)	Normal stresses up to 10,000 psf (480 kPa). Low strength not used in Figure 6.7, corresponds to shredded waste.
Richardson & Reynolds (1991)	Large direct shear tests performed in-situ	$\phi = 18°$ to $43°$, $c = 210$ psf (10 kPa)	Normal stresses range from 300 to 800 psf (14 to 38 kPa). Unit weight of waste and cover estimated as 95 lb/ft^3 (15 kN/m^3).

Used with permission of ASCE.

sion for a frictionless ($\phi = 0$) material and a vertical slopes ($\beta = 90°$) is given by the following equation:

$$c_{\min}|_{\phi=0,\,\beta=90°} \geq H \cdot \gamma / 4$$

Based on this observation and the data plotted in Figure 6.8, a Mohr-Coulomb strength envelope consisting of $\phi = 0$ with $c = 24$ kPa (500 psf) at normal stress below 30 kPa (625 psf) and $\phi = 33°$ with $c = 0$ at higher normal stresses was developed by Kavazanjian et al. (1995) for use in stability analyses of municipal solid waste landfills.

For the long-term stability, any change in shear parameters will depend on the nature of the waste. For municipal solid waste, there is no direct evidence that shear strength parameters change significantly with time. On the other hand, it is reasonable to believe that shear strength should change as the waste degrades and decomposes. The overall shear strength could decrease, for example, if there is substantial local decomposition that leaves weak zones or cavities, but this type of deterioration could not be detected easily through relatively small-scale laboratory shear tests.

TABLE 6.11 Back-Analysis of Existing Landfill Slopes (Kavazanjian et al., 1995)

Landfill	Average Slope		Maximum Slope		Waste Strength, ϕ		
	Height (m)	Slope (H:V)	Height (m)	Slope (H:V)	$FS = 1.0$	$FS = 1.1$	$FS = 1.2$
Lopez Canyon, CA	120	2.5:1	35	1.7:1	25°	27°	29°
OII, CA	75	2:1	20	1.6:1	28°	30°	34°
Babylon, NY	30	1.9:1	10	1.25:1	30°	34°	38°
Private Landfill, OH	40	2:1	10	1.2:1	30°	34°	37°

Note: FS = Factor of safety for back-analysis assuming $c = 5$ kPa (100 psf).
Used with permission of ASCE.

198 Chapter 6 Engineering Properties of Municipal Solid Waste

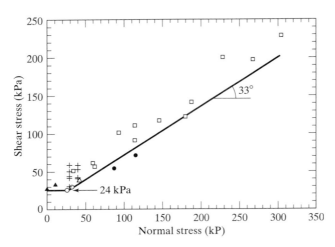

FIGURE 6.8 Shear Strength of Municipal Solid Waste from Different Locations (modified from Kavazanjian et al., 1995).
Used with permission of ASCE.

The strength of municipal solid waste appears to increase with increasing normal stress (confining pressure), in much the same way as a soil. On the other hand, its high organic and "fiber" content immediately after placement make it behave more like a fibrous peat than a mineral soil (Landva and La Rochelle, 1983). The fiber content of municipal solid waste will remain relatively constant, but the putresible organic content will decrease with time. Therefore, some changes in geotechnical properties must be expected. Also, municipal solid waste can be extremely heterogeneous, so that a sample taken at one location may not be indicative of the "average" properties (Howland and Landva, 1992).

The date of waste placement may also be a factor as well. Consider, for example, the increased use of plastics for packaging in the 1960's. Landva and Clark (1990) note that the friction angle between plastic bags is approximately 9 degrees (Figure 6.6). While plastics in the waste stream do not appear to reduce the average bulk strength of municipal solid waste to 9 degrees, Landva and Clark (1990) measured a friction angle of 24 degrees for a shredded refuse with a high plastic content (Figure 6.5).

Municipal solid waste placed in the 1990's again may behave differently than municipal solid waste of previous eras. Permitted landfill air space has become increasingly valuable and, therefore increased compactive effort is seen as a way to place more waste into the same space. Also, stricter enforcement of daily soil cover requirements will tend to increase the average density of the landfill mass. It is unknown whether the effect of increased density of the layered waste and daily cover on stability will be offset by a corresponding increase in strength. Increased recycling also will have an uncertain effect on waste composition and strength. Conversely, the use of alternative daily cover materials (ADCMs), in place of the customary 6 inches (150 mm) of soil, would greatly decrease the overall unit weight. Pohland and Graven (1993) present the following categories of ADCMs, all of which have the effect of decreasing the overall unit weight of MSW:

   (i) Polymer foams,
   (ii) Slurry sprays (paper or woodchip based),

- (iii) Sludges and indigenous materials, and
- (iv) Reusable geosynthetics (geotextiles or geomembranes).

Perhaps the greatest uncertainties regarding shear strength determinations of municipal solid waste have to do with fundamental principles; namely, is the linear Mohr-Coulomb theory appropriate for use with the waste? Municipal solid waste can under go large deformations without failing. If so, at what strains are the strength data relevant? Failure is conventionally defined as either sudden rupture with all loss of strength or alternatively by increasing strain without limit at essentially constant stress. At what strain should failure be declared?

Singh and Murphy (1990) reported triaxial testing of municipal solid waste performed on Shelby tube samples. After undergoing strains greater than 30%, the stress continued to increase with no indication of reaching an asymptotic value. During load tests conducted in Monterey Park, California, a surcharged landfill slope underwent large deformations but no failure plane was apparent (Singh and Murphy, 1990). Based on this and other information, Singh and Murphy concluded that a Mohr-Coulomb characterization of the waste strength may be inappropriate and that slope failure (through the waste material) may not be a critical aspect of landfill design. This situation could differ drastically when liquids are involved. Koerner and Soong (2000) present case histories of 10 massive landfill failures. All were associated with liquids either within the waste mass, between liner materials, or in the foundation soils beneath the waste. Failure surfaces were generally partially within the waste mass.

In general, however, landfills tend to involve slippage along interfaces within the liner system or within weak underlying soils. Therefore, while it is necessary to estimate the strength properties of waste when conducting stability analyses, it is more important (Fassett et al., 1994) to evaluate weak interfaces and/or poor foundation materials properly, as well as strain compatibility between dissimilar materials.

6.8 COMPRESSIBILITY OF MUNICIPAL SOLID WASTE

The compressibility of municipal solid waste has been studied for several decades. Early work focussed on the behavior and suitability of landfills for future construction sites. As the practice of sanitary landfilling increased and landfill sites became more scarce the focus shifted to improving the efficiency of waste placement to improve unit waste volume (Fassett et al., 1994). The general findings of earlier research can be summarized as follows:

- (i) The majority of the settlement occurred quickly;
- (ii) Increased compaction can reduce total settlement; and
- (iii) Settlement under loads decreases with age and depth of municipal solid waste (Fassett et al., 1994).

Two main factors affect settlement of municipal solid waste, namely, the initial density of the waste and compaction effort at placement (which affect mechanical compression); and the moisture content, depth, waste composition, pH, and temperature, which affect physico-chemical and biochemical alteration of the waste (Wallis, 1991).

Traditional soil mechanics theories of compressibility have been applied to municipal solid waste. Therefore, the same considerations and parameters have been used. Note, however, that these theories are based on fully saturated soils and not on partially saturated (i.e., $S \ll 100\%$) MSW materials. Settlement is generally considered to consist of three components:

$$\Delta H_{\text{total}} = \Delta H_i + \Delta H_c + \Delta H_\alpha \qquad (6.11)$$

Where ΔH_{total} = total settlement;
ΔH_i = immediate settlement;
ΔH_c = consolidation settlement; and
ΔH_α = secondary compression, or creep.

Initial settlement of municipal solid waste resulting from increased loads occurs typically within the first three months (Bjarngard and Edgers, 1990). Waste settlement is similar to that of peat soil in which, after rapid immediate and consolidation settlement, additional settlement is accompanied by little or no excess pore pressure buildup and is primarily due to long-term secondary compression (Sowers, 1973). Because the consolidation phase is completed so rapidly, it is generally lumped together with the immediate settlement and called "primary settlement" (Lukas, 1992). However, unlike peat deposits, the secondary compression of municipal solid waste includes an important decomposition component.

The parameters commonly used to estimate the primary settlement of municipal solid waste resulting from an increase in vertical stress include the primary compression index (C_c) and the modified primary compression index (C'_c). These parameters have been defined by Fassett et al. (1994) and are written as

$$C_c = \frac{\Delta e}{\log(\sigma_1/\sigma_0)} \qquad (6.12)$$

and

$$C'_c = \frac{\Delta H}{H_0 \cdot \log(\sigma_1/\sigma_0)} = \frac{C_c}{1 + e_0} \qquad (6.13)$$

where Δe = change in void ratio;
e_0 = initial void ratio;
σ_0 = initial vertical effective stress;
σ_1 = final vertical effective stress;
H_0 = original thickness of waste layer; and
ΔH = change in thickness of waste layer.

There are several problems with this approach (Edil et al., 1990). First, the initial void ratio (e_0) or initial height of waste (H_0), especially for old landfills, is often not known. Second, the effective stress is a function of MSW unit weight (and level of leachate in the landfill) which is generally not accurately known. Third, the e-$\log(\sigma)$

relationship is often not linear; therefore, C_c and C'_c vary with the initial stresses within the landfill and as these stresses change with time.

The secondary compression index (C_α) or the modified secondary compression index (C'_α) are used to estimate the settlement that occurs after completion of the primary settlement (Fassett et al., 1994). This usually occurs while the waste is subjected to a constant load. The indexes are

$$C_\alpha = \frac{\Delta e}{\log(t_2/t_1)} \qquad (6.14)$$

and

$$C'_\alpha = \frac{\Delta H}{H_0 \cdot \log(t_2/t_1)} = \frac{C_\alpha}{1 + e_0} \qquad (6.15)$$

where t_1 = starting time of secondary settlement; and
          t_2 = ending time of secondary settlement.

These parameters tend to change during creep deformation and with chemical or biogradation of the waste. The most important cause of secondary settlement is most likely volume reduction due to decomposition of organic matter. This view is not substantiated, however, by field evidence and test results at this time.

In order to measure the compressibility resulting from an increase in load, plate load tests (Landva et al., 1984), pressuremeters (Steinberg and Lucas, 1984), and odometers (Bjarngard and Edgers, 1990; Chen et al., 1977; Landva et al., 1984; Oakley, 1990) have been used in various investigations. To measure the rate of settlement under constant load, surveying methods have been widely used. Techniques include aerial photo comparisons over time (Druschel and Wardwell, 1991), benchmark surveys on the landfill surface (Steinberg and Lucas, 1984; Dodt et al., 1987; Wallis, 1991; Druschel and Wardwell, 1991), and settlement platforms placed below earth embankments constructed on landfills (Sheurs and Khera, 1980). An additional technique utilizes telescoping inclinometers (Siegel et al., 1990; Galante et al., 1991). These devices allow measurements of settlement at various depths under both increases in load and constant load. Frequent readings during landfilling are needed (Fassett et al., 1994) to back-calculate separate values of C_c and C_α.

The high compressibility of waste fill is evident from Figure 6.9, where consolidation results from five locations are plotted (Landva and Clark, 1990). These tests were all done in a 470-mm diameter apparatus. The samples were placed in the container in about 50-mm lifts and lightly compacted. The gradient of the log pressure versus strain is the modified primary compression index (C'_c). The range of C'_c in Figure 6.9 is 0.2 to 0.5. This value is high in comparison with soils, even with organic soils. Keene (1977) installed nine settlement platforms at various elevations of a sanitary landfill to investigate the landfill settlement for five years. The settlement readings indicated that the primary compression was about 3% and occurred rapidly. Primary settlement essentially ended at one-half to one month after completion of filling. Sowers (1973) pointed out that the compression index, C_c, is related to the initial void ratio, e_0, as shown in Figure 6.10. There is a considerable variation in C_c for any value of e. The higher values are for fills containing large amounts of garbage, wood, brush, and tin

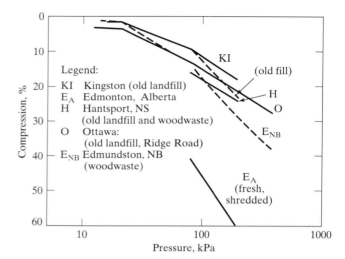

FIGURE 6.9 Compressive Strain versus Log Pressure for Various Landfills in Canada (Landva and Clark, 1990)

KI: $C_C' = 0.17$ ($p = 20 - 200$ kPa)   O: $C_C' = 0.21$ ($p = 100 - 400$ kPa)
E_A: $C_C' = 0.35$ ($p = 80 - 200$ kPa) E_{NB}: $C_C' = 0.36$ ($p = 100 - 400$ kPa)
H: $C_C' = 0.22$ ($p = 80 - 200$ kPa)

cans; the lower values are for the less resilient materials. The maximum C_c for peat is about one-third greater than the maximum observed for waste fills.

Landva and Clark (1990) found that the coefficient of secondary consolidation, C_α, (the gradient of the compression versus log time relationship) was in the range 0.2 to 3.0 percent per log cycle time, depending on the type of waste involved. Field testing results using a settlement platform (Keene, 1977) showed that the coefficient of secondary consolidation, C_α, varies between 0.014 and 0.034. Too few tests have been carried out for any firm relationship to be established between the value of C_α and the type of waste, but it does appear that C_α increases with increasing organic content. Sowers (1973) pointed that the coefficient of secondary consolidation, C_α, is also a

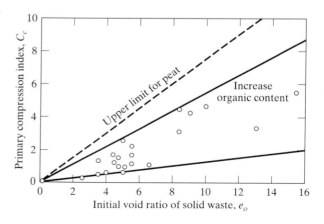

FIGURE 6.10 Compressibility of MSW Landfills (Sowers, 1973)

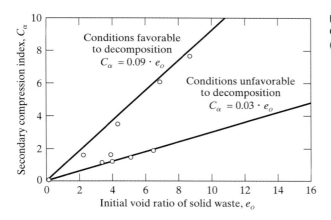

FIGURE 6.11 Secondary Compression of MSW Landfills (Sowers, 1973)

function of the void ratio, as shown in Figure 6.11. For any given void ratio, there is a large range in C_α, related to the potential for physico-chemical and bio-chemical decay. The value is high if the organic content subject to decay is large and the environment is favorable: namely, warm, moist, with fluctuating water table that pumps fresh air into the fill. The value is low for more inert materials and an unfavorable environment. More research and data are necessary before this relationship can be defined more closely.

The most widely reported compressibility parameter is the modified secondary compression index (C'_α). The reported values of C'_α range from 0.001 to 0.59. The lowest value represents the compressibility of a landfill that had been subjected to dynamic compaction. For typical landfills the lower limit of C'_α is generally around 0.01 to 0.03. This compares to 0.005 to 0.02 for common clays (Holtz and Kovacs, 1981). Fasset et al. (1994) observed that the typical upper limit of C'_α appears to be approximately 0.1.

According to Yen and Scanlon (1975), the settlement rate of waste increases with depth, hence larger values of C'_α should be associated with thicker fills. They observed that this effect leveled off at about 90 ft. and suggested that conditions within the landfill at great depths limit the biological activity to anaerobic decomposition, which is much slower than the aerobic decomposition believed to occur in shallower fills.

The values of C_α and C'_α, like C_c and C'_c, are dependent on the values used for e_0 or H_0. The value of C'_α is also dependent on stress level, time, and on how the origin of time is selected. The waste placement or filling period for landfills is often long and should be taken into consideration for settlement rate analyses (Yen and Scanlon, 1975). The zero time selection has a large impact on C'_α particularly during earlier phases of a landfill (Fassett et al., 1994)

An additional problem with determining C'_α is the fact that this parameter is generally not constant. Edgers (1992) presents settlement log-time data from 22 case histories (shown in Figure 6.12). The majority of the curves show a relatively flat slope (i.e. low C'_α values) at small times, but at larger times the slope greatly increases (Figure 6.13). They attributed the higher slopes in the later stages of compression to increasing decomposition, but it may simply be an artifact of the log-time scale. It is

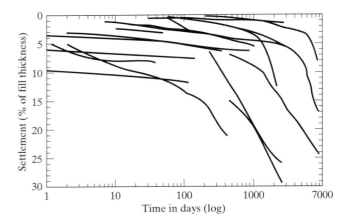

FIGURE 6.12 Landfill Settlement versus Log Time (Bjarnagard and Edgers, 1990)

highly likely that decomposition of municipal solid waste will affect its compressibility. Fassett et al. (1994) have concluded that the relative contribution of mechanical compression, thermal effects, and biological decomposition to the total settlement has not been adequately addressed. What is critical to note in Figure 6.12 is that the settlement of MSW landfills can be enormous. Thirty (30) percent of total height settlement has huge implications in cover soil design and related landfill planning. This issue will be revisited in subsequent chapters of the book.

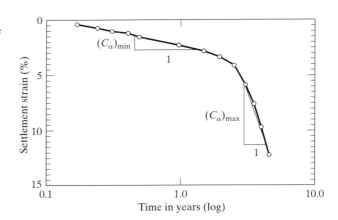

FIGURE 6.13 Idealized Plot of Landfill Settlement versus Log Time (Bjarnagard and Edgers, 1990)

PROBLEMS

6.1 Explain why it is difficult to determine the engineering properties of municipal solid waste.

6.2 Which waste property or parameter is the most useful in landfill design and performance?

6.3 List the factors that influence the unit weight of solid waste.

6.4 How many methods can be used to measure the unit weight of municipal solid waste? What methods do you think are relatively reliable and what methods are the least reliable to obtain accurate unit weight results?

6.5 What is the difference between dry gravimetric and volumetric moisture contents?

6.6 The unit weights of two types of municipal solid waste are 60 lb/ft^3 (9.4 kN/m^3) and 70 lb/ft^3 (11.0 kN/m^3), respectively. The volumetric moisture content of both municipal solid wastes is 30%. What is the dry gravimetric moisture content of each type of waste?

6.7 List the factors that affect the initial moisture content of municipal solid waste.

6.8 Assume that the dry gravimetric water contents of residential, commercial, and industrial waste are 30.7%, 23.2%, and 16.2%, respectively, and average unit weights of residential, commercial, and industrial waste are 65 lb/ft^3 (10.2 kN/m^3), 68 lb/ft^3 (10.7 kN/m^3), and 73 lb/ft^3 (11.5 kN/m^3). Convert dry gravimetric water contents to volumetric water contents.

6.9 The dry gravimetric moisture content and weight percent of various components of a typical municipal solid waste are listed in Table 6.12. Calculate the average dry gravimetric moisture content of this waste. If the unit weight of this waste is 62 lb/ft^3 (9.7 kN/m^3), what is the average volumetric moisture content of this waste?

6.10 What is an average, order-of-magnitude value for the hydraulic conductivity of municipal solid waste?

6.11 The properties of a particular type of solid waste are listed as follows:
Unit weight of the waste: 67 lb/ft^3 (10.5 kN/m^3),
Dry gravimetric moisture content of the waste: 22.5%,
Field capacity of the waste: 29.2%.
Calculate how many gallons of extra water can be absorbed by a cubic yard of this waste without producing any secondary leachate.

6.12 What factors influence the shear strength of municipal solid waste?

6.13 What methods can be used to estimate solid waste strength? What is an average shear strength value for municipal solid waste based on results of past investigations?

TABLE 6.12 Municipal Solid Waste Composition and Moisture Content for Problem 6.9

Component	Percent by Wet Weight	Dry Gravimetric Moisture Content
Organics		
Food Waste	9.0%	60.0%
Garden Waste	18.5%	50.0%
Paper Waste	39.5%	20.0%
Plastics/Rubber	8.0%	10.0%
Textile	2.0%	15.0%
Wood	2.5%	15.0%
Inorganics		
Metals	8.5%	2.0%
Glass/Ceramics	8.0%	2.0%
Ash/Dirt/Rock	2.5%	8.0%
Other	1.5%	3.0%

6.14 What salient compressibility characteristics does solid waste exhibit?

6.15 List the factors affecting the settlement of municipal solid waste.

6.16 What are the main differences in the settlement characteristics of natural clay soils compared to municipal solid waste?

REFERENCES

Bartos, M. J. and Palermo, M. R., (1977) "Physical and Engineering Properties of Hazardous Industrial Wastes and Sludges," Technical Resource Document, EPA-600/2-77-139, U.S. Army Engineer Waterways Experiment Station, Vicksburg, MS.

Bjarngard, A. B. and Edgers, L., (1990) "Settlement of Municipal Solid Waste Landfills," *Proceedings of the 13th Annual Madison Waste Conference,* University of Wisconsin, Madison, WI, September, pp. 192–205.

Blight, G. E., Ball, J. M., and Blight, J. J., (1992) "Moisture and Suction in Sanitary Landfills in Semiarid Areas," *Journal of Environmental Engineering,* ASCE, Vol. 118, No. 6, pp. 865–877.

Brakensiek, D. L., Rawls, W. J., and Stephenson, G. R., (1984) "Modifying SCS Hydrologic Soil Groups and Curve Numbers for Rangeland Soils," American Society of Agricultural Engineers Paper No. PNR-84-203, St. Joseph, MI.

CalRecovery, Inc., (1993) *Handbook of Solid Waste Properties,* Published by Governmental Advisory Associates, Inc., New York, NY.

Chen, W. W. H., Zimmerman, R. E., and Franklin, A. G., (1977) "Time Settlement of Milled Urban Refuse," *Proceedings of the Conference on Geotechnical Practice for Disposal of Solid Waste Materials '77,* Ann Arbor, MI, June, pp. 136–152.

Dodt, M. E., Sweatman, M. B., and Bergstrom, W. R., (1987) "Field Measurements of Landfill Surface Settlements," *Geotechnical Practice for Waste Disposal,* ASCE, Ann Arbor, MI, pp. 406–417.

Druschel, S. J. and Wardwell, R. E., (1991) "Impact of Long Term Landfill Deformation," *Proceedings of Geotechnical Engineering Congress,* Boulder, CO, pp. 1268–1279.

Divinoff, A. H. and Munion, D. W., (1986) "Stability Failure of A Sanitary Landfill," *International Symposium on Environmental Geotechnology,* H. Y. Fang, ed., Lehigh University Press, Bethlehem, PA, pp. 25–35.

Earth Technology, (1988) "In-Place Stability of Landfill Slopes, Puente Hills Landfill, Los Angeles, California," Report No. 88-614-1, prepared for the Sanitation Districts of Los Angeles County, The Earth Technology Corp., Long Beach, CA.

Edgers, L., Noble, J. J., and Williams E., (1992) "A Biologic Model for Long Term Settlement in Landfills," *Environmental Geotechnology,* Proceedings of the Mediterranean Conference on Environmental Geotechnology, A. A. Balkema Publishers, Rotterdam, Netherlands, pp. 177–184.

Edil, T. B., Ranguette, V. J., and Wuellner, W. W., (1990) "Settlement of Municipal Refuse," *Geotechnics of Waste Fills—Theory and Practice,* ASTM STP 1070, Arvid Landva and G. David Knowles, Eds., Philadelphia, PA, pp. 225–239.

Fang, H. Y., Slutter, R. G., and Koerner, R. M., (1977) "Load Bearing Capacity of Compacted Waste Disposal Materials," *Proceedings of the Specialty Session on Geotechnical Engineering and Environmental Control,* 9th ICSMFE, Tokyo, Japan.

Fassett, J. B., Leonards, G. A., and Repetto, P. C., (1994) "Geotechnical Properties of Municipal Solid Wastes and Their Use in Landfill Design," *Waste Tech'94,* Landfill Technology, Technical Proceedings, Charleston, SC, January 13–14.

Fungaroli, A. A. and Steiner, R. L., (1979) "Investigation of Sanitary Landfill Behavior," Vol. 1, Final Report, U.S. Environmental Protection Agency, Cincinnati, Ohio, EPA-600/2-79/053a, p. 331.

Galante, V. N., Eith, A. W., Leonard, M. S. M., and Finn, P. S., (1991) "An Assessment of Deep Dynamic Compaction as a Means to Increase Refuse Density for an Operating Municipal Waste Landfill," *Proceedings of Conference on the Planning and Engineering of Landfills,* Midland Geotechnical Society, University of Birmingham, Birmingham, England, July.

Hentrich, R. L., Swartzbaugh, J. T., and Thomas, J. A., (1979) "Influence of MSW Processing on Gas and Leachate Production," *Proceedings of the Fifth Annual Research Symposium Municipal Solid Waste Land Disposal,* EPA-600/9-79-023a, U.S. Environmental Protection Agency, Cincinnati, OH.

Holmes, R., (1983) "The Absorptive Capacity of Domestic Refuse from a Full-Scale, Active Landfill," *Waste Management,* Vol. 73, No. 11, Atlanta, GA, p. 581.

Holtz, R. D. and Kovacs, W. D., (1981) "An Introduction to Geotechnical Engineering," Prentice-Hall. Englewood Cliffs, NJ.

Howland, J. D. and Landva, A. O., (1992) "Stability Analysis of a Municipal Solid Waste Landfill," *Proceedings of ASCE Specialty Conference on Stability and Performance of Slope and Embankments—II,* Berkeley, CA, June 28–July 1, pp. 1216–1231.

Kavazanjian, S., Jr., Matasovic, N., Bonaparte, R., and Schmertmann, G. R., (1995) "Evaluation of MSW Properties for Seismic Analysis," *Proceedings of GeoEnvironment 2000,* Geotechnical Special Publication No. 46, ASCE, New Orleans, LA, February 24–26, pp. 1126–1141.

Keene, P., (1977) "Sanitary Landfill Treatment, Interstate Highway 84," *Proceedings of the Conference on Geotechnical Practice for Disposal of Solid Waste Materials '77,* Ann Arbor, MI, June, pp. 632–644.

Koerner, R. M. and Soong, T-. Y., (2000) "Stability Assessment of Ten Large Landfill Failures," Advances in Transportation and Geoenvironmental Systems using Geosynthetics, Proceedings of Sessions of GeoDenver 2000, Denver, CD, ASCE Geotechnical Special Publication No. 103, pp. 1–38.

Landva, A. O., Clark, J. I., Weisner, W. R., and Burwash, W. J., (1984) "Geotechnical Engineering and Refuse Landfills," *Sixth National Conference on Waste Management in Canada,* Vancouver, BC, Canada.

Landva, A. O. and Clark, J. I., (1986) "Geotechnical Testing of Waste Fill," *Proceedings of Canadian Geotechnical Conference,* Ottawa, Ontario, pp. 371–385.

Landva, A. O. and Clark, J. I., (1990) "Geotechnics of Waste Fill," *Geotechnics of Waste Fills—Theory and Practice,* ASTM STP 1070, Arvid Landva and G. David Knowles, Eds., Philadelphia, pp. 86–103.

Landva, A. O. and La Rochelle, P., (1983) "Compressibility and Shear Characteristics of Radforth Peats," *Testing of Peats and Organic Soils,* P. M. Jarrett, ed., ASTM, STP 820, Philadelphia, PA.

McAneny, C. C., Tucker, P. G., Morgan, J. M., Lee, C. R., Kelley, M. F., and Horz, R. C., (1985) "Covers for Uncontrolled Hazardous Waste Sites," Technical Resource Document, EPA/540/2-85/002, U.S. Army Engineer Waterways Experiment Station, Vicksburg, MS.

McBean, E. A., Rovers, F. A., and Farquhar, G. J., (1995) "Solid Waste Landfill Engineering and Design," Prentice Hall PTR, Englewood Cliffs, NJ.

Mitchell, R. A. and Mitchell, J. K., (1992) "Stability Evaluation of Waste Landfills," *Proceedings of ASCE Specialty Conference on Stability and Performance of Slope and Embankments—II,* Berkeley, CA, June 28–July 1, pp. 1152–1187.

Oakley, R. E., (1990) "Case History: Use of the Cone Penetrometer to Calculate the Settlement of a Chemically Stabilized Landfill," *Geotechnics of Waste Fills—Theory and Practice,* ASTM STP 1070, Arvid Landva and G. David Knowles, Eds., Philadelphia, PA, pp. 345–357.

Oweis, I. and Khera, R., (1986) "Criteria for Geotechnical Construction of Sanitary Landfills," *International Symposium on Environmental Geotechnology,* H. Y. Fang, ed., Lehigh University Press, Bethlehem, PA, Vol. 1, pp. 205–222.

Oweis, I. S., Smith, D. A., Allwood, R. B., and Greene, D. S, (1990) "Hydraulic Characteristics of Municipal Refuse," *Journal of Geotechnical Engineering,* ASCE, Vol. 116, No. 4, pp. 539–553.

Perry, J. S. and Schultz, D. I., (1977) "Disposal and Alternate Uses of High Ash Papermill Sludge," *Proceedings of the 1977 National Conference on Treatment and Disposal of Industrial Wastewaters and Residues,* University of Houston, Houston, TX.

Pohland, F. and Graven, J.P. (1993). "The Use of Alternative Materials for Daily Cover at Municipal Solid Waste Landfills," EPA 600/R-93/172, U.S. Environmental Protection Agency, Cincinnati, OH.

Poran, C. J. and Ahtchi-Ali, F., (1989) "Properties of Solid Waste Incinerator Fly Ash," *Journal of Geotechnical Engineering,* ASCE, Vol. 115, No. 8, pp. 1118–1133.

Qian, Xuede, (1994) "Analysis of Allowable Reintroduction Rate for Landfill Leachate Recirculation," Michigan Department of Environmental Quality, Waste Management Division, Lansing, MI, November.

Qian, Xuede and Guo, Zhiping, (1998) "Engineering Properties of Municipal Solid Waste," *Chinese Journal of Geotechnical Engineering,* China, Vol. 20, No.5, pp. 1–6.

Reinhart, D. R. and Townsend, T. G., (1998) "Landfill Bioreactor Design and Operation," CRC Press LLC, 2000 Corporate Blvd., N.W., Boca Raton, FL, 33431.

Richardson, G. and Reynolds, D., (1991) "Geosynthetic Considerations in a Landfill on Compressible Clays," *Proceedings of Geosynthetics'91,* Vol. 2, Industrial Fabrics Association International, St. Paul, MN.

Rovers, F. A. and Farquhar, G. J., (1973) "Infiltration and Landfill Behavior," *Journal of Environmental Engineering,* ASCE, Vol. 99, No. 5, pp. 671–690.

Schroeder, P. R., Morgan, J. M., Walski, T. M., and Gibson, A. C., (1984a) "The Hydrologic Evaluation of Landfill Performance (HELP) Model, Volume I, User's Guide for Version I," Technical Resource Document, EPA/530-SW-84-009, U.S. Environmental Protection Agency, Cincinnati, OH, June.

Schroeder, P. R., Gibson, A. C., and Smolen, M.D., (1984b) "The Hydrologic Evaluation of Landfill Performance (HELP) Model, Volume II, Documentation for Version I," Technical Resource Document, EPA/530-SW-84-010, U.S. Environmental Protection Agency, Cincinnati, OH, June.

Schroeder, P. R., Lloyd, C. M., Zappi, P. A., and Aziz, N. M., (1994a) "The Hydrological Evaluation of Landfill Performance (HELP) Model, User's Guide for Version 3," EPA/600/R-94/168a, Risk Reduction Engineering Laboratory, Office of Research and Development, U.S. Environmental Protection Agency, Cincinnati, OH, September.

Schroeder, P. R., Dozier, T. S., Zappi, P. A., McEnore, B. M., Sjostrom, J. W., and Peyton, R. L., (1994b) "The Hydrological Evaluation of Landfill Performance (HELP) Model, Engineering Documentation for Version 3," EPA/600/R-94/168b, Risk Reduction Engineering Laboratory, Office of Research and Development, U.S. Environmental Protection Agency, Cincinnati, OH, September.

Sharma, H. D., Dukes, M. T., and Olsen, D. M., (1990) "Field Measurements of Dynamic Moduli and Poisson's Ratios of Refuse and Underlying Soils at a Landfill Site," *Geotechnics of Waste Fills—Theory and Practice,* ASTM STP 1070, Arvid Landva and G. David Knowles, Eds., Philadelphia, PA, pp. 57–70.

Sharma, H. D. and Lewis, S. P., (1994) "Waste Containment System, Waste Stabilization, and Landfills," John Wiley & Sons, Somerset, NJ.

Sheurs, R. E. and Khera, R. P., (1980) "Stabilization of a Sanitary Landfill to Support a Highway," Transportation Research Record 754, TRB, Washington, DC.

Siegel, R. A., Robertson, R. J., and Anderson, D. G., (1990) "Slope Stability at a Landfill in Southern California," *Geotechnics of Waste Fills—Theory and Practice,* ASTM STP 1070, Arvid Landva and G. David Knowles, Eds., Philadelphia, PA, pp. 259–284.

Singh, S. and Murphy, B., (1990) "Evaluation of the Stability of Sanitary landfills," *Geotechnics of Waste Fills—Theory and Practice,* ASTM STP 1070, Arvid Landva and G. David Knowles, Eds., Philadelphia, PA, pp. 240–258.

Sowers, G. F., (1968) "Foundation Problem in Sanitary LandFills," *Journal of Sanitary Engineering,* ASCE, Vol. 94, No. 1, pp. 103–116.

Sowers, G. F., (1973) "Settlement of Waste Disposal Fills," *Proceedings of the 8th International Conference on Soil Mechanics and Foundation Engineering,* Moscow, USSR, Vol. 1, pp. 207–210.

Steinberg, S. B. and Lucas, R. G., (1984) "Densifying a Landfill for Commercial Development," *Proceedings of International Conference on Case Histories in Geotechnical Engineering,* Vol. 3, University of Missouri-Rolla, Rolla, MO.

Swain, A., (1979) "Field Studies of Pulverized Fuel Ash in Partially Submerged Conditions," *Proceedings of the Symposium of the Engineering Behavior of Industrial and Urban Fill,* The Midland Geotechnical Society, University of Birmingham, Birmingham, England, pp. D49–D61.

Tchobanoglous, T., Theisen, H., and Vigil, S., (1993) "Integrated Solid Waste Management, Engineering Principles and Management Issues," McGraw-Hill, Inc., Hightstown, NJ.

USEPA, (1999) "Characterization of Municipal Solid Waste in the United States: 1998 Update," EPA530-R-99, U.S. Environmental Protection Agency, Municipal and Industrial Solid Waste Division, Office of Solid Waste, Prepared by Franklin Associates, Prairie Village, KS, July.

Wallis, S., (1991) "Factors Affecting Settlement at Landfill Sites," *Proceedings of Conference on the Planning and Engineering of Landfills,* Midland Geotechnical Society, University of Birmingham, Birmingham, England, July.

Walsh, J. J. and Kinman, R. N., (1979) "Leachate and Gas Production under Controlled Moisture Conditions," *Land Disposal: Municipal Solid Waste,* D. W. Shultz ed., 5th Annual Research, Symposium, U. S. Environmental Protection Agency, Cincinnati, Ohio, EPA-600/9-79/023, pp. 41–57.

Walsh, J. J. and Kinman, R. N., (1981) "Leachate and Gas from Municipal Solid Waste Landfill Simulators," *Land Disposal: Municipal Solid Waste,* D. W. Shultz ed., 7th Annual Research Symposium, U. S. Environmental Protection Agency, Cincinnati, OH, EPA-600/9-81/002a, pp. 67–93.

Wigh, R. J., (1979) "Boone County Field Site Interim Report," U. S. Environmental Protection Agency, Cincinnati, OH, EPA-600/2-79/580.

Yen, B. C. and Scanlon, B., (1975) "Sanitary Landfill Settlement Rates," *Journal of Geotechnical Engineering,* ASCE, Vol. 101, No. 5, pp. 475–487.

Zornberg, J. G., Jernigan, B. L., Sanglerat, T. R., and Cooley, B. H., (1999) "Retention of Free Liquids in Landfills Undergoing Vertical Expansion," *Journal of Geotechnical and Geoenvironmental Engineering,* ASCE, Vol. 125, No. 7, pp. 583–594.

CHAPTER 7

Leachate Generation and Evaluation in MSW Landfills

7.1 MSW LEACHATE CHARACTERIZATION
7.2 FACTORS AFFECTING LEACHATE QUANTITY
7.3 ESTIMATION OF LEACHATE PRODUCTION RATE IN AN ACTIVE CONDITION
    7.3.1 PRECIPITATION
    7.3.2 SQUEEZED WASTE LIQUID
    7.3.3 EVAPORATION
    7.3.4 WASTE MOISTURE ABSORPTION
7.4 ESTIMATION OF LEACHATE PRODUCTION RATE IN A POSTCLOSURE CONDITION
    7.4.1 SNOWMELT INFILTRATION
    7.4.2 SURFACE RUNOFF
    7.4.3 EVAPOTRANSPIRATION
    7.4.4 SOIL MOISTURE STORAGE
    7.4.5 LATERAL DRAINAGE
    7.4.6 MOISTURE EXTRACTION FROM WASTE
7.5 HYDROLOGIC EVALUATION OF LANDFILL PERFORMANCE (HELP) MODEL
    7.5.1 VERSIONS OF HELP MODEL
    7.5.2 DATA GENERATION AND DEFAULT VALUES
    7.5.3 LANDFILL PROFILE AND LAYER DESCRIPTIONS
    7.5.4 MODELING PROCEDURE
    7.5.5 PROGRAM INPUT
    7.5.6 PROGRAM OUTPUT
    7.5.7 LIMITS OF APPLICATION
PROBLEMS
REFERENCES

Leachate in municipal solid waste (MSW) landfills are contaminated liquids that contain a number of dissolved or suspended materials. Leachate is generated as a result of the expulsion of liquid from the waste due to its own weight or compaction loading (termed *primary leachate*) and the percolation of water through a landfill (termed *secondary leachate*). The source of percolating water could be precipitation, irrigation, groundwater, or leachate recirculated through the landfill. Percolating water plays a significant role in leachate generation. During the percolation process, the percolating water dissolves some of the chemicals in the waste through chemical reaction; it may also mix with the liquid that is squeezed out due to weight of the waste or compaction load. It should be noted that even if no water is allowed to percolate through the waste, a small volume of contaminated liquid is always expected to form due to biological and chemical reactions. Both the quality and quantity of produced leachate are important issues for landfill design. While this chapter focuses on MSW leachate, the concepts, designs and materials used are the same for any solid or quasi-solid waste material, such as hazardous, radioactive, ash-monofils, construction, and demolition wastes.

7.1 MSW LEACHATE CHARACTERIZATION

Leachate from a decomposing landfill contains a range of inorganic and organic chemicals, and its composition and characteristics are complex. The basic processes of waste decomposition affect the quality of leachate. McBean et al. (1995) have identified three stages in the decomposition of solid waste:

Stage I: Aerobic decomposition occurs rapidly, typically for a duration of less than one month. Once available oxygen within the waste is used up (except in the vicinity of the surface), this phase of decomposition terminates.

Stage II: Anaerobic and facultative organisms (acetogenic bacteria) hydrolyze and ferment cellulose and other putrescible materials, producing simpler, soluble compounds such as volatile fatty acids which produce a high biochemical oxygen demand (BOD) value and ammonia.

Stage III: Slower-growing methanogenic bacteria gradually become established and start to consume simple organic compounds, producing the mixture of carbon dioxide and methane (plus various trace constituents) that constitute landfill gas. This phase is more sensitive than Stage II.

Stage II can last for years, or even decades. Leachate produced during this stage is characterized by high BOD value (commonly greater than 10,000 mg/L) and high ratio of BOD to chemical oxygen demand (COD) (commonly greater than 0.7), indicating that a high proportion of soluble organic materials are readily biodegradable (McBean et al., 1995). Other typical characteristics of Stage II leachates are acidic pH levels (typically 5 to 6), strong, unpleasant smells, and high concentrations of ammonia (often 500 to 1,000 mg/L). The acidity and aggressive chemical nature of this leachate assists in dissolution of other components of the waste, which typically results in high levels of iron, manganese, zinc, calcium, and magnesium in the leachate.

According to McBean et al. (1995), the transition from Stage II to Stage III can take many years and may not be completed for decades. Sometimes it is never completed. However, some wastes have been known to reach Stage III in a few months. In Stage III, bacteria gradually become established and are able to remove the soluble organic compounds (mainly fatty acids) that are largely responsible for the characteristics of Stage II leachates. Leachates generated during Stage III are often referred to as "stabilized," but at this stage the landfill is biologically at its most active level. A dynamic equilibrium is eventually established between acetogenic and methanogenic bacteria, and wastes continue to actively decompose. Leachates produced during Stage III are characterized by relatively low BOD values and low ratios of BOD to COD. However, ammonia nitrogen continues to be released by the first-stage acetogenic process and is present at high levels in the leachate. Inorganic substances such as iron, sodium, potassium, sulfate, and chloride may continue to dissolve and leach from the landfill for many years.

Because of the sequential nature of biochemical reactions in a MSW landfill, the leachate coming from a single location is highly variable over time. Likewise, leachate varies greatly from location to location as well. Some locations will be at one phase of decomposition, while others will be at a very different stage. The leachate at the bottom of the waste is to some degree a result of the processes that have occurred in the waste above (McBean et al., 1995).

Although leachate quality differs from one municipal landfill to another, common factors affect the composition of leachate. These include the following (Ehrig, 1989; Lu et al., 1984; McBean et al., 1995; McGinley and Kmet, 1984; Qasim and Burchinal, 1970; Straub and Lynch, 1982):

  (i) Solid waste composition;
 (ii) Depth of the solid waste;
(iii) Age of the landfill;
(iv) Final cover condition;
 (v) Operation of the landfill such as water addition, leachate recirculation, waste compaction, thickness of dumped layers, and rate of placement;
(vi) Climate variables such as annual rainfall and ambient temperature;
(vii) Hydrogeologic conditions in the vicinity of the landfill site; and
(viii) Conditions within the landfill such as chemical and biological activities, moisture content, temperature, pH, and the degree of stabilization.

Young landfills typically have leachates high in biodegradable organics. As a landfill ages, its contents degrade and produce more complex organics and inorganics, that are not so readily amenable to biodegradation. Characteristics of leachate produced, as well as differences in the quality of leachate generated, by municipal, codisposal, and hazardous waste landfills have been documented (USEPA, 1988). In general, the collected data shows that although the same chemicals are routinely detected at both municipal and hazardous waste landfills, considerably higher concentrations of many chemicals are found in the leachate of hazardous waste facilities. In particular, chemicals such as 1,1,1-trichloroethane, trichloroethene, vinyl chloride, chloroform, pesticides, and PCBs occur with greater frequency and at higher concentrations in leachates at hazardous waste landfills than at municipal facilities. Typical chemical constituents in leachate from MSW landfills (USEPA, 1988) are shown in Tables 7.1 and 7.2.

7.2 FACTORS AFFECTING LEACHATE QUANTITY

The quantity of leachate that is generated is both site specific and waste specific. The amount of leachate produced is affected by the following factors: precipitation, type of site, groundwater infiltration, surface water infiltration, waste composition and moisture content, preprocessing of the waste (baling or shredding), density of the waste,

TABLE 7.1 Municipal Landfill Leachate Data—Indicator Parameters and Inorganic Compounds (USEPA, 1988)

Indicator Parameters	Municipal Solid Waste Landfills Leachate Concentration Reported (ppm)	
	Minimum	Maximum
Alkalinity	470	57,850
Ammonia	0.39	1,200
Biological oxygen demand	7	29,200
Calcium	95.9	2,100
Chemical oxygen demand	42	50,450
Chloride	31	5,475
Fluoride	0.11	302
Iron	0.22	2,280
Phosphorus	0.29	117.18
Potassium	17.8	1,175
Sulfate	8	1,400
Sodium	12	2,574
Total dissolved solids	390	31,800
Total suspended solids	23	17,800
Total Organic carbon	20	14,500
Inorganic Compounds (ppm)		
Aluminum	0.01	5.8
Antimony	0.0015	47
Arsenic	0.0002	0.982
Barium	0.08	5
Beryllium	0.001	0.01
Cadmium	0.0007	0.15
Chromium (total)	0.0005	1.9
Cobalt	0.04	0.13
Copper	0.003	2.8
Cyanide	0.004	0.3
Lead	0.005	1.6
Manganese	0.03	79
Magnesium	74	927
Mercury	0.0001	0.0098
Nickel	0.02	2.227
Vanadium	0.009	0.029
Zinc	0.03	350

thickness of the waste, climate, evaporation, evapotranspiration, gas production, final cover design, and surface flow pattern (Koerner and Daniel, 1997; Reinhart and Townsend, 1998). Continuous production of leachate will occur once the absorptive capacity of waste has been exhausted. The main factors affecting leachate quantity are as follows:

Precipitation. Precipitation represents the largest single contribution to the production of leachate. The amount of rain and snow falling on a landfill influences leachate quantity significantly. As with all cases of infiltration, the most critical

TABLE 7.2 Municipal Landfill Leachate Data—Organic Compounds (USEPA, 1988)

Indicator Parameters	Municipal Solid Waste Landfills Leachate Concentration Reported (ppm)	
	Minimum	Maximum
Acetone	8	11,000
Acrolein	270	270
Aldrin	NA	NA
α-Chlordane	NA	NA
Aroclor-1242	NA	NA
Aroclor-1254	NA	NA
Benzene	4	1,080
Bromomethane	170	170
Butanol	10,000	10,000
1-Butanol	320	360
2-Butanone (methyl ethyl ketone)	110	27,000
Butyl benzyl phenol	21	150
Carbazole	NA	NA
Carbon tetrachloride	6	397.5
4-Chloro-3-methylphenol	NA	NA
Chlorobenzene	1	685
Chloroethane	11.1	860
Bis(2-chloroethyoxy) methane	18	25
2-Chloroethyl vinyl ether	2	1,100
Chloroform	7.27	1,300
Chloromethane	170	400
Bis(chloromethyl)ether	250	250
2-Chloronaphthalene	46	46
p-Cresol	45.2	5,100
2,4-D	7.4	220
4,4'-DDE	NA	NA
4,4-DDT	0.042	0.22
Dibromomethane	5	5
Di-N-butyl phthalate	12	150
1,2-Dichlorobenzene	3	21.9
1,4-Dichlorobenzene	1	52.1
3,3-Dichlorobenzidine	NA	NA
Dichlorodifluoromethane	10.3	450
1,1-Dichloroethane	4	44,000
1,2-Dichloroethane	1	11,000
1,2-Dichloroethylene (Total)	NA	NA
cis-1,2-Dichloroethylene	190	470
trans-1,2-Dichloroethylene	2	4,800
1,2-Dichloropropane	0.03	500
1,3-Dichloropropane	18	30
Diethyl phthalate	3	330
2,4-Dimethyl phenol	10	28
Dimethyl phthalate	30	55
Endrin	0.04	50
Endrin ketone	NA	NA
Ethanol	23,000	23,000

TABLE 7.2 (*continued*)

Indicator Parameters	Municipal Solid Waste Landfills Leachate Concentration Reported (ppm)	
	Minimum	Maximum
Ethyl acetate	42	130
Ethyl benzene	6	4,900
Ethylmethacrylate	NA	NA
Bis(2-ethylhexyl) phthalate	16	750
2-Hexanone (methyl butyl ketone)	6	690
Isophorone	4	16,000
Lindane	0.017	0.023
4-Methyl-2-pentanone (methyl isobutyl ketone)	10	710
Methylene chloride (dichloromethane)	2	220,000
2-Methylnaphthalene	NA	NA
2-Methylphenol	NA	NA
4-Methylphenol	NA	NA
Methoxychlor	NA	NA
Naphthalene	2	202
Nitrobezene	4	120
4-Nitrophenol	17	17
Pentachlorophenol	3	470
Phenanthrene	NA	NA
Phenol	7.3	28,000
1-Propanol	11,000	11,000
2-Propanol	94	26,000
Styrene	NA	NA
1,1,2,2,-Tetrachloroethane	210	210
Tetrachloroethylene	2	620
Tetrahydrofuran	18	1,300
Toluene	5.55	18,000
Toxaphene	1	1
2,4,6-Tribromophenol	NA	NA
1,1,1-Trichloroethane	1	13,000
1,1,2,-Trichloroethane	30	630
Trichloroethylene	1	1,300
Trichlorofluormethane	4	150
1,2,3-Trichloropropane	230	230
Vinyl chloride	8	61
Xylenes	32	310

situation occurs during periods of light rainfall over a long lapse of time; short bursts of heavy rainfall during a storm result in a quick saturation of the cover material, the remainder is shed as runoff, so there is little net infiltration (Canziani and Cossu, 1989). Precipitation depends on geographical location (Bagchi, 1994).

Waste Condition. Waste condition can include waste composition, waste moisture content, preprocessing of the waste (baling or shredding), thickness of the waste, and density of the waste. Leachate quantity will increase if the waste releases pore water when squeezed because of increasing waste filling depth and density of the waste. Unsaturated waste continues to absorb water until it reaches field capacity as described in Chapter 6. Thus, waste in a state lower than its field capacity will reduce leachate formation. Sludge residues from sanitary treatment facilities, combined sewer systems, industrial filter materials, and other quasi-solids are being permitted by many regulatory groups for disposal in MSW landfills. Depending upon their moisture content and quantity, sludges have a major effect on leachate quality and quantity.

Final Cover Implications. Leachate volume is reduced significantly after a landfill is closed and finally covered because of two reasons—vegetation grown in the topsoil of a final cover reduces infiltration significantly by evapotranspiration, and the low permeability barrier reduces percolation. A significantly lesser amount of water will infiltrate into a landfill if it is covered with a composite cap, such as using geomembrane over compacted clay (GM/CCL) or geomembrane over a geosynthetic clay liner (GM/GCL). A properly designed final cover will eventually halt leachate quantity after landfill closure (Bagchi, 1994).

Leachate quantity is site specific and ranges from zero to 100% of precipitation during active landfill operation. Leachate generation rates in arid climates may be very low, even zero. In moist climates, however, leachate generation rates can be quite high. Leachate production from new landfills occurs at relatively low rates, then increases as more waste is placed and larger areas are exposed to precipitation. Leachate production reaches a peak just before closure and then declines significantly with the provision of surface grading and interim or final cover. The trend in leachate generation rates can be seen in Figure 7.1, which presents data for a double-lined municipal solid waste landfill having a geomembrane in the final cover. Monthly average leachate generation rates during the period of cell filling were up to 360 gallons/acre/day [3,400 liters/hectare/day (lphd)]. Rates for the first three years of the post-closure period

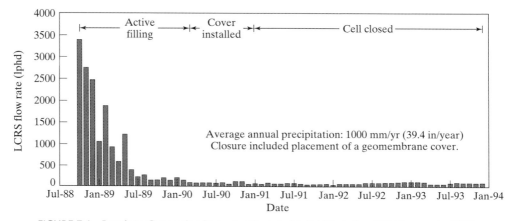

FIGURE 7.1 Leachate Generation Rate at a Municipal Solid Waste Landfill (Bonaparte, 1995). Used with permission of ASCE.

have averaged only 7.4 gallons/acre/day (70 lphd) (Bonaparte, 1995). Leachate quantities, however, vary over extremely wide limits and no single value can be considered as being "typical". For example, average leachate generation rates of all landfills in New York State are from 2,000 to 3,000 gallons/acre/year (20,000 to 30,000 liters/hectare/year). Conversely, landfills in Arizona and Nevada will rarely result in any leachate after the first few lifts of waste are placed in the facility.

7.3 ESTIMATION OF LEACHATE PRODUCTION RATE IN AN ACTIVE CONDITION

The leachate rates in a landfill for active and postclosure conditions vary significantly and the methods used to calculate them also differ. The leachate generation rate is higher during the active period of a landfill and is reduced significantly after construction of a qualified final cover. Procedures for estimating leachate quantity during the active and postclosure conditions are described in this section and next section, respectively.

Leachate is generated primarily as a result of percolating precipitation and pore liquid squeezed from waste placed in a landfill. The leachate generation rate in the active condition is denoted by

$$L_A = P + S - E - WA \qquad (7.1)$$

where L_A = leachate rate in the active condition;
P = precipitation;
S = pore squeeze liquid from waste;
E = moisture lost through evaporation; and
WA = waste moisture absorption.

Figure 7.2 shows how leachate is generated in an active landfill at a municipal solid waste landfill in active condition. Decomposition of putrescible waste mass can also release liquid. For practical design purposes the volume of leachate generated due to decomposition from waste is negligible. Surface run-on water may also cause an increase in leachate quantity; however, in a properly designed landfill surface, water is not allowed to move into an active cell thereby contacting the waste. Hence, this issue is not addressed here. However, surface run-on water is unavoidable in the case of older landfills. In this case, the volume of run-on water must be estimated using principles of local site hydrology, landfill-specific geometry, and site surface flow pattern.

The variables in the Equation 7.1 are discussed in the subsections that follow.

7.3.1 Precipitation

Precipitation includes all water that falls from the atmosphere to the area under consideration. It may occur in a variety of forms, including rainfall, snow, hail, and sleet. Once precipitation strikes the ground, it will produce surface runoff, evaporation, and percolation (Qasim and Chiang, 1994).

Precipitation varies geographically and seasonally. The amounts of precipitation can vary considerably within a short distance. Therefore, reliable precipitation data for

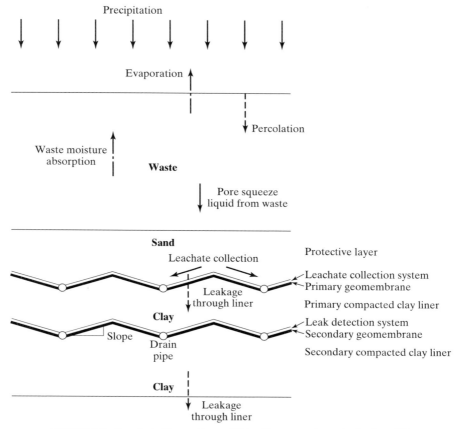

FIGURE 7.2 Leachate Generation or Water Movement in an Active Landfill

the local area under consideration must be developed. The National Weather Service and various other Federal and state agencies that keep historical precipitation data are important sources of rainfall data. The mean annual precipitation in the United States is shown in Figure 7.3.

7.3.2 Squeezed Waste Liquid

The volume of liquid squeezed from the pores of a waste material is directly affected by waste composition, waste moisture content, thickness of the waste, and density of the waste, as well as by operational measures such as water leachate recirculation, waste compaction, lift thickness of waste filling, and landfill build speed. When sludge is disposed in a landfill, the liquid within the pores of the sludge layer is released due to the weight of the sludge and the weight of the layer above it. The pore water is released because of both primary and secondary consolidation of the sludge that can take place. Although secondary consolidation may be high for putrescible waste—partly due to the creep of fibrous material and partly to the microbial decomposition

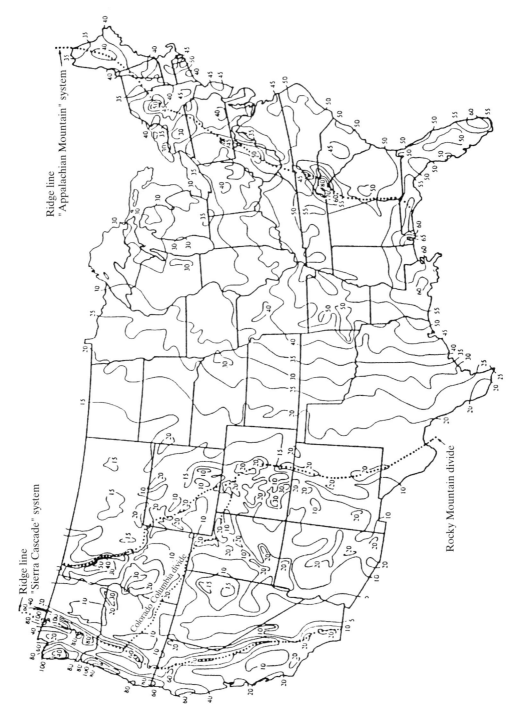

FIGURE 7.3 Mean Annual Precipitation in inches, 1899 to 1938 (U.S. Department of Agriculture, Soil Conservation Service)

of organic matter present in the waste—the total volume of liquid drained due to squeezing is not expected to be high (Bagchi, 1994). Usually, primary consolidation accounts for the majority of the squeezed liquid, and this volume can be predicted reasonably well using laboratory values (Charlie and Wardwell, 1979).

7.3.3 Evaporation

The moisture already present in a landfill as a result of precipitation may evaporate under favorable conditions. Evaporation depends on factors such as ambient temperature, wind velocity, difference of vapor pressure between the evaporating surface and air, atmospheric pressure, and the specific gravity of the evaporating liquid (Bagchi, 1994). A 1% decrease in evaporation rate due to each 1% rise in specific gravity of the evaporating liquid has been reported (Veihmeyer and Henderickson, 1955; Chow, 1964). Soil tends to bind water molecules by an attractive force that depends on the moisture content of the soil and its characteristics. The evaporation rate of unsaturated soils is almost constant over a range of moisture content of the soil. A shallow surface layer of soil [approximately 4 inches (100 mm) for clays and 8 inches (200 mm) for sand] will continue to evaporate until the layer reaches a permanent wilting point (the point at which the moisture content of the soil prevents the soil from supplying water at a sufficient rate to maintain turgor). Evaporation from deeper soil is negligible (Chow, 1964).

The water budget method, energy budget method, and mass transfer method have all been used to predict evaporation from open water bodies (Viessman et al., 1977). Evaporation potential depends on the availability of water. It is 100% from saturated soil, but nearly zero from dry soil (Bagchi, 1994). The mean annual evaporation from shallow lakes and reservoirs in the United States is shown in Figure 7.4.

7.3.4 Waste Moisture Absorption

Waste in a landfill may absorb some moisture before it can percolate through. Internal storage of leachate within the landfill is an important concept both to the water balance used to calculate leachate generation rates and to the success of a leachate recirculation system. Internal storage of leachate is possible because the moisture content of the waste "as received" is generally below the maximum absorptive capacity of the waste (i.e., its field capacity). The term of $(\theta_{FC} - \theta)$ can be defined as moisture absorptive capacity of the waste, in which θ_{FC} is the field capacity of the waste, and θ is initial moisture content of the waste placed in the landfill. Field capacity is the maximum amount of water that the soil or waste can retain in a gravitational field without percolation.

In theory, leachate can develop only after the field capacity moisture content is exceeded. The moisture content of municipal solid waste at placement varies from 10 to 20% by volume (Fenn et al., 1975). If the field capacity of the waste is 30% by volume, then the waste can hold an additional 10 to 20% by volume without leachate generation. However, some leachate will develop at a moisture content less than the

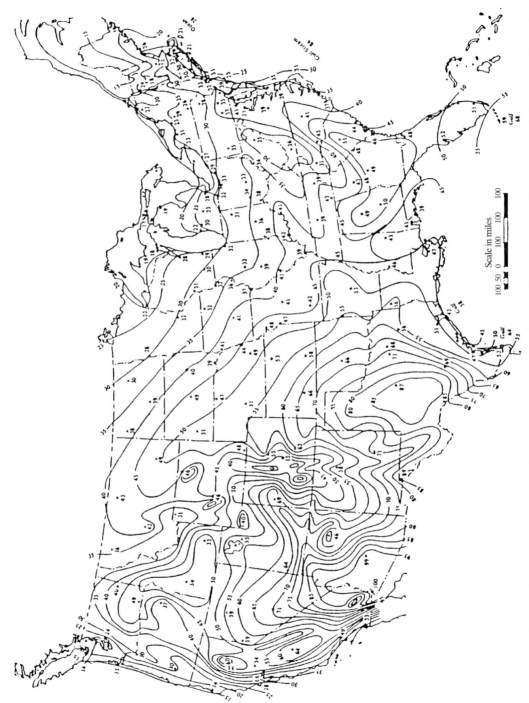

FIGURE 7.4 Mean Annual Evaporation from Shallow Lakes and Reservoirs, in inches (U.S. Department of Agriculture, Soil Conservation Service)

Section 7.4   Estimation of Leachate Production Rate in a Postclosure Condition   223

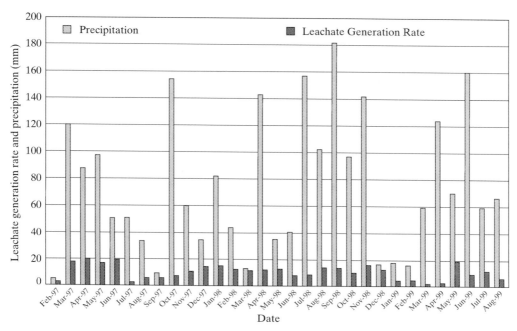

FIGURE 7.5   Precipitation and Leachate Generation Rate at a Municipal Solid Waste Landfill in Active Condition

field capacity because of secondary conductivity (channeling) of the waste (Oweis and Khera, 1998). The moisture content of municipal solid waste was presented in Table 6.5. The field capacity of municipal solid waste was summarized in Table 6.8. The field capacities for various waste materials were listed in Table 6.9.

Figure 7.5 shows a comparison between precipitation and leachate generation rates for a municipal solid waste landfill in active condition.

7.4   ESTIMATION OF LEACHATE PRODUCTION RATE IN A POSTCLOSURE CONDITION

After completion of the final cover for a closed landfill, only the precipitation that can infiltrate through the final cover percolates through the waste to generate leachate. The potential pathways for water movement onto and through a cover are summarized in Figure 7.6. The input of water is precipitation, and output is drainage (percolation) of water out of the cover. Within the cover, water can be stored, drained laterally, or be returned to the atmosphere via evapotranspiration. To conserve mass, the quantity of water that flows into the cover must equal the quantity of flow out of the cover plus the change in amount of water stored within the cover. This principle of conservation of mass is the basis for the term water balance (Koerner and Daniel, 1997). The

224 Chapter 7 Leachate Generation and Evaluation in MSW Landfills

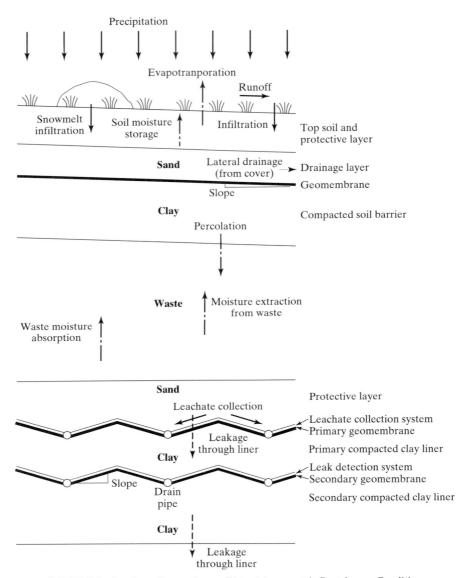

FIGURE 7.6 Leachate Generation or Water Movement in Postclosure Condition

analysis of water routing in covers is called *water balance analysis*. For landfill postclosure condition, the water balance equation can be written as

$$L_C = P + SM - RF - ET - \Delta S - Q - WA - ME \tag{7.2}$$

where L_C = leachate rate in the postclosure condition;
P = precipitation;
SM = snowmelt infiltration;

RF = surface runoff;
ET = moisture lost through evapotranspiration;
ΔS = soil moisture storage;
Q = lateral drainage in the cap drainage layer;
WA = waste moisture absorption; and
ME = moisture extraction from the waste.

Precipitation can fall on the cover in the form of either rain or snow. Rain that reaches the surface of the cover can either run off or infiltrate the subsoil. Snow can accumulate on the surface and later melt. Final covers are usually designed to minimize the amount of percolation of water out the base of the cover. Water percolation is minimized by maximizing runoff, lateral drainage, evapotranspiration, and physically blocking downward infiltration by including one or more hydraulic barriers in the cover cross section. The moisture content of the waste placed in a landfill may be lower than its field capacity after closure, especially for landfills located in semiarid and arid areas. In that case the waste still has capacity to absorb some extra moisture before it percolates through. The terms P and WA in Equation 7.2 are the same as in Equation 7.1. The other variables in Equation 7.2 are discussed in the subsections that follow.

7.4.1 Snowmelt Infiltration

Snow that falls to the surface of the cover is assumed to accumulate on the surface until it either evaporates due to sublimation; melts due to temperature rising above freezing (32°F or 0°C); or melts due to rainfall. Sublimation is typically ignored because it usually plays a minor role in the water balance. Schroeder et al. (1994a & b) provide recommended procedures for calculating snowmelt from the other two mechanisms. Direct (non-rain) snowmelt depends primarily on the surface air temperature, but is affected by the amount of runoff to melted snow (vs. storage within the snow pack) and the tendency of water from snow melt to refreeze. Snowmelt caused by rainfall is analyzed based on a heat balance. The warmer the temperature of the rain, and the greater the amount of rain, the greater the snowmelt.

Infiltration from snow depends on the condition of the ground (frozen or unfrozen), ambient temperature and its duration (snowmelt will depend on whether a temperature of 32°F [0°C] and above prevails for 1 day or several days), radiation energy received (more snowmelt occurs on sunny days than on a cloudy day), rainfall during snow melting (rainfall accelerates the snowmelting process), and so on (Bagchi, 1994). Because of the variables involved, it is difficult to predict snowmelt runoff or infiltration. Two methods are usually used: the *degree-day method* and the U. S. Army Corps of Engineers equation. Only the degree-day method is discussed here because of its relative simplicity. A detailed discussion of the Corps of Engineers method can be found elsewhere (Chow, 1964; Lu et al., 1985). The degree-day method for estimating snowmelt infiltration (SCS, 1975) can be expressed as

$$SM = K \cdot (T - 32°F) \tag{7.3}$$

where SM = potential daily snow melt infiltration, inches of water;
  K = constant that depends on the watershed condition (Table 7.3); and
  T = ambient temperature above 32°F, °F [note: °F = (9/5)°C + 32].

TABLE 7.3 Values of K for the Degree Day Equation Relevant to Landfill Cover Design (SCS, 1972)

Watershed Condition	K
Average Heavily Forested Area	
North Facing Slopes	0.04 to 0.06
South Facing Slopes	0.06 to 0.08
High Runoff Potential	0.03

°F = (9/5)°C + 32

$T - 32°F$ is the number of degree days per day. As Bagchi (1994) notes, the total predicted snowmelt infiltration must not exceed the total water equivalent of precipitated snow (note: 1 inch of water = 10 inches of snow).

7.4.2 Surface Runoff

A portion of incident precipitation runs off the site and is converted to overland flow before it has a chance to infiltrate. The amount of surface runoff depends upon factors such as intensity and duration of storm, the surface slope, the permeability of the soil cover, and the amount and type of vegetation (Qasim and Chiang, 1994). Several methods have been proposed to estimate the runoff from, or percolation into, a sanitary landfill. These methods include field measurements and empirical relationships. The most commonly used relationship is the *rational method* (Qasim and Chiang, 1994), which can be expressed as

$$RF = C \cdot I \cdot A_s \tag{7.4}$$

where RF = peak surface runoff, ft³/sec (1 ft³/sec = 0.0283 m³/sec);
  C = runoff coefficient;
  I = uniform precipitation rate, inches/hour (1 inch/hour = 25.4 mm/hour); and
  A_s = area of the landfill surface, acres (1 acre = 0.4 hectare).

The runoff Coefficient C is defined as the ratio of runoff to precipitation. For example, if $C = 0.1$, then 10% of the precipitation is assumed to run off and 90% is assumed to infiltrate the soil. Runoff is one of the most difficult parameters to determine accurately because very little information is available on actual runoff rates from landfill covers. Two approaches are used for estimating the runoff coefficient. The simplest approach is to estimate a value based on the type of soil and average angle of the slope. Guidance provided by Fenn et al. (1975), which is summarized in Table 7.4, is recommended if no better information is available for a specific site.

The procedure recommended by Schroeder et al. (1994a & b) and used in the Hydrologic Evaluation of Landfill Performance (HELP) model is more complicated and involves use of the Soil Conservation Service (SCS) *curve number method*, proposed by the Soil Conservation Service of the United States and used to predict surface runoff from agricultural land (SCS, 1975). In addition to rainfall volume, soil type,

Section 7.4 Estimation of Leachate Production Rate in a Postclosure Condition

TABLE 7.4 Suggested Runoff Coefficients, C (Fenn et al., 1975)

Description of Soil	Slope	Runoff Coefficient
Sandy soil	Flat ($\leq 2\%$)	0.05–0.10
	Average (2–7%)	0.10–0.15
	Steep ($\geq 7\%$)	0.15–0.20
Clayey Soil	Flat ($\leq 2\%$)	0.13–0.17
	Average (2–7%)	0.18–0.22
	Steep ($\geq 7\%$)	0.25–0.35

and land cover, the method accounts for land use and antecedent moisture conditions. The antecedent moisture condition is first divided into three groups based on season (dormant and growing) and a 5-day total antecedent rainfall in inches. Soil is grouped into four different types based on ability to cause runoff (e.g., a clayey soil has high runoff potential and sand or gravel has low runoff potential; all other soil types are classified between these two extremes). The land use and land cover type is then determined. The weighted curve number is next established using tables developed for this purpose (Bagchi, 1994). The direct runoff can then be estimated for different rainfall using the equation

$$R_i = \frac{\{W_p - 0.2[(1000/CN) - 10]\}^2}{W_p + 0.8[(1000/CN) - 10]} \tag{7.5}$$

where R_i = surface runoff, inches;
W_p = rainfall, inches/hour; and
CN = curve number.

The curve number method, which is applicable for large storms on small watersheds, was developed by plotting measured runoff in streams versus rainfall. The typical trend that has been observed is shown in Figure 7.7. Initially, no runoff occurs, but with time and continuing rainfall, the slope of the diagram approaches 45 degrees,

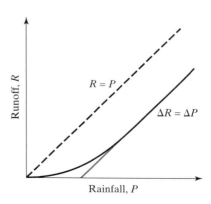

FIGURE 7.7 Typical Relationship between Runoff and Rainfall for Large Storms in Small Watersheds (Schroeder et al., 1994a & b)

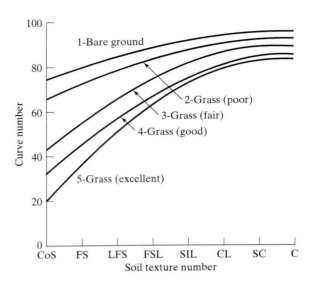

FIGURE 7.8 Relationship between SCS Curve Number and Soil Texture Number for Various Levels of Vegetation (Schroeder, et al., 1994a & b)

indicating that all of the rainfall is running off. A series of curves were developed (hence the name, SCS curve-number method), which are used along with information on soil moisture to compute runoff. Schroeder et al. (1994a & b) describe the procedure used in the computer program as well as a method used to correct the curve numbers for steeply sloping landfill covers. The curve number CN used by Schroeder et al. (1994a & b) in HELP model for mild slope is given in Figure 7.8. The CN number from Figure 7.8 is adjusted for various slopes (0.04 to 0.5 ft/ft or m/m) and slope lengths (50 to 500 ft or 15 to 150 m) using the equation

$$CN_S = 100 - (100 - CN)(L^2/S)^m \tag{7.6}$$

where CN_S = curve number adjusted for slope angle and slope length;
CN = curve number from Figure 7.8 (standard for slope of 4%, 500 ft long) (1 ft = 0.3048 m);
L = nondimensional length which is equal to actual slope length (ft)/500 ft;
S = nondimensional slope which is equal to actual slope (ft/ft)/0.04 (if the actual slope is 3(H):1(V), then $S = 0.33/0.04 = 8.25$); and
$m = (CN)^{-0.81}$.

7.4.3 Evapotranspiration

When water infiltrates a cover, the water may be stored or it may be routed elsewhere. Near the ground surface, water can evaporate directly to the atmosphere. Within the root zone of plants, water can be withdrawn from the soil by the plants and returned to the atmosphere by the process of transpiration. *Evapotranspiration* is a term that combines evaporation and transpiration. The second process (transpiration) is by far the more significant because evaporation is limited to the surface of the soil, but transpiration occurs within the entire root depth of plants. The amount of evapotranspiration increases with increasing water content of the soil, surface temperature, wind speed, plant density, and depth of plant root penetration; it increases with decreasing relative humidity in the atmosphere (Koerner and Daniel, 1997).

The estimation of evapotranspiration is probably the most complicated part of the water balance analysis. Many empirical equations and methods have been proposed for estimating the potential evapotranspiration. The procedures provided by Thornthwaite and Mather (1955, 1957) are the most commonly applied for all these equations and methods. Fenn et al. (1975) analyzed evapotranspiration using the relatively simple and largely observational-based method of Thornthwaite (Thornthwaite and Mather, 1955). Schroeder et al. (1994b) used a more complex computerized method involving an energy balance in the Hydrologic Evaluation of Landfill Performance (HELP) model (Ritchi, 1972).

The quantity of evapotranspiration is approximately equal to the frequently reported "lake evapotranspiration," which is about 0.7 times the pan evaporation level (Lutton et al., 1979).

7.4.4 Soil Moisture Storage

The amount of water that can be stored in a soil mainly depends on the type, destiny, and initial moisture content of the soil, as well as the total thickness of the soil layer of the final cover. The moisture content in the soil continually changes—increasing due to infiltration and decreasing due to evapotranspiration. The depletion of moisture due to evapotranspiration is limited to an upper soil zone defined by the effective root zone depth. In the water balance method, it is important to account for the change in the moisture storage within the soil cover. The maximum moisture that a soil can hold against the pull of gravitational force is the field capacity. The minimum moisture a soil loses due to vegetation is its moisture content at wilting point. The difference between the two limits is the moisture holding capacity available to the plants from the soil. The higher the available moisture holding capacity of the soil, the more moisture it may lose to the atmosphere by evapotranspiration. The soil within the root zone can be dried by evapotranspiration to moisture contents considerably below the field capacity. At the permanent wilting point, the remaining moisture is essentially unavailable for withdrawal by plants (Qasim and Chiang, 1994). The water holding characteristics for different soil types was shown in Figure 6.4.

7.4.5 Lateral Drainage

If a lateral drainage layer is present, the maximum rate of flow within the layer is typically calculated using Darcy's formula, namely,

$$q = k \cdot (\Delta H/L) \cdot A = k \cdot (\Delta H/L)(t)(1) \qquad (7.7)$$

where q = the rate of flow (volume/time);
        k = the hydraulic conductivity (length/time) of the material comprising the drainage layer;

L = distance along the path of flow;
ΔH = the head loss (length) over distance L along the path of flow;
A = the cross sectional area of flow, which is equal to the thickness of the drainage layer times a unit distance of length; and
t = the thickness of the drainage layer.

The transmissivity θ of a layer is defined as the hydraulic conductivity times the saturated thickness that conducts seepage (t):

$$\theta = k \cdot t \tag{7.8}$$

Thus, if transmissivity is used rather than hydraulic conductivity, Darcy's formula can be rewritten as

$$q = \theta \cdot (\Delta H/L) \tag{7.9}$$

Because Equation (7.9) is a simpler formulation than Equation (7.7), transmissivity is often used rather than hydraulic conductivity. Viewed in more basic terms, the flow rate is proportional to hydraulic conductivity times saturated thickness, i.e., transmissivity. The Depuit-Forcheimer assumptions are normally employed, which means that the line of seepage is assumed to be parallel to the slope of the drainage layer, and the hydraulic gradient throughout the flow regime is assumed to be constant and equal to the sine of the angle of the slope (Koerner and Daniel, 1997).

7.4.6 Moisture Extraction from Waste

Moisture extraction from the waste is the removal of moisture by the landfill gas collection system (e.g., gas extraction wells and trenches by a vacuum source) and by biodegradation processes within the waste. Degradation of the waste is a time-dependent process that not only reduces the waste moisture with time, but may also change the waste moisture retention properties (e.g., field capacity) (Zornberg, 1999). The details of waste biodegradation processes with time are still not clear.

Temperatures of landfill gas within the landfill and at the wellhead usually range from 80 to 125°F (27 to 52°C) depending upon a number of factors (LandTech, 1994; Prosser and Janechek, 1995). As landfill gas moves through refuse, it is typically saturated with water vapor at these temperatures. During the gas collection stage, part of the moisture in the waste can be extracted in the form of water vapor and transported through the gas collection system, where it gradually cools and converts to a liquid. This liquid is known as landfill gas condensate. For every million cubic feet of gas that is generated, about 50 to 600 gallons (190 to 2,300 liters) of condensate can be generated, depending on the vacuum extraction pressure of the system and the moisture content of the waste. Detailed procedures for estimating condensate production in the landfill gas are described in Section 10.8.6.

7.5 HYDROLOGIC EVALUATION OF LANDFILL PERFORMANCE (HELP) MODEL

The Hydrologic Evaluation of Landfill Performance (HELP) computer program. Versions 1, 2, and 3, were developed by the U.S. Army Engineer Waterways Experiment Station (WES) for the U.S. Environmental Protection Agency (EPA) in response to requirements established in the Resource Conservation and Recovery Act (RCRA) and the Comprehensive Environmental Response, Compensation and Liability Act (CERCLA or Superfund). The primary purpose of the model is to assist in the comparison of landfill design alternatives on the basis of water balances and hydrologic performance (Schroeder et al., 1994a & 1994b).

The importance of the HELP model in landfill practice cannot be overemphasized. It is used to predict leachate generation by all states in the USA and in most countries of the world. The model is well configured and is very "user friendly." The salient features are presented in this text. If greater detail and explanation are required, then the HELP model user's manual should be consulted.

The model accepts weather, soil, and design data and uses solution techniques that account for the effects of surface storage, snowmelt, runoff, infiltration, evapotranspiration, vegetative growth, soil moisture storage, lateral subsurface drainage, leachate recirculation, unsaturated vertical drainage, and leakage through soil, geomembrane, or composite liners. Landfill systems, including various combinations of vegetation, cover soils, waste cells, lateral drain layers, low permeability barrier soils, and synthetic geomembrane liners, can be simulated in the model. Results are expressed as daily, monthly, annual and long-term average water budgets.

7.5.1 Versions of HELP Model

The HELP model is a quasi-two-dimensional, deterministic, water-routing model for determining water balances. The model was adapted from the Hydrologic Simulation Model for Estimating Percolation at Solid Waste Disposal Sites (HSSWDS model) of the U.S. Environmental Protection Agency (Perrier and Gibson, 1980; Schroeder and Gibson, 1982; Schroeder et al., 1994b) and various models of the U.S. Agricultural Research Service (ARS), including the Chemical Runoff and Erosion from Agricultural Management Systems (CREAMS) model (Knisel, 1980), the Simulator for Water Resources in Rural Basins (SWRRB) model (Arnold et al., 1989), the SNOW-17 routine of the National Weather Service River Forecast System (NWSRFS) Snow Accumulation and Ablation Model (Anderson, 1973), and the WGEN synthetic weather generator (Richardson and Wright, 1984).

7.5.1.1 HELP Version 1. Version 1 of the HELP model (Schroeder et al., 1984a & 1984b) represented a major advance beyond the Hydrologic Simulation Model for Estimating Percolation at Solid Waste Disposal Sites (HSSWDS model) (Perrier and Gibson, 1980; Schroeder and Gibson, 1982), which was also developed at the U.S. Army Engineer Waterways Experiment Station. The HSSWDS model simulated only the cover system, did not model lateral flow through drainage layers, and handled vertical drainage only in a rudimentary manner. The infiltration, percolation and evapotranspiration routines were almost identical to those used in the Chemical Runoff and Erosion from Agricultural Management Systems (CREAMS) model that was developed by Knisel (1980) for the U.S. Department of Agriculture (USDA). The runoff and infiltration routines relied heavily on the Hydrology of the National Engineering Handbook (USDA, Soil Conservation Service, 1985). Version 1 of the HELP model incorporated a lateral subsurface drainage model and improved unsaturated drainage and liner leakage model into the HSSWDS model. In addition, the

HELP model provided simulation of the entire landfill including leachate collection and liner systems.

Version 1 of the HELP program was tested extensively using both field and laboratory data. HELP Version 1 simulation results were compared to field data for 20 landfill cells from seven sites (Schroeder and Peyton, 1987a). The lateral drainage component of HELP Version 1 was tested against experimental results from two large-scale physical models of landfill liner and drain systems (Schroeder and Peyton, 1987b). The results of these tests provided motivation for some of the improvements incorporated into HELP Version 2 (Schroeder et al., 1994b).

7.5.1.2   HELP Version 2.   Version 2 of the HELP model (Schroder et al., 1988a & 1988b) greatly enhanced the capabilities of the HELP model. The WGEN synthetic weather generator developed by the USDA Agricultural Research Service (Richardson and Wright, 1984) was added to the model to yield daily values of precipitation, temperature, and solar radiation. This replaced the use of normal mean monthly temperature and solar radiation values and improved the modeling of snow and evapotranspiration. Also, a vegetative growth model from the Simulator for Water Resources in Rural Basins (SWRRB) model developed by the USDA Agricultural Research Service (Arnold et al., 1989) was merged into the HELP model to calculate daily leaf area indices. Modeling of unsaturated hydraulic conductivity and flow and lateral drainage computations were improved. Default soil data were improved also, and the model permitted use of more layers and initialization of soil moisture content.

7.5.1.3   HELP Version 3.   Version 3 of the HELP model was presented in 1994 (Schroder et al., 1994a & 1994b). Additional improvements were incorporated in the Version 3 model. The number of layers that can be modeled was increased and the default soil/material texture list was expanded to account for additional waste materials, geomembranes, geosynthetic drainage nets and compacted soils. The model also permits the use of a user-built library of soil textures. Computations of leachate recirculation and groundwater drainage into the landfill have been added. Moreover, HELP Version 3 accounts for leakage through geomembrane due to manufacturing defects (pinholes) and installation defects (punctures, tears and seaming flaws) and by vapor diffusion through the liner based on the equations compiled by Giroud et al. (1989, 1992). The estimation of runoff from the surface of the landfill has been improved to account for large landfill surface slopes and slope lengths. The snowmelt model has been replaced with an energy-based model; the Priestly-Taylor potential evapotranspiration model has been replaced with a Penman method, incorporating wind and humidity effects as well as long wave radiation losses (heat loss at night). A frozen soil model has been added to improve infiltration and runoff predictions in clod regions. The unsaturated vertical drainage model has also been improved to aid in storage computations. Input and editing have been further simplified with interactive, full-screen, menu-driven input techniques.

7.5.2   Data Generation and Default Values

The HELP model requires general climate data for computing potential evapotranspiration, daily climatologic data, soil characteristics, and design specifications to perform

Section 7.5 Hydrologic Evaluation of Landfill Performance (HELP) Model 233

the analysis (Schroeder et al., 1994b). The required general climate data include growing season, average annual wind speed, average quarterly relative humidities, normal mean monthly temperatures, maximum leaf area index, evaporative zone depth, and latitude. Default values for these parameters were compiled or developed from the "Climates of the States" (Ruffner, 1985) and "Climatic Atlas of the United States" (National Oceanic and Atmospheric Administration, 1974) for 183 U.S. cities. Daily climatologic (weather) data requirements include precipitation, mean temperature, and total global solar radiation. Daily rainfall data may be input by the user, generated stochastically, or taken from the model's historical database. The model contains parameters for generating synthetic precipitation for 139 U.S. cities. The historical database contains five years of daily precipitation data for 102 U.S. cities. Daily temperature and solar radiation data are generated stochastically or may be input by the user (Schroeder et al., 1994a & 1994b).

Necessary soil data include porosity, field capacity, wilting point, saturated hydraulic conductivity, initial moisture storage, and Soil Conservation Service (SCS) runoff curve number for antecedent moisture condition II (Schroeder et al., 1994b). The HELP model provides default values for the total porosity, field capacity, wilting point, and saturated hydraulic conductivity of numerous soil and waste materials, as well as for geosynthetic materials. The model contains default soil characteristics for 42 material types for use when measurements or site-specific estimates are not available. The soil texture types are classified according to two standard systems, the U.S. Department of Agriculture (USDA) texture classification system and the Unified Soil Classification System (USCS). Applicable USDA and USCS soil texture abbreviations are provided in Table 7.5.

The default characteristics compiled by Schroeder et al., (1994a and 1994b) for 42 soil/material types are shown in Table 7.6. The default characteristics of Types 1

TABLE 7.5 Default Soil Texture Abbreviations (Schroeder et al., 1994a & 1994b)

U.S. Department of Agriculture	Definition
G	Gravel
S	Sand
Si	Silt
C	Clay
L	Loam (sand, silt, clay and humus mixture)
Co	Coarse
F	Fine

Unified Soil Classification System	Definition
G	Gravel
S	Sand
M	Silt
C	Clay
P	Poorly Graded
W	Well Graded
H	High Plasticity or Compressibility
L	Low Plasticity or Compressibility

TABLE 7.6 Default Soil, Waste, and Geosynthetics Characteristics (Schroeder et al., 1994a & 1994b)

HELP	Classification USDA	Classification USCS	Total Porosity (vol/vol)	Field Capacity (vol/vol)	Wilting Point (vol/vol)	Hydraulic Conductivity (cm/sec)
1	CoS	SP	0.417	0.045	0.018	1.0×10^{-2}
2	S	SW	0.437	0.062	0.024	5.8×10^{-3}
3	FS	SW	0.457	0.083	0.033	3.1×10^{-3}
4	LS	SM	0.437	0.105	0.047	1.7×10^{-3}
5	LFS	SM	0.457	0.131	0.058	1.0×10^{-3}
6	SL	SM	0.453	0.190	0.085	7.2×10^{-4}
7	FSL	SM	0.473	0.222	0.104	5.2×10^{-4}
8	L	ML	0.463	0.232	0.116	3.7×10^{-4}
9	SiL	ML	0.501	0.284	0.135	1.9×10^{-4}
10	SCL	SC	0.398	0.244	0.136	1.2×10^{-4}
11	CL	CL	0.464	0.310	0.187	6.4×10^{-5}
12	SiCL	CL	0.471	0.342	0.210	4.2×10^{-5}
13	SC	SC	0.430	0.321	0.221	3.3×10^{-5}
14	SiC	CH	0.479	0.371	0.251	2.5×10^{-5}
15	C	CH	0.475	0.378	0.265	1.7×10^{-5}
16	Barrier Soil		0.427	0.418	0.367	1.0×10^{-7}
17	Bentonite Mat (0.6 cm)		0.750	0.747	0.400	3.0×10^{-9}
18	Municipal Waste (900 lb/yd³ or 312 kg/m³)		0.671	0.292	0.077	1.0×10^{-3}
19	Municipal Waste (channeling and dead zones)		0.168	0.073	0.019	1.0×10^{-3}
20	Drainage Net (0.5 cm)		0.850	0.010	0.005	10
21	Gravel		0.397	0.032	0.013	3.0×10^{-1}
22	L*	ML	0.419	0.307	0.180	1.9×10^{-5}
23	SiL*	ML	0.461	0.360	0.203	9.0×10^{-6}
24	SCL*	SC	0.365	0.305	0.202	2.7×10^{-6}
25	CL*	CL	0.437	0.373	0.266	3.6×10^{-6}
26	SiCL*	CL	0.445	0.393	0.277	1.9×10^{-6}
27	SC*	SC	0.400	0.366	0.288	7.8×10^{-7}
28	SiC*	CH	0.452	0.411	0.311	1.2×10^{-6}
29	C*	CH	0.451	0.419	0.332	6.8×10^{-7}
30	Coal-Burning Electric Plant Fly Ash*		0.541	0.187	0.047	5.0×10^{-5}
31	Coal-Burning Electric Plant Bottom Ash*		0.578	0.076	0.025	4.1×10^{-3}
32	Municipal Incinerator Fly Ash*		0.450	0.116	0.049	1.0×10^{-2}
33	Fine Copper Slag*		0.375	0.055	0.020	4.1×10^{-2}
34	Drainage Net (0.6 cm)		0.850	0.010	0.005	33
35	High Density Polyethylene (HDPE)					2.0×10^{-13}
36	Low Density Polyethylene (LDPE)					4.0×10^{-13}
37	Polyvinyl Chloride (PVC)					2.0×10^{-11}
38	Butyl Rubber					1.0×10^{-12}
39	Chlorinated Polyethylene (CPE)					4.0×10^{-12}
40	Hypalon or Chlorosulfonated Polyethylene (CSPE)					3.0×10^{-12}
41	Ethylene-Propylene Diene Monomer (EPDM)					2.0×10^{-12}
42	Neoprene					3.0×10^{-12}

*Moderately Compacted. Note: USDA denotes U.S. Department of Agriculture; USCS denotes Unified Soil Classification System.

through 15 in the table are typical of surficial and disturbed agricultural soils, which may be less consolidated and more aerated than soils typically placed in landfills (Breazeale and McGeorge, 1949; England, 1970; Lutton et al., 1979; Rawls et al., 1982). Clays and silts in landfills would generally be compacted except within the vegetative layer, which might be tilled to promote vegetative growth. Untilled vegetative layers may be more compacted than the loams listed in Table 7.6. Soil texture Types 22 through 29 are compacted soils. Type 18 is representative of typical municipal solid waste that has been compacted; Type 19 is the same waste but it accounts for 65 percent of the waste being in dead zones not contributing to drainage and storage. Soil Types 16 and 17 denote very well compacted clay soils that might be used for barrier soil liners.

The porosity, field capacity, wilting point, and saturated hydraulic conductivity are used to estimate the soil water evaporation coefficient and Brooks-Corey soil moisture retention parameters. The porosity, field capacity, and wilting point are all dimensionless numbers between 0 and 1. The porosity must be greater than the field capacity, which in turn must be greater than the wilting point. The wilting point must be greater than zero. The values for porosity, field capacity, and wilting point are not used for liners, except for initializing the soil water storage of liners to the porosity value (Schroeder et al., 1994a).

Landfill design requirements for the HELP model include such items as the slope and maximum drainage distance for lateral drainage layers, as well as layer thicknesses, layer description, area, leachate recirculation procedure, subsurface inflows, surface characteristics, and geomembrane characteristics.

7.5.3 Landfill Profile and Layer Descriptions

The HELP Version 3 program may be used to model landfills with up to 20 layers of materials—soils, geosynthetics, wastes or other materials (Schroeder et al., 1994b). Figure 7.9 shows a typical landfill profile with eleven layers. The top portion of the profile (Layers 1 through 4) is the cap or cover. The bottom portion of the landfill is a double liner system (Layers 6 through 11), in this case composed of a geomembrane and a GM/CCL composite liner. Immediately above the bottom composite liner is a secondary leachate collection layer (leakage detection drainage layer) to collect leakage from the primary liner, in this case a geomembrane. Above the primary liner is a geosynthetic drainage net and a sand layer that serve as drainage layers for leachate collection. The drain layers composed of sand are typically at least 1-ft thick and have suitably spaced perforated or open joint drain pipe embedded below the surface of the liner. The leachate collection drainage layer serves to collect any leachate that may percolate through the waste layers. In the event the liner is solely a geomembrane, a drainage net may be used to rapidly drain leachate from the liner, avoiding a significant buildup of head and limiting leakage. The liners are sloped to prevent ponding by encouraging leachate to flow toward the drains. The net effects are that very little leachate should leak through the primary liner and virtually no migration of leachate through the bottom composite liner to the natural formation below. Taken as a whole, the drainage layers, geomembrane and compacted clay liners may be referred to as a double liner system.

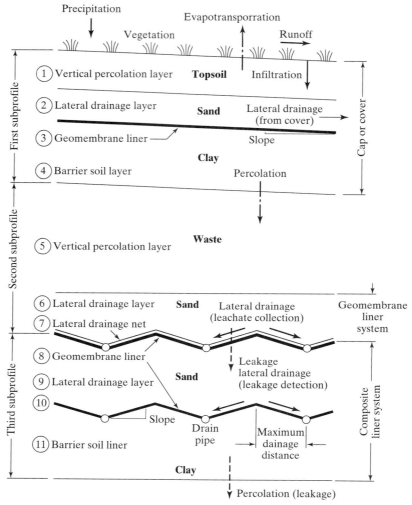

FIGURE 7.9 Typical Municipal Solid Waste Landfill Profile (Schroeder, et al., 1994a & b)

The layers shown in Figure 7.9 can be assigned as follows—4 in the cover or cap, 1 as the waste layer, 3 in the primary leachate collection and removal system, and 3 in the secondary leachate collection and removal system (leakage detection). These 11 layers also constitute three subprofiles or modeling units. A subprofile consists of all layers between (and including) the landfill surface and the bottom of the top liner system, between the bottom of one liner system and the bottom of the next lower liner system, or between the bottom of the lowest liner system and the bottom of the lowest soil layer modeled. In the sketch, the top subprofile contains the cover layers, the middle subprofile contains the waste, drain and liner system for leachate collection, and the bottom subprofile contains the drain and liner system for leakage detection. Six subprofiles in a single landfill profile may be simulated by the model.

Schroeder et al. (1994) point out that layers in a landfill can also be characterized or typed by the hydraulic function that they perform. They have identified four types of layers—vertical percolation layers, lateral drainage layers, barrier soil liners, and geomembrane liners. These layer types are illustrated in Figure 7.9. The topsoil and waste layers are generally vertical percolation layers. Sand layers above liners are typically lateral drainage layers; compacted clay layers are typically barrier soil liners. Geomembranes are typed as geomembrane liners. Composite liners are modeled as two layers. Geotextiles are not considered as layers unless they perform a unique hydraulic function.

Flow in a vertical percolation layer (e.g., Layers 1 and 5 in Figure 7.9) is either downward due to gravity drainage or extracted upwards by evapotranspiration. Unsaturated vertical drainage is assumed to occur by gravity drainage whenever the soil moisture is greater than the field capacity (greater than the wilting point for soils in the evaporative zone) or when the soil suction of the layer below the vertical percolation layer is greater than the soil suction in the vertical percolation layer. The rate of gravity drainage (percolation) in a vertical percolation layer is assumed to be a function of the soil moisture storage and largely independent of conditions in adjacent layers. The rate can be restricted when the layer below is saturated and drains more slowly than the vertical percolation layer. Layers, whose primary hydraulic function is to provide storage of moisture and detention of drainage, should normally be designated as vertical percolation layers. Waste layers and layers designed to support vegetation should be designated as vertical percolation layers, unless the layers provide lateral drainage to collection systems.

Lateral drainage layers (e.g., Layers 2, 6, 7, and 9 in Figure 7.9) are layers that promote lateral drainage to collection systems at or below the surface of liner systems. Vertical drainage in a lateral drainage layer is modeled in the same manner as for a vertical percolation layer, but saturated lateral drainage is allowed. The saturated hydraulic conductivity of a lateral drainage layer generally should be greater than 1.0×10^{-3} cm/sec for significant lateral drainage to occur. A lateral drainage layer may be underlain by only a liner or another lateral drainage layer. The slope of the bottom of the layer may vary from 0 to 40 percent.

Barrier soil liners (e.g., Layers 4 and 11 in Figure 7.9) are intended to restrict vertical flow. These layers should have hydraulic conductivities substantially lower than those of the other types of layers, typically below 1.0×10^{-6} cm/sec. The program allows only downward flow in barrier soil liners. Thus, any water moving into a liner will eventually percolate through it. The leakage (percolation) rate depends upon the depth of water-saturated soil (head above the base of the layer, the thickness of the liner and the saturated hydraulic conductivity of the barrier soil. Leakage occurs whenever the moisture content of the layer above the liner is greater than the field capacity of the layer. The program assumes that barrier soil liner is permanently saturated and that its properties do not change with time.

Geomembrane liners (e.g., Layers 3, 8, and 10 in Figure 7.9) are layers of nearly impermeable material that restrict significant leakage to small areas around defects. Leakage (percolation) is computed to result from three sources—vapor diffusion, manufacturing flaws (pinholes), and installation defects (punctures, cracks, tears, and bad seams). Leakage by vapor diffusion is computed to occur across the entire area of the liner as a function of the head on the surface of the liner, the thickness of the geomembrane, and its vapor diffusivity. Leakage through pinholes and installation defects is computed as a function of head on the liner, size of hole, and the saturated hydraulic conductivity of the soils or materials adjacent to the geomembrane liner. Furthermore, the rate of leakage in the wetted area is computed as a function of the head, thickness of soil and membrane, and the saturated hydraulic conductivity of the soils or materials adjacent to the geomembrane liner.

7.5.4 Modeling Procedure

The HELP program simulates daily liquid movement into, through, and out of a landfill. In general, the hydrologic processes modeled by the program can be divided into two categories: surface processes and subsurface processes (Schroeder et al., 1994). The surface processes modeled are snowmelt, interception of rainfall by vegetation, surface runoff, evaporation of water, and interception and snow from the surface. The subsurface processes modeled are evaporation of water from the soil, plant transpiration, vertical unsaturated drainage, geomembrane liner leakage, barrier soil liner percolation, and lateral saturated drainage. Vegetative growth and frozen soil models are also included in the program to aid modeling of the water routing processes.

Daily infiltration into the landfill is determined indirectly from a surface-water balance. Each day, infiltration is assumed to equal the sum of rainfall and snowmelt, minus the sum of runoff, surface storage, and surface evaporation. No liquid water is held in surface storage from one day to the next, except in the snow cover. The daily process of surface-water accounting is described next. Snowfall and rainfall are added to the surface snow storage, if present, and the snowmelt plus excess storage of rainfall is computed. The total outflow from the snow cover is then treated as rainfall in the absence of a snow cover for the purpose of computing runoff. A rainfall-runoff relationship is used to determine the runoff. Surface evaporation is then computed. Surface evaporation is not allowed to exceed the sum of surface snow storage and intercepted rainfall. Interception is computed only for rainfall, not for outflow from the snow cover. The snowmelt and rainfall that does not run off or evaporate is assumed to infiltrate into the landfill. Computed infiltration is excess of the storage and drainage capacity of the soil is routed back to the surface and is added to the runoff of held as surface storage.

The first subsurface processes considered are evaporation from the soil and plant transpiration from the evaporative zone of the upper subprofile. These are computed on a daily basis. The evapotranspiration demand is distributed among the seven modeling segments in the evaporative zone.

The other subsurface processes are modeled one subprofile at a time, from top to bottom, using a design-dependent time step, varying from 30 minutes to 6 hours. Unsaturated vertical drainage is computed for each modeling segment starting at the top of the subprofile, proceeding downward to the liner system or bottom of the subprofile. The program performs a water balance on each segment to determine the water storage and drainage for each segment, accounting for infiltration or drainage from above, subsurface inflow, leachate recirculation, moisture content and material characteristics. If the subprofile contains a liner, water-routing or drainage from the segment directly above the liner is computed as leakage or percolation through the liner, and lateral drainage to the collection system, if present. The sum of the lateral drainage and leakage/percolation is first estimated to compute the moisture storage and head on the liner. Using the head, the leakage and lateral drainage is computed and compared with their initial guesses. If the sum of these two outflows is not sufficiently close to the initial estimate, new estimates are generated and the procedure is repeated until acceptable convergence is achieved. The moisture storage in liner systems is assumed to be constant; therefore, any drainage into a liner results in an equal drainage out of the liner. If the subprofile does not contain a liner, the lateral drainage is zero and the vertical drainage from the bottom subprofile is computed in the same manner as the upper modeling segments.

7.5.5 Program Input

Version 3 of the HELP program is started by typing "HELP3" from the DOS prompt in the directory where the program resides (Schroeder et al., 1994a). The program starts by displaying a title screen, a preface, a disclaimer, and then the main menu. The user moves from the title screen to the main menu by striking any key, such as the space bar. Upon reaching the main menu, the user can select any of the following options:

1. Enter/Edit Weather Data,
2. Enter/Edit Soil and Design Data,
3. Execute Simulation,
4. View Results,
5. Print Results,
6. Display Guidance, and
7. Quit.

The program automatically solicits input from the user based on the option selected. In general, the HELP model requires the following data, some of which may be selected from the default values (Schroeder et al., 1994a):

1. Units,
2. Location,
3. Weather data file names,
4. Evapotranspiration information,
5. Precipitation data,
6. Temperature data,
7. Solar radiation data,
8. Soil and design data file name,
9. General landfill and site information,
10. Landfill profile and soil/waste/geomembrane data, and
11. SCS runoff curve number information.

An example of HELP model input for active condition is shown in Appendix I and an example of HELP model input for postclosure condition is shown in Appendix II.

7.5.6 Program Output

The HELP program always produces output consisting of the identifying labels and input data (except daily precipitation) supplied by the user, and a summary of the simulation results. Daily, monthly, and yearly output may be obtained at the option of the user.

7.5.6.1 Summary Output. Basic program output composed of all default and manual input information except daily precipitation data, and summary results are always reported. The information described herein was extracted directly from the User's Guide for the HELP Model Version 2 and Version 3 compiled by Schroeder et al. (1988a and 1994a). The reader should consult this guide for further details.

Following the input data summary, the program produces a table of the daily results, a table of the monthly totals, and a table of the annual totals for each year of simulation if these options are used. If a different set of climatologic data were used for each year of simulation, these input values other than precipitation data would be printed before the results of each year of simulation. After the results for all years are printed, the program produces a summary of the output. The summary includes average monthly totals, average annual totals, and peak daily values for various simulation variables.

The program reports average monthly totals for precipitation, runoff, evapotranspiration (total of evaporation from the surface and soil, and plant transpiration), percolation through the base of each subprofile, and lateral drainage from each subprofile. These results are reported in inches. The output values indicate averages of the monthly totals for a particular month of all years of simulation. For example, if five years of simulation were run, the reported average monthly precipitation total for March would be the average of the five monthly totals for March precipitation.

The next table in the summary output is a listing of the average annual totals for the simulation period. Average annual values for precipitation, runoff, evapotranspiration, percolation through the base of each subprofile, and lateral drainage from each subprofile are reported in terms of inches, cubic feet, and as a percent of the total average annual precipitation.

In the last summary table, peak daily values for precipitation, runoff, percolation through the base of each subprofile, and precipitation accumulation on the surface in the form of snow are reported in terms of inches and cubic feet. These values represent the maximum values that occurred on any day during the simulation period. The table also contains the maximum head on the barrier soil layer of each subprofile and the maximum and minimum soil moisture content of the evaporative zone. These variables are reported in inches.

The daily values reported in these summary tables are sufficient for rapidly screening alternative designs and roughly sizing drainage and leachate collection and treatment systems in most cases. However, more detailed information showing trends and variability in the results may be obtained by requesting annual, monthly or daily output (Schroeder et al., 1988a).

7.5.6.2 Annual, Monthly and Daily Output. The Users Manual (Schroeder et al. (1988a) also explains how the program can be interrogated for detailed output. The program will print, for example, annual totals of precipitation, runoff, evapotranspiration, percolation through the base of each subprofile, and lateral drainage from each subprofile for each year of simulation. Each of these output variables are reported in terms of inches, cubic feet and as a percent of the total annual precipitation. The program also lists the soil moisture contents and snow accumulations at the start and end of the year in inches and cubic feet.

If the user requests monthly output, the program produces tables that report monthly totals in inches for precipitation, runoff, evapotranspiration, percolation through the base of each subprofile, and lateral drainage from each subprofile for each year of simulation.

If daily output is requested, the program prints a table containing the Julian data, and the daily values of precipitation, runoff, evapotranspiration, head on the soil barrier layer at the base of the cover, percolation through the base of the cover, total lateral drainage from all subprofiles in the cover, head on the soil barrier layer at the base of the landfill, percolation through the base of the landfill, total lateral drainage from all subprofiles below the cover, and the soil moisture content of the evaporative zone. Where applicable, the units of the variables are in inches, except for the soil moisture content that is reported in dimensionless form (volume of water/volume of soil). The program prints an asterisk after the Julian data for dates when the mean temperature is below freezing (32°F or 0°C).

An example of HELP model output for active condition is shown in Appendix I, and an example of HELP model output for postclosure condition is shown in Appendix II.

7.5.7 Limits of Application

The HELP model can simulate water routing through, or storage in, up to 20 layers of soil, waste, geosynthetics or other materials. As many as five liner systems, either barrier soil, geomembrane or composite liners, can be used. The simulation period can range from 1 to 100 years. The model cannot simulate a capillary break or unsaturated lateral drainage (Schroeder et al., 1994a).

Flow across a geomembrane is very subjective in that the number and size of holes (called *defects* in the program) must be estimated. Since the designer has no idea as to the qualification of the installer, or the degrees and sophistication of the inspection at this point in the process, the program's default values are often used. These default values were estimated based on geomembrane installation practice in the late 1980s to early 1990s. Recent data, as shown in Figure 5.28, suggests that the default values may be quite overestimated.

The HELP model has limits on the order that layers can be arranged in the landfill profile. Each layer must be described as being one of four operational types: vertical percolation layer, lateral drainage layer, barrier soil liner, or geomembrane liner. The model does not permit a vertical percolation layer to be placed directly below a lateral drainage layer. A barrier soil liner may not be placed directly below another barrier soil liner. A geomembrane liner may not be placed directly below another geomembrane liner. Three or more liners—barrier soil or geomembrane—cannot be placed adjacent to each other. The top layer may not be a barrier soil or geomembrane liner.

The HELP program performs water balance analysis for a minimum period of one year. All simulations start on January 1 and end on December 31. Surface run-on from adjacent areas is not permitted. The condition of the landfill, soil properties, thicknesses, geomembrane hole density, maximum level of vegetation, etc., are assumed to be constant throughout the simulation period. The program cannot simulate the actual filling operation of an active landfill. Active landfills are modeled a year at a time, adding a yearly lift of material and updating the initial moisture of each layer for each year of simulation.

A limitation of the HELP program for design of drainage layers in cover soils is that the minimum time step is one day—24 hours is simply too long a time period to assess an intense, short duration storm and its potential infiltration into a relatively thin cover soil. The situation can be serious if the cover soil has a high hydraulic conductivity. The minimum time step in such cases should be one hour. See Koerner and Soong (1998) for case history failures, and Koerner and Daniel (1997) for a spread sheet calculation procedure.

PROBLEMS

7.1 What is landfill leachate? What is the difference between primary and secondary leachate?

7.2 List the factors affecting the composition (quality) of leachate.

7.3 List the factors affecting the generation rate of leachate.

7.4 List the factors affecting both the composition (quality) and amount (quantity) of leachate.

7.5 Explain what factors tend to increase, and what factors tend to decrease, leachate quantity and their relative importance during the following landfill stages:
1. Active filling period, and
2. Postclosure period.

7.6 What are the differences in the rate of leachate generation, and what mainly controls this rate during active and postclosure conditions, respectively?

7.7 What is the difference between evaporation and evapotranspiration? Explain under what condition one should be used as opposed to the other.

7.8 What affects or controls the amount of evapotranspiration?

7.9 What are the definitions or meanings of "field capacity" and "wilting point" for a soil?

7.10 Based on their hydraulic function, how many "types of layers" are identified in the HELP model? What hydraulic function or purpose does each of these layers serve?

7.11 Use the HELP model to evaluate a primary leachate collection system design for Cells 1 and 2. Check whether the design restricts the leachate head development on the primary liner to no more than 12 inches (300 mm)—the maximum allowable. If not, revise the design so that it meets the requirements (note: don't change the liner system materials). Use a five-year precipitation period and landfill locations at Providence, Rhode Island and Olympia, Washington, respectively, to check the design. All information needed for running the HELP model is listed in Tables 7.7 and 7.8. Use synthetically generated rainfall data, bare ground, and default soil characteristics.

7.12 Describe limitations in the use of the HELP model. Can the HELP model simulate actual day-by-day filling operation of an active landfill? Why?

TABLE 7.7 Liner System Information for Problem 7.11

Layer Description	Layer Thickness (inches)		Hydraulic Conductivity
	inch	mm	
Solid Waste	120	3,600	
Drainage Sand	24	600	$k = 1.0 \times 10^{-2}$ cm/sec
Clay with FML	36	900	
Drainage Sand	12	300	$k = 1.0 \times 10^{-2}$ cm/sec
Clay with FML	36	900	

TABLE 7.8 Landfill Cell Information for Problem 7.11

Cell Dimension	Cell 1	Cell 2
Maximum Drainage Distance	100 ft (30 m)	300 ft (90 m)
Width of Landfill Cell	1,000 ft (300 m)	1,000 (300 m)
Slope at the Base (%)	2	25
Total Area of Cell Surface	100,000 ft^2 (9,000 m^2)	300,000 ft^2 (27,000 m^2)

REFERENCES

Anderson, E., (1973) "National Weather Service River Forecast System—Snow Accumulation and Ablation Model," Hydrologic Research Laboratory, National Oceanic and Atmospheric Administration, Silver Spring, MD.

Arnold, J. G., Williams, J. R., Nicks, A. D., and Sammons, N. B., (1989) "SWRRB, A Simulator for Water Resources in Rural Basins," Agricultural Research Service, USDA, Texas A&M University Press, College Station, TX.

Bagchi, A., (1994) "Design, Construction, and Monitoring of Sanitary Landfill," John Wiley & Sons, Inc., 605 Third Avenue, New York, NY 10158-0012.

Bonaparte, R., (1995) "Long-Term Performance of Landfills," *Proceedings of GeoEnvironment 2000,* Geotechnical Special Publication No. 46, ASCE, New Orleans, LA, February 24–26, pp. 514–553.

Breazeale, E. and McGeorge, W. T., (1949) "A New Technique for Determining Wilting Percentage of Soil," *Soil Science* 68, pp. 371–374.

Canziani and Cossu, (1989) "Landfill Hydrology and Leachate Production," *Sanitary Landfilling: Process, Technology and Environmental Impact,* Edited by T. H. Christensen, R. Cossu, and R. Stegmann, Academic Press Limited, London, UK, pp. 185–212.

Charlie, W. A. and Wardwell, R. E., (1979) "Leachate Generation from Sludge Disposal Area," *Journal of Environmental Engineering,* ASCE, Vol. 105, No. 5, pp. 947–960.

Chow, V. T., (1964) *Handbook of Applied Hydrology,* McGraw-Hill, New York.

Ehrig, H. J., (1989) "Leachate Quality," *Sanitary Landfilling: Process, Technology and Environmental Impact,* Edited by T. H. Christensen, R. Cossu, and R. Stegmann, Academic Press Limited, London, UK, pp. 213–229.

England, C. B., (1970) "Land Capability: A Hydrologic Response Unit in Agricultural Watersheds," ARS 41-172, USDA Agricultural Research Service, Washington, DC.

Fenn, D. G., Hanley, K. J., and DeGeare, T. V., (1975) "Use of the Water Balance for Predicting Leachate Concentration from Solid Waste Disposal Sites," EPA/530-SW-168, U.S. Environmental Protection Agency, Washington, DC.

Giroud, J. P. and Badu-Tweneboah, K., (1992) "Rate of Leakage through a Composite Liner due to Geomembrane Defects," *Geotextiles and Geomembranes,* Vol. 11(1), Elsevier Science Publishers Ltd., Oxford, UK, pp. 1–28.

Giroud, J. P., Khatami, A., and Badu-Tweneboah, K., (1989) "Evaluation of Leakage through Composite Liners," *Geotextiles and Geomembranes,* Vol. 8(4), Elsevier Science Publishers Ltd., Oxford, England, UK, pp. 337–340.

Knisel, W. G., ed., (1980) "CREAMS, A Field Scale Model for Chemical Runoff and Erosion from Agricultural Management Systems, Volumes I, II, and III," Conservation Report 26, USDA-SEA, Washington DC.

Koerner, R. M. and Daniel, D. E., (1997) "Final Covers for Solid Waste Landfills and Abandoned Dumps," ASCE Press, Reston, VA, and Thomas Telford, London, UK.

Koerner, R. M. and Soong, T-. Y., (1998) "Analysis and Design of Veneer Cover Soils," Proceedings 6th Intl. Geosynthetics Conference, Atlanta, GA, IFAI, pp. 1–23.

LandTech, (1994) "Landfill Gas System Engineering—Practical Approach," Landfill Gas System Engineering Design Seminar, Landfill Control Technologies, Commerce, CA.

Lu, J. C. S., Morrison, R. D., and Stearns, R. J., (1984) "Production and Management of Leachate from Municipal Landfills: Summary and Assessments," U.S. Environmental Protection Agency, EPA-600/2-84-092, May, Cincinnati, OH

Lu, J. C. S., Eichenberger, B., and Stearns, R. J., (1985) "Leachate from Municipal Landfills, Production and Management," Noyes Publ., Park Ridge, NJ, pp. 109–121.

Lutton, R. J., Regan, G. L., and Jones, L. W., (1979) "Design and Construction of Covers for Solid Waste Landfills," EPA-600/2-79-165, U. S. Environmental Protection Agency, Cincinnati, OH, August.

McBean, E. A., Rovers, F. A., and Farquhar, G. J., (1995) "Solid Waste Landfill Engineering and Design," Prentice Hall PTR, Englewood Cliffs, NJ.

McGinley, P. M. and Kmet P., (1984) "Formation, Characteristics, Treatment and Disposal of Leachate from Municipal Solid Waste Landfills," Wisconsin Department of Natural Resources, Special Report, August, Madison, WI.

National Oceanic and Atmospheric Administration, (1974) "Climatic Atlas of the United States," U.S. Department of Commerce, Environmental Science Services Administration, Nation Climatic Center, Ashville, NC.

Oweis, I. S. and Khera, R. P., (1998) "Geotechnology of Waste Management," 2nd Edition, PWS Publishing Company, 20 Park Plaza, Boston, MA 02116-4324.

Perrier, E. R. and Gibson, A. C., (1980) "Hydrologic Simulation on Solid Waste Disposal Sites," Technical Resource Document, EPA-SW-868, U.S. Environmental Protection Agency, Cincinnati, OH.

Prosser, R. and Janechek, A., (1995) "Landfill Gas and Groundwater Contamination," *Landfill Closures—Environmental Protection and Land Recovery,* ASCE, Geotechnical Special Publication, No. 53, R. Jeffrey Dunn and Udai P. Singh, Eds. New York, NY, pp. 258–271.

Qasim, S. R. and Burchinal, J. C., (1970) "Leaching from Simulated Landfills," *Journal of Water Pollution Control Federation,* Vol. 42, pp. 371–379.

Qasim, S. R. and Chiang, W., (1994) "Sanitary Landfill Leachate," Technomic Publishing Company, Inc., 851 New Holland Avenue, Box 3535, Lancaster, PA, 17604.

Reinhart, D. R. and Townsend, T. G., (1998) "Landfill Bioreactor Design and Operation," CRC Press LLC, 2000 Corporate Blvd., N.W., Boca Raton, FL, 33431.

Richardson, C. W. and Wright, D. A., (1984) "WGEN: A model for Generating Daily Weather Variables," ARS-8, Agricultural Research Service, USDA, Washington, DC.

Ritchie, J. T., (1972) "A Model for Predicting Evaporation from a Row Crop with Incomplete Cover," Water Resources Research, Vol. 8, No. 5, pp. 1204–1213.

Ruffner, J. A., (1985) "Climates of the States," National Oceanic and Atmospheric Administration Narrative Summaries, Tables and Maps for Each State, Volume 1: Alabama–New Mexico, and Volume 2: New York - Wyoming and Territories, Gale Research Company, Detroit, MI.

Schroeder, P. R. and Gibson, A. C., (1982) "Supporting Documentation for the Hydrologic Simulation Model for Estimating Percolation at Solid Waste Disposal Sites (HSSWDS)," Draft Report, U.S. Environmental Protection Agency, Cincinnati, OH.

Schroeder, P. R., Morgan, J. M., Walski, T. M., and Gibson, A. C., (1984a) "The Hydrologic Evaluation of Landfill Performance (HELP) Model, Volume I, User's Guide for Version I," Technical Resource Document, EPA/530-SW-84-009, U.S. Environmental Protection Agency, Cincinnati, OH, June.

Schroeder, P. R., Gibson, A. C., and Smolen, M. D., (1984b) "The Hydrologic Evaluation of Landfill Performance (HELP) Model," Vol. II, *Documentation for Version I,* Technical Resource Document, EPA/530-SW-84-010, U.S. Environmental Protection Agency, Cincinnati, OH, June.

Schroeder, P. R. and Peyton, R. L., (1987a) "Verification of the Hydrologic Evaluation of Landfill Performance (HELP) Model Using Field Data," Technical Resource Document, EPA 600/2-87-050, U.S. Environmental Protection Agency, Cincinnati, OH.

Schroeder, P. R. and Peyton, R. L., (1987b) "Verification of the Lateral Drainage Component of the HELP Model Using Physical Models," Technical Resource Document, EPA 600/2-87-049, U.S. Environmental Protection Agency, Cincinnati, OH.

Schroeder, P. R., Peyton, R. L., McEnore, B. M., and Sjostrom, J. W., (1988a) "The Hydrologic Evaluation of Landfill Performance (HELP) Model: Volume III, User's Guide for Version 2," Internal Working Document EL-92-1, Report 1, U.S. Army Engineer Waterways Experiment Station, Vicksburg, MS.

Schroeder, P. R., McEnore, B. M., Peyton, R. L., and Sjostrom, J. W., (1988b) "The Hydrologic Evaluation of Landfill Performance (HELP) Model: Volume IV, Document for Version 2," Internal Working Document EL-92-1, Report 2, U.S. Army Engineer Waterways Experiment Station, Vicksburg, MS.

Schroeder, P. R., Lloyd, C. M., Zappi, P. A., and Aziz, N. M., (1994a) "The Hydrological Evaluation of Landfill Performance (HELP) Model, User's Guide for Version 3," EPA/600/R-94/168a, Risk Reduction Engineering Laboratory, Office of Research and Development, U.S. Environmental Protection Agency, Cincinnati, OH, September.

Schroeder, P. R., Dozier, T. S., Zappi, P. A., McEnore, B. M., Sjostrom, J. W., and Peyton, R. L., (1994b) "The Hydrological Evaluation of Landfill Performance (HELP) Model, Engineering Documentation for Version 3," EPA/600/R-94/168b, Risk Reduction Engineering Laboratory, Office of Research and Development, U.S. Environmental Protection Agency, Cincinnati, OH, September.

SCS (Soil Conservation Service), (1972) "Procedure for Computing Sheet and Rill Erosion on Project Areas," Release No. 51, U.S. Department of Agriculture, Engineering Division, Washington, DC.

SCS (Soil Conservation Service), (1975) "Urban Hydrology for Small Watersheds," Technical Release No. 55, U. S. Department of Agriculture, Engineering Division, Washington, DC.

Straub, N. A. and Lynch, D. R., (1982) "Models of Landfill Leaching: Organic Strength," *Journal of Environmental Engineering,* ASCE, Vol. 108, April, No. 2, pp. 251–268.

Thornthwaite, C. W. and Mather, J. R., (1955) "The Water Balance," Publications in Climatology, Drexel Institute of Technology, Vol. 8, No. 1, Centerton, NJ.

Thornthwaite, C. W. and Mather, J. R., (1957) "Instructions and Tables for Computing Potential Evapotranspiration and the Water Balance," Publications in Climatology, Drexel Institute of Technology, Centerton, NJ. Vol. 10, No. 3, pp.185–311.

USEPA, (1988) "Summary of Data on Municipal Solid Waste Landfill Characteristics—Criteria for municipal Solid Waste Landfills (40CFR Part 258)," OSWER, U.S. Environmental Protection Agency, EPA/530-SW-88-038, Washington. DC.

USDA, Soil Conservation Service, (1985) *National Engineering Handbook,* Section 4, "Hydrology," U.S. Government Printing Office, Washington, DC.

Veihmeyer, F. J. and Henderickson, A. H., (1955) "Rate of Evaporation from Wet and Dry Soils and Their Significance," *Soil Science,* Vol. 80, pp. 61–67.

Viessman, W., Jr., Knapp, J. W., Lewis, G. L., and Harbaugh, T. E., (1977) *Introduction to Hydrology,* Harper & Row, New York, NY, pp. 43–87.

Zornberg, J. G., Jernigan, B. L., Sanglerat, T. R., and Cooley, B. H., (1999) "Retention of Free Liquids in Landfills Undergoing Vertical Expansion," *Journal of Geotechnical and Geoenvironmental Engineering,* ASCE, Vol. 125, No. 7, pp. 583–594.

CHAPTER 8

Liquid Drainage Layer

8.1 PROFILE OF LEACHATE DRAINAGE LAYER
8.2 SOIL DRAINAGE AND FILTRATION LAYER
8.3 GEOTEXTILE DESIGN FOR FILTRATION
    8.3.1 GEOTEXTILE OVERVIEW
    8.3.2 ALLOWABLE VERSUS ULTIMATE GEOTEXTILE PROPERTIES
    8.3.3 CROSS-PLANE PERMEABILITY
    8.3.4 SOIL RETENTION
    8.3.5 LONG TERM COMPATIBILITY
8.4 GEONET DESIGN FOR LEACHATE DRAINAGE
    8.4.1 GEONET OVERVIEW
    8.4.2 HYDRAULIC PROPERTIES OF GEONETS
    8.4.3 ALLOWABLE GEONET FLOW RATE
    8.4.4 DESIGNING WITH GEONETS FOR DRAINAGE
8.5 ESTIMATION OF MAXIMUM LEACHATE HEAD IN THE DRAINAGE LAYER
    8.5.1 METHODS FOR ESTIMATING MAXIMUM LIQUID HEAD
    8.5.2 COMPARISON OF VARIOUS CALCULATION METHODS
    PROBLEMS
    REFERENCES

All landfills must have at their base a leachate drainage layer consisting of natural soils (sands or gravels) and/or a geosynthetic drainage material. In addition, municipal solid waste landfills with a double-composite liner system are required to have both primary and secondary leachate drainage layers (i.e., leachate collection and leak detection layers). The most essential requirement for a landfill leachate drainage layer is that it has adequate drainage capacity to handle the maximum leachate flow produced during landfill operations. The leachate head build-up in the drainage layer must be less than 12 inches (0.3 m), based on requirements of Federal and state regulations in the United States and many other countries as well.

    The purpose of this chapter is to present various drainage materials, both natural soils and geosynthetics, in order to establish a foundation and lead-in to the next chapter on the complete leachate collection and removal system. In addition to discussing drainage materials, this chapter will also address filtration.

8.1 PROFILE OF LEACHATE DRAINAGE LAYER

A liner system consists of a combination of one or more leachate drainage layers and low-permeability barriers (i.e., liners). The liners and drainage layers complement one another. The liner impedes the migration of leachate and gas out of the landfill and improves the performance of any associated drainage layer. The drainage layer limits the buildup of hydraulic head on the underlying liner and conveys liquid that percolates into the drainage layer to a network of perforated leachate collection pipes. In the case of landfill covers, drainage layers above the liner serve a similar function while drainage layers beneath the liner system limit the buildup of gas pressure beneath the overlying liner and conveys the gas to a venting system.

A double composite liner system, widely used at present in municipal solid waste landfills, has both primary and secondary leachate drainage layers (i.e., leachate collection and leak detection layers). Figures 8.1 to 8.6 illustrate the components of various types of double composite liner system with both primary and secondary leachate drainage layers. From top to bottom, these components include the following:

Protective Layer. Consists of a minimum 2-foot (0.6-m) sand, gravel, or other permeable material (Figures 8.3 and 8.4). Its function is to prevent solid waste from damaging and clogging the leachate drainage layer, yet not impede leachate flow.

Primary Leachate Drainage Layer. Consists of a permeable soil (i.e., sand or gravel, shown in Figures 8.1, 8.2, 8.5, and 8.6) or geosynthetic drainage material (i.e., geotextile bonded to one or both surfaces of a geonet, called a geocomposite, shown in Figures 8.3 and 8.4). Its function is to limit the buildup of leachate head on the primary liner and to collect and convey leachate from the solid waste to a perforated leachate collection pipe.

Primary Composite Liner. Consists of a geomembrane atop a low-permeability soil. The low-permeability soil component may be constructed of compacted clay or a geosynthetic clay liner (GCL). The geomembrane and soil components of a composite liner are designed to maintain "intimate" contact with one another.

Secondary Leachate Drainage Layer. Consists of a permeable soil (i.e., sand or gravel, shown in Figures 8.1 and 8.2) or geosynthetic drainage material (i.e., geotextile and geonet or geocomposite, shown in Figures 8.3, 8.4, 8.5, and 8.6). Its function is to collect and remove the leachate that has passed through the primary liner system. It is often called the *leak detection* layer.

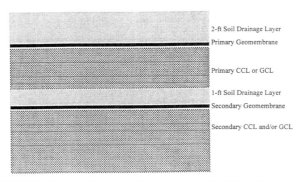

FIGURE 8.1 Double Composite Liner System with Soil as Primary and Secondary Leachate Drainage Layers

Section 8.1 Profile of Leachate Drainage Layer

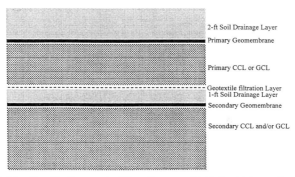

FIGURE 8.2 Double Composite Liner System with Soil as Primary Leachate Drainage Layer and Geotextile and Soil as Secondary Leachate Drainage Layer

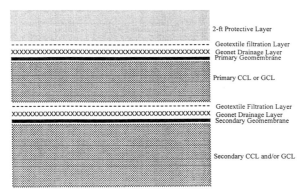

FIGURE 8.3 Double Composite Liner System with Geotextile and Geonet as Primary and Secondary Leachate Drainage Layers for Flat Slopes

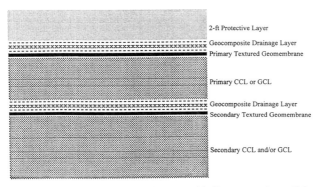

FIGURE 8.4 Double Composite Liner System with Geocomposite as Primary and Secondary Leachate Drainage Layers for Steep Slopes

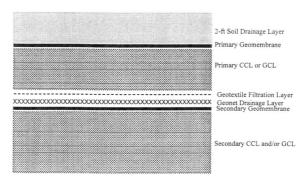

FIGURE 8.5 Double Composite Liner System with Soil as Primary Leachate Drainage Layer and Geotextile and Geonet as Secondary Leachate Drainage Layers for Flat Slopes

Secondary Composite Liner. Consists of a geomembrane atop a low-permeability soil. The low-permeability soil component may be constructed of natural clay soil, compacted clay liner (CCL) and/or geosynthetic clay liner (GCL). The geomembrane and soil components of a composite liner are designed to maintain "intimate" contact with one another.

In earlier times, sand and other granular materials were usually used for the leachate drainage layer. A two-foot- (0.6-meter)-thick sand was used as the primary leachate drainage layer and one-foot- (0.3-meter)-thick sand as the secondary leachate drainage layer (Figure 8.1). The hydraulic conductivity of the sand or other granular materials that are used as the leachate drainage layer must be greater than 1.0×10^{-2} cm/sec. The sand which is used as leachate drainage layer should be free of organic material, and should have less than 5% of the material, by weight, pass the #200 sieve and 100% pass the 3/8-inch (9.5-mm) sieve.

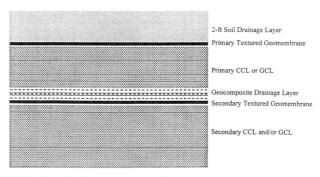

FIGURE 8.6 Double Composite Liner System with Soil as Primary Leachate Drainage Layer and Geocomposite as Secondary Leachate Drainage Layer for Steep Slopes

In order to prevent clay particles from the primary compacted clay liner from extruding into the secondary leachate drainage layer and thereby reducing the hydraulic conductivity of the sand, a layer of geotextile can be placed between the primary clay liner and the secondary sand drainage layer (Figure 8.2).

More recently, geotextiles and geonets have been widely used as landfill leachate drainage materials (Figures 8.3 to 8.6). The hydraulic transmissivity of a geonet is much greater than that of sand. Thus, a very thin geonet can be used in lieu of several feet of sand as the leachate drainage layer to reduce the total thickness of liner system so as to increase the waste filling space. When a geotextile and geonet are used as the primary leachate drainage layer, two-foot- (0.6-meter)-thick sand layer must be placed over it as a protective blanket. The hydraulic conductivity of the sand protective layer must be greater than 1.0×10^{-4} cm/sec.

It is important in landfill design to prevent a liner system from sliding when placed on steep side slopes. Interface friction angles between geomembranes and geonets are very low, which can be a problem when a geonet is used as the leachate drainage system on a steep side slope. In order to increase the interface friction angle between the geomembrane and geonet drainage layer, a geocomposite drainage material is usually used (Figure 8.4). A drainage geocomposite consists of a HDPE geonet sandwiched between one or two layers of nonwoven geotextiles. The geotextiles are heat bonded to the geonet in the factory to form a strong, coherent, multilayer product. A geocomposite has almost the same hydraulic properties as a geonet, but the interface friction angle between the geomembrane and geocomposite is higher than that between the geomembrane and geonet. This greatly improves the liner stability conditions on a side slope. This is particularly the case when the geomembrane is textured on the surface facing the geotextile. This produces what is often called a Velcro effect with very high friction behavior.

It should be noted that there are other types of geocomposite drainage materials than geonet-related drainage products. Drainage cores consisting of polymer columns, nubs, cuspations and 3-D fibrous networks are also available. In the United States, however, either biplanar or triplanar geonet drainage cores are used almost exclusively.

Figures 8.5 and 8.6 show two other configurations of leachate drainage layers, which have been used widely in landfill design in recent years. In these two examples, a two-foot- (0.6-meter)-thick sand is used as both the primary leachate drainage and a protective layer. The hydraulic conductivity of the sand must be greater than 1.0×10^{-2} cm/sec. Geotextile and geonet or geocomposites are used as the secondary leachate drainage layer. The cross section shown in Figure 8.5 is suitable for use on flat areas in a landfill cell. The cross section shown in Figure 8.6 is suitable for use on side slope areas particularly when using textured geomembrane.

8.2 SOIL DRAINAGE AND FILTRATION LAYER

Natural soils (sands and gravels) are used extensively in landfill units. The following are the most common uses, according to the U.S. Environmental Protection Agency(1993):

(i) as a leachate collection layer in liner systems to remove leachate for treatment and to remove precipitation from the landfill or cell in areas where waste has not yet been placed;

(ii) as a leak detection layer in a double composite liner system to monitor performance of the primary liner and to serve as a secondary leachate collection layer;
(iii) as a gas collection layer in a final cover system to channel gas to vents for controlled removal of potentially dangerous landfill gases;
(iv) as a drainage layer in a final cover system to reduce the hydraulic head on the underlying barrier layer and to enhance slope stability by reducing seepage forces in the cover system; and
(v) as a drainage trench to collect horizontally flowing fluids (e.g., groundwater and gas).

Soil drainage layers are constructed of materials that have high hydraulic conductivity and good internal stability. The drainage material must also maintain a high hydraulic conductivity over time and resist plugging or clogging. The hydraulic conductivity of drainage materials depends primarily on the grain size of the finest particles present in the soil. Hazen's formula is an equation that is used occasionally to estimate hydraulic conductivity of granular materials; it is written as

$$k = C \cdot (D_{10})^2 \tag{8.1}$$

where k = hydraulic conductivity, cm/sec;
D_{10} = equivalent grain diameter at which 10% of the soil is finer by weight, mm; and
C = constant depending on tortuosity and particle shape.

Laboratory tests have shown that the percentage of fines in a soil controls its hydraulic conductivity. For example, the data in Table 8.1 illustrates the influence of a small amount of fines upon the hydraulic conductivity of a filter sand. The addition of just a few percent of fine material to a drainage material can reduce the hydraulic conductivity of the drainage material by 100 times or more. Design and construction specifications usually stipulate a minimum hydraulic conductivity for the drainage layer. The minimum value specified varies considerably. A recent survey of regulations in the 50 U.S. states shows the following results (Koerner et al, 1998):

10 states: $k \geq 1 \times 10^{-2}$ cm/sec   1 state: $k \geq 1 \times 10^{-4}$ cm/sec
14 states: $k \geq 1 \times 10^{-3}$ cm/sec   25 states: k is site specific

TABLE 8.1 Effect of Fines on Hydraulic Conductivity of a Washed Filter Aggregate (Cedergren, 1989)

Percent Passing No.100 Sieve	Hydraulic Conductivity	
	(cm/sec)	(ft/day)
0	3.0×10^{-2} to 1.1×10^{-1}	80 to 300
2	4.0×10^{-3} to 4.0×10^{-2}	10 to 100
4	7.0×10^{-4} to 2.0×10^{-2}	2 to 50
6	2.0×10^{-4} to 7.0×10^{-3}	0.5 to 20
7	7.0×10^{-5} to 1.0×10^{-3}	0.2 to 3

Note: Opening size of No.100 sieve is 0.15 mm.

Two comments are offered regarding this information. First, the values—particularly those less than 0.01 cm/sec—are very low. Second, regulations never state that the minimum hydraulic conductivity must be at the end of service life, only at the beginning, although regulations are invariably interpreted as being the initial hydraulic conductivity. Clearly, such low values will not function if the landfill is being considered as a bioreactor system. (This issue will be further discussed in Chapter 15.)

Drainage materials may also be required to serve as filters. For instance, as shown in Figure 8.7, a filter layer may be needed to protect a drainage layer from plugging. This requirement may be critical if coarse drainage materials, (i.e., gravels with $k \gg 1 \times 10^{-2}$ cm/sec) are being used. The filter layer must serve three functions (USEPA, 1993):

(i) The filter must prevent passage of significant amounts of soil through the filter (i.e., the filter must retain soil).

(ii) The filter must have a relatively high hydraulic conductivity (e.g., the filter should be more permeable than the material above it).

(iii) The soil particles within the filter must not migrate significantly into the adjacent drainage layer.

Filter specifications vary somewhat, but the design procedures are similar. Requirements for a filter material are determined as follows (USEPA, 1993):

(i) The grain size distribution curve of the soil to be retained (protected) is determined following procedures outlined in particle size analysis of soil (ASTM D422). The size of the protected soil at which 15% is finer $[(D_{15})_{soil}]$ and 85% is finer $[(D_{85})_{soil}]$ is determined.

(ii) Experience shows that the particles of the protected soil will not significantly penetrate into the filter if the size of the filter at which 15% is finer $[(D_{15})_{filter}]$ is less than 4 to 5 times D_{85} that of the protected soil; that is,

$$(D_{15})_{filter} \leq (4 \text{ to } 5)(D_{85})_{soil} \tag{8.2}$$

(iii) Experience also shows that the hydraulic conductivity of the filter will be significantly greater than that of the protected soil if the following criterion is satisfied:

$$(D_{15})_{filter} \geq 4 \cdot (D_{15})_{soil} \tag{8.3}$$

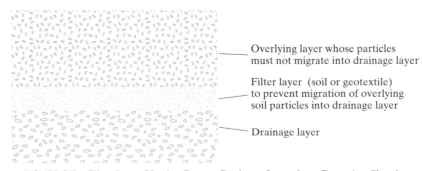

FIGURE 8.7 Filter Layer Used to Protect Drainage Layer from Excessive Clogging

(iv) To ensure that the particles within the filter do not tend to migrate excessively into the drainage layer, the following criterion may be applied:

$$(D_{15})_{drain} \leq (4 \text{ to } 5)(D_{15})_{filter} \tag{8.4}$$

(v) The hydraulic conductivity of the drain will be significantly greater than that of the filter if the following criterion is satisfied:

$$(D_{15})_{drain} \geq 4 \cdot (D_{15})_{filter} \tag{8.5}$$

Filter design is complicated significantly by the presence of biodegradable waste materials (e.g., municipal solid waste) being placed directly on top of the filter. In such circumstances, the usual filter criteria may be modified to satisfy site-specific requirements. Some degree of reduction in hydraulic conductivity of the filter layer may be acceptable, as long as the reduction does not impair the ability of the drainage system to serve its intended function. A laboratory test method ASTM D1987 can be used to quantify the hydraulic properties of both soil and geotextile filters that are exposed to leachate. However, regardless of specific design criteria, the gradational characteristics of the filter material control the behavior of the filter.

8.3 GEOTEXTILE DESIGN FOR FILTRATION

Geotextiles, which some refer to as filter fabrics or construction fabrics, consist of polymeric yarns (fibers) made into woven or nonwoven textile sheets and supplied to the job site in large rolls. When ready for placement, the rolls are removed from their protective covering, properly positioned and unrolled over the substrate materials. The substrate upon which the geotextile is placed is usually a geonet, geocomposite, drainage soil or other soil material. The roll edges and ends are either overlapped for a specified distance, or are sewn together. The geotextile is then covered with the overlying material. Depending on site-specific conditions, this overlying material can be a geomembrane, geosynthetic clay liner, compacted clay liner, geonet, or drainage soil.

8.3.1 Geotextile Overview

The primary functions of geotextiles are separation, reinforcement, filtration, and drainage. Within these functions, however, there are a large number of applications or use areas. The main use areas of geotextiles include separation of dissimilar materials, reinforcement of weak soils and other materials, filtration (cross-plane flow), and drainage (in-plane flow).

In landfill drainage situations, geotextiles are used as filters over various drainage materials, such as soil materials, geonets, geonet/geotextile geocomposites, leachate collection trenches, leachate collection pipes, leachate collection sumps, gas extraction wells, and so on.

Nonwoven needle punched geotextiles are commonly used in landfill projects as a filtration layer. Typical physical properties of two nonwoven needle punched geotextiles that are widely used in landfills are listed in Tables 8.2 and 8.3.

Depending on the site-specific situation, open woven monofilament geotextiles have also been used. Table 8.4 presents various styles of a woven, monofilament, polypropylene geotextile.

TABLE 8.2 Typical Properties of a Nonwoven, Needle-Punched Polyester Geotextile

Property	Style (a)	Style (b)	Style (c)	Style (d)	Style (e)	Style (f)
Fabric Weight (oz/yd^2)	4.2	6.0	7.5	10.5	13.5	16.5
(g/m^2)	143	204	255	356	458	560
Thickness (mils)	70	90	110	140	170	210
(mm)	1.75	2.25	2.75	3.5	4.25	5.25
Permeability (cm/sec)	0.45	0.52	0.56	0.57	0.58	0.57
Permittivity (sec^{-1})	2.54	2.27	2.01	1.60	1.34	1.07
AOS (sieve size)	70–100	70–100	70–100	100–120	120–140	140–170
O_{95} (mm)	.210–.149	.210–.149	.210–.149	.149–.125	.125–.106	.106–.088
Porosity (%)	92.8	93.0	93.0	92.7	92.0	92.1
Roll Width (feet)	12.5 & 15	12.5 & 15	12.5 & 15	12.5 & 15	12.5 & 15	12.5 & 15
(m)	3.8 & 4.6	3.8 & 4.6	3.8 & 4.6	3.8 & 4.6	3.8 & 4.6	3.8 & 4.6
Roll Length (feet)	300	300	300	300	300	300
(m)	90	90	90	90	90	90

TABLE 8.3 Typical Properties of a Nonwoven, Needle-Punched Polypropylene Geotextile

Property	Style (a)	Style (b)	Style (c)	Style (d)	Style (e)	Style (f)	Style (g)
Fabric Weight (oz/yd^2)	3.3	4.0	5.7	7.1	10.0	13.0	16.0
(g/m^2)	112	136	193	241	339	441	543
Thickness (mils)	50	55	75	95	125	150	185
(mm)	1.25	1.375	1.875	2.375	3.125	3.75	4.625
Permeability (cm/sec)	0.26	0.28	0.33	0.35	0.34	0.31	0.25
Permittivity (sec^{-1})	2.07	2.01	1.74	1.47	1.07	0.80	0.53
AOS (sieve size)	50	50	70	70	70	100	100
O_{95} (mm)	0.297	0.297	0.210	0.210	0.210	0.149	0.149
Porosity (%)	90+	90+	90+	90+	90+	90+	90+
Roll Width (feet)	12.5 & 15	12.5 & 15	12.5 & 15	12.5 & 15	12.5 & 15	12.5 & 15	12.5 & 15
(m)	3.8 & 4.6	3.8 & 4.6	3.8 & 4.6	3.8 & 4.6	3.8 & 4.6	3.8 & 4.6	3.8 & 4.6
Roll Length (feet)	400	400	300	300	300	300	300
(m)	120	120	90	90	90	90	90

TABLE 8.4 Typical Properties of a Woven, Monofilament, Polypropylene Geotextile

Property	Style (a)	Style (b)	Style (c)	Style (d)	Style (e)
Fabric Weight (oz/yd^2)	5.8	5.3	5.4	6.4	6.0
Permittivity (sec^{-1})	0.28	0.95	1.36	0.51	2.14
Flow Rate (gal/min-ft^2)	730	2850	4070	1430	5900
POA (%)	5–6	10–12	12–15	15–20	20–30
AOS (sieve size)	70	40	40	50	40
O_{95} (mm)	0.212	0.425	0.425	0.300	0.425
Roll Width (feet)	8 & 12	8 & 12	8 & 12	8 & 12	8 & 12
Roll Length (feet)	300	300	300	250	250

8.3.2 Allowable versus Ultimate Geotextile Properties

It is important to recognize that many of the geotextile laboratory test properties represent ideal conditions and thereby result in artificially high numeric values (called "ultimate" values) when used in design. The data of Tables 8.2, 8.3, and 8.4 is of this type. In the design-by-function concept, the factor of safety should be based on an "allowable" value and expressed as

$$FS = \frac{\text{Allowable Property}}{\text{Required Property}}$$

where, Allowable Property = a value based on a laboratory test that models the actual situation and Required Property = a value based on a design method that models the actual situation.

Thus, a laboratory test value cannot generally be used directly and must be modified for in-situ conditions. This modification could technically be incorporated into the test procedure (i.e., by conducting a true performance test), but in many cases, this is not practical. Such considerations as test specimen size, long-term creep testing, use of site-specific liquids, and representation of in-situ pore water and total stresses are often not feasible in laboratory testing or simulations. To compensate for the difference between a laboratory-measured test value and the true performance value, two approaches can be taken:

(i) use a higher than usual factor of safety at the end of a design, or
(ii) incorporate reduction factors in the laboratory generated test value to make it into a site-specific allowable value.

The approach taken herein will be to refer to the laboratory-obtained value as an "ultimate" value and to modify it by reduction factors to an "allowable" value. For problems dealing with flow through or within a geotextile, this approach takes the form of the equation that follows, with typical values given in Table 8.5. Note that these values must be tempered with site-specific conditions and consideration of the product-specific material (Koerner, 1998).

TABLE 8.5 Recommended Reduction Factor Values for Geotextiles Used in Filtration and Drainage (Koerner, 1998)

Application	Range of Reduction Factors				
	RF_{SCB}	RF_{CR}	RF_{IN}	RF_{CC}[1]	RF_{BC}[2]
Retaining Wall Filters	2.0 to 4.0	1.5 to 2.0	1.0 to 1.2	1.0 to 1.2	1.0 to 1.3
Underdrain Filters	5.0 to 10.0	1.0 to 1.5	1.0 to 1.2	1.2 to 1.5	2.0 to 4.0
Erosion Control Filters	2.0 to 10.0	1.0 to 1.5	1.0 to 1.2	1.0 to 1.2	2.0 to 4.0
Landfill Filters	5.0 to 10.0	1.5 to 2.0	1.0 to 1.2	1.2 to 1.5	5.0 to 10.0
Gravity Drainage	2.0 to 4.0	2.0 to 3.0	1.0 to 1.2	1.2 to 1.5	1.2 to 1.5
Pressure Drainage	2.0 to 3.0	2.0 to 3.0	1.0 to 1.2	1.1 to 1.3	1.1 to 1.3

[1] Values can be higher particularly for high alkalinity groundwater.
[2] Values can be higher for microorganism contents greater than 2,500 mg/l.

$$q_{\text{allow}} = \frac{q_{\text{ult}}}{RF_{\text{SCB}} \times RF_{\text{CR}} \times RF_{\text{IN}} \times RF_{\text{CC}} \times RF_{\text{BC}}} \qquad (8.6)$$

where q_{allow} = allowable flow rate;
q_{ult} = ultimate flow rate;
RF_{SCB} = reduction factor for soil clogging and blinding;
RF_{CR} = reduction factor for creep reduction of void space;
RF_{IN} = reduction factor for adjacent materials intruding into geosynthetic's void space;
RF_{CC} = reduction factor for chemical clogging; and
RF_{BC} = reduction factor for biological clogging.

8.3.3 Cross-Plane Permeability

Geotextile filtration involves the movement of liquid through the fabric (i.e., across its manufactured plane). At the same time, the fabric serves the purpose of retaining the soil on its upstream side. Both adequate permeability (requiring an open fabric structure) and soil retention (requiring a tight fabric structure) are required simultaneously. In addition, long-term soil-to-fabric flow compatibility is required so that it will not excessively clog during the lifetime of the system. A satisfactory filtration system can be defined as an equilibrium fabric-to-soil system that allows for free liquid flow (but without soil loss) across the plane of the fabric over an indefinitely long period of time. The filtration function is extremely important.

Fabric permeability in this case refers to cross-plane permeability when liquid flow is perpendicular to the plane of the fabric. Some of the fabrics used for this purpose are relatively thick and compressible. For this reason, the thickness is included in the permeability coefficient and is referred to as a "permittivity" instead. Permittivity is defined as

$$\psi = k/t \qquad (8.7)$$

where ψ = permittivity;
k = cross-plane permeability coefficient; and
t = thickness at a specified normal pressure.

The required permittivity ψ_{reqd} or required permeability k_{reqd} of geotextiles for a specific project can be calculated from

$$q = k \cdot i \cdot A = k \cdot (\Delta h/t) \cdot A = (k/t) \cdot \Delta h \cdot A = \psi \cdot \Delta h \cdot A \qquad (8.8)$$
$$q = r \cdot A \qquad (8.9)$$
$$\psi_{\text{reqd}} = q/(\Delta h \cdot A) = r/\Delta h \qquad (8.10)$$
$$k_{\text{reqd}} = q/(i \cdot A) = r/i \qquad (8.11)$$

where q = cross-plane flow rate of geotextile, ft^3/sec or m^3/sec;
ψ = permittivity of geotextile, sec^{-1};
k = cross-plane permeability coefficient, ft/sec or m/sec;
r = inflow rate, ft^3/sec/ft^2 or m^3/sec/m^2;

A = filtration area, usually use 1 ft², 1 acre (1 acre = 43,560 ft²), or 1 m², 1 hectare (1 ha = 10,000 m²);
ψ_{reqd} = required permittivity of geotextile, sec^{-1};
k_{reqd} = required cross-plane permeability coefficient of geotextile, ft/sec or m/sec;
i = hydraulic gradient;
Δh = liquid head from bottom of geotextile, ft or m;

and

$$\Delta h = H - t \tag{8.12}$$

where H = liquid head on the geomembrane liner, ft or m; and
t = thickness of drainage layer, ft or m.

In landfill design, the liquid head H on the geomembrane liner may be known based on hydrological analysis or regulatory requirement. Then, the liquid head from the bottom of geotextile, Δh, and the required permittivity of geotextile ψ_{reqd} can be determined using the preceding equations.

Using data from Table 8.5, the allowable permittivity or allowable permeability can be calculated from the following equations:

$$\psi_{allow} = \frac{\psi_{ult}}{RF_{SCB} \times RF_{CR} \times RF_{IN} \times RF_{CC} \times RF_{BC}} \tag{8.13}$$

$$k_{allow} = \frac{k_{ult}}{RF_{SCB} \times RF_{CR} \times RF_{IN} \times RF_{CC} \times RF_{BC}} \tag{8.14}$$

The values of ultimate permittivity, ψ_{ult}, and ultimate permeability, k_{ult}, are obtained from the approximate method. Then, the factor of safety that is used to evaluate the filtration capacity of the selected geotextile can finally be obtained:

$$FS = \psi_{allow}/\psi_{reqd} = k_{allow}/k_{reqd} \tag{8.15}$$

Equations 8.10 to 8.15 can be used to calculate the cross-plane flow capacity for the selected geotextile.

8.3.4 Soil Retention

To allow for the required flow of water through a geotextile, the void spaces in it must be adequately large. However, there is a limit, namely, when the upstream soil or fine waste fragment particles start to pass through the fabric voids along with the flowing liquid. This leads to an unacceptable situation called "piping", in which the finer soil particles are carried through the fabric, leaving large residual voids behind. The liquid velocity then increases, accelerating the whole process, until the upgradient structure begins to collapse. This collapse often leads to minute sinkhole-type patterns that grow larger with time.

This entire process can be prevented by making the geotextile voids small enough to retain the soil on the upstream side of the fabric. The coarser soil fraction must initially be retained; this is the targeted soil size in the design process. These particles eventually block the finer sized particles by establishing a bridging mechanism. Fortunately, filtration concepts are well established in the design of soil filters, and those same ideas can be used to design an adequate geotextile filter.

There are a number of approaches for achieving soil retention, all of which use particle sizes that are measured by sieving and are compared with the 0_{95} size (the apparent opening size) of the geotextile. The test for apparent opening size (AOS) was developed by the U.S. Army Corps of Engineers to evaluate woven fabrics. The test has since been extended to cover all fabrics, including nonwoven types. The apparent opening size (AOS) or equivalent opening size (EOS) are defined in CW-02215 as the U.S. standard sieve number having openings closest in size to the openings in the fabric. Note that AOS and EOS are equivalent terms. The equivalent ASTM test is designated D4751. The test uses known-sized glass beads designated by number and determines AOS by dry sieving. Sieving is done by using beads of successively smaller size until 5% or less pass through the fabric. The AOS or EOS of the fabric specimen is the "retained on" U.S. standard sieve number of the size. It is sometimes given as the equivalent sieve size opening in millimeters, and when done so is referred to as the 95% opening size or 0_{95}. Thus, AOS, EOS, and 0_{95} all refer to the same thing, the difference being that AOS and EOS are sieve numbers and 0_{95} is the corresponding sieve opening size in millimeters. Table 8.6 gives the interrelated values. Note that as the AOS sieve number increases, the 0_{95} particle size value decreases (i.e., they are inversely related to one another).

TABLE 8.6 Conversion of U. S. Standard Sieve Sizes to Equivalent Square Opening Sizes

Sieve Size (no.)	Opening Size	
	mm	mil
4	4.750	187.0
6	3.350	132.0
8	2.360	93.7
10	2.000	78.7
12	1.700	66.1
16	1.180	46.9
20	0.850	33.1
30	0.590	23.4
40	0.425	16.5
50	0.297	11.7
60	0.250	9.8
70	0.212	8.3
80	0.180	7.0
100	0.150	5.9
120	0.125	4.9
140	0.106	4.1
170	0.090	3.5
200	0.075	2.9
270	0.053	2.1
400	0.038	1.5

Note: 1 mil = 0.001 in.

Three approaches for determining soil retention opening size are described below in the subsections that follow.

8.3.4.1 Task Force 25 Method. The simplest of these methods examines the percentage of soil passing the No. 200 sieve (= 0.075 mm). The Task Force 25 (1983) makes the following recommendations:

1. Particles < 50% passing the No. 200 sieve
   AOS of the fabric > No. 30 sieve (i.e., O_{95} < 0.59 mm)
2. Particles > 50% passing the No. 200 sieve
   AOS of the fabric > No. 50 sieve (i.e., O_{95} < 0.297 mm)

8.3.4.2 Carroll Method. Slightly more restrictive is the recommendation of Carroll (1983) for the O_{95} size in millimeters, which is

$$O_{95} < (2 \text{ or } 3)d_{85}$$

where d_{85} = the particle size in millimeters for which 85% of sample is finer.

8.3.4.3 Giroud Method. The most conservative method is proposed by Giroud (1982), who presents a method for recommended O_{95} values (i.e., the opening size in millimeters corresponding to the AOS value) in terms of relative density (D_R), coefficient of uniformity (CU), and average particle size (d_{50}) for granular soils and in terms of Plasticity Index (PI) for fine-grained soils. It is presented below for steady-state flow conditions. There is a parallel set of criteria for dynamic-flow conditions, but these are not likely to occur in landfill filtration design situations.

Granular Soils. The definition of granular soils for the Giroud Method is the soil which contains less than 10% fines (less than #200 sieve opening size) or contains more than more than 10% fines, but the Plasticity Index (PI) is less than 5.

Case 1: For Loose Granular Soils (D_R < 50%)

    If $1 < CU < 3$, $O_{95} < (CU)(d_{50})$.
    If $CU > 3$, $O_{95} < (9 \cdot d_{50})/CU$.

Case 2: For Intermediate Granular Soils (50% < D_R < 80%)

    If $1 < CU < 3$, $O_{95} < 1.5 \cdot (CU)(d_{50})$.
    If $CU > 3$, $O_{95} < (13.5 \cdot d_{50})/CU$.

Case 3: For Dense Granular Soils (D_R > 80%)

    If $1 < CU < 3$, $O_{95} < 2 \cdot (CU)(d_{50})$.
    If $CU > 3$, $O_{95} < (18 \cdot d_{50})/CU$.

In all cases,

    O_{95} = apparent opening size of geotextile, mm (if data is not given by the manufacturer, this value is approximately the AOS sieve value in millimeters);

CU = coefficient of uniformity ($= d_{60}/d_{10}$);
d_{10} = soil particle size corresponding to 10% finer, mm;
d_{60} = soil particle size corresponding to 60% finer, mm; and
d_{50} = soil particle size corresponding to 50% finer, mm.

Fine-Grained Soils. If the soil contains more than 10% fine-size particles and the Plasticity Index is larger than 5 (thus lending cohesion to the soil structure), the soil is considered fine-grained for the Giroud Method. If the soil is determined to be nondispersive by the Double Hydrometer Method (i.e., DHR < 0.5), then O_{95} must be less than 0.21 mm (i.e., less than #70 sieve) opening size. In other words,

$$O_{95} < 0.21 \text{ mm (for fine-grained soils)}.$$

Note that any soil that is used to construct a compacted clay liner must be nondispersive soil.

8.3.5 Long Term Compatibility

Perhaps the most frequently asked question associated with the use of geotextiles in hydraulic related systems is, "Will it clog?" Undoubtedly, some soil particles will embed themselves within the fabric structure. A more relevant question is, "Will it excessively clog?"

A reasonable answer or response to the clogging question is simply to avoid situations that have been known to lead to severe clogging problems. To minimize the risk of clogging, the following precautions are recommended when using geotextiles:

1. Use the largest available opening size geotextile satisfying the retention criteria;
2. For nonwoven geotextiles (recall Tables 8.2 and 8.3): porosity > 40% under the actual stress conditions that the geotextile is serving;
3. For woven geotextiles (recall Table 8.4): percent open area (POA) > 6%.

The porosity of a nonwoven geotextile can be calculated using

$$n = 1 - \mu/(t_g \cdot \rho) \times 100\% \qquad (8.16)$$

where n = geotextile porosity or planar porosity, expressed as a percentage;
μ = geotextile mass per unit area;
t_g = geotextile thickness; and
ρ = density of filaments.

Percent open area (POA) is a fabric property that has applicability only for woven fabrics, and even then primarily for monofilament woven fabric. POA is a comparison of the total open area (the void spaces between adjacent fibers) to the total specimen area. Woven monofilament fabrics vary from essentially a closed structure (POA ≈ 0) to some that are quite open (POA = 36%). Many commercial woven geotextiles are in the range of 4 to 20%.

Other situations that have caused excessive clogging problems of geotextiles are filtration of very high alkalinity groundwater. For high pH liquids, the retardation of flow at the fabric interface can cause a calcium, sodium, or magnesium precipitate to be deposited, thereby blinding the fabric's upstream surface. The potential of biological clogging has often been considered, but for groundwater the likelihood is relatively remote. Conversely, for municipal landfill leachate the likelihood is relatively high. Obviously, landfill leachates (particularly for MSW landfills) are a concern. G. R. Koerner (1993) has investigated ten geotextile filters and two soil filters, with three different leachates, and compared their responses to water as a permeant. His general finding was that leachates with more than 2,500 g/ml of TSS or BOD required laboratory simulation to assess the severity of the clogging.

This discussion of soil-to-fabric compatibility assumes the establishment of a set of mechanisms that are in equilibrium with the flow regime being imposed on the system. Some insights into possible mechanisms have been suggested (McGown, 1978), including upstream soil filter, blocking, arching, and partial clogging. Most likely, a number of them are working together simultaneously, and just what mechanism dominates under what conditions of soil type and flow regime is still an issue that needs further research.

EXAMPLE 8.1

Use Task Force 25, Carroll ($O_{95} < 2 \cdot d_{85}$), and Giroud methods, respectively, to design a needle-punched nonwoven geotextile used in the primary leachate drainage layer (use average values from Table 8.5). The geotextile is placed between protective sand blanket and geonet as a filtration layer (shown in Fig. 8.3). The thickness of geonet is 0.25 inch (6.4 mm). The maximum inflow rate is 0.20 inch/hour (5 mm/hour). The liquid head over the geomembrane liner is assumed to be 1.5 inch (38 mm). The soil placed over the geotextile is a clean sand with $d_{85} = 0.95$ mm, $d_{50} = 0.19$ mm, $CU = 5.5$, $D_R = 65\%$, and 9% soil particles passing #200 sieve. The typical properties of commercially available, needle-punched nonwoven geotextile products are listed in Table 8.3.

Solution:
Permeability:

$$r = 0.20 \text{ in/hr} = 1.67 \times 10^{-2} \text{ ft/hr} = 4.630 \times 10^{-6} \text{ ft/sec } (1.4 \times 10^{-3} \text{ mm/sec})$$

$$\Delta h = H - D = 1.5 - 0.25 = 1.25 \text{ inch} = 0.104 \text{ ft } (31.8 \text{ mm})$$

$$\psi_{\text{reqd}} = r/\Delta h = (4.630 \times 10^{-6})/0.104 = 4.452 \times 10^{-5} \text{ sec}^{-1} = 0.160 \text{ hr}^{-1}$$

Soil Retention:

1. Task Force 25:
   9% soil particles passing the No. 200 sieve $< 50\%$
   So, AOS of geotextile must be greater than #30 sieve (i.e., $O_{95} < 0.59$ mm).
   Select Geotextile Style (b) ($O_{95} = 0.297$ mm, AOS #50) from Table 8.3.
2. Carroll Method ($O_{95} < 2.5 \cdot d_{85}$):
   $d_{85} = 0.95$ mm;
   $O_{95} < 2.5 \cdot d_{85} = 2.5 \times 0.95 = 2.4$ mm.
   Select Geotextile Style (b) ($O_{95} = 0.297$ mm, AOS #50) from Table 8.3.
3. Giroud Method:
   9% soil particles passing the No. 200 sieve $< 10\%$: Granular Soil.
   Granular Soils:
   $50\% < D_R = 65\% < 80\%$: Intermediate Granular Soil;

$CU = d_{60}/d_{10} = 5.5 > 3;$
$d_{50} = 0.19$ mm;
$O_{95} < (13.5 \cdot d_{50})/CU = (13.5 \times 0.19)/5.5 = 0.466$ mm.
Select Geotextile Style (b) ($O_{95} = 0.297$ mm, AOS #50) from Table 8.3.

Long-Term Compatibility:
Style (b) is a needle-punched nonwoven geotextile, and its porosity can be found in Table 8.3:

$$n = 90\% > 40\% \text{ (OK)}.$$

Thus, Style (b) geotextile is selected by all three methods. Check whether Style (b) geotextile has adequate filtration capacity.

Style (b) geotextile, Permittivity $\psi_{ult} = 2.01$ sec^{-1}, and using average values of reduction factors from the "landfill filters" row in Table 8.5,

$$\psi_{allow} = \frac{\psi_{ult}}{RF_{SCB} \times RF_{CR} \times RF_{IN} \times RF_{CC} \times RF_{BC}} \tag{8.13}$$

$$= (2.01)/(7.5 \times 1.75 \times 1.1 \times 1.35 \times 7.5)$$
$$= (2.01)/(146)$$
$$= 0.014 \text{ sec}^{-1} = 50 \text{ hr}^{-1}.$$

$$FS = \psi_{allow}/\psi_{reqd} \tag{8.15}$$
$$= (50)/(0.160)$$
$$= 313 \text{ (OK)}.$$

Thus, Style (b) geotextile can meet all requirements of permeability, soil retention, and long-term compatibility.

8.4 GEONET DESIGN FOR LEACHATE DRAINAGE

Biplanar and triplanar geonets are both members of the geosynthetic family. They are indeed net-like materials and are used exclusively for their in-plane (or transmissivity) drainage function. All of the geonets are formed from polyethylene in the specific gravity range of 0.935 to 0.950 gm/cc. Thus they cover the range from medium-to-high density polyethylene (i.e., MDPE-to-HDPE). However, when formulated with carbon black and antioxidants, they are all higher than 0.941 gm/cc and thus in the HDPE category.

8.4.1 Geonet Overview

Geonets are unitized sets of repeating parallel ribs positioned in layers such that liquids can be transmitted within their open spaces. Thus their primary function is drainage. There are two variations of geonets. *Biplanar* geonets have two parallel sets of ribs overlapping one another. Flow is approximately the same in all directions, although slightly greater in the direction of the sets of ribs. *Triplanar* geonets have three parallel sets of ribs. The main (and largest) ribs are in the center where the flow rate is the greatest. The upper and lower sets of ribs are significantly smaller and flow in these directions is proportionately lower.

Geonets are always covered with either a geomembrane or a geotextile on their upper and lower surfaces—they are never in direct contact with soil cover since the soil particles would fill the apertures of the geonet rendering it useless. Many geonets have a geotextile bonded to one or both surfaces. They are then referred to as *drainage composites* or *geocomposites*. The geotextile is usually bonded on the surface of the geonet by heat fusion to form an interface that is usually stronger than the geotextile itself.

Geonets are used almost exclusively for their drainage capability. As such, they are single-function geosynthetics. Geonets are used for their in-place drainage capability, and are not used for reinforcement. They do have considerable strength (particularly when confined in soil), but this attribute is normally discounted. Current uses of geonets in the landfill projects include leachate drainage of landfill side slopes, leachate drainage of landfill base areas, leak detection layers between liner systems, surface water drainage of landfill caps, and groundwater interception and drainage around landfills.

8.4.2 Hydraulic Properties of Geonets

The in-plane hydraulic test used to determine planar flow rate, or transmissivity, of geonets can be performed using ASTM D4716. A schematic diagram of an in-plane flow-rate testing device is shown in Figure 8.8. The relationship between flow rate and normal stress with various hydraulic gradients for a 0.25-inch (6.3-mm) biplanar geonet sandwiched between two 60-mil (1.5-mm) HDPE geomembranes is shown in Figure 8.9. Flow rate decreases with increasing normal stress with the effect

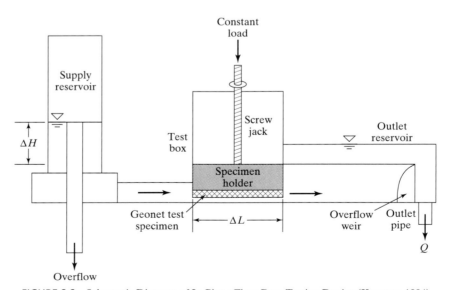

FIGURE 8.8 Schematic Diagram of In-Plane Flow-Rate Testing Device (Koerner, 1994)

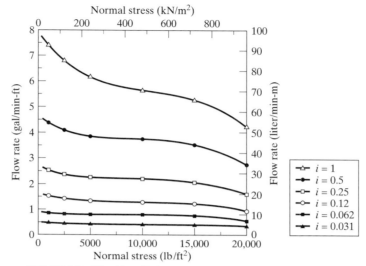

FIGURE 8.9 Flow Rate Behavior of a 0.25-inch (6.3-mm) Geonet Sandwiched between Two 60-mil (1.5-mm) HDPE Geomembranes (USEPA, 1989)

being most pronounced at large values of gradient, namely, $i \geq 0.25$. A transmissivity value can be calculated from this data assuming that the system is saturated and under laminar flow. When one considers that these flow rates are extremely high in comparison to flow rates in soil, one should exercise caution when using transmissivity. For comparison purposes, 12 inches (0.3 m) of sand having a coefficient of permeability of 0.2 ft/min (about 1 mm/sec) at a hydraulic gradient of 1.0 can carry only 1.5 gal/min-ft (19 liter/min-m). Thus, geonets can handle significantly larger flow rates compared with soil, and the flow within their apertures are probably turbulent (Koerner, 1994).

The data shown in Figure 8.9 are sometimes called *index* data since site-specific conditions have not been included. The values also can be called *ultimate*, hence "q_{ult}." More realistic simulations can be incorporated into the procedure to make the test yield "performance" data, the goal being the allowable value, or "q_{allow}." To do so one must have representative conditions above and below the test specimen, and liquids of the type (and sometimes temperature) to be conveyed in the actual system. Figure 8.10 illustrates the impact of one such variation. Here a cross-sectional profile consisting of a layer of clay (Kaolinite clay at 15% water content), a 16 oz/yd^2 (540 g/m^2) needle-punched nonwoven continuous filament polyester geotextile, a geonet, and a 60 mil (1.5 mm) HDPE geomembrane was used. Since the geonet was the same type as that which produced the data of Figure 8.9, the results can be compared directly, the only difference being the geotextile/clay over the geonet.

The marked decrease in flow rates of Figure 8.10 compared with Figure 8.9 comes from intrusion of the geotextile/clay into the core space of the geonet. Flow reductions are compared numerically in Table 8.7. The reduction in flow rate is largest at higher hydraulic gradients. In addition to this short-term intrusion effect, the

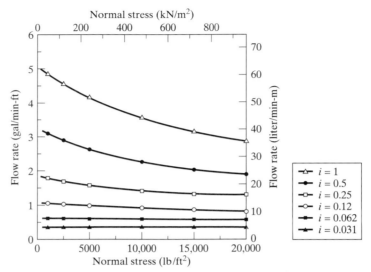

FIGURE 8.10 Flow Rate Behavior of a 0.25-inch (6.3-mm) Geonet Sandwiched between a 16-oz/yd² (540-g/m²) Nonwoven Needle-Punched Geotextile with Clay above and a 60-mil (1.5-mm) HDPE Geomembrane below (USEPA, 1989)

geotextile-geonet system also must be capable of sustaining these loads over time, suggesting that long-term tests (or reduction factors) be required to adequately assess such situations as well.

8.4.3 Allowable Geonet Flow Rate

An allowable flow rate must be extracted from hydraulic testing of the type just described. Accordingly, it is necessary to assess how realistic the test setup is in contrast to the actual field conditions. If it does not model site-specific conditions adequately, then some adjustments to the laboratory value must be made. This is often the case. As mentioned previously, the laboratory-generated index value is treated as an ultimate value, which means that when using ASTM D4716 for flow rate determination, the value must be reduced before used in design. In other words,

$$q_{\text{allowable}} < q_{\text{ultimate}}$$

The recommended way of doing this is to prescribe reduction factors on each of the items not assessed adequately in the laboratory test. For example (Koerner, 1998),

$$q_{\text{allow}} = \frac{q_{\text{ult}}}{RF_{\text{IN}} \times RF_{\text{CR}} \times RF_{\text{CC}} \times RF_{\text{BC}}} \tag{8.17}$$

where q_{ult} = ultimate flow rate determined from ASTM D4716 for short-term index tests between solid plates using water as the transported liquid under laboratory test temperatures;
q_{allow} = allowable flow rate for final design purposes;

TABLE 8.7 Flow Rates and Reduction from Curves of Figures 8.9 and 8.10 (modified from Koerner, 1994)

Normal Stress	Cross Section			Hydraulic Gradient (i)					
				0.03	0.06	0.12	0.25	0.50	1.00
1,000 lb/ft² (48 kN/m²)	HDPE (both sides)	gal/min-m		0.5	0.9	1.5	2.5	4.4	7.5
		liter/min-m		6.2	11.2	18.6	31.1	54.6	93.2
	GT/Clay (one side)	gal/min-m		0.4	0.7	1.1	1.8	3.2	5.0
		liter/min-m		5.0	8.7	13.7	22.4	39.7	62.1
	Difference	gal/min-m		0.1	0.2	0.4	0.7	1.2	2.5
		liter/min-m		1.2	2.5	4.9	8.7	14.9	31.1
	Reduction			20%	22%	30%	28%	28%	33%
5,000 lb/ft² (240 kN/m²)	HDPE (both sides)	gal/min-m		0.5	0.8	1.4	2.3	3.9	6.2
		liter/min-m		6.2	9.9	17.4	28.6	48.4	77.0
	GT/Clay (one side)	gal/min-m		0.4	0.6	1.0	1.6	2.7	4.3
		liter/min-m		5.0	7.5	12.4	19.9	33.5	53.4
	Difference	gal/min-m		0.1	0.2	0.4	0.7	1.2	1.9
		liter/min-m		1.2	2.4	5.0	8.7	14.9	23.6
	Reduction			20%	25%	29%	30%	31%	31%
10,000 lb/ft² (480 kN/m²)	HDPE (both sides)	gal/min-m		0.4	0.8	1.3	2.2	3.8	5.7
		liter/min-m		5.0	9.9	16.1	27.3	47.2	70.8
	GT/Clay (one side)	gal/min-m		0.3	0.6	0.9	1.4	2.3	3.6
		liter/min-m		3.7	7.5	11.2	17.4	28.6	44.7
	Difference	gal/min-m		0.1	0.2	0.4	0.8	1.5	2.1
		liter/min-m		1.3	2.4	4.9	9.9	18.6	26.1
	Reduction			25%	25%	31%	36%	39%	37%
20,000 lb/ft² (960 kN/m²)	HDPE (both sides)	gal/min-m		0.4	0.6	0.9	1.6	2.7	4.3
		liter/min-m		5.0	7.5	11.2	19.9	33.5	53.4
	GT/Clay (one side)	gal/min-m		0.3	0.5	0.8	1.3	1.9	2.8
		liter/min-m		3.7	6.2	9.9	16.1	23.6	34.8
	Difference	gal/min-m		0.1	0.1	0.1	0.3	0.8	1.5
		liter/min-m		1.3	1.3	1.3	3.8	9.9	18.6
	Reduction			25%	17%	11%	19%	30%	35%

RF_{IN} = reduction factor for elastic deformation, or intrusion, of the adjacent geosynthetics into the geonet's core space;

RF_{CR} = reduction factor for creep deformation of the geonet and adjacent geosynthetics into geonet's core space;

RF_{CC} = reduction factor for chemical clogging and/or precipitation of chemicals in the geonet's core space; and

RF_{BC} = reduction factor for biological clogging in the geonet's core space.

Some guidelines as to various reduction factors to be used in different situations are given in Table 8.8. Note that these values are based on preliminary and relatively sparse information. Other reduction factors, such as installation damage, viscosity effects and temperature effects, could also have been incorporated. If needed, they can be included on a site-specific basis. An example problem follows, which illustrates the use of geonets and points out that high factors of safety are warranted in critical situations.

EXAMPLE 8.2

What is the allowable geonet flow rate to be used in the design of a secondary leachate collection (i.e., leak detection) system? Assume that laboratory testing at proper design load and proper hydraulic gradient gave a short-term between-rigid-plate index value of 1.2 gal/min-ft (14.9 liter/min-m).

Solution: Average values from Table 8.8 are used (however, note the large resulting reduction):

$$q_{allow} = \frac{q_{ult}}{RF_{IN} \times RF_{CR} \times RF_{CC} \times RF_{BC}} \quad (8.17)$$

$$= \frac{1.2}{1.75 \times 1.7 \times 1.75 \times 1.75}$$

$$= \frac{1.2}{9.11} = 0.13 \text{ gal/min-ft (1.6 liter/min-m)}$$

TABLE 8.8 Recommended Preliminary Reduction Factors for Determining Allowable Flow Rate or Transmissivity of Biplanar Geonets (Koerner, 1998)

Application Area	Reduction Factor Values			
	RF_{IN}	RF_{CR}*	RF_{CC}	RF_{BC}
Sport fields	1.0 to 1.2	1.0 to 1.5	1.0 to 1.2	1.1 to 1.3
Capillary break	1.1 to 1.3	1.0 to 1.2	1.1 to 1.5	1.1 to 1.3
Roof and plaza decks	1.2 to 1.4	1.0 to 1.2	1.0 to 1.2	1.1 to 1.3
Retaining walls, seeping rock and soil slopes	1.3 to 1.5	1.2 to 1.4	1.1 to 1.5	1.0 to 1.5
Drainage blankets	1.3 to 1.5	1.2 to 1.4	1.0 to 1.2	1.0 to 1.2
Surface water drains for landfill caps	1.3 to 1.5	1.1 to 1.4	1.0 to 1.2	1.2 to 1.5
Secondary leachate collection (landfill)	1.5 to 2.0	1.4 to 2.0	1.5 to 2.0	1.5 to 2.0
Primary leachate collection (landfill)	1.5 to 2.0	1.4 to 2.0	1.5 to 2.0	1.5 to 2.0

*These values are sensitive to the density of the resin used in the geonet's manufacture. The higher the density, the lower the reduction factor. Creep of the covering geotextile(s) is a product-specific issue.

8.4.4 Designing with Geonets for Drainage

Several design problems and their solutions are presented in this section. The necessary theory for proper understanding of the example problems is summarized at the beginning.

8.4.4.1 Flow Rate Factor of Safety.
"Design by Function" requires the formulation of a factor of safety as follows:

$$FS = \frac{\text{Allowable (test) value}}{\text{Required (design) value}}$$

For geonets serving as a drainage medium, the targeted value is flow rate and the above equation can be written as

$$FS = q_{\text{allow}}/q_{\text{reqd}} \tag{8.18}$$

where q_{allow} = the allowable flow rate; and
q_{reqd} = the required flow rate.

8.4.4.2 Transmissivity.
For saturated systems under laminar flow conditions, Darcy's formula can be used to arrive at an alternative to flow rate, namely, the transmissivity. Darcy's formula states that

$$Q = k \cdot i \cdot A = k \cdot i \cdot (w \cdot t) \tag{8.19}$$

$$Q = (k \cdot t) \cdot i \cdot w = \theta \cdot i \cdot w \text{ and} \tag{8.20}$$

$$\theta = k \cdot t = Q/(i \cdot w) \tag{8.21}$$

where Q = the volumetric flow rate, cm³/sec;
k = the coefficient of permeability, cm/sec;
i = the hydraulic gradient,
A = the flow cross-sectional area, cm²;
θ = the transmissivity, cm²/sec;
w = the width, cm; and
t = the thickness, cm.

In a similar manner as the preceding, a factor of safety can be formulated using transmissivity (viz., $FS = \theta_{\text{allow}}/\theta_{\text{reqd}}$). Thus, Q/w and transmissivity carry the same units and are directly related to one another by means of the hydraulic gradient i. At a hydraulic gradient of 1.0, they are numerically identical. At other values of hydraulic gradient, they are not equal. Also note that the system must be saturated and flow must be laminar in order to use transmissivity. When in doubt, it is usually best to use flow rate per unit width.

8.4.4.3 Required Flow Rate.
The required flow rate for the leachate drainage layer can be calculated from the equations

$$q_{\text{reqd}} = (r \cdot w \cdot L_H)/w \quad \text{(suitable only for square or rectangular cell)} \tag{8.22}$$

and

$$q_{\text{reqd}} = [r \cdot (L_H)_{\text{max}} \cdot dw]/dw \tag{8.23}$$

where q_{reqd} = required flow rate of geonet outlet, ft³/sec/ft or m³/sec/m;
 r = inflow rate, ft³/sec/ft² or m³/sec/m;
 w = total width of cell along leachate collection trench direction, ft or m;
 L_H = horizontal distance perpendicular to the leachate collection pipe for the rectangular cell, ft or m;
 $(L_H)_{max}$ = maximum horizontal distance perpendicular to the leachate collection pipe of the landfill cell, ft or m; and
 dw = 1 ft or 1 m.

8.4.4.4 Normal Pressure Acting on Geonet. The total overburden pressure over a geonet is

$$W = \left(\sum \gamma_i \cdot H_i\right) \cdot L_H \tag{8.24}$$

The total normal force acting on the geonet is

$$N = \left(\sum \gamma_i \cdot H_i\right) \cdot L_H \cdot \cos\beta \tag{8.25}$$

The normal pressure acting on the geonet is

$$\sigma_n = \left(\sum \gamma_i \cdot H_i\right) \cdot L_H \cdot \cos\beta / L$$

where

$$L = L_H / \cos\beta$$

Thus,

$$\sigma_n = \left(\sum \gamma_i \cdot H_i\right) \cdot L_H \cdot \cos\beta / (L_H / \cos\beta)$$

and it follows that

$$\sigma_n = \left(\sum \gamma_i \cdot H_i\right) \cdot \cos^2\beta \tag{8.26}$$

where W = total weight of solid waste and soil over geonet, lb/ft or kN/m;
 N = total normal force acting on geonet, lb/ft or kN/m;
 σ_n = normal pressure acting on geonet, lb/ft² or kN/m²;
 γ_i = unit weight of solid waste or soil, lb/ft³ or kN/m³;
 H_i = depth of solid waste or soil over geonet, ft or m;
 L_H = horizontal distance from top to bottom of cell subgrade or side slope, ft or m;
 L = actual distance cell subgrade or side slope, ft or m; and
 β = angle of cell subgrade or side slope, degrees.

Typical hydraulic properties of a biplanar geonet drainage composite and a triplanar geonet that are widely used in landfills are listed in Tables 8.9 and 8.10, respectively.

TABLE 8.9 Ultimate Flow Rate and Transmissivity of a Biplanar Geonet Drainage Composite

Load	Gradient	Ultimate Flow Rate		Transmissivity		
		gal/min-ft	liter/min-m	gal/min-ft	m²/sec ($\times 10^{-3}$)	ft²/sec ($\times 10^{-3}$)
500 lb/ft² (24 kN/m²)	0.02	0.46	5.71	22.98	47.58	51.21
	0.10	1.28	15.90	12.80	26.49	28.52
	0.25	2.53	31.42	10.14	20.98	22.58
	0.50	3.97	49.30	7.94	16.43	17.69
	0.75	5.28	65.57	7.04	14.58	15.69
	1.00	6.28	77.99	6.28	13.01	14.00
1,000 lb/ft² (48 kN/m²)	0.02	0.37	4.59	18.36	38.01	40.91
	0.10	1.21	15.03	12.05	24.95	26.86
	0.25	2.33	28.93	9.31	19.27	20.74
	0.50	3.79	47.06	7.57	15.67	16.87
	0.75	4.91	60.97	6.65	13.55	14.59
	1.00	5.78	71.78	5.78	11.97	12.89
2,000 lb/ft² (96 kN/m²)	0.02	0.34	4.22	16.81	34.81	37.46
	0.10	1.07	13.29	10.72	22.20	23.89
	0.25	2.02	25.08	8.07	16.70	17.97
	0.50	3.23	40.11	6.46	13.36	14.39
	0.75	4.22	52.40	5.62	11.64	12.53
	1.00	5.10	63.33	5.10	10.55	11.35
5,000 lb/ft² (240 kN/m²)	0.02	0.28	3.48	14.05	29.08	31.30
	0.10	0.81	10.06	6.14	16.85	18.14
	0.25	1.51	18.75	6.05	12.52	13.47
	0.50	2.33	28.93	4.65	9.64	10.37
	0.75	2.98	37.01	3.97	8.22	8.85
	1.00	3.61	44.83	3.61	7.48	8.05
10,000 lb/ft² (480 kN/m²)	0.02	0.19	2.36	9.70	20.07	21.61
	0.10	0.64	7.95	6.38	13.22	14.22
	0.25	1.13	14.03	4.51	9.34	10.05
	0.50	1.80	22.35	3.60	7.46	8.03
	0.75	2.27	28.19	3.03	6.28	6.76
	1.00	2.72	33.78	2.72	5.63	6.06
15,000 lb/ft² (720 kN/m²)	0.02	0.15	1.86	7.72	15.98	17.20
	0.10	0.40	4.97	3.97	8.23	8.86
	0.25	0.77	9.56	3.07	6.35	6.83
	0.50	1.25	15.53	2.51	5.19	5.58
	0.75	1.64	20.37	2.18	4.62	4.87
	1.00	1.98	24.59	1.98	4.09	4.41
20,000 lb/ft² (960 kN/m²)	0.02	0.09	1.12	4.67	9.66	10.40
	0.10	0.28	3.48	2.82	5.84	6.29
	0.25	0.53	6.58	2.11	4.36	4.69
	0.50	0.85	10.56	1.71	3.63	3.80
	0.75	1.07	13.29	1.43	2.96	3.12
	1.00	1.33	16.52	1.33	2.75	2.96

TABLE 8.10 Ultimate Flow Rate and Transmissivity of a Triplanar Geonet

Load	Gradient	Ultimate Flow Rate		Transmissivity		
		gal/min-ft	liter/min-m	gal/min-ft	m^2/sec ($\times 10^{-3}$)	ft^2/sec ($\times 10^{-3}$)
500 lb/ft^2 (24 kN/m^2)	0.01	0.24	3.00	80.2	16.6	17.9
	0.10	0.66	8.25	22.6	4.68	5.04
	0.30	1.21	15.00	13.3	2.76	2.97
	0.50	1.62	20.13	10.6	2.20	2.37
	1.00	2.33	29.00	7.63	1.58	1.70
1,000 lb/ft^2 (48 kN/m^2)	0.01	0.17	2.10	64.3	13.3	14.3
	0.10	0.69	8.63	22.8	4.72	5.08
	0.30	1.22	15.13	13.4	2.78	2.99
	0.50	1.60	19.88	10.8	2.23	2.40
	1.00	2.31	28.75	7.63	1.58	1.70
5,000 lb/ft^2 (240 kN/m^2)	0.01	0.16	1.96	56.0	11.6	12.5
	0.10	0.59	7.38	19.7	4.07	4.38
	0.30	1.10	13.63	12.0	2.48	2.67
	0.50	1.45	18.00	9.61	1.99	2.14
	1.00	2.10	26.13	6.91	1.43	1.54
10,000 lb/ft^2 (480 kN/m^2)	0.01	0.14	1.79	49.3	10.2	11.0
	0.10	0.59	7.38	19.5	4.04	4.35
	0.30	1.03	12.75	11.3	2.34	2.52
	0.50	1.35	16.75	8.94	1.85	1.99
	1.00	1.96	24.38	6.43	1.33	1.43
15,000 lb/ft^2 (720 kN/m^2)	0.01	0.11	1.37	38.7	8.01	8.62
	0.10	0.54	6.75	18.0	3.73	4.01
	0.30	0.95	11.75	10.4	2.15	2.31
	0.50	1.24	15.38	8.21	1.70	1.83
	1.00	1.81	22.50	5.94	1.23	1.32
20,000 lb/ft^2 (960 kN/m^2)	0.01	0.15	1.92	34.2	7.07	7.60
	0.10	0.45	5.63	15.0	3.11	3.35
	0.30	0.85	10.50	9.42	1.95	2.10
	0.50	1.11	13.75	7.44	1.54	1.66
	1.00	1.62	20.13	5.31	1.10	1.18
25,000 lb/ft^2 (1,200 kN/m^2)	0.01	0.10	1.24	32.9	6.81	7.33
	0.10	0.44	5.50	14.5	3.01	3.24
	0.30	0.78	9.63	8.70	1.80	1.94
	0.50	1.03	12.75	6.72	1.39	1.50
	1.00	1.47	18.25	4.83	1.00	1.08

EXAMPLE 8.3

What is the factor of safety of a biplanar geonet drainage composite used as a primary leachate drainage layer in a landfill cell (use average reduction factors from Table 8.8)? The drainage composite lies between a two-foot protective sand layer (γ_{sand} = 115 lb/ft^3 or 18 kN/m^3) and a

Section 8.4 Geonet Design for Leachate Drainage

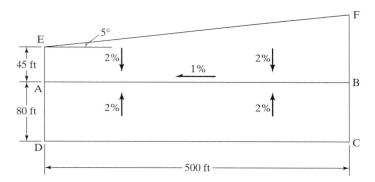

FIGURE 8.11 Landfill Cell Plan for Example 8.3

60-mil (1.5-mm) HDPE geomembrane, and the design inflow rate is 3,000 gal/acre/day (28,400 liter/ha/day). The slope of the landfill bottom floor is 2%. The landfill when filled completely will be 250 ft (76 m) high with a unit weight of waste of 70 lb/ft³ (11 kN/m³). The ultimate flow rate of the geonet is listed in Table 8.9. The plan view of the landfill cell is shown in Figure 8.11. AB in the plan view is a leachate collection pipe and flow is assumed to be perpendicular to it.

Solution:

1. Determine the maximum horizontal distance $(L_H)_{max}$:
   For the lower portion of the cell,
   $$[(L_H)_{max}]_{lower} = AD = BC = 80 \text{ ft } (24.4 \text{ m})$$
   For the upper portion of the cell,
   $$[(L_H)_{max}]_{upper} = BF = 45 + (500 \times \tan 5°) = 45 + 43.7 = 88.7 \text{ ft } (27.0 \text{ m})$$
   But
   $$[(L_H)_{max}]_{upper} > [(L_H)_{max}]_{lower}$$
   $$(L_H)_{max} = [(L_H)_{max}]_{upper} = 88.7 \text{ ft } (27.0 \text{ m})$$

2. Calculate the required flow rate q_{reqd}:
   $$r = 3{,}000 \text{ gal/acre/day} \div 43{,}560 \text{ ft}^2/\text{acre} \div 1{,}440 \text{ min/day}$$
   $$= 4.783 \times 10^{-5} \text{ gal/ft}^2/\text{min } (1.949 \times 10^{-3} \text{ liter/m}^2/\text{min})$$
   $$q_{reqd} = [r \cdot (L_H)_{max} \cdot dw]/dw \quad (8.23)$$
   $$= (4.783 \times 10^{-5})(88.7)(1)/(1)$$
   $$= 4.24 \times 10^{-3} \text{ gal/min-ft } (5.27 \times 10^{-2} \text{ liter/min-m})$$

3. Calculate the normal pressure acting on the geonet, σ_n:
   For 2% slope, $\beta = 1.15°$, $\cos \beta = \cos(1.15°) = 0.9998$;
   $$\sigma_n = (\gamma_{sand} \cdot H_{sand} + \gamma_{waste} \cdot H_{waste}) \cdot \cos^2 \beta \quad (8.26)$$
   $$= (115 \times 2 + 70 \times 250) \times (0.9998)^2$$
   $$= (230 + 17500) \times 0.9996 = 17{,}730 \text{ lb/ft}^2 \text{ (849 kN/m}^2)$$

4. Determine the ultimate flow rate of the geonet, q_{ult}:
   From Table 8.9,

   $i = 0.02$, $\sigma_n = 15{,}000$ lb/ft² (720 kN/m²), $q_{ult} = 0.15$ gal/min-ft (1.86 liter/min-m)

   and

   $i = 0.02$, $\sigma_n = 20{,}000$ lb/ft² (960 kN/m²), $q_{ult} = 0.09$ gal/min-ft (1.12 liter/min-m)

   For this particular geonet, $i = 0.02$, $\sigma_n = 17{,}730$ lb/ft² (849 kN/m²); thus, by interpolation,

   $q_{ult} = 0.09 + [(0.15 - 0.09)/(20{,}000 - 15{,}000)] \times (20{,}000 - 17{,}730)$

   $= 0.117$ gallon/min-ft (1.45 liter/min-m)

5. Calculate the allowable flow rate q_{allow}:
   Using average data from Table 8.8,

   $$q_{allow} = \frac{q_{ult}}{RF_{IN} \times RF_{CR} \times RF_{CC} \times RF_{BC}} \tag{8.17}$$

   $$= \frac{0.117}{1.75 \times 1.7 \times 1.75 \times 1.75}$$

   $= (0.117)/(9.11) = 12.84 \times 10^{-3}$ gal/min-ft (0.16 liter/min-m)

6. Calculate the factor of safety:

   $$FS = q_{allow}/q_{reqd} \tag{8.18}$$

   $= (12.84 \times 10^{-3})/(4.24 \times 10^{-3}) = 3.01$

   Thus, the biplanar geonet drainage composite can meet the design requirement.

8.5 ESTIMATE OF MAXIMUM LIQUID HEAD IN A DRAINAGE LAYER

The maximum liquid head over a barrier must be estimated in two distinct locations during the design of a landfill. The first requirement is to calculate the maximum leachate head over the base, or primary, liner. This liquid head must not exceed 300 mm based on Federal and state regulations in the United States and as well as many other countries. The second is to calculate the maximum saturated depth in the final cover system above the barrier. This saturated depth is one of the most important parameters affecting the stability of the final cover system.

To properly design a leachate drainage system for a landfill liner, the design engineer must be able to estimate the maximum saturated depth over the barrier for any proposed configuration. Most regulations (Federal, state and international) limit the maximum leachate head over the liner to 12 inches (300 mm) for municipal solid waste landfills (USEPA, 1991; MDEQ, 1999). Factors affecting this maximum saturated depth include the inflow rate into the drainage layer, the hydraulic conductivity of the leachate drainage layer, the leachate flow distance from the upstream boundary to the leachate collection pipe, the slope of the landfill bottom liner, and the hydraulic condition at the downstream end of the drainage layer.

8.5.1 Methods for Estimating Maximum Liquid Head

In general, the hydraulic conductivity of the drainage layer, the drainage slope, and the drainage distance from the upstream boundary to the drainage outlet are relatively easy to determine for the landfill design. They are constants for a specific landfill. But, the rate of inflow into the landfill leachate drainage layer or final cover drainage layer is variable. The flow conditions in the drainage layers for both leachate collection system and final cover system are in an unsteady drainage state. Calculation of liquid head over the barriers for the unsteady flow is very complex. In order to simplify the calculations and still obtain reliable results, the flow in the drainage layers of the leachate collection system and final cover system will be assumed to be in a steady flow state. In that case, the inflow rate will be constant and is assumed to be equal to the maximum inflow rate. Thus, the maximum liquid head over the barriers can be calculated based on the maximum inflow rate using a steady-state assumption. If the calculated maximum leachate head over the bottom liner is not greater than 300 mm under these worst case conditions, the leachate head over the liner will always meet the regulatory requirements in any other case. Thus, this method will provide conservative results (Qian, 1994).

Four methods are currently used to calculate the maximum leachate head over the landfill liner or the maximum saturated depth in the final cover system. Two of these methods were proposed by Moore, and the other two were proposed by Giroud et al. and McEnroe, respectively.

8.5.1.1 Moore's 1980 Method. An explicit formula for estimating the maximum liquid head over a sloping barrier (see Figure 8.12) can be found in several technical guidance documents of the U.S. Environmental Protection Agency (USEPA, 1980; USEPA, 1989). The maximum liquid head can be determined by the following expression:

$$y_{max} = L \cdot (r/k)^{1/2}[(k \cdot S^2/r) + 1 - (k \cdot S/r)(S^2 + r/k)^{1/2}] \tag{8.27}$$

where y_{max} = maximum liquid head on the landfill barrier, in. or mm;
L = horizontal drainage distance, in. or mm;
r = inflow rate (i.e., rate of vertical inflow to the drainage layer per unit horizontal area), in./day or mm/sec;

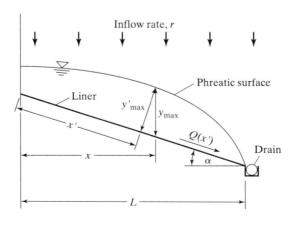

FIGURE 8.12 Phreatic Surface in Landfill Drainage Layer

k = hydraulic conductivity of the drainage layer, in./day or mm/sec;
S = slope of the drainage layer, $S = \tan \alpha$; and
α = slope angle of drainage layer, measured from horizontal, degrees.

This formula was first presented by C. A. Moore in 1980 without derivation or explanation of its origin or limitations.

8.5.1.2 Moore's 1983 Method. Moore presented another formula for estimating the maximum liquid head over a sloping barrier in 1983 (USEPA, 1983). This formula is expressed as:

$$y_{\max} = L \cdot [(r/k + S^2)^{1/2} - S] \tag{8.28}$$

where $y_{\max}$ = maximum liquid head on the landfill liner, in. or mm;
L = horizontal drainage distance, in. or mm;
r = inflow rate, in./day or mm/sec;
k = hydraulic conductivity of the drainage layer, in./day or mm/sec;
S = slope of the drainage layer, $S = \tan \alpha$; and
α = slope angle of drainage layer, measured from horizontal, degrees.

This formula is simpler than Moore's 1980 formula. But again, neither derivation nor explanation of its origin or limitations was included in the 1983 report.

8.5.1.3 Giroud's 1992 Method. Giroud et al. presented a different formula for estimating the maximum liquid head over a sloping liner based on a simplifying assumption and numerical methods in 1992 (Giroud et al., 1992; Giroud and Houlihan, 1995). This formula is expressed as

$$y_{\max} = j \cdot L \cdot [(4 \cdot r/k + S^2)^{1/2} - S]/(2 \cdot \cos \alpha) \tag{8.29}$$

A parameter j in the formula can be calculated as

$$j = 1 - 0.12 \cdot \exp\{-[\log(1.6 \cdot r/k/S^2)^{5/8}]^2\} \tag{8.30}$$

where $y_{\max}$ = maximum liquid head on the landfill liner, in. or mm;
L = horizontal drainage distance, in. or mm;
r = inflow rate, in./day or mm/sec;
k = hydraulic conductivity of the drainage layer, in./day or mm/sec;
S = slope of the drainage layer, $S = \tan \alpha$; and
α = slope angle of drainage layer, measured from horizontal, degrees.

8.5.1.4 McEnroe's 1993 Method. Based on the standard Dupuit assumptions, McEnroe presented a graphic method in 1989 (McEnroe, 1989) to estimate the maximum leachate head over the liner. It is only suitable for slopes of less than 10%. McEnroe presented another set of formulas for estimating the maximum saturated depth over a sloping liner in 1993 (McEnroe, 1993). In the derivation of these formulas, the lateral drainage over the liner was described by an extended form of the Dupuit discharge formula (Harr, 1962; Childs, 1971; Chapman, 1980).

According to McEnroe (1993), the explicit formulas for estimating the maximum liquid head over a landfill liner that is draining freely, with no backwater effect from the collection trough, are expressed as

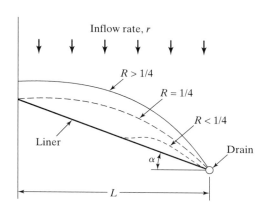

FIGURE 8.13 Phreatic Surfaces for Different R-Values

$$R = r/(k \cdot \sin^2 \alpha) \quad (8.31)$$
$$A = (1 - 4 \cdot R)^{1/2} \quad (8.32)$$
$$B = (4 \cdot R - 1)^{1/2} \quad (8.33)$$

If $R < 1/4$ (Figure 8.13),

$$y_{max} = L \cdot S \cdot (R - R \cdot S + R^2 \cdot S^2)^{1/2} \cdot \{[(1 - A - 2 \cdot R)(1 + A - 2 \cdot R \cdot S)] / [(1 + A - 2 \cdot R)(1 - A - 2 \cdot R \cdot S)]\}^{1/(2A)} \quad (8.34)$$

If $R = 1/4$ (Figure 8.13),

$$y_{max} = L \cdot S \cdot R \cdot (1 - 2 \cdot R \cdot S)/(1 - 2 \cdot R) \cdot \exp\{2 \cdot R \cdot (S - 1) / [(1 - 2 \cdot R \cdot S)(1 - 2 \cdot R)]\} \quad (8.35)$$

If $R > 1/4$ (Figure 8.13),

$$y_{max} = L \cdot S \cdot (R - R \cdot S + R^2 \cdot S^2)^{1/2} \cdot \exp\{(1/B) \cdot \tan^{-1}[(2 \cdot R \cdot S - 1)/B] - (1/B) \cdot \tan^{-1}[(2 \cdot R - 1)/B]\} \quad (8.36)$$

where y_{max} = maximum liquid head on the landfill liner, in. or mm;
 L = horizontal drainage distance, in. or mm;
 r = inflow rate, in./day or cm/sec;
 k = hydraulic conductivity of the drainage layer, in./day or mm/sec;
 S = slope of the drainage layer, $S = \tan \alpha$; and
 α = slope angle of drainage layer, measured from horizontal, degrees.

If the drainage system is working properly, i.e., not excessively clogged, the liquid level in the drainage trench will be below the upper edge of the trench, and will have no effect on the saturated-depth profile over the liner. This has been termed the "free drainage condition." All methods described herein are only suitable in the "free drainage condition."

Based on typical regulatory requirements (see for example MDEQ, 1999) the maximum leachate head on the primary liner (y_{max}) must be less than 12 inches (300 mm). The required saturated hydraulic conductivity of the leachate drainage layer (k) must not be less than 0.01 cm/sec. In addition, the minimum slope of the leachate collection pipe is 1% and the minimum bottom liner grade perpendicular to the leachate collection pipe is 2%.

EXAMPLE 8.4

A landfill cell is constructed on a pipe slope of 1% and a bottom liner grade perpendicular to the pipe of 2%. The horizontal distance from upstream to pipe is 100 ft (30 m). The granular soil drainage layer is 24 inches (0.3 m) thick. The hydraulic conductivity of the drainage material is 0.01 cm/sec. The amount of leachate inflow rate is estimated to be 3,000 gallons/acre/day (1,024 mm/year). Estimate the maximum leachate head over the primary liner.

Solution The actual landfill base grade is shown as Figure 8.14. Liquids always flow along the maximum grade. The largest grade in Figure 8.14 is not perpendicular to the leachate collection pipes. Thus, the maximum leachate flow distance from upstream boundary to the leachate collection pipe should be larger than 100 feet (30 m). The actual leachate flow grade, S, and the maximum leachate flow distance from the upstream boundary to the leachate collection pipe, L, can be calculated as follows:

Assume that
   a. the slope of the bottom liner grade perpendicular to the leachate collection pipe is S_1;
   b. the slope of the leachate collection pipe is S_2.

Then

$$S_1 = a/b \quad \text{and} \quad S_2 = c/x$$

In Figure 8.14, it can be seen that the landfill base grade varies with change of x. The landfill base grade, $S(x)$, can be expressed as follows:

$$S(x) = \frac{m}{n}$$

where $\quad m = S_1 \cdot b + S_2 \cdot x \quad$ and $\quad n = (b^2 + x^2)^{0.5}$

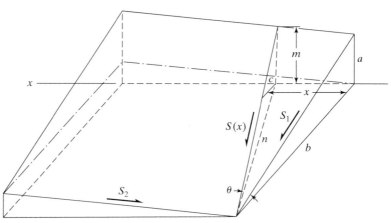

FIGURE 8.14   Landfill Cell Floor Grade for Example 8.4

Section 8.5 Estimate of Maximum Liquid Head in a Drainage Layer

Thus,

$$S(x) = \frac{S_1 \cdot b + S_2 \cdot x}{(b^2 + x^2)^{0.5}}$$

and

$$\begin{aligned}\frac{dS(x)}{dx} &= -\frac{S_1 \cdot b \cdot x}{(b^2 + x^2)^{1.5}} + \frac{S_2}{(b^2 + x^2)^{0.5}} - \frac{S_2 \cdot x^2}{(b^2 + x^2)^{1.5}} \\ &= \frac{-S_1 \cdot b \cdot x + S_2 \cdot b^2 + S_2 \cdot x^2 - S_2 \cdot x^2}{(b^2 + x^2)^{1.5}} \\ &= \frac{S_2 \cdot b^2 - S_1 \cdot b \cdot x}{(b^2 + x^2)^{1.5}}\end{aligned}$$

The landfill base grade is a maximum in which $dS(x)/dx = 0$:

$$\frac{S_2 \cdot b^2 - S_1 \cdot b \cdot x}{(b^2 + x^2)^{1.5}} = 0$$

Because $b^2 + x^2$ can never be equal to 0,

$$S_2 \cdot b^2 - S_1 \cdot b \cdot x = 0$$

and

$$x = (S_2/S_1) \cdot b.$$

For the maximum landfill base grade, S_{max},

$$m = S_1 \cdot b + S_2 \cdot x = S_1 \cdot b + S_2 \cdot (S_2/S_1) \cdot b = \frac{(S_1^2 + S_2^2) \cdot b}{S_1}$$

and

$$n = (b^2 + x^2)^{0.5} = [b^2 + (S_2/S_1)^2 \cdot b^2]^{0.5} = b \cdot [1 + (S_2/S_1)^2]^{0.5} = \frac{(S_1^2 + S_2^2)^{0.5} \cdot b}{S_1}$$

Thus,

$$S_{max} = \frac{m}{n} = \frac{(S_1^2 + S_2^2) \cdot b}{(S_1^2 + S_2^2)^{0.5} \cdot b}$$

or

$$S_{max} = (S_1^2 + S_2^2)^{0.5} \tag{8.37}$$

where S_{max} = maximum landfill base grade (i.e., actual leachate flow grade);
S_1 = slope of the bottom liner grade perpendicular to the leachate collection pipe; and
S_2 = slope of the leachate collection pipe.

The horizontal distance from upstream boundary to the leachate collection pipe along the maximum drainage grade is

$$L_{max} = b \cdot [1 + (S_2/S_1)^2]^{0.5} \tag{8.38}$$

where L_{max} = horizontal distance from upstream boundary to the leachate collection pipe along the maximum drainage grade, ft or m;
b = horizontal distance from upstream boundary to the leachate collection pipe, which is perpendicular to the leachate collection pipe, ft or m;
S_1 = slope of the bottom liner grade perpendicular to the leachate collection pipe; and
S_2 = slope of the leachate collection pipe.

For the example problem in which $S_1 = 2\%$ and $S_2 = 1\%$, the maximum base grade (i.e., actual leachate flow grade) is

$$S_{max} = (S_1^2 + S_2^2)^{0.5} \tag{8.37}$$
$$= [(0.02)^2 + (0.01)^2]^{0.5}$$
$$= 0.0224$$

The horizontal distance from the upstream boundary to the leachate collection pipe along the maximum drainage grade is

$$L_{max} = b \cdot [1 + (S_2/S_1)^2]^{0.5} \tag{8.38}$$
$$= 100 \times [1 + (0.01/0.02)^2]^{0.5}$$
$$= 112 \text{ feet} = 1,341 \text{ inches } (33.6 \text{ m})$$

Based on a given inflow rate,

$r = 3,000$ gallons/acre/day $= 40.33$ inches/year $= 0.1105$ inch/day (2.81×10^{-3} m/day)

Moore's 1980 Method:

$$y_{max} = L \cdot (r/k)^{1/2} [k \cdot S^2/r + 1 - (k \cdot S/r)(S^2 + r/k)^{1/2}] \tag{8.27}$$
$$= (1,341.6)(0.1105/340.16)^{1/2}[340.16 \times 0.0224^2/0.1105 + 1$$
$$\quad - (340.16 \times 0.0224/0.1105)(0.0224^2 + 0.1105/340.16)^{1/2}]$$
$$= (1,341.6)(0.0180)[1.545 + 1 - (68.956)(0.0288)]$$
$$= (1,341.6)(0.0180)(0.559)$$
$$= 13.4 \text{ inches} > 12.0 \text{ inches } (336 \text{ mm} > 300 \text{ mm})$$

Moore's 1983 Method:

$$y_{max} = L \cdot [(r/k + S^2)^{1/2} - S] \tag{8.28}$$
$$= (1,341.6)[(0.1105/340.16 + 0.0224^2)^{1/2} - 0.0224]$$
$$= (1,341.6)[(0.000325 + 0.000502)^{1/2} - 0.0224]$$
$$= (1,341.6)(0.0288 - 0.0224)$$
$$= (1,341.6)(0.0064)$$
$$= 8.6 \text{ inches} < 12.0 \text{ inches } (216 \text{ mm} < 300 \text{ mm})$$

Giroud's 1992 Method:

$$j = 1 - 0.12 \cdot \exp\{-[\log(1.6 \cdot r/k/S^2)^{5/8}]^2\} \tag{8.30}$$
$$= 1 - 0.12 \cdot \exp\{-[\log(1.6 \times 0.1105/340.16/0.0224^2)^{5/8}]^2\}$$
$$= 1 - 0.12 \cdot \exp\{-[\log(1.0358646)^{5/8}]^2\}$$

$$= 1 - 0.12 \cdot \exp[-(\log 1.0222671)^2]$$
$$= 1 - 0.12 \cdot \exp(-0.00091477)$$
$$= 0.8800$$

$$y_{max} = j \cdot L \cdot [(4 \cdot r/k + S^2)^{1/2} - S]/(2 \cdot \cos\alpha) \quad (8.29)$$
$$= (0.8800)(1{,}341.6)[(4 \times 0.1105/340.16 + 0.0224^2)^{1/2} - 0.0224]/(2 \times 0.9997)$$
$$= (0.8800)(1{,}341.6)(0.04244 - 0.0224)/(2 \times 0.9997)$$
$$= (0.8800)(1{,}341.6)(0.02004)/(2 \times 0.9997)$$
$$= 11.8 \text{ inches} < 12.0 \text{ inches } (299.8 \text{ mm} < 300 \text{ mm})$$

McEnroe's 1993 Method:

$$R = r/(k \cdot \sin^2\alpha) \quad (8.31)$$
$$= 0.1105/(340.16 \times 0.0224^2) = 0.6474 > 1/4$$

Because $R > 1/4$, we use Equation 8.36 to calculate y_{max}:

$$B = (4 \cdot R - 1)^{1/2} \quad (8.33)$$
$$= (4 \times 0.6474 - 1)^{1/2} = 1.2608$$

$$y_{max} = L \cdot S \cdot (R - R \cdot S + R^2 \cdot S^2)^{1/2} \cdot \exp\{(1/B) \cdot \tan^{-1}[(2 \cdot R \cdot S - 1)/B]$$
$$- (1/B) \cdot \tan^{-1}[(2 \cdot R - 1)/B]\} \quad (8.36)$$
$$= (1{,}341.6)(0.0224)[0.6474 - 0.6474 \times 0.0224 + (0.6474)^2(0.0224)^2]^{1/2}$$
$$\cdot \exp\{(1/1.2608)\tan^{-1}[(2 \times 0.6474 \times 0.0224 - 1)/1.2608]$$
$$- (1/1.2608)\tan^{-1}[(2 \times 0.6474 - 1)/1.2608]\}$$
$$= (1{,}341.6)(0.0224)(0.7957) \cdot \exp(-0.5205 - 0.1822)$$
$$= (1{,}341.6)(0.0224)(0.7959) \cdot \exp(-0.7027)$$
$$= (1{,}341.6)(0.0224)(0.7959)(0.4952)$$
$$= 11.8 \text{ inches} < 12.0 \text{ inches } (299.8 \text{ mm} < 300 \text{ mm})$$

EXAMPLE 8.5

A final cover for a municipal solid waste landfill is constructed on a slope of 10% with a maximum horizontal distance of 450 feet (135 m). The granular soil drainage layer is 24 inches (0.3 m) thick. The amount of rainfall percolating though the cover is estimated to be 30 inches/year (762 mm/year). The hydraulic conductivity of the drainage material is 0.01 cm/sec. Estimate the maximum saturated depth over the cover liner.

Solution:

$$k = 0.01 \text{ cm/sec} = 340.16 \text{ in/day } (0.1 \text{ mm/day});$$
$$L = 450 \text{ ft} = 5{,}400 \text{ in } (135 \text{ m});$$
$$r = 30 \text{ in/year} = 0.0822 \text{ in/day } (2.1 \text{ mm/day});$$
$$S = \tan\alpha = 0.1;$$
$$\alpha = 5.7°, \sin\alpha = 0.0995, \cos\alpha = 0.9950.$$

282 Chapter 8 Liquid Drainage Layer

Moore's 1980 Method:

$$y_{max} = L \cdot (r/k)^{1/2} \cdot [k \cdot S^2/r + 1 - (k \cdot S/r)(S^2 + r/k)^{1/2}] \quad (8.27)$$
$$= (5400)(0.0822/340.16)^{1/2}[340.16 \times 0.1^2/0.0822 + 1$$
$$\quad - (340.16 \times 0.1/0.0822)(0.1^2 + 0.0822/340.16)^{1/2}]$$
$$= (5400)(0.0155)[41.382 + 1 - (413.820)(0.101)]$$
$$= (5400)(0.0155)(0.586)$$
$$= 49.0 \text{ inches } (1,245 \text{ mm})$$

Moore's 1983 Method:

$$y_{max} = L \cdot [(r/k + S^2)^{1/2} - S] \quad (8.28)$$
$$= (5400)[(0.0822/340.16 + 0.1^2)^{1/2} - 0.1]$$
$$= (5400)[(0.000242 + 0.01)^{1/2} - 0.1]$$
$$= (5400)(0.00120)$$
$$= 6.5 \text{ inches } (165 \text{ mm})$$

Giroud's 1992 Method:

$$j = 1 - 0.12 \cdot \exp\{-[\log(1.6 \cdot r/k/S^2)^{5/8}]^2\} \quad (8.30)$$
$$= 1 - 0.12 \cdot \exp\{-[\log(1.6 \times 0.0822/340.16/0.1^2)^{5/8}]^2\}$$
$$= 1 - 0.12 \cdot \exp\{-[\log(0.0386642)^{5/8}]^2\}$$
$$= 1 - 0.12 \cdot \exp[-(\log 0.13094)^2]$$
$$= 1 - 0.12 \cdot \exp(-0.77956)$$
$$= 0.9450$$

$$y_{max} = j \cdot L \cdot [(4 \cdot r/k + S^2)^{1/2} - S]/(2 \cdot \cos\alpha) \quad (8.29)$$
$$= (0.9450)(5,400)[(4 \times 0.0822/340.16 + 0.1^2)^{1/2} - 0.1]/(2 \times 0.9950)$$
$$= (0.9450)(5,400)(0.10472 - 0.1)/(2 \times 0.9950)$$
$$= (0.9450)(5,400)(0.00472)/(2 \times 0.9950)$$
$$= 12.1 \text{ inches } (307 \text{ mm})$$

McEnroe's 1993 Method:

$$R = r/(k \cdot \sin^2\alpha) \quad (8.31)$$
$$= 0.0822/(340.16 \times 0.0995^2) = 0.0244 < 1/4$$

Because $R < 1/4$, use Equation 8.34 to calculate y_{max}:

$$A = (1 - 4 \cdot R)^{1/2} \quad (8.32)$$
$$= (1 - 4 \times 0.0244)^{1/2} = 0.902^{1/2} = 0.950$$

$$y_{max} = L \cdot S \cdot (R - R \cdot S + R^2 \cdot S^2)^{1/2} \cdot \{[(1 - A - 2 \cdot R)(1 + A - 2 \cdot R \cdot S)] \quad (8.34)$$
$$\qquad /[(1 + A - 2 \cdot R)(1 - A - 2 \cdot R \cdot S)]\}^{1/(2A)}$$
$$= (5400 \times 0.1)(0.0244 - 0.0244 \times 0.1 + 0.0244^2 \times 0.1^2)^{1/2}$$
$$\quad \{[(1 - 0.950 - 2 \times 0.0244)(1 + 0.950 - 2 \times 0.0244 \times 0.1)]$$
$$\quad /[(1 + 0.950 - 2 \times 0.0244)(1 - 0.950 - 2 \times 0.0244 \times 0.1)]\}^{1/(2 \times 0.950)}$$

$$= (5400 \times 0.1)(0.0220)^{1/2}\{[(0.0012)(1.945)]/[(1.901)(0.0451)]\}^{0.526}$$
$$= 5400 \times 0.1 \times 0.148 \times (0.0272)^{0.526}$$
$$= 5400 \times 0.1 \times 0.148 \times 0.150$$
$$= 12.0 \text{ inches (305 mm)}$$

8.5.2 Comparison of Various Calculation Methods

The derivation of McEnroe's 1993 method is based on the extended Dupuit assumptions, whereas Giroud's 1992 method is based on simplified assumptions and numerical methods. No derivations or explanations for Moore's 1980 or 1983 methods can be found from the U.S. EPA documents that recommend these two methods. Theoretically, McEnroe's 1993 method is felt to be the preferred method to estimate the maximum leachate head on the landfill liner and the maximum saturated depth in the final cover system.

8.5.2.1. Landfill Leachate Drainage Layer. Assume a landfill cell has a hydraulic conductivity of the leachate drainage layer k of 0.01 cm/sec, the leachate drainage slope S of 2%, the horizontal drainage distance from most upstream to leachate collection pipe L of 25 m, and the inflow rate r of 3 mm/day. The maximum liquid head over the liner y_{max} calculated by these different methods are as follows:

Moore's 1980 Method: $y_{max} = 269.0$ mm
Moore's 1983 Method: $y_{max} = 183.4$ mm
Giroud's 1992 Method: $y_{max} = 245.6$ mm
McEnroe's 1993 Method: $y_{max} = 246.1$ mm

Compared with McEnroe's 1993 method, the above results show that Moore's 1980 method overestimates the maximum leachate head over the liner. In contrast, Moore's 1983 method greatly underestimates the maximum leachate head over the liner for this specific landfill cell. Moore's 1980 method overestimates the maximum leachate head 9%, and Moore's 1983 method underestimates the maximum leachate head 26% in the example shown previously. The result from Giroud's 1992 method is almost the same as that from McEnroe's 1993 method. The difference between these two methods is only 0.2%.

In order to conduct further comparisons among these four methods for various design conditions of landfill cells, detailed calculations were conducted for a landfill cell with different leachate drainage distances and slopes, various values of hydraulic conductivity of the drainage layer, and different inflow rates for all four methods. Figure 8.15 shows the relationship between the maximum leachate head and the horizontal drainage distance for $r = 3$ mm/day, $k = 0.01$ cm/sec, and $S = 2\%$. Figure 8.16 shows the relationship between the maximum leachate head and the drainage slope for $r = 3$ mm/day, $k = 0.01$ cm/sec, and $L = 25$ m. Figure 8.17 shows the relationship between the maximum leachate head and the inflow rate for $L = 25$ m, $k = 0.01$ cm/sec, and $S = 2\%$. Figure 8.18 shows the relationship between

284 Chapter 8 Liquid Drainage Layer

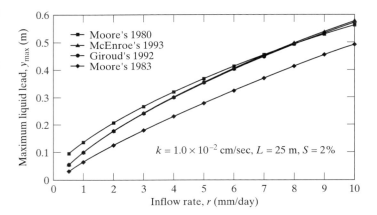

FIGURE 8.15 Relationship between Maximum Liquid Head and Horizontal Drainage Distance for Landfill Cell from Different Calculation Methods

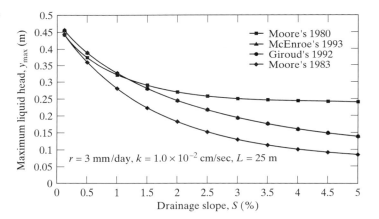

FIGURE 8.16 Relationship between Maximum Liquid Head and Drainage Slope for Landfill Cell from Different Calculation Methods

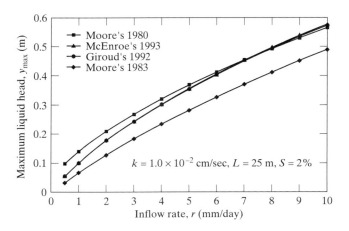

FIGURE 8.17 Relationship between Maximum Liquid Head and Inflow Rate for Landfill Cell from Different Calculation Methods

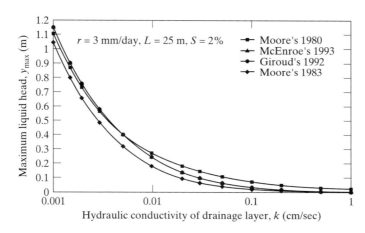

FIGURE 8.18 Relationship between Maximum Liquid Head and Hydraulic Conductivity of Drainage Layer for Landfill Cell from Different Calculation Methods

the maximum leachate head and the hydraulic conductivity of the drainage layer for $r = 3$ mm/day, $L = 25$ m, and $S = 2\%$.

A comparison of results shown in Figure 8.15 shows that McEnroe's 1993 method tends to overestimate Moore's 1980 method, whereas it underestimates Moore's 1983 method for calculating the maximum leachate head in a landfill cell with $r = 3$ mm/day, $k = 0.01$ cm/sec, and $S = 2\%$. The discrepancy between McEnroe's 1993 method and Moore's 1980 or 1983 method increases with an increase in the horizontal drainage distance. Figure 8.15 also shows that the maximum liquid heads calculated from Giroud's 1992 method and McEnroe's 1993 method are almost the same; the difference between these two methods does not exceed 0.2% for various drainage distances with $r = 3$ mm/day, $k = 0.01$ cm/sec, and $S = 2\%$.

Figure 8.16 shows that Moore's 1980 method underestimates the maximum leachate head when $S < 1.3\%$ and overestimates the maximum leachate head when $S > 1.3\%$ for a landfill cell with $r = 3$ mm/day, $k = 0.01$ cm/sec, and $L = 25$ m. Overall, Moore's 1983 method tends to underestimate the maximum leachate head. The difference between McEnroe's 1993 method and Moore's 1980 (when $S > 1.3\%$) or 1983 method increases with an increase in the drainage slope. The results from Moore's 1980 and 1983 methods do not approach those of McEnroe's 1993 method until the drainage slope reduces to 0.2%. The use of such flat slopes is not allowed in landfill subbase design. The results from Giroud's 1992 method and McEnroe's 1993 method are approximately the same; the difference between these two methods is less than 1% for various drainage slopes with $r = 3$ mm/day, $k = 0.01$ cm/sec, and $L = 25$ m.

Figure 8.17 shows that Moore's 1980 method overestimates the maximum leachate head when $r < 7$ mm/day and underestimates the maximum leachate head when $r > 7$ mm/day for a landfill cell with $k = 0.01$ cm/sec, $L = 25$ m, and $S = 2\%$. Overall Moore's 1983 method still underestimates the maximum leachate head. The difference between the results of McEnroe's 1993 method and Moore's 1983 method increase with an increase in the inflow rate. The results from McEnroe's 1993 method and Moore's 1980 method only approach one another when the inflow rate lies between 4 and 10 mm/day. On the other hand, the results from Giroud's 1992 method and McEnroe's 1993 method are still almost the same; the maximum difference between these two methods is only about 0.5% for various inflow rates, with $k = 0.01$ cm/sec, $L = 25$ m, and $S = 2\%$.

Finally, Figure 8.18 shows that Moore's 1980 method underestimates the maximum leachate head when $k < 5.0 \times 10^{-3}$ cm/sec and overestimates the maximum leachate head when $k > 5.0 \times 10^{-3}$ cm/sec for a landfill cell with $r = 3$ mm/day, $L = 25$ m, $S = 2\%$. Overall, Moore's 1983 method tends to underestimate the maximum leachate head. The results from Giroud's 1992 method are still approximately the same as those from McEnroe's 1993 method; the difference between these two methods is less than 1% for various values of hydraulic conductivity of the drainage layer with $r = 3$ mm/day, $L = 25$ m, $S = 2\%$.

A detailed comparison of the various methods shows that Moore's 1983 and Moore's 1980 method tend to underestimate or overestimate the maximum leachate head depending on the drainage slope, inflow rate, and hydraulic conductivity of the drainage layer. Moore's 1983 method always underestimates the maximum leachate head on the liner compared to the other methods. The results from Giroud's 1992 method compare very closely to those from McEnroe's 1993 method; the difference between these two methods is less than 1% for all cases examined.

8.5.2.2 Landfill Cover System. Assume a landfill cover has the hydraulic conductivity of a sand drainage layer of 0.01 cm/sec, a drainage layer slope of 25%, a horizontal distance from most upstream to toe drain of 100 m and an inflow rate of 3 mm/day. The maximum liquid head over the liner from three different methods can be calculated as follows:

Moore's 1980 Method: $y_{max} = 933.0$ mm
Moore's 1983 Method: $y_{max} = 69.3$ mm
Giroud's 1992 Method: $y_{max} = 139.1$ mm
McEnroe's 1993 Method: $y_{max} = 144.0$ mm

The preceding analysis shows that Moore's 1980 method greatly overestimates the maximum saturated depth in the final cover system, compared with either McEnroe's 1993 method or Giroud's 1992 method, respectively. Conversely, Moore's 1983 method greatly underestimates the maximum liquid head relative to these other two methods. The result from Moore's 1980 method is almost 6.5 times that of McEnroe's 1993 method, whereas Moore's 1983 method yields a result less than 50% of that from either McEnroe's 1993 method or Giroud's 1992 method. The depth calculated by these latter two methods differs by only 3.4%.

In order to conduct further comparisons for these four methods for various design conditions of a landfill cover system, detailed calculations were also carried out using all four methods for a landfill cover with different drainage distances and slopes, various values of hydraulic conductivity of the drainage layer, and for different inflow rates. Figure 8.19 shows the relationship between the maximum saturated depth and the horizontal drainage distance for $r = 3$ mm/day, $k = 0.01$ cm/sec, and $S = 25\%$. Figure 8.20 shows the relationship between the maximum saturated depth and the drainage slope for $r = 3$ mm/day, $k = 0.01$ cm/sec, and $L = 100$ m. Figure 8.21 shows the relationship between the maximum saturated depth and the inflow rate for

Section 8.5 Estimate of Maximum Liquid Head in a Drainage Layer

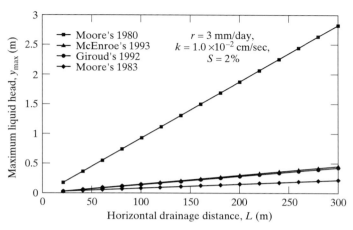

FIGURE 8.19 Relationship between Maximum Liquid Head and Horizontal Drainage Distance for Landfill Cover from Different Calculation Methods

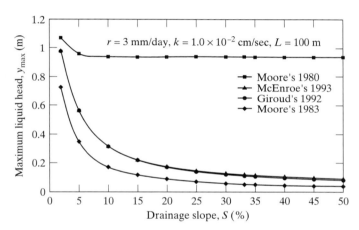

FIGURE 8.20 Relationship between Maximum Liquid Head and Drainage Slope for Landfill Cover from Different Calculation Methods

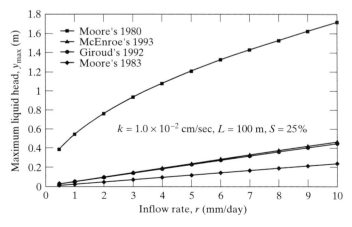

FIGURE 8.21 Relationship between Maximum Liquid Head and Inflow Rate for Landfill Cover from Different Calculation Methods

$k = 0.01$ cm/sec, $L = 100$ m, and $S = 25\%$. Figure 8.22 shows the relationship between the maximum liquid head and the hydraulic conductivity of the drainage layer for $r = 3$ mm/day, $L = 100$ m, and $S = 25\%$.

A detailed comparison of the results in Figures 8.19 to 8.22 shows that Moore's 1980 method always greatly overestimates, whereas Moore's 1983 method always underestimates the maximum saturated depth in the final cover system, compared with the other two methods for various values of drainage distances and slopes, inflow rates, and hydraulic conductivity of the drainage layer. The difference between the results of Giroud's 1992 method and McEnroe's 1993 method is less than 4% with a 25% drainage slope, about 5.4% with a 3(H):1(V) drainage slope, and up to 10.6% with a 2(H):1(V) drainage slope for various values of drainage distances, inflow rates, and hydraulic conductivity of the drainage layer.

8.5.2.3 Concluding Comments about Various Calculation Methods. Several conclusions can be drawn from a comparison of four currently available methods for calculating the maximum liquid head over landfill barriers. From a theoretical standpoint, McEnroe's 1993 method appears the best way for estimating the maximum leachate head on a landfill liner and the maximum saturated depth in the final cover system. Moore's 1980 method either underestimates or overestimates the maximum leachate head when the drainage slope is flat, the horizontal drainage distance is short (e.g., for a leachate drainage layer), or hydraulic conductivity of the drainage layer is low (e.g., $k < 5.0 \times 10^{-3}$ cm/sec). Moore's 1980 method always greatly overestimates the maximum liquid head when the drainage layer has a steep slope and long horizontal drainage distance (e.g., for a final cover system). Moore's 1983 method always underestimates the maximum liquid head under any conditions. In general, the results calculated from Giroud's 1992 method are slightly lower than those calculated from McEnroe's 1993 method. The drainage slope is a major factor that affects the difference in the results between Giroud's 1992 and McEnroe's 1993 method. The difference between the results of these two methods increases with an increase in the slope of the drainage layer. This difference between the results of these two methods is about 1%

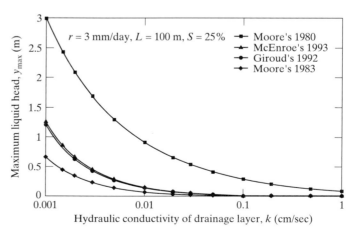

FIGURE 8.22 Relationship between Maximum Liquid Head and Hydraulic Conductivity of Drainage Layer for Landfill Cover from Different Calculation Methods

when the drainage slope is less than 10% and less than 5.5% for a 3(H):1(V) drainage slope. McEnroe's 1993 method can be viewed as a theoretically precise or "exact" method and is recommended for the general analysis and design of a drainage system for landfill covers and bottom liners. Giroud's 1992 method is a numerical approximation method. The method is simpler than McEnroe's 1993 method and provides accurate results for general landfill design conditions. Giroud's 1992 method is recommended for analysis and design of drainage systems for landfill covers and bottom liners with a drainage slope of less than 3(H):1(V). Finally, it should be noted that the methods described in this section are only suitable for "free drainage conditions." This implies that the liquid level in the drainage trench is always below the upper edge of the trench, and has no effect on the saturated-depth profile over the liner.

PROBLEMS

8.1 Compare the functions, specifications, and uses of the top layer of sand shown in Figure 8.4 and 8.6.

8.2 Where are natural soil drainage materials (e.g., sand and gravel) mainly used in landfill projects? What are the basic requirements for a soil that is proposed as a drainage material?

8.3 What are the main functions for a soil filter layer?

8.4 If the flow rate of a geotextile predicted from an index-type test is 1.7 gal/min-ft (21.1 liter/min-m), what would be the average allowable flow rate for design purposes according to Table 8.5 for:

(a) Underdrain filters?

(b) Landfill filters?

(c) Gravity drainage?

8.5 Use Task Force 25, Carroll ($O_{95} < 2.5 \cdot d_{85}$), and Giroud methods respectively to design a needle-punched, nonwoven geotextile for use in a primary leachate drainage layer (use average data from Table 8.5). The geotextile is placed between a protective sand blanket and geonet as a filtration layer like that shown in Figure 8.3. Typical properties of needle-punched nonwoven geotextiles are shown in Tables 8.2 and 8.3. The hydraulic gradient for filtration is 1.5. The maximum leachate percolation rate at the time of landfill opening is 40 inches per year. The soil placed over the geotextile is a well-graded sand (SW) in a medium-dense state ($D_R = 70\%$). Information about the soil gradation is as follows:

$$13\% \text{ soil particles passing the No. 200 sieve;}$$
$$d_{60}/d_{10} = 6.6;$$
$$d_{85} = 1.2 \text{ mm;}$$
$$d_{50} = 0.275 \text{ mm.}$$

8.6 Use Task Force 25, Carroll ($O_{95} < 2 \cdot d_{85}$), and Giroud methods, respectively, to design a needle-punched nonwoven geotextile for use in a lateral drainage layer of a landfill final cover system (use average data for "*Underdrain Filters*" from Table 8.5). The geotextile is placed between a soil protective layer and a geonet as a filtration layer. Typical properties of needle-punched nonwoven geotextiles are shown in Tables 8.2 and 8.3. The thickness of geonet is 0.2 inch (5 mm). Assume the maximum inflow rate is 0.14 inch/hour

(3.6 mm/hour). The liquid head over the geomembrane liner is 2 inches (50 mm). The index properties of the soil placed over the geotextile are as follows:

21% soil particles passing the No. 200 sieve;

$d_{85} = 3.7$ mm;

PI = 16.

8.7 Use of a geonet beneath a frost protection layer in a landfill cap requires an allowable flow rate of 2.25 gal/min-ft (28 liter/min-m). What is the necessary ultimate flow rate based on average data from Table 8.8?

8.8 The ultimate flow rate of a geonet being considered for a primary leachate collection system on a landfill side slope is 3.5 gal/min-ft (43.5 liter/min-m). Using the maximum values in Table 8.8, what is the allowable flow rate?

8.9 Is a drainage composite in Table 8.9 adequate for a primary leachate collection system (use average data from Table 8.8)? The drainage composite in Table 8.9 consists of a geotextile bonded to one side of a bi-planar drainage net. The drainage composite lies between a two-foot- (0.6-m)-thick protective layer ($\gamma_{sand} = 115$ lb/ft^3 or 18 kN/m^3) and a HDPE geomembrane. The design inflow rate is 3,500 gal/acre/day (33,000 liter/ha/day). The slope of the bottom is 2%. The horizontal distance of the slope is 100 feet (30 m). The landfill when completed will be 200 ft (60 m) high with a unit weight of waste of 65 lb/ft^3 (10.2 kN/m^3) (The required factor of safety must be greater than 2).

8.10 What is the factor of safety of a biplanar geonet drainage composite in Table 8.11 when used as a primary leachate drainage layer in a valley fill type landfill (use average data from Table 8.8). The drainage composite in Table 8.11 consists of a geotextile bonded to both sides of a biplanar drainage net. The drainage composite lies between a two-foot (0.6-meter) protective layer ($\gamma_{sand} = 110$ lb/ft^3 or 17.3 kN/m^3) and a HDPE geomembrane. The bottom slope is 10%. The horizontal distance of the slope is 250 feet (75 m). The landfill when completed will be 350 ft (105 m) high with a unit weight of waste of 70 lb/ft^3 (11 kN/m^3). Assume the maximum inflow rate is 0.20 inch/day (5 mm/day). The ultimate flow rate of the drainage geocomposite is listed in Table 8.11. If the factor of

TABLE 8.11 Ultimate Flow Rate and Transmissivity of a Commercially Available Biplanar Geonet Drainage Composite Product for Problem 8.10

Load	Gradient	Ultimate Flow Rate			Transmissivity	
		gal/min-ft	liter/min-m	gal/min-ft	m^2/sec($\times 10^{-4}$)	ft^2/sec ($\times 10^{-3}$)
20,000 lb/ft^2 (960 kN/m^2)	0.10	0.29	3.60	2.94	6.09	6.55
	0.25	0.53	6.58	2.12	4.38	4.72
	0.50	0.81	10.06	1.63	3.37	3.63
	0.75	1.02	12.67	1.36	2.82	3.03
	1.00	1.21	15.03	1.21	2.50	2.70
25,000 lb/ft^2 (1,200 kN/m^2)	0.10	0.14	1.74	1.40	2.91	3.13
	0.25	0.25	3.11	0.98	2.03	2.19
	0.50	0.35	4.35	0.70	1.44	1.55
	0.75	0.45	5.59	0.60	1.25	1.35
	1.00	0.54	6.71	0.54	1.12	1.21

TABLE 8.12 Physical Properties of Commercially Available Geonet Product for Problem 8.12

Property	Style (a)	Style (b)	Style (c)	Style (d)
Thickness (inches)	0.160	0.200	0.200	0.270
(mm)	4	5	5	7
Ultimate Transmissivity (m^2/sec)	1×10^{-3} (at 2,000 lb/ft^2 or 96 kN/m^2)	1×10^{-3} (at 15,000 lb/ft^2 or 718 kN/m^2)	1×10^{-3} (at 4,000 lb/ft^2 or 192 kN/m^2)	1×10^{-3} (at 15,000 lb/ft^2 or 718 kN/m^2)
Tensile Strength (lb/in)	30	42	23	44
(kN/m)	5.25	7.35	4.03	7.7
Width × Length (feet)	7.54 × 300	7.54 × 300 14.5 × 300	7.54 × 300	7.54 × 220
(m)	2.3 × 90	2.3 × 90 4.4 × 90	2.3 × 90	2.3 × 66

safety is less than 2.0, change to a triplanar geonet drainage composite listed in Table 8.10 to check whether the factor of safety can be greater than 2.0.

8.11 Consider a proposed landfill design with a side slope of 33% [3(H):1(V)] and a maximum height of 100 ft (30 m). The primary leachate drainage is a 24-inch- (0.6-meter)-thick sand layer with a hydraulic conductivity of 1.0×10^{-2} cm/sec. Assume the leachate inflow rate is 100 inch/year (2,540 mm/year) at the landfill site. Use Giroud's 1992 and McEnroe's 1993 methods to estimate the maximum leachate head over the bottom liner.

8.12 A final cover for a municipal solid waste landfill is constructed on a slope of 10% with a maximum horizontal drainage distance of 500 feet (150 m). A commercially available geonet, Style (b), is used as the lateral drainage layer. Typical physical properties of the geonet are listed in Table 8.12. Use average values of reduction factors in Table 8.8 for determining allowable transmissivity of the geonet. Assume the maximum inflow rate is 22,000 gallon/acre/day (208,000 liter/ha/day). Use Giroud's 1992 and McEnroe's 1993 methods to estimate the maximum saturated depth over the cover liner.

8.13 With leachate collection pipes of 1% slope at 200-ft (60-m) intervals, a percolation rate of 4.5 inch/month (114.3 mm/month), and a 2-ft primary leachate drainage layer with hydraulic conductivity of 5.0×10^{-2} cm/sec, find the maximum leachate head acting on a liner using Giroud's 1992 and McEnroe's 1993 methods, assuming a bottom slope of 2%. Recalculate the maximum head acting on the liner, assuming a bottom slope of 1%.

REFERENCES

Carroll, R. G., Jr., (1983) "Geotextile Filter Criteria," TRR 916, Engineering Fabrics in Transportation Construction, Washington, DC., pp. 46–53.

Cedergren, H. R., (1989) " Seepage, Drainage, and Flow Nets," 3 rd Edition, John Wiley & Sons, Inc., New York, NY.

Chapman, T. C., (1980) "Modeling Groundwater Flow over Sloping Beds," *Water Resources Research*, Vol. 16, No. 6, pp. 1114–1118.

Childs, E. C., (1971) "Drainage of Groundwater Resting on a Sloping Bed," *Water Resources Research*, Vol. 7, No. 5, pp. 1256–1263.

GeoSyntec, (1996) "Final Report: Interface Direct Shear Testing Tenax CE-9," GeoSyntec Consultants, Atlanta, GA.

Giroud, J. P., (1982) "Filter Criteria for Geotextile," *Proceedings of 2nd International Conference on Geotextiles*, Las Vegas, NV, August, Vol. 1, pp. 103–109.

Giroud, J. P., Badu-Tweneboah, K., and Bonaparte, R., (1992) "Rate of Leakage through a Composite Liner due to Geomembrane Defects," *Geotextiles and Geomembranes,* Vol. 11, No. 1, Elsevier Science Publishers Ltd., Oxford, England, UK, pp. 1–28.

Giroud, J. P. and Houlihan, M. F., (1995) "Design of Leachate Collection Layers," Proceedings of the Fifth International Landfill Symposium, Sardinia, Italy, October, Vol. 2, pp. 613–640.

Harr, M. E., (1962) "Groundwater and Seepage," McGraw-Hill Book Co., New York, NY, pp. 210–226.

Koerner, R. M., (1994) *Designing with Geosynthetics,* 3rd Edition, Prentice Hall Inc., Englewood Cliffs, NJ, 07632.

Koerner, R. M., (1998) "Designing with Geosynthetics," 4th Edition, Prentice Hall Inc., Upper Saddle River, NJ, 07458.

Koerner, R. M. and Hwu, B. -L., (1989) "Behavior of Double Geonet Systems," *Geotechnical Fabrics Report,* Industrial Fabrics Association International (IFAI), Roseville, MN, September/October, pp. 39–42.

Koerner, J. R., Soong T-. Y., and Koerner, R. M., (1998) "A Survey of Solid Waste Landfill Liner and Cover Systems: Part 1—USA States," GRI Report #21, GSI, Folsom, PA, 211 pp.

Koerner, G. R., (1993) "Performance Evaluation of Geotextile Filters Used in Leachate Collection of Solid Waste Landfills," Ph.D. Dissertation, Drexel University, Philadelphia, PA, 219 pp.

McEnroe, B. M., (1989) "Steady Drainage of Landfill Covers and Bottom Liners," *Journal of Environmental Engineering,* ASCE, Vol. 115, December, No. 6, pp. 1114–1122.

McEnroe, B. M., (1993) "Maximum Saturated Depth over Landfill Liner," *Journal of Environmental Engineering*, ASCE, Vol. 119, March/April, No. 2, pp. 262–270.

McGown, A., (1978) "The Properties of Nonwoven Fabrics Presently Identified as Being Important in Public Works Applications," *Index 78 Programme,* University of Strathclyde, Glasgow, Scotland, UK.

MDEQ, (1999) "Solid Waste Management Act Administrative Rules Promulgated Pursuant to Part 115 of the Natural Resources and Environmental Protection Act, 1994 PA 451, as amended (Effective April 12, 1999)," Michigan Department of Environmental Quality, Waste Management Division, Lansing, MI.

Qian, Xuede, (1994) "Estimation of Maximum Saturated Depth on Landfill Liner," Michigan Department of Environmental Quality, Waste Management Division, Lansing, MI, September.

Task Force No. 25, (1983) "Report on Task Force," Joint Committee Report of AASHTO-AGC-ARTBA, December 2, Washington, D.C.

USEPA, (1980) "Landfill and Surface Impoundment Performance Evaluation," EPA/536/SW-869-C, U. S. Environmental Protection Agency, Washington, D.C., September.

USEPA, (1983) "Landfill and Surface Impoundment Performance Evaluation," SW-869, Revised Edition, Office of Solid Waste and Emergency Response, U.S. Environmental Protection Agency, Washington, D.C., April.

USEPA, (1989) "Requirements for Hazardous Waste Landfill Design, Construction, and Closure," EPA/625/4-89/022, Center for Environmental Research Information, Office of Research and Development, U. S. Environmental Protection Agency, Cincinnati, OH, August.

USEPA, (1991) "Federal Register, Part II, 40 CFR Parts 257 and 258, Solid Waste Disposal Facility Criteria; Final Rule," U. S. Environmental Agency, Washington, DC, October 9.

USEPA, (1993) "Quality Assurance and Quality Control for Waste Containment Facilities," Technical Guidance Document, EPA/600/R-93/182, U.S. Environmental Protection Agency, Office of Research and Development, Washington, DC, September.

Williams, N., Giroud, J. P., and Bonaparte, R., (1984) "Properties of Plastics Nets for Liquid and Gas Drainage Associated with Geomembranes," *Proceedings of International Conference on Geomembranes,* Denver, CO.

CHAPTER 9

Leachate Collection and Removal Systems

9.1 SUBBASE GRADING
9.2 LEACHATE COLLECTION TRENCHES
9.3 SELECTION OF LEACHATE COLLECTION PIPE
    9.3.1 TYPE OF PIPE MATERIAL
    9.3.2 PIPE DESIGN ISSUES
    9.3.3 PIPE PERFORATIONS
9.4 DEFORMATION AND STABILITY OF LEACHATE COLLECTION PIPE
    9.4.1 PIPE DEFLECTION
    9.4.2 PIPE WALL BUCKLING
9.5 SUMP AND RISER PIPES
9.6 LEACHATE REMOVAL PUMPS
    PROBLEMS
    REFERENCES

A leachate collection and removal system is designed and constructed to collect leachate and convey the leachate out of the landfill for treatment or reinjection depending on the liquid management strategy. Most regulations stipulate that the system must ensure that less than 12 inches (0.3 m) of leachate accumulates over the liner to minimize leakage through the liner and possible contamination of the underlying groundwater. The leachate collection and removal system consists of a leachate filter and drainage layer (which has been discussed in Chapter 8), a perforated collection pipe network, sumps, riser pipes or manholes, cleanout ports, pumps, and leachate storage tanks. A leachate storage tank may not be necessary for a site where the leachate can be directly discharged to a sewer system and then pumped to a treatment facility. These components must be designed to handle large leachate flows associated with initial operations and to resist problems that can degrade the long-term flow capacity of the system. For double lined systems, a leak detection drainage system is necessary. It must have similar components to a leachate collection system.

9.1 SUBBASE GRADING

In order to avoid the accumulation of leachate at the bottom of a landfill, the base is generally graded into a series of sloped terraces (Figure 9.1). Note that there are many options available in this regard. The profile and configuration of the bottom of a land-

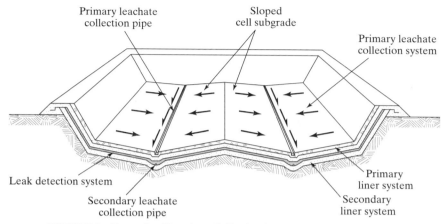

FIGURE 9.1 Example of Leachate Collection System with Sloped Subgrade

fill must be such that gravitational flow to a low point always exists. This requirement must be satisfied for both primary and secondary leachate collection systems. Thus, accurate grading of the bottom of the landfill (and every cell within a landfill) is very important. The consequence of improper design, localized low points, subsidence of subsoil, poor construction quality control and assurance, etc., is that leachate will pond above the geomembrane and create leakage above that anticipated, rather than being properly removed and treated.

Based on Federal and some state regulations, the landfill base floor must be graded a minimum slope of 2% in directions perpendicular to the leachate collection pipes to promote drainage and prevent ponding above the liner. The leachate collection pipes should be laid on a slope of 1 percent or more in a direction to intercept liquid flow. The low point of the leachate collection system must terminate at a sump with a riser pipe or manhole rising from this location through the waste as it is placed. Older designs of penetrations through the liner systems (so that gravity flow is continued) are not recommended. The difficulty of making leak-free connections at this low point in the entire facility is formidable and has been the source of leakage in the past.

For sites where removal pipe systems within the leachate sand drainage layer must be periodically inspected, cleaned, and flushed, both access and egress must be provided. This will require careful planning and can completely dictate the nature of the grading plan.

9.2 LEACHATE COLLECTION TRENCHES

Leachate collection pipes are sometimes installed in gravel-filled trenches. The trenches (not the pipe) should be lined with a geotextile to minimize entry of fines from the liner into the trench and eventually into the leachate collection pipe. An example of a specific design is shown in Figure 9.2. It is recommended that a deeper compacted clay liner be placed below the collection trench so that the compacted clay liner has the same minimum design thickness below the trench as above it.

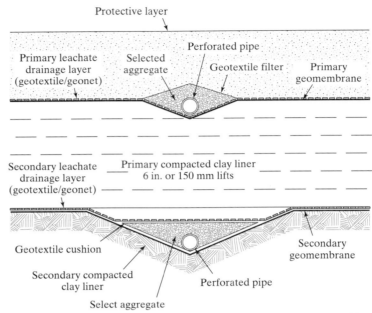

FIGURE 9.2 Example of Leachate Collection Pipe and Trench for Double Geomembrane/Compacted Clay Composite Liner System

The gravel used in the trench should be mounded to distribute the load of compaction machinery and thereby provide more protection for the pipe against crushing. The geotextile, which acts as a filter, should be folded over the gravel. Alternatively, a graded sand filter may be designed to minimize the infiltration of fines into the trench from the waste (Bagchi, 1994). Designs for both soil and geotextile filters are discussed below.

Granular Soil Filter. Several criteria that differ only slightly from one another are available for the design of granular soil filters. These design criteria were presented and discussed in Section 8.2 (i.e., Equations 8.2 to 8.5).

The criteria are focused to prevent migration of overlying soils into the filter layer and also to ensure sufficient hydraulic conductivity of the filter layer to maintain proper flow. A critique of several design criteria can be found in Sherard et al. (1984a and 1984b), along with their suggestions for alternative approaches.

Geotextile Filter. Design criteria for geotextiles have been discussed in Chapter 8. The approach for filter fabric design primarily consists of comparing the soil particle size characteristics with the apparent opening size (AOS, also called the equivalent opening size or EOS) of the filter geotextile and being assured of adequate flow through the geotextile. The design process of a geotextile filter follows that of a granular soil filter. The entire design process was presented in Section 8.2.

9.3 SELECTION OF LEACHATE COLLECTION PIPE

A leachate collection pipe may fail due to clogging, crushing, or faulty installation. Design specifications for leachate pipes for each of these situations should include the following:

 (i) Type of pipe material;
 (ii) Design issues (e.g., diameter and wall thickness of the pipes or equivalent SDR);
 (iii) Size and distribution of slots and perforations on the pipes;
 (iv) Type of pipe bedding material and required compaction used to support the pipes.

9.3.1 Type of Pipe Material

There are many choices of pipe materials for leachate collection and leak detection systems. Of these, polymeric pipes are by far the most common. Within the available types of versions, HDPE and PVC (both are available in profiled or smooth wall conditions) are used almost exclusively. Fortunately, both pipe materials have well crafted material specifications available from ASTM (e.g., D3350 for HDPE and D1784 for PVC).

Regarding the HDPE pipe, it should be recognized that the density is higher than HDPE geomembrane (e.g., 0.95 g/cc versus 0.92 g/cc, respectively). Also to be noted is that PVC pipe (unlike PVC geomembrane) contains no plastizers.

The profiled or corrugated version of HDPE drainage pipe is currently undergoing a tremendous growth period. For example, use of HDPE profiled pipe in New York has gone from 1% to 60% market share in the last six years.

9.3.2 Pipe Design Issues

The design of perforated leachate collection pipes should consider the following factors:

 (i) The required flow, using estimated percolation impingement rates, maximum drainage slope, and maximum pipe spacing.
 (ii) Pipe size, using required flow and minimum pipe slope.
 (iii) The structural strength of the pipe.

The pipe spacing may be determined by using the HELP model or the Giroud's 1992 and McEnroe's 1993 methods as discussed in Chapter 8. In using the HELP model or Giroud's 1992 and McEnroe's 1993 methods to calculate the pipe spacing, the maximum height of liquid between two parallel perforated drainage pipes should be less than 12 inches (0.3 m).

The amount of unit area leachate production (or percolation) which is used to design the size of pipes can be estimated in the following two ways:

 1. HELP model, considering 25-year, 24-hour storm (e.g., for Michigan, 0.14 inch/hour or 3.6 mm/hour) and using the maximum peak daily leachate flow rate.

2. Percolation rate given by the appropriate rules, such as 40 inches/year or 1,020 mm/year, if there are such regulations.

The required flow rate used to determine the pipe size can be calculated with the equation

$$Q_{reqd} = q_{max} \times A_{cell} \tag{9.1}$$

where Q_{reqd} = required leachate flow rate, ft³/sec or m³/sec;
q_{max} = maximum unit area leachate production, ft³/sec/ft² or m³/sec/m² (obtained from one of the two methods cited previously); and
A_{cell} = cell area served by a leachate collection pipe, ft² or m².

If the base of the leachate drainage layer of a landfill is graded in a V-shape and the leachate collection pipe is placed at the bottom of the V-shape, it means that the pipe will take on leachate from both sides of the pipe. In that case, the cell area, A_{cell}, in Equation 9.1 should be equal to the summation of the cell areas of both sides of the pipe.

Once the required leachate flow rate, pipe slope, and material of the pipe are known, the size of the pipe can be determined by a trial-and-error procedure using Manning's equation. To determine the suitable pipe size, assume a pipe size first and then calculate the flow rate of the pipe based on the assumed pipe size using Manning's equation. The calculated flow rate from Manning's equation must be greater than or equal to the required leachate flow rate. If not, another pipe size must be tried and the flow rate calculation repeated until the calculated flow rate from Manning's equation is greater than or equal to the required leachate flow rate.

Manning's Equation (in U.S. units):

$$Q = (1.49/n) \cdot A \cdot r_h^{2/3} \cdot S^{1/2} \tag{9.2}$$

where Q = flow rate of pipe, ft³/sec;
n = Manning's roughness coefficient, $n \approx 0.009$ for PVC and $n \approx 0.011$ for HDPE pipe;
A = area in flow, ft²;
S = slope of the pipe; and
r_h = hydraulic radius, ft.

Manning's Equation (in SI units):

$$Q = (1/n) \cdot A \cdot r_h^{2/3} \cdot S^{1/2} \tag{9.3}$$

where Q = flow rate of pipe, m³/sec;
n = Manning's roughness coefficient, $n \approx 0.009$ for PVC and $n \approx 0.011$ for HDPE pipe;
A = area in flow, m²;
S = slope of the pipe; and
r_h = hydraulic radius, m.

Hydraulic Radius:

$$r_h = A/P_w \qquad (9.4)$$

where A = area in flow, ft² or m²;
r_h = hydraulic radius, ft or m; and
P_w = wetted perimeter, ft or m.

For full pipe flow,

$$r_h = D_{in}/4 \qquad (9.5)$$

where D_{in} = inside diameter of pipe, ft or m.

Standard Dimension Ratio:

$$SDR = D_o/t \qquad (9.6)$$

where SDR = standard dimension ratio, the same as the dimension ratio, DR;
D_o = outside diameter of pipe, in or m; and
t = thickness of pipe, in or m.

9.3.3 Pipe Perforations

The most important parameter for determining the size and distribution of slots or perforations in the leachate collection pipes is the maximum leachate inflow per unit length of the pipe. The maximum leachate inflow per unit length of the pipe, which mainly depends on the maximum unit area leachate production and the maximum servicing unit cell area per foot of pipe, can be calculated from the equation

$$Q_{in} = q_{max} \times A_{unit} \qquad (9.7)$$

where Q_{in} = maximum leachate inflow per unit length of pipe, ft³/sec/ft or m³/sec/m;
q_{max} = maximum unit area leachate production, ft³/sec/ft² or m³/sec/m², obtain from HELP model (peak daily flow rate) or percolation rate given by regulations; and
A_{unit} = maximum servicing unit cell area per foot of pipe, ft²/ft or m²/m.

Note that

$$A_{unit} = (L_H)_{max} \times dw \qquad (9.8)$$

where $(L_H)_{max}$ = maximum horizontal distance of leachate flow, ft or m; and
dw = unit width at the area of the maximum horizontal distance of leachate flow.

As stated previously, if the base of the leachate drainage layer of a landfill is graded in a V-shape and the leachate collection pipe is placed at the bottom of the V-shape, it means that the pipe will take on leachate from both sides of the pipe. In that case, the maximum horizontal distance of leachate flow $(L_H)_{max}$ in Equation 9.8 should be equal to the maximum summation of the horizontal leachate flow distances of both sides of the pipe.

The inflow capacity per orifice can be calculated from the Bernoulii equation if the size and shape of the orifice are known.

Bernoulii Equation:

$$Q_b = C \cdot A_b \cdot (2 \cdot g \cdot \Delta h)^{0.5} \qquad (9.9)$$

where Q_b = inflow capacity per orifice (per hole or slot), ft³/sec or m³/sec;
C = discharge coefficient = 0.62;
A_b = cross-sectional area of a slot or hole on the selected perforated pipe, ft² or ft²;
g = gravitational constant (g = 32.2 ft/sec² = 9.81 m/sec²); and
Δh = liquid head, ft or m.

In Equation 9.9,

$$(2 \cdot g \cdot \Delta h)^{0.5} = v_{ent} \qquad (9.10)$$

where v_{ent} = limit leachate entrance velocity, ft/sec or m/sec.

Equation 9.9 also can be written as

$$Q_b = C \cdot A_b \cdot v_{ent} \qquad (9.11)$$

From Equation 9.11, it can be observed that the inflow capacity per hole or slot mainly depends on the size and shape of the opening and the limiting leachate entrance velocity. The limiting leachate entrance velocity v_{ent} is usually assumed to be equal to 0.1 ft/sec (0.03 m/sec) in the calculation of the pipe perforation (Driscoll, 1986). The diameter d of a perforation hole of 1/4 or 3/8 inches (6 or 10 mm) is a typical value for leachate collection pipes.

Once the maximum leachate inflow per unit length of pipe and the inflow capacity per opening are known, the number of the perforated holes per unit length of pipe can be calculated using the equation

$$N = Q_{in}/Q_b \qquad (9.12)$$

where N = number of perforation holes per foot of pipe or number of perforation holes per meter of pipe;
Q_{in} = maximum leachate inflow per foot pipe, ft³/sec/ft or m³/sec/m; and
Q_b = maximum inflow capacity per orifice (per hole or slot), ft³/sec or m³/sec.

The holes in a perforated solid wall leachate collection pipe should be made as shown in Figure 9.3. To maintain the lowest possible leachate head, the pipe should be laid such that the holes are located along the lower half of the pipe, but not directly at the invert. Holes close to the spring line reduce the strength of the pipe and hence should be avoided. An alternative to perforated solid wall pipe is slotted profiled pipe.

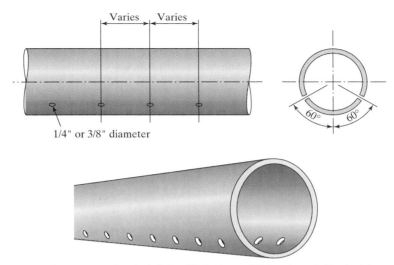

FIGURE 9.3 Perforations in Solid Wall HDPE or PVC Leachate Collection Pipes

Most slotted profiled pipe is made from HDPE, but the wall structure is completely different from solid wall pipe. (See Figure 9.4 for slotted profiled pipe.)

When perforated pipes are embedded in granular materials that are used as filters, no unplugged ends should be allowed and the filter materials in contact with pipes must be coarse enough not to enter joints, holes, or slots. The U.S. Army Corps of

FIGURE 9.4 Profiled or Corrugated HDPE Pipe with Perforations (i.e., Slots in Valleys)

Engineers (1955) and the U.S. Army et al. (1971) use the following criteria for gradation of filter materials in relation to holes and slots:
For Circular Holes,

$$\frac{85\% \text{ Size of Filter Material}}{\text{Hole Diameter}} > 1.0 \qquad (9.13)$$

For Slots,

$$\frac{85\% \text{ Size of Filter Material}}{\text{Slot Width}} > 1.2 \qquad (9.14)$$

The U.S. Bureau of Reclamation (1973) uses the following criterion for grain size of filter materials in relation to openings in pipes:

$$\frac{D_{85} \text{ of the Filter nearest the Pipe}}{\text{Maximum Opening of Pipe Drain}} = 2 \text{ or more} \qquad (9.15)$$

Formula 9.13, 9.14, and 9.15 represent a reasonable range over which satisfactory performance can be expected (Cedergren, 1989).

EXAMPLE 9.1

A leachate collection pipe is placed in the middle of a rectangular landfill cell. Upgradient of the leachate collection pipe is adjacent to a 3(H):1(V) side slope with a height of 65 ft (19.5 m). The downgradient side of the leachate collection pipe rests on a bottom floor sloping at 2% with a width of 100 ft (30 m). The total length of this leachate collection pipe is 1,000 ft (300 m). The peak leachate generation rate at the side slope area is 0.0786 ft³/day/ft² (0.0240 m³/day/m²), and the peak leachate generation rate at the bottom floor area is 0.0355 ft³/day/ft² (0.0108 m³/day/m²). Design a perforated solid wall pipe to meet both drainage and perforation requirements.

Solution:
1) *Determine Amount of Leachate Generation*
Area of the upgradient portion of the landfill cell:

$$A_u = 1{,}000 \times (3 \times 65) = 195{,}000 \text{ ft}^2 \, (18{,}000 \text{ m}^2)$$

Maximum leachate flow generated for the upgradient portion of the cell:

$$(Q_u)_{max} = (q_u)_{max} \times A_u$$
$$= 0.0786 \times 195{,}000$$
$$= 15{,}327 \text{ ft}^3/\text{day} = 0.1774 \text{ ft}^3/\text{sec} \, (5.023 \times 10^{-3} \text{ m}^3/\text{sec})$$

Area of the downgradient portion of the landfill cell:

$$A_d = 1{,}000 \times 100 = 100{,}000 \text{ ft}^2 \, (9{,}000 \text{ m}^2)$$

Maximum leachate flow generated at the downgradient portion of the cell:

Section 9.3 Selection of Leachate Collection Pipe

$$(Q_d)_{max} = (q_d)_{max} \times A_d$$
$$= 0.0355 \times 100,000$$
$$= 3,550 \text{ ft}^3/\text{day} = 0.0411 \text{ ft}^3/\text{sec} (1.164 \times 10^{-3} \text{ m}^3/\text{sec})$$

Maximum leachate flow generated for the entire cell:

$$Q_{max} = (Q_u)_{max} + (Q_d)_{max}$$
$$= 0.1774 + 0.0411 = 0.2185 \text{ ft}^3/\text{sec} (6.187 \times 10^{-3} \text{ m}^3/\text{sec})$$

2) Selection of Pipe Size
Select 6-in (150-mm) SDR 11 HDPE pipe:

$$SDR = D_o/t \tag{9.6}$$

So
$$t = D_o/SDR = 6/11 = 0.55 \text{ in } (0.014 \text{ m})$$
$$D_i = D_o - 2 \cdot t = 6 - 2 \times 0.55 = 4.9 \text{ in} = 0.408 \text{ ft } (0.124 \text{ m})$$
$$r_h = D_i/4 = 0.408/4 = 0.102 \text{ ft } (0.031 \text{ m}) \tag{9.5}$$
$$A = \pi \cdot (D_i/2)^2 = 3.1416 \times (0.408/2)^2 = 0.131 \text{ ft}^2 (1.217 \times 10^{-2} \text{ m}^2)$$

For HDPE pipe, $n \approx 0.011$.

Manning's Equation:

$$Q = (1.49/n) \cdot A \cdot r_h^{2/3} \cdot S^{1/2} \tag{9.2}$$
$$= (1.49/0.011) \times 0.131 \times (0.102)^{2/3} \times (0.01)^{1/2}$$
$$= 135.455 \times 0.131 \times 0.218 \times 0.1$$
$$= 0.387 \text{ ft}^3/\text{sec} (10.96 \times 10^{-3} \text{ m}^3/\text{sec}) > Q_{max} = 0.2185 \text{ ft}^3/\text{sec} (6.187 \times 10^{-3} \text{ m}^3/\text{sec}) (OK)$$

3) Pipe Perforations
Maximum Leachate Inflow per Unit Length of Pipe

$$Q_{in} = (q_u)_{max} \times (A_u)_{unit} + (q_d)_{max} \times (A_d)_{unit}$$
$$= 0.0786 \times 195 + 0.0355 \times 100$$
$$= 15.327 + 3.55$$
$$= 18.877 \text{ ft}^3/\text{day/ft} = 0.0002184 \text{ ft}^3/\text{sec/ft} (2.029 \times 10^{-5} \text{ m}^3/\text{sec/m})$$

Assume that the diameter of a perforation hole, d, in the pipe is 0.25 inch (6 mm). Then the cross-sectional area of a hole on the perforated pipe is

$$A_b = \pi \cdot (d/2)^2$$
$$= 3.1416 \times (0.25/12/2)^2 = 0.000341 \text{ ft}^2 (0.0000316 \text{ ft}^2)$$

Also,
Discharge coefficient $C = 0.62$;
Limiting leachate entrance velocity $v_{ent} = 0.1$ ft/sec (0.03 m/sec).
Bernoulii Equation:

$$Q_b = C \cdot A_b \cdot v_{ent} \tag{9.11}$$
$$= 0.62 \times 0.000341 \times 0.1$$
$$= 0.00002114 \text{ ft}^3/\text{sec} (5.986 \times 10^{-7} \text{ m}^3/\text{sec})$$

Number of Perforation Holes:

$$N = Q_{in}/Q_b \qquad (9.12)$$
$$= 0.0002184/0.00002114$$
$$= 10.35 \text{ holes/ft } (34 \text{ holes/m})$$

So, use 12 holes/ft (40 holes/m); that is 6 holes per foot (20 holes per meter) each side as shown in Figure 9.3.

9.4 DEFORMATION AND STABILITY OF LEACHATE COLLECTION PIPE

All components of the leachate collection and removal system must have sufficient strength to support the weight of the overlying waste, cover system, and post-closure loadings, as well as the stresses from operating equipment. The component that is perhaps the most vulnerable to compressive strength failure is the drainage layer piping. Leachate collection and removal system piping can fail by excessive deflection, which may lead to buckling or collapsing. Pipe strength calculations should include resistance to pipe deflection and critical buckling pressure. This situation is heightened by the current tendency to create extremely large landfills, sometimes called "megafills."

9.4.1 Pipe Deflection

Leachate collection pipes may excessively deform during construction, during the active life of the landfill or under the post-closure loading. This deformation may lead to buckling and eventual collapse. Thus, leachate pipes should be handled carefully and brought on site only when the trench is ready. Passage of heavy equipment directly over a pipe must be avoided. A pipe can be installed in either a positive or negative projection mode. However, every effort should be made to install it in a negative projection mode (Figure 9.2), although at times it may be necessary to install a pipe in a positive projecting mode (Figure 9.5). The essential difference between these two con-

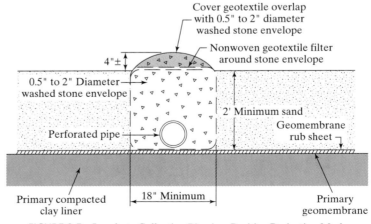

FIGURE 9.5 Leachate Collection Pipe in a Positive Projection Mode

cepts is that a negative projection allows for soil arching which limits the load on the pipe. Conversely, positive projection can actually add load to the pipe. Spangler (1960), among others, explains these concepts for deeply buried pipelines. The design of a pipe must be checked to ascertain whether it will be able to withstand the load during both preconstruction and postconstruction periods. Usually one of two types of pipes are used, HDPE or PVC. These are considered as flexible type pipes. This infers that they do not rupture or break under excessive load, they deform, and if excessively, buckle and/or collapse. The basic design approach consists of calculating the deflection of the pipe, which should not exceed the allowable value. The following formula, commonly known as the Modified Iowa formula, can be used to estimate pipe deflection (Spangler and Handy, 1973; Moser, 1990).

Modified Iowa Formula:

$$\Delta X = \frac{D_L \cdot K \cdot W_c \cdot r^3}{E \cdot I + 0.061 E' \cdot r^3} \tag{9.16}$$

where ΔX = horizontal deflection, in or m (Figure 9.6);
K = bedding constant, its value depending on the bedding angle (see Table 9.1 and Figure 9.7); also, as a general rule, a value of $K = 0.1$ is assumed;
D_L = deflection lag factor (see Table 9.2);
W_c = vertical load per unit length of the pipe, lb/in or kN/m;
r = mean radius of the pipe, $r = (D_o - t)/2$, in or m;
E = elastic modulus of the pipe material, lb/in² or kN/m²;
I = moment of inertia of the pipe wall per unit length, $I = t^3/12$, in⁴/in = in³ or m⁴/m = m³;
t = thickness of pipe, in or m; and
E' = soil reaction modulus, lb/in² or kN/m², see Table 9.3.

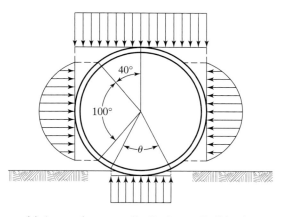

(a) Assumed pressure distribution on flexible pipe

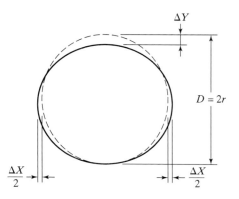

(b) Pipe deflection under pressure

FIGURE 9.6 Buried Flexible Pipe

TABLE 9.1 Values of Bedding Constant, K

Bedding Angle, θ (degree)	Bedding Constant, K
0	0.110
30	0.108
45	0.105
60	0.102
90	0.096
120	0.090
180	0.083

The deflection of the pipe, ΔX, calculated from Equation 9.16 is the deflection in the horizontal direction, as shown in Figure 9.6. When the deflection of pipe is not large (e.g., less than 10%), the vertical deflection of pipe, ΔY, is usually assumed to be approximately equal to the horizontal deflection of pipe, ΔX.

Vertical Load per Unit Length of Pipe:
For Solid Pipe,

$$W_c = (\Sigma \gamma_i \cdot H_i) \cdot D_o \qquad (9.17)$$

where W_c = vertical load per unit length of the pipe, lb/in or kN/m;
γ_i = unit weight of material i on the pipe (sand, clay or solid waste), lb/in³ or kN/m³;
H_i = thickness of material i, in or m; and
D_o = outside diameter of the pipe, in or m.

For Perforated Pipe,

$$W_c = \frac{(\Sigma \gamma_i \cdot H_i) \cdot D_o}{(1 - n \cdot d/12)} \qquad (9.18)$$

where W_c = vertical load per unit length of the pipe, lb/in or kN/m;
γ_i = unit weight of material i (soils or solid waste), lb/in³ or kN/m³;
H_i = thickness of material i, in or m;
D_o = outside diameter of the pipe, in or m;
d = diameter of perforated hole or width of perforated slot on the pipe, in or m; and
n = number of perforated holes or slots per row per foot of pipe.

FIGURE 9.7 Pipe Bedding Angle

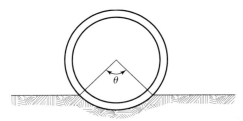

TABLE 9.2 Approximate Range of Values of D_L

Variable	Range	Remarks
D_L	1.5 to 2.5	If the soil in the trench is not compacted, then the higher value of D_L should be used.
	1.0	When deflection calculations are based on prism loads.

TABLE 9.3 Average Values of Soil Reaction Modulus, E' (for Short Term Flexible Pipe Deflection) (Howard, 1977)

Soil type-pipe bedding material (United Classification System)[a]	E' for degree of compaction of bedding			
	Dumped	Slight, < 85% Proctor, < 40% relative density	Moderate, 85%–95% Proctor, 40%–70% relative density	High, > 95% Proctor, > 70% relative density
Fine-grained soils (LL > 50)[b] Soils with medium to high plasticity CH, MH, CH-MH	No data available; consult a competent soils engineer; Otherwise use $E' = 0$			
Fine-grained soils (LL < 50) Soils with medium to no plasticity CL, ML, ML-CL, with less than 25% coarse-grained particles	50 lb/in² (345 kN/m²)	200 lb/in² (1,380 kN/m²)	400 lb/in² (2,760 kN/m²)	1,000 lb/in² (6,900 kN/m²)
Fine-grained soils (LL < 50) Soils with medium to no plasticity CL, ML, ML-CL, with more than 25% coarse-grained particles Coarse-grained soils with fines GM, GC, SM, SC contains more than 12% fines	100 lb/in² (690 kN/m²)	400 lb/in² (2,760 kN/m²)	1,000 lb/in² (6,900 kN/m²)	2,000 lb/in² (13,800 kN/m²)
Coarse-grained soils with little or no fines GW, GP, SW, SP[c] contains less than 12% fines	200 lb/in² (1,380 kN/m²)	1,000 lb/in² (6,900 kN/m²)	2,000 lb/in² (13,800 kN/m²)	3,000 lb/in² (20,700 kN/m²)
Crushed rock	1,000 lb/in² (6,900 kN/m²)	3,000 lb/in² (20,700 kN/m²)	3,000 lb/in² (20,700 kN/m²)	3,000 lb/in² (20,700 kN/m²)
Accuracy in term of percentage deflection[d]	±2	±2	±1	±0.5

[a] ASTM Designation D2487, USBR Designation E-3
[b] LL = Liquid Limit
[c] or any borderline soil beginning with one of these symbols (i.e., GM-GC, GC-SC)
[d] for ±1% accuracy and predicted deflection of 3%, actual deflection would be between 2% and 4%

Note: Values applicable only for soil fills less than 50 ft (15 m). Table does not include any safety factor. For use in predicting initial deflections only—appropriate deflection lag factor must be applied for long-term deflections. If bedding falls on the borderline between two compaction categories, select lower E' value or average the two values. Percentage Proctor based on laboratory maximum dry density from test standards using about 12,500 ft-lb/ft³ (600 m-kN/m³) (ASTM D698, AASHO T-99, USBR Designation E-11).
Used with permission of ASCE.

The parameter that controls the pipe deformation is known as the deflection ratio. The deflection ratio of a pipe is defined as the ratio of the vertical deflection of pipe and the mean diameter of the pipe.

Deflection Ratio:

$$\text{Deflection Ratio}(\%) = (\Delta Y/D) \times 100\% \qquad (9.19)$$

where ΔY = vertical deflection of pipe, $\Delta Y \approx \Delta X$ when the deflection is less than 10%, in or m; and
D = mean diameter of pipe, in or m.

Mean Diameter of Pipe:

$$D = (D_o + D_i)/2 = D_o - t = D_i + t \qquad (9.20)$$

where D = mean diameter of pipe, in or m;
D_o = outside diameter of pipe, in or m;
D_i = inside diameter of pipe, in or m; and
t = thickness of pipe, in or m.

There is another formula that can be used to estimate the deflection of the pipe. It is essentially an alternative version of the Modified Iowa formula and has been widely used in the engineering field. This formula is

$$\Delta X = \frac{D_L \cdot K \cdot W_c}{0.149 \cdot PS + 0.061 \cdot E'} \qquad (9.21)$$

where ΔX = horizontal deflection, in or m (Figure 9.6);
K = bedding constant, its value depending on the bedding angle (see Table 9.1 and Figure 9.7); as a general rule, a value of $K = 0.1$ is assumed;
D_L = deflection lag factor, see Table 9.2;
W_c = vertical load per unit length of the pipe, lb/in or kN/m;
PS = pipe stiffness, lb/in^2 or kN/m^2; and
E' = soil reaction modulus, lb/in^2 or kN/m^2.

The vertical pressure on solid pipe is given by

$$P_{tp} = \sum \gamma_i \cdot H_i \qquad (9.22)$$

The vertical pressure on perforated pipe is given by

$$P_{tp} = \frac{\sum \gamma_i \cdot H_i}{(1 - n \cdot d/12)} \qquad (9.23)$$

where P_{tp} = vertical pressure on the pipe, $P_{tp} = W_c/D_o$, lb/in^2 or kN/m^2;
γ_i = unit weight of material i on the pipe (sand, clay or solid waste), lb/in^3 or kN/m^3;
H_i = thickness of material i, in or m;

d = diameter of perforated hole or width of perforated slot on the pipe, in or m; and

n = number of perforated holes or slots per row per foot of pipe.

Pipe stiffness is measured according to ASTM D2412 (Standard Test Method for External Loading Properties of Plastic Pipe by Parallel-Plate Loading). The elastic modulus of the pipe material depends on the type of resin and formulation being used. Three formulas that can be used to calculate pipe stiffness are

$$PS = \frac{E \cdot I}{0.149 \cdot r^3} \quad (9.24)$$

$$PS = 0.559 \cdot E \cdot (t/r)^3 \quad (9.25)$$

and

$$PS = 4.47 \cdot \frac{E}{(SDR - 1)^3} \quad (9.26)$$

where PS = pipe stiffness, lb/in² or kN/m²;
E = elastic modulus of the pipe material, lb/in² or kN/m²;
I = moment of inertia of the pipe wall per unit length,
  $I = t^3/12$, in⁴/in = in³ or m⁴/m = m³;
r = mean radius of pipe, in or m;
t = wall thickness of pipe, in or m; and
SDR = standard dimension ratio, the same as the dimension ratio.

The allowable deflection ratios for a typical commercial polyethylene pipe are listed in Table 9.4.

Deflections of buried flexible pipe are commonly calculated using Equation 9.16 or 9.21. These equations use the soil reaction modulus, E', as a surrogate parameter for soil stiffness. It should be noted that the values of E' in Table 9.3 only apply for soil fills of less than 50 ft (15 m). However, megafills built over leachate collection pipes often exceed 150 ft (46 m) in height. The soil reaction modulus is not a directly measurable soil parameter; instead it must be determined by back-calculation using observed pipe deflections. Research by Selig (1990) showed that E' is a function of the bedding condition and overburden pressure. Selig's studies were carried out to seek a correlation between the soil reaction modulus and soil stiffness parameters such as

TABLE 9.4 Allowable Deflection Ratio of Polyethylene Pipe

SDR	Allowable Deflection Ratio
11	2.7%
13.5	3.4%
15.5	3.9%
17	4.2%
19	4.7%
21	5.2%
26	6.5%
32.5	8.1%

Young's modulus of soil, E_s, and the constrained modulus of soil, M_s, where E_s and D_s are related through Poisson's ratio of soil, v_s, by

$$M_s = \frac{E_s \cdot (1 - v_s)}{(1 + v_s)(1 - 2 \cdot v_s)} \tag{9.27}$$

where M_s = constrained modulus of soil, lb/ft² or kN/m²;
E_s = elastic modulus of soil, lb/ft² or kN/m²; and
v_s = Poisson's ratio of soil.

The studies and analyses by Neilson (1967), Allgood and Takahashi (1972), and Hartely and Duncan (1987) indicated that for

$$E' = k \cdot M_s \tag{9.28}$$

the value of k may vary from 0.7 to 2.3. Using $k = 1.5$ as a representative value and $v_s = 0.3$, in addition to combining Equations 9.27 and 9.28 yields the following relationship between the elastic modulus of the pipe and soil (Selig, 1990):

$$E' = 2 \cdot E_s \tag{9.29}$$

The values of elastic parameters, E_s and v_s, can be found in Table 9.5 according to different percents of density from a standard Proctor compaction test (ASTM D698).

TABLE 9.5 Elastic Soil Parameters (Selig, 1990)

Soil Type	Stress Level		85% Standard Density			95% Standard Density		
			E_s			E_s		
	psi	kPa	psi	MPa	v_s	psi	MPa	v_s
SW, SP, GW, GP	1	7	1,300	9	0.26	1,600	11	0.40
	5	35	2,100	14	0.21	4,100	28	0.29
	10	70	2,600	18	0.19	6,000	41	0.24
	20	140	3,300	23	0.19	8,600	59	0.23
	40	280	4,100	28	0.23	13,000	90	0.25
	60	420	4,700	32	0.28	16,000	110	0.29
GM, SM, ML, and GC, SC with < 20% fines	1	7	600	4	0.25	1,800	12	0.34
	5	35	700	5	0.24	2,500	17	0.29
	10	70	800	6	0.23	2,900	20	0.27
	20	140	850	6	0.30	3,200	22	0.29
	40	280	900	6	0.38	3,700	25	0.32
	60	420	1,000	7	0.41	4,100	28	0.35
CL, MH, GC, SC	1	7	100	1	0.33	400	3	0.42
	5	35	250	2	0.29	800	6	0.35
	10	70	400	3	0.28	1,100	8	0.32
	20	140	600	4	0.25	1,300	9	0.30
	40	280	700	5	0.35	1,400	10	0.35
	60	420	800	6	0.40	1,500	10	0.38

9.4.2 Pipe Wall Buckling

Buckling can occur because of insufficient stiffness. Buckling may govern design of flexible pipes subjected to internal vacuum, external hydrostatic pressure, or high soil pressures in compacted soil (Figure 9.8). As Moser (1990) notes the more flexible the conduit (e.g., high values of SDR), the more unstable the wall structure will be in resisting buckling.

Most conduits are buried in a soil medium that does offer considerable shear resistance. An exact rigorous solution to the problem of buckling of a cylinder in an elastic medium entails some advanced mathematics (Moser, 1990). However, because of uncertainties in the behavior and performance of the surrounding soil, an exact solution is not necessary. Meyerhof and Baike (1963) developed the following empirical formula for computing the critical buckling pressure in a buried circular conduit:

$$P_{cr} = 2 \cdot \{[E'/(1-\mu^2)](E \cdot I/r^3)\}^{1/2} \tag{9.30}$$

Where,

P_{cr} = critical buckling pressure, lb/in² or kN/m²;
E' = modulus of soil reaction, lb/in² or kN/m², see Table 9.3;
μ = Poisson's ratio of pipe material;
E = modulus of elasticity of the pipe material, lb/in² or kN/m²;
I = moment of inertia of the pipe wall per unit length, in⁴/in = in³ or m⁴/m = m³, $I = t^3/12$; and
r = mean radius of the pipe, in or m.

Because $I = t^3/12$ and $r = D/2$, Equation 9.30 can be rewritten as

$$P_{cr} = 2 \cdot (G_b \cdot E')^{1/2} \tag{9.31}$$

where

$$G_b = \frac{2 \cdot E}{3 \cdot (1-\mu^2)} \cdot (t/D)^3 \tag{9.32}$$

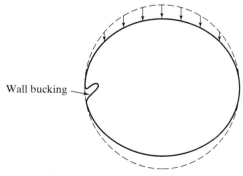

FIGURE 9.8 Localized Wall Buckling

in which

t = thickness of pipe, in or m and
D = mean diameter of pipe, in or m

The factor of safety for pipe wall buckling can be determined by

$$FS = P_{cr}/P_{tp} \qquad (9.33)$$

where P_{tp} = actual vertical pressure at the top of the pipe, obtained from Equation 9.22 or 9.23, lb/in² or kN/m².

In both Equations 9.30 and 9.31 initial out-of-roundness is neglected but the reduction in P_{cr} because of this has been assumed to be no greater than 30% (Moser, 1990). As a result, a factor of safety ≥ 2 is recommended for use with Equation 9.33 in the design of a flexible conduit to resist buckling.

EXAMPLE 9.2

An 8-inch (200-mm) SDR 11 HDPE perforated pipe with 8, 0.25-inch (6-mm) holes per foot (i.e., 4 holes per side per foot) is selected as a primary leachate collection pipe. The maximum load acting on the pipe includes a 2-ft (0.6-m) protective sand layer (γ_{sand} = 115 lb/ft³ or 18 kN/m³), 100-ft (30-m) solid waste (γ_{waste} = 60 lb/ft³ or 9.4 kN/m³), 12-inch (0.3-m) gas venting layer (γ_{sand} = 115 lb/ft³ or 18 kN/m³), 18-inch (0.45-m) compacted clay layer (γ_{clay} = 110 lb/ft³ or 17.3 kN/m³), 24-inch (0.6-m) drainage and protective layer (γ_{silt} = 110 lb/ft³ or 17.3 kN/m³), and 6-inch (0.15-m) topsoil (γ_{top} = 90 lb/ft³ or 14 kN/m³). Assume bedding angle θ = 0°, deflection lag factor D_L = 1.0, elastic modulus of the pipe material for 50 years at 73°F (23°C) temperature E = 28,200 lb/in², (194,000 kN/m²), Poisson's ratio of pipe material μ = 0.3. The bedding material of the pipe is poorly graded gravel (GP) with 85% standard density. What will be the deflection ratio (%) and critical buckling pressure of the pipe?

Solution: The maximum load applied on the pipe is given by

$$W_c = \frac{\left(\sum \gamma_i \cdot H_i\right) \cdot D_o}{(1 - n \cdot d/12)} \qquad (9.18)$$

$$= \frac{[(115)(2) + (60)(100) + (115)(1) + (110)(3.5) + (90)(0.5)] \times 8/12}{(1 - 4 \times 0.25/12)}$$

$$= \frac{(230 + 6{,}000 + 115 + 385 + 45) \times 8/12}{0.917}$$

$$= \frac{6{,}775 \times 8/12}{0.917}$$

$$= 4{,}925 \text{ lb/ft} = 410 \text{ lb/in } (72 \text{ kN/m})$$

The maximum pressure applied on the pipe can be obtained from

$$P_{tp} = W_c/D_o = 410/8 = 51.3 \text{ lb/in}^2 \text{ (354 kN/m}^2)$$

From Table 9.5,

$$P_{tp} = 40 \text{ lb/in}^2, E_s = 4{,}100 \text{ lb/in}^2$$

and
$$P_{tp} = 60 \text{ lb/in}^2, \; E_s = 4{,}700 \text{ lb/in}^2$$

For $P_{tp} = 51.3 \text{ lb/in}^2$,
$$E_s = 4{,}100 + (51.3 - 40)(4{,}700 - 4{,}400)/20 = 4{,}100 + 339 = 4{,}439 \text{ lb/in}^2$$

The soil reaction modulus is given by
$$E' = 2 \cdot E_s = 2 \times 4{,}439 = 8{,}878 \text{ lb/in}^2 \; (61{,}200 \text{ kN/m}^2) \tag{9.29}$$

The thickness of pipe is given by
$$t = D_o/SDR \tag{9.6}$$
$$= 8/11 = 0.73 \text{ in } (0.0185 \text{ m})$$

The mean diameter of pipe is
$$D = D_o - t \tag{9.20}$$
$$= 8 - 0.73 = 7.27 \text{ in } (0.1847 \text{ m})$$

Also,

Deflection lag factor, $D_L = 1.0$;
Bedding angle $\theta = 0°$, $K = 0.11$;
Mean radius of the pipe, $r = 3.635 \text{ in } (0.0923 \text{ m})$;
Elastic modulus of the pipe material, $E = 28{,}200 \text{ lb/in}^2 \; (194{,}000 \text{ kN/m}^2)$;
Soil reaction modulus, $E' = 8{,}878 \text{ lb/in}^2 \; (61{,}200 \text{ kN/m}^2)$;
and
Inertia moment of the pipe wall per unit length, $\text{in}^4/\text{in} = \text{in}^3$, given by
$$I = t^3/12 = (0.73)^3/12 = 0.389/12 = 0.0324 \text{ in}^3 \; (5.276 \times 10^{-7} \text{ m}^3)$$

Modified Iowa Formula:
$$\Delta X = \frac{D_L \cdot K \cdot W_c \cdot r^3}{E \cdot I + 0.061 E' \cdot r^3} \tag{9.16}$$
$$= \frac{(1.0)(0.11)(410)(3.635)^3}{(400{,}000)(0.0324) + (0.061)(1{,}000)(3.635)^3}$$
$$= \frac{(1.0)(0.11)(410)(48.03)}{(28{,}200)(0.0324) + (0.061)(8{,}878)(48.03)}$$
$$= \frac{2{,}166}{914 + 26{,}011}$$
$$= 0.08 \text{ in } (2.0 \text{ mm})$$

Deflection Ratio:
$$\text{Deflection Ratio} = (\Delta Y/D) \times 100\% \tag{9.19}$$
$$= (0.08/7.27) \times 100\%$$
$$= 1.1\% < 2.7\% \text{ (ok, as shown in Table 9.4)}$$

Wall Buckling of Pipe:
Modulus of soil reaction, $E' = 8{,}878 \text{ lb/in}^2$, $(61{,}200 \text{ kN/m}^2)$;

Poisson's ratio of pipe material, $\mu = 0.3$;
Modulus of elasticity of the pipe material, $E = 28{,}200$ lb/in^2 (194,000 kN/m^2);
Moment of inertia of the pipe wall per unit length, $I = 0.0324$ in^3 (5.276 × 10^{-7} m^3);
Mean radius of the pipe, $r = 3.635$ in (0.0923 m).
Thus,

$$\begin{aligned} P_{cr} &= 2 \cdot \{[E'/(1-\mu^2)](E \cdot I/r^3)\}^{1/2} \\ &= 2 \times \{[8{,}878/(1-0.3^2)][(28{,}200 \times 0.0324)/(3.635)^3]\}^{1/2} \\ &= 2 \times [9{,}756 \times (913.68/48.03)]^{1/2} \\ &= 2 \times (185{,}589)^{1/2} \\ &= 2 \times 431 \\ &= 862 \text{ lb/in}^2 \text{ (5,943 kN/m}^2\text{)} \end{aligned} \qquad (9.30)$$

The factor of safety for pipe wall buckling is, then,

$$FS = P_{cr}/P_{tp} = 862/51.3 = 16.8 > 2 \text{ (OK)} \qquad (9.33)$$

9.5 SUMP AND RISER PIPES

Leachate collection sumps are low points in the landfill liner constructed to collect and removal leachate. The sumps are filled with gravel to provide the maximum space (volume) for leachate accumulation, as well as to support the weight of the overlying waste, cover system, and post-closure loadings. Commonly, the composite liner system is slightly depressed or indented to create these sumps (shown in Figures 9.9 and 9.10). The absence of sketches illustrating continued gravity flow of leachate beyond the limits of the cells and/or landfill using liner penetrations is intentional. The authors do not recommend such practice due to the difficulty of making liner seams in this remote of all locations. With double liner systems, the situation is even more difficult. Even with the sketches of Figures 9.9 and 9.10 it is difficult to test the geomembrane seaming in such sumps because of the slope and corners at which the seams occur. Because of the difficulty in seam testing sumps, sump areas often are designed with an additional layer of geomembrane. Sulfates are one of the most common and abundant constituents in landfill leachate. Accordingly, all concrete components in a sump (e.g., riser pipe and foundation pad) must be constructed using low water/cement ratios and sulfate resistant, Class V Portland cement (ACI, 1998). Failure to observe this precaution can lead to sulfate attack and disintegration of the concrete. Sulfate attack occurs when calcium, alumina, and sulfate combine to form the mineral ettringite ($3CaO \cdot Al_2O_3 \cdot 32H_2O$) in the cement matrix. The volume of ettringite is over 200% that of the original constituents, which can result in massive swelling and cracking when sufficient ettringite forms by the sulfation of alumina. Alternatively, many sumps now are being constructed using premanufactured units made of HDPE, with large-diameter HDPE pipe or HDPE manholes. Although more costly, the factory manufactured sumps can be thoroughly tested and installed as a unit.

Figure 9.9 shows details of vertical riser (manhole) removal designs for primary and secondary leachate collection systems. The manhole riser extends vertically

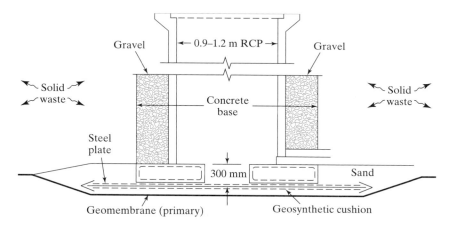

(a) Low-volume primary leachate collectiom sump and manhole

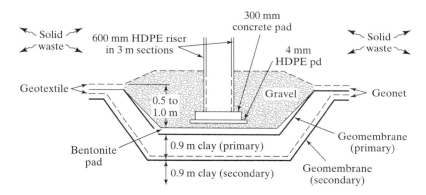

(b) High-volume primary leachate collection sump and manhole

FIGURE 9.9 Various Removal Designs for Primary Leachate Removal System (Koerner, 1998)

through the waste as it is being placed and eventually penetrates the landfill cover when the site is closed. Leachate is removed using submersible pumps until it is no longer generated. A related aspect of this leachate removal manhole is the downward force exerted on the outside of the pipe risers by the subsiding waste mass. Called negative skin friction, or downdrag, by geotechnical engineers subsiding soil around bearing piles, piers, and caissons can create very large pulldown (downdrag) forces. In some cases, the downward drag force could be excessive and cause foundation failure. Hence, the importance of paying careful attention to detail beneath the footing in both sketches of Figure 9.9 is emphasized (Koerner, 1998).

The negative skin friction on the manhole riser can be calculated using

$$f_n = K_o \cdot \sigma_v \cdot \tan \delta \tag{9.34}$$

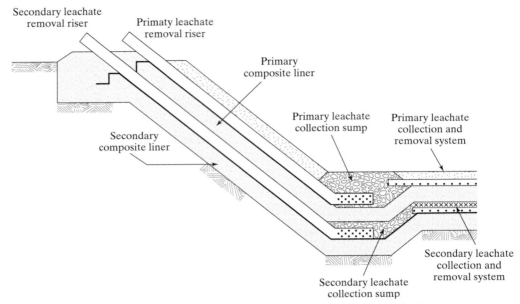

FIGURE 9.10 Sideslope Riser Pipe to Remove Liquid from Leachate Collection Sump

where f_n = negative skin stress, lb/ft² or kN/m²;
K_o = coefficient of at-rest solid waste pressure, $K_o = 1 - \sin\phi_w$;
ϕ_w = internal friction angle of solid waste, degree;
σ_v = vertical effective stress at any depth, $\sigma_v = \gamma_{waste} \cdot z$, lb/ft² or kN/m²;
γ_{waste} = unit weight of solid waste, lb/ft³ or kN/m³;
z = depth of solid waste, ft or m; and
δ = interface friction angle between manhole riser wall and solid waste, degree.

The total downward drag force on a manhole riser is

$$Q_n = \int_0^{H_f} (p \cdot K_o \cdot \gamma_{waste} \cdot \tan\delta) \cdot z \cdot dz \qquad (9.35)$$

Therefore,

$$Q_n = 0.5 \cdot p \cdot K_o \cdot \gamma_{waste} \cdot H_f^2 \cdot \tan\delta \qquad (9.36)$$

where Q_n = total downward drag force on the manhole riser, lb or kN;
p = perimeter of the manhole riser, ft or m;
K_o = coefficient of at-rest solid waste pressure, $K_o = 1 - \sin\phi_w$;
ϕ_w = internal friction angle of solid waste, degree;
γ_{waste} = unit weight of solid waste, lb/ft³ or kN/m³;
H_f = height of solid waste fill, ft or m; and
δ = interface friction angle between manhole riser wall and solid waste, degree.

EXAMPLE 9.3

Refer to Figure 9.9(a). The primary leachate removal HDPE manhole riser is circular in cross section with an inside diameter of 3 ft (0.9 m) and an outside diameter of 3.75 ft (1.13 m). The height of solid waste surrounding the manhole riser is given by $H_f = 210$ ft (63 m); the unit weight of solid waste is given by $\gamma_{waste} = 70$ lb/ft^3 (11 kN/m^3); the internal friction angle of solid waste is given by $\phi_w = 33°$; and the interface friction angle between the concrete manhole riser wall and solid waste is $\delta = 24°$. Determine the total downdrag force, stress on the concrete, and factor of safety if the compressive strength of concrete is 5,000 lb/in^2 (34,500 kN/m^2).

Solution:

$$p = \pi(3.75) = 11.78 \text{ ft } (3.5 \text{ m})$$
$$K_o = 1 - \sin\phi_w = 1 - \sin 33° = 1 - 0.545 = 0.455$$
$$\tan\delta = \tan 24° = 0.445$$

a. The downdrag force is

$$Q_n = 0.5 \cdot p \cdot K_o \cdot \gamma_{waste} \cdot H_f^2 \cdot \tan\delta \qquad (9.36)$$
$$= (0.5)(11.78)(0.455)(70)(210)^2(0.445)$$
$$= 3,681,500 \text{ lb } (16,400 \text{ kN})$$

b. The downdrag stress is

$$\sigma_n = Q_n/A_{concrete}$$
$$= Q_n/[(\pi/4)(D_o^2 - D_i^2)]$$
$$= 3,681,500/[(\pi/4)(3.75^2 - 3.0^2)]$$
$$= 3,681,500/3.98$$
$$= 925,000 \text{ lb/ft}^2$$
$$= 6,424 \text{ lb/in}^2 (44,300 \text{ kN/m}^2)$$

c. Finally, the factor of safety against concrete overstress is

$$FS = \sigma_{allow}/\sigma_n$$
$$= 5,000/6,424$$
$$= 0.78 < 1.0 \text{ (No Good)}$$

Figure 9.11 shows that a concrete manhole riser was collapsed by downdrag force due to waste settlement. To relieve the downdrag somewhat, the outside of the concrete riser sections extending through the waste is sometimes made from HDPE as in Figure 9.9(b) or wrapped in a low-friction material. Other downdrag-reducing methods such as bitumen slip layers and telescoping risers are also possible.

This downdrag problem can be largely eliminated by using a side wall riser scheme for leachate removal as shown in Figure 9.10. This design eliminates many operational problems associated with vertical risers. Here a larger diameter HDPE or PVC pipe, typically 24-inch (0.6-m) diameter, is brought from the sump area up the side slope above the primary liner into an enclosure or shed. A submersible pump is

FIGURE 9.11   Concrete Manhole Riser Collapse Caused by Downdrag Force due to Waste Settlement

lowered in the pipe on a sled down into the sump area, where it terminates at a T (with perforations in the pipe section forming the T). The pump can be withdrawn for maintenance or if problems arise.

The key steps for designing a leachate collection sump and riser pipe are listed as follows:

1. *Assume dimension of the sump.*

The assumption of the dimension of a leachate collection sump is based on the total area of a landfill cell served by this sump. The following data are commonly used in the sump design:

Bottom dimension:	15 ft × 15 ft or 20 ft × 20 ft (4.5 m × 4.5 m or 6 m × 6 m)
Depth:	1.5 ~ 4.5 ft (0.45 ~ 1.35 m)
Sump slope:	3(H):1(V)

2. *Calculate total volume of the sump.*

Once the dimension of the sump is assumed, the total volume of the sump can be determined. Note that the top elevation for calculating storage volume of the sump is equal to the bottom elevation of the leachate collection trench.

3. *Calculate volume of riser pipe section placed in the sump.*

The section of the riser pipe located in the bottom of the sump is perforated and is used to contain a submersible pump (Figure 9.10).

4. *Calculate void volume of the sump.*

Because the sump is filled by gravel and a part of the riser pipe also lies at the bottom of the sump, the void volume of the sump mainly depends on the porosity of the fill material and the volume occupied by the riser pipe in the sump. That is,

$$V_{void} = n \cdot (V_{total} - V_{pipe}) + V_{pipe} \tag{9.37}$$

where V_{void} = void volume of sump, ft³ or m³;
V_{total} = total volume of sump, ft³ or m³;
V_{pipe} = volume of pipe section in sump, ft³ or m³; and
n = porosity of gravel filled in sump, $n = 30 \sim 40\%$.

5. *Determine pump on/off level.*

The sump typically houses a submersible pump which is positioned close to the sump floor to pump the leachate and to maintain a 12-inch (0.3-m) maximum leachate head on the bottom liner. Low-volume sumps can present operational problems. Because they may run dry frequently, there is an increased probability of the submersible pumps burning out. For this reason, some landfill operators prefer to have sumps placed at depths between 1.5 and 3 feet (0.45 and 0.9 meters) and keep a 12-inch (0.3-m) minimum liquid depth in the sump to avoid the submersible pump burning out. The sump is designed with level controls to control initiation and shut-off of the pumping sequence.

Pump Off Level = Sump Floor + 1 ft (0.3 m), or = Pump Top + 0.5 ft (0.15 m);
Pump On Level = Bottom of Leachate Trench.

6. *Calculate sump storage volume available for pumping.*

Let V_s = sump storage volume.
This volume is also equal to the volume stored in the voids, therefore,

V_s = Void Volume between Pump On Level and Pump Off Level.

7. *Calculate riser pipe perforations and strength.*

The flow capacity of the riser pipe perforations must be greater than the pump rate. The methods to calculate the riser pipe perforations and strength are the same as the methods described in Sections 9.3 and 9.4.

Figure 9.12 shows a cross section of a landfill leachate collection and removal system including a leachate drainage layer, trench, perforated pipe, sump, riser pipe, and

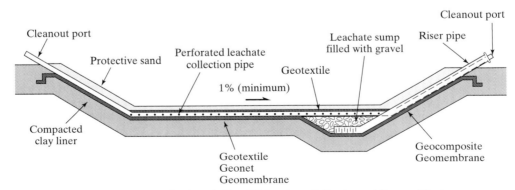

FIGURE 9.12 Cross Section of a Landfill Leachate Collection and Removal System

320 Chapter 9 Leachate Collection and Removal Systems

FIGURE 9.13 Sideslope Riser Pipes and Cleanout Port

cleanout port. Figure 9.13 is a photograph showing the primary and secondary riser pipes and a cleanout port next to the primary riser pipe.

9.6 LEACHATE REMOVAL PUMPS

The capacity of leachate removal pumps must be carefully calculated for proper functioning. The pumps used to remove leachate from the sumps should be sized to ensure removal of leachate at the maximum rate of generation. These pumps also should have a sufficient operating head to lift the leachate to the required height from the sump to the access port. Figure 9.10 illustrated that a submersible pump can be installed in the riser pipe extending down to the leachate collection sump. Usually automatic submersible pumps are used to pump the leachate in the sumps. The leachate removal pumps are installed in riser pipes or manholes. The shut off switch must be located at least 6 inches (0.15 m) higher than the top of the pump to keep the pump under liquid and prevent the pump from burning out. A schematic diagram of the installation of a leachate removal pump in a sideslope riser pipe is shown in Figure 9.14. Figure 9.15 is a photograph showing a submersible pump installed in the riser pipe.

Most pumps work more efficiently if run continuously; however, since the leachate generation rate varies, intermittent operation of the pump becomes necessary. A 12-minute cycle is considered satisfactory (Bureau of Reclamation, 1978), a

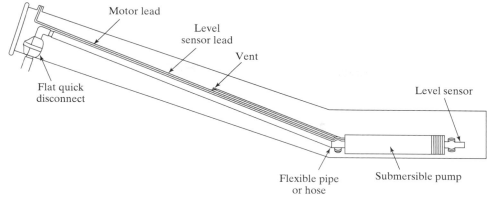

FIGURE 9.14 Schematic Diagram of Installation of a Leachate Collection Pump in Sideslope Riser Pipe

specification which should be verified from prospective pump manufacturers. The sump size can also be estimated based on the volume of leachate inflow in one-half the cycling time (Bagchi, 1994); this would mean equal on and off time. Another approach is to assume that pump discharge flow is double that of the leachate inflow.

FIGURE 9.15 A Submersible Pump Installed in Sideslope Riser Pipe

The total discharge head of pump is given by

$$TDH = H_s + H_f + H_m \tag{9.39}$$

where H_s = static head, ft or m;
H_f = friction head loss, ft or m; and
H_m = minor head loss, ft or m.

The static head is given by

H_s = Elevation of top of side slope or leachate storage tank if a storage tank is used − Elevation of bottom of sump

where H_s = static head, ft or m.

The friction head loss (in U.S. units) is given by

$$H_f = \frac{3.022 \cdot v^{1.85} \cdot L}{C^{1.85} \cdot D_i^{1.165}} \tag{9.40}$$

or

$$H_f = \frac{10.44 \cdot Q^{1.85} \cdot L}{C^{1.85} \cdot d_i^{4.87}} \tag{9.41}$$

where H_f = friction head loss, ft;
v = flow velocity, ft/sec
L = flow length, ft;
C = Hazen-Williams coefficients, see Table 9.6;
D_i = inside diameter of pipe, ft;
Q = discharge flow rate, gallon/min; and
d_i = inside diameter of pipe, inches.

The friction head loss (in SI units) is given by

$$H_f = \frac{6.821 \cdot v^{1.85} \cdot L}{C^{1.85} \cdot D_i^{1.165}} \tag{9.42}$$

where H_f = friction head loss, m;
v = flow velocity, m/sec

TABLE 9.6 Hazen-Williams Coefficients

Type of Pipe	C
Pipes extremely straight and smooth	140
Pipes very smooth	130
Smooth wood, smooth masonry	120
New riveted steel, vitrified clay	110
Old cast iron, ordinary brick	100
Old riveted steel	95
Old iron in bad condition	60 ~ 80

L = flow length, m;
C = Hazen-Williams coefficients, see Table 9.6; and
D_i = inside diameter of pipe, m.

Head losses that are caused by localized disturbances to a flow, such as by valves or pipe fittings, are traditionally termed minor losses. Since they are not necessarily negligible, a more precise term would be transition losses, because they typically occur at transitions from one fully developed flow to another. As localized disturbances, they do not vary with the conduit length but are instead described by an overall transition loss coefficient K. The greater the flow disturbance, the larger the value of K. For example, as a valve is gradually closed, the associated value of K for the valve increases. Similarly, K for different pipe fittings will vary somewhat with pipe size, decreasing with increasing pipe size, since fittings are generally not geometrically similar for different pipe sizes. Values of transition loss coefficient K for various types of transitions are given in Tables 9.7, 9.8, and 9.9. The total minor head loss can be calculated using either of the equations that follow.

The minor head loss is given by either

$$H_m = \sum K_i \frac{v_i^2}{2 \cdot g} \tag{9.43}$$

$$H_m = \sum K_i \frac{Q_i^2}{2 \cdot g \cdot A^2} \tag{9.44}$$

where H_m = minor head loss, ft or m;
V_i = flow velocity, ft/sec or m/sec;
g = gravitational acceleration, 32.2 ft/sec² or 9.81 m/sec²;
K_i = transition loss coefficient for each valve or fitting;
Q_i = flow in the pipe, ft³/sec or m³/sec; and
A = area in flow, ft² or m².

Pump Type Selection

Pump manufacturer's catalogs can be used to select a suitable type and model of submersible pump according to the design discharge flow rate and calculated total discharge head (*TDH*). The relationships between pump capacity (pump rate) and total discharge head (*TDH*) for two models (i.e., Style "2" and Style "12") of one manufacturer's sideslope riser pumps are illustrated in Figures 9.16 and 9.17, respectively. The inside diameters of outlet pipe for Style "2" pump and Style "12" pump are 1.25 inches

TABLE 9.7   Transition Loss Coefficient of Fittings

Fitting	Transition Loss Coefficient, K
Square-Edged Entrance	0.5
Well-Rounded Entrance	0.03
Outlet	1.0
General Contraction (30° included angle)	0.02
General Contraction (70° included angle)	0.07

TABLE 9.8 Transition Loss Coefficient of Bends and Elbows

	Miter Bend								
Angle	10°	20°	30°	40°	50°	60°	70°	80°	90°
K	0.04	0.10	0.20	0.30	0.40	0.55	0.70	0.90	1.10

	Elbow											
Angle	20°	30°	40°	50°	60°	70°	80°	90°	100°	120°	140°	160°	180°
K	0.40	0.55	0.65	0.75	0.83	0.88	0.95	1.00	1.05	1.13	1.20	1.27	1.33

(31 mm) and 2.0 inches (50 mm), respectively. Such submersible pumps are made of stainless steel and teflon. The wheeled design of this type of pump allows for ease of installation and removal in the sideslope riser pipe. A pressure transducer for level sensing is incorporated in the pump to detect the leachate level in the sump.

The pump-off time can be obtained by

$$T_{off} = V_s/Q_{in} \tag{9.45}$$

where T_{off} = pump off time, min;
V_s = sump storage volume, ft³ or m³; and
Q_{in} = inflow = the rate of leachate flow into sump, ft³/min or m³/min.

The pump-on time can be obtained by

$$T_{on} = V_s/(Q_p - Q_{in}) \tag{9.46}$$

where T_{on} = pump on time, min;
V_s = sump storage volume, ft³ or m³;
Q_p = pump rate, ft³/min or m³/min; and
Q_{in} = inflow, i.e., the rate of leachate flow into sump, ft³/min or m³/min.

Finally, the cycle time is obtained from

$$T_{cycle} = T_{off} + T_{on} \tag{9.47}$$

where T_{cycle} = pump cycle time, min.

TABLE 9.9 Transition Loss Coefficient of Valves

Type of Valve		Screwed			Flanged		
		1"	2"	4"	2"	4"	8"
Globe Valve	fully open	8.2	6.9	5.7	8.5	6.0	5.8
	half open	20	17	14	21	15	14
	one-quarter open	57	48	40	60	42	41
Angle Valve (fully open)		4.7	2.0	1.0	2.4	2.0	2.0
Swing Valve (fully open)		2.9	2.1	2.0	2.0	2.0	2.0
Gate Valve (fully open)		0.24	0.16	0.11	0.35	0.16	0.07

Section 9.6 Leachate Removal Pumps 325

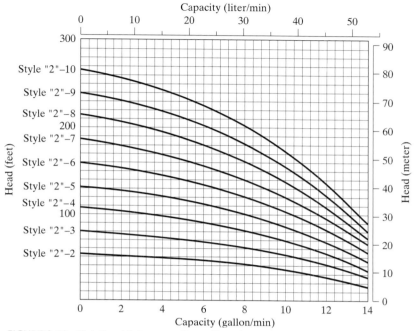

FIGURE 9.16 Relationship between Pump Capacity and Discharge Head for Style "2" Pump (adapted from EPG Sure Pump™ Co.)

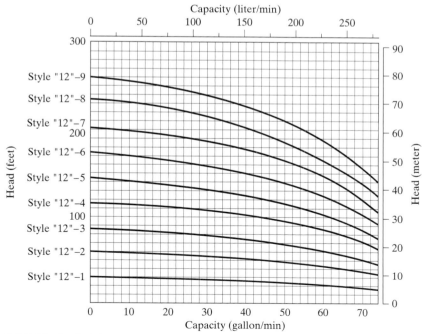

FIGURE 9.17 Relationship between Pump Capacity and Discharge Head for Style "12" Pump (adapted from EPG Sure Pump™ Co.)

EXAMPLE 9.4

The bottom dimension of a primary leachate collection sump is 15 ft × 15 ft (4.5 m × 4.5 m). The depth of the sump is 4.5 ft (1.35 m), and sump slope is 3(H):1(V). The depth of the primary leachate trench is 6 inches (0.15 m). The porosity of the gravel filled in the sump is 40%. The leachate inflow is 35 gallon/min (132 liter/min). A 12-inch (300-mm) diameter SDR 17 riser pipe is placed on the side slope with a slope of 3(H):1(V) and height of 50 ft (15 m). The length of the riser pipe placed at the bottom of the sump is 15 ft (4.5 m). The length of the pump pipe extended at the top of the slope is 12 ft (3.6 m). A swing valve with screwed connector is installed at the top of the slope. Assume Hazen-Williams coefficient of the pump pipe is 130. A Style "12" pump having 2-inch (50-mm) inside diameter of outlet pipe and the characteristics presented in Figure 9.17 is installed in the riser pipe at the bottom of the sump to pump the leachate. A schematic diagram of pump installation is shown in Figure 9.13. Assume the leachate in the sump is required to pump to the top of the side slope and then it can gravitationally drain to a sewer system.

1. Calculate the sump storage volume available for pumping;
2. Select the model of pump;
3. Calculate the pump cycle time for the selected pump.

Solution:

1. *Calculate Sump Storage Volume Available for Pumping*

Sump Depth = 4.5 ft (1.35);

Leachate Trench Depth = 0.5 ft (0.15 m);

Pump On Level = Bottom of Leachate Trench;

Pump Off Level = Sump Floor + 1 ft (0.3 m).

Sump Volume for Pumping:

$$V = 3 \times [(21 + 39)/2]^2 = 3 \times 30^2 = 2{,}700 \text{ ft}^3 \ (76.45 \text{ m}^3)$$

Volume Occupied by Riser Pipe:

$$V_{pipe} = 3.14 \times (0.5)^2 \times (9^2 + 3^2)^{1/2}$$
$$= 3.14 \times 0.25 \times 9.49$$
$$= 7.45 \text{ ft}^3 \ (0.21 \text{ m}^3)$$

Thickness of Pipe, $t = D_o/SDR = 12/17 = 0.71$ in (0.018 m)

Inside Radius of Pipe, $r_{in} = 6 - 0.71 = 5.29$ in $= 0.44$ ft (0.134 m)

Effective Storage Volume of Pipe Section in Sump:

$$(V_{pipe})_{eff} = 3.14 \times (0.44)^2 \times 9.49 = 5.77 \text{ ft}^3 \ (0.16 \text{ m}^3)$$

Sump Volume for Filling Gravel:

$$V_f = V - V_{pipe}$$
$$= 2{,}700 - 7.45 = 2{,}693 \text{ ft}^3 \ (76.24 \text{ m}^3)$$

Gravel Void Volume for Leachate Storage:

$$V_v = n \cdot V_f$$
$$= 0.4 \times 2{,}693$$
$$= 1{,}077 \text{ ft}^3 \ (30.50 \text{ m}^3)$$

Sump Storage Volume Available for Pumping:

$$V_s = V_v + (V_{\text{pipe}})_{\text{eff}}$$
$$= 1{,}077 + 5.77$$
$$= 1{,}083 \text{ ft}^3 = 8{,}099 \text{ gallons } (30.66 \text{ m}^3)$$

2. Select Pump Model

Static Head:

$$H_s = 4.5 + 50 = 54.5 \text{ ft } (16.35 \text{ m})$$

Assume that the pump discharge flow is 2 times the rate of the leachate inflow; then

$$Q = 2 \cdot Q_{\text{in}} = 2 \times 35 = 70 \text{ gallons/min } (264 \text{ liter/min})$$

Flow Length:

$$L = 15 + \{(4.5 + 50)^2 + [(4.5 + 50) \times 3]^2\}^{1/2} + 12$$
$$= 199.3 \text{ ft } (59.8 \text{ m})$$

The Hazen-Williams coefficient $C = 130$;

The inside diameter of the outlet pipe for Style "12" pump, $d_i = 2$ in (50 mm).

Friction Head Loss:

$$H_f = \frac{10.44 \cdot Q^{1.85} \cdot L}{C^{1.85} \cdot d_i^{4.87}} \tag{9.41}$$

$$= \frac{10.44 \times 70^{1.85} \times 199.3}{130^{1.85} \times 2^{4.87}}$$

$$= \frac{10.44 \times 2590.8 \times 199.3}{8143.2 \times 29.2}$$

$$= 22.7 \text{ ft } (6.81 \text{ m})$$

Entrance, $K_{\text{in}} = 0.5$ (Table 9.7);

18.4° miter bend, $K_{18.4°} = 0.10$ (Table 9.8);

90° miter bend, $K_{18.4°} = 1.10$ (Table 9.8);

Swing valve with "2" screwed connector (fully open), $K_{\text{valve}} = 2.1$ (Table 9.9);

Outlet, $K_{\text{out}} = 1.0$ (Table 9.7);

Area in Flow, $A = 3.1416 \times (1/12)^2 = 0.022 \text{ ft}^2 \ (2.044 \times 10^{-3} \text{ m}^3)$;

Flow in Pipe, $Q = 70 \text{ gallons/min} = 0.156 \text{ ft}^3/\text{sec } (4.417 \times 10^{-3} \text{ m}^3/\text{sec})$.

Minor Head Loss:

$$H_m = \sum K_i \cdot \frac{Q_i^2}{2 \cdot g \cdot A^2} \tag{9.44}$$

$$= (0.5 + 0.10 + 1.10 + 2.1 + 1.00)[0.156^2/(2 \times 32.2 \times 0.022^2)]$$

$$= (4.8)(0.0243/0.0312)$$

$$= 3.7 \text{ ft } (1.11 \text{ m})$$

Total Discharge Head of Pump:

$$TDH = H_s + H_f + H_m \tag{9.39}$$

$$= 54.5 + 22.7 + 3.7$$

$$= 80.9 \text{ ft} \approx 81 \text{ ft } (24.5 \text{ m})$$

From Figure 9.17, select the Style "12"-5 pump model.

For this particular model, when discharge head, $TDH = 81$ ft (24.5 m), discharge flow, $Q_p = 72$ gallon/min (273 liter/min) from Figure 9.17. This value is quite close to the assuming discharge flow rate of 70 gallon/min (265 liter/min). If not, another discharge flow should be assumed to repeat the preceding calculation until the assuming discharge flow is equal to the discharge flow of the selected pump from Figures 9.16 or 9.17.

3. *Calculate Pump Cycle Time*

Pump Off Time:

$$T_{off} = V_s/Q_{in} \tag{9.45}$$

$$= 8,099/35$$

$$= 231 \text{ min} = 3.85 \text{ hrs}$$

Pump On Time:

$$T_{on} = V_s/(Q_p - Q_{in}) \tag{9.46}$$

$$= 8,099/(72 - 35)$$

$$= 219 \text{ min} = 3.65 \text{ hrs}$$

Pump Cycle Time:

$$T_{cycle} = T_{off} + T_{on} \tag{9.47}$$

$$= 3.85 + 3.65 = 7.5 \text{ hrs}$$

PROBLEMS

9.1 What components compose a leachate collection system?

9.2 What factors should be considered when selecting the leachate collection pipes?

9.3 Describe procedures of design of leachate collection pipes.

9.4 A leachate collection pipe is placed in the middle of a rectangular landfill cell. A 3.5(H):1(V) side slope with a height of 60 ft (18 m) is located on the left side of the collection pipe. A 2.5% bottom floor with a width of 120 ft (36 m) is located to the right side. The total length of this leachate collection pipe is 800 ft (240 m). The peak leachate generation rate at the side slope area is 0.0615 ft^3/day/ft^2 (0.0187 m^3/day/m^2), and the peak leachate generation rate at the bottom floor area is 0.0428 ft³/day/ft² (0.0130 m³/day/m²). This pipe is also used to carry a leachate flow rate of 0.284 ft³/sec (8.042 × 10⁻³ m³/sec) from an upstream cell. Design a perforated pipe to meet both drainage and perforation requirements.

9.5 A plan of a landfill cell is shown in Figure 8.11, in which AB is a leachate collection pipe. The bottom of the cell has is a 2% slope with a peak leachate generation rate of 0.943 in/day (24 mm/day). The slope of leachate collection pipe is 1%. Design an HDPE perforated pipe to meet both drainage and perforation requirements. The diameter of perforation hole is 0.3 inch (7.5 mm). The limiting leachate entrance velocity is 0.1 ft/sec (0.03 m/sec). Arrange the perforation holes along the pipe. What is the distance between two holes?

9.6 A 10-inch (250-mm) SDR 11 PVC perforated pipe with 12, 0.25-inch holes per foot (or 6-mm holes per meter) is selected as a primary leachate collection pipe. The maximum load applied on the pipe includes surcharge from a 2-ft (0.6-m) protective sand layer (γ_{sand} = 110 lb/ft³ or 17.3 kN/m³), 300-ft (90-m) of solid waste (γ_{waste} = 70 lb/ft³ or 11 kN/m³), a 12-inch (0.3-m) gas venting layer (γ_{sand} = 110 lb/ft³), an 18-inch (0.45-m) compacted clay layer (γ_{clay} = 115 lb/ft³ or 18 kN/m³), a 24-inch (0.6-m) drainage and protective layer (γ_{silt} = 110 lb/ft³ or 17.3 kN/m³), and a 6-inch (0.15-m) topsoil (γ_{top} = 90 lb/ft³ or 14 kN/m³). Assume bedding angle $\theta = 0°$, deflection lag factor $D_L = 1.0$, elastic modulus of the pipe material for 50 years at 73°F (23°C) temperature E = 28,200 lb/in², (194,000 kN/m²), Poisson's ratio of pipe material, $\mu = 0.3$. The bedding material of the pipe is poorly graded gravel (GP) with 85% standard density. What will be the deflection ratio (%) and critical buckling pressure of the pipe? Recalculate the deflections (%) and critical buckling pressures of the pipe if SDR of the pipe is changed to 15.5.

9.7 A vertical concrete manhole was used in a landfill. The manhole riser is circular in cross section with an outside diameter of 4.5 ft (1.35 m). The height of solid waste surrounding the manhole riser is 320 ft (96 m). The unit weight of solid waste is 65 lb/ft³ (10.2 kN/m³). The internal friction angle of the solid waste is assumed to be 35°. The interface friction angle between manhole riser wall and solid waste is approximately 26°. Determine the total downdrag force on the riser.

9.8 The concrete manhole in Problem 9.7 is replaced by a 3-ft (0.9-m) diameter HDPE manhole with an interface friction angle between manhole wall and solid waste of 18°. Recalculate the total downdrag force on the riser.

9.9 The bottom dimension of a primary leachate collection sump is 20 ft × 20 ft (6 m × 6 m). The depth of the sump is 4 ft (1.2 m), and sump slope is 3(H):1(V). The depth of the primary leachate trench is 6 inches (0.15 m). The porosity of the gravel fill in the base of the sump is 35%. The leachate inflow is 30 gallon/min (114 liter/min). A 12-inch (300-mm) SDR 15.5 riser pipe is placed on the side slope which has an inclination of 3.5(H):1(V) and height of 40 ft (12 m). The length of the riser pipe placed at the bottom of the sump is 20 ft (6 m). The length of the riser pipe extension at the top of the slope is 10 ft (3 m). A quick disconnect valve is installed at the top of the slope. Assume the Hazen-Williams coefficient of the pump pipe is 130, and the minor loss factor for a 16° elbow is 0.20. A Style "12" pump having 2-inch (50-mm) inside diameter of outlet pipe and the characteristics presented in Figure 9.17 is installed in the riser pipe at the bottom of sump to pump out and remove the leachate. Assume the leachate in the sump is required to pump to the top of the side slope and then it can gravitationally drain to a sewer system.

(1) Calculate the sump storage volume available for pumping;
(2) Select the model of the pump;
(3) Calculate the pump cycle time for the selected pump.

REFERENCES

ACI (1998). Guide to Durable Concrete. ACI 201.2R-92, ACI Manual of Concrete Practice, Vol. 1, American Concrete Institute, Skokie, IL.

Allgood, J. F. and Takahashi, H., (1972) "Balanced Design and Finite Element Analysis of Culverts," Highway Research Record No. 413, Washington, DC, pp. 44–56.

Bagchi, A., (1994) *Design, Construction, and Monitoring of Sanitary Landfill,* 2nd Edition, John Wiley & Sons, Inc., New York, NY.

Bonaparte, R., (1995) "Long-Term Performance of Landfills," *Proceedings of GeoEnvironment 2000,* Geotechnical Special Publication No. 46, ASCE, New Orleans, LA, U.S.A., February 24–26, pp. 514–553.

Bureau of Reclamation, (1978) "Drainage Manual," First Edition, Bureau of Reclamation, Eng. Res. Cent., Denver Federal Center, Denver, CO.

Cedergren, H. R., (1977) *Seepage, Drainage, and Flow Nets,* Second Edition, John Wiley & Sons, New York, NY.

Cedergren, H. R., (1989) *Seepage, Drainage, and Flow Nets,* 3rd Edition, John Wiley & Sons, Inc., New York, NY.

Corps of Engineers, (1977) "Civil Works Construction Guide Specification CW 02215," U.S. Department of the Army, Washington, DC.

Driscoll, F. G., (1986) "Groundwater and Wells," Second Edition, Johnson Division, St. Paul, MN, p. 997.

Ford, H. W., (1974) "Low Pressure Jet Cleaning of Plastic Drains in Sandy Soil," *Trans.* ASAE, Vol. 17 No. 5, pp. 895–897.

Hartley, J. P. and Duncan, J. M., (1987) "E' and its Variation with Depth," *Journal of Transportation Engineering,* Vol. 113, No. 5, pp. 538–553.

Howard, A. K., (1977) "Modulus of Soil Reaction Values for Buried Flexible Pipe," *Journal of Geotechnical Engineering,* ASCE, Vol. 103, No. 1, pp. 33–43.

Koerner, R. M., (1986) *Designing with Geosynthetics,* Prentice Hall Inc., Englewood Cliffs, NJ.

Koerner, R. M., (1994) *Designing with Geosynthetics,* 3rd Edition, Prentice Hall Inc., Englewood Cliffs, NJ.

Koerner, R. M., (1998) *Designing with Geosynthetics,* 4th Edition, Prentice Hall Inc., Upper Saddle River, NJ.

MDEQ, (1999) "Solid Waste Management Act Administrative Rules Promulgated Pursuant to Part 115 of the Natural Resources and Environmental Protection Act, 1994 PA 451, as amended (Effective April 12, 1999)," Michigan Department of Environmental Quality, Waste Management Division, Lansing, MI.

Meyerhof, G. G. and Baike, L. D., (1963) "Strength of Steel Culverts Sheets Bearing against Compacted sand Backfill," Highway Research Board Proceeding, Vol. 30, Washington, DC.

Moser, A. P., (1990) *Buried Pipe Design,* McGraw-Hill, Inc., New York, NY.

Neilson, F. D., (1967) "Modulus of Soil Reaction as Determined from Triaxial Shear Test," Highway Research Record No. 185, Washington, DC, pp. 80–90.

Selig, E. T., (1990) "Soil Properties for Plastic Pipe Installations," *Buried Plastic Pipe Technology,* ASTM STP 1093, George S. Buczala and Michael J. Cassady, Eds., American Society for Testing and Materials, Philadelphia, pp. 141–158.

Sherard, J. L., Dunnigan, L. P., and Talbot, J. R., (1984a) "Basic Properties of Sand Gravel Filters," *Journal of Geotechnical Engineering,* ASCE, Volume 110, No. 6, pp. 684–700.

Sherard, J. L., Dunnigan, L. P., and Talbot, J. R., (1984b) "Filters for Silts and Clays," *Journal of Geotechnical Engineering,* ASCE, Volume 110, No. 6, pp. 701–717.

Spangler, M. G., (1960) "Soil Engineering," 3rd Edition, International Textbook Co., Scranton, PA, 483 pages.

Spangler, M. G. and Handy, R. L., (1973) "Soil Engineering," 3rd Edition, Intext Education Publishers, 257 Park Avenue South, New York, NY.

Tchobanoglous, T., Theisen, H., and Vigil, S., (1993) *Integrated Solid Waste Management, Engineering Principles and Management Issues,* McGraw-Hill, Inc., Hightstown, NJ.

U.S. Army Corps of Engineers, (1955) "Drainage and Erosion Control—Subsurface Drainage Facilities for Airfields," *Part XIII, Chapter 2, Engineering Manual,* Military Construction, Washington, DC, June 1955.

U.S. Army, U.S. Navy, and U.S. Air Force, (1971) "Dewatering and Groundwater Control for Deep Excavation," TM 5-818-5, NAVFAC P-418, AFM 88-5, Chapter 6, April, Government Printing Office, Washington, DC.

U.S. Bureau of Reclamation, (1973) *Design of Small Dams,* 2nd Edition, U.S. Government Printing Office, Washington, DC.

CHAPTER 10

Gas Collection and Control Systems

10.1 GAS GENERATION
10.2 GAS COMPOSITION
10.3 FACTORS AFFECTING GAS GENERATION
10.4 GAS GENERATION RATE
10.5 GAS MIGRATION
10.6 TYPES AND COMPONENTS OF GAS COLLECTION SYSTEMS
    10.6.1 PASSIVE GAS COLLECTION SYSTEM
    10.6.2 ACTIVE GAS COLLECTION SYSTEM
10.7 GAS CONTROL AND TREATMENT
    10.7.1 GAS FLARING
    10.7.2 GAS PROCESSING AND ENERGY RECOVERY
10.8 DESIGN OF GAS COLLECTION SYSTEM
    10.8.1 CALCULATION OF NMOC EMISSION RATE
    10.8.2 ESTIMATION OF GAS GENERATION RATE
    10.8.3 GAS EXTRACTION WELL SYSTEM LAYOUT AND SPACING
    10.8.4 GAS FLOW GENERATED FROM EACH EXTRACTION WELL OR COLLECTOR
    10.8.5 COLLECTION PIPE SYSTEM LAYOUT AND ROUTING
    10.8.6 ESTIMATION OF CONDENSATE PRODUCTION
    10.8.7 HEADER PIPE SIZING AND PRESSURE LOSS CALCULATIONS
    10.8.8 VALVE AND FITTING PRESSURE LOSS CALCULATIONS
    PROBLEMS
    REFERENCES

A solid waste landfill can be conceptualized as a relatively long-term biochemical reactor, with solid waste and water as the major inputs, and with landfill gas and leachate as the principal outputs. Material stored in the landfill includes partially biodegraded organic material and the other inorganic waste materials originally placed in the landfill. Landfill gas is a normal by-product of biological anaerobic decomposition of organic materials in landfills. A landfill gas collection and control system is employed to prevent unwanted movement of landfill gas into the atmosphere or the lateral and vertical movement through the surrounding soil. Recovered landfill gas can be used to produce energy or can be flared under controlled conditions to eliminate the discharge of harmful constituents to the atmosphere.

10.1 GAS GENERATION

Municipal solid waste (MSW) can generate tremendous quantities of gas during its decomposition. Landfill gas generation is a biological process in which microorganisms decompose organic waste to produce carbon dioxide, methane, and other gases. Landfill gas is composed of a number of gases that are present in large amounts (the principal gases) and a number of gases that are present in very small amounts (the trace gases). The principal gases are produced from the decomposition of the organic fraction of municipal solid waste. A simple equation for the production of the principal landfill gas from cellulose can be expressed as follows:

$$C_6H_{10}O_5 + H_2O \rightarrow 3CH_4 + 3CO_2$$

Some of the trace gases, although present in small quantities, can be toxic and could present risks to public health.

Landfill gas generation occurs in several phases, as shown in Figure 10.1. Initially, the distribution of gases in the landfill is representative of the distribution of gases in the atmosphere—about 80% nitrogen and 20% oxygen, with some carbon dioxide and other compounds. Aerobic decomposition begins soon after the waste is placed in a landfill and continues until all of entrained oxygen is depleted from the voids in the waste and from within the organic material itself. Aerobic bacteria produce a gaseous product characterized by relatively high temperatures (130 to 160°F or 54 to 71°C), high carbon dioxide content, and no methane content. Other by-products include water, residual organics, and heat. Aerobic decomposition may continue (EMCON, 1998) from 6 to as long as 18 months in the case of waste placed in the bottom of the landfill, although it may last only 3 to 6 months in the upper lifts if methane-rich landfill gas from below flushes oxygen from voids in the disposed waste.

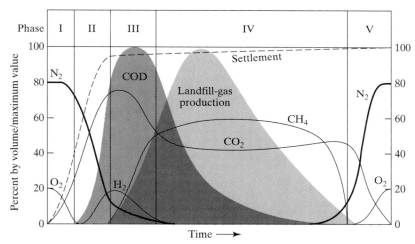

FIGURE 10.1 Changes in Landfill Gas Composition over Time (UKDOE, 1993)

Waste decomposition undergoes several distinct changes with time. After all entrained oxygen is depleted, decomposition enters a transitional phase in which acid-forming bacteria begin to hydrolyze and ferment the complex organic compounds in the waste. Decomposition then enters an anaerobic phase, during which methane-forming bacteria, which thrive in an oxygen deficient environment, become dominant. Studies have shown (EMCON, 1998) that anaerobic gas production is typified by somewhat lower temperatures (100 to 130°F or 38 to 54°C), significantly higher methane concentrations (45 to 57 percent) and lower carbon dioxide concentrations (40 to 48 percent). Anaerobic gas production will continue until all of the carbonaceous material is depleted or until oxygen is re-introduced into the waste, which would then return the decomposition process to aerobic conditions. A return to aerobic decomposition does not stop landfill gas production, but it will retard the process until anaerobic conditions resume (EMCON, 1998). A summary of the five phases of municipal solid waste (MSW) landfill gas generation is presented in Table 10.1. The total time duration of gas generation for a landfill can be 10 to 80 or more years. The lower number is felt to be representative of bioreactor landfills (which will be discussed in Chapter 15), while the higher number is probably indicative of conventional landfill practice.

10.2 GAS COMPOSITION

Landfill gas is commonly said to be composed of roughly equal quantities of methane (CH_4) and carbon dioxide (CO_2). It would be more accurate to state that landfill gas, as observed in most landfill gas samples, typically is composed of many more constituents. Typical concentrations of these constituents are listed in Table 10.2. A discussion of individual landfill gas components follows (modified from EMCON, 1998).

Methane (CH_4)

Methane is a by-product of anaerobic decomposition. It is a colorless, odorless, tasteless gas that is lighter than air, relatively insoluble, and can be highly explosive in concentrations of 5 to 25 percent in air (the "explosive range").

TABLE 10.1 Summary of MSW Landfill Gas Generation Phases (EMCON, 1998)

Phase	Name	Primary Activity Signaling the End of Phase
I	Aerobic	No oxygen in the landfill gas (several hours to 1 week)
II	Aerobic/Acid Generation	Formation of free fatty acids is at its peak and methane generation begins (1 to 6 months)
III	Transition to Anaerobic	Methane and carbon dioxide concentrations stabilize and no nitrogen in the landfill gas (3 months to 3 years)
IV	Anaerobic	Methane and carbon dioxide concentrations begin to reduce and some nitrogen (air) returns to the system (8 to 40 years)
V	Transition to Stabilization	Gas is primary air and all anaerobic decomposition is complete (1 to 40 or more years)

TABLE 10.2 Typical Constituents in Municipal Solid Waste Landfill Gas (EMCON, 1998)

Component	Percent
Methane (CH_4)	45 to 58
Carbon Dioxide (CO_2)	35 to 45
Nitrogen (N_2)	<1 to 20
Oxygen (O_2)	<1 to 5
Hydrogen (H_2)	<1 to 5
Water Vapor (H_2O)	1 to 5
Trace Constituents*	<1 to 3

*NMOCs are among the trace constituents

Carbon Dioxide (CO_2)

Carbon dioxide is a by-product of both the aerobic and anaerobic phases of decomposition. It is also colorless and odorless, but is heavier than air, noncombustible, and highly soluble in water. Carbon dioxide will increase water hardness and decrease water pH.

Nitrogen (N_2) *and Oxygen* (O_2)

Nitrogen and oxygen are usually found in landfill gas samples. They are not typically found in landfill gas as it is being generated. Their presence in landfill gas is the result of air intrusion through the cover of the landfill, air leaks into a landfill gas recovery or control system, air leaks in the sampling train during collection of landfill gas samples, or as a result of leachate removal activities such as insertion of removal pumps. Typically, the combined volumes of nitrogen and oxygen in landfill gas are less than 10 percent and their ratios are the same as in air. If their volumes increase, it may indicate an air leak in the collection system.

Hydrogen (H_2)

In landfills, hydrogen is typically produced only during aerobic decomposition and the earliest states of anaerobic decomposition. If hydrogen is present in anything more than trace concentrations in a mature landfill, it may indicate that areas of the site have not attained the mature landfill gas generation phase for one reason or another.

Water Vapor (H_2O)

Landfill gas is typically saturated with water vapor. The water vapor in landfill gas comes from water in the landfill that becomes entrained in the gas. Water vapor that condenses from landfill gas is the primary component of condensate. Consideration must always be given to proper handling and disposing of condensate as part of any landfill gas management effort.

Trace Constituents

Landfill gas also contains small quantities (usually less than 1 percent) of non-methane organic components (NMOCs), as well as various other trace compounds.

The presence of trace compounds in landfill gas is most likely caused by the disposal of waste containing these compounds into the landfill, although some also may be present as a result of natural decomposition processes within the landfill [e.g., hydrogen sulfide (H_2S) from decomposition of gypsum board in a landfill].

As many as 150 different gases, mostly in the parts per million (ppm) or parts per billion (ppb) range, have been identified in landfill gas, although not all landfills will have all of these compounds in their landfill gas (EMCON, 1998). These gases may include harmful, toxic, or even carcinogenic compounds, such as vinyl chloride, benzene, toluene, xylene, perchloroethlyne, carbonyl sulfide, and various other chlorinated and fluorinated hydrocarbons. Other trace compounds found in landfill gas include hydrogen sulfide and mercaptans, which carry the distinctive odor associated with landfill gas.

Trace constituents typically enter the landfill as part of a particular waste disposed there (such as the perchloroethylene contained in the fluff from dry cleaning establishments) and are picked up in minute quantities by the landfill gas. Because the sources, and therefore, quantities of nonmethane organic components (NMOCs) and other trace constituents, are so unpredictable in both type and quantity, there is no direct or predictable relationship between the amount of methane and the amount of trace constituents (including NMOCs) contained in landfill gas.

10.3 FACTORS AFFECTING GAS GENERATION

The ability of a landfill to generate gas depends on many factors, including waste composition, moisture content, waste particle size, age of waste, pH, temperature, and others. Decomposition and gas production can be expected to continue, theoretically, for up to 30 to 100 years, but in practice, they occur at a high level for a much shorter period of time (McBean et al., 1995; EMCON, 1998). The factors that affect gas generation are discussed as follows (modified from EMCON, 1998; and McBean et al., 1995).

Waste Composition

The majority of residential and commercial waste placed in a typical municipal solid waste landfill is decomposable. The remaining portion typically consists of various inert materials such as concrete, ash, soil, metals, plastics, and other nondecomposable materials. The more easily the organic fraction of the waste decomposes, the faster will be the landfill gas generation rate. Food wastes typically fall into this category. Thus, a high percentage of food wastes in a landfill likely will lead to a faster landfill gas generation rate. Some decomposable wastes, such as large pieces of wood, are not inert, but decompose so slowly that for most practical purposes they do not contribute significantly to landfill gas generation.

Moisture Content of Waste

For most landfills, after waste composition, the moisture content of waste is the most significant factor in the production rate of landfill gas. The higher the moisture content, the greater the gas generation rate. This aspect will be emphasized further in Chapter 15, which is on bioreactor landfills. Moisture content in a conventional landfill will change over time. Changes in landfill moisture content may result from changes in surface water infiltration and/or groundwater inflow, release of water as a result of waste decomposition, seasonal variations in the moisture content of the waste, and managed additions of liquids. Theoretically, the optimum condition for gas generation is total waste saturation. On the other hand, if the waste were saturated, extraction of the landfill gas would be extremely difficult, if not impossible.

Particle Size of Waste

The smaller the disposed waste units or particles, the larger its specific surface area. A particle of waste with a larger specific surface area will decompose faster than a particle with a smaller one. For example, a disposed tree stump will decompose more quickly if it is ground into wood chips, than if disposed whole. Therefore, a landfill that accepts shredded waste will have a faster overall decomposition rate (i.e., faster gas generation rate) than a landfill that accepts only nonshredded waste.

Age of Waste

Landfill gas (methane) generation has two primary time-dependent variables: lag time and conversion time. Lag time is the period from waste placement to the start of methane generation (start of Phase III). Conversion time is the period from waste placement to the end of methane generation (end of Phase V). For example, yard waste has very short lag and conversion times, while leather and plastic have very long lag and conversion times.

pH

The optimum pH range for most anaerobic bacteria is 6.7 to 7.5 or close to neutral [i.e., pH = 7.0] (McBean et al., 1995). Within the optimum pH range, methanogens grow at a high rate so that methane production is maximized. Outside the optimum range—below a pH of 6 or above 8—methane production is severely limited. Most landfills tend to have slightly acidic environments (EMCON, 1998).

Temperature

Temperature conditions within a landfill influence the type of bacteria that are predominant and the level of gas production. The optimum temperature range for mesophilic bacteria is 30 to 35°C (86 to 95°F), whereas the optimum for thermophilic bacteria is 45 to 65°C (113 to 149°F). Thermophiles generally produce higher gas generation rates; however, most landfills exist in the mesophilic range. Landfill temperatures often reach a maximum within 45 days after placement of wastes as a result of the aerobic microbial activity. Landfill temperature then decreases once anaerobic conditions develop. Greater temperature fluctuations are typical in the upper zones of a landfill as a result of changing ambient air temperature. Landfill waste at a depth of 15 m (50 ft) or more is relatively unaffected by ambient air temperatures. Temperatures as high as 70°C (158°F) have been observed (McBean et al., 1995). Elevated gas temperatures within a landfill are a result of biological activity. Landfill gas temperatures typically are reported to be in the range from 30 to 60°C (86 to 140°F) (EMCON, 1980 and 1981). Optimum temperatures range from 30 to 40°C (86 to 104°F), whereas temperatures below 15°C (59°F) severely limit methanogenic activity (McBean et al., 1995). Liner temperatures, on the other hand, lie in the range 20 to 25°C (68 to 77°F), attesting to the fact that heat rises within the waste mass (G. R. Koerner et al., 1996).

Other Factors

Other factors that may influence landfill gas generation rate include nutrient content, bacterial content, oxidation-reduction potential, density of gas production, waste compaction, landfill dimensions (area and depth), and landfill operation and waste processing variables (Zehnder, 1978; Fungaroli and Steiner, 1979; McBean et al., 1995).

10.4 GAS GENERATION RATES

All landfills containing organic, decomposable materials will generate gas although the total volume or incremental rate of production may vary widely over time. The total volume of landfill gas generated over the entire decompositional life of the landfill is mainly a direct function of the total quantity of organic material contained in the landfill, albeit some components decompose rapidly, some at a moderate rate, and some over a much longer period of time. Therefore, the quantity of the waste available for decomposition is the primary factor (EMCON, 1998) in determining the total volume of landfill gas that will be generated.

In a mature landfill, which approaches fully anaerobic, steady-state conditions, potential landfill gas generation rates may range from 0.04 ft^3 of landfill gas (LFG) per pound of waste per year [2.50 m^3 (LFG)/Mg (waste)/yr] to a theoretical maximum of 0.14 ft^3 (LFG)/lb (waste)/yr [8.74 m^3 (LFG)/Mg (waste)/yr]. Empirical evidence (EMCON, 1998) suggests that the gas generation rate at most landfills may actually be at or below 0.10 ft^3 (LGF)/lb (waste)/yr [6.24 m^3 (LFG)/Mg (waste)/yr].

The incremental rate at which landfill gas is produced is primarily a function of the type of waste involved. Food waste, for example, decomposes much more rapidly than paper or cardboard. The overall rate of decomposition for all waste components in a given section of the landfill also is influenced by a variety of other factors (EMCON, 1998), such as moisture content, waste particle size, site configuration, temperature, pH, and others. If the landfill is very dry, little gas will be generated. In some landfills, seasonal temperature changes also might influence gas generation. In the colder months, the rate of gas generation can decrease significantly in shallow landfills. Basically, the better the conditions within a landfill are for the anaerobic bacteria, the faster decomposition will take place, resulting in a faster overall gas generation rate.

There is no specific standard at present for the length of landfill gas generation. EMCON (1998) quotes several sources that claim a 20-year life for generation of the majority of landfill gas from disposed waste as the norm. It should be noted, however, that this figure is a theoretical estimate since the first U.S. landfill gas recovery facility began operation only 18 years ago, and presently only a few such facilities have accumulated more than 10 years of operating data. Many older landfills are still going strong as landfill gas producers, and some newer sites appear depleted after only a few years. Dry landfills in arid climates could generate a minimal amount of gas for a very long time, perhaps 100 years. In landfills where there is considerable moisture added to the organic matter (as in bioreactor landfills), landfill gas generation is very rapid for 8 to 15 years and then is anticipated to decline to a minimal level.

10.5 GAS MIGRATION

Because landfill gas is primarily generated by the decomposition of solids in the waste, the resulting gas occupies a substantially larger volume than the waste. This causes pressure to build within the landfill as the gas tries to expand to its equilibrium volume. The gas pressure at the bottom of a landfill has been measured at levels ranging from less than one inch of water column to as high as 4 atmospheres (Prosser and Janechek, 1995). It is this pressure that drives the gas from a landfill into the atmosphere, into the soil, and, potentially, into surrounding structures. Landfill gas migration from landfills occurs via two mechanisms—convection and diffusion. *Convection* is transport induced by pressure gradients formed by gas production in layers surrounded by low hydraulic conductivity or saturated layers. Convection also results from buoyancy forces because methane is lighter than carbon dioxide and air.

Diffusion is the transport of substances induced by concentration gradients. Anaerobic decomposition produces a gas mixture with concentrations of methane and carbon dioxide that are much greater than those found in the surrounding air. Therefore, molecules of methane and carbon dioxide will diffuse from the landfill gas to the air. Diffusion plays a much smaller role in gas migration than convection.

Landfill gas migration occurs in response to pressure, concentration, and possibly temperature gradients. Because gas generation causes gas pressure to build, a gradient is established that seeks to equalize itself. Gas migration in the landfill follows the path of least resistance. The degree to which gas migrates vertically or horizontally depends on many factors, including the nature of the landfill design, surrounding soils, type of waste, degree of waste segregation in the landfill, the type of daily or final cover used at the facility. With a sand and gravel soil cover of relatively high permeability, gas tends to vent equally and vertically, perhaps through the cover onto the surface, and at a relatively uniform rate. With a low-permeability cover (geomembrane and/or GCL), gas tends to migrate horizontally. If a low-permeability cover is placed on the landfill, the gas no longer has a pathway for upward release. This situation can also arise when a soil layer at the surface becomes frozen. If lateral resistance to migration is less than vertical resistance, the gas tends to move laterally where it can collect in low spots, such as manholes and pump stations adjacent to the landfill, resulting in oxygen-depleted and potentially explosive environments. Any structures on or near the landfill (such as offices, homes, basements, etc.) must be monitored to ensure that gas is not accumulating in that area. These migration concepts are shown in Figures 10.2 and 10.3 (USEPA, 1994).

Gas generally is transported through the unsaturated portion of the soil (i.e., that portion not filled with water) or fractures in rock. But because gas is soluble, it can migrate through saturated soil under sufficient pressure. For example, in some large old landfills, relatively impermeable waste often was placed directly on the ground with no leachate collection system or drainage beneath the waste. These landfills can generate significant amounts of gas. At one facility, the pressure was so high that it drove the gas into the ground water. The gas then exited from the groundwater into unsaturated soil at the site perimeter, where pressure was lower (USEPA, 1994).

Many factors affect gas migration. Some of the more important factors are in the landfill design, including waste cell construction, final cover design, and incorporation

340 Chapter 10 Gas Collection and Control Systems

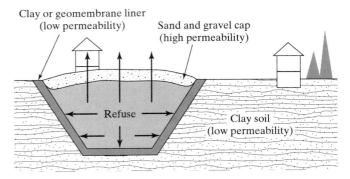

FIGURE 10.2 Landfill Condition Resulting in Vertical Gas Migration (USEPA, 1994)

of gas migration control measures. Low hydraulic conductivity soil layers and geomembranes are very effective barriers to gas migration. Sand and gravel layers and void spaces provide effective corridors for channeling gas migration. Other channels affecting migration are cracks and fissures between and in lifts of waste or soil due to differential settlement and subsidence.

Other factors affecting gas migration include the gas production rate, the presence of natural and artificial conduits and barriers adjacent to the landfill, and climatic and seasonal variations in site conditions. High gas production rates increase migration. Corridors at the site adjacent to landfill, such as water conduits, drain culverts, buried lines, and sand and gravel lenses, promote uncontrolled migration from the site. Barriers can include clay deposits, high or perched water tables, roads, and compacted, low hydraulic conductivity soils. Environmental variations can result from the intermittent occurrence of saturated or frozen surface soils, which seal the surface and promote lateral migration. Barometric pressure changes also affect the rate of gas release to the surface. Seasonal changes in moisture content can change the gas production rate and, therefore, the extent and quantity of migration.

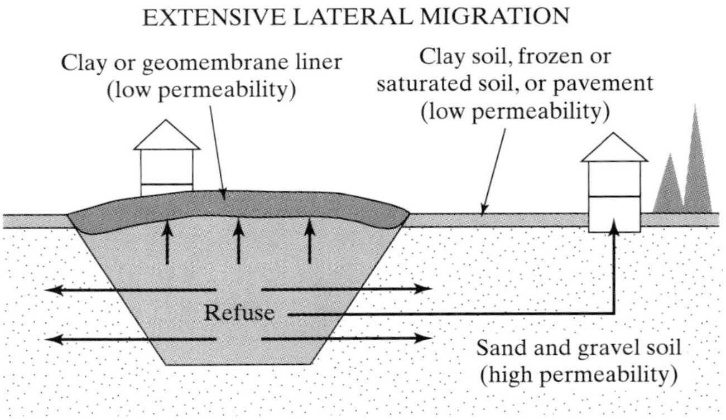

FIGURE 10.3 Landfill Condition Resulting in Lateral Gas Migration (USEPA, 1994)

10.6 TYPES AND COMPONENTS OF GAS COLLECTION SYSTEMS

The Resource Conservation and Recovery Act (RCRA) requires that landfill gas be controlled so that the concentration of methane at a landfill's property line is less than 5% by volume (40 CFR Part 256.23). Two different systems can be used to collect vented gas: passive and active collection systems. With either type, redundancy in the gas collection system is important for ensuring continued operation of the system. Redundancy protects against the loss of system components caused by settlement and failure of the entire system from a single malfunctioning component. Redundancy can include additional gas extraction wells and header pipes.

10.6.1 Passive Gas Collection System

Passive gas collection systems allow gas to be released without using mechanical devices such as blowers or pumps. The systems can be used outside or within the landfill. Perimeter trenches and pipes vented to the atmosphere can act as a passive system by intercepting lateral migration of gas through the soil. A trench is dug around the landfill to the depth of the water table (if shallow), and is backfilled with pervious stone and pipes, which act as a passive barrier. Depending on the types of soil at the facility, a more solid and less permeable barrier might be needed on the trench side away from the landfill to improve passive venting within the trench. If the soil is sandy with a permeability similar to that of the trench, a geomembrane placed on the outside of the trench will help stop gas migration and allow the gas to pass up through the vent. For facilities with a deeper water table, a slurry wall (perhaps including a geomembrane within it) can be used as a remedial measure to stop gas migration.

Figure 10.4 presents a typical passive vent used at landfill with its final cover system. A perforated collection pipe is placed in a granular or geocomposite layer above the waste. Typically, coarse sand is used for the vent layer, but geotextiles and geonets can be combined as an alternative. This perforated pipe is connected to a vertical riser pipe, which is connected to a 90-degree elbow (gooseneck) through which gas is vented. A barrier layer placed above the vent layer causes gases to stop at the geomembrane or the clay surface and migrate laterally to the pipes and up to the atmosphere. Vents can be installed independently of one another or connected in a

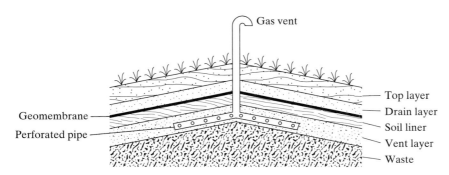

FIGURE 10.4 Typical Passive Gas Collection System for Venting of Landfill Gas (USEPA, 1992)

system of lateral header pipes. Piping should be buried deep enough to prevent frost heaving. Care must also be taken to protect these vents; if they are broken the piping will provide a conduit for surface water to enter the waste.

The only advantage of passive gas collection systems is that they are relatively inexpensive and require little maintenance. However, air intrusion is usually a problem, due to the shallow construction of these vents, thus limiting the effectiveness of the vent for collecting gas. The major disadvantage of a passive gas collection system is that landfill gases are vented directly into the atmosphere. This is a source of air pollution and must be challenged insofar as sound environmental practice is concerned. Thus for environmental concerns and/or if a passive system is not working properly and vents are connected with a header pipe in a portion of the landfill, the system can be converted to an active gas collection system (USEPA, 1994).

10.6.2 Active Gas Collection System

Based on the New Source Performance Standards/Emission Guidelines (NSPS/EG) requirements promulgated on March 12, 1996, an active gas collection system must be designed if a municipal solid waste landfill has a design capacity equal to or greater than 2.5 million megagrams (2.8 million U.S. tons) or 2.5 million cubic meters (3.3 million cubic yards) and the calculated nonmethane organic compounds (NMOC) emission rate is equal to or greater than 50 megagrams per year (55 U.S. tons per year). Most active gas collection systems use negative pressure and apply a vacuum to pull gas out of the waste in the landfill via vertical extraction wells, horizontal extraction trenches, or a venting layer beneath the cover barrier system.

A profile of an active gas collection system is shown in Figure 10.5. The main components of an active gas collection system includes gas extraction well or extrac-

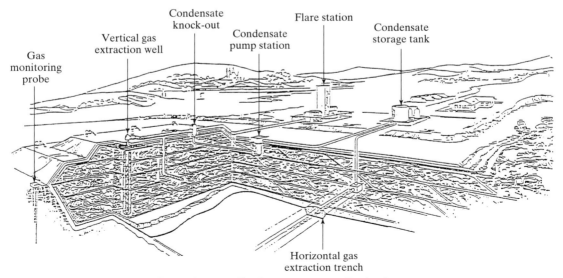

FIGURE 10.5 Profile of an Active Gas Collection System

tion trench, gas extraction wellhead, gas collection header pipe, condensate collection and pump station, condensate storage tank, vacuum source (blower), and gas migration probe. Active gas collection systems are the direction of the future not only from an environmentalist perspective, but also from an economic perspective. For a large landfill, the energy provided by captured landfill gas represents an excellent financial incentive to the owner/operator of the facility.

10.6.2.1 Gas Extraction Well Landfill gas is extracted from landfills using vertical wells or horizontal trenches. Vertical extraction wells are drilled into the landfill (Figures 10.5 and 10.6). Horizontal extraction trenches must be installed as lifts (layers) of waste are placed in the landfill (Figure 10.5). A partially perforated pipe is placed in the well or trench that is then backfilled with gravel, forming the gas collection zone. The pipe (well casing) terminates at the surface to a wellhead that includes a flow control valve and may include a flow measurement element and gas sampling ports. Wells are interconnected with pipe forming a landfill gas extraction system.

Figure 10.7 illustrates a vertical gas extraction well that can be used within the landfill or around the perimeter. Gas extraction wells are typically drilled (EMCON, 1998) with a 3-foot (0.3-m) diameter bucket auger to within 25 feet (7.5 m) of the landfill base or until the leachate level is encountered, whichever occurs first (Michels, 1996). Wells are usually installed with a 6-inch diameter schedule 80 PVC casing with the upper one-third solid pipe and bottom two-thirds slotted or perforated. The boreholes are backfilled with 1- to 2-inch (25- to 50-mm)-diameter washed stone, and a surface seal typically consisting of fine soil (stone screenings) and bentonite.

FIGURE 10.6 Vertical Gas Extraction Well and Wellhead

344 Chapter 10 Gas Collection and Control Systems

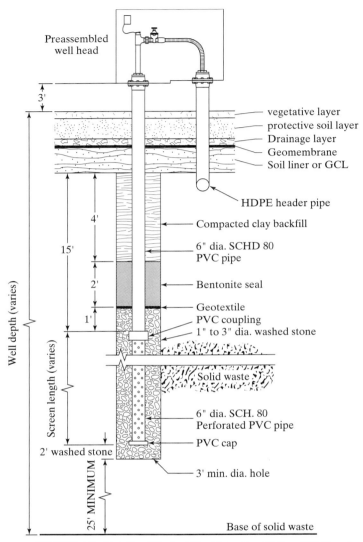

FIGURE 10.7 Details of Vertical Gas Extraction Well and Wellhead

Gas wells or header systems should be equipped with a valve to regulate flow and serve as a sampling port (USEPA, 1994). Such a valve is important because gas generation can vary throughout different parts of the landfill. Additionally, over time, the flow in certain areas might need to be adjusted. By monitoring gas quality (e.g., methane content) and measuring gas pressures, the operator can assess more readily the seasonal and long-term changes in gas production and distribution within the landfill and make appropriate adjustments.

Gas extraction wells should be sealed to minimize atmospheric releases. Depending on the age of the landfill and the location of the wells, differential settlement can occur, leading to well damage. Effort to design the extraction system with flexible connections and materials capable of withstanding deformations will help maintain system integrity. In areas of seismic activity, flexibility of the entire system is an overriding concern.

10.6.2.2 Gas Extraction Wellhead Designs of gas extraction wellheads vary greatly. These wellheads typically have built-in gas flow metering and gas temperature gauges, PVC flow control valve, quick-connect gas pressure and sampling ports, flex hose for connection to header, with diameter and length varying and depending upon the wellhead design (Figure 10.7).

Buried/vaulted wellheads are sometimes installed for aesthetic and safety purposes. This design makes it less likely that visitors to the landfill will either see or run into projecting wellheads above the landfill cap. Sometimes buried/vaulted wellheads are also used to minimize vandalism (EMCON, 1998). Postclosure uses of a municipal solid waste landfill will be affected by design layout wellheads. (See Chapter 18 for details.)

10.6.2.3 Gas Collection Header Pipe Vacuum for gas extraction and gas flow is conveyed to the gas extraction wells through a network of buried piping. The main collection header is designed as a looped network, as shown in Figure 10.8, to provide for distribution of gas flows and lower overall system pressure drops due to variance of actual field flows. The buried header piping is sloped and installed at controlled grade with engineered low points to provide for gravity collection of condensate and minimize blockages from differential settlement. The diameter of the pipe may be increased to reduce head loss due to friction. These pipes are embedded in trenches filled with sand. PVC or HDPE pipes are used for header pipes. The header pipe must

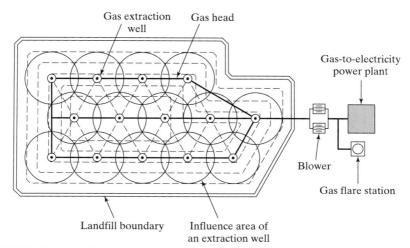

FIGURE 10.8 Equilateral Triangular Distribution for Vertical Gas Extraction Wells and a Network of Gas Collection Header Pipes

not be perforated. Online header shut-off valves are installed at different points along the header piping to allow for the isolation of different sections of the wellfield during system maintenance or expansion.

A SDR 17 HDPE pipe is typically used as gas collection header pipe with pipe diameter from 4 inches to 18 inches (100 to 450 mm) (EMCON, 1998). Pipe sections are fusion welded and placed in a trench at an adequate slope to allow condensate drainage to the pump station. With differential settlement, low spots in the connection or header pipes can develop, and the pipe can fill with gas condensate, which effectively plugs the pipe. The header piping must be sufficiently sloped to allow for differential settlement and still provide for adequate drainage in the future. A 3% gradient should be the absolute minimum, and 6 to 12% is preferred for shorter piping runs.

In buried systems, PVC piping joints and connections do not stand up well to landfill differential settlement and frequently exhibit breakage from stress due to their inflexibility. This is often observed at piping elbows and connections. When PVC is buried, flex connectors are used to compensate for some of the likely settlement. For flexible piping, pipe stiffness rather than crush strength is usually the controlling property. One that should be considered is nonperforated corrugated HDPE pipe in such conditions and it is being used regularly.

10.6.2.4 Condensate Collection and Pump Station Temperatures of landfill gas within the landfill and at the wellhead usually range from 80 to 125°F (27 to 52°C) depending upon a number of factors (LandTech, 1994; Prosser and Janechek, 1995). As landfill gas moves through waste, it is typically saturated with water at these temperatures. As gas is extracted from the landfill and transported through the gas collection system, it gradually cools and a liquid is formed. This liquid is known as landfill gas condensate. In an uninsulated, above ground, collection piping system the gas temperature approaches that of the ambient temperature. The same cooling occurs in a buried system, although more slowly, because the piping system is more insulated.

Landfill gas condensate may have similarities to the composition of leachate (for a particular site), but due to its nature, it will also have some differences. This is because a fraction of condensate is distilled from the leachate (as the gas is formed, becomes saturated, and cools). The composition of leachate is representative of the components leached from the waste by water moving through and formed within the landfill.

Controlling and removing condensate from the gas flow is important for the efficient operation of the gas collection system. Landfill gas condensate is collected in the low spots of the gas collection system, cutting off the vacuum to the wells and impairing system operation. A condensate knock-out (or condensate trap) eliminates this problem by promoting the formation of liquid droplets and separating them from the gas flow in a controlled manner and location. Condensate traps, spaced 200 to 500 feet (60 to 150 m) apart, should be included in the design of the gas collection system. These traps will allow the condensate, which migrates with gas, to drop out of the gas collection pipe, thus preventing pipe plugging. For every million cubic feet of gas generated, about 50 to 600 gallons (190 to 2,300 liters) of condensate might be generated, depending on the vacuum pressure of the system and the moisture content of the waste.

Once the condensate has accumulated in a knock-out reservoir or low point in the gas collection system, it is forced to drain into a pumping station reservoir. There must be a complete pipe network connecting all condensate traps and leading the liquid to the aforementioned reservoir. The pipe networks are often smooth wall HDPE pipe or nonperforated/nonslotted corrugated HDPE pipe. The condensate is then pumped to a storage tank or sewer line for treatment and disposal. The number of pumping stations required at each landfill will depend on the number of gas extraction low points and installed condensate knock-outs. Figure 10.9 shows the details of a condensate collection and pump station.

10.6.2.5 Condensate Storage Tank It is recommended that a condensate storage tank be a double wall tank, like a leachate storage tank with a secondary containment system. It is constructed of HDPE, fiberglass, or steel, and can be installed either underground or aboveground (Figure 10.5). If the storage tank is installed in an area of high groundwater, a concrete ballast should be installed to prevent buoyant forces from "popping" the tank out of the ground (EMCON, 1998).

10.6.2.6 Vacuum Source (Blower) The blowers should be installed in a shed at an elevation slightly higher than the end of the header pipe to facilitate condensate dripping. The blowers are usually housed in the landfill gas-to-electricity plant or flare station (see Figure 10.8). The blowers create a vacuum on the extraction system bringing the landfill gas to the landfill gas-to-electricity plant or flare station.

The size of the blower and the amount of head are design parameters that should be based on total negative head and volume of gas to be extracted. Blower capacity should be matched against future needs for gas management as a facility expands over time or as needs change, such as when landfill cells are added to or disconnected from the gas recovery system.

Blowers used to pull gas from the landfill generally operate from 300 to 2,000 ft^3/min (8.5 to 57 m^3 min). The total head of a landfill gas collection system blower is on the order of 45 to 60 inches (1,143 to 1,524 mm) of water column (W.C.) so that suction vacuum on the wellfield usually runs 10 to 50 inches (254 to 1,270 mm) of water column (W.C.). In extreme climates or in areas where sound deadening is required, blowers are typically housed in prefabricated sheds. These sheds are normally insulated (EMCON, 1998) to reduce noise levels, but are not heated, and have a concrete floor, lighting, and a methane alarm system to warn if methane leaks into the building.

10.6.2.7 Gas Migration Probe If the gas extraction wellfield is not properly adjusted, landfill gas can migrate and leave the landfill. Since landfill gas can be explosive, gas migration probes are often placed along the perimeter of the landfill, in native soil, to detect escaping methane before it poses a threat to the nearby community (recall Figure 10.5).

Bore holes for gas migration probes are typically drilled with a hollow stem auger, and extend to just below the groundwater table or 5 feet (1.5 m) below the base of the landfill. A 1-inch (25-mm) diameter schedule 40 or schedule 80 polyvinyl chloride (PVC) casing is commonly used for sampling. The bore hole is normally

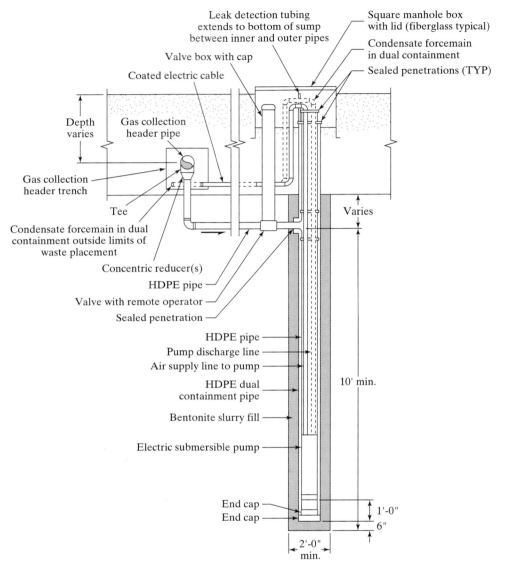

FIGURE 10.9 Example of Condensate Collection and Pump Station

backfilled with pea gravel and various sealing materials (including bentonite). A 6-inch- (150-mm)-diameter steel riser with locking cap is also installed at the surface to protect the PVC casing (EMCON, 1998). Figure 10.10 shows the details of a landfill gas migration probe.

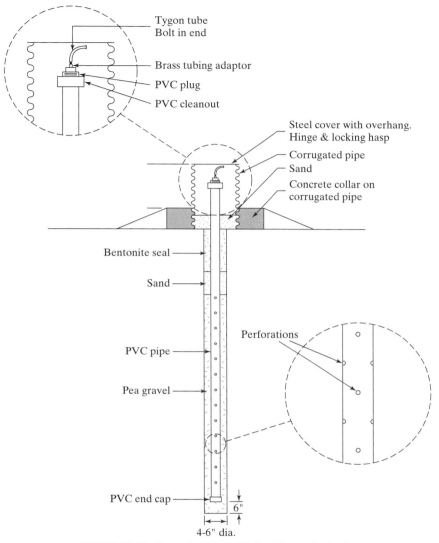
FIGURE 10.10 Example of Landfill Gas Monitoring Probe

10.7 GAS CONTROL AND TREATMENT

If landfill gas is collected in an active gas collection system, the collected gas must be treated before releasing to atmosphere. Gas treatment methods include gas flaring and gas processing and energy recovery.

10.7.1 Gas Flaring

A flare is sometimes referred to as a "controlled combustion unit." Flaring is a common treatment method when enough methane (e. g., greater than approximately 20 percent by volume) is present in the gas. Flaring reduces odors and is a much more effective and environmentally friendly method for odor control than passive venting. Most flares designed today are enclosed flares, which allow longer residence times, elevated combustion temperatures, and greater thermal destruction efficiency than open flares. (See Figure 10.11.)

Generally, gas enters the flare system from the landfill through a valve located upstream of the blower (Figure 10.11). The blower outlet exits through a pipe to the flare stack, which contains instruments to verify temperature and flame presence and to prevent burnback of gas into the blower. These instruments use passive safety mechanisms, such as flame arresters and liquid-filled flashback units, or active protection systems, such as thermocouples (to detect combustion flashback), self-actuated valves (to shut off gas entry), and auto-shutdown sensors. If for any reason the flame goes out, a flame detector will immediately sense that no flame is present and will shut down the self-actuated valve, thus preventing uncombusted gas from escaping into the atmosphere. A flare should include (USEPA, 1994) both passive and active safety systems; if one of these systems malfunctions, the other system can take over. The stack is generally purged before flare startup, and typically, propane is used for ignition and pilot fuel. The flare stack also can include equipment for monitoring air quality exiting the system, which might be required by some regulatory permits.

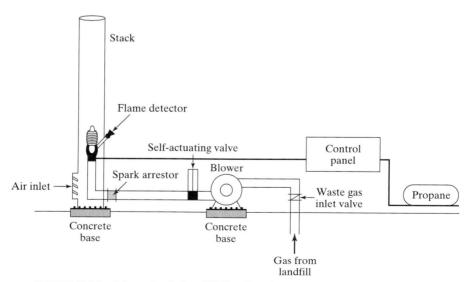

FIGURE 10.11 Schematic of a Landfill Gas Flare System with Blower (USEPA, 1994)

10.7.2 Gas Processing and Energy Recovery

Landfill gas also can be processed by removing water and impurities, including carbon dioxide. The heat value of unprocessed landfill gas is about 500 Btu per cubic foot (18,000 Btu per cubic meter). This heat value is about half that of natural gases, primarily because only about half of landfill gas is methane. Processing increases the heat value of the gas to approximately 1,000 Btu per cubic foot (35,000 Btu per cubic meter). At this level, gas can be used on-site or directed into a pipeline and sold to a utility as natural gas.

Energy recovery is being used at many landfills, particularly larger landfills where the size of the operation and the potential life of the project make energy recovery economically viable. Whether energy can be recovered at a reasonable cost depends on the quality and volume of the gas. At a small landfill, gas with a heat value of 500 Btu per cubic foot (18,000 Btu per cubic meter) can be used to run a modified internal combustion engine or a generator to convert gas to electrical energy. At a larger landfill, moisture and carbon dioxide removal (through scrubbing and gas polishing with carbon or polymer adsorption) makes the gas usable enough to run boilers and turbine generators for energy recovery. Figure 10.12 shows a landfill gas-to-electricity power plant.

Generally, landfills closed for fewer than 5 years are the best candidates for energy recovery because over time, even with proper conditions, the ability of a landfill to generate gas decreases. Under optimum conditions, however, a landfill might

FIGURE 10.12 Landfill Gas-to-Electricity Power Plant

produce gas for 15 years or more, depending on the rate of gas generation, the water content of the waste, and the manner of landfill closure. Modern closure requirements for landfills are intended to limit moisture infiltrating the landfill. The extent to which these requirements will affect long-term gas generation is unknown, but they will likely shorten the duration of gas generation after closure (USEPA, 1994).

10.8 DESIGN OF GAS COLLECTION SYSTEM

On March 12, 1996, the U.S. Environmental Protection Agency (EPA) promulgated the New Source Performance Standards/Emission Guidelines (NSPS/EG) for Municipal Solid Waste Landfills (USEPA, 1997). Based on NSPS/EG requirements, if a municipal solid waste landfill has a design capacity equal to or greater than 2.5 million megagrams (2.8 million U.S. tons) or 2.5 million cubic meters (3.3 million cubic yards), and the calculated nonmethane organic compounds (NMOC) emission rate is equal to or greater than 50 megagrams per year (55 U.S. tons), the landfill owner or operator must submit a gas collection and control system design plan to the regulatory agency within one year and install a gas collection and control system within 18 months of the submittal of the design plan. An active gas collection system must be designed to handle the maximum expected gas flow rate from the entire area of the landfill and to minimize off-site migration of subsurface gas. An active gas collection system must be able to collect gas from each area, cell, or group of cells in the landfill and also collect gas at a sufficient extraction rate. A designed gas collection and control system should be operated so that the methane concentration is less than 500 parts per million (ppm) above background at the surface of the landfill (USEPA, 1997).

Key factors for designing a landfill gas collection system are as follows:

(i) Estimation of maximum gas generation rate;
(ii) Well or collector layout and spacing;
(iii) Gas flow generated from each well or collector;
(iv) Collection piping layout and routing;
(v) Condensate liquid handling;
(vi) Pipe sizing and pressure loss calculations; and
(vii) Value and fitting pressure loss calculations.

10.8.1 Calculation of NMOC Emission Rate

If the actual year-to-year solid waste acceptance rate for a landfill is known, the NMOC emission rate can be calculated using the equation (USEPA, 1997)

$$M_{\text{NMOC}} = \sum_{i=1}^{n} 2 \cdot k \cdot L_o \cdot M_i \cdot (e^{-k \cdot t_i})(C_{\text{NMOC}})(3.6 \times 10^{-9}) \qquad (10.1)$$

where M_{NMOC} = total NMOC emission rate from the landfill, megagrams per year (1 Mg = 2,205 lb);
k = methane generation rate constant, yr^{-1};

L_o = methane generation potential, cubic meters per megagram solid waste;
M_i = mass of solid waste in the ith section, megagrams;
t_i = age of the ith section, years;
C_{NMOC} = concentration of NMOC, parts per million by volume as hexane;
3.6×10^{-9} = conversion factor.

If the actual year-to-year solid waste acceptance rate is unknown, the equation that should be used to calculate the NMOC emission rate is (USEPA, 1997)

$$M_{NMOC} = 2 \cdot k \cdot L_o \cdot m_o \cdot (e^{-k \cdot c} - e^{-k \cdot t})(C_{NMOC})(3.6 \times 10^{-9}) \qquad (10.2)$$

where
M_{NMOC} = total NMOC emission rate from the landfill, megagrams per year;
k = methane generation rate constant, yr^{-1};
L_o = methane generation potential, cubic meters per megagram solid waste;
m_o = average annual solid waste acceptance rate, megagrams per year;
c = time since closure, years (for active landfill, $c = 0$ and $e^{-k \cdot c} = 1$);
t = age of landfill, years;
C_{NMOC} = concentration of NMOC, parts per million by volume as hexane;
3.6×10^{-9} = conversion factor.

If a landfill has been closed, the time since closure can be calculated from

$$c = t - t_a \qquad (10.3)$$

where
c = time since closure, years,
t = age of the landfill, years;
t_a = total years of active period of the landfill, years.

For more convenient use of Equation 10.2, it can be divided into two equations according to different periods of landfill operation:

For an active landfill,

$$M_{NMOC} = 2 \cdot k \cdot L_o \cdot m_o \cdot (1 - e^{-k \cdot t})(C_{NMOC})(3.6 \times 10^{-9}) \qquad (10.4)$$

For a closed landfill,

$$M_{NMOC} = 2 \cdot k \cdot L_o \cdot m_o \cdot (e^{k \cdot t_a} - 1) \cdot e^{-k \cdot t} \cdot (C_{NMOC})(3.6 \times 10^{-9}) \qquad (10.5)$$

where
M_{NMOC} = total NMOC emission rate from the landfill, megagrams per year;
k = methane generation rate constant, yr^{-1};
L_o = methane generation potential, cubic meters per megagram solid waste;

m_o = average annual solid waste acceptance rate, megagrams per year;
t = age of landfill, years;
t_a = total years of active period of the landfill, years;
C_{NMOC} = concentration of NMOC, parts per million by volume as hexane;
3.6×10^{-9} = conversion factor.

The values to be used in Equations 10.4 and 10.5 are 0.05 per year for k, 170 cubic meters per megagram for L_o, and 4,000 parts per million by volume as hexane for the C_{NMOC}. The mass of nondegradable solid waste may be subtracted from the total mass of solid waste in a particular section of the landfill when calculating the value for M_i when using Equation 10.1. The mass of nondegradable solid waste may also be subtracted from the average annual acceptance rate when calculating a value of m_o when using Equation 10.2.

10.8.2 Estimation of Gas Generation Rate

For a new landfill gas collection system, the NSPS/EG §60.759(c)(2) (USEPA, 1997) requires that the gas collection and control system be sized for the maximum flow rate in accordance with EPA's landfill gas generation modeling equations.

For the purposes of determining the maximum expected gas generation flow rate from the landfill, the total gas generation from the landfill for each year during either active period or closure period should be calculated based on each year's waste mass and waste age. Then, the maximum annual gas generation rate can be found by comparing each year's amount of gas generation.

For a landfill with a constant or unknown year-to-year solid waste acceptance rate, the annual gas generation rate can be calculated using EPA's modeling equation (USEPA, 1997)

$$Q_t = 2 \cdot L_o \cdot m_o \cdot (e^{-k \cdot c} - e^{-k \cdot t}) \tag{10.6}$$

where Q_t = expected gas generation rate in the tth year, ft³/yr or m³/yr;
L_o = methane generation potential, ft³/lb or m³/Mg;
m_o = constant or average annual solid waste acceptance rate, lb/yr or Mg/yr;
k = methane generation rate constant, yr^{-1};
t = age of the landfill, yr;
c = time since closure, yr. For an active landfill, $c = 0$ and $e^{-k \cdot c} = 1$.

As with Equation 10.2, Equation 10.6 can also be divided into two equations according to different periods of landfill operation:

For an active landfill,

$$Q_t = 2 \cdot L_o \cdot m_o \cdot (1 - e^{-k \cdot t}) \tag{10.7}$$

For a closed landfill,

$$Q_t = 2 \cdot L_o \cdot m_o \cdot (e^{k \cdot t_a} - 1) \cdot e^{-k \cdot t} \tag{10.8}$$

where Q_t = expected gas generation rate in the tth year, ft³/yr or m³/yr;
       L_o = methane generation potential, ft³/lb or m³/Mg;
       m_o = constant or average annual solid waste acceptance rate, lb/yr or Mg/yr;
       k = methane generation rate constant, yr^{-1};
       t = age of the landfill, yr;
       t_a = total years of active period of the landfill, yr.

For a landfill with a known and changed year-to-year solid waste acceptance rate, if the active life of a landfill is assumed to be 15 years and this landfill will be closed after 15 years, the waste filling procedures and the age changes of the waste filled in each year during both active and closure periods are shown in Figure 10.13. In Figure 10.13, M_1 through M_{15} represent the waste mass filled in each active year, respectively. M_1 represents the waste mass filled in the first year, M_2 represents the waste mass filled in the second year, and so on. The number shown in each block in Figure 10.13 represents the age of a certain waste mass. For example, for the waste mass filled in the 5th year, M_5, its age is 4 years old in the 8th year of the landfill, 11 years old in the 15th year of the landfill, and 21 years old in the 25th year of the landfill. Thus, the age of the waste mass, M_i, in the tth year can be determined from

$$t_i = t - i + 1 \tag{10.9}$$

where t_i = age of the waste mass, M_i, in the tth year, yr;
       t = age of the landfill, yr;
       i = time filled waste mass, M_i, yr.

The expected gas generation rate of any waste mass, M_i, in the tth year can be calculated using

$$(Q_i)_t = 2 \cdot k \cdot L_o \cdot M_i \cdot e^{-k \cdot t_i} \tag{10.10}$$

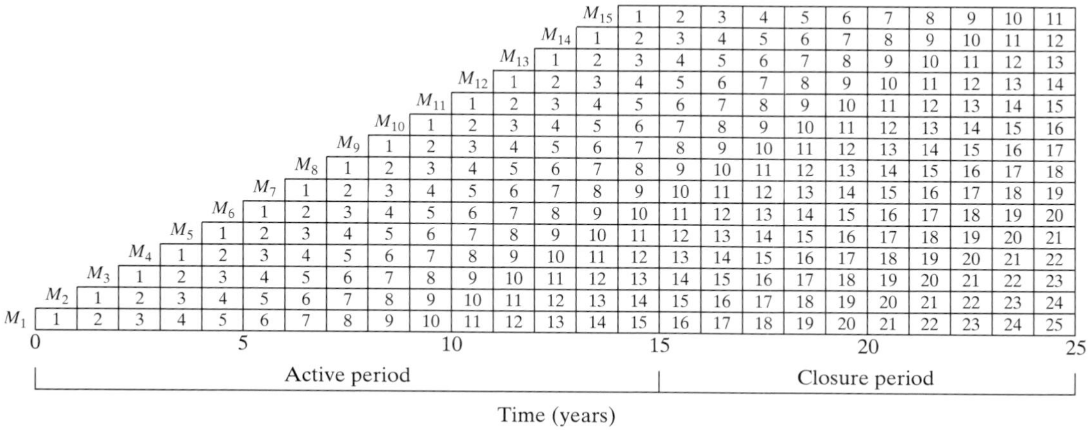

FIGURE 10.13 Age Changes of Annually Filled Solid Waste during Active and Closure Periods

where $(Q_i)_t$ = expected gas generation rate for waste mass, M_i, in the tth year, ft³/yr or m³/yr;
k = methane generation rate constant, yr⁻¹;
L_o = methane generation potential, ft³/lb or m³/Mg;
M_i = mass of solid waste filled in the ith year, lb or Mg;
t_i = age of the waste mass, M_i, in the tth year, yr.

For a landfill with a known and changed year-to-year solid waste acceptance rate, the annual gas generation rate can be calculated using EPA's modeling equation (USEPA, 1997)

$$Q_t = \sum_{i=1}^{n} 2 \cdot k \cdot L_o \cdot M_i \cdot (e^{-k \cdot t_i}) \quad (n \leq t_a) \quad (10.11)$$

where Q_t = expected gas generation rate in the tth year, ft³/yr or m³/yr;
k = methane generation rate constant, yr⁻¹;
L_o = methane generation potential, ft³/lb or m³/Mg;
M_i = mass of solid waste filled in the ith year, lb or Mg;
t_i = age of the waste mass, M_i, in the tth year, yr;
t_a = total years of active period of the landfill, yr.

The total years of active period for a landfill shown in Figure 10.13 are 15 years, (i.e., t_a = 15 years). The range and suggested values for methane generation potential, L_o, and methane generation rate constant, k, can be found in Table 10.3.

If the year-to-year waste acceptance rates are known and the waste filling sequence is like that shown in Figure 10.13, the year-to-year gas generation rate can be calculated as follows using Equation 10.11:

Gas generation rate in the 1st year,

$$Q_1 = 2 \cdot k \cdot L_o \cdot M_1 \cdot (e^{-k \cdot 1})$$

Add appropriate terms to Equation 10.11 for the gas generation rate in the 2nd year to the 4th year.

Gas flow rate in the 5th year,

$$Q_5 = 2 \cdot k \cdot L_o \cdot (M_1 \cdot e^{-k \cdot 5} + M_2 \cdot e^{-k \cdot 4} + M_3 \cdot e^{-k \cdot 3} + M_4 \cdot e^{-k \cdot 2} + M_5 \cdot e^{-k \cdot 1})$$

TABLE 10.3 Range and Suggested Values for L_o and k (LandTech, 1994)

Parameter		Range		Suggested Values
L_o	ft³/lb	0 ~ 5		2.25 ~ 2.88
	m³/Mg	0 ~ 310		140 ~ 180
k	yr⁻¹	0.003 ~ 0.40	Wet Climate	0.10 ~ 0.35
			Medium Moisture Climate	0.05 ~ 0.15
			Dry Climate	0.02 ~ 0.10

Add appropriate terms to Equation 10.11 for the gas generation rate in the 6th year to the 14th year.

Gas flow rate in the 15th year (the last year in active period),

$$Q_{15} = 2 \cdot k \cdot L_o \cdot (M_1 \cdot e^{-k \cdot 15} + M_2 \cdot e^{-k \cdot 14} + M_3 \cdot e^{-k \cdot 13} + M_4 \cdot e^{-k \cdot 12} + M_5 \cdot e^{-k \cdot 11}$$
$$+ M_6 \cdot e^{-k \cdot 10} + M_7 \cdot e^{-k \cdot 9} + M_8 \cdot e^{-k \cdot 8} + M_9 \cdot e^{-k \cdot 7} + M_{10} \cdot e^{-k \cdot 6}$$
$$+ M_{11} \cdot e^{-k \cdot 5} + M_{12} \cdot e^{-k \cdot 4} + M_{13} \cdot e^{-k \cdot 3} + M_{14} \cdot e^{-k \cdot 2} + M_{15} \cdot e^{-k \cdot 1})$$

Gas flow rate in the 16th year (the 1st year after landfill closure),

$$Q_{16} = 2 \cdot k \cdot L_o \cdot (M_1 \cdot e^{-k \cdot 16} + M_2 \cdot e^{-k \cdot 15} + M_3 \cdot e^{-k \cdot 14} + M_4 \cdot e^{-k \cdot 13} + M_5 \cdot e^{-k \cdot 12}$$
$$+ M_6 \cdot e^{-k \cdot 11} + M_7 \cdot e^{-k \cdot 10} + M_8 \cdot e^{-k \cdot 9} + M_9 \cdot e^{-k \cdot 8} + M_{10} \cdot e^{-k \cdot 7}$$
$$+ M_{11} \cdot e^{-k \cdot 6} + M_{12} \cdot e^{-k \cdot 5} + M_{13} \cdot e^{-k \cdot 4} + M_{14} \cdot e^{-k \cdot 3} + M_{15} \cdot e^{-k \cdot 2})$$

Add appropriate terms to Equation 10.11 for the gas generation rate in the 17th year to the 24th year.

Gas flow rate in the 25th year (the 10th year after landfill closure),

$$Q_{25} = 2 \cdot k \cdot L_o \cdot (M_1 \cdot e^{-k \cdot 25} + M_2 \cdot e^{-k \cdot 24} + M_3 \cdot e^{-k \cdot 23} + M_4 \cdot e^{-k \cdot 22} + M_5 \cdot e^{-k \cdot 21}$$
$$+ M_6 \cdot e^{-k \cdot 20} + M_7 \cdot e^{-k \cdot 19} + M_8 \cdot e^{-k \cdot 18} + M_9 \cdot e^{-k \cdot 17} + M_{10} \cdot e^{-k \cdot 16}$$
$$+ M_{11} \cdot e^{-k \cdot 15} + M_{12} \cdot e^{-k \cdot 14} + M_{13} \cdot e^{-k \cdot 13} + M_{14} \cdot e^{-k \cdot 12} + M_{15} \cdot e^{-k \cdot 11})$$

After obtaining the gas generation rate for each year in both active and closure periods from Equations 10.7 and 10.8, or Equation 10.11, the maximum expected gas generation rate used for the gas collection and control system can be determined.

If a portion of gas collection and control system has been installed, actual gas flow data measured from the existing portion of gas collection system may be used in design of a new portion of gas collection system. If the landfill is still accepting waste, the actual measured flow data will not equal the maximum expected gas generation rate. So, Equations 10.7 and 10.8 or Equation 10.11 still should be used to predict the maximum expected gas generation rate over the intended period of use of the gas control system equipment.

EXAMPLE 10.1

A landfill receives 235,000 U.S. tons/year (213,000 Mg/year) of solid waste. The total active period of this landfill is 15 years. The landfill will be closed after 15 years. Assume the methane generation potential, L_o, is 2.50 ft^3/lb (156 m^3/Mg), and the methane generation rate constant, k, is 0.12 yr^{-1}. Estimate changes of landfill gas generation rate with time, and determine the maximum gas generation rate.

Solution:
Annual waste acceptance rate:

$$235{,}000 \text{ tons/yr} = 4.7 \times 10^8 \text{ lb/yr } (2.13 \times 10^5 \text{ Mg/yr})$$

The waste filling sequence and the age changes of annually filled waste is shown in Figure 10.13.

Gas generation rate in active period:

$$Q_t = 2 \cdot L_o \cdot m_o \cdot (1 - e^{-kt}) \tag{10.7}$$
$$= 2 \times 2.50 \times 4.7 \times 10^8 \times (1 - e^{-0.12 \cdot t})$$
$$= 23.5 \times 10^8 \times (1 - e^{-0.12 \cdot t})$$

Gas generation rate in the 1st year,

$$Q_1 = 23.5 \times 10^8 \times (1 - e^{-0.12})$$
$$= 2.657 \times 10^8 \text{ ft}^3/\text{yr} = 506 \text{ ft}^3/\text{min } (14 \text{ m}^3/\text{min})$$

Gas generation rate in the 5th year,

$$Q_5 = 23.5 \times 10^8 \times (1 - e^{-0.60})$$
$$= 10.603 \times 10^8 \text{ ft}^3/\text{yr} = 2{,}017 \text{ ft}^3/\text{min } (57 \text{ m}^3/\text{min})$$

Gas generation rate in the 10th year,

$$Q_{10} = 23.5 \times 10^8 \times (1 - e^{-1.20})$$
$$= 16.422 \times 10^8 \text{ ft}^3/\text{yr} = 3{,}124 \text{ ft}^3/\text{min } (88 \text{ m}^3/\text{min})$$

Gas generation rate in the 15th year (the last year in active period),

$$Q_{15} = 23.5 \times 10^8 \times (1 - e^{-1.80})$$
$$= 19.615 \times 10^8 \text{ ft}^3/\text{yr} = 3{,}732 \text{ ft}^3/\text{min } (106 \text{ m}^3/\text{min})$$

Gas generation rate in postclosure period:

$$t_a = 15 \text{ yr},$$
$$Q_t = 2 \cdot L_o \cdot m_o \cdot (e^{k \cdot t_a} - 1) \cdot e^{-k \cdot t} \tag{10.8}$$
$$= 2 \times 2.50 \times 4.7 \times 10^8 \times [e^{(0.12)(15)} - 1] \cdot e^{-0.12 \cdot t}$$
$$= 2 \times 2.50 \times 4.7 \times 10^8 \times (e^{1.80} - 1) \cdot e^{-0.12 \cdot t}$$
$$= 1.187 \times 10^{10} \times e^{-0.12 \cdot t}$$

Gas generation rate in the 16th year,

$$Q_{16} = 1.187 \times 10^{10} \times e^{-1.92}$$
$$= 17.397 \times 10^8 \text{ ft}^3/\text{yr} = 3{,}310 \text{ ft}^3/\text{min } (94 \text{ m}^3/\text{min})$$

Gas generation rate in the 20th year,

$$Q_{20} = 1.187 \times 10^{10} \times e^{-2.40}$$
$$= 10.765 \times 10^8 \text{ ft}^3/\text{yr} = 2{,}048 \text{ ft}^3/\text{min } (58 \text{ m}^3/\text{min})$$

Gas generation rate in the 25th year,

$$Q_{25} = 1.187 \times 10^{10} \times e^{-3.00}$$
$$= 5.908 \times 10^8 \text{ ft}^3/\text{yr} = 1{,}124 \text{ ft}^3/\text{min } (32 \text{ m}^3/\text{min})$$

The calculated annual gas generation rates are listed in Table 10.4. The maximum gas generation rate for this landfill is 3,732 ft^3/min (106 m^3/min) that occurs in the 15th year.

TABLE 10.4 Annual Landfill Gas Generation Rate

Year	Acceptance Rate		Gas Generation Rate	
	U.S. ton/yr	Mg/yr	ft³/min	m³/min
1	235,000	213,000	506	14
2	235,000	213,000	954	27
3	235,000	213,000	1,352	38
4	235,000	213,000	1,704	48
5	235,000	213,000	2,017	57
6	235,000	213,000	2,295	65
7	235,000	213,000	2,541	72
8	235,000	213,000	2,759	78
9	235,000	213,000	2,953	84
10	235,000	213,000	3,124	88
11	235,000	213,000	3,277	93
12	235,000	213,000	3,412	97
13	235,000	213,000	3,532	100
14	235,000	213,000	3,638	103
15	235,000	213,000	3,732	106
16	0	0	3,310	94
17	0	0	2,936	83
18	0	0	2,604	74
19	0	0	2,309	65
20	0	0	2,048	58
21	0	0	1,817	51
22	0	0	1,611	46
23	0	0	1,429	40
24	0	0	1,267	36
25	0	0	1,124	32

10.8.3 Gas Extraction Well System Layout and Spacing

Spacing of gas extraction wells is an important step in the design of a gas collection system. A sufficient density of gas extraction wells will be capable of efficiently controlling and extracting landfill gas from all portions of the landfill to meet Federal and state operational requirements. Gas extraction wells should be spaced such that their zone of influence overlaps, as shown in Figure 10.8. The most efficient well layout configuration is normally a triangular layout with equilateral spacing between wells, as shown in Figure 10.14. The distance between wells for a triangulated layout is determined by the formula

$$x = 2 \cdot r \cdot \cos 30° \qquad (10.12)$$

where x = distance between triangulated wells,
 r = specified or desired planning "radius of influence."

The design of gas vertical extraction wells and their spacing combine to form a science requiring considerable engineering judgement. Many consultants and

FIGURE 10.14 Landfill Gas Well Layout Configuration

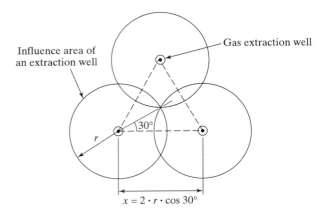

owners/operators recognize there is no precise model that provides a rational basis for selecting well spacing and layout. Yet, most recognize that there is a reasonable spacing that will generally gather the majority of all landfill gas generated within a specific landfill. A summary of suggested collector density is provided in Table 10.5.

10.8.4 Gas Flow Generated from Each Extraction Well or Collector

Gas flow generated from each vertical extraction well or horizontal collector is a very important value for sizing gas collection finger pipes, calculating piping pressure loss, and estimating condensate volume. The method to determine the gas flow rate for each vertical extraction well or horizontal collector is similar to the method to determine the expected gas generation rate for the whole landfill presented in Section 10.8.1.

For an extraction well or collector with a constant year-to-year solid waste acceptance rate, the year-to-year gas generation rate from a vertical extraction well or horizontal collector can be calculated using the following equations:

For an active landfill,

$$q_t = 2 \cdot L_o \cdot w_o \cdot (1 - e^{-k \cdot t}) \tag{10.13}$$

TABLE 10.5 Summary of Suggested Gas Collector Density (EMCON, 1998)

Type of Landfill Condition	Approximate Spacing of Vertical Collectors*
Migration control for adjacent structures	100 to 200 feet (30 to 60 meters)
Migration control for deep vadose zone	200 to 300 feet (60 to 90 meters)
Landfill with wet waste	250 to 300 feet (75 to 90 meters)
Landfill with dry waste	350 to 400 feet (105 to 120 meters)
Horizontal collectors	30 to 50 feet (9 to 15 meters) vertically and 150 to 300 feet (45 to 90 meters) horizontally

*Use the lower value in wet waste or critical areas.

For a closed landfill,

$$q_t = 2 \cdot L_o \cdot w_o \cdot (e^{k \cdot t_a} - 1) \cdot e^{-k \cdot t} \tag{10.14}$$

where q_t = gas generation rate in an extraction well or collector in the tth year, ft³/yr or m³/yr;
  L_o = methane generation potential, ft³/lb or m³/Mg;
  w_o = constant or average annual solid waste acceptance rate in the influence area of an extraction well or collector, lb/yr or Mg/yr;
  k = methane generation rate constant, yr⁻¹;
  t = age of the influence area of an extraction well or collector, yr;
  t_a = total years of active period in the influence area of an extraction well or collector, yr.

For an extraction well or collector with known and changed year-to-year solid waste acceptance rate, the annual gas flow rate generated from a vertical extraction well or a horizontal collector can be calculated using

$$q_t = \sum_{i=1}^{n} 2 \cdot k \cdot L_o \cdot W_i \cdot (e^{-k \cdot t_i}) \quad \text{(for } n \leq t_a\text{)} \tag{10.15}$$

where q_t = gas generation rate in an extraction well or collector in t^{th} year, ft³/yr or m³/yr;
  k = methane generation rate constant, yr⁻¹;
  L_o = methane generation potential, ft³/lb or m³/Mg;
  W_i = mass of solid waste placed in the influence area of an extraction well or collector in the ith year, lb or Mg;
  t_i = age of the waste mass, W_i, in the tth year, determined from Equation 10.9, yr;
  t_a = total years of active period in the influence area of an extraction well or collector, yr.

The total years of active period in the influence area of an extraction well or collector may not be equal to the total active years of the landfill. The active years for an extraction well or collector should be counted from the time the waste is first placed in the influence area of the well or collector, and ended when waste placement stops in the influence area in this well or collector. The range and suggested values for methane generation potential, L_o, and methane generation rate constant, k, used in the equations discussed earlier also can be selected from Table 10.3. After obtaining the gas generation rate from each well or collector for each year in both active and closure periods from Equations 10.13 and 10.14, or Equation 10.15, the maximum gas generation rate for each well or collector can be determined.

The mass of solid waste covered by an extraction gas well or collector can be calculated using the following equations:

For an extraction well or collector with constant year-to-year solid waste acceptance rate,

$$w_o = \pi \cdot r^2 \cdot d_o \cdot \gamma_{waste} \tag{10.16}$$

362 Chapter 10 Gas Collection and Control Systems

where w_o = annual solid waste acceptance rate in the influence area of an extraction well or collector, lb/yr or Mg/yr;
r = influence radius of an extraction well or collector, ft or m;
d_o = annual filling depth of the solid waste in the influence area of an extraction well or collector, ft or m;
γ_{waste} = unit weight of the solid waste, lb/ft^3 or Mg/m^3.

For an extraction well or collector with known and changed year-to-year solid waste acceptance rate,

$$W_i = \pi \cdot r^2 \cdot d_i \cdot \gamma_{waste} \tag{10.17}$$

where W_i = mass of solid waste placed in the influence area of an extraction well or collector in the ith year, lb or Mg;
r = influence radius of an extraction well or collector, ft or m;
d_i = filling depth of the solid waste in the influence area of an extraction well or collector in the ith year, ft or m;
γ_{waste} = unit weight of the solid waste, lb/ft^3 or Mg/m^3.

For a triangulated well layout, the influence radius for an extraction well can calculated (recall Equation 10.12) from the equation

$$r = x/(2 \cdot \cos 30°) \tag{10.18}$$

where r = influence radius of a well or collector, ft or m;
x = distance between triangulated wells, ft or m.

EXAMPLE 10.2

The influence radius of a vertical gas extraction well is 145 feet (43.5 m). The year-to-year filling depths of the solid waste in the influence area of this gas extraction well are listed in Table 10.6.

TABLE 10.6 Annual Waste Filling Depths in the Well Influence Area

Year	Filling Depth (feet)	
	feet	meters
1	15	4.5
2	18	5.4
3	25	7.5
4	22	6.6
5	20	6.0
6	0	0
7	0	0
8	0	0
9	21	6.3
10	23	6.9
11	16	4.8

Section 10.8 Design of Gas Collection System

FIGURE 10.15 Age Changes of Annually Filled Solid Waste in the Influence Area of an Extraction Well

No waste is placed in the influence area of this well after 11th year. Assume the density of the solid waste, γ_{waste}, is 65 lb/ft^3 (1.04 Mg/m^3); the methane generation potential, L_o, is 2.80 ft^3/lb (175 m^3/Mg), and the methane generation rate constant, k, is 0.22 yr^{-1}. Estimate changes of landfill gas generation rate from this well with time, and determine the maximum gas generation rate of this well.

Solution: The waste filling sequence and the age changes of the solid waste that is annually placed in the influence radius of the gas extraction well is shown in Figure 10.15.

Calculate year-to-year waste mass placed in the influence area of the gas extraction well using Equation 10.17:

$$W_i = \pi \cdot r^2 \cdot d_i \cdot \gamma_{waste} \tag{10.17}$$

$W_1 = 3.14 \times (145)^2 \times 15 \times 65 = 6.440 \times 10^7$ lb (29,000 Mg)

$W_2 = 3.14 \times (145)^2 \times 18 \times 65 = 7.728 \times 10^7$ lb (35,000 Mg)

$W_3 = 3.14 \times (145)^2 \times 25 \times 65 = 10.733 \times 10^7$ lb (49,000 Mg)

$W_4 = 3.14 \times (145)^2 \times 22 \times 65 = 9.445 \times 10^7$ lb (43,000 Mg)

$W_5 = 3.14 \times (145)^2 \times 20 \times 65 = 8.587 \times 10^7$ lb (39,000 Mg)

$W_6 = 0$

$W_7 = 0$

$W_8 = 0$

$W_9 = 3.14 \times (145)^2 \times 21 \times 65 = 9.016 \times 10^7$ lb (41,000 Mg)

$W_{10} = 3.14 \times (145)^2 \times 23 \times 65 = 9.875 \times 10^7$ lb (45,000 Mg)

$W_{11} = 3.14 \times (145)^2 \times 16 \times 65 = 6.869 \times 10^7$ lb (31,000 Mg)

Calculate gas generation rate from the gas extraction well using Equation 10.15:

$$q_t = \sum_{i=1}^{n} 2 \cdot k \cdot L_o \cdot W_i \cdot (e^{-k \cdot t_i}) \quad (\text{for } n \leq t_a) \tag{10.15}$$

$$= \sum_{i=1}^{n} 2 \times 0.22 \times 2.80 \cdot W_i \cdot (e^{-0.22 \cdot t_i})$$

$$= \sum_{i=1}^{n} 1.232 \cdot W_i \cdot (e^{-0.22 \cdot t_i})$$

Gas generation rate in the 1st year,

$$q_1 = 2 \cdot k \cdot L_o \cdot W_1 \cdot (e^{-k \cdot 1})$$
$$= 1.232 \times 6.440 \times 10^7 \times e^{-0.22}$$
$$= 6.367 \times 10^7 \text{ ft}^3/\text{yr} = 121 \text{ ft}^3/\text{min } (3.4 \text{ m}^3/\text{min})$$

Gas generation rate in the 5th year,

$$q_5 = 2 \cdot k \cdot L_o \cdot (W_1 \cdot e^{-k \cdot 5} + W_2 \cdot e^{-k \cdot 4} + W_3 \cdot e^{-k \cdot 3} + W_4 \cdot e^{-k \cdot 2} + W_5 \cdot e^{-k \cdot 1})$$
$$= 1.232 \times 10^7 \times (6.440 \cdot e^{-1.10} + 7.728 \cdot e^{-0.88} + 10.733 \cdot e^{-0.66} + 9.445 \cdot e^{-0.44} + 8.587 \cdot e^{-0.22})$$
$$= 29.409 \times 10^7 \text{ ft}^3/\text{yr} = 560 \text{ ft}^3/\text{min } (15.9 \text{ m}^3/\text{min})$$

Gas generation rate in the 6th year,

$$q_6 = 2 \cdot k \cdot L_o \cdot (W_1 \cdot e^{-k \cdot 6} + W_2 \cdot e^{-k \cdot 5} + W_3 \cdot e^{-k \cdot 4} + W_4 \cdot e^{-k \cdot 3} + W_5 \cdot e^{-k \cdot 2})$$
$$= 1.232 \times 10^7 \times (6.440 \cdot e^{-1.32} + 7.728 \cdot e^{-1.10} + 10.733 \cdot e^{-0.88} + 9.445 \cdot e^{-0.66} + 8.587 \cdot e^{-0.44})$$
$$= 23.601 \times 10^7 \text{ ft}^3/\text{yr} = 449 \text{ ft}^3/\text{min } (12.7 \text{ m}^3/\text{min})$$

Gas generation rate in the 8th year,

$$q_8 = 2 \cdot k \cdot L_o \cdot (W_1 \cdot e^{-k \cdot 8} + W_2 \cdot e^{-k \cdot 7} + W_3 \cdot e^{-k \cdot 6} + W_4 \cdot e^{-k \cdot 5} + W_5 \cdot e^{-k \cdot 4})$$
$$= 1.232 \times 10^7 \times (6.440 \cdot e^{-1.76} + 7.728 \cdot e^{-1.54} + 10.733 \cdot e^{-1.32} + 9.445 \cdot e^{-1.10} + 8.587 \cdot e^{-0.88})$$
$$= 15.200 \times 10^7 \text{ ft}^3/\text{yr} = 289 \text{ ft}^3/\text{min } (8.9 \text{ m}^3/\text{min})$$

Gas generation rate in the 9th year,

$$q_9 = 2 \cdot k \cdot L_o \cdot (W_1 \cdot e^{-k \cdot 9} + W_2 \cdot e^{-k \cdot 8} + W_3 \cdot e^{-k \cdot 7} + W_4 \cdot e^{-k \cdot 6} + W_5 \cdot e^{-k \cdot 5} + W_9 \cdot e^{-k \cdot 1})$$
$$= 1.232 \times 10^7 \times (6.440 \cdot e^{-1.98} + 7.728 \cdot e^{-1.76} + 10.733 \cdot e^{-1.54} + 9.445 \cdot e^{-1.32} + 8.587 \cdot e^{-1.10} + 9.016 \cdot e^{-0.22})$$
$$= 21.113 \times 10^7 \text{ ft}^3/\text{yr} = 402 \text{ ft}^3/\text{min } (11.4 \text{ m}^3/\text{min})$$

Gas generation rate in the 11th year,

$$q_{11} = 2 \cdot k \cdot L_o \cdot (W_1 \cdot e^{-k \cdot 11} + W_2 \cdot e^{-k \cdot 10} + W_3 \cdot e^{-k \cdot 9} + W_4 \cdot e^{-k \cdot 8} + W_5 \cdot e^{-k \cdot 7} + W_9 \cdot e^{-k \cdot 3} + W_{10} \cdot e^{-k \cdot 2} + W_{11} \cdot e^{-k \cdot 1})$$
$$= 1.232 \times 10^7 \times (6.440 \cdot e^{-2.42} + 7.728 \cdot e^{-2.20} + 10.733 \cdot e^{-1.98} + 9.445 \cdot e^{-1.76} + 8.587 \cdot e^{-1.54} + 9.016 \cdot e^{-0.66} + 9.875 \cdot e^{-0.44} + 6.869 \cdot e^{-0.22})$$
$$= 28.224 \times 10^7 \text{ ft}^3/\text{yr} = 537 \text{ ft}^3/\text{min } (15.2 \text{ m}^3/\text{min})$$

Gas generation rate in the 12th year,

$$q_{12} = 2 \cdot k \cdot L_o \cdot (W_1 \cdot e^{-k \cdot 12} + W_2 \cdot e^{-k \cdot 11} + W_3 \cdot e^{-k \cdot 10} + W_4 \cdot e^{-k \cdot 9} + W_5 \cdot e^{-k \cdot 8} + W_9 \cdot e^{-k \cdot 4} + W_{10} \cdot e^{-k \cdot 3} + W_{11} \cdot e^{-k \cdot 2})$$
$$= 1.232 \times 10^7 \times (6.440 \cdot e^{-2.64} + 7.728 \cdot e^{-2.42} + 10.733 \cdot e^{-2.20} + 9.445 \cdot e^{-1.98} + 8.587 \cdot e^{-1.76} + 9.016 \cdot e^{-0.88} + 9.875 \cdot e^{-0.66} + 6.869 \cdot e^{-0.44})$$
$$= 22.651 \times 10^7 \text{ ft}^3/\text{yr} = 431 \text{ ft}^3/\text{min } (12.2 \text{ m}^3/\text{min})$$

Gas generation rate in the 15th year,

$$q_{15} = 2 \cdot k \cdot L_o \cdot (W_1 \cdot e^{-k \cdot 15} + W_2 \cdot e^{-k \cdot 14} + W_3 \cdot e^{-k \cdot 13} + W_4 \cdot e^{-k \cdot 12} + W_5 \cdot e^{-k \cdot 11} + W_9 \cdot e^{-k \cdot 7}$$
$$+ W_{10} \cdot e^{-k \cdot 6} + W_{11} \cdot e^{-k \cdot 5})$$
$$= 1.232 \times 10^7 \times (6.440 \cdot e^{-3.30} + 7.728 \cdot e^{-3.08} + 10.733 \cdot e^{-2.86}$$
$$+ 9.445 \cdot e^{-2.64} + 8.587 \cdot e^{-2.42} + 9.016 \cdot e^{-1.54} + 9.875 \cdot e^{-1.32}$$
$$+ 6.869 \cdot e^{-1.10})$$
$$= 11.707 \times 10^7 \text{ ft}^3/\text{yr} = 223 \text{ ft}^3/\text{min} \ (6.3 \text{ m}^3/\text{min})$$

The calculated year-to-year waste masses and gas generation rates are listed in Table 10.7. From Table 10.7, it can be found that the maximum gas generation rate from the gas extraction well is 560 ft^3/min (15.9 m^3/min) that occurs in its 5th year.

10.8.5 Collection Piping System Layout and Routing

The main goal of a gas collection piping line is to transport gas from extraction wells or horizontal collectors to gas treatment devices (i.e., flare stations or gas-to-electricity power plant) in an efficient steady-state manner. The gas collection piping layout can be characterized as having a main line and branch headers or piping "legs". Pressure loss calculations are then performed on each leg or branch (EMCON, 1998) working backward from the furthest reaches of the piping system. A gas collection and control system should operate as close to steady-state conditions as possible because this will reduce the possibility of air intrusion.

TABLE 10.7 Year-to-Year Waste Filling Depth, Waste Mass and Gas Generation Rate

Year	Filling Depth		Waste Mass		Gas Generation Rate	
	feet	meter	× 10^7 lb	Mg	ft^3/min	m^3/min
1	15	4.5	6.440	29,000	121	3.4
2	18	5.4	7.728	35,000	243	6.9
3	25	7.5	10.733	49,000	397	11.2
4	22	6.6	9.445	43,000	496	14.0
5	20	6.0	8.587	39,000	**560**	**15.9**
6	0	0	0	0	449	12.7
7	0	0	0	0	360	10.2
8	0	0	0	0	289	8.2
9	21	6.3	9.016	41,000	402	11.4
10	23	6.9	9.875	45,000	508	14.4
11	16	4.8	6.869	31,000	537	15.2
12	0	0	0	0	431	12.2
13	0	0	0	0	346	9.8
14	0	0	0	0	278	7.9
15	0	0	0	0	223	6.3

Note: Bold face numbers represent the maximum values.

The gas collection pipe system must consider the slope (gradient) of the pipelines to prevent condensate blocking the flow of landfill gas. Important guidelines for establishing piping gradients include the following (modified from EMCON, 1998):

(i) The header piping must be sufficiently sloped to allow for total and/or differential settlement and still provide for adequate drainage in the future.

(ii) A 3% gradient when located within the limits of waste is a suggested minimum, more if the gas header is buried below a geomembrane cap.

(iii) In instances where the desirable gradient cannot be achieved (e.g., when trench depth is limited due to long piping runs), a designer should strive for as much gradient as is practical in combination with shortened distances between drain points.

10.8.6 Estimation of Condensate Production

Satisfactory design must account for drainage and collection of landfill gas condensate throughout the gas collection system. A gas collection piping system works best when it is free draining and clear of liquid. Build-up of condensate in the gas collection system can result in pipe or blower surging and liquid blockage that will restrict gas flow and increase pressure loss. The effects of this buildup may make collection and interpretation of monitoring data difficult and thus will hamper system adjustment, operation, and control.

The following caveats and precautions should be observed to ensure that condensate is properly collected and drained (modified from EMCON, 1998):

(i) Condensate drainage inside the gas collection piping should run with, and not counter to, the gas flow whenever possible to avoid unnecessary pressure losses and surging.

(ii) An adequate number of condensate drain/collection points should be incorporated into the design.

(iii) Condensate drain points should be provided for efficient removal under the worst-case conditions (largest volume) and with sufficient handling capacity and redundancy so that the system will operate smoothly even when a few points fail.

(iv) The minimum recommended header slope is 3%. If enough slope cannot be achieved, the distance between drain points should be shortened and the number of drainouts increased.

Special considerations must be made in seasonally cold and freezing regions to prevent collected gas condensate from freezing. As long as landfill gas is being continuously collected, the warm gas should keep condensate freely flowing to drain points. A problem can occur if the system is shut down for even a brief time. When condensate is collected and stored under freezing conditions, the following guidelines apply (modified from EMCON, 1998):

(i) The gas collection systems must be buried below the local frost penetration depth.

(ii) For short above ground piping, pipe insulation must be used. Insulation materials may be geofoam (sheet or blanket) or soil.

(iii) If heat tapes or wraps are used, they must be properly designed not to melt thermoplastic pipe, of which HDPE has the lowest melt temperature (110 to 140°C or 230 to 284°F).

The amount of landfill gas condensate that will drop out should be determined for design of a landfill gas collection system. The quantity of gas condensate in saturated landfill gas at any point in a gas collection system is temperature dependent. The total condensate that will be collected at that point also depends on the volume of gas passing through that point during the period in question. A network analysis (LandTech, 1994) can be performed to determine the quantities of condensate collected during a given period throughout the entire landfill gas collection system.

The analysis basically assumes that all condensate is water. To calculate the amount of condensate which can accumulate in a network, branch, or header per flowing quantity of landfill gas, the procedures described next are required. Extreme values for a worst-case scenario and conservative average values should be used for a reasonable prediction. Calculation for both summer and winter conditions is recommended.

The step-by-step procedures to estimate the amount of condensate produced in a gas collection system are as follows (modified from LandTech, 1994):

(i) *Gas Temperature at Initial Point*
Determine the initial point from which the landfill gas originates and the gas temperature at that point.

(ii) *Gas Temperature at Final Point*
Determine the final point the gas temperature (for the branch or header) where the condensate will drop out and accumulate or be collected.

(iii) *Gas Pressure in Branch*
Determine the landfill gas pressure in a collection branch or system, (i.e., average pipeline pressure).

(iv) *Fraction of Liquid Gas Condensate*
Calculate the percent condensate water fraction in the landfill gas at each point, using

$$F_{cond} = \frac{P_{water}}{P_{LFG}} = \frac{V_{cond}}{V_{LFG}} \quad (10.19)$$

where F_{cond} = fraction of liquid gas condensate (assume to be primary aqueous water);
P_{water} = vapor (partial) pressure of water at landfill gas temperature determined, lb/in^2 or kN/m^2;
P_{LFG} = total pressure of landfill gas (i.e., average pipeline pressure), lb/in^2 or kN/m^2;
V_{cond} = volume of landfill gas condensate, ft^3 or m^3;
V_{LFG} = volume of landfill gas, ft^3 or m^3.

The values of vapor pressure of water at different temperatures are listed in Table 10.8.

(v) *Fraction of Water Vapor Dropped Out*
Subtract the fraction of liquid gas condensate at the final point from the fraction of liquid gas condensate at the initial point to obtain the fraction of water vapor that is dropped out; that is,

$$(F_{cond})_{drop} = (F_{cond})_{initial} - (F_{cond})_{final} \qquad (10.20)$$

where $(F_{cond})_{drop}$ = fraction of water vapor dropped out, [i.e., cubic foot of water vapor per cubic foot of landfill gas, ft^3/ft^3 (or cubic meter of water vapor per cubic meter of landfill gas, m^3/m^3)];
$(F_{cond})_{initial}$ = fraction of liquid gas condensate at the initial point;
$(F_{cond})_{final}$ = fraction of liquid gas condensate at the final point.

(vi) *Volume of Water Condensed from a Cubic Foot of Landfill Gas*
At 14.7 lb/in^2 (1.01 bar) air pressure and 60°F (15.5°C), the density of water vapor is 0.04757 lb/ft^3 (7.473 N/m^3), and the density of water is 8.34 lb/gal (9.81 N/liter).

U.S. Units

$$1 \text{ ft}^3(\text{water vapor})/\text{ft}^3(\text{LFG}) = \frac{1 \text{ ft}^3/\text{ft}^3 \times 0.04757 \text{ lb/ft}^3}{8.34 \text{ lb/gal}}$$
$$= 0.005704 \text{ gal(water)}/\text{ft}^3(\text{LFG})$$

Thus, 1 cubic foot of water vapor can condense 0.005704 gallon of water.

TABLE 10.8 Values of Vapor Pressure of Water at Different Temperatures

Temperature		Vapor Pressure of Water			
°C	°F	lb/in^2	kN/m^2	mmHg	bar
0	32	0.088	0.607	4.537	0.006
10	50	0.176	1.214	9.126	0.012
20	68	0.336	2.317	17.394	0.023
30	86	0.611	4.213	31.594	0.042
40	104	1.063	7.329	54.958	0.073
50	122	1.778	12.259	91.951	0.123
60	140	2.872	19.802	148.532	0.198
70	158	4.494	30.986	232.414	0.310
80	176	6.832	47.107	353.312	0.471
90	194	10.116	69.750	523.168	0.698
100	212	14.625	100.839	756.337	1.008
110	230	20.685	142.623	1069.743	1.426

°F = (9/5)°C + 32
1 lb/in^2 = 6.895 kN/m^2 = 51.7149 mmHg = 0.06895 bar

SI Units

$$1 \text{ m}^3(\text{water vapor})/\text{ft}^3(\text{LFG}) = \frac{1 \text{ m}^3/\text{m}^3 \times 7.473 \text{ N/m}^3}{9.81 \text{ N/liter}}$$
$$= 0.762 \text{ liter}(\text{wter})/\text{m}^3(\text{LFG})$$

Thus, 1 cubic meter of water vapor can condense 0.762 liter of water.

U.S. Units

$$\omega_{cond} = 0.005704 \times (F_{cond})_{drop} \qquad (10.21a)$$

SI Units

$$\omega_{cond} = 0.762 \times (F_{cond})_{drop} \qquad (10.21b)$$

where ω_{cond} = volume of water condensed from a unit volume of landfill gas, gal/ft³, or liter/m³;
$(F_{cond})_{drop}$ = fraction of water vapor dropped out, unit volume of water vapor per unit volume of landfill gas, ft³/ft³ or m³/m³.

(vii) *Volume of Liquid Condensate Produced in a Branch*
With a known value of ω_{cond} and the volume of landfill gas generated in a branch gas collection line or the whole gas collection system, the volume of liquid condensate can be calculated using the equation

$$V_{cond} = \omega_{cond} \cdot V_{LFG} \qquad (10.22)$$

where V_{cond} = volume of liquid condensate produced in a branch gas collection line or the whole gas collection system, gal/min or liter/min;
V_{LFG} = volume of landfill gas generated in a branch gas collection line or the whole gas collection system, ft³/min or m³/min.

EXAMPLE 10.3

A branch gas collection line carries 1,520 ft³/min (43 m³/min) of landfill gas for 580 ft (174 m) where condensate will drain and be collected. Average initial temperature of the gas at the beginning of the line is 50°C (122°F). Temperature at drop out point is 20°C (68°F). The branch is at 30 inches (762 mm) W.C. vacuum (average). What is the quantity of landfill gas condensate that would be condensed out in the branch header line?

Solution: The differential temperature between the beginning and the end of the line is 30°C (54°F). The vapor pressure (lb/in²) of water at the initial and drop out temperature points can be found in Table 10.5.

At 50°C, the vapor pressure, $(P_{water})_{initial}$, of water is approximately 1.778 lb/in² (0.123 bar), and at 15°C, the vapor pressure, $(P_{water})_{final}$, of water is approximately 0.336 lb/in² (0.0232 bar).

Because 1 inch (25.4 mm) W.C. = 0.0361 lb/in² (0.00249 bar), and 1 atmosphere pressure = 14.7 lb/in² (1.01 bar),

30 inches (762 mm) W.C. vacuum = 14.7 − (0.0361 × 30) = 13.617 lb/in² (0.939 bar)

The gas pressure (P_{LFG}) in the branch is 13.617 lb/in² (0.939 bar).

Using Equation 10.13, the fractions of liquid gas condensate at the initial and final points of the line can be calculated as the follows:

$$(F_{cond})_{initial} = \frac{(P_{water})_{initial}}{P_{LFG}} = \frac{1.778}{13.617} = 0.1306$$

$$(F_{cond})_{final} = \frac{(P_{water})_{final}}{P_{LFG}} = \frac{0.336}{13.617} = 0.0247$$

Using Equation 10.20,

$$(F_{cond})_{drop} = (F_{cond})_{initial} - (F_{cond})_{final} \qquad (10.20)$$
$$= 0.1306 - 0.0247$$
$$= 0.1059$$
$$= 10.59\%$$

Using Equation 10.21 yields

$$\omega_{cond} = 0.005704 \times (F_{cond})_{drop} \qquad (10.21a)$$
$$= 0.005704 \times 0.1059$$
$$= 6.041 \times 10^{-4} \text{ gal/ft}^3 \text{ (0.081 liter/m}^3\text{)}$$

Calculate the volume of liquid condensate using Equation 10.22:

$$V_{cond} = \omega_{cond} \cdot V_{LFG} \qquad (10.22)$$
$$= 6.041 \times 10^{-4} \times 1520$$
$$= 0.918 \text{ gal/min}$$
$$= 1,322 \text{ gal/day (5,000 liter/day)}$$

10.8.7 Header Pipe Sizing and Pressure Loss Calculations

A key principle for header pipe sizing and pressure loss calculations is to confirm that a conservative approach is used in the calculations. Landfill gas header pipe sizing by a designer is an iterative process that is usually done as follows (modified from EMCON, 1998):

(i) Estimate maximum gas flow rate from each gas extraction well using the method presented in Section 10.8.3.
(ii) Determine a design flow rate for each branch of gas collection pipeline.
(iii) Select a trial pipe using Table 10.9. When landfill gas and condensate flow in the same direction, the maximum gas flow velocity in the pipe should be between 40

TABLE 10.9 HDPE Pipe Diameter versus Landfill Gas Flow Rate, ft³/min (m³/min) (EMCON, 1998)

Pipe Diameter inch (mm)	Gas Flow Velocity = 20 ft/sec (6 m/sec) Countercurrent LFG & Condensate			Gas Flow Velocity = 40 ft/sec (12 m/sec) Concurrent LFG & Condensate			Head Loss = 1.0 inch (25.4 mm) W.C. per 100 feet (30 m) of header		
	SDR 21	SDR 17	SDR 11	SDR 21	SDR 17	SDR 11	SDR 21	SDR 17	SDR 11
4 (100)	108 (3.1)	102 (2.9)	87 (2.5)	215 (6.1)	204 (5.8)	174 (4.9)	130 (3.7)	120 (3.4)	98 (2.8)
6 (150)	233 (6.6)	221 (6.3)	189 (5.4)	466 (13.2)	466 (13.2)	466 (13.2)	359 (10.2)	335 (9.5)	273 (7.7)
8 (200)	395 (11.2)	375 (10.6)	320 (9.1)	791 (22.4)	750 (21.2)	640 (18.1)	724 (20.5)	675 (19.1)	547 (15.5)
10 (250)	614 (17.4)	583 (16.5)	497 (14.1)	1225 (34.7)	1165 (33.0)	995 (28.2)	1290 (36.5)	1205 (34.1)	980 (27.8)
12 (300)	864 (24.5)	820 (23.2)	700 (19.8)	1727 (48.9)	1639 (46.4)	1399 (39.6)	2030 (57.5)	1890 (53.5)	1540 (43.6)
16 (400)	1360 (38.5)	1291 (36.6)	1102 (31.2)	2720 (77.0)	2581 (73.1)	2204 (62.4)	3690 (104)	3450 (97.7)	2800 (79.3)
18 (450)	1721 (48.7)	1634 (46.3)	1394 (39.5)	3443 (97.5)	3267 (92.5)	2789 (79.0)	5040 (143)	4700 (133)	3830 (108)
24 (600)	3060 (86.7)	2904 (82.2)	2479 (70.2)	6120 (173)	5808 (164)	4958 (140)	10800 (306)	10100 (286)	8200 (232)

Caution: This chart should only be used as a general guide to in landfill gas header pipes. Analysis that includes loop routines may allow smaller header selection. Also, this analysis does not account for condensate or leachate that may be flowing in the header

and 50 ft/sec (12 and 15 m/sec) without causing water hammer for condensate water. When landfill gas and condensate flow in opposite directions, the maximum gas flow velocity in the pipe should be between 20 and 30 ft/sec (6 and 9 m/sec) without causing water hammer for condensate water.

(iv) Calculate pressure loss along the pipeline. Never exceed one-inch (25.4-mm) water column head loss per 100 feet (30 m) of gas header pipe.

(v) Add the four pressures together for the blower requirement, which are pressure losses along the header pipes, pressure losses at the blower station, pressure required by the gas flare manufacturer, and residual vacuum required at the wellhead.

Generally, a looped piping system provides a more even distribution of vacuum and additional redundancy in vacuum distribution (EMCON, 1998) and thus provides easier and more efficient gas collection and control system operation. Use of looping and branch inter-ties provides more operating flexibility.

Landfill gas in an active extraction system is almost always in fully turbulent flow conditions. The size of the branch header pipe is usually 4 to 8 inches (100 to 200 mm) nominal. The landfill gas flow velocity in the header pipe can be calculated using the continuity equation

$$V = Q/A \tag{10.23}$$

where V = landfill gas flow velocity in the header pipe, ft/min or m/min;
Q = design gas flow rate in the header pipe, ft³/min or m³/min;
A = cross-sectional area of the header pipe, ft² or m².

The design gas flow rate for the main pipe that carries gas flow generated from the whole landfill is equal to the maximum gas generation rate estimated using the methods presented in Section 10.8.1. The design gas flow rate for the finger pipes that only carry the gas flow generated from a single extraction well or collector is equal to the maximum gas generation rate in each well or collector, which can be estimated using the methods presented in Section 10.8.3. The design gas flow rate for a branch header line can be estimated using either of the following two methods:

(i) Calculate the maximum gas generation rate for each well or collector connected to this branch line using the methods presented in Section 10.8.3. Summarize the maximum gas flow rate for each well or collector to obtain the design gas flow rate of the branch header pipe. This is a very conservative method because the maximum gas generation rate for each well or collector may not occur at the same time.

(ii) Calculate the year-to-year gas generation rate for each well or collector connected to this branch line using the methods presented in Section 10.8.3, such as Example 10.2. List the year-to-year gas flow rate for each well or collector in a table. Summarize the gas flow rate for each well or collector year by year, then find a maximum value and use this value as a design gas flow rate for the branch header pipe. This is a more reasonable way to determine the design gas flow rate for a branch line.

There are two equations that can be used to calculate the piping pressure loss along the gas collection pipe. They are the Darcy-Weisbach Equation and the Low-Pressure Mueller Equation.

Darcy-Weisbach Equation

Calculate the Reynolds number using

$$N_{Re} = D \cdot V \cdot \gamma_g / \mu_g \tag{10.24}$$

where N_{Re} = Reynolds number (dimensionless);
D = internal pipe diameter, ft or m;
V = mean velocity of gas flow, ft/sec or m/sec;
γ_g = unit weight of landfill gas, 0.0612 lb/ft³ or 9.615 N/m³;
μ_g = absolute viscosity of landfill gas, 8.14×10^{-6} lb/ft-sec or 1.19×10^{-4} N/m-sec.

Calculate the Darcy friction factor (empirical formula) using

$$f = 0.0055 + 0.0055 \cdot [(20{,}000 \cdot \varepsilon/D) \cdot (1{,}000{,}000/N_{Re})]^{1/3} \tag{10.25}$$

where f = Darcy friction factor;
ε = absolute roughness, ft or m, use 0.000005 ft or 0.0000015 m for PVC pipe and 0.00007 ft or 0.000021 m for HDPE pipe;
D = internal pipe diameter, ft or m;
N_{Re} = Reynolds number (dimensionless).

The Darcy-Weisbach equations for pressure loss are given next.

U.S. Units

$$\Delta P = 27.73 \cdot \frac{f \cdot \gamma_g \cdot L \cdot V^2}{144 \cdot (2 \cdot g) \cdot D} \tag{10.26a}$$

where ΔP = pressure loss, inches of water column (inch W.C.);
f = Darcy friction factor;
γ_g = unit weight of landfill gas, 0.0612 lb/ft^3;
L = length of piping, ft;
V = gas flow velocity, ft/sec;
D = internal pipe diameter, ft;
g = gravitational constant, 32.2 ft/sec^2.

Note: The value of 27.73 in Equation 10.26a is the factor to convert from pressure loss in lb/in^2 to inches W.C. [i.e., 1 lb/in^2 = 27.73 inches W.C. (68°F)].

SI Units

$$\Delta P = 0.102 \cdot \frac{f \cdot \gamma_g \cdot L \cdot V^2}{(2 \cdot g) \cdot D} \tag{10.26b}$$

where ΔP = pressure loss, millimeters of water column (mm W.C.);
f = Darcy friction factor;
γ_g = unit weight of landfill gas, 9.62 N/m^3;
L = length of piping, m;
V = gas flow velocity, m/sec;
D = internal pipe diameter, m;
g = gravitational constant, 9.81 m/sec^2.

Note: The value of 0.102 in Equation 10.26b is the factor to convert from pressure loss in N/m^2 to mm W.C. [i.e., 1 N/m^2 = 0.102 mm W.C. (20°C)].

Low-Pressure Mueller Equation

U.S. Units

$$\Delta P = L \cdot \left[\frac{60 \cdot Q \cdot G^{0.425}}{2971 \cdot d^{2.725}} \right]^{1.739} \tag{10.27a}$$

SI Units

$$\Delta P = (2.103 \times 10^8) \cdot L \cdot \left[\frac{Q \cdot G^{0.425}}{d^{2.725}} \right]^{1.739} \tag{10.27b}$$

Where ΔP = pressure loss, inch W.C. or mm W.C.;
L = length of pipe, feet or m;
Q = gas flow rate in pipe, ft³/min or m³/min;
G = specific density of the gas, use 0.87 for landfill gas;
d = internal pipe diameter, inch or mm.

EXAMPLE 10.4

A branch gas collection line carries 500 ft³/min (14 m³/min) of landfill gas for 580 feet (174 m). Assume the gas header is a 4-inch (100-mm) nominal Schedule 40 PVC pipe [internal diameter = 4.026 inches (102 mm)]. Use the Darcy-Weisbach Equation and the Low-Pressure Mueller Equation to calculate pressure loss per 100 feet (30 m), respectively.

Solution: Calculate the cross-sectional area of the header pipe:

Internal Diameter = 4.026 in = 0.336 ft (0.102 m)

$$A = \pi \cdot r^2 = 3.1416 \times (0.336/2)^2 = 0.088 \text{ ft}^2 \ (0.0082 \text{ m}^2)$$

Calculated the gas flow velocity in the header pipe using Equation 10.23:

$$V = Q_{max}/A \tag{10.23}$$
$$= 500/0.088 = 5{,}656 \text{ ft/min} = 94.27 \text{ ft/sec} \ (28.7 \text{ m/sec})$$

Calculate the Reynolds number using Equation 10.24:

$$N_{Re} = D \cdot V \cdot \gamma_g / \mu_g \tag{10.24}$$
$$= 0.336 \times 94.27 \times 0.0612/(8.14 \times 10^{-6})$$
$$= 237{,}790$$

Calculate the Darcy friction factor using Equation 10.25:

$$f = 0.0055 + 0.0055 \times [(20{,}000 \cdot \varepsilon/D)(1{,}000{,}000/N_{Re})]^{1/3} \tag{10.25}$$
$$= 0.0055 + 0.0055 \times [(20{,}000 \times 0.000005/0.3355)(1{,}000{,}000/237{,}790)]^{1/3}$$
$$= 0.0055 + 0.0055 \times (1.2536)^{1/3}$$
$$= 0.0055 + 0.005931$$
$$= 0.0114$$

Calculate the pressure loss per 100 feet (30 m) using the Darcy-Weisbach Equation:

$$\Delta P = 27.73 \cdot \frac{f \cdot \gamma_g \cdot L \cdot V^2}{144 \cdot (2 \cdot g) \cdot D} \tag{10.26a}$$
$$= 27.73 \times \frac{(0.0114)(0.0612)(100)(94.27)^2}{(144)(2)(32.2)(0.336)}$$
$$= 5.54 \text{ in. W.C. } (141 \text{ mm W.C.})$$

Calculate pressure loss per 100 feet (30 m) using the Low-Pressure Mueller Equation:

$$\Delta P = L \cdot \left[\frac{60 \cdot Q \cdot G^{0.425}}{2971 \cdot d^{2.725}} \right]^{1.739} \tag{10.27a}$$

$$= 100 \times \left[\frac{60 \times 500 \times 0.87^{0.425}}{2971 \times 4.026^{2.725}} \right]^{1.739}$$

$$= 100 \times (0.214)^{1.739}$$

$$= 6.84 \text{ in. W.C. (174 mm W.C.)}$$

Because the head loss per 100 feet (30 m) of gas head pipe exceeds one-inch (25.4-mm) water column, use a 6-inch (150-mm) Schedule 40 PVC pipe [internal diameter = 6.065 inches (154 mm)] to calculate pressure loss again.

Calculate the cross-sectional area of the header pipe for 6-inch (150-mm) pipe:

Internal Diameter = 6.065 in = 0.505 ft (0.15 m)

$$A = \pi \cdot r^2 = 3.1416 \times (0.505/2)^2 = 0.201 \text{ ft}^2 \text{ (0.0177 m}^2\text{)}$$

Calculate the gas flow velocity in the header pipe using Equation 10.23:

$$V = Q_{max}/A \tag{10.23}$$

$$= 500/0.201$$

$$= 2{,}493 \text{ ft/min} = 41.55 \text{ ft/sec (12.7 m/sec)}$$

Calculate the Reynolds number using Equation 10.24:

$$N_{Re} = D \cdot V \cdot \gamma_g / \mu_g \tag{10.24}$$

$$= 0.505 \times 41.55 \times 0.0612/(8.14 \times 10^{-6})$$

$$= 157{,}900$$

Calculate the Darcy friction factor using Equation 10.25:

$$f = 0.0055 + 0.0055 \cdot [(20{,}000 \cdot \varepsilon/D)(1{,}000{,}000/N_{Re})]^{1/3} \tag{10.25}$$

$$= 0.0055 + 0.0055 \cdot [(20{,}000 \times 0.000005/0.505)(1{,}000{,}000/157{,}882)]^{1/3}$$

$$= 0.0055 + 0.0055 \times (1.2534)^{1/3}$$

$$= 0.0055 + 0.005930$$

$$= 0.0114$$

Calculate the pressure loss per 100 feet (30 m) using the Darcy-Weisbach Equation:

$$\Delta P = 27.73 \cdot \frac{f \cdot \gamma_g \cdot L \cdot V^2}{144 \cdot (2 \cdot g) \cdot D} \tag{10.26a}$$

$$= 27.73 \times \frac{(0.0114)(0.0612)(100)(41.55)^2}{(144)(2)(32.2)(0.505)}$$

$$= 0.715 \text{ in. W.C. (18 mm W.C.)}$$

Calculate the pressure loss per 100 feet (30 m) using the Low-pressure Mueller Equation

$$\Delta P = L \cdot \left[\frac{60 \cdot Q \cdot G^{0.425}}{2971 \cdot d^{2.725}} \right]^{1.739} \tag{10.27}$$

$$= 100 \times \left[\frac{60 \times 500 \times 0.87^{0.425}}{2971 \times 6.065^{2.725}} \right]^{1.739}$$

$$= 100 \times (0.07003)^{1.739}$$
$$= 0.982 \text{ in. W.C. } (25 \text{ mm W.C.})$$

The calculated head losses from both equations are less than one inch (25.4 mm) W.C. per 100 feet (30 m) for 6-inch (150-mm) pipe. Thus, a 6-inch (150-mm) Schedule 40 PVC pipe should be selected for this gas line.

10.8.8 Valve and Fitting Pressure Loss Calculations

Valve and fitting selection and design sometimes require calculation of the pressure losses across the valves and fittings. Most manufacturers' literature provides information for valves and fittings in liquid service, but is either silent or unclear about gas service. The use of a transition loss coefficient, K, is a convenient way to rate most valves and fittings supplied by manufacturers. These should be used in conjunction with a flow coefficient, C_v, either calculated theoretically or provided directly by the manufacturers. The flow coefficient, C_v, normally assumes water as the flowing fluid, but it can be adapted to gases as well by adjusting for density. The flow coefficient, C_v, is expressed in terms of flow rate of water in gallons per minute at 60°F (15.5°C) with 1 lb/in^2 (0.07 bar) of pressure loss across the valve and fitting (LandTech, 1994).

The values of transition loss coefficient, K, can be calculated using the following equations for sudden enlargement and contraction connectors between two different diameter pipes.

Sudden Enlargement:

$$K = (1 - A_1/A_2)^2 \qquad (10.28)$$

where K = transition coefficient for sudden enlargement;
A_1 = cross-sectional area of the small pipe;
A_2 = cross-sectional area of the large pipe.

For circular pipe,

$$K = [1 - (D_1/D_2)^2]^2 \qquad (10.29)$$

where K = transition coefficient for sudden enlargement;
D_1 = internal diameter of the small circular pipe;
D_2 = internal diameter of the large circular pipe.

Sudden Contraction:

$$K = 0.5 \cdot (1 - A_1/A_2)^2 \qquad (10.30)$$

where K = transition coefficient for sudden contraction;
A_1 = cross-sectional area of the small pipe;
A_2 = cross-sectional area of the large pipe.

For the circular pipe,

$$K = 0.5 \cdot [1 - (D_1/D_2)^2]^2 \qquad (10.31)$$

where K = transition coefficient for sudden contraction;
D_1 = internal diameter of the small circular pipe;
D_2 = internal diameter of the large circular pipe.

The values of transition coefficient, K, will be normally supplied by the valve and fitting manufacturers. In the absence of better estimates for K, the values of transition loss coefficient for various types of transitions can be used from Table 9.5, 9.6, and 9.7.

Flow Coefficient for Valve or Fitting:

U.S. Units:

$$C_v = 29.9 \cdot d^2 / K^{0.5} \qquad (10.32a)$$

SI Units:

$$C_v = 0.0463 \cdot d^2 / K^{0.5} \qquad (10.32b)$$

where C_v = valve or fitting flow coefficient;
d = internal pipe diameter, inches or mm;
K = transition loss coefficient of the valve or fitting.

Pressure Loss across the Valve or Fitting:

U.S. Units:

$$\Delta P = (\gamma_g / \gamma_w) \cdot (7.48 \cdot Q / C_v)^2 \qquad (10.33a)$$

where ΔP = pressure loss across the valve or fitting, lb/in²;
γ_g = density of landfill gas, 0.0612 lb/ft³;
γ_w = density of water, 62.4 lb/ft³;
Q = gas flow rate through the valve or fitting, ft³/min;
C_v = valve or fitting flow coefficient.

SI Units:

$$\Delta P = (6.895) \cdot (\gamma_g / \gamma_w) \cdot (264.2 \cdot Q / C_v)^2 \qquad (10.33b)$$

where ΔP = pressure loss across the valve or fitting, kN/m²;
γ_g = density of landfill gas, 0.00962 kN/m³;
γ_w = density of water, 9.81kN/m³;
Q = gas flow rate through the valve or fitting, m³/min;
C_v = valve or fitting flow coefficient.

EXAMPLE 10.5

A landfill gas collection branch line is shown in Figure 10.16. There are five vertical extraction wells connected to this branch line. They are named W-1, W-2, W-3, W-4, and W-5, respectively. The individual lengths of the feeder pipes, AF, BG, CH, DI, and EJ, between wells and main header pipe are each 300 feet (90 m). The section lengths of the main header pipe between the feeders, FG, GH, HI, and JI, are also 300 feet (90 m). The distance of pipe section JK is 400 feet (120 m). The feeder pipelines are inclined at an angle of 60 degrees to the header pipe. The

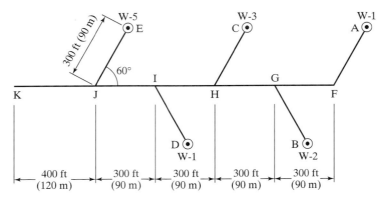

FIGURE 10.16 Schematic of a Gas Collection Header Line for Example 10.5

header pipe slopes in the direction from F to K. The year-to-year gas generation rates from each extraction well are listed in Table 10.10. Conduct the following calculations:

1. Determine the design gas flow rate in each header and feeder pipe.
2. Select diameters of the feeder pipes and the different sections of the header pipe using the Darcy-Weisbach Equation.
3. Assume the minimum vacuum pressure at the wellhead for each extraction well is -25 inches W.C. (-635 mm W.C.) and the transition loss coefficient of the joints G, H, I, and J is equal to 0.55. Determine the minimum required vacuum pressure at the end of the header pipe.
4. Check the vacuum pressures at the wellhead of each extraction well to make sure the pressures at the wellhead of each extraction well do not exceed -25 inches W.C. (-635 mm W.C.).

Solution:

1. *Determine the design gas flow rate in each header and feeder pipe*

The design gas flow rates in each gas extraction well should be equal to the maximum gas generation rate from each gas extraction well. The design gas flow rates in each feeder pipe, AF, BG, EH, DI, and EJ, are listed as follows:

$$Q_{AF} = 563 \text{ ft}^3/\text{min } (15.9 \text{ m}^3/\text{min})$$
$$Q_{BG} = 591 \text{ ft}^3/\text{min } (16.7 \text{ m}^3/\text{min})$$
$$Q_{CH} = 545 \text{ ft}^3/\text{min } (15.4 \text{ m}^3/\text{min})$$
$$Q_{DI} = 576 \text{ ft}^3/\text{min } (16.3 \text{ m}^3/\text{min})$$
$$Q_{EJ} = 534 \text{ ft}^3/\text{min } (15.1 \text{ m}^3/\text{min})$$

Because the gas flow rate varies along the main header pipe, FGHILK, the header pipe should be divided into several sections to determine the design gas flow rate for each section. The pipe

Section 10.8 Design of Gas Collection System

TABLE 10.10 Year-to-Year Gas Flow Rates Generated from Each Extraction Well

	Gas Generation Rate (ft^3/min)									
	W-1		W-2		W-3		W-4		W-5	
Year	ft^3/min	m^3/min	ft^3/min	m^3/min	ft^3/min	m^3/min	ft^3/min	m^3/min	ft^3/min	m^3/min
1	121	3.4	101	2.9	88	2.5	0	0	0	0
2	243	6.9	191	5.4	168	4.8	72	2.0	0	0
3	397	11.2	270	7.6	213	6.0	136	3.9	0	0
4	462	13.1	341	9.7	349	9.9	193	5.5	115	3.3
5	563	15.9	403	11.4	306	8.7	243	6.9	231	6.5
6	449	12.7	459	13.0	403	11.4	288	8.2	297	8.4
7	360	10.2	508	14.4	495	14.0	328	9.3	382	10.8
8	289	8.2	552	15.6	434	12.3	363	10.3	460	13.0
9	402	11.4	591	16.7	351	9.9	394	11.2	375	10.6
10	508	14.4	521	14.8	437	12.4	422	11.9	303	8.6
11	537	15.2	460	13.0	482	13.6	446	12.6	401	11.4
12	430	12.2	410	11.6	545	15.4	487	13.8	465	13.2
13	346	9.8	363	10.3	477	13.5	515	14.6	493	14.0
14	298	8.4	322	9.1	428	12.1	542	15.3	534	15.1
15	224	6.3	284	8.0	364	10.3	576	16.3	468	13.3
16	179	5.1	253	7.2	310	8.8	493	14.0	540	15.3
17	143	4.0	225	6.4	233	6.6	419	11.9	477	13.5
18	114	3.2	186	5.3	189	5.4	357	10.1	402	11.4
19	91	2.6	157	4.4	147	4.2	281	8.0	323	9.1
20	74	2.1	119	3.4	112	3.2	208	5.9	271	7.7

section FG will carry the gas generated only from one well (i.e., W-1). The design gas flow rate for section FG should be equal to the maximum gas flow rate generated from W-1:

$$Q_{FG} = 563 \text{ ft}^3/\text{min} \ (15.9 \text{ m}^3/\text{min})$$

The pipe section GH will carry the gas generated from two wells (i.e., W-1 and W-2). The year-to-year gas flow rates in the section GH are equal to the sum of year-to-year gas generation rates from W-1 and W-2. The year-to-year gas flow rates in the section GH are listed in Table 10.11. The design gas flow rate for section GH should be equal to the maximum gas flow rate in section GH:

$$Q_{GH} = 1,029 \text{ ft}^3/\text{min} \ (29.1 \text{ m}^3/\text{min})$$

The year-to-year gas flow rates in the section HI are equal to the sum of year-to-year gas generation rates from W-1, W-2, and W-3. The year-to-year gas flow rates in the section HI are listed in Table 10.11. The design gas flow rate for section HI should be equal to the maximum gas flow rate in section HI:

$$Q_{HI} = 1,479 \text{ ft}^3/\text{min} \ (41.9 \text{ m}^3/\text{min})$$

The pipe section IJ will carry the gas generated from four wells, (i.e., W-1, W-2, W-3, and W-4). The year-to-year gas flow rates in the section IJ are equal to the sum of year-to-year gas generation rates from W-1, W-2, W-3, and W-4. The year-to-year gas flow rates in the section IJ are

TABLE 10.11 Year-to-Year Gas Flow Rates in Feeder Pipes and Header Pipes

Year	Unit	Gas Flow Rate in Feeder Pipes (ft³/min)					Gas Flow Rate in Header Pipe (ft³/min)				
		AF	BG	CH	DI	EJ	FG	GH	HI	IJ	JK
1	ft³/min	121	101	88	0	0	121	222	310	310	310
	m³/min	3.4	2.9	2.5	0	0	3.4	6.3	8.8	8.8	8.8
2	ft³/min	243	191	168	72	0	243	434	359	431	431
	m³/min	6.9	5.4	4.8	2.0	0	6.9	12.3	10.2	12.2	12.2
3	ft³/min	397	270	213	136	0	397	667	880	1016	1016
	m³/min	11.2	7.6	6.0	3.9	0	11.2	18.9	24.9	28.8	28.8
4	ft³/min	462	341	349	193	115	462	803	1152	1345	1460
	m³/min	13.1	9.7	9.9	5.5	3.3	13.1	22.7	32.6	38.1	41.3
5	ft³/min	**563**	403	306	243	231	**563**	966	1272	1515	1746
	m³/min	**15.9**	11.4	8.7	6.9	6.5	**15.9**	27.4	36.0	42.9	49.4
6	ft³/min	449	459	403	288	297	449	908	1311	1599	1896
	m³/min	12.7	13.0	11.4	8.2	8.4	12.7	25.7	37.1	45.3	53.7
7	ft³/min	360	508	495	328	382	360	868	1363	1691	2073
	m³/min	10.2	14.4	14.0	9.3	10.8	10.2	24.6	38.6	47.9	58.7
8	ft³/min	289	552	434	363	460	289	841	1275	1638	2098
	m³/min	8.2	15.6	12.3	10.3	13.0	8.2	23.8	36.1	46.4	59.4
9	ft³/min	402	**591**	351	394	375	402	993	1344	1738	2113
	m³/min	11.4	**16.7**	9.9	11.2	10.6	11.4	28.1	38.1	49.2	59.8
10	ft³/min	508	521	437	422	303	508	**1029**	1466	1888	2191
	m³/min	14.4	14.8	12.4	11.9	8.6	14.4	**29.1**	41.5	53.5	62.0
11	ft³/min	537	460	482	446	401	537	997	**1479**	**1925**	2326
	m³/min	15.2	13.0	13.6	12.6	11.4	15.2	28.2	**41.9**	**54.5**	65.9
12	ft³/min	430	410	**545**	487	465	430	840	1385	1872	**2337**
	m³/min	12.2	11.6	**15.4**	13.8	13.2	12.2	23.8	39.2	53.0	**66.2**
13	ft³/min	346	363	477	515	493	346	709	1186	1701	2194
	m³/min	9.8	10.3	13.5	14.6	14.0	9.8	20.1	33.6	48.2	62.1
14	ft³/min	298	322	428	542	**534**	298	620	1048	1590	2124
	m³/min	8.4	9.1	12.1	15.3	**15.1**	8.4	17.6	29.7	45.0	60.1
15	ft³/min	224	284	364	**576**	468	224	508	872	1448	1916
	m³/min	6.3	8.0	10.3	**16.3**	13.3	6.3	14.4	24.7	41.0	54.3
16	ft³/min	179	253	310	493	540	179	432	742	1235	1775
	m³/min	5.1	7.2	8.8	14.0	15.3	5.1	12.2	21.0	35.0	50.3
17	ft³/min	143	225	233	419	477	143	368	601	1020	1497
	m³/min	4.0	6.4	6.6	11.9	13.5	4.0	10.4	17.0	28.9	42.4
18	ft³/min	114	186	189	357	402	114	300	489	724	1126
	m³/min	3.2	8.1	5.4	10.1	11.4	3.2	8.5	13.8	20.5	31.9
19	ft³/min	91	157	147	281	323	91	248	395	676	999
	m³/min	2.6	4.4	4.2	8.0	9.1	2.6	7.0	11.2	19.1	28.3
20	ft³/min	74	119	112	208	271	74	193	305	513	784
	m³/min	2.1	3.4	3.2	5.9	7.7	2.1	5.5	8.6	14.5	22.2

Note: Bold face numbers represent the maximum values in each pipe.

listed in Table 10.11. The design gas flow rate for section IJ should be equal to the maximum gas flow rate in section IJ:

$$Q_{IJ} = 1{,}925 \text{ ft}^3/\text{min } (54.5 \text{ m}^3/\text{min})$$

The pipe section JK will carry the gas generated from all five wells, (i.e., W-1, W-2, W-3, W-4, and W-5). The year-to-year gas flow rates in the section JK are equal to the sum of year-to-year gas generation rates from W-1, W-2, W-3, W-4, and W-5. The year-to-year gas flow rates in the section JK are listed in Table 10.11. The design gas flow rate for section JK should be equal to the maximum gas flow rate in the section JK:

$$Q_{JK} = 2{,}337 \text{ ft}^3/\text{min } (66.2 \text{ m}^3/\text{min})$$

2. Select diameters of the feeder pipes and the different sections of the header pipe

An important criterion for sizing gas pipes is never to exceed a one-inch (25.4-mm) water column head loss per 100 feet (30 m) of gas header pipe.

Selection of Feeder Pipes

From Table 10.9, an 8-inch (200-mm) SDR17 HDPE pipe may be selected for all of the well feeder pipes.

Check pressure loss per 100 feet (30 m) for selected pipe:

The maximum flow rate for 5 feeder pipes is 591 ft^3/min (16.7 m^3/min). The internal diameter of 8-inch (200-mm) SDR17 HDPE pipe is 7.059 inches (179 mm).

Calculate the cross-sectional area of an 8-inch (200-mm) SDR17 HDPE pipe:

Internal Diameter of feeder pipe $= 7.059 \text{ in} = 0.588 \text{ ft } (0.179 \text{ m})$

$$A = \pi \cdot r^2 = 3.1416 \times (0.588/2)^2 = 0.272 \text{ ft}^2 \ (0.0252 \text{ m}^2)$$

Calculate the gas flow velocity in the pipe using Equation 10.23:

$$V = Q/A = 591/0.272 = 2{,}173 \text{ ft/min} = 36.22 \text{ ft/sec } (11.0 \text{ m/sec})$$

Calculate the Reynolds number using Equation 10.24:

$$N_{Re} = D \cdot V \cdot \gamma_g/\mu_g \tag{10.24}$$
$$= 0.588 \times 36.22 \times 0.0612/(8.14 \times 10^{-6})$$
$$= 160{,}123$$

Calculate the Darcy friction factor using Equation 10.25:

$$f = 0.0055 + 0.0055 \cdot [(20{,}000 \cdot \varepsilon/D) \cdot (1{,}000{,}000/N_{Re})]^{1/3} \tag{10.25}$$
$$= 0.0055 + 0.0055 \cdot [(20{,}000 \times 0.00007/0.588) \cdot (1{,}000{,}000/160{,}123)]^{1/3}$$
$$= 0.0055 + 0.01353$$
$$= 0.01903$$

Calculate the pressure loss per 100 feet using the Darcy-Weisbach Equation:

$$\Delta P = 27.73 \cdot \frac{f \cdot \gamma_g \cdot L \cdot V^2}{144 \cdot (2 \cdot g) \cdot D} \tag{10.26}$$

$$= 27.73 \times \frac{(0.01903)(0.0612)(100)(36.22)^2}{(144)(2)(32.2)(0.588)}$$

$$= 0.777 \text{ in. W.C. } (19.7 \text{ mm W.C.})$$

The pressure loss per 100 feet (30 m) of gas head pipe is less than one-inch (25.4-mm) water column for the maximum gas flow rate from all five feeder pipes. So, an 8-inch (200-mm) SDR17 HDPE pipe will be suitable for all five feeder pipes.

Selection of Header Pipe

Section FG:

Because the design gas flow rate of section FG is 563 ft³/min (15.9 m³/min), which is less than 591 ft³/min (16.7 m³/min) used to design the feeder pipes in the preceding calculation, an 8-inch (200-mm) SDR17 HDPE pipe will be also suitable for section FG of the main header.

Section GH:

The design gas flow rate of section GH, Q_{GH}, is 1,029 ft³/min (29.1 m³/min). From Table 10.9, a 10-inch (250-mm) SDR17 HDPE pipe may be selected for section GH.

Check pressure loss per 100 feet (30 m) for a 10-inch (250-mm) SDR17 HDPE pipe:

The internal diameter of a 10-inch (250-mm) SDR17 HDPE pipe is 8.824 inches (224 mm).

Calculate the cross-sectional area of a 10-inch (250-mm) SDR17 HDPE pipe:

Internal Diameter of the pipe = 8.824 in = 0.735 ft (0.224 m)

$$A = \pi \cdot r^2 = 3.1416 \times (0.735/2)^2 = 0.424 \text{ ft}^2 \; (0.0394 \text{ m}^2)$$

Calculate the gas flow velocity in the pipe using Equation 10.23:

$$V = Q_{GH}/A = 1{,}029/0.424 = 2{,}427 \text{ ft/min} = 40.45 \text{ ft/sec } (12.3 \text{ m/sec})$$

Calculate the Reynolds number using Equation 10.24:

$$N_{Re} = D \cdot V \cdot \gamma_g / \mu_g \tag{10.24}$$

$$= 0.735 \times 40.45 \times 0.0612/(8.14 \times 10^{-6})$$

$$= 223{,}528$$

Calculate the Darcy friction factor using Equation 10.25:

$$f = 0.0055 + 0.0055 \cdot [(20{,}000 \cdot \varepsilon/D) \cdot (1{,}000{,}000/N_{Re})]^{1/3} \tag{10.25}$$

$$= 0.0055 + 0.0055 \cdot [(20{,}000 \times 0.00007/0.735) \cdot (1{,}000{,}000/223{,}528)]^{1/3}$$

$$= 0.0055 + 0.01123$$

$$= 0.01673$$

Calculate the pressure loss per 100 feet using the Darcy-Weisbach Equation:

$$\Delta P = 27.73 \cdot \frac{f \cdot \gamma_g \cdot L \cdot V^2}{144 \cdot (2 \cdot g) \cdot D} \tag{10.26}$$

$$= 27.73 \times \frac{(0.01673)(0.0612)(100)(40.45)^2}{(144)(2)(32.2)(0.735)}$$

$$= 0.682 \text{ in. W.C. } (17.3 \text{ mm W.C.})$$

The head loss per 100 feet (30 m) of gas head pipe is less than 1-inch (25.4-mm) water column for the design gas flow rate in section GH. Therefore, a 10-inch (250-mm) SDR17 HDPE pipe will be suitable for section GH of the header.

Section HI:

The design gas flow rate of section HI, Q_{HI}, is 1,479 ft³/min (41.9 m³/min). From Table 10.9, a 12-inch (300-mm) SDR17 HDPE pipe may be selected for section HI.

Check pressure loss per 100 feet (30 m) for a 12-inch (300-mm) SDR17 HDPE pipe:

The internal diameter of a 12-inch (300-mm) SDR17 HDPE pipe is 10.588 inches (269 mm).

Calculate the cross-sectional area of a 12-inch (300-mm) SDR17 HDPE pipe:

$$\text{Internal Diameter of the pipe} = 10.588 \text{ in} = 0.882 \text{ ft } (0.269 \text{ m})$$
$$A = \pi \cdot r^2 = 3.1416 \times (0.882/2)^2 = 0.611 \text{ ft}^2 (0.0568 \text{ m}^2)$$

Calculate the gas flow velocity in the pipe using Equation 10.23:

$$V = Q_{HI}/A = 1{,}479/0.611 = 2{,}421 \text{ ft/min} = 40.35 \text{ ft/sec } (12.3 \text{ m/sec})$$

Calculate the Reynolds number using Equation 10.24:

$$N_{Re} = D \cdot V \cdot \gamma_g / \mu_g \tag{10.24}$$
$$= 0.882 \times 40.35 \times 0.0612/(8.14 \times 10^{-6})$$
$$= 267{,}570$$

Calculate the Darcy friction factor using Equation 10.25:

$$f = 0.0055 + 0.0055 \cdot [(20{,}000 \cdot \varepsilon/D) \cdot (1{,}000{,}000/N_{Re})]^{1/3} \tag{10.25}$$
$$= 0.0055 + 0.0055 \cdot [(20{,}000 \times 0.00007/0.882) \cdot (1{,}000{,}000/267{,}571)]^{1/3}$$
$$= 0.0055 + 0.009957$$
$$= 0.01546$$

Calculate the pressure loss per 100 feet using the Darcy-Weisbach Equation:

$$\Delta P = 27.73 \cdot \frac{f \cdot \gamma_g \cdot L \cdot V^2}{144 \cdot (2 \cdot g) \cdot D} \tag{10.26}$$
$$= 27.73 \times \frac{(0.01546)(0.0612)(100)(40.35)^2}{(144)(2)(32.2)(0.882)}$$
$$= 0.522 \text{ in. W.C. } (13.3 \text{ mm W.C.})$$

The head loss per 100 feet (30 m) of gas head pipe is less than 1-inch (25.4-mm) water column for the design gas flow rate in section HI. Accordingly, a 12-inch (300-mm) SDR17 HDPE pipe will be suitable for section HI of the header pipe.

Section IJ:

The design gas flow rate of section IJ, Q_{IJ}, is 1,925 ft³/min (54.5 m³/min). From Table 10.9, a 14-inch (350-mm) SDR17 HDPE pipe may be selected for section IJ.

Check pressure loss per 100 feet (30 m) for a 14-inch (350-mm) SDR17 HDPE pipe:

The internal diameter of a 14-inch (350-mm) SDR17 HDPE pipe is 12.353 inches (314 mm).

Calculate the cross-sectional area of a 14-inch (350-mm) SDR17 HDPE pipe:

Internal diameter of the pipe = 12.353 in = 1.029 ft (0.314 m)

$$A = \pi \cdot r^2 = 3.1416 \times (1.029/2)^2 = 0.832 \text{ ft}^2 \; (0.0774 \text{ m}^2).$$

The calculated gas flow velocity in the pipe, using Equation 10.23, is

$$V = Q_{IJ}/A = 1{,}925/0.832 = 2{,}314 \text{ ft/min} = 38.57 \text{ ft/sec} \; (11.8 \text{ m/sec})$$

Calculate the Reynolds number using Equation 10.24:

$$N_{Re} = D \cdot V \cdot \gamma_g / \mu_g \qquad (10.24)$$
$$= 1.029 \times 38.57 \times 0.0612 / (8.14 \times 10^{-6})$$
$$= 298{,}395$$

Calculate the Darcy friction factor using Equation 10.25:

$$f = 0.0055 + 0.0055 \cdot [(20{,}000 \cdot \varepsilon/D) \cdot (1{,}000{,}000/N_{Re})]^{1/3} \qquad (10.25)$$
$$= 0.0055 + 0.0055 \cdot [(20{,}000 \times 0.00007/1.029) \cdot (1{,}000{,}000/298{,}395)]^{1/3}$$
$$= 0.0055 + 0.009120$$
$$= 0.01462$$

Calculate the pressure loss per 100 feet using the Darcy-Weisbach Equation:

$$\Delta P = 27.73 \cdot \frac{f \cdot \gamma_g \cdot L \cdot V^2}{144 \cdot (2 \cdot g) \cdot D} \qquad (10.26)$$
$$= 27.73 \times \frac{(0.01462)(0.0612)(100)(38.57)^2}{(144)(2)(32.2)(1.029)}$$
$$= 0.387 \text{ in. W.C.} \; (9.8 \text{ mm W.C.})$$

The pressure loss per 100 feet (30 m) of gas head pipe in the case is much less than 1-inch (25.4-mm) water column for the design gas flow rate in section IJ using a 14-inch (350-mm) SDR17 HDPE pipe. Therefore, try a smaller diameter 12-inch (300-mm) SDR17 HDPE pipe for section IJ.

Check the pressure loss per 100 feet (30 m) for a 12-inch (300-mm) SDR17 HDPE pipe:

The internal diameter of a 12-inch (300-mm) SDR17 HDPE pipe is 10.588 inches (269 mm).

Calculate the cross-sectional area of a 12-inch (300-mm) SDR17 HDPE pipe:

Internal diameter of the pipe = 10.588 in = 0.882 ft (0269 m)

$$A = \pi \cdot r^2 = 3.1416 \times (0.882/2)^2 = 0.611 \text{ ft}^2 \; (0.0568 \text{ ft}^2)$$

Calculate the gas flow velocity in the pipe using Equation 10.23:

$$V = Q_{IJ}/A = 1{,}925/0.611 = 3{,}151 \text{ ft/min} = 52.52 \text{ ft/sec} \; (16.0 \text{ m/sec})$$

Calculate the Reynolds number using Equation 10.24:

$$N_{Re} = D \cdot V \cdot \gamma_g / \mu_g \qquad (10.24)$$
$$= 0.882 \times 52.52 \times 0.0612 / (8.14 \times 10^{-6})$$
$$= 348{,}273$$

Calculate the Darcy friction factor using Equation 10.25:

$$f = 0.0055 + 0.0055 \cdot [(20{,}000 \cdot \varepsilon/D) \cdot (1{,}000{,}000/N_{Re})]^{1/3} \quad (10.25)$$
$$= 0.0055 + 0.0055 \cdot [(20{,}000 \times 0.00007/0.882)(1{,}000{,}000/348{,}273)]^{1/3}$$
$$= 0.0055 + 0.009119$$
$$= 0.01462$$

Calculate the pressure loss per 100 feet using the Darcy-Weisbach Equation:

$$\Delta P = 27.73 \cdot \frac{f \cdot \gamma_g \cdot L \cdot V^2}{144 \cdot (2 \cdot g) \cdot D} \quad (10.26)$$
$$= 27.73 \times \frac{(0.01462)(0.0612)(100)(52.52)^2}{(144)(2)(32.2)(0.882)}$$
$$= 0.837 \text{ in. W.C. } (21.3 \text{ mm W.C.})$$

The pressure loss per 100 feet (30 m) of gas head pipe is still less than 1-inch (25.4-mm) water column for the design gas flow rate in section IJ. Therefore, a 12-inch (300-mm) SDR17 HDPE pipe will satisfy requirements for section IJ of the header pipe.

Section JK:

The design gas flow rate of section JK, Q_{JK}, is 2,337 ft³/min (66.2 ft²/min). From Table 10.9, a 14-inch (350-mm) SDR17 HDPE pipe may be selected for section JK.

Check pressure loss per 100 feet (30 m) for a 14-inch (350-mm) SDR17 HDPE pipe:

The internal diameter of a 14-inch (350-mm) SDR17 HDPE pipe is 12.353 inches (314 mm).

Calculate the cross-sectional area of a 14-inch (350-mm) SDR17 HDPE pipe:

$$\text{Internal diameter of the pipe} = 12.353 \text{ in} = 1.029 \text{ ft } (0.314 \text{ m})$$
$$A = \pi \cdot r^2 = 3.1416 \times (1.029/2)^2 = 0.832 \text{ ft}^2 \ (0.0774 \text{ m}^2)$$

Calculate the gas flow velocity in the pipe using Equation 10.23:

$$V = Q_{JK}/A = 2{,}337/0.832 = 2{,}809 \text{ ft/min} = 46.82 \text{ ft/sec } (14.3 \text{ m/sec})$$

Calculate the Reynolds number using Equation 10.24:

$$N_{Re} = D \cdot V \cdot \gamma_g / \mu_g \quad (10.24)$$
$$= 1.029 \times 46.82 \times 0.0612/(8.14 \times 10^{-6})$$
$$= 362{,}221$$

Calculate the Darcy friction factor using Equation 10.25:

$$f = 0.0055 + 0.0055 \cdot [(20{,}000 \cdot \varepsilon/D) \cdot (1{,}000{,}000/N_{Re})]^{1/3} \quad (10.25)$$
$$= 0.0055 + 0.0055 \cdot [(20{,}000 \times 0.00007/1.029)(1{,}000{,}000/362{,}221)]^{1/3}$$
$$= 0.0055 + 0.008549$$
$$= 0.01404$$

Calculate the pressure loss per 100 feet using the Darcy-Weisbach Equation:

$$\Delta P = 27.73 \cdot \frac{f \cdot \gamma_g \cdot L \cdot V^2}{144 \cdot (2 \cdot g) \cdot D} \quad (10.26)$$

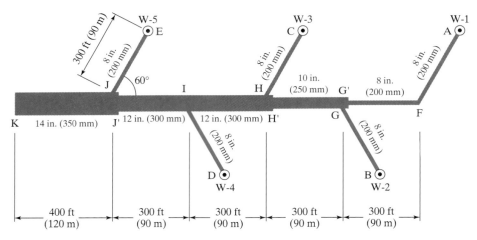

FIGURE 10.17 Detailed Drawing of a Gas Collection Header Liner for Example 10.5

$$= 27.73 \times \frac{(0.01404)(0.0612)(100)(46.82)^2}{(144)(2)(32.2)(1.029)}$$

$$= 0.547 \text{ in. W.C. (13.9 mm W.C.)}$$

The pressure loss per 100 feet (30 m) of gas head pipe is again sufficiently less than 1-inch (25.4-mm) water column for the design gas flow rate in section JK. Therefore, a 14-inch (350-mm) SDR17 HDPE pipe will meet requirements for section JK of the header pipe. Figure 10.17 is a schematic diagram that shows the pipe diameters for each feeder pipe and header pipe section and the three-way joints and enlargement connectors.

3. Determine the minimum required vacuum pressure at the end of the header pipe

Figure 10.17 shows that well W-1 (point A) is the farthest well from the end of the main header pipe (point L). If a -25-inch (-635-mm) water column vacuum pressure can be maintained at the wellhead of well W-1, it should be possible to keep the minimum vacuum pressures of -25-inch (-635-mm) water column at the wellheads of the other wells.

Determine the friction pressure losses along the pipes

Calculate the pressure loss for feeder pipe AF:

The design flow rate for feeder pipe AF is 563 ft^3/min (15.9 m^3/min). The feeder pipe AF is an 8-inch (200-mm) SDR17 HDPE pipe with a length of 300 feet (30 m). The internal diameter of an 8-inch (200-mm) SDR17 HDPE pipe is 7.059 inches or 0.588 feet (0.179 m). The cross-sectional area of an 8-inch (200-mm) SDR17 HDPE pipe is 0.272 ft^2 (0.0252 m^2).

Calculate the gas flow velocity in the pipe using Equation 10.23:

$$V = Q_{AF}/A = 563/0.272 = 2{,}070 \text{ ft/min} = 34.50 \text{ ft/sec (10.5 m/sec)}$$

Calculate the Reynolds number using Equation 10.24:

$$N_{Re} = D \cdot V \cdot \gamma_g / \mu_g \qquad (10.24)$$

$$= 0.588 \times 34.50 \times 0.0612/(8.14 \times 10^{-6})$$

$$= 152{,}519$$

Calculate the Darcy friction factor using Equation 10.25:

$$f = 0.0055 + 0.0055 \cdot [(20{,}000 \cdot \varepsilon/D) \cdot (1{,}000{,}000/N_{Re})]^{1/3} \qquad (10.25)$$
$$= 0.0055 + 0.0055 \cdot [(20{,}000 \times 0.00007/0.588)(1{,}000{,}000/152{,}519)]^{1/3}$$
$$= 0.0055 + 0.01375$$
$$= 0.01925$$

Calculate the pressure loss per 100 feet using the Darcy-Weisbach Equation:

$$\Delta P = 27.73 \cdot \frac{f \cdot \gamma_g \cdot L \cdot V^2}{144 \cdot (2 \cdot g) \cdot D} \qquad (10.26)$$
$$= 27.73 \times \frac{(0.01925)(0.0612)(100)(34.50)^2}{(144)(2)(32.2)(0.588)}$$
$$= 0.713 \text{ in. W.C. (18.1 mm W.C.)}$$
$$\Delta P_{AF} = (0.713/100) \times 300 = 2.14 \text{ in. W.C. (54.4 mm W.C.)}$$

Calculate the pressure loss for header pipe section FG:

The pipe of section FG has the same diameter and length and the same design gas flow rate as the feeder pipe AF; therefore, the pressure loss of section FG should be same as that of feeder pipe AF:

$$\Delta P_{FG} = \Delta P_{AF} = 2.14 \text{ in. W.C. (54.4 mm W.C.)}$$

Calculate the pressure loss in header pipe section GH:

The length of section GH is 300 feet (90 m). The pressure loss per 100 feet (30 m) of section GH is 0.682 in. W.C. (17.3 mm W.C.):

$$\Delta P_{GH} = (0.682/100) \times 300 = 2.05 \text{ in. W.C. (52.1 mm W.C.)}$$

Calculate the pressure loss in header pipe section HI:

The length of section HI is 300 feet (90 m). The pressure loss per 100 feet (30 m) of section HI is 0.522 in. W.C. (13.3 mm W.C.):

$$\Delta P_{HI} = (0.522/100) \times 300 = 1.57 \text{ in. W.C. (39.9 mm W.C.)}$$

Calculate the pressure loss in header pipe section IJ:

The length of section IJ is 300 feet (90 m). The pressure loss per 100 feet (30 m) of section IJ is 0.837 in. W.C. (21.3 mm W.C.):

$$\Delta P_{IJ} = (0.837/100) \times 300 = 2.51 \text{ in. W.C. (63.8 mm W.C.)}$$

Calculate the pressure loss in header pipe section JK:

The length of section JK is 400 feet (120 m). The pressure loss per 100 feet (30 m) of section JK is 0.547 in. W.C. (13.9 mm W.C.):

$$\Delta P_{JK} = (0.547/100) \times 400 = 2.19 \text{ in. W.C. (55.6 mm W.C.)}$$

Calculate the total friction pressure loss from A to K:

$$\Delta P_{\text{friction}} = \Delta P_{AF} + \Delta P_{FG} + \Delta P_{GH} + \Delta P_{HI} + \Delta P_{IJ} + \Delta P_{JK}$$
$$= 2.14 + 2.14 + 2.05 + 1.57 + 2.51 + 2.19$$
$$= 12.60 \text{ in. W.C. (320.0 mm W.C.)}$$

Determine the fitting pressure losses

Determine the transition loss coefficient for a 60° miter bend F (Table 9.6):

$$K_F = 0.55$$

Determine the transition loss coefficient for three-way joints G, H, I, and J:

$$K_G = K_H = K_I = K_J = 0.55$$

Determine the transition loss coefficient for sudden enlargement connectors:

$$K = [1 - (D_{small})^2/(D_{large})^2]^2 \tag{10.29}$$

Transition loss coefficient of connector G′:

$$\begin{aligned} K_{G'} &= [1 - (D_{FG})^2/(D_{GH})^2]^2 \\ &= [1 - (7.059)^2/(8.824)^2]^2 \\ &= 0.130 \end{aligned}$$

Transition loss coefficient of connector H′:

$$\begin{aligned} K_{H'} &= [1 - (D_{GH})^2/(D_{HI})^2]^2 \\ &= [1 - (8.824)^2/(10.588)^2]^2 \\ &= 0.0933 \end{aligned}$$

Transition loss coefficient of connector J′:

$$\begin{aligned} K_{J'} &= [1 - (D_{IJ})^2/(D_{JK})^2]^2 \\ &= [1 - (10.588)^2/(12.353)^2]^2 \\ &= 0.0704 \end{aligned}$$

Calculate the flow coefficient for each fitting:

$$C_v = 29.9 \cdot d^2/K^{0.5} \tag{10.32}$$

Flow coefficient of 60° miter bend F:

$$(C_v)_F = 29.9 \times (7.059)^2/(0.55)^{0.5} = 2{,}009$$

Flow coefficient of three-way joint connection G:

$$(C_v)_G = 29.9 \times (8{,}824)^2/(0.55)^{0.5} = 3{,}139$$

Flow coefficient of three-way joint connection H:

$$(C_v)_H = 29.9 \times (10.588)^2/(0.55)^{0.5} = 4{,}520$$

Flow coefficient of three-way joint connection I:

$$(C_v)_I = 29.9 \times (10.588)^2/(0.55)^{0.5} = 4{,}520$$

Flow coefficient of three-way joint connection J:

$$(C_v)_J = 29.9 \times (12.353)^2/(0.55)^{0.5} = 6{,}152$$

Flow coefficient of connector G′:

$$(C_v)_{G'} = 29.9 \times (7.059)^2/(0.130)^{0.5} = 4{,}132$$

Flow coefficient of connector H′:

$$(C_v)_{H'} = 29.9 \times (8.824)^2/(0.0933)^{0.5} = 7{,}622$$

Flow coefficient of connector J′:

$$(C_v)_{J'} = 29.9 \times (10.588)^2/(0.0704)^{0.5} = 12{,}633$$

Calculate the pressure loss across each fitting:

$$\Delta P = (\gamma_g/62.4) \cdot (7.48 \cdot Q/C_v)^2 \qquad (10.33)$$
$$= (0.0612/62.4)(7.48)^2 (Q/C_v)^2$$
$$= 0.0549 \cdot (Q/C_v)^2$$

Pressure loss across 60° miter bend F:

$$\Delta P_F = 0.0549 \cdot [Q_{AF}/(C_v)_F]^2$$
$$= 0.0549 \cdot (563/2{,}009)^2$$
$$= 0.00431 \text{ psi} = 0.120 \text{ in. W.C.} \ (3.0 \text{ mm W.C.})$$

Pressure loss across three-way joint G:

$$\Delta P_G = 0.0549 \cdot [Q_{GH}/(C_v)_G]^2$$
$$= 0.0549 \cdot (1{,}029/3{,}139)^2$$
$$= 0.00102 \text{ psi} = 0.164 \text{ in. W.C.} \ (4.2 \text{ mm W.C.})$$

Pressure loss across three-way joint H:

$$\Delta P_H = 0.0549 \cdot [Q_{HI}/(C_v)_H]^2$$
$$= 0.0549 \cdot (1{,}479/4{,}520)^2$$
$$= 0.00588 \text{ psi} = 0.163 \text{ in. W.C.} \ (4.1 \text{ mm W.C.})$$

Pressure loss across three-way joint I:

$$\Delta P_I = 0.0549 \cdot [Q_{IJ}/(C_v)_I]^2$$
$$= 0.0549 \cdot (1{,}925/4{,}520)^2$$
$$= 0.00996 \text{ psi} = 0.276 \text{ in. W.C.} \ (7.0 \text{ mm W.C.})$$

Pressure loss across three-way joint J:

$$\Delta P_J = 0.0549 \cdot [Q_{JK}/(C_v)_J]^2$$
$$= 0.0549 \cdot (2{,}337/6{,}152)^2$$
$$= 0.00792 \text{ psi} = 0.220 \text{ in. W.C.} \ (5.6 \text{ mm W.C.})$$

Pressure loss across connector G′:

$$\Delta P_{G'} = 0.0549 \cdot [Q_{FG}/(C_v)_{G'}]^2$$
$$= 0.0549 \cdot (563/4{,}132)^2$$
$$= 0.00102 \text{ psi} = 0.028 \text{ in. W.C.} \ (0.7 \text{ mm W.C.})$$

Pressure loss across connector H′:

$$\Delta P_{H'} = 0.0549 \cdot [Q_{GH}/(C_v)_{H'}]^2$$
$$= 0.0549 \cdot (1{,}029/7{,}622)^2$$
$$= 0.00101 \text{ psi} = 0.028 \text{ in. W.C. } (0.7 \text{ mm W.C.})$$

Pressure loss across connector J′:

$$\Delta P_{J'} = 0.0549 \cdot [Q_{IJ}/(C_v)_{J'}]^2$$
$$= 0.0549 \cdot (1{,}925/12{,}633)^2$$
$$= 0.00127 \text{ psi} = 0.035 \text{ in. W.C. } (0.9 \text{ mm W.C.})$$

Calculate the total fitting pressure loss from A to K:

$$\Delta P_{\text{fitting}} = \Delta P_F + \Delta P_G + \Delta P_H + \Delta P_I + \Delta P_J + \Delta P_{G'} + \Delta P_{H'} + \Delta P_{J'}$$
$$= 0.120 + 0.164 + 0.163 + 0.276 + 0.220 + 0.028 + 0.028 + 0.035$$
$$= 1.034 \text{ in. W.C. } (26.3 \text{ mm W.C.})$$

Determine the total pressure loss from A to K:

$$\Delta P_{AK} = \Delta P_{\text{friction}} + \Delta P_{\text{fitting}}$$
$$= 12.60 + 1.034$$
$$= 13.63 \text{ in. W.C. } (346.2 \text{ mm W.C.})$$

Determine the minimum required vacuum pressure at the end of the header pipe:

$$P_K = -25.0 - \Delta P_{AK}$$
$$= -25.0 - 13.63$$
$$= -38.63 \text{ in. W.C. } (-981.2 \text{ mm W.C.})$$

4. *Estimate the pressures at the wellhead of each extraction well*

Determine the pressure losses for each feeder pipe:

Calculate the pressure loss for feeder pipe CH:

The design flow rate for feeder pipe CH is 545 ft³/min (15.4 m³/min). The feeder pipe CH is an 8-inch (200-mm) SDR17 HDPE pipe with a length of 300 feet (90 m). The internal diameter of an 8-inch (200-mm) SDR17 HDPE pipe is 7.059 in. or 0.588 ft (0.179 m). The cross-sectional area of an 8-inch SDR17 HDPE pipe is 0.272 ft² (0.0252m²).

Calculate the gas flow velocity in the pipe using Equation 10.23:

$$V = Q_{CH}/A = 545/0.272 = 2{,}004 \text{ ft/min} = 33.40 \text{ ft/sec } (10.2 \text{ m/sec})$$

Calculate the Reynolds number using Equation 10.24:

$$N_{Re} = D \cdot V \cdot \gamma_g / \mu_g \quad (10.24)$$
$$= 0.588 \times 33.40 \times 0.0612/(8.14 \times 10^{-6})$$
$$= 147{,}666$$

Calculate the Darcy friction factor using Equation 10.25:

$$f = 0.0055 + 0.0055 \cdot [(20{,}000 \cdot \varepsilon/D) \cdot (1{,}000{,}000/N_{Re})]^{1/3} \quad (10.25)$$

$$= 0.0055 + 0.0055 \cdot [(20{,}000 \times 0.00007/0.588)(1{,}000{,}000/147{,}666)]^{1/3}$$
$$= 0.0055 + 0.01389$$
$$= 0.01939$$

Calculate the pressure loss per 100 feet (30 m) using the Darcy-Weisbach Equation:

$$\Delta P = 27.73 \cdot \frac{f \cdot \gamma_g \cdot L \cdot V^2}{144 \cdot (2 \cdot g) \cdot D} \tag{10.26}$$

$$= 27.73 \times \frac{(0.01939)(0.0612)(100)(33.40)^2}{(144)(2)(32.2)(0.588)}$$

$$= 0.674 \text{ in. W.C. } (17.1 \text{ mm W.C.})$$

$$\Delta P_{CH} = (0.674/100) \times 300 = 2.02 \text{ in. W.C. } (51.3 \text{ mm W.C.})$$

Calculate the pressure loss for feeder pipe DI:

The design flow rate for feeder pipe DI is 576 ft³/min (16.3 m³/min). The feeder pipe DI is an 8-inch (200-mm) SDR17 HDPE pipe with a length of 300 feet (90 m). The internal diameter of an 8-inch (200-mm) SDR17 HDPE pipe is 7.059 in or 0.588 ft (0.179 m). The cross-sectional area of an 8-inch (200-mm) SDR17 HDPE pipe is 0.272 ft² (0.0252 ft²).

Calculate the gas flow velocity in the pipe using Equation 10.23:

$$V = Q_{DI}/A = 576/0.272 = 2{,}118 \text{ ft/min} = 35.30 \text{ ft/sec } (10.8 \text{ m/sec})$$

Calculate the Reynolds number using Equation 10.24:

$$N_{Re} = D \cdot V \cdot \gamma_g / \mu_g \tag{10.24}$$
$$= 0.588 \times 35.30 \times 0.0612/(8.14 \times 10^{-6})$$
$$= 156{,}055$$

Calculate the Darcy friction factor using Equation 10.25:

$$f = 0.0055 + 0.0055 \cdot [(20{,}000 \cdot \varepsilon/D) \cdot (1{,}000{,}000/N_{Re})]^{1/3} \tag{10.25}$$
$$= 0.0055 + 0.0055 \cdot [(20{,}000 \times 0.00007/0.588)(1{,}000{,}000/156{,}055)]^{1/3}$$
$$= 0.0055 + 0.01364$$
$$= 0.01914$$

Calculate the pressure loss per 100 feet (30 m) using the Darcy-Weisbach Equation:

$$\Delta P = 27.73 \cdot \frac{f \cdot \gamma_g \cdot L \cdot V^2}{144 \cdot (2 \cdot g) \cdot D} \tag{10.26}$$

$$= 27.73 \times \frac{(0.01914)(0.0612)(100)(35.30)^2}{(144)(2)(32.2)(0.588)}$$

$$= 0.743 \text{ in. W.C. } (18.9 \text{ mm W.C.})$$

$$\Delta P_{DI} = (0.743/100) \times 300 = 2.23 \text{ in. W.C. } (56.6 \text{ mm W.C.})$$

Calculate the pressure loss for feeder pipe EJ:

The design flow rate for feeder pipe EJ is 534 ft³/min (15.1 m³/min). The feeder pipe EJ is an 8-inch (200-mm) SDR17 HDPE pipe with a length of 300 feet (90 m). The internal diameter of an 8-inch (200-mm) SDR17 HDPE pipe is 7.059 in. or 0.588 ft (0.179 m). The cross-sectional area of an 8-inch (200-mm) SDR17 HDPE pipe is 0.272 ft² (0.0252 m²).

Calculate the gas flow velocity in the pipe using Equation 10.23:

$$V = Q_{EJ}/A = 534/0.272 = 1,963 \text{ ft/min} = 32.72 \text{ ft/sec } (10.0 \text{ m/sec})$$

Calculate the Reynolds number using Equation 10.24:

$$N_{Re} = D \cdot V \cdot \gamma_g / \mu_g \qquad (10.24)$$
$$= 0.588 \times 32.72 \times 0.0612/(8.14 \times 10^{-6}) = 144,650$$

Calculate the Darcy friction factor using Equation 10.25:

$$f = 0.0055 + 0.0055 \cdot [(20,000 \cdot \varepsilon/D) \cdot (1,000,000/N_{Re})]^{1/3} \qquad (10.25)$$
$$= 0.0055 + 0.0055 \cdot [(20,000 \times 0.00007/0.588)(1,000,000/144,650)]^{1/3}$$
$$= 0.0055 + 0.01399$$
$$= 0.01949$$

Calculate the pressure loss per 100 feet (30 m) using the Darcy-Weisbach Equation:

$$\Delta P = 27.73 \cdot \frac{f \cdot \gamma_g \cdot L \cdot V^2}{144 \cdot (2 \cdot g) \cdot D} \qquad (10.26)$$
$$= 27.73 \times \frac{(0.01949)(0.0612)(100)(32.72)^2}{(144)(2)(32.2)(0.588)}$$
$$= 0.649 \text{ in. W.C. (16.5 mm W.C.)}$$
$$\Delta P_{EJ} = (0.649/100) \times 300 = 1.95 \text{ in. W.C. (49.5 mm W.C.)}$$

Pressure loss for feeder pipe AF:

$$\Delta P_{AF} = 2.14 \text{ in. W.C. (54.4 mm W.C.)}$$

Pressure loss for feeder pipe BG:

$$\Delta P_{BG} = (0.777/100) \times 300 = 2.33 \text{ in. W.C. (59.2 mm W.C.)}$$

Estimate the pressures at the wellhead of each well

Estimate the pressure at the wellhead of well W-1:

$$P_{W-1} = -25.0 \text{ in. W.C. } (-635 \text{ mm W.C.})$$

Estimate the pressure at the wellhead of well W-2:

$$P_{W-2} = P_K + (\Delta P_{JK} + \Delta P_{IJ} + \Delta P_{HI} + \Delta P_{GH} + \Delta P_{BG} + \Delta P_J + \Delta P_I + \Delta P_H$$
$$+ \Delta P_G + \Delta P_{J'} + \Delta P_{H'})$$
$$= -38.63 + (2.19 + 2.51 + 1.57 + 2.05 + 2.33 + 0.220 + 0.276 + 0.163$$
$$+ 0.164 + 0.035 + 0.028)$$
$$= -38.63 + 11.54 = -27.09 \text{ in. W.C. } (-688.1 \text{ mm W.C.})$$

Estimate the pressure at the wellhead of well W-3:

$$P_{W\text{-}3} = P_K + \Delta P_{JK} + \Delta P_{IJ} + \Delta P_{HI} + \Delta P_{CH} + \Delta P_J + \Delta P_I + \Delta P_H + \Delta P_{J'})$$
$$= -38.63 + (2.19 + 2.51 + 1.57 + 2.02 + 0.220 + 0.276 + 0.163 + 0.035)$$
$$= -38.63 + 8.98 = -29.65 \text{ in. W.C.} (-753.1 \text{ mm W.C.})$$

Estimate the pressure at the wellhead of well W-4:

$$P_{W\text{-}4} = P_K + (\Delta P_{JK} + \Delta P_{IJ} + \Delta P_{DI} + \Delta P_J + \Delta P_I + \Delta P_{J'})$$
$$= -38.63 + (2.19 + 2.51 + 2.23 + 0.220 + 0.276 + 0.035)$$
$$= -38.63 + 7.46 = -31.17 \text{ in. W.C.} (-791.7 \text{ mm W.C.})$$

Estimate the pressure at the wellhead of well W-5:

$$P_{W\text{-}5} = P_K + (\Delta P_{JK} + \Delta P_{EJ} + \Delta P_J)$$
$$= -38.63 + (2.19 + 1.95 + 0.220)$$
$$= -38.63 + 4.36 = -34.27 \text{ in. W.C.} (-870.5 \text{ mm W.C.})$$

The pressures at the wellhead of each extraction well are equal to or higher than the minimum required vacuum pressure of -25.0 in. W.C. (-635 mm W.C.). Accordingly, the minimum vacuum pressure requirement is observed throughout the entire system.

Summary: The calculated design flow rates, pipe diameters, and friction pressure losses for the finger pipes and different sections of header pipe are listed in Table 10.12. The calculated fitting

TABLE 10.12 Design Flow Rates, Pipe Lengths, Pipe Diameters, and Pressure Losses for Feeder Pipes and Different Sections of Main Header Pipe

Pipe	Feeder Pipe					Header Pipe Section				
	AF	BG	CH	DI	EJ	FG	GH	HI	IJ	JK
Design Flow Rate										
ft³/min	563	591	545	576	534	563	1,029	1,479	1,925	2,337
m³/min	15.9	16.7	15.4	16.3	15.1	15.9	29.1	41.9	54.5	66.2
Pipe Length										
feet	300	300	300	300	300	300	300	300	300	400
m	90	90	90	90	90	90	90	90	90	120
SDR 17 Pipe Diameter										
in.	8	8	8	8	8	8	10	12	12	14
mm	200	200	200	200	200	200	250	300	300	350
Pressure Loss										
in. W.C./100 ft	0.713	0.777	0.674	0.743	0.649	0.713	0.682	0.522	0.837	0.547
mm W.C./30 m	18.1	19.7	17.1	18.9	16.5	18.1	17.3	13.3	21.3	13.9
Pressure Loss										
in. W.C.	2.14	2.33	2.02	2.23	1.95	2.14	2.05	1.57	2.51	2.19
mm W.C.	54.4	59.2	51.3	56.6	49.5	54.4	52.1	39.9	63.8	55.6

TABLE 10.13 Fitting Pressure Losses across a Bend, Joints, and Connectors

Fitting	60° Bend	Three-Way Joint				Enlargement Connector		
	F	G	H	I	J	G′	H′	J′
Pressure Loss								
in. W.C.	0.120	0.164	0.163	0.276	0.220	0.028	0.028	0.035
mm W.C.	3.0	4.2	4.1	7.0	5.6	0.7	0.7	0.9

TABLE 10.14 Vacuum Pressures at the Wellhead of Each Well and the End of the Header Line

Well and Branch Line	Gas Extraction Well					End of Header Line
	W-1	W-2	W-3	W-4	W-5	K
Pressure						
in. W.C.	−25.0	−27.09	−29.65	−31.17	−34.27	−38.63
mm W.C.	−635.0	−688.1	−753.1	−791.7	−870.5	−981.2

pressure losses across the bend and the different joins and enlargement connectors are listed in Table 10.13. The calculated vacuum pressures at the wellhead of each extraction well and at the end of the header pipe are listed in Table 10.14.

PROBLEMS

10.1 What are the main constituents of municipal solid waste landfill gas?

10.2 Is landfill gas a clean gas or a polluted gas? Why?

10.3 List the factors influencing landfill gas generation. What constituents in the waste itself affect gas generation?

10.4 How can the gas generation rate from a landfill be increased without changing waste composition?

10.5 How long does the gas generation phase last? What factors affect the duration of the gas generation phase?

10.6 What are two mechanisms of gas migration from landfills? How do these two mechanisms differ from one another?

10.7 What are the major engineering measures to prevent gas migration from landfills?

10.8 What are the main advantages and disadvantages of a passive gas collection system?

10.9 What are the main differences between passive and active gas collection systems?

10.10 What components make up an active gas collection system?

10.11 For an active gas collection system, an adequate vacuum pressure must be applied to the gas collection header pipe to extract the landfill gas. Why must the header pipes still be sloped sufficiently?

10.12 What are key factors that must be considered for designing a landfill gas collection system?

10.13 A landfill receives 188,500 U.S. ton/year (171,200 Mg/year) of solid waste. The total active period of this landfill is 12 years. The landfill will be closed after 12 years. Assume the methane generation potential, L_o, is 2.65 ft³/lb (165 m³/Mg), and the methane generation rate constant, k, is 0.31 year^{-1}. Estimate changes of landfill gas generation rate with time, and determine the maximum gas generation rate.

10.14 The total active period of a landfill is 10 years. The landfill will be closed after 10 years. The year-to-year waste acceptance tonnage is listed in Table 10.15. Assume the methane generation potential, L_o, is 2.25 ft³/lb (140 m³/Mg), and the methane generation rate constant, k, is 0.18 year^{-1}. Estimate changes in the landfill gas generation rate with time, and determine the maximum gas generation rate.

10.15 The influence radius of a vertical gas extraction well is 120 feet (36 m). The average annual waste filling depth in the influence area of this gas extraction well is 18 feet (5.4 m). No waste is placed in the influence area of this well after the 10th year. Assume the density of the solid waste, γ_{waste}, is 70 lb/ft³ (1.12 Mg/m³); the methane generation potential, L_o, is 2.25 ft³/lb (140 m³/Mg), and the methane generation rate constant, k, is 0.15 year^{-1}. Estimate changes in the landfill gas generation rate from this well with time, and determine the maximum gas generation rate of this well.

10.16 The influence radius of a vertical gas extraction well is 120 feet (36 m). The year-to-year waste filling depths in the influence area of the gas extraction well are listed in Table 10.16. No waste is placed in the influence area after the 10th year. Assume the density of the solid waste, γ_{waste}, is 70 lb/ft³ (1.12 Mg/m³); the methane generation potential, L_o, is 2.25 ft³/lb (140 m³/Mg), and the methane generation rate constant, k, is 0.15 year^{-1}. Estimate changes in the landfill gas generation rate from this well with time, and determine the maximum gas generation rate of this well.

10.17 A gas collection header line carries 1,400 ft³/min (40 m³/min) of landfill gas over a distance of 450 ft (135 m) where condensate will drain and be collected. Average initial temperature of the gas at the beginning of the line is 62°C (144°F). The temperature at drop-out point is 25°C (77°F). The header is at 30 inches (762 mm) W.C. vacuum (average). What is the quantity of landfill gas condensate that would be condensed out in the header pipe?

TABLE 10.15 Annual Waste Acceptance Tonnage for Problem 10.14

Year	Acceptance Tonnage	
	U.S. ton	Mg
1	175,000	158,900
2	230,000	208,840
3	258,000	234,260
4	229,000	207,932
5	201,000	182,508
6	187,000	169,796
7	311,000	282,388
8	250,000	227,000
9	212,000	192,496
10	198,000	179,784

TABLE 10.16 Annual Waste Filling Depths in the Well Influence Area for Problem 10.16

Year	Waste Filling Depth	
	Feet	m
1	21	6.3
2	28	8.4
3	0	0
4	22	6.6
5	17	5.1
6	14	4.2
7	29	8.7
8	0	0
9	27	8.1
10	22	6.6
Total	180	54

10.18 A landfill gas collection header line is shown in Figure 10.16. There are five vertical extraction wells connected to this header line. They are named W-1, W-2, W-3, W-4, and W-5, respectively. The length of each of the feeder pipes, AF, BG, EH, DI, and EJ, between wells and header pipe, is 300 feet (90 m). The section lengths, FG, GH, HI, and IJ, of the header pipe are also 300 feet (90 m). The distance of pipe section JK is 400 feet (120 m). The angle between well feeder pipes and the header pipe is 60 degrees. The feeder pipes slope from the well to the header line, and the header slopes from F to K. The year-to-year filling depths of the solid waste in the influence area of each gas extraction well are listed in Table 10.17. No waste is placed in the influence area of the 5 wells after the 20th year. Assume the density of the solid waste, γ_{waste}, is 65 lb/ft^3 (1.04 Mg/m^3); the methane generation potential, L_o, is 2.25 ft^3/lb (140 m^3/Mg); and the methane generation rate constant, k, is 0.30 year^{-1}. Determine the following:

1. The gas generation rate from each well with time.
2. The design gas flow rate in each feeder and header pipe.
3. Select the diameters of the feeder pipes and the different sections of the header pipe using the Darcy-Weisbach Equation.
4. Assume the minimum vacuum pressure at the wellhead for each extraction well is −25 in. (−635 mm) W.C. and the transition loss coefficient of the joints G, H, I, and J is equal to 0.55. Determine the minimum required vacuum pressure at the end of the header pipe.
5. Check the vacuum pressures at the wellhead of each extraction well to make sure the pressures at the wellhead of each extraction well do not exceed −25 inches W.C. (−635 mm W.C.).

REFERENCES

EMCON Associates, (1980) "Methane Generation and Recovery from Landfills," Ann Arbor Science, Ann Arbor, MI.

EMCON Associates, (1981) "State-of-the-Art of Methane Gas Enhancement in Landfills," Report to Argonne National Laboratory, ANL/CNSV-23, June, Argonne, IL.

TABLE 10.17 Year-to-Year Gas Flow Rates Generated from Each Extraction Well for Problem 10.18

Year	Waste Filling Depth									
	W-1		W-2		W-3		W-4		W-5	
	ft	m	ft	m	ft	m	ft	m	ft	m
1	15	4.5	6	1.8	25	7.5	10	3	0	0
2	15	4.5	27	8.1	25	7.5	24	7.2	0	0
3	15	4.5	29	8.7	25	7.5	31	9.3	0	0
4	15	4.5	31	9.3	25	7.5	27	8.1	0	0
5	15	4.5	23	6.9	25	7.5	30	9	0	0
6	15	4.5	0	0	25	7.5	24	7.2	20	6
7	15	4.5	15	4.5	0	0	12	3.6	20	6
8	15	4.5	26	7.8	0	0	0	0	20	6
9	15	4.5	24	7.2	0	0	0	0	20	6
10	15	4.5	21	6.3	0	0	28	8.4	20	6
11	15	4.5	0	0	0	0	22	6.6	20	6
12	15	4.5	0	0	0	0	10	3	20	6
13	15	4.5	14	4.2	0	0	0	0	20	6
14	15	4.5	22	6.6	0	0	20	6	20	6
15	15	4.5	25	7.5	25	7.5	34	10.2	20	6
16	15	4.5	20	6	25	7.5	18	5.4	20	6
17	15	4.5	17	5.1	25	7.5	0	0	20	6
18	15	4.5	0	0	25	7.5	0	0	20	6
19	15	4.5	0	0	25	7.5	10	3	20	6
20	15	4.5	0	0	25	7.5	0	0	20	6
Total	300	90	300	90	300	90	300	90	300	90

EMCON Associates, (1998) "Municipal Solid Waste Landfill Gas Design Plan Review," Student Manual, APTI Workshop T018, First Edition, North Carolina State University, College of Engineering, Industrial Extension Service, Environmental Programs, Raleigh, NC.

Fungaroli, A. and Steiner, R., (1979) "Investigation of Sanitary Landfill Behavior," Volume 2, Final Report, U.S. Environmental Protection Agency, EPA-600/2-79-053a, Cincinnati, OH.

Koerner, G. R., Yazdani, R. and Mackey, R. E., (1996) "Long Term Temperature Monitoring of Landfill Geomembranes," *Proceedings, SWANA Conference on Landfills*, Nov. 4–6, 1996, Publ. GR D0401, Silver Spring, MD, pp. 61–63.

LandTech, (1994) "Landfill Gas System Engineering—Practical Approach," Landfill Gas System Engineering Design Seminar, Landfill Control Technologies, Commerce, CA.

McBean, E. A., Rovers, F. A., and Farquhar, G. J., (1995) "Solid Waste Landfill Engineering and Design," Prentice Hall PTR, Englewood Cliffs, NJ.

Michels, M. S., (1996) "Landfill Gas Collection System Components," *Solid Waste Technologies*, Vol. 10, No.1, Adams/Green Industry Publishing, Inc., January/February, Riverton, NJ, pp. 20–30.

Prosser, R. and Janechek, A., (1995) "Landfill Gas and Groundwater Contamination," *Landfill Closures—Environmental Protection and Land Recovery*, ASCE, Geotechnical Special Publication, No. 53, R. Jeffrey Dunn and Udai P. Singh, Eds. New York, NY, pp. 258–271.

UKDOE, (1993) "The Technical Aspects of Controlled Waste Management—Understanding Landfill Gas," Report No. CWM/040/92, Department of the Environment, London, UK.

USEPA, (1992) "Seminars: Design, Operation, and Closure of Municipal Solid Waste Landfills," U.S. Environmental Protection Agency, Office of Research and Development, Washington, DC, EPA/600/K-92/002, April.

USEPA, (1994) "Design, Operation, and Closure of Municipal Solid Waste Landfills," U.S. Environmental Protection Agency, Office of Research and Development, Washington, DC, 20460, EPA/625/R-94/008, September.

USEPA, (1997) "Code of Federal Regulations, 40 CFR Part 60," Revised as of July 1, 1997, U. S. Environmental Agency, Washington, DC.

Zehnder, A. J. B., (1978) "Ecology of Methane Formation," Water Pollution Microbiology 2, pp. 349–376.

CHAPTER 11

Final Cover System

11.1 COMPONENTS OF FINAL COVER SYSTEM
    11.1.1 EROSION CONTROL LAYER
    11.1.2 PROTECTION LAYER
    11.1.3 DRAINAGE LAYER
    11.1.4 HYDRAULIC BARRIER LAYER
    11.1.5 GAS VENT LAYER
    11.1.6 FOUNDATION LAYER
11.2 ALTERNATIVE LANDFILL COVERS
    11.2.1 WATER BALANCE OF EARTHEN COVERS
    11.2.2 CAPILLARY BARRIER
    11.2.3 MONOLAYER BARRIER
11.3 FIELD STUDY OF LANDFILL COVERS
11.4 SOIL EROSION CONTROL
    11.4.1 NATURE OF SOIL EROSION
    11.4.2 SOIL LOSS PREDICTION
    11.4.3 LIMITATIONS OF UNIVERSAL SOIL LOSS EQUATION
    11.4.4 EROSION CONTROL PRINCIPLES
    11.4.5 MANUFACTURED EROSION CONTROL MATERIALS
11.5 EFFECTS OF SETTLEMENT AND SUBSIDENCE
11.6 DIFFERENTIAL SUBSIDENCE CASE HISTORY
    PROBLEMS
    REFERENCES

When municipal solid waste and hazardous waste landfills are filled to capacity, they are capped with a final cover that keeps out infiltration and keeps in gases and volatile components. These final covers are multilayered systems that provide multiple functions.

The regulations dealing with final covers for municipal solid waste (MSW) landfills in the United States are found in Title 40, Part 258 of the Resource Conservation and Recovery Act (RCRA). The basic requirement of the final cover system for MSW landfills is set forth as follows (USEPA, 1995):

"Owners or operators of all MSW landfill units must install a final cover system that is designed to minimize infiltration and erosion. The final cover system must be comprised of an erosion layer underlain by an infiltration layer as follows:

(i) The infiltration layer must be comprised of a minimum of 18 inches (450 mm) of earthen material that has a permeability less than or equal to the permeability of any bottom liner system or natural subsoils present, or a permeability no greater than 1.0×10^{-5} cm/sec, whichever is less; and

(ii) The erosion layer must consist of a minimum of 6 inches (150 mm) of earthen material that is capable of sustaining native plant growth."

The regulations permit the director of an approved state to approve an alternative final cover design that includes an equivalent infiltration layer and erosion layer, and also require postclosure care and maintenance for at least 30 years, unless a different period is approved by the director of an approved state. It should be noted, however, that cover regulations in other countries vary considerably from the preceding and from one another as well. See Koerner et al. (1999) for a worldwide review of cover and liner regulations.

11.1 COMPONENTS OF FINAL COVER SYSTEM

Covers placed over landfills are multicomponent cover systems that are constructed directly on top of the waste shortly after a specific unit or cell has been filled to capacity. As illustrated in Figure 11.1, the usual components within a final cover system are the erosion control layer, protection layer, drainage layer, hydraulic barrier layer, gas vent layer, and foundation layer. Not all components are needed for all final covers. For example, a gas vent layer may be required for some covers but not others, depending upon whether the waste is producing gases that require collection and management. In addition, some of the layers may be combined. For instance, the gas

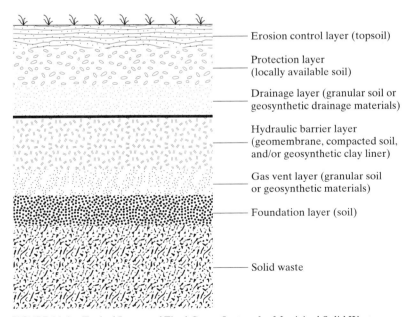

FIGURE 11.1 Typical Layers of Final Cover System for Municipal Solid Waste Landfill

collection layer can be combined as a single layer with the foundation layer. The materials that are typically used for the various components are noted in Figure 11.1 (Daniel, 1995; Koerner and Daniel, 1997).

11.1.1 Erosion Control Layer

The most commonly used material for the erosion control layer is fertile topsoil. Such vegetated topsoil helps to minimize erosion and promote transpiration of water back to the atmosphere. Vegetation also provides a leaf cover or canopy above the soil that reduces rainfall impact and decreases wind velocity at the soil surface.

Selection of plant species is an important consideration in the establishment of vegetation on the surface of the final cover system. The use of shrubs and trees is usually inappropriate because the root systems extend to a depth that would normally invade the drainage layer and the barrier layer if it is low-permeability soil. Most cover systems are seeded with grasses. Suitable plant species, such as grasses and herbaceous plants, are available for various climates. The timing of seeding is also important to successful establishment of vegetation.

The thickness of the topsoil layer depends on many site-specific factors, including climate, plant species to be grown on the cover, and soil materials to be used. As a minimum, the topsoil layer should be at least 6 inches (150 mm), because this is typically the minimum depth of root penetration of most plant species.

Erosion can be controlled temporarily by placing a geosynthetic erosion control material such as erosion control blankets on the surface to hold the soil in place until vegetation can become established. Figure 11.2 shows a landfill cover with some erosion damages. Figure 11.3 shows a landfill cover with well established vegetation.

In arid sites, it may be very difficult to establish and to maintain consistent vegetation on the surface of the final cover. Soil with a sparse cover of vegetation is vulnerable to excessive erosion from water, wind, or both. Cobbles or gravel have been used to replace the topsoil as an erosion control layer in arid sites. Depending on their size, cobbles and gravel are highly resistant to erosion from both wind and water.

11.1.2 Protection Layer

The protection layer lies directly beneath the erosion control layer, and in some cases can be combined with it. The protection layer minimizes frost penetration into the compacted soil layer, and protects a geomembrane from accidental intrusion, burrowing animals, and other deleterious effects. Locally available native soil is the material that is most commonly used for the protection layer.

The minimum required thickness of the protection layer depends on many site-specific factors (Koerner and Daniel, 1997), including the need to:

(i) support growth of vegetation;
(ii) avoid frost penetration;
(iii) protect underlying layers from desiccation; and
(iv) prevent accidental human intrusion, penetration by burrowing animals, or root penetration into underlying materials.

FIGURE 11.2 A Landfill Cover with Erosion Damage

FIGURE 11.3 A Landfill Cover with Well Established Grass

The protection layer is often designed primarily to prevent underlying layers from freezing. As discussed in Chapter 3, freeze-thaw cycles damage compacted clay materials to varying degrees. Thus, a compacted clay liner should be placed below the depth of maximum frost penetration. Figure 11.4 shows contours of maximum frost penetration depths in the continental U.S. and state averages. It may be prudent to prevent the drainage layer from freezing as well. The primary use of an underlying drainage layer in many final cover systems is to dissipate pore water pressures that tend to cause slope instability. If the drainage layer freezes, its function is destroyed for part of the year. During the thaw period, it is particularly important that both the drainage layer and the protection layer function properly.

Within the protection layer, a designer can incorporate a layer of cobbles, commonly called a *biobarrier*. The purpose is to prevent burrowing animals from reaching the underlying drainage and barrier layers. The size of the cobbles is related to the digging power of the animal species of concern. The range is typically from 4.0 in. (100 mm) to 12.0 (300 mm) in size. Filtration layers are not necessary unless one is designing a capillary break barrier, a topic described in Section 11.2.2.

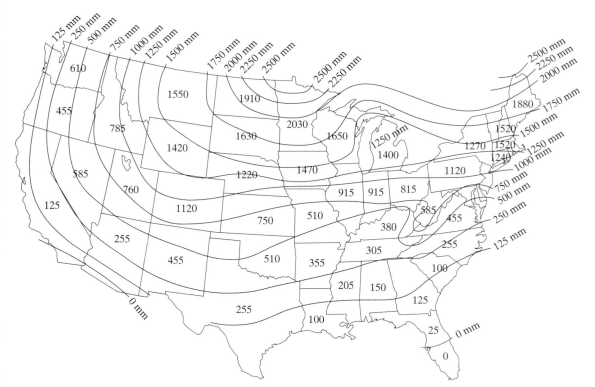

FIGURE 11.4 Contours of Maximum Frost Penetration Depths in Continental U.S. and State Averages (USEPA, 1989)

11.1.3 Drainage Layer

Water that penetrates through the cover soil should be removed from the cover system using a drainage layer. By draining water from the slope in a drainage layer, the stability of the slope is greatly enhanced. The stabilizing impact of drainage in a slope can be dramatic, causing a major difference in the factor of safety (i.e., doubling or halving the value, depending on whether drainage is or is not controlled) (Koerner and Daniel, 1997). A drainage layer is placed below the protection layer and above the hydraulic barrier layer. A drainage layer serves three principal functions:

(i) To reduce the head of water on the barrier layer, thus minimizing infiltration;
(ii) To drain water from the overlying soil, allowing it to absorb and retain additional water; and
(iii) To reduce and control pore water pressures at the interface to the underlying barrier layer and thus improve slope stability.

The third function is often the most important. In areas that receive sufficient rainfall to soak the protection layer to a significant depth, a drainage layer is usually essential to maintain stability of slopes. If the cover soil becomes saturated and the water table rises to the surface of the cover system, the factor of safety against slope failure reduces to about half the value for a nonsaturated cover.

The materials for use in the drainage layer are either cohesionless soils or drainage geosynthetics. Granular materials used for drainage layer are almost always sand or gravel. If a drainage layer is composed of granular material, minimum thickness of 12 inches (300 mm) and minimum slope of 4% at the bottom of the layer are recommended. Hydraulic conductivity of drainage material should be no less than 1.0×10^{-2} cm/sec. Sometimes the drainage material inherently provides adequate protection against migration of soil particles from the adjacent soil and a separate filter is not required. However, the drainage material does usually not meet filter criteria described in Chapter 8 and a filter (soil or geotextile) is almost invariably required. An adequate filter must be provided and an applicable filter design criteria must be followed. Inadequate filtration and subsequent excessive clogging of drainage soils is one of the most common causes of final cover failure, a problem easily solved by a properly designed and constructed filter.

As an alternative to natural drainage soils, a geotextile/geonet or geocomposite can be placed between the protective layer and the hydraulic barrier as a drainage layer. If composed of geosynthetic materials, the drainage layer material should meet the following specifications (Koerner and Daniel, 1997):

(i) Hydraulic transmissivity of a geonet or other type of drainage core that is based on a site-specific design, but that is never less than 3.0×10^{-5} m^2/sec under anticipated overburden during the design life;
(ii) Inclusion of a geotextile filter layer above drainage material to prevent intrusion and clogging by the overlying protective soil; and
(iii) Inclusion of geotextile cushion layer bedding beneath the drainage layer, if necessary, to increase friction and minimize slippage between the drainage core and the underlying geomembrane, and to prevent intrusion, by deformation, of the geomembrane into the geonet or drainage core of the drainage layer.

The method for designing a drainage layer using either granular materials or geosynthetics in the final cover system is the same as the method for designing a leachate drainage layer discussed in Chapter 8. Controlling the discharge of water from the drainage layer is an important detail. Water must be allowed to discharge freely from the drainage layer at the base of the cover system, through what is often called a "toe drain" (Figure 11.5). If the toe drain excessively clogs, freezes, or is of inadequate capacity, the slope will become unstable. Drainage pipes within the toe must have adequate capacity and freeze protection. Adequate filtration around the toe drain is crucial. The HELP model is generally not acceptable for estimating the maximum flow rate in the drainage layer for a landfill in a closed condition. The reason that the HELP model is not acceptable for this situation is that the program's necessary time step of 24 hours is too long to assess an intense rainfall. A 24-hour averaging time decreases the intensity accordingly and results in an underdesign of the drainage layer and its outlet. Needed for design instead is a site-specific local storm intensity and then incremental calculations based on a water balance procedure (see Koerner and Daniel, 1997).

11.1.4 Hydraulic Barrier Layer

The hydraulic barrier layer can be considered the most critical component of an engineered final cover system. The function of the hydraulic barrier layer is to minimize percolation of water through the cover system by directly blocking water and by indirectly promoting drainage or storage of water in the overlying layers. Water that is

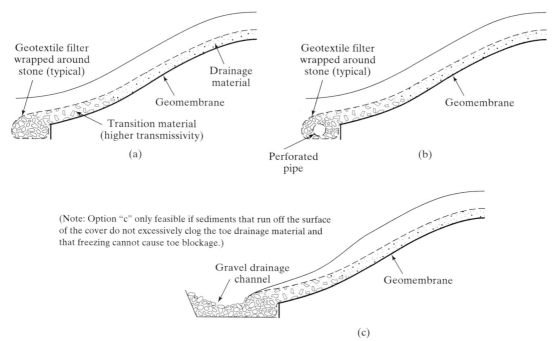

FIGURE 11.5 Various Designs Allowing for Free Drainage at Toe of Final Cover (Soong and Koerner, 1996)

temporarily stored is eventually removed by runoff, evapotranspiration, or internal drainage. Furthermore, the hydraulic barrier layer prevents landfill gases from escaping into the atmosphere. Such gases have been shown to be a major source of air pollution and ozone depletion.

The hydraulic barrier layer for municipal solid waste landfills has consisted typically of a composite layer made up of a geomembrane overlying a compacted clay liner (CCL), the latter having a thickness of 18 inches (450 mm) and a hydraulic conductivity not greater than 1.0×10^{-5} cm/sec. In recent years, geomembrane and/or geosynthetic clay liners (GCLs) have been used to replace compacted clay liners in the final cover system.

Perhaps the most critical issue with respect to the hydraulic barrier layer is maintenance of its integrity and function over time. All of the barrier materials can provide extremely high impedance to downward percolation of water if they are properly installed. Similarly, they can be compromised by desiccation, puncture, or other types of damage. Compacted clay liners are especially vulnerable because they have low resistance to wet-dry cycles (which causes desiccation cracking), freeze-thaw cycles (which increases hydraulic conductivity), and distortion caused by differential settlement (which causes tensile cracks to form). Thin liners, such as geomembranes and geosynthetic clay liners, are vulnerable to accidental puncture, but an occasional puncture is not nearly so damaging in a final cover system as it can be at a more critical location, such as the sump in a bottom liner.

The hydraulic barrier layer is also intended to stop the upward migration of gases, if they are present. Some materials are better barriers to gas movement than others. Most types of geomembranes, for instance, are highly impermeable to gases, while the gas permeability of compacted clay liners is extremely sensitive to the water content and degree of cracking in the compacted clay liner. A saturated, crack-free compacted clay liner makes an excellent barrier to gas migration, but a relative dry or significantly cracked clay liner makes a very poor barrier.

The use of thin liners, such as geomembranes and geosynthetic clay liners, creates a potential problem with interface shear between their surfaces and both underlying and overlying components. The potential for interface shear should be addressed during the design process by determining interface shear parameters for site-specific conditions and project-specific materials followed by use of appropriate methods of slope stability analysis. The use of published values of interface shear without confirmation for the project-specific materials during the design or construction phase is inappropriate because of the inherent variability and changes of both geosynthetic and natural materials. Both site-specific and product- or material-specific conditions must be modeled and assessed accordingly. Chapter 13 discusses these slope stability issues in detail.

Geomembranes

Geomembranes form an essential part of most barrier layers. The most common types of geomembranes being used in final covers are high density polyethylene (HDPE), linear low density polyethylene (LLDPE), chlorosulfonated polyethylene (CSPE), polyvinyl chloride (PVC), and flexible polypropylene (fPP) geomembranes. All of these geomembranes are available with smooth and textured surfaces for increased friction and shear strength when used on steep side slopes. They are also available as single-sided textured geomembranes, the other surface being smooth. Occasionally, scrim reinforced geomembranes (fPP-R and CSPE-R) have been used as well.

The geomembrane material used for the final cover system must be long lasting and must tolerate anticipated subsidence-induced strains. Stress situations such as bridging over subsidence and friction between the geomembrane and other cover components (i.e., compacted soil, geosynthetic drainage material, granular drainage materials, etc.), especially on side slopes, will require special laboratory tests to ensure the design has incorporated site-specific materials. Typical three-dimensional axisymmetric stress-strain curves for various geomembranes are shown in Figure 11.6. The accommodation of different geomembranes to differential settlement is poorest for scrim reinforced chlorosulphonated polyethylene (CSPE-R) and HDPE, and best for VLDPE, LLDPE, and PVC.

Compacted Soil Layer

A compacted soil layer usually with clay material should be at least 18 inches thick and have a hydraulic conductivity no greater than 1.0×10^{-5} cm/sec. The soil layer should be free of detrimental rock, clods and other soil debris and should be located below the maximum frost line. A compacted clay soil layer is constructed in lifts that typically have a thickness after compaction of 6 inches (150 mm). The surface of the compacted soil layer should be smooth so that no small-scale stress points are created for the geomembrane.

In designing the compacted soil layer, various causes of failure (e.g., subsidence, desiccation cracking, and freeze-thaw cycling) must be considered. Some of the settlement will have taken place by the time the cover is put into place, but the majority of subsidence still remains to take place. Although estimating this potential is difficult, information about voids and compressible materials in the underlying waste will aid in calculating subsidence.

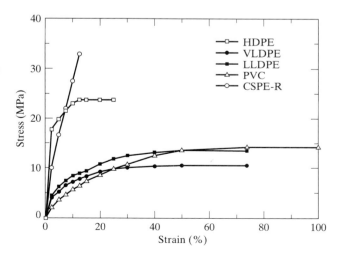

FIGURE 11.6 Three-Dimensional Axisymmetric Stress-versus Strain Response Curves of Various Types of Geomembranes (Koerner and Daniel, 1997). Used with permission of ASCE.

Because freeze-thaw conditions can cause soils to crack, lessen soil density, and lessen soil strength, this entire low hydraulic conductivity soil/geomembrane composite layer should be below the depth of the maximum frost penetration. In northern areas, the maximum depth of the topsoil and protection layer will often exceed the recommended minimum of 30 inches (750 mm).

Compacted soil liners used in final cover systems must be sufficiently ductile to accommodate differential settlement and must be resistant to cracking from moisture variations (e.g., desiccation). Sandy-clay mixtures are acceptable if resistance to shrinkage and desiccation-induced cracking are important (Daniel and Wu, 1993). Ductility is enhanced by avoiding the use of dense, dry soils, which tend to be brittle.

Geosynthetic Clay Liner

A geosynthetic clay liner can replace the compacted clay soil layer because the former has both lower permeability and higher allowable out-of-plane tensile strain. A higher allowable out-of-plane tensile strain can better accommodate differential settlement of the final cover. The allowable tensile strain of compacted clay is less than 1%, while the allowable tensile strain of geosynthetic clay liners varies from 7% to 20% (LaGatta et al., 1996; and Koerner et al., 1996). If a geosynthetic clay liner is used to replace the compacted soil layer, the geosynthetic clay liner should be underlain by approximately 18 inches (450 mm) of earthen material to protect the liner from waste and minimize the effect of settlement. (The properties and behaviors of geosynthetic clay liners (GCLs) are described in Chapter 5.)

Geosynthetic clay liners compare favorably to compacted clay liners and especially compare to compacted soil liners on the basis of advective flow rate calculations that were described in detail in Chapter 5. This comparison shows that, in general, geosynthetic clay liners provide equivalent or superior performance compared with compacted soil layers in final covers. One of the main design questions still outstanding with regard to performance of geosynthetic clay liners is their interface shear strength when they are placed on steep slopes. There are three interface surfaces of concern: upper, internal, and lower. The mobilization of interface friction or shear strength on these surfaces affects sliding resistance and stability of the final cover. In order to address this issue, 14 full-scale test plots were constructed in Cincinnati, Ohio at 3(H)-to-1(V) and 2(H)-to-1(V) slopes. The cross section replicated final cover systems of the type described in Section 11.1 and on Figure 11.1. After five years, no interface problems were encountered on any of the reinforced GCLs. Two slides were encountered at the interface between a covering geomembrane and the woven geotextile on the surface of the GCL (see Daniel et al., 1998). The way to avoid such a problem is to use nonwoven geotextiles on both surfaces of GCLs when placed on steep side slopes.

Penetrations of all of these low hydraulic conductivity layers with gas vents or leachate drainage risers should be kept to a minimum. Where a vent or riser is necessary, there should be a secure, gas- and liquid-tight seal between the vent and the geomembrane. If settlement or subsidence is a major concern, this seal must be designed for flexibility to allow for vertical movement.

11.1.5 Gas Vent Layer

The purpose of a gas vent layer is to vent gases generated from the decomposition of putrescible waste or volatile organics from underlying wastes. The gas vent layer may be constructed of sand, gravel, geonet composite, geotextile, geocomposite, or other gas-transmitting material. Gas is captured in the gas vent layer and flows from it into periodically spaced collection pipes or risers. Flow of gas may occur passively under a natural pressure gradient within the closed landfill or may flow actively in response to a vacuum system located at the surface (as was described in Chapter 10). A gas vent layer is necessary for those solid waste materials that produce gas or volatiles that must be subsequently treated or released in a controlled manner.

The gas vent layer must have a high in-plane transmissivity and must resist clogging by fine-grained materials that are located above or beneath. When using natural soil (such as sand or gravel), the gas vent layer should be a minimum of 12-inches (300-mm)-thick. Geosynthetic materials used for a gas vent layer can be selected from thick needle-punched nonwoven geotextiles, geonet composites, or related drainage geocomposites. Filters may be needed to prevent fine materials from adjacent layers (above and/or below) from clogging the gas vent layer.

11.1.6 Foundation Layer

The lowest layer in the vertical cross section of a multilayer final cover system is a soil foundation layer. The main function of the foundation layer is to minimize differential settlement of the final cover. The foundation layer is often the interim soil cover layer that has already been placed, perhaps with some additional compaction before the final cover is constructed. If there is no interim cover, if the interim cover is too thin, or if differential settlement has created an irregular surface, then additional soil is typically brought in, spread, and compacted to form the foundation layer. As often is the case, when the upper 12 inches (300 mm) of the foundation layer is a granular soil, it can also serve as the gas vent layer.

Irrespective of the site-specific situation, the foundation layer represents the last time that mechanical compaction can take place so as to minimize final cover settlements from occurring. For this reason, the foundation layer is always heavily proof rolled with large compactors. As many load repetitions as practical are used so that stresses are transmitted as deeply as possible into the underlying waste mass. In the extreme, deep dynamic compaction (DDC) can be used to densify the waste thereby minimizing both total and differential settlement. A decision to adopt this procedure depends upon individual, site-specific situations, but may be necessary if a postclosure use of the site is anticipated.

Composite final cover systems have been widely used in municipal solid waste landfills. Various types of final cover systems that are widely used in the design of municipal solid waste landfills are shown in cross sections in Figures 11.7 to 11.12. To prevent the ponding of water on a completed fill surface, the grading contours of the final cover should be sufficient to prevent the development of local depressions due to post-construction settlement. Some states require that slopes of the final cover must not be less than 4% at any location, but no more than 25% (MDEQ, 1999).

410 Chapter 11 Final Cover System

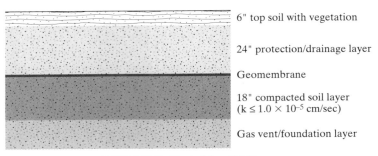

FIGURE 11.7 Typical Landfill Cover

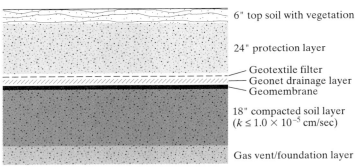

FIGURE 11.8 Landfill Cover with Geotextile and Geonet Drainage Layer

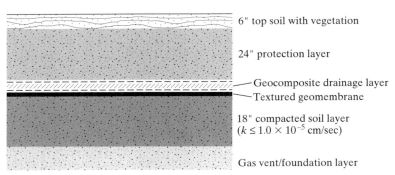

FIGURE 11.9 Landfill Cover with Geocomposite Drainage Layer for Steep Slopes

Section 11.1 Components of Final Cover System

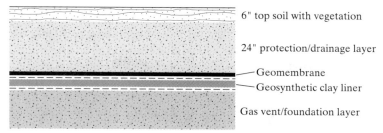

FIGURE 11.10 Landfill Cover with Geosynthetic Clay Liner

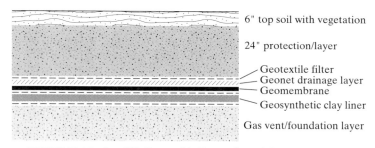

FIGURE 11.11 Landfill Cover with Geotextile and Geonet Drainage Layer and Geosynthetic Clay Liner

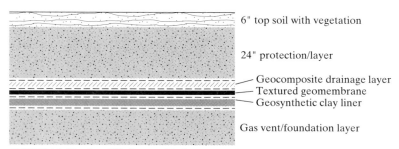

FIGURE 11.12 Landfill Cover with Geocomposite Drainage Layer and Geosynthetic Clay Liner for Steep Slopes

11.2 ALTERNATIVE LANDFILL COVERS

Most landfill covers in the U.S. must meet the minimum regulatory performance standards set forth under the EPA's Resource Conservation and Recovery Act (RCRA) in Subtitle D for municipal landfills (soil cover) and Subtitle C for hazardous waste landfills (compacted clay cover). Although Subtitles D and C describe particular cover designs in detail, landfill design engineers are not required to use them. The regulations allow the governing regulatory agency to consider and approve an alternative final cover as long as it meets general performance standards (Dwyer, 1998).

Most landfills in the U.S. implement the soil cover and compacted clay cover designs of Subtitles D and C, often with little regard for regional conditions. These designs are used even though the EPA design guidelines state that a barrier layer composed of clay under Subtitle C is "not very effective" in arid regions. This claim is based on the observation that the clay soil is compacted with more moisture than is needed to achieve optimum density. When a clay soil dries under arid conditions, it tends to shrink or contract in volume. This shrinkage can result in cracking that greatly increases hydraulic conductivity of the clay mass as was explained in Chapter 3. In this regard, the EPA guidelines state, "This traditional cover not only is inherently problematic, but is also very expensive and difficult to construct. Furthermore, the basic soil cover barrier layer specified in Subtitle D is subject to deterioration due to freeze-thaw cycles as well." Fortunately, in semiarid and arid regions, economical alternative covers constructed solely from earth materials can prove as effective in limiting percolation as the prescriptive cover. Although these covers may not be as effective in limiting gas migration, the volume of gas generated at these sites may be sufficiently low that management of landfill gas is not as problematic. This section describes the principles behind two types of alternative earthen covers: *capillary barriers* and *monolayer barriers*.

11.2.1 Water Balance of Earthen Covers

Alternative earthen covers generally exploit the unique characteristics of unsaturated flow, the storage capacity of fine-grained soils, and the natural capacity of plants to remove water entering the cover during wet periods. As explained by Benson and Khire (1995), these factors are linked by an input/output water balance, which accounts for movement of water into, within, and out of a final cover. The water balance consists of precipitation in the form of rain, snow, or ice, surface runoff, soil moisture storage, evapotranspiration—including evaporation from the surface and transpiration by vegetation—lateral drainage in cover drainage layer, and percolation from the cover base. In algebraic form, the water balance can be described by

$$PR = P - RF - \Delta S - ET - Q \qquad (11.1)$$

where PR = percolation from the cover base;
P = precipitation;
RF = surface runoff;
ΔS = soil moisture storage;
ET = moisture lost through evapotranspiration;
Q = lateral drainage in the cap drainage layer.

Inspection of Equation 11.1 indicates that percolation can be minimized by enhancing surface runoff, soil moisture storage, evapotranspiration, or lateral drainage. Of these four factors, soil moisture storage and lateral drainage are the easiest to optimize during design. For example, soil moisture storage can be increased by selecting surficial soils containing a greater percentage of fines or by increasing the thickness of the cover, whereas lateral drainage can be increased by adding a wicking layer (Yeh et al., 1994) or by employing a layer with anisotropic hydraulic conductivity (Stormont, 1995a). Evaporation can be enhanced by selecting soils having unsaturated hydraulic conductivity (k_ψ) that changes gradually with matric suction (ψ). Fine-grained soils typically have this type of unsaturated hydraulic conductivity function (Hillel, 1980). Transpiration can be enhanced by careful selection of vegetation and manipulating the extent and density of the plant canopy (Rockhold et al., 1995).

Accordingly, design of an alternative earthen cover for a semiarid or arid environment entails manipulating soil water moisture storage capacity, lateral drainage, and evapotranspiration in the cover such that an acceptable value for percolation occurs in a worst-case design year (e.g., a year in which rainfall or snowfall is abnormally high and temperature and solar radiation are abnormally low). Note that the terms *arid* and *semiarid* are quantifiable indexes. Thornthwaite (1948) uses a moisture index defined as annual precipitation minus evapotranspiration in units of inches per year. In doing so, an arid climate is between -60 and -40. A semiarid climate is between -40 and -20. This definition restricts these areas of the United States to the southwest for arid areas, and most of the west (except for the Pacific coast) for semiarid areas. In particular, the cover is designed (Benson and Khire, 1995) such that it has adequate capacity to store or divert water that infiltrates during late fall and winter and sufficient vegetation and evaporative potential to remove the stored water during spring, summer, and early fall. Two types of earthen covers that can be designed on this principle are a *capillary* barrier and *monolayer* barrier. It is important to note that neither uses a geomembrane; thus, these strategies depart from a composite liner and the general concept that the cover must be equally as impermeable as the liner beneath the waste.

11.2.2 Capillary Barrier

The use of a layer of fine-textured soil overlying a layer of coarse-textured soil to form a "capillary barrier" is a fairly recent development. Field studies have suggested that capillary barriers can be used for restricting percolation in semiarid and arid climates (Nyhan et al., 1993; Stormont, 1995; Gee and Ward, 1997; Nyhan et al., 1997). Capillary barriers are constructed in various forms, ranging from a simple design consisting of two layers to more complex designs that include layers of finer-grained and coarser-grained soils (e.g., Stormont, 1995b; Morris and Stormont, 1997; Stormont and Morris, 1997; Nyhan et al., 1997). In its basic form, however, a capillary barrier consists of a fine-grained layer overlying a coarse-grained layer [Figure 11.13(a)] (Benson and Khire, 1995). The contrast in particle size limits downward migration of water by (1) storing water in the upper fine-grained layer until it can be later removed by evaporation and transpiration, or (2) diverting the water laterally in the surface layer

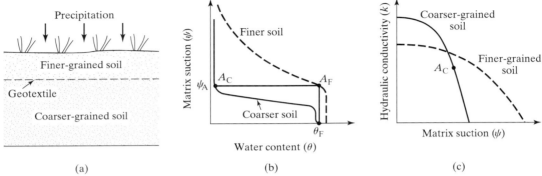

FIGURE 11.13 Capillary Barrier (Khire et al., 1999). Used with permission of ASCE.
(a) Schematic of Capillary Barrier
(b) Unsaturated Hydraulic Conductivity Functions for Finer- and Coarser-Grained Soils
(c) Soil-Water Characteristic Curves for Finer- and Coarser-Grained Soils

(e.g., Stormont, 1995b) (Khire et al., 1999). When the annual precipitation is low (e.g., < 30 cm) and the fine-grained layer is silty or clayey, storage in the fine-grained layer and subsequent evapotranspiration is often considered the primary means of preventing downward movement of water (Meyer et al., 1996).

A capillary barrier limits downward migration of water because the fine-grained surface layer must become nearly saturated before water will enter the coarse-grained soil. Water will enter the coarser soil when the matric suction at the surface of the coarser layer decreases to the value corresponding to the rapid change in slope near residual water content in the soil-water characteristic curve (Stormont and Anderson, 1999). This point is noted as A_C in Figure 11.13(b), and the corresponding matric suction is ψ_A. Since continuity in pore-water pressure requires that the matric suction be equal at the interface between the two layers, the matric suction in the finer layer at the interface must equal ψ_A before water will enter the coarser layer. This water content is noted as θ_F in Figure 11.13(b); θ_F is near saturation and corresponds to point A_F on the soil-water characteristic curve for the finer layer. Even when A_F is reached, water still enters the coarse-grained layer slowly because the saturation and consequently the relative hydraulic conductivity of the coarse-grained layer is still low at A_C [Figure 11.13(c)], and is generally lower than that of the fine-grained layer (Khire et al., 1999).

Thus, if the subsoil remains unsaturated, a fine-textured soil overlying a coarse-textured soil will tend to function as a moisture reservoir that retains nearly all the soil moisture and the underlying layer will behave as a barrier to water percolation due to its dryness. The contact surface between the fine- and coarse-textured soils can be sloped, much like at the interface between a drainage layer and underlying barrier layer. However, in a capillary barrier, lateral movement of water in the fine-textured soil occurs in an unsaturated state. The layer is sometimes called a "wicking layer." The concept has been shown to work in pilot-scale experiments (Nyhan et al., 1990; Fayer et al., 1992). The water in the fine-textured soil must be drained as the slope increases, otherwise it will become saturated and move into the coarse-textured soil. Steep side slopes can be a problem in this regard.

There are two main concerns with a capillary barrier. One is that the fine-textured soil must not be allowed to migrate or drift over time into the underlying coarse-grained soil. A geotextile, used as a separator, must be considered for placement at the interface between the fine-textured soil and the coarse-textured soil. For extremely long service lifetimes, fiberglass geotextiles have been considered for this application. The second concern is associated with periods of extremely high (relatively speaking) precipitation. In such cases, the capillary barrier concept may cease to function, at least temporarily, as the coarse-textured soil becomes moist and loses its water-impeding capability (Koerner and Daniel, 1997). More elaborate designs employing layers having contrasting grain sizes are also possible. These covers employ the capillary barrier principle to divert infiltrating water via lateral flow, while ensuring that deep percolation does not occur (Nyhan et al., 1993; Yeh et al., 1994).

Field studies reported by Benson and Khire (1995) have shown that capillary barriers in their simplest form (two layers) are effective final covers in semiarid and arid regions. In addition, more complex designs having more than two layers have been found to be effective in humid regions (Benson and Khire, 1995). Nyhan et al. (1990) and Khire et al. (1994 a, 1994b) have shown that capillary barriers can be more effective than prescriptive final covers constructed from earthen materials. The capillary barrier covers can be easier to construct and can be less costly than the prescriptive final covers.

Capillary barriers perform better if the saturated hydraulic conductivity of the surface layer is low. Nyhan et al. (1993) reported that percolation from a capillary barrier constructed with a clay-loam surface layer had 11 times less percolation than an identical capillary barrier constructed with a loam surface layer. A capillary barrier constructed with a clay-loam surface also sheds more water as surface runoff (Benson and Khire, 1995).

Capillary barriers perform more effectively if a uniform coarse material is used in the lower soil layer. Stormont and Anderson (1999) conducted a series of laboratory infiltration tests on the capillary barriers composed of (i) silty sand over pea gravel, (ii) concrete sand over pea gravel, and (iii) silty sand over concrete sand. They found that the barrier formed with the concrete sand as the coarse layer permitted breakthrough at a greater suction head than did the barrier with the pea gravel, indicating that the more uniform and coarser the lower soil layer, the more effective the capillary barrier will be.

If the surface layer of a capillary barrier cracks, its effectiveness will be compromised. Accordingly, caution must be exercised when selecting soils for the surface layer. Suitable candidate soils include clayey silts, silty sands, and some sandy clays. Clay-rich soils should be avoided if possible, because clays shrink and crack when dried (Montgomery and Persons, 1990). If clays are used, they should be placed at low water contents to minimize desiccation cracking (Kleppe and Olsen, 1984; Daniel and Wu, 1993). Furthermore, if significant potential exists for the surface layer to shrink and crack on drying, then a three-layer capillary barrier may prove to be a more effective alternative (Yeh et al., 1994).

Barriers constructed with thick layers of fine-grained soil must be protected from intrusion by biota. Burrows or tunnels created by animals (e.g., woodchucks or ground squirrels) or residual holes left by roots of woody plants can become preferential flow paths (Khire et al., 1994b). Thus, a biota barrier should be included in any design where preferential flow through a fine-grained layer will compromise performance of the cover (Benson and Khire, 1995).

11.2.3 Monolayer Barrier

Monolayer barriers are covers that include a thick layer of fine-grained soil with a layer of vegetated topsoil. This cover encourages water storage and enhanced evapotranspiration year-round, rather than just during the growing seasons. The soil allows water storage, which, when combined with the vegetation, will increase evapotranspiration. Monolayer barriers exploit two characteristics of fine-grained soils: (i) their large soil moisture storage capacity when unsaturated, and (ii) their low saturated hydraulic conductivity relative to coarse-grained soils (Morrison-Knudsen, 1993). Their low saturated conductivity limits infiltration through the surface during rainfall or snowmelt. Their high moisture storage capacity makes them capable of storing water that does infiltrate until it can later be removed by evapotranspiration. The barrier must be sufficiently thick, however, such that changes in water content (θ) do not occur near its base (i.e., all changes in soil moisture storage tend to occur in the upper portion of the barrier, as in Figure 11.14). Otherwise, water will percolate into the underlying waste. The necessary thickness is a function of the type of precipitation received, the unsaturated hydraulic properties of the soil, and the rate at which water can be removed by evapotranspiration. Monolayer barriers are constructed from silty sands, silts, and clayey silts. These barriers are cost-effective when large quantities of fine-grained soil requiring little processing is available on site (Benson and Khire, 1995).

Geologic Associates (1993) described a field study conducted to assess the performance of a thick soil barrier used as a final cover for a landfill in Southern California. The barrier was 2-m (6.6-ft)-thick and was constructed from a clayey silt. Water movement was limited to the upper 0.6 m of soil; no changes in water content were observed at the base of the barrier. The data indicated that the water content of the upper soil layers increased rapidly after rainfall and then decreased as water was removed by evapotranspiration.

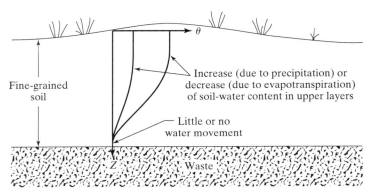

FIGURE 11.14 Monolayer Barrier (Benson and Khire, 1995).
Used with permission of ASCE.

Preferential flow pathways can also develop in monolayer barriers as a result of desiccation cracking and by intrusion of biota. Sufficient data about the performance of monolayer barriers have not been gathered so far from which to judge their reliability and effectiveness in this regard. Field tests including large-scale measurements of percolation are needed before definitive conclusions regarding monolayer barriers can be drawn (Benson and Khire, 1995).

11.3 FIELD STUDY OF LANDFILL COVERS

In response to the need for more effective landfill cover designs, Sandia National Laboratories initiated an alternative landfill cover demonstration project at Kirtland Air Force Base in Albuquerque, New Mexico. The demonstration project (Dwyer, 1998) is a large-scale field test comparing the performance of alternative landfill cover technologies of varying costs and complexities for interim stabilization or final closure of landfills in arid and semiarid environments. Six landfill test covers have been constructed side by side for direct comparison of their performance, cost, and ease of construction. Two of the covers are conventional designs that act as baseline covers, and four are alternative designs.

The two conventional covers are an RCRA Subtitle D soil cover and an RCRA Subtitle C compacted clay cover. The four alternative covers are a geosynthetic clay liner (GCL) cover, a capillary barrier, an anisotropic barrier, and a monolayer barrier cover (or evapotranspiration cover). The test covers are each 13-m wide by 100-m (328-ft) long. The 100-m (328-ft) length is crowned at the middle, with half of the length sloping to the east and the other half to the west. All covers were constructed with a 5% grade in all layers. The eastern slope of each cover was installed with a sprinkler system to facilitate stress testing of the covers. The western slope, on the other hand, is subjected only to ambient conditions. Both slopes are continuously monitored for water balance data (Dwyer, 1998). The purpose of this project is to provide comparative performance and cost data to weigh the merits of alternative, lower-cost landfill covers. A summary of the cross section, percolation, and construction cost for each cover is presented in Table 11.1.

This study has provided some useful information about performance of different cover designs in arid and semiarid climates. Unfortunately, experimental conditions in some of the field covers are not representative of standard cover designs. For example, a purposely designed number of holes were cut in the geomembrane in the simulated RCRA Subtitle C cover and geosynthetic clay liner (GCL), rather than using an intact geomembrane. The cutting of holes in the geomembrane will bias results because, in reality, there would be no leakage at all with an intact geomembrane.

11.4 SOIL EROSION CONTROL

Soil erosion is the removal of surface layers of soil by the agencies of water, wind, and ice. Soil erosion involves a process of both particle detachment and transport by these agencies. Erosion is initiated by drag, impact, or tractive forces acting on individual particles of soil at the surface. Weathering processes, such as frost action and wet-dry

TABLE 11.1 Comparison for Various Landfill Covers

Type of Cover	Subtitle D Cover	Subtitle C Cover	GCL Cover	Capillary Barrier Cover	Anisotropic Barrier Cover	Monolayer Barrier Cover
Cross Section	Topsoil: 6" (150 mm), Barrier soil (10^{-5} cm/s): 18" (450 mm)	Topsoil: 6" (150 mm), Native soil: 18" (450 mm), Geotextile, Sand: 12", (300 mm) GM: 40 mil (1 mm), Clay (10^{-7} cm/s): 24" (600 mm)	Topsoil: 6" (150 mm), Native soil: 18" (450 mm), Geotextile, Sand: 12" (300 mm), GM: 40 mil (1 mm), GCL	Topsoil: 12" (300 mm), Sand: 3" (75 mm), Gravel: 9" (225 mm), Barrier soil: 18" (450 mm), Sand: 12" (300 mm)	Topsoil: 6" (150 mm), Native soil: 24" (600 mm), Fine sand: 6" (150 mm), Pea gravel: 6" (150 mm)	Topsoil: 6" (150 mm), Native soil: 36" (900 mm)
Percolation	1,776 gallons (6,724 liters)	12 gallons (46 liters)	151 gallons (572 liters)	212 gallons (804 liters)	17 gallons (63 liters)	21 gallons (80 liters)
Cost	$4.77/ft² ($51.40/m²)	$14.64/ft² ($157.54/m²)	$8.36/ft² ($89.99/m²)	$8.61/ft² ($92.64/m²)	$6.99/ft² ($75.26/m²)	$6.86/ft² ($73.89/m²)

cycling, set the stage for erosion by breaking up rock into smaller particles and weakening bonds between particles.

11.4.1 Nature of Soil Erosion

Rainfall erosion is the most common type of erosion. Rainfall erosion starts with particle detachment caused by falling raindrops themselves. When these drops impact on bare or fallow ground, they can dislodge and transport soil particles a relatively large distance. At the onset of runoff, water collects into small rivulets which may erode very small channels called *rills*. These rills may eventually combine into larger and deeper channels called *gullies*. Gullying is a complex and destructive process; once started, gullies are difficult to stop. Bare or unprotected earth surfaces are the most vulnerable to all forms of surficial erosion.

Erosion is basically a twofold process that involves: (i) particle detachment and (ii) particle transportation (Gray and Sotir, 1996).

Erosion is caused by:

$$\text{Drag or tractive forces (Water, Wind, Ice)} = f(\text{Velocity, Discharge, Shape and Roughness})$$

Erosion is resisted by:

$$\text{Inertial, frictional, and cohesive forces} = f(\text{Basic soil properties, Soil structure, Physicochemical interaction})$$

Erosion protection essentially consists of (i) decreasing drag or tractive forces by decreasing the velocity of water flowing over the surface or by dissipating the energy of the water in a defended area, and (ii) increasing resistance to erosion by protecting/reinforcing the surface with a suitable cover or by increasing interparticle bond strength.

Rainfall erosion is controlled by four basic factors: climate, soil type, topography, and vegetation. This relationship can be expressed schematically as follows:

$$\textit{Rainfall Erosion} = f(\textit{Climate, Soil, Topography, Vegetation})$$

Climate: storm intensity and duration;
Soil: inherent erodibility;
Topography: length and steepness of slope;
Vegetation: type and extent of cover.

The rainfall erosion factors can be expressed in terms of identifiable and measurable parameters as noted. These parameters in turn can be used to estimate or predict rainfall erosion losses from a site as explained in the next section.

The most important climatic parameters controlling rainfall erosion are intensity and duration of precipitation. Wischmeier and Smith (1958) have shown that the most important single measure of the erosion-producing power of a rainstorm is the product of rainfall energy times the maximum 30-minute rainfall intensity. Raindrops impacting on bare soil not only cause erosion, but they also tend to compact the soil and decrease its infiltration capacity.

The susceptibility of a soil to erosion is known as its *erodibility*. Some soils (e.g., silts) are inherently more erodible than others (e.g., well-graded sand, gravels, and most clays). In general, increasing the organic content and clay size fraction of a soil decreases erodibility. Erodibility also depends upon such parameters as soil texture, antecedent moisture content, void ratio, pH, and composition or ionic strength of the eroding water. The dependence of soil erodibility on all these variables can be summed up as follows (Gray and Leiser, 1982):

Soil erodibility trends:

 (i) Are low in well-graded gravels;
 (ii) Are high in uniform silts and fine sands;
 (iii) Decrease with increasing clay and organic content;
 (iv) Decrease with low void ratios and high antecedent moisture content; and
 (v) Increase with increasing sodium adsorption ratio and decreasing ionic strength of water.

There is not yet a simple and universally accepted erodibility index for soils. Instead, various tests have been proposed for this purpose, including the SCS dispersion test (Volk, 1937), crumb test (Emerson, 1967), and pinhole test (Sherard et al., 1976). A suggested hierarchy of erodibility based on the Unified Soil classification system is

$$\text{Most Erodible} \rightarrow \text{Least Erodible}$$
$$\text{ML} > \text{SM} > \text{SP} > \text{SW} > \text{SC} > \text{MH} > \text{OL}$$
$$\gg$$
$$\text{CL} > \text{CH} > \text{GM} > \text{GP} > \text{GW}$$

where GW = well graded gravel;
       GP = poorly graded gravel;
       GM = silty gravel;
       SW = well graded sand;
       SP = poorly graded sand;
       SM = silty sand;
       SC = clayey sand;
       ML = low plasticity silt;
       MH = high plasticity silt;
       CL = low plasticity clay;
       CH = high plasticity clay;
       OL = low plasticity organic soil.

This erodibility hierarchy is simple, but it is based on gradation and plasticity indices of remolded soils. It fails to take into account the effects of soil structure, void ratio, and antecedent moisture content. Wischmeier et al. (1971) published an erodibility nomograph for use with the Universal Soil Loss Equation that is based on easily measured soil properties.

Topographic variables influencing rainfall erosion are (i) slope angle, (ii) length of slope, and (iii) size and shape of watershed. The influence or importance of length tends to increase as slopes become steeper. For instance, a doubling of slope length from 100 to 200 ft (30 to 60 m) will only increase soil loss by 29% in a 6% slope, whereas the same doubling of slope length in a 20% slope will result in a 49% increase in soil loss. This is one of the reasons for benching or terracing and for contour wattling on long, steep slopes.

Vegetation plays an extremely important role in controlling rainfall erosion. Removal or stripping of vegetation by either human or natural agencies (e.g., wildfires) often results in accelerated erosion.

11.4.2 Soil Loss Prediction

A semiempirical equation known as the Universal Soil Loss Equation (USLE) was developed by the USDA Agricultural Research Service in the early 1960s. Regulatory agencies require the use of the USLE in determining maximum soil loss for a given area of final cover. The USLE takes into account all the factors known to affect rainfall erosion (viz., climate, soil, topography, and vegetation). It is based on a statistical analysis of erosion measured in the field on scores of test plots under natural and simulated rainfall. The annual soil loss from a site is predicted according to the relationship

$$A = R \cdot K \cdot LS \cdot C \cdot P \quad (11.2)$$

where A = computed soil loss (dry weight), U.S. tons/acre/year (note: 1 U.S. ton/acre/year = 2.3 metric tons/hectare/year);
       R = rainfall energy factor;
       K = soil erodibility factor;
       LS = slope-length factor;
       C = cropping management (vegetation) factor;
       P = erosion control practice factor, $P = 1$ in landfill design.

11.4.2.1 Rainfall Energy Factor. The rainfall energy factor, R, is also known as the *rainfall erosion index*. As noted previously, the single most important measure of the erosion producing power of a rainstorm is the product of the rainfall energy times the maximum 30-minute rainfall intensity. The rainfall index for a single storm is thus defined as

$$R = E \cdot I/100 \tag{11.3}$$

where E = total kinetic energy of a given storm, ft-U.S. ton/acre (note: 1 ft-U.S. ton/acre = 0.69 m-metric ton/hectare);
 I = maximum 30-minute rainfall in the area, inch/hour (note: 1 inch/hour = 25.4 mm/hour).

The rainfall energy factor can also be expressed as function of rainfall intensity alone:

$$R = (916 + 331 \cdot \log I) \cdot I/100 \tag{11.4}$$

The records of individual storms are summed over a given time interval to obtain cumulative R-values for other periods of time (e.g., a month or a year). The annual R-factors for approximately 2,000 locations in the United States were summarized in the form of "isoerodent" maps by Wischmeier and Smith (1965). Annual R-factor values vary from a low of approximately 50 in the northern Great Plains to a high 600 in the Gulf Coast region.

Studies by the USDA Soil Conservation Service (1972) have established a relationship between Type II, two-year frequency, six-hour duration rainfall and the average annual rainfall energy factor. This particular duration and frequency storm can be considered a typical "average" storm because it can be expected to occur 50% of the time, and the six-hour duration has been found by the Soil Conservation Service to be the most frequently occurring storm length. The relationship between annual rainfall energy factor and Type II rainfall is shown graphically in Figure 11.15, together with a similar curve for Type I rainfall. Type I and II refer to rainfall characteristics in different regions or zones of the United States, as shown in Figure 11.16. The two-year frequency, six-hour duration rainfall depths for various parts of the United States are also superposed on Figure 11.16.

Thus, both the type and depth for a two-year, six-hour rainfall for any location under study can be obtained from the map shown in Figure 11.16. With this information, the average annual erosion energy factor can be determined from the curves in Figure 11.15. Alternatively, the rainfall for a particular location can be determined from weather records published by the U.S. Weather Bureau (1963). The contours of average annual rainfall energy factor, R, in continental U.S. are shown in Figure 11.17.

11.4.2.2 Soil Erodibility Factor. The soil erodibility factor, K, represents the inherent susceptibility to erosion of the soil; it is governed by textural and gradation properties of the soil, discussed previously. The soil erodibility factor, K, can be estimated on the U.S. Department of Agriculture (USDA) Textural Classification, as shown in Table 11.2. The soil texture classification, depending on percentage of sand, silt, and clay, can be determined using Figure 11.18.

422 Chapter 11 Final Cover System

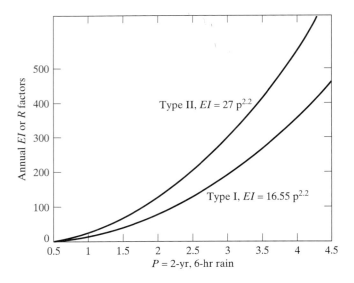

FIGURE 11.15 Relationship between Annual Average Rainfall Energy Factor and the two-year, six-hour Rainfall Depth for Two Rainfall Types

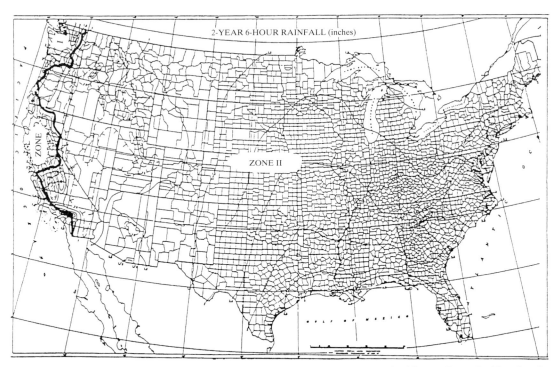

FIGURE 11.16 Depth of the two-year, six-hour Rainfall in Various Parts of the United States, Zones for Type I and Type II rainfall are also shown. (U.S. Weather Bureau, 1963)

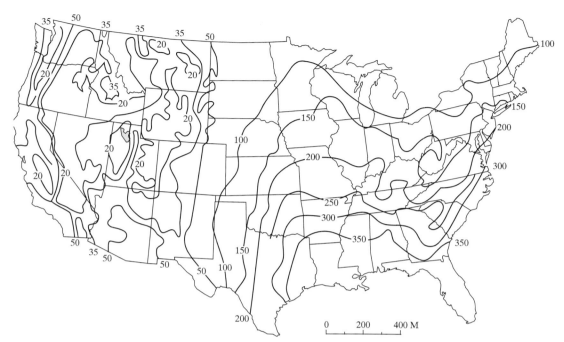

FIGURE 11.17 Contours of Average Rainfall Energy Factor, R, in Continental U.S. (adapted from U.S. Weather Bureau, 1963)

TABLE 11.2 Erodibility K-Values for USDA Texture Classification

Texture Classification	Soil Erodibility Factor, K		
	Organic Matter Content		
	< 0.5%	2%	4%
Sand	0.05	0.03	0.02
Fine Sand	0.16	0.14	0.10
Very Fine Sand	0.42	0.36	0.28
Loamy Sand	0.12	0.10	0.08
Loamy Fine Sand	0.24	0.20	0.16
Loamy Very Fine Sand	0.44	0.38	0.30
Sandy Loam	0.27	0.24	0.19
Fine Sandy Loam	0.35	0.30	0.24
Very Fine Sandy Loam	0.47	0.41	0.33
Loam	0.38	0.34	0.29
Silt Loam	0.48	0.42	0.33
Silt	0.60	0.52	0.42
Sandy Clay Loam	0.27	0.25	0.21
Clay Loam	0.28	0.25	0.21
Silty Clay Loam	0.37	0.32	0.26
Sandy Clay	0.14	0.13	0.12
Silty Clay	0.25	0.23	0.19
Clay		$0.13 \sim 0.29$	

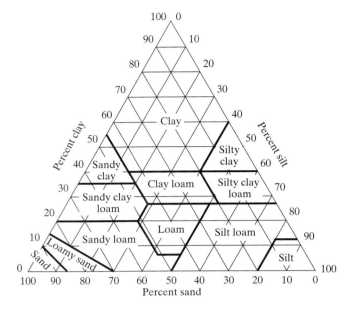

FIGURE 11.18 U.S. Department of Agriculture Textural Classification

Wischmeier et al. (1971) have also published a convenient nomograph (Figure 11.19) that can be used to determine erodibility K-values of soils. The nomograph is valid for exposed subsoil at construction sites as well as farmlands. Only five soil parameters are required:

(i) Percent silt and very fine sand (0.002 ~ 0.10 mm);
(ii) Percent sand (0.10 ~ 2.0 mm);
(iii) Percent organic matter;
(iv) Structure; and
(v) Permeability

The first three parameters will often suffice to provide a reasonable approximation of the erodibility. This approximation can be refined by including information on permeability and soil structure as indicated on the nomograph (Figure 11.19).

11.4.2.3 Slope-Length Factor. The slope-length factor, LS, is the ratio of soil loss per unit area from a given site to that from a unit plot having 9% slope and 72.6-ft (22-m) length. The slope-length factor, LS, can be computed from an empirical equation, which is graphed in Figure 11.20.

The slope-length factor, LS, has been extended by the U.S. Soil Conservation Service (1972) to cover slope length up to 1,600 ft and for slope steepness up to 100% (equivalent to 1H:1V). Figure 11.20 shows extensions of the original chart beyond the 400-ft (120-m) length and 20% slope, the extent of physical data on which the Universal Soil Loss Equation (USLE) was based. These extensions and additions (shown as dashed lines) are extrapolations beyond confirmed data; therefore, they

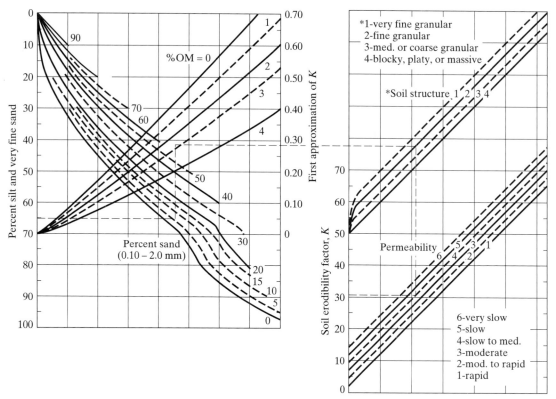

FIGURE 11.19 Soil Erodibility Nomograph for Determining K-Values (Wischmeier et al., 1971)

should be treated as speculative estimates. The slope-length factor LS-values are also tabulated in Table 11.3.

11.4.2.4 Cropping Management Factor. The cropping management factor, C, is defined as the ratio of soil loss from land cropped under specific conditions to the corresponding loss from tilled, continuous fallow (bare) land. In physical terms, it describes the protective effects of vegetation against erosion.

Vegetation or cropping management affects erosion via three separate and distinct, but interrelated zones of influence (viz., canopy cover, vegetative cover in direct contact with the soil, and crop residue at or beneath the surface). The effects of these three constituent influences can be defined separately, but for practical purposes, they are represented by a single value of the C-factor.

For completely bare or fallow ground, the C-factor is unity (1.0). Some cropping management factor C-values are tabulated in Table 11.4. Information in Table 11.4 reveals the benefit of vegetation or plant cover for reducing erosion. Factor C-values range as low as 0.003 for well-established plant cover. This corresponds to almost a

426　Chapter 11　Final Cover System

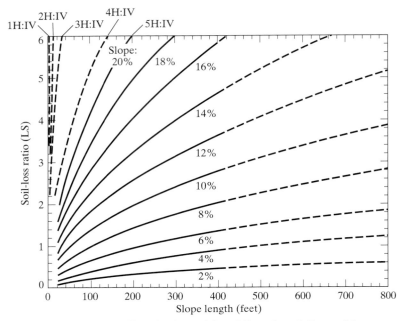

FIGURE 11.20　Chart for determining of Slope-Length Factor, LS

TABLE 11.3　Slope-Length Factor, LS

Slope Length feet (m)	LS Value (% Slope)													
	4	6	8	10	12	14	16	18	20	25	30	35	40	45	50
50 (15)	0.3	0.5	0.7	1.0	1.3	1.6	2.0	2.4	3.0	4.3	6.0	7.9	10.1	12.6	15.4
100 (30)	0.4	0.7	1.0	1.4	1.8	2.3	2.8	3.4	4.2	6.1	8.5	11.2	14.4	17.9	21.7
150 (45)	0.5	0.8	1.2	1.6	2.2	2.8	3.5	4.2	5.1	7.5	10.4	13.8	17.6	21.9	26.6
200 (60)	0.6	0.9	1.4	1.9	2.6	3.3	4.1	4.8	5.9	8.7	12.0	15.9	20.3	25.2	30.7
250 (75)	0.7	1.0	1.6	2.2	2.9	3.7	4.5	5.4	6.6	9.7	13.4	17.8	22.7	28.2	34.4
300 (90)	0.7	1.2	1.7	2.4	3.1	4.0	5.0	5.9	7.2	10.7	14.7	19.5	24.9	30.9	37.6
350 (105)	0.8	1.2	1.8	2.6	3.4	4.3	5.4	6.4	7.8	11.5	15.9	21.0	26.9	33.4	40.6
400 (120)	0.8	1.3	2.0	2.7	3.6	4.6	5.7	6.8	8.3	12.3	17.0	22.5	28.7	35.7	43.5
450 (135)	0.9	1.4	2.1	2.9	3.8	4.9	6.1	7.2	8.9	13.1	18.0	23.8	30.5	37.9	46.1
500 (150)	0.9	1.5	2.2	3.1	4.0	5.2	6.4	7.6	9.3	13.7	19.0	25.1	32.1	39.9	48.6
550 (165)	1.0	1.6	2.3	3.2	4.2	5.4	6.7	8.0	9.8	14.4	19.9	26.4	33.7	41.9	50.9
600 (180)	1.0	1.6	2.4	3.3	4.4	5.7	7.0	8.3	10.2	15.1	20.8	27.5	35.2	43.7	53.2
650 (195)	1.1	1.7	2.5	3.5	4.6	5.9	7.3	8.7	10.6	15.7	21.7	28.7	36.6	45.5	55.4
700 (210)	1.1	1.8	2.6	3.6	4.8	6.1	7.6	9.0	11.1	16.3	22.5	29.7	38.0	47.2	57.5
750 (225)	1.1	1.8	2.7	3.7	4.9	6.3	7.9	9.3	11.4	16.8	23.3	30.8	39.3	48.9	59.5
800 (240)	1.2	1.9	2.8	3.8	5.1	6.5	8.1	9.6	11.8	17.4	24.1	31.8	40.6	50.5	61.4
900 (270)	1.2	2.0	3.0	4.1	5.4	6.9	8.6	10.2| 12.5	18.5	25.5	33.7	43.1	53.5	65.2
1,000 (300)	1.3	2.1	3.1	4.3	5.7	7.3	9.1	10.8| 13.2	19.5	26.9	35.5	45.4	56.4	68.7

TABLE 11.4 Cropping Management Factor, C (adapted from USDA Soil Conservation Service, 1978)

Ground Cover	Cropping Management Factor, C
95% ~ 100%	
as Grasses	0.003
as Weeds	0.010
80%	
as Grasses	0.010
as Weeds	0.040
60%	
as Grasses	0.040
as Weeds	0.090
High Productivity	
Grass and Legume Mix	0.004
Moderate Productivity	
Grass and Legume Mix	0.010

thousandfold reduction in erosion losses, compared with the continuous-fallow or bare-ground cases. Mulching also can be considered as a form of cropping management. The cropping management factor C-values for mulching with various types of organic mulches (straw, hay, woodchips, etc.) are listed in Table 11.5.

11.4.2.5 Erosion Control Practice Factor. The erosion control practice factor, P, is a parameter representing the reduction of soil loss resulting from soil conservation measures, such as contour tillage, contour strip cropping, terracing, and stabilized waterways. Factor P-values for standard erosion control practices are tabulated in Table 11.6. Values of P range from 0.95 for contouring on steep slopes (18 to 24%) to 0.25 for contour strip cropping on gentle slopes. Terracing effectively reduces the length of slope from that of the entire site to the horizontal distance between terraces. The methods of determining P for a given conservation practice and, alternatively, the

TABLE 11.5 Cropping Management Factor, C, for Mulching Sites (adapted from USDA Soil Conservation Service, 1978)

Type of Cover	Cropping Management Factor, C
Hay Rate of Application:	
0.5 U.S. ton/acre (1.1 ton/hectare)	0.25
1.0 U.S. ton/acre (2.3 ton/hectare)	0.13
1.5 U.S. ton/acre (3.4 ton/hectare)	0.07
2.0 U.S. ton/acre (4.5 ton/hectare)	0.02
Small Grain Straw: 2 U.S. ton/acre (4.5 ton/hectare)	0.02
Wood Chips: 6 U.S. ton/acre (13.6 ton/hectare)	0.06
Wood Cellulose: 1.75 U.S. ton/acre (4.0 ton/hectare)	0.10
Fiberglass: 1.5 U.S. ton/acre (3.4 ton/hectare)	0.05

428 Chapter 11 Final Cover System

TABLE 11.6 Factor *P*-Values for Standard Erosion Control Practices (from USDA Soil Conservation Service, 1978)

Slope (%)	Up and Down Hill	Cross Slope Farming without Strips	Contour Tillage	Cross Slope Farming with Strips	Contour Stripcropping
2.0 ~ 7.0	1.0	0.75	0.50	0.37	0.25
7.1 ~ 12.0	1.0	0.80	0.60	0.45	0.30
12.1 ~ 18.0	1.0	0.90	0.80	0.60	0.40
18.1 ~ 24.0	1.0	0.95	0.90	0.67	0.45

selection of a conservation practice, using the Universal Soil Loss Equation, have been described by Wischmeier and Smith (1965). A factor *P*-value of 1.0 should be used for a landfill with a smooth surface cover.

A number of erosion control practices, such as structural, mechanical, chemical, vegetative, or combinations thereof, can be used to reduce soil losses from disturbed slopes or construction sites. Because the allowable soil loss given by the U.S. and many state regulations must not exceed 2 U.S. tons per acre per year (4.5 metric tons per hectare per year) at the landfill area, meeting this level of erosion control requires careful design of slope gradient and length plus other mitigating measures.

Water-related erosion can be controlled not only by vegetation, but also by hardened covers using stones or riprap. Such hardened covers allow more water to infiltrate than vegetative covers because no vegetative evapotranspiration occurs. Hardened covers increase the need for a barrier layer, but they reduce long-term maintenance.

EXAMPLE 11.1

A landfill is located in Southeastern Michigan. The landfill final cover has slope of 5(H):1(V) and length of 250 feet (75 m). Topsoil is loam with 4% organic matter content. Ground cover as grasses is 80%. Use Universal Soil Loss Equation to calculate the annual soil loss.

Solution:

Southeastern Michigan (Figure 11.17): $R = 100$;
Soil erodibility factor for loam with 4% organic matter content (Table 11.2): $K = 0.29$;
Slope-length factor for $S = 20\%$ and $L = 250$ ft (75 m) (Table 11.3): $LS = 6.6$;
Cropping management factor (Table 11.4): $C = 0.010$; and
Erosion control practice factor: $P = 1$ (for landfill design).

Universal Soil Loss Equation

$$A = R \cdot K \cdot LS \cdot C \cdot P \qquad (11.2)$$
$$= 100 \times 0.29 \times 6.6 \times 0.010 \times 1$$
$$= 1.91 \text{ U.S. ton/acre/yr (4.33 metric ton/ha/year)} < 2 \text{ ton/acre/yr (4.5 metric ton/ha/yr)}$$

11.4.3 Limitations of Universal Soil Loss Equation

The Universal Soil Loss Equation (USLE) is a semiempirically-based equation that predicts sheet and rill erosion from relatively small areas. Although it explicity takes into account all the factors known to affect rainfall erosion and is widely employed around the world, it has limitations (Gray and Sotir, 1996):

(i) *It is semiempirical.* Considerable judgment is required in assigning correct values to some of the factors in certain situations. This is particularly true in the case of the vegetation or cover factor.

(ii) *It predicts average annual soil loss.* The rainfall factor is based on the two-year, six-hour rainfall index. Unusual storms or weather events during a particular year could produce more sediment than predicted.

(iii) *It does not predict gully erosion.* The USLE predicts soil loss from sheet and rill erosion, not erosion resulting from concentrated flow in large channels or gullies.

(iv) *It does not predict sediment delivery.* The USLE predicts soil loss, not sediment deposition. Soil loss from the upper portions of a slope or watershed will not automatically end up as sediment in a body of water. Instead, it may collect on the lower portion of a slope or watershed.

(v) *The influence of erosion control products on soil loss predictions is still the subject of on going research.* These products will be described in Section 11.4.5.

In spite of its limitations, the USLE provides a simple, straightforward method of estimating soil loss, identifying critical areas, and evaluating the effectiveness of soil loss reduction measures. Significantly, the USLE provides an idea of the range of variability of each of the parameters, their relative importance in affecting erosion, and the extent to which each can be changed or managed to limit soil losses.

11.4.4 Erosion Control Principles

Soil erosion can be controlled or prevented by observing some basic principles. These principles are universally applicable; they should be observed regardless of whether conventional or other soil erosion treatments are contemplated. Erosion control principles are based on common sense, but are frequently violated in site development work. Many erosion control measures and products have been introduced over the years. They are most effective when applied in conjunction with the following principles (Gray and Sotir, 1996):

(i) Fit the development plan to the site. Avoid extensive grading and earthwork in erosion-prone areas.
(ii) Install hydraulic conveyance facilities to handle increased runoff.
(iii) Keep runoff velocities low.
(iv) Divert runoff away from steep slopes and denuded areas by constructing interceptor drains and berms.
(v) Save native site vegetation whenever possible.

(vi) If vegetation must be removed, clear the site in small, workable increments. Limit the duration of exposure.

(vii) Protect cleared areas with mulches and temporary, fast-growing herbaceous covers.

(viii) Construct sediment basins to prevent eroded soil or sediment from leaving the site.

(ix) Install erosion-control measures as early as possible.

(x) Inspect and maintain control measures.

Observance of these 10 basic principles will greatly minimize erosion losses. Some additional discussion is warranted here on the impact of grading practices on soil erosion. A linear or planar slope will exhibit higher soil loss than a slope with a concave or decreasing gradient near the toe. Likewise, drainage channels brought down and across a slope in a curvilinear manner that lengthens the flow path and reduces the gradient are less susceptible to erosion than exposed channels brought directly down the face of a slope.

11.4.5 Manufactured Erosion Control Materials

Over the past 10 to 15 years, a wide range and variety of manufactured erosion control products have been developed. These products, which often are referred to as *rolled erosion control products* (RECPs), are generally manufactured in rolls and laid on the prepared soil to be protected. Edges and ends of the rolls are overlapped and staked to the soil using long nails or staples ("U" shaped pins). A classification of the materials by categories is given in Table 11.7.

TABLE 11.7 Classification of Erosion Control Materials and Products (modified from Theisen, 1992)

TERMs	PERMs	
	Soft-armor	Hard-armor
Straw, hay, and hydraulic mulches	UV-stabilized fiber roving systems (FRs)	Aggregate filled, geocellular containment systems (GCSs)
Tackifiers and soil stabilizers	Erosion control revegetation matrices (ECRMs)	Fabric formed revetments (FFRs)
Hydraulic mulch geofibers	Turf reinforcement mats (TRMs)	Articulated, vegetated, concrete block systems
Degradable, erosion control meshes and nets (ECMNs)	Discrete length geofibers	Articulated concrete block systems (ACBs)
Degradable, erosion control blankets (ECBs)	Earth filled, vegetated, geo-cellular containment systems (GCSs)	Vegetated, stone riprap
UV-degradable, fiber roving systems (FRs)		Vegetated, gabion mattresses

The preceding table is generalized because there are many individual products in each of the specification areas. In general, the materials and products categorized in Table 11.7 have been successful in preventing or greatly reducing erosion losses on slopes. Considerable research activity is focused at present on adapting the USLE's variables to the various erosion-control materials and products. It is important to note that the performance of rolled erosion control products (RECPs) is limited by the effectiveness of their attachment to the ground surface. Insufficient use of point attachments (such as pins, pegs, and staples) can allow rills and small gullies to form at the ground interface between attachment points. Burying a blanket or mat at periodic intervals in shallow trenches dug on contours across a sloping surface can minimize this problem. [See Gray and Sotir (1996) for further discussion of this problem and for a description of various biotechnical groundcover alternatives for erosion control.]

11.5 EFFECTS OF SETTLEMENT AND SUBSIDENCE

Two types of settlement are of concern with respect to landfill covers: total settlement and differential settlement. Total settlement of the surface of a cover is the total downward movement of a fixed point on the surface. Total settlement of municipal solid waste can be enormous. As seen in Figure 11.21, total settlement of 10 to 15% of the thickness is generally to be expected, and 20 to 30% also is possible. The contours of the final cover must take such anticipated settlement into account.

Differential settlement is even more insidious and problematic. Differential settlement is always measured between two points and is defined as the difference between the total settlements at these two points; that is,

$$\Delta Z_{i,i+1} = Z_{i+1} - Z_i \tag{11.5}$$

where $\Delta Z_{i,i+1}$ = differential settlement between points i and $i + 1$,
  Z_i = total settlement of point i.
  Z_{i+1} = total settlement of point $i + 1$.

Distortion is defined as the differential settlement between two points divided by the distance along the ground surface between the two points, or

$$\psi_{i,i+1} = \frac{\Delta Z_{i,i+1}}{L_{i,i+1}} \tag{11.6}$$

where $\psi_{i,i+1}$ = distortion between points i and $i+1$,
  $\Delta Z_{i,i+1}$ = differential settlement between points i and $i+1$,
  $L_{i,i+1}$ = distance between points i and $i+1$.

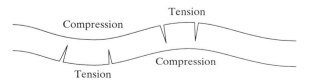

FIGURE 11.21 Tension Cracking of Landfill Cover due to Differential Settlement

Excessive differential settlement of underlying waste can damage a cover system. If differential settlement occurs, tensile strains develop in cover materials as a result of bending stresses and/or elongation. Tensile strain is defined as the amount of stretching of an element divided by the original length of the element, written as

$$\varepsilon_{i,i+1} = \frac{(L_{i,i+1})_{Fnl} - (L_{i,i+1})_{Int}}{(L_{i,i+1})_{Int}} \times 100\% \qquad (11.7)$$

where
$\varepsilon_{i,i+1}$ = tensile strain between points i and $i+1$,
$(L_{i,i+1})_{Int}$ = distance between points i and $i+1$ in their initial positions,
$(L_{i,i+1})_{Fnl}$ = distance between points i and $i+1$ in their post-settlement positions.

When the cover settles differentially, some part of it will be subjected to tension and will undergo tensile strain. Tensile strains are of concern because the larger the stretching (tensile strain), the greater the possibility that the soil will crack and that a geomembrane or geosynthetic clay liner will rupture.

Bending stresses—stresses that occur when an object is bent—result when covers settle differentially; part of the bent cover is in tension and part is in compression. Bending stresses are of concern because the tensile stresses associated with bending may be large enough to cause the soil to crack (Figure 11.21). Geomembranes can withstand far larger tensile strains without failing than soils (recall Figure 11.6). Geomembranes have the ability to elongate (stretch) a great deal without rupturing or breaking. On the contrary, compacted clay is very weak in tension; it cracks at tensile strain of less than 1%. Geosynthetic clay liners are intermediate between these two extremes.

Gilbert and Murphy (1987) discuss the prediction and mitigation of subsidence damage to the landfill covers. Gilbert and Murphy developed a relationship between tensile strain in a cover and distortion. This relationship is shown in Figure 11.22. As the distortion increases, the tensile strain in the cover soils increases.

Minor cracking to topsoil or drainage layers as a result of tensile stresses is of little concern. However, cracking of a hydraulic barrier, such as a layer of low hydraulic con-

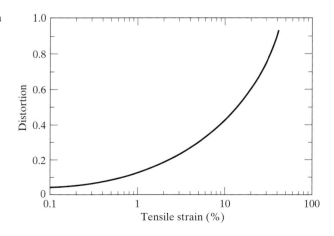

FIGURE 11.22 Relationship between Distortion and Tensile Strain in a Cover (Gilbert and Murphy, 1987)

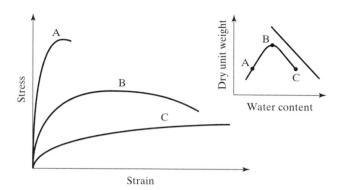

FIGURE 11.23 Relationship between Shearing Characteristics of Compacted Soils and Conditions of Compaction (USEPA, 1991)

ductivity soil, is of great concern because the hydraulic integrity of the barrier layer is compromised if it is cracked. The amount of strain that a low hydraulic conductivity, compacted soil can withstand prior to cracking depends significantly upon the water content of the soil. As shown in Figure 11.23, soils compacted wet of optimum are more ductile than soils compacted dry of optimum. For cover systems, ductile soils that can withstand significant strain without cracking are preferred. For this reason, as well as the hydraulic conductivity consideration, it is preferable to compact low hydraulic conductivity soil layers wet of optimum. The soil must then be kept safe from drying out and cracking. One way of accomplishing this is to cover the clay with a geomembrane that acts as a vapor trap in addition to its own intrinsic barrier capabilities.

Gilbert and Murphy (1987) summarize information concerning tensile strain at failure for compacted, clayey soils. The available data show that such soils can withstand maximum tensile strains of 0.1 to 1%. If the lower limit (0.1%) is used for design, the maximum allowable value of distortion is approximately 0.05%.

EXAMPLE 11.2

A circular depression with a radius (R) of 10 feet (3 m) develops in a landfill cover with a compacted clay liner as the barrier. The maximum allowable distortion (ψ) is 0.05%. What is the maximum allowable settlement at the center of the depression to avoid cracking from excessive tensile strain?

Solution: The maximum allowable differential settlement,

$$\Delta Z = \psi \cdot R$$
$$= 0.05 \times 10$$
$$= 0.5 \text{ ft} = 6 \text{ inches (150 mm)}$$

Therefore, settlement at center of depression $\leq$ 6 inches (150 mm).

Some wastes (such as loose municipal solid waste or unconsolidated sludge of varying thickness) are so compressible that constructing a cover system above the waste will almost certainly produce distortions that are far larger than 0.05. The hydraulic integrity of a low hydraulic conductivity layer of compacted soil is likely to be seriously damaged by the distortion caused by large differential settlement. If the waste is continuing to settle (e.g., as a result of decomposition), it may be prudent to place a temporary cover on the waste and wait for settlement to take place prior to constructing the final cover system. Alternatives for stabilizing the waste include deep dynamic compaction, soil preloading, and the use of wick drains to consolidate sludges. A case history of differential subsidence of a municipal solid waste landfill cover will serve to illustrate the magnitude that tensile strains can reach.

11.6 DIFFERENTIAL SUBSIDENCE CASE HISTORY

Shown in Figure 11.24 is the plan view of a 65-acre (26-ha) municipal solid waste landfill. It was progressively filled from 1969 through 1978. It was abandoned at that time and the owners filed for bankruptcy. No activity was performed at the site for approximately 10 years. Funds became available in 1988, and by 1990, the landfill was proof-rolled and an 18-inch (450-mm) compacted clay liner with 18 inches (450-mm) of protective soil, and 6 inches (150 mm) of topsoil completed the cross section.

In 1997, a survey was taken which resulted in a large "horseshoe" pattern of subsidence and (more importantly) differential subsidence in many areas. The photographs of Figure 11.25 illustrate typical subsidence areas.

Measurements at the site were taken of the seven main location areas that resulted in the information provided in Figure 11.26. The resulting tensile strains vary from 1.8% to 27.4%. Clearly, all of the values far exceed the capability of compacted

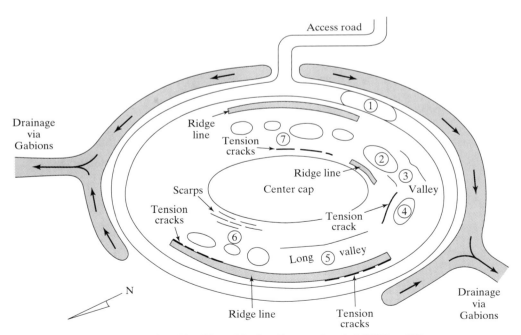

FIGURE 11.24 Plan View of Surface Topography of a MSW Landfill

FIGURE 11.25 Differential Subsidence "Craters" in MSW Landfill Covers

clay liners (recall Example 11.2), and the larger values exceed the tensile strain capability of geosynthetic clay liners and some geomembranes. As seen in Figure 11.6, only VLDPE, LLDPE and PVC geomembranes can accommodate such high values of out-of-plane tensile strain. Evidence that the compacted clay liner at this site had failed was clear, due to the strong methane odor over the entire site.

In closing, it should be mentioned that this degree of differential subsidence is probably an extreme case. There was no compaction of the waste during its placement (i.e., it was a classic "dump"). However, differential subsidence can be an issue and the designer must carefully weigh the site-specific situation in light of the liner system (and related components) of the final cover.

PROBLEMS

11.1 What is the main purpose for constructing a final cover system atop a landfill?

11.2 Describe the basic requirements for an RCRA Subtitle D final cover?

11.3 What components or layers comprise a final cover system? Explain the function of each component.

11.4 Discuss the reason and basis for designing an alternative landfill cover.

11.5 Describe the "water-holding" function for a capillary barrier type cover.

11.6 Describe the function of minimizing percolation for a monolayer barrier type cover.

11.7 Explain what soil erosion is and how it differs from mass movement. What causes soil erosion? What forms does it take?

11.8 What basic approaches can be employed to prevent or resist soil erosion?

11.9 List the basic factors affecting rainfall erosion and explain how each factor affects soil losses.

11.10 How can soils be selected or modified to minimize erodibility? What is the most erodible and least erodible type of soil based on Unified Soil Classification?

11.11 A landfill is located in Detroit, Michigan. The landfill final cover has a slope of 4(H):1(V) and length of 450 feet (135 m). Topsoil is fine, sandy loam with 4% organic matter content. The top surface cover is grass. The grass cover is 80% at the end of the first year and increases to 95% the second year. Use the Universal Soil Loss Equation

436 Chapter 11 Final Cover System

Location	Description	Approximate dimensions (ft)	Max. strain (%)
1	Road subsidence	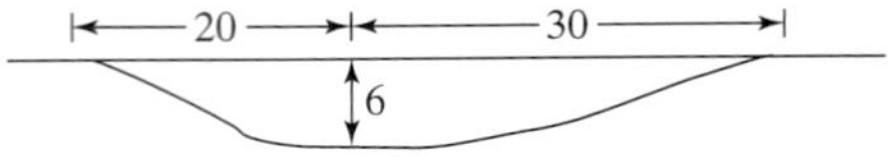 20 / 30 / 6	5.9
2	Major crater	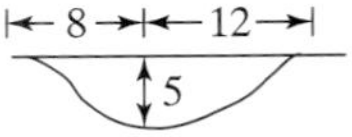 8 / 12 / 5	24.3
3	100-ft long vally	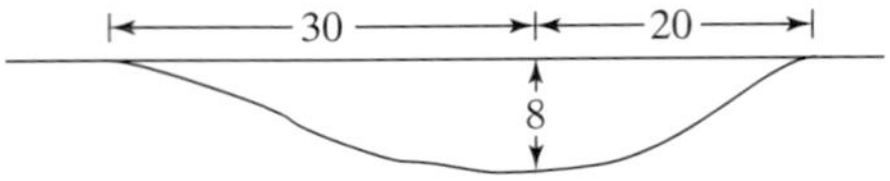 30 / 20 / 8	10.4
4	Large crater	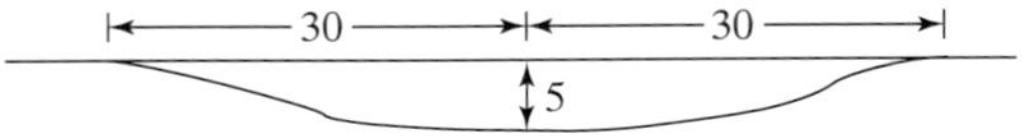 30 / 30 / 5	1.8
5	350-ft long vally	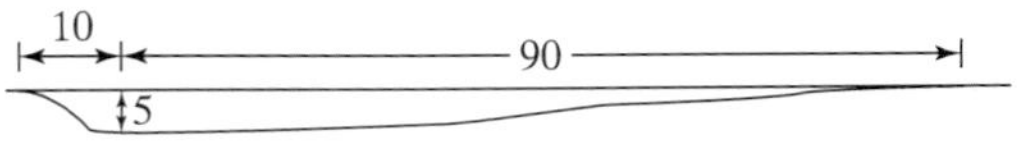 10 / 90 / 5	15.9
6	Three craters	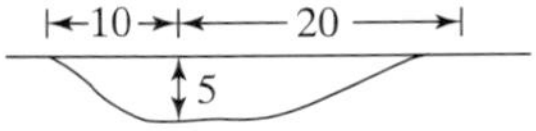 10 / 20 / 5	15.9
6	Three craters	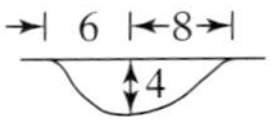 6 / 8 / 4	27.4
6	Three craters	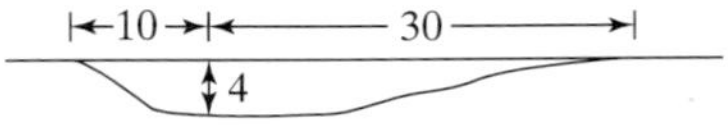 10 / 30 / 4	10.4
7	Four craters	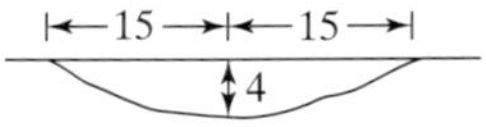 15 / 15 / 4	4.7
7	Four craters	 5 / 5 / 3	22.5
7	Four craters	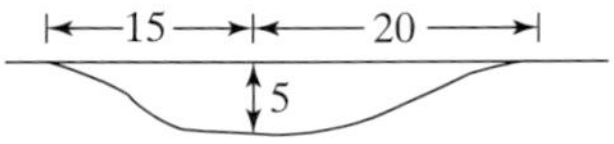 15 / 20 / 5	7.3
7	Four craters	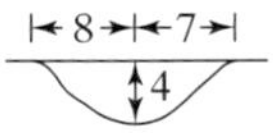 8 / 7 / 4	15.9

FIGURE 11.26 Various Differential Subsidence Patterns at MSW Landfill

(USLE) to calculate the annual soil loss. If the calculated annual soil loss exceeds 2 U.S. tons/acre/year (4.5 tons/hectare/year), suggest an effective way to reduce the annual soil loss to less than 2 U.S. tons/acre/year (4.5 tons/hectare/year).

11.12 Explain the limitations of the Universal Soil Loss Equation.

11.13 If the estimated soil loss for a landfill cover cannot meet the rule requirement, what measures can be used to control soil erosion and reduce soil losses?

REFERENCES

Benson, C. H. and Khire, M. V., (1995) "Earthen Covers for Semiarid and Arid Climates," *Landfill Closures—Environmental Protection and Land Recovery,* ASCE, Geotechnical Special Publication, No. 53, R. Jeffrey Dunn and Udai P. Singh, Eds., New York, NY, pp. 201–217.

Daniel, D. E., (1995) "Soil Barrier Layers versus Geosynthetic Barrier in Landfill Cover Systems," *Landfill Closures—Environmental Protection and Land Recovery,* ASCE, Geotechnical Special Publication, No. 53, R. Jeffrey Dunn and Udai P. Singh, Eds., New York, NY, pp. 1–18.

Daniel, D. E., Koerner, R. M., Bonaparte, R., Landreth, R. E., Carson, D. A., and Scranton, H. B., (1998) "Slope Stability of Geosynthetic Clay Liner Test Plots," *Journal of Geotechnical and Geoenvironmental Engineering,* ASCE, Vol. 124, No. 7, pp. 628–637.

Daniel, D. E. and Wu, Y. -K., (1993) "Compacted Clay Liners and Covers for Arid Sites," *Journal of Geotechnical Engineering,* ASCE, Volume 119, No. 2, pp. 223–237.

Dwyer, S. F., (1995) "Alternative Landfill Cover Demonstration," *Landfill Closures—Environmental Protection and Land Recovery,* ASCE, Geotechnical Special Publication, No. 53, R. Jeffrey Dunn and Udai P. Singh, Eds., New York, NY, pp. 19–34.

Dwyer, S. F., (1998) "Alternative Landfill Covers Pass the Test," *Civil Engineering* Magazine, ASCE, September, pp. 50–52.

Emerson, W. W., (1967) "A Classification of Soil Aggregates Based on their Coherence in Water," *Australian J. Soil Res.,* 2: pp. 211–217.

Fayer, M., Rockhold M., and Campbell, M., (1992) "Hydrologic Modeling of Protective Barriers: Comparison of Field Data and Simulation Results," *Soil Science Society of America Journal,* Vol. 56, pp. 690–700.

Gee, G. and Ward, A., (1997) "Still in Quest of the Perfect Cap," *Proceedings of Landfill Capping in the Semiarid West: Problems, Perspectives, and Solutions,* Environmental Science and Research Foundation, Idaho Falls, ID, pp. 145–164.

GeoLogic Associates (1993) "Evaluation of Unsaturated Fluid Flow, Coastal Sage Scrub Habitat Area, Coyote Canyon Final Cover System, Orange County, CA," Report prepared by GeoLogic Associates for Orange County Integrated Waste Management Department.

Gilbert, P. A. and Murphy, W. L., (1987) "Prediction/Mitigation of Subsidence Damage to Hazardous Waste Landfill Covers," U. S. Environmental Protection Agency, EPA/600/2-87/025 (PB87-175386), Cincinnati, OH.

Gray, D. H. and Leiser, A. T., (1982) "Biotechnical Slope Protection and Erosion Control," Van Nostrand-Reinhold, NY.

Gray, D. H. and Sotir, R. B., (1996) "Biotechnical and Soil Bioengineering Slope Stabilization," John Wiley & Sons, Inc., 605 Third Avenue, New York, NY 10158-0012.

Hillel, D., (1980) "Fundamentals of Soil Physics," Academic Press, Inc.

Khire, M. V., Benson, C. H., Bosscher, P. J., and Pliska, R., (1994a) "Field-Scale Comparison of Capillary and Resistive Landfill Covers in a Arid Climate," *Fourteenth Annual Hydrology Days Conference,* Fort Collins, CO, pp. 195–209.

Khire, M. V., Benson, C. H., and Bosscher, P. J., (1994b) "Final Cover Hydrologic Evaluation—Phase III," Environmental Geotechnics Report 94-4, Department of Civil and Environmental Engineering, University of Wisconsin, Madison, WI.

Khire, M. V., Benson, C. H., and Bosscher, P. J., (1999) "Field Data from a Capillary Barrier and Model Predictions with UNSAT-H," *Journal of Geotechnical and Geoenvironmental Engineering,* ASCE, Vol. 125, No. 6, pp. 518–527.

Koerner, R. M. and Daniel, D. E., (1997) "Final Covers for Solid Waste Landfills and Abandoned Dumps," ASCE Press, Reston, VA, and Thomas Telford, London, UK.

Koerner, J. R. and Koerner, R. M., (1999) "A Survey of Solid Waste Landfill Liners and Cover Regulations: Part II—Worldwide Status," GRI Report #23, Folsom, PA, 177 pp.

Kleppe, J. H. and Olson, R. E., (1984) "Desiccation Cracking of Soil Barriers," Hydraulic Barriers in Soil and Rock, ASTM 874, pp. 263–275.

LaGatta, M. D., (1992) "Hydraulic Tests on Geosynthetic Clay Liners Subjected to Differential Settlement," MSCE Thesis, University of Texas, Austin, TX.

MDEQ, (1999) "Solid Waste Management Act Administrative Rules Promulgated Pursuant to Part 115 of the Natural Resources and Environmental Protection Act, 1994 PA 451, as amended (Effective April 12, 1999)," Michigan Department of Environmental Quality, Waste Management Division, Lansing, MI.

Meyer, P., Rockhold, M., Nichols, W., and Gee, G., (1996) "Hydrologic Evaluation Methodology for Estimating Water Movement through the Unsaturated Zone at Commercial Low-Level Radioactive Waste Disposal Sites," Pacific Northwest Laboratory, Richland, WA.

Montgomery, R. and Parsons, L., (1990) "The Omega Hills Cover Test Plot Study: Fourth Year Data Summary," *Proceedings of the 22nd Mid-Atlantic Industrial Waste Conference,* Drexel University, Philadelphia, PA, pp. 43–56.

Morris, C. E. and Stormont, J. C., (1997) "Capillary Barriers and Subtitle D Covers: Estimating Equivalency," *Journal of Environmental Engineering,* ASCE, Vol. 123, No. 1, pp. 3–10.

Morrison-Knudsen, (1993) "White Paper, Implementation of Soil/Vegetative Covers for Final Remediation of the Rocky Mountain Arsenal," prepared by Morrison-Knudsen Corporation, Denver, CO, for Shell Oil Company, December.

Nyhan, J., Langhorst, G., Martin, C., Martinez, J., and Schofield, T., (1993) "Hydrologic Studies of Multilayered Landfill Closure of Waste Landfills at Los Alamos," *Proceedings of 1993 DOE Environmental Remediation Conference "ER" 93,* October 1993, Augusta, GA.

Nyhan, J., Hakonson, T., and Drennon, B., (1990) "A Water Balance Study of Two Landfill Cover Designs for Semiarid Regions," *Journal of Environmental Quality,* Vol. 19, pp. 281–288.

Nyhan, J., Schofield, T., and Starmer, R., (1997) "A Water balance Study of Four Landfill Cover Designs Varying in Slope for Semiarid Regions," *Journal of Environmental Quality,* Vol. 26, pp. 1385–1392.

Oweis, I. S. and Khera, R. P., (1998) "Geotechnology of Waste Management," 2nd Edition, PWS Publishing Company, 20 Park Plaza, Boston, MA.

Rockhold, M., Fayer, M., Kincaid, C., and Gee, G., (1995) "Estimation of Natural Ground Water Recharge for the Performance Assessment of a Low-Level Waste Disposal Facility at the Hanford Site," PNL-10508, Pacific Northwest Laboratory, Richland, WA.

Sherard, J. L., et al., (1976) "Pinhole Test for Identifying Dispersive Soils," *Journal of Geotechnical Engineering,* ASCE, Volume 102, No. 1, pp. 69–85.

Soong, T. -Y. and Koerner, R. M., (1996) "Seepage Induced Slope Instability," *Proceedings of GRI-9 Conference on Geosynthetics in Infrastructure Enhancement and Remediation,* GII Publications, Philadelphia, PA, pp. 245–265.

Stormont, J. C., (1995a) "The Effect of Constant Anisotropy on Capillary Barrier Performance," *Water Resources Research,* Vol. 31, pp. 783–786.

Stormont, J. C., (1995b) "The Performance of Two Capillary Barriers during Constant Infiltration," *Landfill Closures—Environmental Protection and Land Recovery,* ASCE, Geotechnical Special Publication, No. 53, R. Jeffrey Dunn and Udai P. Singh, Eds., New York, NY, pp. 77–92.

Stormont, J. C. and Anderson, C. E., (1999) "Capillary Barrier Effect from Underlying Coarser Soil Layer," *Journal of Geotechnical and Geoenvironmental Engineering,* ASCE, Volume 125, No. 8, pp. 641–648.

Stormont, J. C. and Morris, C. E., (1997) "Unsaturated Drainage Layers for Diversion of Infiltrating Water," Journal of Irrigation and Drainage Engineering, ASCE, Vol. 123, No. 5, pp. 364–366.

Theisen, M. S., (1992) "The Role of Geosynthetics in Erosion Control: An Overview," *Journal of Geotextiles and Geomembranes,* Vol. 11, Nos. 4–6, pp. 192–214.

Thornthwaite, C. W., (1948) "An Approach toward a Rational Classification of Climate," Geographic Review, Vol. 38, pp. 89–99.

USDA Soil Conservation Service, (1972) "Procedures for Computing Sheet and Rill Erosion on Project Areas," Technical Release, No. 51, Washington, DC.

USDA Soil Conservation Service, (1978) "Predicting Rainfall Erosion Losses: A Guide to Conservation Planning," *USDA Agric. Handbook,* No. 537, Washington, DC.

USEPA, (1989) "Technical Guidance Document: Final Covers on Hazardous Waste Landfills and Surface Impoundments," U.S. Environmental Protection Agency, Office of Solid Waste and Emergency Response, Washington, DC., EPA/530-SW-89-047.

USEPA, (1991) "Design and Construction of RCRA/CERCLA Final Covers (Seminar Publication)," U.S. Environmental Protection Agency, Office of Research and Development, Washington, DC., EPA/625/4-91/025, May.

USEPA, (1995) "Code of Federal Regulations, 40 CFR Parts 190 to 259," Revised as of July 1, 1995, U. S. Environmental Agency, Washington, DC.

U.S. Weather Bureau, (1963) "Rainfall Frequency Atlas for the USA for Durations from 30 minutes to 24 hours and Return Periods from 1 to 100 years," Technical Paper, No. 40, Washington, DC.

Volk, G. M., (1937) "Method of Determining the Degree of Dispersion of the Clay Fraction of Soils," *Proc. Soil Sci. Soc. Amer.,* pp. 432–445.

Wischmeier, W. H. and Smith D. D., (1958) "Rainfall Energy and Its Relationship to Soil Loss," *Trans. Amer. Geophysical Union,* Vol. 39, No. 2, pp. 285–291.

Wischmeier, W. H. and Smith D. D., (1965) "Predicting Rainfall-Erosion Losses from Cropland East of the Rocky Mountains," *Agriculture Handbook,* No. 282, U.S. Government Printing Office, Washington, DC.

Wischmeier, W. H., Johnson, C. B., and Cross, B. V., (1971) "A Soil Erodibility Monograph for Farmland and Construction Sites," *Journal of Soil Water Conservation,* Vol. 26, No. 5, pp. 189–193.

Yeh, T., Guzman, A., Srivastava, R., and Gagnard, P., (1994) "Numerical Simulation of the Wicking Effect in Liner Systems," *Ground Water,* Vol. 32, No. 1, pp. 2–11.

CHAPTER 12

Landfill Settlement

12.1 MECHANISM OF SOLID WASTE SETTLEMENT
12.2 EFFECT OF DAILY COVER
12.3 LANDFILL SETTLEMENT RATE
12.4 ESTIMATION OF LANDFILL SETTLEMENT
    12.4.1 SETTLEMENT OF NEW SOLID WASTE
    12.4.2 SETTLEMENT OF EXISTING SOLID WASTE
12.5 EFFECT OF WASTE SETTLEMENT ON LANDFILL CAPACITY
12.6 OTHER METHODS FOR ESTIMATING LANDFILL SETTLEMENT
    12.6.1 EMPIRICAL FUNCTIONS
    12.6.2 APPLICATION OF EMPIRICAL FUNCTIONS TO FIELD CASE STUDY
    12.6.3 SUMMARY COMMENTS ABOUT THE EMPIRICAL FUNCTIONS
12.7 ESTIMATION OF LANDFILL FOUNDATION SETTLEMENT
    12.7.1 TOTAL SETTLEMENT OF LANDFILL FOUNDATION
    12.7.2 DIFFERENTIAL SETTLEMENT OF LANDFILL FOUNDATION
    PROBLEMS
    REFERENCES

Settlement is an important concern in the management of municipal solid waste landfills. Landfill settlement continues over an extended period of time, with a final settlement that can approach 30% of the initial fill height as shown in Figure 12.1 from Spikula (1997). From an operator's viewpoint, landfill capacity will increase if most of the settlement occurs during the initial or early filling stages. Accordingly, waste placement or management strategies that maximize the rate/amount of early settlement are desirable from an economic standpoint. On the other hand, a large postclosure settlement is undesirable from a maintenance point of view, since it may lead to surface ponding, development of cracks in soil materials (such as compacted clay liners), tearing of the geomembrane, and damage to the geocomposite drainage layer. In addition, ancillary landfill facilities, such as gas collection and drainage pipes and leachate injection pipes (as in bioreactor landfills), may be damaged as a result of large differential postclosure settlement.

12.1 MECHANISM OF SOLID WASTE SETTLEMENT

The settlement of landfills affects the design of protection systems such as covers, barriers, and drains. Landfill storage capacity, and the cost and feasibility of using the underlying refuse for the support of buildings, pavements, and utilities will also be affected. Excessive settlements may cause ponding and even fracture of covers and

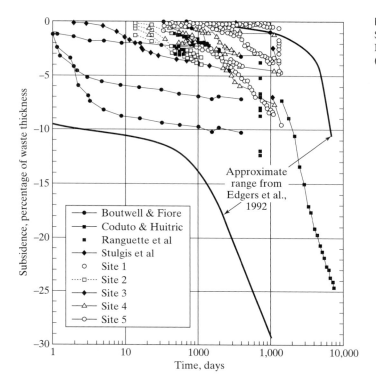

FIGURE 12.1 Landfill Subsidence, New and Previously Published Data (after Spikula, 1997)

drains. The latter outcome may increase the amount of moisture entering the landfill, which, in turn, will produce more leachate.

Settlement of municipal solid waste often begins rapidly as load is placed and continues to occur for long periods thereafter. The mechanisms of refuse settlement are complex, even more so than for soil, because of the extreme heterogeneity of waste fill and the presence of large voids. The main mechanisms involved in waste settlement are the following (Sowers, 1973; Murphy and Gilbert, 1985; Edil et al., 1990; Edgers et al., 1992):

(i) *Mechanical Compression:* Densification, distortion, bending, crushing, and reorientation; similar to consolidation of organic soils. Compression caused by the self-weight of the landfill and imposed loads, occurs in the from of initial, and/or primary consolidation, and/or secondary (delayed) compression.

(ii) *Raveling:* The movement of finer particles into larger voids or cavities within the fill. It is usually difficult to distinguish this mechanism from others.

(iii) *Physical-Chemical Change:* The deterioration and volume loss of waste products by corrosion, oxidation, and combustion.

(iv) *Bio-Chemical Decomposition:* The reduction of waste mass by fermentation and decay, both aerobic and anaerobic processes. This mechanism will be discussed at length in Chapter 15 on bioreactor landfills.

Many factors affect the magnitude of the settlement, and several of them are interrelated: (i) initial density or void ratio of the solid waste, including the types and amount of daily cover used; (ii) waste compaction effort and placement sequence; (iii) content of the decomposable materials in the waste; (iv) overburden pressure and stress history, such as conducting vertical expansion to overfill over an old landfill; (v) leachate level and fluctuations in landfills; (vi) landfill operation methods, such as leachate recirculation can accelerate waste biodegradation; and (vii) environmental factors, such as moisture content, oxygen which reaches the waste, temperature within the landfill, and gases present or generated within the landfill (Edil et al., 1990).

Settlement can occur jointly with evolution and release of large quantities of landfill gas. A detailed geotechnical mechanism or explanation for settlement caused by or associated with gas generation is not available at present. Similarly, the role of moisture equilibration (Noble et al, 1989) and hydrodynamic effects in the compression of partially saturated materials also is not well understood.

It should be noted that waste settles substantially both under its own weight and under the weight of a new load (for example, the placement of new waste over existing waste as in vertical expansions, which will be covered in Chapter 14). The introduction of cover soil to or on top of the waste fill complicates the computation of stresses due to these weights. As a result, two types of waste unit weight can be defined: (i) Actual waste unit weight (weight of refuse per unit volume of refuse); and (ii) Effective waste unit weight (weight of waste plus cover per unit volume of landfill) (Ham, et al., 1978). Waste unit weights are highly erratic, typically varying within a landfill from 32 to 70 lb/ft^3 (5 to 11 kN/m^3). Moisture contents typically range from 10 to 50% on a dry-weight basis (Sowers, 1968, 1973; Ham et al., 1978).

Settlement of waste fill is characteristically irregular. Initially, there is a large settlement within one or two months after completing construction, followed by a substantial amount of secondary compression over an extended period of time (recall Figure 12.1). The magnitude of settlement decreases over time and with increasing depth below the surface of the landfill. Waste settlement under its own weight typically ranges from 5 to 30% of the original thickness, with most of the settlement occurring in the first year or first two years (Edil et al., 1990).

12.2 EFFECT OF DAILY COVER

Complications arising from the settlement behavior of the daily cover itself are normally ignored when estimating landfill settlement. Placement of a daily cover of inorganic soil over waste fill is standard practice at most landfill sites; it is done to keep waste from blowing away, to restrict access to rodents, birds, and insects, and to provide additional overburden pressure. Typical procedures consist of placing 2 feet (0.60 m) of compacted waste and 6 inches (0.15 m) of soil cover. A simple settlement analysis assumes that this intermediate zone of soil material would settle as an independent layer between the much thicker layers of waste, while remaining largely intact and undergoing some consolidation settlement of its own.

This conceptual model of cover soil behavior does not accurately simulate actual behavior. Although the inert soil component initially occupies approximately 20% of

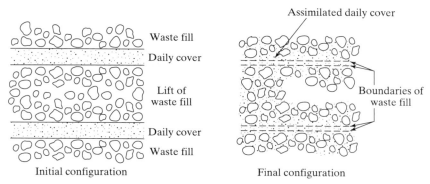
FIGURE 12.2 Absorption of Daily Cover into Waste Fill (Morris and Woods, 1990)

the total fill volume, observations at operating landfills indicate that the proportion becomes significantly reduced as settlement proceeds. This reduction is caused to some extent by compression of the soil under self-weight, but more importantly, it is caused by migration of the soil into the voids in the adjacent solid waste (Morris and Woods, 1990), as shown in Figure 12.2. The net result is that this soil layer can compress over time to less than one-quarter of the original volume. The resultant density of the fill is therefore considerably more than would be expected if this effect was ignored.

The soil particle migration effect has implications also for daily operation and design. One suggested way of increasing the storage efficiency of a landfill is to increase waste lift thickness relative to daily cover thickness. The suggestion is superficially appealing as it appears to minimize dilution of valuable storage volume with natural soil. Furthermore, the decreased overall density achieved by eliminating some or all of the denser cover soil (which is typically much denser than solid waste by a factor of 2 to 4) should reduce overall settlement of the final landfill and make design of final closure and abandonment easier. As pointed out by Morris and Woods (1990), these arguments are flawed for the following reasons:

Effect on Storage Volume: The net loss of storage due to the presence of the extra daily cover of soil is not large in practice for the reasons described next. A soil layer that may initially have occupied 20% of the overall disposal volume will occupy only 5% of the overall fill volume at depth after assimilation into adjacent fill. This minor loss of waste storage volume no longer represents a serious economic penalty for the operator. Indeed, the corresponding increase in overall fill density will assist in self-weight compression of the landfill as a whole.

Effect on Settlement: The second argument that reduction of daily cover will decrease post-construction settlement by reducing self-weight densities is true, but its applicability is debatable. Overall fill settlement may be less as a result of lower self-weight stresses, but not by as much as might be supposed, because the absence of soil "infilling" of the voids in the solid waste will cause the overall compressibility to be increased. Furthermore, the time taken for settlement will generally be increased such that a large proportion of the settlement is likely to take place after closure. If this is the case, then it is possible that postclosure set-

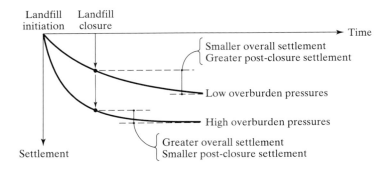

FIGURE 12.3 Possible Settlement Curves for Dense and Light Fills (Morris and Woods, 1990)

tlement of the landfill may actually be more than before, although preclosure settlement may have been reduced somewhat, as illustrated in Figure 12.3.

Morris and Woods (1990) conclude from the results of these conflicting tendencies or outcomes that recommended practice should specify fairly generous layers of daily cover soil. This reduces fire hazard, blowing, vermin problems, etc., and greatly improves landfill management at the expense of relatively minor (at worst) reductions in storage capacity. However, another strategy is to use alternative daily cover materials (ADCMs), as described by Pohland and Graven (1993).

12.3 LANDFILL SETTLEMENT RATE

Yen and Scanlon (1975) studied field settlement records extending up to nine years for three completed sanitary landfills located in Los Angeles County, California, to establish a general trend of landfill settlement rates. The areas of the three filled sites are 19 acres (7.6 ha), 80 acres (32 ha), and 22 acres (8.8 ha); the maximum height of fill was approximately 125 feet (38 m).

A sketch illustrating the parameters/variables used in the settlement rate analyses is shown in Figure 12.4. The settlement rate is defined as

$$m = \frac{\text{(Change in elevation of survey monument, in feet or meter)}}{\text{(Elapsed time between surveys, in months)}} \quad (12.1)$$

The time variable used here is the estimated "median age of a fill column" as defined by Yen and Scanlon (1975). This is measured as the elapsed time between the date of the settlement survey and that time when the fill column is half completed. The term "fill column" is defined (Yen and Scanlon, 1975) as the column of earth and waste fill lying directly below a survey monument, and extending to the natural soil. Other variables used were the completed fill column depth, H_f, and the total fill construction time t_c, in months. These parameters are chosen because the construction or filling period of a sanitary landfill is usually long and should be taken into consideration for settlement rate analyses. From Figure 12.4 and the data available, the median age of the fill column is estimated as

$$t_1 = t - t_c/2 \quad (12.2)$$

where t = the total elapsed time since the beginning of construction.

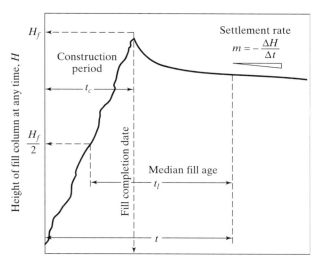

FIGURE 12.4 Diagram Showing Notations Used in Analysis (Yen and Scanlon, 1975). Used with permission of ASCE.

Yen and Scanlon (1975) then analyze the data to identify the relationships between median fill age (t_1), rate of settlement m, depth of fill H_f, and construction time t_c. Since the fill column, H_f, is a measure of average stress acting upon and within the waste, the fill columns were divided into subgroups based on ranges of H_f and construction time t_c. Thus, within each subgroup, the variation in H_f and t_c is restricted, and the relationship between elapsed time t_1 and settlement rate m can be seen more clearly.

Plots of m versus $\log(t_1)$ for fills between 40 feet and 80 feet (12 m and 24 m), between 80 feet and 100 feet (24 m and 30 m), and greater than 100 feet (30 m) are shown in Figures 12.5, 12.6 and 12.7. The linear relationship between m and t_1 is obtained by least-squares fitting with the regression coefficient shown. These three plots are only for the fill column subgroups with t_c between 70 months and 82 months. Table 12.1 summarizes results of other subgroups in which the construction period, t_c, ranges from less than one year to nearly seven years. The mean values of settlement rate, m, for the fill column of different depth ranges are shown in Table 12.1. Values of mean m were computed only in those intervals that contained at least three field survey measurements.

Effect of Median Fill Age on Settlement Rates as Function of Fill Depth

The decreasing trend of m with respect to t_1 can be seen in each of the referenced figures and in Table 12.1. However, after a period of 6 years from the completion of a fill (70 months to 120 months of median fill age), settlement rates on the order of 0.02 ft/month (6 mm/month) may still be expected. It is also interesting to note the range of post-construction settlement for the sites investigated. An estimate of the average total post-construction settlement was computed (Yen and Scanlon, 1975) by integrating the m and $\log(t_1)$ functions of Figures 12.5, 12.6, and 12.7 between the fill completion date and the extrapolated value of t_1 for which $m = 0$. The results show

446 Chapter 12 Landfill Settlement

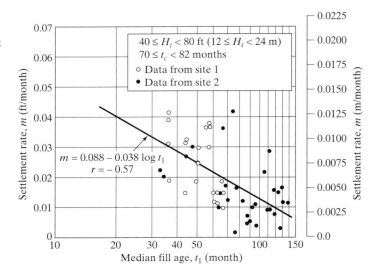

FIGURE 12.5 Settlement Rates versus Time Elapsed for Fill Depths between 40 ft and 80 ft (12 m and 24 m) (Yen and Scanlon, 1975). Used with permission of ASCE.

that post-construction settlement ranges between 4.5% and 6% of total fill depth, i.e., about 4.5 feet to 6 feet (1.4 m to 1.8 m) of past construction settlement would be expected for a 100 feet (30 m) of fill without subjecting the fill to any external load other than its own weight and biodegradation.

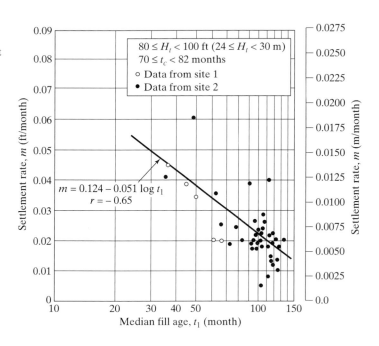

FIGURE 12.6 Settlement Rates versus Time Elapsed for Fill Depths between 80 ft and 100 ft (24 m and 31 m) (Yen and Scanlon, 1975). Used with permission of ASCE.

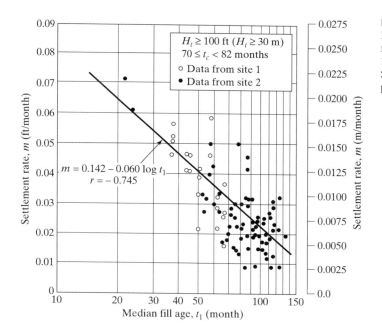

FIGURE 12.7 Settlement Rates versus Time Elapsed for Fill Depths Greater than 100 ft (31 m) (Yen and Scanlon, 1975). Used with permission of ASCE.

Effect of Depth on Rate of Settlement

As can be seen from Figures 12.5 to 12.7 and from Table 12.1, deeper fill depth generally shows a faster rate of settlement regardless of construction period, t_c, and median fill age t_1. However, the differences are not linearly proportional to the depth of fill, H_f. The results from the three sites studied indicate that although the rate of settlement increases with depth, this effect levels off for depths greater than 90 feet (27 m). The settlement rates for fill depth of more than 100 feet (30 m) are substantially the same after five years as those with H_f between 80 feet to 100 feet (24 m to 30 m). This finding is not surprising when considering that waste at greater depth has been subjected to a higher number of compaction cycles and prior densification.

Effect of Construction Period on Settlement Rate

The length of the construction period, t_c, also affects settlement rate. In Table 12.2, the time required for the completion of settlement (i.e., when $m = 0$) was computed by Yen and Scanlon (1975) using least-squares fitted $m - \log(t_1)$ functions (Figures 12.5, 12.6, and 12.7). The data in Table 12.2 indicate that a faster rate of the construction (smaller t_c values) will result in a much shorter time for the fill to complete its settlement. This effect is more pronounced for the shallower fills. This finding suggests it may be advantageous to construct a sanitary landfill as fast as possible in order to accelerate settlement.

Length of Post-Construction Settlement

The fill column subgroups with $70 \leq t_c \leq 82$ months had the largest data populations with the widest coverage in terms of t_1. For some of the fill columns in this group,

Table 12.1 Mean Values of Settlement Rate, m, in feet/month (millimeter/month) (Yen and Scanlon, 1975)

Completed Fill Height H_f, feet (meter)	Construction Period $t_c \leq 12$ months	Construction Period $24 \leq t_c \leq 50$ months	Construction Period $70 \leq t_c \leq 82$ months
(a) For Median Fill Age, $t_1 \leq 40$ months			
40 (12)	×	×	×
41 to 80 (12.3 to 24)	0.030 (9.14)	×	0.030 (9.14)
81 to 100 (24.3 to 30)	0.050 (15.24)	×	×
> 100 (> 30)	×	×	0.057 (17.37)
(b) For Median Fill Age, $40 \leq t_1 \leq 60$ months			
40 (12)	0.016 (4.88)	0.016 (4.88)	0.015 (4.57)
41 to 80 (12.3 to 24)	0.010 (3.05)	0.026 (7.92)	0.029 (8.84)
81 to 100 (24.3 to 30)	0.030 (9.14)	×	0.040 (12.19)
> 100 (> 30)	×	×	0.041 (12.50)
(c) For Median Fill Age, $60 \leq t_1 \leq 80$ months			
40 (12)	0.016 (4.88)	0.010 (3.05)	0.009 (2.74)
41 to 80 (12.3 to 24)	0.009 (2.74)	0.012 (3.66)	0.016 (4.88)
81 to 100 (24.3 to 30)	0.036 (10.97)	×	0.025 (7.62)
> 100 (> 30)	×	×	0.025 (7.62)
(d) For Median Fill Age, $80 \leq t_1 \leq 100$ months			
40 (12)	0.008 (2.44)	0.012 (3.66)	×
41 to 80 (12.3 to 24)	×	0.012 (3.66)	0.008 (2.44)
81 to 100 (24.3 to 30)	×	×	0.022 (6.71)
> 100 (> 30)	×	×	0.025 (7.62)
(e) For Median Fill Age, $100 \leq t_1 \leq 120$ months			
40 (12)	×	×	×
41 to 80 (12.3 to 24)	×	×	0.015 (4.57)
81 to 100 (24.3 to 30)	×	×	0.020 (6.19)
> 100 (> 30)	×	×	0.020 (6.19)

"×" indicates that less than three field settlement survey data were available in that particular H_f, t_1, t_c interval, and therefore, no mean value of settlement rate m was computed.
Data in Column 2 from Site 2; data in Column 3 from Sites 2 and 3; data in Column 4 from Site 1 and 2.
Used with permission of ASCE.

the settlement observations extended over a nine-year period. Figures 12.5, 12.6, and 12.7 show the scatter diagrams for these subgroups along with their regression coefficient r. Extrapolation of these functions suggests that, even without the addition of surface loads, the settlement process may last over 250 months before $m = 0$. This represents a very long time for completion of settlement from a geotechnical engineering viewpoint, and appears to be one of the major limitations of using deep landfill sites as load-bearing fills.

Table 12.2 Comparison of Settlement and Construction Period (Yen and Scanlon, 1975)

Range of Fill Depth H_f, feet, (meter)	Average Construction Period, t_c (month)	Total Time Required for Construction and Settlement (months)	Approximate Time Required for Settlement to Complete (month)
40 to 80 (12 to 24)	12	113	101
40 to 80 (12 to 24)	72	324	252
80 to 100 (24 to 30)	12	245	233
80 to 100 (24 to 30)	72	310	238

Used with permission of ASCE.

12.4 ESTIMATION OF LANDFILL SETTLEMENT

The usual laboratory tests for soil consolidation testing are not well suited for obtaining accurate consolidation parameters for solid waste that has a heterogeneous composition and extremely large particle sizes. By analyzing the field settlement data from some large-scale pilot landfill cells, Sowers (1973) proposed an alternative method to estimate the amount of the landfill settlement. In recent years, this method has been revised and refined several times by other investigators.

The settlement of solid waste includes primary settlement and long-term secondary compression. The total amount of settlement is given by the expression

$$\Delta H = \Delta H_c + \Delta H_\alpha \tag{12.3}$$

where ΔH = total settlement of solid waste;
ΔH_c = primary settlement of solid waste;
ΔH_α = long-term secondary settlement of solid waste.

12.4.1 Settlement of New Solid Waste

Based on the procedure proposed by Sowers (1973), the equations that follow can be used to calculate the settlement for new landfilled solid waste. The *Initial primary settlement* is given by

$$\Delta H_c = C_c \cdot \frac{H_o}{1 + e_o} \cdot \log \frac{\sigma_i}{\sigma_o} \tag{12.4}$$

or

$$\Delta H_c = C_c' \cdot H_o \cdot \log \frac{\sigma_i}{\sigma_o} \tag{12.5}$$

where ΔH_c = primary settlement;
e_o = initial void ratio of the waste layer before settlement;
H_o = initial thickness of the waste layer before settlement;
C_c = primary compression index (recall Figure 6.10);
C_c' = modified primary compression index, $C_c' = 0.17 \sim 0.36$;
σ_o = previously applied pressure in the waste layer (assumed equal to the compaction pressure, σ_o = 1,000 lb/ft² or 48 kN/m²);
σ_i = total overburden pressure applied at the mid level of the waste layer.

The previous compaction pressure applied on the solid waste layer during placement with compaction equipment is assumed to be 1,000 lb/ft² (48 kN/m²) based on 1973 compaction efforts for municipal solid waste landfills. In other words, the waste that has been placed in the landfill is essentially incompressible at normal pressure below 1,000 lb/ft² (48 kN/m²) due to the preconsolidation effect caused by previous compaction of the material. The value of the previously applied pressure, σ_o, should be changed during estimation of settlement if the compaction effort is much lower or higher than 1,000 lb/ft² (48 kN/m²) for a specific landfill project. Indeed, current practices of using waste compactors in the 100 to 150 U.S. tons (900 to 1,300 kN) range will significantly increase the value of σ_o.

The *long-term secondary settlement* can be obtained from

$$\Delta H_\alpha = C_\alpha \cdot \frac{H_o}{1 + e_o} \cdot \log \frac{t_2}{t_1} \tag{12.6}$$

or

$$\Delta H_\alpha = C'_\alpha \cdot H_o \cdot \log \frac{t_2}{t_1} \tag{12.7}$$

where ΔH_α = long-term secondary settlement;
e_o = initial void ratio of the waste layer before settlement;
H_o = initial thickness of the waste layer before settlement;
C_α = secondary compression index (recall Figure 6.11);
C'_α = modified secondary compression index, $C'_\alpha = 0.03 \sim 0.1$;
t_1 = starting time of the time period for which long-term settlement of the layer is desired, $t_1 = 1$ month;
t_2 = ending time of the time period for which long-term settlement of the layer is desired.

Because a standard consolidation test method for solid waste has not yet been developed, the selection of waste compression indices are mainly based on experience and limited field data. The value of the primary compression index C_c can be selected from Figure 6.10 based on the initial void ratio and organic content of the solid waste. The value of the secondary compression index C_α can be selected from Figure 6.11 based on the initial void ratio of the waste and the decomposition conditions.

Generally, the initial void ratio of municipal solid waste placed in a landfill after compaction is quite difficult to determine, and hence the values of the primary compression index C_c and the secondary compression index C_α cannot be estimated readily for settlement analysis. Accordingly, an alternative approach has been used in engineering practice—namely, the use of a "modified" primary compression index C'_c and a "modified" secondary compression index C'_α. Based on experience, the value of the modified primary compression index C'_c varies from 0.17 to 0.36, and the value of the modified secondary compression index C'_α varies from 0.03 to 0.1 for municipal solid waste (depending on the initial compaction effort and composition of the solid waste). The value of the modified secondary compression index C'_α for common clay ranges from 0.005 to 0.02. Therefore, the secondary settlement for municipal solid waste is approximately five to six times that of common clay.

12.4.2 Settlement of Existing Solid Waste

The following equations can be used to calculate the settlement of an existing solid waste landfill caused by vertical expansion (Chapter 14) or other additional extra loading, such as a light structure on a raft foundation.

The *primary settlement* is obtained by

$$\Delta H_c = C_c \cdot \frac{H_o}{1 + e_o} \cdot \log \frac{\sigma_o + \Delta \sigma}{\sigma_o} \tag{12.8}$$

or

$$\Delta H_c = C'_c \cdot H_o \cdot \log \frac{\sigma_o + \Delta \sigma}{\sigma_o} \tag{12.9}$$

where ΔH_c = primary settlement;
 e_o = initial void ratio of the waste layer before settlement;
 H_o = initial thickness of the waste layer of the existing landfill;
 C_c = primary compression index;
 C'_c = modified primary compression index, $C'_c = 0.17 \sim 0.36$;
 σ_o = existing overburden pressure acting at the mid level of the waste layer;
 $\Delta \sigma$ = increment of overburden pressure due to vertical expansion or other extra load.

The *long-term secondary settlement* is given by

$$\Delta H_\alpha = C_\alpha \cdot \frac{H_o}{1 + e_o} \cdot \log \frac{t_2}{t_1} \tag{12.10}$$

or

$$\Delta H_\alpha = C'_\alpha \cdot H_o \cdot \log \frac{t_2}{t_1} \tag{12.11}$$

where ΔH_α = secondary settlement;
 e_o = initial void ratio of the waste layer before starting secondary settlement;
 H_o = initial thickness of the waste layer before starting secondary settlement;
 C_α = secondary compression index;
 C'_α = modified secondary compression index, $C'_\alpha = 0.03 \sim 0.1$;
 t_1 = starting time of the secondary settlement. It is assumed to be equal to the age of the existing landfill for vertical expansion project;
 t_2 = ending time of the secondary settlement.

EXAMPLE 12.1

The filling procedure of a new municipal solid waste landfill is listed in Table 12.3.

Table 12.3 Solid Waste Filling Record

Time Period	Height of Solid Waste Filled	
	feet	meter
1st Month	12	3.6
2nd month	18	5.4
3rd month	16	4.8
4th month	10	3.0
5th month	14	4.2

Assume the following:
   Unit weight of solid waste, $\gamma_{waste} = 70$ lb/ft^3 (11 kN/m^3);
   Original applied pressure for solid waste, $\sigma_o = 1000$ lb/ft^2; (48 kN/m^2);
   Modified primary compression index, $C_c' = 0.26$;
   Modified secondary compression index, $C_\alpha' = 0.07$; and
   Secondary settlement starting time, $t_1 = 1$ month.
Calculate the total settlement of the top of the landfill at the end of the 5th month.

Solution:

$$\Delta H_c = C_c' \cdot H_o \cdot \log \frac{\sigma_i}{\sigma_o} \tag{12.5}$$

$$\Delta H_\alpha = C_\alpha' \cdot H_o \cdot \log \frac{t_2}{t_1} \tag{12.7}$$

$$\Delta H = \Delta H_c + \Delta H_\alpha \tag{12.3}$$

Calculate the solid waste depth over the mid-level of each waste layer:

$H_1 = (0.5)(12) + 18 + 16 + 10 + 14 = 64$ ft (19.2 m)
$H_2 = (0.5)(18) + 16 + 10 + 14 = 49$ ft (14.7 m)
$H_3 = (0.5)(16) + 10 + 14 = 32$ ft (9.6 m)
$H_4 = (0.5)(10) + 14 = 19$ ft (5.7 m)
$H_5 = (0.5)(14) = 7$ ft (2.1 m)

Calculate the total overburden pressure acting on the mid-level of each waste layer:

$\sigma_1 = \gamma_{waste} \cdot H_1 = 70 \times 64 = 4480$ lb/ft^2 (215 kN/m^2)
$\sigma_2 = \gamma_{waste} \cdot H_2 = 70 \times 49 = 3430$ lb/ft^2 (164 kN/m^2)
$\sigma_3 = \gamma_{waste} \cdot H_3 = 70 \times 32 = 2240$ lb/ft^2 (107 kN/m^2)
$\sigma_4 = \gamma_{waste} \cdot H_4 = 70 \times 19 = 1330$ lb/ft^2 (64 kN/m^2)
$\sigma_5 = \gamma_{waste} \cdot H_5 = 70 \times 7 = 490$ lb/ft^2 (23 kN/m^2) $< \sigma_o = 1000$ lb/ft^2 (48 kN/m^2)

Calculate the settlement of each waste layer:

$$\Delta H_{ci} = C'_c \cdot H_{oi} \cdot \log \frac{\sigma_i}{\sigma_o} \qquad (12.5)$$

$$\Delta H_{\alpha i} = C'_\alpha \cdot H_{oi} \cdot \log \frac{t_2}{t_1} \qquad (12.7)$$

$$\Delta H_i = \Delta H_{ci} + \Delta H_{\alpha i} \qquad (12.3)$$

First Layer:

$\Delta H_{c1} = 0.26 \times 12 \times \log(4480/1000) = 0.26 \times 12 \times 0.651 = 2.03$ ft (0.62 m)
$\Delta H_{\alpha 1} = 0.07 \times 12 \times \log(4.5/1) = 0.07 \times 12 \times 0.653 = 0.55$ ft (0.17 m)
$\Delta H_1 = \Delta H_{c1} + \Delta H_{\alpha 1} = 2.03 + 0.55 = 2.58$ ft (0.79 m)

Second Layer:

$\Delta H_{c2} = 0.26 \times 18 \times \log(3430/1000) = 0.26 \times 18 \times 0.535 = 2.51$ ft (0.77 m)
$\Delta H_{\alpha 2} = 0.07 \times 18 \times \log(3.5/1) = 0.07 \times 18 \times 0.544 = 0.69$ ft (0.21 m)
$\Delta H_2 = \Delta H_{c2} + \Delta H_{\alpha 2} = 2.51 + 0.69 = 3.20$ ft (0.98 m)

Third Layer:

$\Delta H_{c3} = 0.26 \times 16 \times \log(2240/1000) = 0.26 \times 12 \times 0.350 = 1.46$ ft (0.45 m)
$\Delta H_{\alpha 3} = 0.07 \times 16 \times \log(2.5/1) = 0.07 \times 16 \times 0.398 = 0.45$ ft (0.21 m)
$\Delta H_3 = \Delta H_{c3} + \Delta H_{\alpha 3} = 1.46 + 0.45 = 1.91$ ft (0.58)

Fourth Layer:

$\Delta H_{c4} = 0.26 \times 10 \times \log(1330/1000) = 0.26 \times 10 \times 0.124 = 0.32$ ft (0.098 m)
$\Delta H_4 = 0.07 \times 10 \times \log(1.5/1) = 0.07 \times 18 \times 0.176 = 0.22$ ft (0.067 m)
$\Delta H_4 = \Delta H_{c4} + \Delta H_{\alpha 4} = 0.32 + 0.22 = 0.54$ ft (0.16 m)

Fifth Layer:

$\Delta H_{c5} = 0$ [because $\sigma_5 = 490$ lb/ft^2 (23 kN/m^2) $< \sigma_o = 1000$ lb/ft^2 (48 kN/m^2)]
$\Delta H_{\alpha 5} = 0$ (because $t_2 = 0.5$ month $< t_1 = 1$ month)
$\Delta H_5 = \Delta H_{c5} + \Delta H_{\alpha 5} = 0 + 0 = 0$

Calculate the total settlement of the landfill at the end of the 5th month:

$$\Delta H_{total} = \Delta H_1 + \Delta H_2 + \Delta H_3 + \Delta H_4 + \Delta H_5$$
$$= 2.58 + 3.20 + 1.91 + 0.54 + 0$$
$$= 8.23 \text{ ft } (2.51 \text{ m})$$
$$\Delta H_{total}/(H_o)_{total} = 8.23/70 = 11.8\%$$

12.5 EFFECT OF WASTE SETTLEMENT ON LANDFILL CAPACITY

Solid waste settlement in a sanitary landfill offers a significant opportunity to a landfill operator for increasing potential disposal capacity. Because the settlement of solid waste is much larger and faster than the settlement of clay, a large settlement can be achieved during the filling process. Use of a suitable model to predict this behavior can help a landfill operator estimate actual remaining space of an existing landfill. Accurate estimation of the exact filling capacity including the extra volume achieved by waste settlement is also very important for local government solid waste planning efforts.

Most solid waste regulations require that once a portion of a landfill is filled to the approved final grade, a temporary or final cover layer must be placed over the top of the waste to close this portion. Additional filling will not be allowed again, even if the top of this portion settles below the approved final grade. Therefore, the usable volume or capacity increase from waste settlement is the additional space achieved during the filling period. In other words, only preclosure settlement volume can be counted into landfill capacity estimates. For example, let us assume that the design height of a landfill cell is 120 feet (36 m), and it takes 10 months to fill it to this height. Furthermore, assume that the waste settlement in 10 months is 21 feet (6.3 m). This 21-foot (6.3-m) space represents the usable volume increase from settlement. As a result, this cell can actually be filled to 141 feet (42.3 m) in 10 months, but the final height of the landfill in this cell will be only 120 feet (36 m) at the end of 10th month. The increment ratio due to the settlement in this case is 17.5%.

Theoretically, the larger the settlement achieved during the filling period, the more extra filling space the landfill operator can utilize. A settlement analysis and estimate of the space increment due to settlement during the waste filling period are presented in the next example.

EXAMPLE 12.2

The dimension of a landfill cross section is shown in Figure 12.8. The filling procedure or sequence used to place the waste is shown in numeric sequence. The filling time for each waste block shown in Figure 12.8 is assumed to be one month. The following are the relevant placement and solid waste parameters:

Thickness of each waste block, H_o = 20 ft (6 m);
Unit weight of solid waste, γ_{waste} = 70 lb/ft³ (11 kN/m³);
Original applied pressure for solid waste, σ_o = 1000 lb/ft²; (48 kN/m²);
Modified primary compression index, C'_c = 0.26;
Modified secondary compression index, C'_α = 0.07;
Secondary settlement starting time, t_1 = 1 month.

Calculate the increment of filling space due to the solid waste settlement.

Solution: The increment of filling space due to waste settlement corresponding to the filling procedure shown in Figure 12.8 can be estimated as follows:

Settlement at Point A

$$\Delta H_A = 0$$

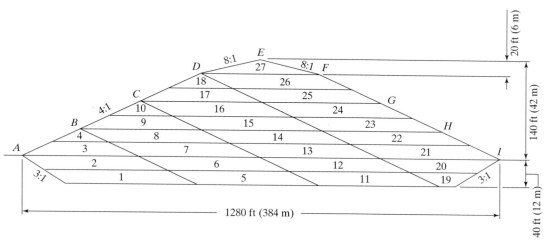

FIGURE 12.8 Landfill Cross Section and Waste Filling Procedure for Use in Example 10.2

Settlement at Point B

Solid waste depth over the mid-level of each waste layer:

$$H_1 = (0.5)(20) + 20 \times 3 = 70 \text{ ft (21 m)}$$
$$H_2 = (0.5)(20) + 20 \times 2 = 50 \text{ ft (15 m)}$$
$$H_3 = (0.5)(20) + 20 = 30 \text{ ft (9 m)}$$
$$H_4 = (0.5)(20) = 10 \text{ ft (3 m)}$$

Total overburden pressure acting on the mid-level of each waste layer:

$$\sigma_1 = \gamma_{waste} \cdot H_1 = 70 \times 70 = 4900 \text{ lb/ft}^2 \text{ (237 kN/m}^2\text{)}$$
$$\sigma_2 = \gamma_{waste} \cdot H_2 = 70 \times 50 = 3500 \text{ lb/ft}^2 \text{ (168 kN/m}^2\text{)}$$
$$\sigma_3 = \gamma_{waste} \cdot H_3 = 70 \times 30 = 2100 \text{ lb/ft}^2 \text{ (101 kN/m}^2\text{)}$$
$$\sigma_4 = \gamma_{waste} \cdot H_4 = 70 \times 10 = 700 \text{ lb/ft}^2 \text{ (34 kN/m}^2\text{)} < \sigma_o = 1000 \text{ lb/ft}^2 \text{ (48 kN/m}^2\text{)}$$

Primary settlement at Point B:

$$\Delta H_{ci} = C'_c \cdot H_{ci} \cdot \log \frac{\sigma_i}{\sigma_o} \tag{12.5}$$

$$\Delta H_{cB} = 0.26 \times 20 \times [\log(4900/1000) + \log(3500/1000) + \log(2100/1000)]$$
$$= 0.26 \times 20 \times 1.556$$
$$= 8.09 \text{ ft (2.47 m)}$$

Time period of secondary settlement for each layer (count from the mid-level of each layer):

$$(t_2)_1 = 3.5 \text{ month,}$$
$$(t_2)_2 = 2.5 \text{ month,}$$
$$(t_2)_3 = 1.5 \text{ month,}$$
$$(t_2)_4 = 0.5 \text{ month} < t_1 = 1 \text{ month}$$

Secondary settlement at Point B at the end of the 4th month:

$$\Delta H_{\alpha i} = C'_\alpha \cdot H_{oi} \cdot \log \frac{t_2}{t_1} \tag{12.7}$$

$$\Delta H_{\alpha B} = 0.07 \times 20 \times [\log(3.5/1) + \log(2.5/1) + \log(1.5/1)]$$
$$= 0.07 \times 20 \times 1.118$$
$$= 1.57 \text{ ft } (0.48 \text{ m})$$

Total settlement at Point B at the end of the 4th month:

$$\Delta H_i = \Delta H_{ci} + \Delta H_{\alpha i} \tag{12.3}$$

$$\Delta H_B = \Delta H_{cB} + \Delta H_{\alpha B}$$
$$= 8.09 + 1.57$$
$$= 9.66 \text{ ft } (2.94 \text{ m})$$

Settlement at Point C

Solid waste depth over the mid-level of each waste layer:

$H_1 = 110$ ft (33 m), $H_2 = 90$ ft (27 m), $H_7 = 70$ ft (21 m), $H_8 = 50$ ft (15 m), $H_9 = 30$ ft (9 m), $H_{10} = 10$ ft (3 m)

Total overburden pressure acting on the mid-level of each waste layer:

$\sigma_1 = 7{,}700$ lb/ft² (369 kN/m²), $\sigma_2 = 6{,}300$ lb/ft² (302 kN/m²), $\sigma_7 = 4{,}900$ lb/ft² (235 kN/m²),
$\sigma_8 = 3{,}500$ lb/ft² (168 kN/m²), $\sigma_9 = 2{,}100$ lb/ft² (101 kN/m²),
$\sigma_{10} = 700$ lb/ft² (34 kN/m²) $< \sigma_o = 1000$ lb/ft² (48 kN/m²)

Primary settlement at Point C:

$$\Delta H_{cC} = 0.26 \times 20 \times [\log(7700/1000) + \log(6300/1000) + \log(4900/1000) + \log(3500/1000) + \log(2100/1000)]$$
$$= 16.86 \text{ ft } (5.14 \text{ m})$$

Time period of secondary settlement for each layer (count from the mid-level of each layer):

$(t_2)_1 = 9.5$ month, $(t_2)_2 = 8.5$ month, $(t_2)_7 = 3.5$ month, $(t_2)_8 = 2.5$ month,
$(t_2)_9 = 1.5$ month, $(t_2)_{10} = 0.5$ month $< t_1 = 1$ month

Secondary settlement at Point C at the end of the 10th month:

$$\Delta H_{\alpha C} = 0.07 \times 20 \times [\log(9.5/1) + \log(8.5/1) + \log(3.5/1) + \log(2.5/1) + \log(1.5/1)]$$
$$= 4.24 \text{ ft } (1.29 \text{ m})$$

Total settlement at Point C at the end of 10th month:

$$\Delta H_c = \Delta H_{cC} + H_{\alpha C}$$
$$= 16.86 + 4.24$$
$$= 21.10 \text{ ft } (6.43 \text{ m})$$

The same procedure can be used to calculate the primary, secondary, and total settlements at each of the remaining points (viz., D, E, F, G, and H) based on the filling sequence shown in

Section 12.5 Effect of Waste Settlement on Landfill Capacity

Table 12.4 Secondary Settlement Time Period (months) of Each Waste Layer at Each Point

B		C		D		E		F		G		H	
Layer	t_2	Layer	t_2	Layer	t_2	Layer	t_2	Layer	t_2	Layer	t_2	Layer	t_2
						27	0.5						
				18	0.5	26	1.5	26	0.5				
				17	1.5	25	2.5	25	1.5				
		10	0.5	16	2.5	16	11.5	24	2.5	24	0.5		
		9	1.5	15	3.5	15	12.5	23	3.5	23	1.5		
4	0.5	8	2.5	8	10.5	14	13.5	14	12.5	22	2.5	22	0.5
3	1.5	7	3.5	7	11.5	13	14.5	13	13.5	21	3.5	21	1.5
2	2.5	2	8.5	6	12.5	6	21.5	12	14.5	12	12.5	20	2.5
1	3.5	1	9.5	5	13.5	5	22.5	11	15.5	11	13.5	19	3.5

Figure 12.8. The secondary time periods (t_2) of each waste layer at each point can likewise be calculated systematically in the manner just shown. The latter are summarized in Table 12.4. The primary, secondary, and total settlements at each point are listed in Table 12.5.

The same procedure can be used to calculate the primary, secondary, and total settlements at each of the remaining points (viz., D, E, F, G, and H) based on the filling sequence shown in Figure 12.8. The secondary time periods (t_2) of each waste layer at each point can likewise be calculated systematically in the manner just shown. The latter are summarized in Table 12.4. The primary, secondary, and total settlements at each point are listed in Table 12.5.

Volume Gained in filling space due to waste settlement

Volume gained in fill spacing due to waste settlement (Figure 12.9):

$$\Delta V = 0.5 \times [(0 + 9.66) + (9.66 + 21.10) + (21.10 + 34.76) + (34.76 + 43.54) \\ + (43.54 + 35.14) + (35.14 + 21.55) + (21.55 + 9.66) + (9.66 + 0)] \times (1280/8)$$

Table 12.5 Summary of Settlement at Each Point

Point	Primary Settlement		Secondary Settlement		Total Settlement	
	feet	meter	feet	meter	feet	meter
A	0	0	0	0	0	0
B	8.09	2.47	1.57	0.48	9.66	2.94
C	16.86	5.14	4.24	1.29	21.10	6.43
D	27.16	8.28	7.60	2.32	34.76	10.60
E	32.75	9.98	10.79	3.29	43.54	13.27
F	27.16	8.28	7.98	2.43	35.14	10.71
G	16.86	5.14	4.69	1.43	21.55	6.57
H	8.09	2.47	1.57	0.48	9.66	2.94
I	0	0	0	0	0	0

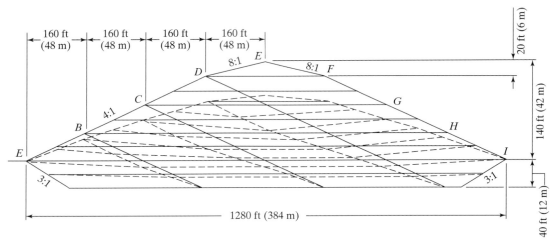

FIGURE 12.9 Landfill Cross Section before and after Solid Waste Settlement Resulting from the Calculations of Example 10.2

$$= (9.66 + 21.10 + 34.76 + 43.54 + 35.14 + 21.55 + 9.66) \times 160$$
$$= 175.45 \times 160$$
$$= 28{,}065 \text{ ft}^3/\text{ft} = 3{,}118 \text{ yd}^3/\text{yd} \; (2{,}607 \text{ m}^3/\text{m})$$

Cross-section volume of landfill:

$$V_o = (0.5 \times 320 \times 20) + 0.5 \times (320 + 1{,}280) \times 120 + 0.5 \times (1{,}280 + 1{,}040) \times 40$$
$$= 3{,}200 + 96{,}000 + 46{,}400$$
$$= 145{,}600 \text{ ft}^3/\text{ft} = 16{,}178 \text{ yd}^3/\text{yd} \; (13{,}527 \text{ m}^3/\text{m})$$

Ratio of volume gained due to waste settlement to total volume of landfill:

$$\Delta V/V_o = 3{,}118/16{,}178$$
$$= 0.193 = 19.3\%$$

The preceding analysis demonstrates that waste settlement significantly increases landfill filling space. According to this example, settlement can increase the available or effective filling space by almost 20%.

12.6 OTHER METHODS FOR ESTIMATING LANDFILL SETTLEMENT

A municipal solid waste fill exhibits heterogeneous and anisotropic material properties that are difficult to characterize. The unit weight and void ratio vary with the types of waste, composition, depth, method of compaction, and rate of decomposition. The rate of decomposition is further complicated by several factors, including the effects of time, temperature, and environmental conditions (Ling et al., 1998). The method described in Section 12.4 is a straight forward way of estimating landfill settlement and is widely used in current landfill designs. This method is basically the same as the conventional method used to estimate soil settlement. Specific characteristics of the solid waste that influence the settlement, such as physical-chemical changes and biochemical decomposition of solid waste with elapsed time, are not explicitly considered in this method. Other approaches reported by Ling et al. (1998) for estimating landfill settlement based on empirical functions are described briefly in this section.

12.6.1 Empirical Functions

The conventional approach to soil compression requires a separation of primary consolidation and secondary compression; and treatment of each with different mathematical expressions. In the long term, secondary compression of solid waste is larger than primary compression (consolidation), and it is often difficult to distinguish between the two. Empirical relationships, such as logarithmic, power, and hyperbolic functions, are often preferred in practice. These functions combine all stages of compression to estimate landfill settlement.

12.6.1.1 Logarithmic Function. Yen and Scanlon (1975) analyzed the settlement rate for three waste fills, 30 m high, with the data recorded over a period of 9 years. The settlement was determined and approximated using the relationship

$$\rho = \frac{dS}{dt} = m' - n' \cdot \log t \tag{12.12}$$

where ρ = settlement rate between time of interval, i.e., $\rho = \rho_i - \rho_o$;
S = settlement between time of interval, i.e., $S = S_i - S_o$;
t = difference between time of interest and time of the start of measurement, i.e., $t = t_i - t_o$;
m' = empirical constant;
n' = empirical constant.

This relationship has been designated by Ling et al. (1998) as a ρ-$\log t$, or logarithmic function. Yen and Scanlon reported a settlement rate of about 0.006 m per month in the first six years. The settlement rate increased with the depth of the fill, until it reached a limit. Sohn and Lee (1994) showed that the settlement rate is linearly proportional to the fill height.

Since Equation 12.12 expresses strain rate-time relationship, it can be integrated with respect to time to yield settlement. That is,

$$S = [m' - n' \cdot (\log t - 1)] \cdot t \tag{12.13}$$

Using $\log t$ (Ling et al., 1998), the settlement may be expressed directly as

$$S = m + n \cdot \log t \tag{12.14}$$

where S = settlement between time of interval, i.e., $S = S_i - S_o$, meter;
t = difference between time of interest and time of the start of measurement, i.e., $t = t_i - t_o$, day;
m = empirical constant;
n = empirical constant.

12.6.1.2 Power Function.
This function has been used by Edil et al. (1990) to relate settlement rate with time. It may be written as

$$\rho = \frac{dS}{dt} = \frac{p'}{t^{q'}} \qquad (12.15)$$

where ρ = settlement rate between time of interval, i.e., $\rho = \rho_i - \rho_o$;
S = settlement between time of interval, i.e., $S = S_i - S_o$;
t = difference between time of interest and time of the start of measurement, i.e., $t = t_i - t_o$;
p' = empirical constant, p' may be defined as the settlement rate at unit time;
q' = empirical constant.

Since Equation 12.15 expresses strain rate-time relationship like Equation 12.12, it can also be integrated with respect to time to yield settlement. That is,

$$S = \frac{p'}{1 - q'} \cdot t^{1-q'} \qquad (12.16)$$

Using the exponent t^q (Ling et, al., 1998), the settlement may be expressed directly as

$$S = p \cdot t^q \qquad (12.17)$$

where S = settlement between time of interval, i.e., $S = S_i - S_o$, meter;
t = difference between time of interest and time of the start of measurement, i.e., $t = t_i - t_o$, day;
p = positive empirical constant, $p = p'/q$;
q = positive empirical constant, $q = 1 - q'$.

Alternative equations 12.14 and 12.17 are advantageous, because an integration does not have to be conducted as in the case of Equations 12.12 and 12.15.

12.6.1.3 Hyperbolic Function.
A hyperbolic function has been used with success to approximate deformation associated with certain geotechnical problems, such as the settlement of an embankment on soft ground (Tan et al., 1991) and the simulation of triaxial test results (Kondner and Zelasko, 1963). For municipal solid waste landfill settlement (Ling et al., 1998), the equivalent hyperbolic expression relating settlement and time may be written as

$$S = \frac{t}{1/\rho_o + t/S_{ult}} \qquad (12.18)$$

where S = difference between settlement at time t_i and that measured at time t_o, i.e., $S = S_i - S_o$, meter;
t = difference between time of interest and time of the start of measurement, i.e., $t = t_i - t_o$, day;
ρ_o = initial rate of settlement at $t = t_o$;
S_{ult} = ultimate settlement, i.e., $t \to \infty$.

The parameters ρ_o and S_{ult} may be determined by transforming Equation 12.18 through t/S versus t relationships and conducting a linear regression analysis, which is written as

$$\frac{t}{S} = \frac{1}{\rho_o} + \frac{t}{S_{ult}} \qquad (12.19)$$

where the reciprocals of intercept and slope give ρ_o and S_{ult}, respectively.

The final settlement in the field will most likely lie between 80 and 95% of the ultimate value. The time taken to reach 95% of this ultimate value is calculated as $t_f = 19S_{ult}/\rho_o$. Notation t_o normally refers to time at $t = 0$, but the hyperbolic function offers the flexibility for referral at any instant of time. This option is particularly useful if there is a change in loading conditions, such as waste surcharging, so that the analysis may be restarted at a later time (Ling et al., 1998).

12.6.2 Application of Empirical Functions to Field Case Study

A total of nine case histories of settlement measurements were used by Ling et al. (1998) to examine the accuracy of the previously described empirical functions. They selected the measurements reported for three landfill sites: Southeastern Wisconsin (Edil et al., 1991), Meruelo landfill, Spain (Sanchez-Alciturri et al., 1995), and Spadra landfill, California (Merz and Stone, 1962) because of their relatively long-term results.

The site reported by Edil et al. (1991) is located in the southeastern part of Wisconsin. This landfill started accepting waste in the early 1970s. The settlement there was primarily due to self-weight loading (i.e., there was no additional placement of waste above the weighing platform during the measurement period). The measurements were taken for a period of 1.5 years, between 1984 and 1986. The platform was embedded under a waste column about 27 meters (90 feet) high, with the age of the waste estimated to lie between 0 and 4 years. Edil et al. (1991) estimated the bulk unit weight of the waste to be 10.7 kN/m^3 (68 lb/ft^3). The moisture content of the waste was not reported.

The Meruelo landfill is located in a valley in Merielo, which is in the northern part of Spain. The results presented by Sanchez-Alciturri et al. (1995) covered a three-year period of monitoring conducted by surveying. Filling started in late 1998 and finished in early 1992. The landfill consists of cells compacted using a sheep-foot roller and having a daily cover of 0.2 m (8 inches). The waste in this landfill was composed mainly of organic material and paper. The three measurement points used in this study are points 26, 27, and 28, which cover a longer period of measurement than other points. The waste thickness was about 15 meters (50 feet) at these three locations. The mean unit weight of the combined waste and soil was 12 kN/m^3 (76 lbft^3), and the water content was 48%.

Spadra landfill number 2 (Merz and Stone, 1962) is located near the city of Pomona, California. This landfill, which was constructed as part of a research project, provided some of the earliest measurements on landfill settlement. The results of surface settlement of five cells, each with an area of 4.65 m^2 (50 ft^2), were analyzed in the cited paper. The wastes were primarily garbage, papers, and grass. The dry unit weight

Table 12.6 Best-Fit Parameters for Logarithmic Function, $S = m + n \cdot \log t$ (adapted from Ling et al., 1998)

Landfill Sites		Parameters		Correlation Coefficient r
		m	n	
Southeastern Wisconsin		−0.927	0.468	0.995
Meruelo	Point 26	−0.478	0.331	0.994
	Point 27	−0.403	0.287	0.992
	Point 28	−0.365	0.242	0.982
Spadra	Cell 1	0.082	0.140	0.993
	Cell 2	0.312	0.115	0.998
	Cell 3	0.408	0.111	0.995
	Cell 4	0.647	0.095	0.996
	Cell 5	0.294	0.142	0.977

Used with permission of ASCE.

of the waste ranged from 3 to 6.8 kN/m³ (19 to 43 lb/ft³). The average depth of the waste in these cells was 6.7 meters (22 feet). Each cell had a different moisture content, ranging from 30% to 90%. The largest settlement was reported to have occurred during the first month of measurement.

Tables 12.6, 12.7 and 12.8 summarize the best-fit parameters obtained by Ling et al. (1998) for three different functions used: logarithmic, power, and hyperbolic. Note that r is the coefficient of correlation. In the table and figures, the settlement is given in meters, while the time is given in days. The settlements fitted using logarithmic, power, and hyperbolic functions are shown in Figures 12.10, 12.11, and 12.12 for each of the landfills. To obtain satisfactory agreement for long-term settlement, some of the initial data were excluded while searching for the best-fit parameters m and n in Equation 12.14. Similarly, for the power function, Equation 12.17, some of the initial data were excluded in order to achieve a better prediction of long-term settlement. The power function provides a better simulation when compared to the logarithmic function.

Table 12.7 Best-Fit Parameters for Power Function, $S = p \cdot t^q$ (adapted from Ling et al., 1998)

Landfill Sites		Parameters		Correlation Coefficient r
		p	q	
Southeastern Wisconsin		0.002	0.884	0.998
Meruelo	Point 26	0.013	0.558	0.853
	Point 27	0.018	0.483	0.986
	Point 28	0.010	0.537	0.996
Spadra	Cell 1	0.170	0.162	0.997
	Cell 2	0.357	0.091	0.999
	Cell 3	0.445	0.075	0.997
	Cell 4	0.667	0.049	0.996
	Cell 5	0.356	0.104	0.984

Used with permission of ASCE.

Table 12.8 Best-Fit Parameters for Hyperbolic Function, $S = t/(1/\rho_o + t/S_{ult})$ (adapted from Ling et al., 1998)

Landfill Sites		Parameters		Correlation Coefficient
		ρ_o	S_{ult}	r
Southeastern Wisconsin		0.001	1.14	0.961
Meruelo	Point 26	0.003	0.62	0.998
	Point 27	0.002	0.58	0.992
	Point 28	0.001	0.51	0.988
Sparda	Cell 1	0.015	0.48	0.9980
	Cell 2	0.040	0.63	0.9997
	Cell 3	0.054	0.72	0.9997
	Cell 4	0.102	0.91	0.9298
	Cell 5	0.036	0.69	0.9998

Used with permission of ASCE.

The results of the settlement prediction using a hyperbolic function in Figure 12.12 show very good agreement between the measured and fitted values, with a coefficient of correlation close to 1.0. The average ultimate settlements S_{ult} for the three landfills were determined as 1.14 m (3.8 ft), 0.57 m (1.9 ft), and 0.69 m (2.3 ft), respectively. These settlements were calculated using the empirical functions and associated empirical parameters in Tables 12.6, 12.7, and 12.8. The time t_f taken to reach 95% ultimate settlement and the magnitude of this final settlement S_f are calculated and summarized in Table 12.9. The t_f value is the largest for the Southeastern Wisconsin landfill (59 years), whereas it lies between 0.5 year and 2 years in the case of the Spadra landfill. Since no actual ultimate settlement results are available for comparison, these estimates serve primarily as a guide. The curve fitting for any site can be refined as additional data become available and the estimate revised if necessary.

12.6.3 Summary Comments about the Empirical Functions

It should be noted that both logarithmic and power functions (Equations 12.14 and 12.17) will produce infinitely large settlement as t becomes extremely large. As a check on the validity of these calculations Ling et al. (1998) recommend that the settlement should be less than approximately 40% of the waste fill thickness. In the process of calculating settlement and time data by means of logarithmic and power functions, some of the initial data had to be excluded to obtain better agreement for long-term settlement prediction (see Figures 12.10 and 12.11). The hyperbolic function, on the other hand, is capable of simulating settlement satisfactorily at any time interval based on the measurements and parameters obtained from the time interval of interest (see Figure 12.12). Ling et al. (1998) also reported some ancillary findings from the settlement case studies. Figure 12.13 shows, for example, that m decreases, while n increases slightly with an increase in the water content of the waste materials for the logarithmic function. The same trend holds for p and q of the power function (Figure 12.14). For

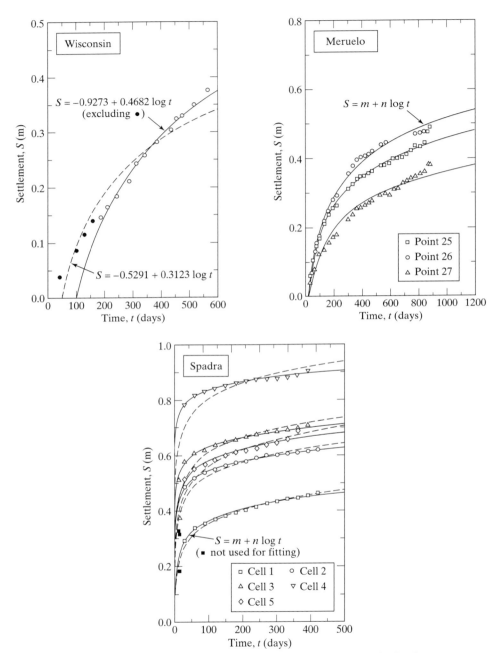

FIGURE 12.10 Settlement Estimation Based on Logarithmic Function for Southeastern Wisconsin, Meruelo, and Spadra Landfills (Ling et al., 1998). Used with permission of ASCE.

Section 12.6 Other Methods for Estimating Landfill Settlement 465

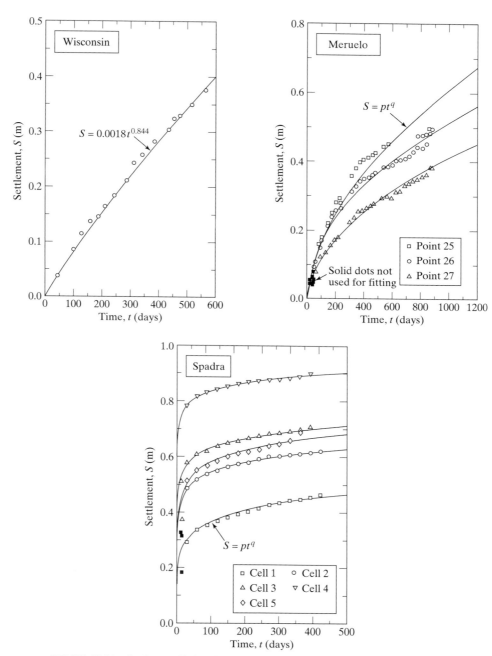

FIGURE 12.11 Settlement Estimation Based on Power Function for Southeastern Wisconsin, Meruelo, and Spadra Landfills (Ling et al., 1998). Used with permission of ASCE.

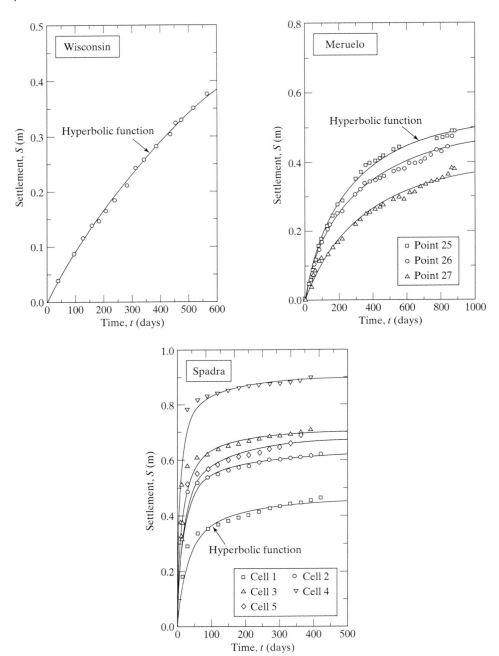

FIGURE 12.12 Settlement Estimation Based on Hyperbolic Function for Southeastern Wisconsin, Meruelo, and Spadra Landfills (Ling et al., 1998). Used with permission of ASCE.

Section 12.6   Other Methods for Estimating Landfill Settlement   467

Table 12.9 Final Settlement and Time Taken for Hyperbolic Function (Ling et al., 1998)

Landfill Sites		$t_f = 19 S_{ult}/\rho_o$ (day)	$S_f = 0.95 S_{ult}$ (meter)
Southeastern Wisconsin		21,660	1.08
Meruelo	Point 26	3,927	0.59
	Point 27	5,510	0.55
	Point 28	9,690	0.48
Spadra	Cell 1	608	0.46
	Cell 2	299	0.60
	Cell 3	253	0.68
	Cell 4	170	0.86
	Cell 5	364	0.66

Used with permission of ASCE.

the hyperbolic function, Figure 12.15 shows that both ρ_o and S_{ult} decrease with an increase in the water content of the waste materials.

The key to using the three empirical functions for estimating landfill settlement is how to predict reliably the values of the two empirical coefficients for each empirical function consistent with the waste properties and landfill dimension. The precise relationship between empirical coefficients for each empirical function and the waste properties (e.g., unit weight, water content, void ratio, and waste composition), landfill dimension (e.g., overburden pressure or landfill height), and environmental factors

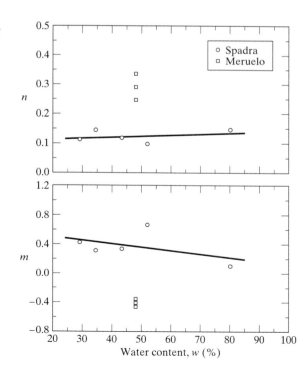

FIGURE 12.13 Variation of Parameters with Water Content of Waste Materials for Logarithmic Function (Ling et al., 1998). Used with permission of ASCE.

468 Chapter 12 Landfill Settlement

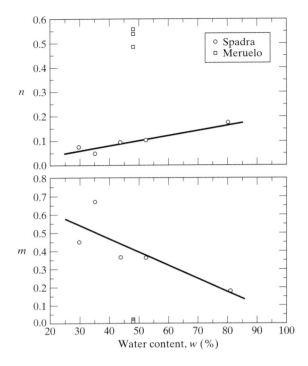

FIGURE 12.14 Variation of Parameters with Water Content of Waste Materials for Power Function (Ling et al., 1998). Used with permission of ASCE.

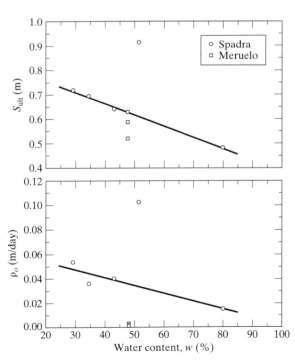

FIGURE 12.15 Variation of Parameters with Water Content of Waste Materials for Hyperbolic Function (Ling et al., 1998). Used with permission of ASCE.

(e.g., temperature within landfill and oxygen reaching the waste) still is not entirely clear. These functions should be used with caution in engineering practice and should be supported by additional testing data and research.

12.7 ESTIMATION OF LANDFILL FOUNDATION SETTLEMENT

If the landfill is underlain by a soil layer, particularly a thick layer of soft, fine-grained soil, consolidation settlements may be large. In these cases, design analyses should consider settlement of the foundation clay layer. Both primary consolidation and long-term secondary settlement should be considered. Calculations are performed using conventional equations from soil mechanics theory and a time frame at least equal to the active life and postclosure care period of the landfill.

Excessive settlement of an underlying foundation clay layer will affect the performance of a landfill liner and leachate collection system. The purposes of analyzing the settlement of a foundation clay layer and overlying landfill liner and leachate collection/removal system are as follows:

(i) Tensile strain induced in the liner system and leachate collection and removal system must be limited to a minimum allowable tensile strain for the components of these two systems. The compacted clay liner usually has the smallest allowable tensile strain value between 0.1% and 1.0% and an average allowable tensile strain of 0.5%.

(ii) Post-settlement grades of the landfill cell subbase and the leachate collection pipes must be sufficient to maintain leachate performance to prevent grade reversal and leachate ponding in accordance with the rule requirements.

12.7.1 Total Settlement of Landfill Foundation

The total settlement of landfill foundation soil can be divided into three portions: elastic settlement, primary consolidation settlement, and secondary consolidation settlement. The settlement of sandy soils includes only elastic settlement. The settlement of clayey soils includes all three types of settlements. The total settlement of clayey soil is equal to the sum of the elastic settlement and the primary and secondary settlements. Because the permeability of clay is quite low, it takes a long time to complete the whole process of consolidation settlement. The settlement of clayey soil is usually much larger than the settlement of sandy soils.

Because the settlement of sandy soils includes only elastic settlement, the settlement of sand layer can be calculated from the Elastic Settlement equation, which is

$$Z_e = (\Delta\sigma/M_s)H_o \tag{12.20}$$

where Z_e = elastic settlement of soil layer, ft or m;
H_o = initial thickness of soil layer, ft or m;
$\Delta\sigma$ = increment of vertical effective stress, lb/ft² or kN/m²;
M_s = constrained modulus of soil, lb/ft² or kN/m².

The constrained modulus is given by

$$M_s = \frac{E_s \cdot (1 - v_s)}{(1 + v_s)(1 - 2 \cdot v_s)} \tag{12.21}$$

where M_s = constrained modulus of soil, lb/ft² or kN/m²;
E_s = elastic modulus of soil, see Table 9.5, lb/ft² or kN/m²;
v_s = Poisson's ratio of soil, see Table 9.5.

The *primary consolidation settlement* is given by

$$Z_c = C_r \cdot \frac{H_{oi}}{1 + e_{oi}} \cdot \log\frac{p_c}{\sigma_o} + C_c \cdot \frac{H_o}{1 + e_{oi}} \cdot \log\frac{\sigma_o + \Delta\sigma}{p_c} \tag{12.22}$$

where Z_c = primary consolidation settlement of clay layer, ft or m;
H_o = initial thickness of clay layer, ft or m;
e_{oi} = initial void ratio of clay layer;
C_r = recompression index;
C_c = primary compression index.
σ_o = initial vertical effective stress, lb/ft² or kN/m²;
p_c = preconsolidation pressure, lb/ft² or kN/m²;
$\Delta\sigma$ = increment of vertical effective stress, lb/ft² or kN/m².

The *secondary compression settlement* is given by

$$Z_\alpha = C_\alpha \cdot \frac{H_{os}}{1 + e_{os}} \cdot \log\frac{t_2}{t_1} \tag{12.23}$$

where Z_α = long-term secondary compression settlement, ft or m;
e_{os} = initial void ratio of clay layer before starting secondary consolidation settlement;
C_α = secondary consolidation compression index;
H_{os} = initial thickness of clay layer before starting secondary consolidation settlement, ft or m;
t_1 = starting time of the time period for which long-term settlement of the layer is desired;
t_2 = ending time of the time period for which long-term settlement of the layer is desired.

The total settlement of clay layer includes three portions: elastic settlement, primary consolidation settlement, and secondary consolidation settlement. These three types of settlement for clayey soil layers can be calculated from Equations 12.20, 12.22, and 12.23, respectively. The total settlement of clayey soil at point i can be determined from the equation

$$Z_i = (Z_e)_i + (Z_c)_i + (Z_\alpha)_i \tag{12.24}$$

where Z_i = total settlement of points i;
$(Z_e)_i$ = elastic settlement of point i;
$(Z_c)_i$ = primary consolidation settlement of point i;
$(Z_\alpha)_i$ = secondary consolidation settlement of point i.

The preceding settlement equations (Equations 12.20 through 12.24) provide a framework and means to account for different types of settlement. Not all foundations on soil settle exactly in the manner previously described. In addition to soil type, the amount of settlement in each category also depends on the degree of saturation, load duration, and load distribution.

The settlement calculations should be performed at discrete points along several selected settlement lines, such as Lines 1, 2, 3, and 4 for a municipal solid waste landfill as shown in Figure 12.16. The following principles should guide the arrangement of the settlement lines for landfill foundation settlement calculations:

(i) Some settlement lines should be set along the leachate collection pipelines (usually 1% slope) to check the grade changes and tensile strains of the leachate collection pipes due to settlement.

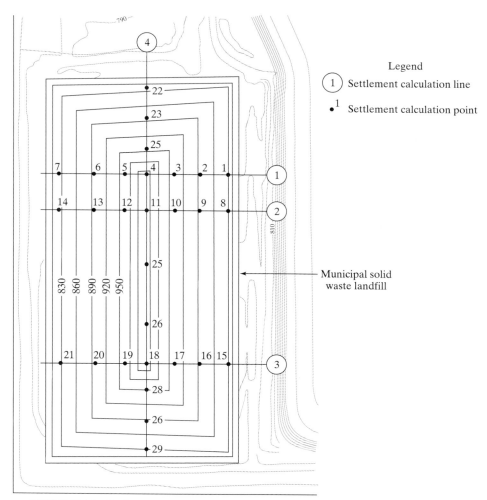

FIGURE 12.16 Settlement Calculation Locations for Landfill Liner System Foundation

(ii) Some settlement lines should be set perpendicular to the leachate collection pipes (i.e., along the direction of leachate flow in the leachate drainage layer—usually 2% slope), to check the slope change of leachate drainage layer.

(iii) Settlement lines are usually set locations where there are the large changes of overburden pressures, which may cause large differential settlements of the subgrade.

At each settlement point shown in Figure 12.16 (i.e., Point 1 to Point 29), the thickness of the various soil units at each point and the thickness of the waste to be placed (i.e., overburden pressure) can be estimated from the engineering plans or cross-sections for the specific projects. The value of the total settlement at each point depends on both the engineering properties of soils and the load due to the waste fill.

12.7.2 Differential Settlement of Landfill Foundation

The differential settlements, tensile strains of liner system materials and leachate collection pipes, and changes of final grades between adjacent settlement points after settlement can be evaluated from the calculated values of the total settlements at various settlement points along each settlement line ranged on the landfill subgrade.

The differential settlement between adjacent points can be calculated using the equation

$$\Delta Z_{i,i+1} = Z_{i+1} - Z_i \qquad (12.25)$$

where $\Delta Z_{i,i+1}$ = differential settlement between points i and $i + 1$;
Z_i = total settlement of point i;
Z_{i+1} = total settlement of point $i + 1$.

The final slope angle between adjacent points after settlement can be calculated using the equation

$$\tan \beta_{Fnl} = \frac{X_{i,i+1} \cdot \tan \beta_{Int} - \Delta Z_{i,i+1}}{X_{i,i+1}} \qquad (12.26)$$

where $X_{i,i+1}$ = horizontal distance between points i and $i + 1$;
$\Delta Z_{i,i+1}$ = differential settlement between points i and $i + 1$;
β_{Int} = initial slope angle between points i and $i + 1$;
β_{Fnl} = final slope angle between points i and $i + 1$ after settlement.

The landfill subgrade changes along each settlement line due to different settlement can be calculated from the Equation 12.26. Figure 12.17 presents the slope changes due to differential settlement along a settlement line. The differential settlement will result in grade reversal between points 3 and 4 as shown in Figure 12.17. As a result, leachate will pond on the liner at this area.

The tensile strains of a liner system and a leachate collection system resulting from the settlements can be estimated using the equation

$$\varepsilon_{i,i+1} = \frac{(L_{i,i+1})_{Fnl} - (L_{i,i+1})_{Int}}{(L_{i,i+1})_{Int}} \times 100\% \qquad (12.27)$$

where $\varepsilon_{i,i+1}$ = tensile strain in liner system between points i and $i + 1$;

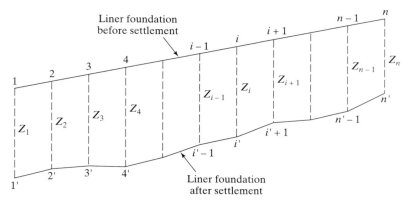

FIGURE 12.17 Subgrade Changes along a Settlement Line due to Differential Settlement

$(L_{i,i+1})_{Int}$ = distance between points i and $i + 1$ in their initial positions;
$(L_{i,i+1})_{Fnl}$ = distance between points i and $i + 1$ in their post-settlement positions.

The distance between points i and $i + 1$ in their initial positions can be calculated using the equation

$$(L_{i,i+1})_{Int} = [(X_{i,i+1})^2 + (X_{i,i+1} \cdot \tan\beta_{Int})^2]^{1/2} \tag{12.28}$$

The distance between points i and $i + 1$ in their post-settlement positions can be calculated using the equation

$$(L_{i,i+1})_{Fnl} = [(X_{i,i+1})^2 + (X_{i,i+1} \cdot \tan\beta_{Int} - \Delta Z_{i,i+1})^2]^{1/2} \tag{12.29}$$

The maximum acceptable tensile strains (i.e., the elongations at yield) of various liner system and leachate collection system components must be obtained from material specific laboratory testing.

PROBLEMS

12.1 What benefit does a landfill operation gain from some settlement in the landfill waste?

12.2 Describe the impact of landfill settlement on landfill design and performance.

12.3 What are the main mechanisms of solid waste settlement?

12.4 List the main factors that influence the amount of landfill settlement.

12.5 What are the main differences of the settlement of solid waste compared with the settlement of clay?

12.6 What are the main advantages and disadvantages of the calculation method for landfill settlement described in Section 12.4?

12.7 The filling sequence for a municipal solid waste landfill is listed in the following table:

Table 12.10 Solid Waste Filling Record for Problem 12.7

Time Period	Height of Solid Waste Filled	
	feet	meter
1st month	12	3.6
2nd month	16	4.8
3rd month	20	6.0
4th month	25	7.5
5th month	22	6.6

Assume the following:
Unit weight of solid waste, γ_{waste} = 65 lb/ft^3 (10.2 kN/m^3);
Original applied pressure on the solid waste, σ_o = 1000 lb/ft^2 (48 kN/m^2);
Modified primary compression index, C_c' = 0.24;
Modified secondary compression index, C_α' = 0.075; and
Secondary settlement starting time, t_1 = 1 month.
Filling or placement of solid waste stops at the end of the 5th month.
1. Calculate the total settlement of the landfill at the end of the 5th month.
2. Calculate the total settlement of the landfill at the end of the 12th month.

12.8 The filling sequence for a municipal solid waste landfill is listed in Table 12.11. Assume the following:
Unit weight of solid waste, γ_{waste} = 65 lb/ft^3 (10.2 kN/m^3);
Original applied pressure on the solid waste, σ_o = 1000 lb/ft^2 (48 kN/m^2);
Modified primary compression index, C_c' = 0.28;
Modified secondary compression index, C_α' = 0.065; and
Secondary settlement starting time, t_1 = 1 month.
Filling or placement of solid waste stops at the end of the 8th month.

Table 12.11 Solid Waste Filling Record for Problem 12.8

Time Period	Height of Solid Waste Filled	
	feet	meter
1st month	25	7.5
2nd month	31	9.3
3rd month	18	5.4
4th month	0	0
5th month	0	0
6th month	8	2.4
7th month	25	7.5
8th month	27	8.1

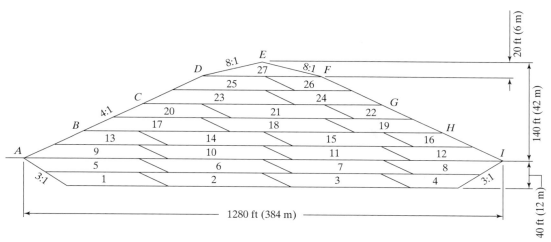

FIGURE 12.18 Landfill Cross Section and Waste Filling Procedure for Problem 12.9

1. Calculate the total settlement of the landfill at the end of the 3rd month.
2. Calculate the total settlement of the landfill at the end of the 5th month.
3. Calculate the total settlement of the landfill at the end of the 8th month.
4. Calculate the total settlement of the landfill at the end of the 18th month.

12.9 The dimension of a landfill cross section is shown in Figure 12.18. The waste filling procedure is also shown in Figure 12.18. The filling time for each waste block shown in Figure 12.18 is assumed to be one month. The following solid waste properties and parameters are applicable:
Thickness of each waste block, H_o = 20 ft (6 m);
Unit weight of solid waste, γ_{waste} = 70 lb/ft³ (11 kN/m³);
Original applied pressure for solid waste, σ_o = 1000 lb/ft² (48 kN/m²);
Modified primary compression index, C'_c = 0.26;
Modified secondary compression index, C'_α = 0.07; and
Secondary settlement starting time, t_1 = 1 month.
Calculate the additional capacity or increment of filling space due to waste settlement corresponding to the stated filling procedure.

12.10 Comment on the three empirical functions used to estimate landfill settlement. Include a discussion of advantages and limitations of each method.

12.11 What are the main purposes for evaluating landfill foundation settlement? How are the settlement lines and points arranged or located in the landfill cells? (You may use a drawing or sketch to help explain the procedure.)

REFERENCES

Edgers, L., Noble, J. J., and Williams E., (1992) "A Biologic Model for Long Term Settlement in Landfills," *Environmental Geotechnology,* Proceedings of the Mediterranean Conference on Environmental Geotechnology, A. A. Balkema Publishers, pp. 177–184.

Edil, T. B., Fox, P. J., and Lan, L. T., (1991) "Observational Procedure for Settlement of Peat," *Proceedings of Geo-Coast 91 Conference,* Port & Harbor Research Inst., Japan, pp. 165–170.

Edil, T. B., Ranguette, V. J., and Wuellner, W. W., (1990) "Settlement of Municipal Refuse," *Geotechnics of Waste Fills—Theory and Practice,* ASTM STP 1070, Arvid Landva and G. David Knowles, Eds., Philadelphia, pp. 225–239.

Ham, R. K., Reinhardt, J. J., and Sevick, G. W., (1978) "Density of Milled and Unprocessed Refuse," *Journal of Environmental Engineering,* ASCE, Vol. 104, No. 1.

Kondner, R. L. and Zelasko, J. S., (1963) "A Stress-Strain Formulation of Sands," *Proceedings of 2nd Pan-America Conference on Soil Mechanics and Foundation Engineering,* Brazil, p. 289.

Ling, H. I., Leshchinsky, D., Mohri, Y., and Kawabata, T., (1998) "Estimation of Municipal Solid Waste Landfill Settlement," *Journal of Geotechnical and Geoenvironmental Engineering,* ASCE, Vol. 124, No. 1, pp. 21–28.

Merz, R. C. and Stone, R., (1962) "Landfill Settlement Rates," *Public Works,* Volume 93, No. 9, pp. 103–106, 210–212.

Morris, D. V. and Woods, C. E., (1990) "Settlement and Engineering Consideration in Landfill and Final Cover Design," *Geotechnics of Waste Fills—Theory and Practice,* ASTM STP 1070, Arvid Landva and G. David Knowles, Eds., Philadelphia, PA, pp. 9–21.

Murphy, W. L. and Gilbert, P. A., (1985) "Settlement and Cover Subsidence of Hazardous Waste Landfills," Final Report to Municipal Environmental Research Laboratory, Office of Research and Development, US Environmental Protection Agency, Cincinnati, OH, Report No. EPA-600/2-85-035.

Noble, J. J., Nair, G. M., and Heestand, J. F., (1989) "Some Numerical Predictions for Moisture Transport in Capped Landfills at Long Times," *Proceedings of the 12th Annual Madison Waste Conference,* University of Wisconsin, Madison, Wisconsin, September, pp. 353–366.

Pohland, F. and Graven, J. P., (1993) "The Use of Alternative Materials for Daily Cover at Municipal Solid Waste Landfills," US EPA Report, EPA/600/R-93/172, US Environmental Protection Agency, Washington, DC.

Sanchez-Alciturri, J. M., Palma, J., Sagaseta, C., and Canizal, J., (1995) "Three Years of Deformation Monitoring at Meruelo Landfill," *Waste Disposal by Landfill—GREEN 3,* Sarsby Ed., A. A. Balkema, Rotterdam, The Netherlands, pp. 365–371.

Spikula, D. R. (1997) "Subsidence Performance of Landfills," Proceedings GRI-10 Conference, GSI, Folsom, Pennsylvania, PP. 237–244.

Sohn, K. C. And Lee, S., (1994) "A Method for Prediction of Long Term Settlement of Sanitary Landfill," *Proceedings of the First International Congress on Environmental Geotechnics,* Edmonton, Canada, pp. 807–811.

Sowers, G. F., (1968) "Foundation Problem in Sanitary Land Fills," *Journal of Sanitary Engineering,* ASCE, Vol. 94, No. 1, pp. 103–116.

Sowers, G. F., (1973) "Settlement of Waste Disposal Fills," *Proceedings of the 8th International Conference on Soil Mechanics and Foundation Engineering,* Moscow, Vol. 1, pp. 207–210.

Tan, T.-S., Inoue, T., and Lee, S.-L., (1991) "Hyperbolic Method for Consolidation Analysis," *Journal of Geotechnical Engineering,* ASCE, Vol. 117, No. 11, pp. 1723–1737.

Yen, B. C. and Scanlon, B., (1975) "Sanitary Landfill Settlement Rates," *Journal of Geotechnical Engineering,* ASCE, Vol. 101, No. 5, pp. 475–487.

CHAPTER 13

Landfill Stability Analysis

13.1 TYPES OF LANDFILL INSTABILITY
    13.1.1 SLIDING FAILURE OF LEACHATE COLLECTION SYSTEM
    13.1.2 SLIDING FAILURE OF FINAL COVER SYSTEM
    13.1.3 ROTATIONAL FAILURE OF SIDEWALL SLOPE OR BASE
    13.1.4 ROTATIONAL FAILURE THROUGH WASTE, LINER, AND SUBSOIL
    13.1.5 ROTATIONAL FAILURE WITHIN THE WASTE MASS
    13.1.6 TRANSLATIONAL FAILURE BY MOVEMENT ALONG LINER SYSTEM
13.2 FACTORS INFLUENCING LANDFILL STABILITY
13.3 SELECTION OF APPROPRIATE PROPERTIES
    13.3.1 GEOSYNTHETIC MATERIALS PROPERTIES
    13.3.2 SOLID WASTE PROPERTIES
    13.3.3 IN-SITU SOIL SLOPE AND SUBSOIL PROPERTIES
13.4 VENEER SLOPE STABILITY ANALYSIS
    13.4.1 COVER SOIL (GRAVITATIONAL) FORCES
    13.4.2 TRACKED CONSTRUCTION EQUIPMENT FORCES
    13.4.3 INCLUSION OF SEEPAGE FORCES
    13.4.4 INCLUSION OF SEISMIC FORCES
    13.4.5 GENERAL RESULTS
13.5 SUBSOIL FOUNDATION FAILURES
    13.5.1 METHOD OF ANALYSIS
    13.5.2 CASE HISTORIES
    13.5.3 GENERAL REMARKS
13.6 WASTE MASS FAILURES
    13.6.1 TRANSLATIONAL FAILURE ANALYSIS
    13.6.2 CASE HISTORIES
    13.6.3 GENERAL REMARKS
13.7 CONCLUDING REMARKS
    PROBLEMS
    REFERENCES

Modern solid waste landfills serve a variety of functions, including maximization of waste storage per unit area, isolation of waste from the surrounding environment, and conversion opportunities to useable land areas after closure. Until recently, attention related to the design, construction, filling, and post-closure monitoring and maintenance of new landfills has focused mainly on the prevention of unacceptable levels of leakage in the surrounding groundwater. Closure plans for old landfills have likewise centered on similar concerns of gas emissions that are fully warranted. However, stability is an issue that has sometimes been overlooked. Several huge failures along liner slopes—through landfill foundations and within the waste mass itself—have occurred.

Along lined slopes, two stability situations can be readily identified: (i) the leachate collection layer along the base liner before waste is placed, and (ii) the final cover system above the waste. While instability of these relatively thin layers can be classified as failures, their impact is usually localized and repairs can sometimes be made at a reasonable cost. They are often referred to as *veneer* failures.

Foundation failures beneath the waste and failures of the waste mass itself are in completely different categories in that implications are generally severe. For example, the 1988 failure of 490,000 m^3 of hazardous waste was very significant due to its repair cost and legal implications, which required a thorough post-failure analysis (Mitchell et al., 1990; Seed et al. 1990; Byrne et al. 1992). This failure was very significant in that it stimulated (i) the introduction of textured geomembranes, (ii) consideration of post-peak soil shear strengths, and (iii) concern over low shear strengths of compacted clay liners placed beneath geomembranes. In addition, waste placement and closure plans for landfills on very soft foundations can be seriously affected by the potential for stability failures.

Consequently, the mass stability of landfills is now a major concern in their design, construction, filling, and closure. The situation is further heightened when bioreactor landfills (with large amounts of liquids) are being considered. (See Chapter 15.) This chapter includes a discussion of potential failure mechanisms, evaluation of relevant material properties, description of stability methods, factor of safety calculations, and description of a number of case histories involving landfill failures.

13.1 TYPES OF LANDFILL FAILURES

Landfills can fail in several ways—during cell excavation, during liner system construction, during waste filling, and after landfill closure. An important feature in the identification and assessment of potential failure mode is the fact that both covers and liners for modern landfills are typically multi-layer composites composed of both soil and geosynthetic materials. A schematic diagram of a typical double composite liner system used for a municipal solid waste landfill was shown in Figure 1.10. The liner system shown in Figure 1.10 contains several interfaces whose resistance against interface shear stresses may be low, and thus act as possible failure surfaces. Additionally, all classical geotechnical failure modes are possible depending upon site-specific conditions (usually involving saturated fine-grained soils) and the placement and geometry of the waste mass. Potential failure modes are summarized schematically in Figure 13.1. A brief description of each situation follows.

13.1.1 Sliding Failure of Leachate Collection System

As seen in Figure 13.1(a), the leachate collection system can slide on the underlying liner system if the slope is too steep or too long. This type of veneer failure has often occurred during heavy rains. It is remedied by pushing the sand or gravel back onto the lined slope. However, if the failure surface is within the liner system, the

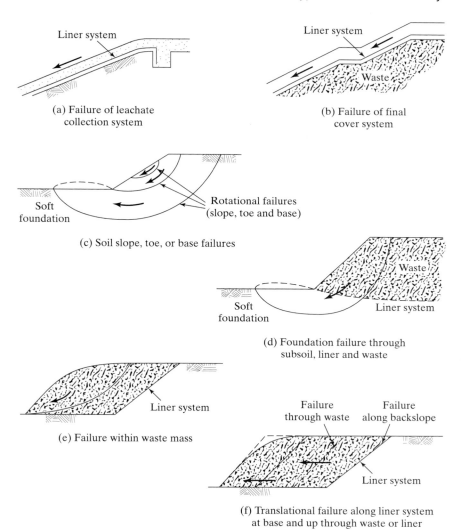

FIGURE 13.1 Various Types of Landfill Failures to be Described/Analyzed in this Chapter

reconstruction will require more effort and cost. Leachate collection soil sliding above or within the liner system is not an uncommon situation.

13.1.2 Sliding Failure of Final Cover System

As seen in Figure 13.1(b), the final cover system (topsoil and protection soil) can slide on the liner system if the slope is too steep or too long. This type of veneer failure often occurs during a heavy rain, and should also be investigated for seismic stability depending on site-specific conditions. If only soil is displaced, the remedy is to replace the soil. However, the question of long-term stability remains. If the failure surface is within the liner system, the implications are more severe. Cover soil failures of this type above the waste mass are not uncommon.

13.1.3 Rotational Failure of Sidewall Slope or Base

As seen in Figure 13.1(c), the soil mass behind the waste repository or beneath the site could be unstable and fail. Failure is usually rotational, emerging along the slope, at the toe, or within the foundation. This is completely a geotechnical problem (i.e., geosynthetics are not included) and applies to steep side slopes and/or soft foundation soils. It is obviously site specific and does not involve liner systems or waste properties. Nevertheless, such situations should be investigated.

13.1.4 Rotational Foundation Failure Through Waste, Liner, and Subsoil

As seen in Figure 13.1(d), a rotational failure can be initiated in a soft foundation soil that can propagate up through the waste mass. If a liner system is present, it offers only negligible resistance and should be discounted in the analysis. Such failures have occurred (in both unlined and lined sites) and a few have been massive (e.g., up to 500,000 m^3).

13.1.5 Rotational Failure within the Waste Mass

As seen in Figure 13.1(e), failure can occur within the waste mass, completely independent of the liner system. It is handled exactly as the geotechnical failures illustrated in Figure 13.1(c), except that the material is solid waste (municipal or hazardous) instead of soil. Such failures are prompted by steep waste slopes, high liquid content, and lack of placement (operations) control.

13.1.6 Translational Failure by Movement along the Liner System

As seen in Figure 13.1(f), a lateral translational failure can occur with the solid waste sliding above, within, or beneath the liner system at the base of the waste mass. The extension of the failure plane back from the toe can propagate up through the waste, or continue in the liner system along the back slope. Such failures have occurred at both clay-lined sites and at geosynthetically-lined sites. They have resulted in the largest failures to date, two of which involved over 1,000,000 m^3 of waste.

13.2 FACTORS INFLUENCING LANDFILL STABILITY

The stability of a landfill is influenced by many variables and considerations in addition to those present in conventional geotechnical analyses. A landfill is a complex system with multiple locations and interfaces, which constitute possible loci for failure. The factors influencing landfill stability can be summarized as follows:

(i) Interface shear strengths between various geosynthetic materials;
(ii) Interface shear strengths between geosynthetics and soil materials;
(iii) Internal shear strengths of compacted clay liners (CCLs);
(iv) Internal shear strength of solid waste;
(v) Internal shear strength of subbase and side slope soils;
(vi) Slope and height of excavated side slopes;
(vii) Waste filling slope and height;
(viii) Landfill subbase slopes;
(ix) Normal stresses;
(x) Pore water pressures acting on base liner;
(xi) Subbase geological profiles;
(xii) Groundwater level;
(xiii) Local site hydrology;
(xiv) Freeze-thaw conditions;
(xv) Construction placement and operations equipment; and
(xvi) Dynamic and/or seismic stresses.

13.3 SELECTION OF APPROPRIATE PROPERTIES

The three types of materials usually encountered in failures of the types addressed in this chapter are geosynthetics, solid wastes, and natural soils. Each will be addressed. However, failures sometimes happen in materials other than the above—for example, one failure occurred within a layer of leachate-saturated wood bark chips. Thus, the situation is ever-evolving and the lessons learned from these unfortunate incidences are important for future projects and their respective designs.

13.3.1 Geosynthetic Materials Properties

Within a multilayer liner system, there are numerous materials of concern with regard to low interface shear strengths. The following lists some of them, along with the current method of lessening the concern:

(i) Low-friction, smooth-geomembrane surfaces, the shear strength of which is greatly enhanced by *texturing*. Recall Section 4.2.4.
(ii) Geotextile-to-geonet surfaces; the shear strength concern of this material is essentially eliminated by thermal bonding in the factory.
(iii) Bentonite containing GCLs; the shear strength of this material is enhanced by needle punching, stitch bonding, or encasement between two geomembranes.
(iv) Bentonite extruding from GCLs; the shear susceptibility of this material is avoided by using nonwoven geotextiles on both surfaces of the GCL.
(v) Geomembrane-to-compacted clay liners; the shear strength of this material is lessened by proper moisture content control when placing the compacted clay.

For all of the listed interfaces, and others as well, the project-specific materials evaluated under site-specific conditions are necessary. By project-specific materials, we mean replicates of the candidate geosynthetics to be used at the site, as well as the associated soils at their targeted density and moisture conditions. By site-specific conditions, we mean normal stresses, strain rates, peak or residual shear strengths, and temperature extremes (high and/or low). Moisture content is also important and anticipated conditions (often saturated) are necessary. Water is usually used, although in some cases, site-specific leachates are required. Note that it is completely inappropriate to use values of interface shear strengths from the literature for final design.

The test for the assessment of shear strength of interfaces involving geosynthetics is the direct shear test that has been utilized in geotechnical engineering for many years. The procedure to follow is ASTM D5321, and a special adaptation when GCLs are involved, namely, ASTM D6243.

In conducting a direct shear test on a specific interface, one typically performs three replicate tests with the only variable being different values of normal stress. The middle value is usually targeted to the anticipated site-specific condition, with a lower and higher value of normal stress covering the range of possible values. The results of these three tests yield a set of shear displacement versus shear stress curves. (See Figure 13.2(a).) From each curve, a peak shear strength (τ_p) and a residual shear strength (τ_r) are obtained. As a next step, these shear-strength values, together with their respective normal stress values, are plotted on Mohr-Coulomb stress space to obtain the shear-strength parameters of friction and adhesion. (See Figure 13.2(b).)

The points are then connected (usually with a straight line), and the two fundamental shear-strength parameters are obtained. These shear-strength parameters are as follows:

δ = the angle of shearing resistance, peak and/or residual, of the two opposing surfaces (often called the interface friction angle); and

c_a = the adhesion of the two opposing surfaces, peak and/or residual (the compliment of cohesion when testing fine-grained soils against one another).

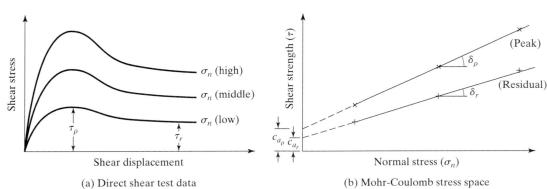

(a) Direct shear test data   (b) Mohr-Coulomb stress space

FIGURE 13.2 Direct Shear Test Results and Method of Analysis to Obtain Shear Strength Parameters

Each set of parameters constitutes the equation of a straight line that is the Mohr-Coulomb failure criterion common to geotechnical engineering. The concept is readily adaptable to geosynthetic materials in the following forms:

$$\tau_p = c_{ap} + \sigma_n \cdot \tan\delta_p \tag{13.1}$$

$$\tau_r = c_{ar} + \sigma_n \cdot \tan\delta_r \tag{13.2}$$

The upper limit of δ when soil is involved as one of the interfaces is ϕ, the angle of shearing resistance of the soil component. The upper limit of the c_a-value is c, the cohesion of the soil component. In the slope stability analyses to follow, the c_a term will be included for the sake of completeness, but then it will be neglected (as being a conservative assumption) in the design graphs and numeric examples. To utilize an adhesion value, there must be a clear physical justification for use of such values when geosynthetics are involved. Only unique situations such as textured geomembranes with physical interlocking of soils having cohesion or the needle-punched fibers of a GCL are valid reasons for including such a term.

Note that residual strengths are always equal to, or lower than, peak strengths. The amount of difference is very dependent on the material and no general guidelines can be given. Clearly, material-specific and site-specific direct shear tests must be performed to determine the appropriate values. Further, each direct shear test must be conducted to a relatively large displacement to determine the residual (or, at least, large displacement) behavior (see Stark and Poeppel, 1994). The decision as to the use of peak, large displacement, or residual strengths in the subsequent analysis is a very subjective one. It is both a materials-specific and site-specific issue that is left up to the designer and/or regulator. Even further, the combination of use of peak values at the crest of a slope, and large displacement or residual values at the toe may be justified. As such, the analyses to follow will use an interface δ-value with no subscript, thereby concentrating on the computational procedures, rather than this particular detail. However, the importance of appropriate and accurate δ-values to be used in the analysis cannot be stressed enough.

Due to the manufactured structure of many geosynthetics, the size of the recommended shear box is quite large. It must be at least 300 mm by 300 mm unless it can be shown that data generated by a smaller device contains no scale or edge effects—that is, that no bias exists when using a smaller shear box. The implications of such a large shear box should not be taken lightly. The following issues should receive particular attention:

(i) Unless it can be justified otherwise, the interface will usually be tested in a saturated state. Thus, complete and uniform saturation over the entire specimen area must be achieved. This is particularly important for CCLs and GCLs (Daniel et al., 1993). Hydration takes relatively long in comparison to soils in conventional (smaller) testing shear boxes.

(ii) Consolidation of soils (including CCLs and GCLs) in larger shear boxes is similarly effected.

(iii) Uniformity of normal stress over the entire area must be maintained during consolidation and shearing to prevent stress concentrations from occurring. This is particularly important when testing at high normal stresses.

(iv) The application of relatively low normal stresses (e.g., 1 to 3 psi or 7 to 20 kPa) simulating typical cover soil thicknesses, challenges the accuracy of some commercially available shear box setups and monitoring systems, especially pressure gages.

(v) Shear rates necessary to attain drained conditions (if this is the desired situation) are extremely slow, requiring long testing times.

(vi) Deformations necessary to attain residual strengths require large relative movement of the two respective halves of the shear box. To prevent travel over the edges of the opposing shear box sections, devices should have the lower shear box significantly longer than 12 inches (300 mm). However, with a lower shear box longer than the upper traveling section, new surface is constantly being added to the shearing plane. This influence is not clear in the material's response or in the subsequent behavior.

(vii) The attainment of a true residual strength is difficult to achieve. ASTM D5321 states that one should "run the test until the applied shear force remains constant with increasing displacement". Many commercially available shear boxes have insufficient travel to reach this condition, and thus most report large-displacement values.

(viii) The ring torsion shearing apparatus is an alternative device that can be used to determine true residual strength values. However, this device has its own problems that are caused by inherent large deformations. See Stark and Poeppel (1994) for information and data using this alternative test method.

13.3.2 Solid Waste Properties

The shear strength of solid waste of all types is high. Indeed near vertical cuts have been made in municipal solid waste with no indications of instability. Hazardous waste, which is either particulate or contains large amounts of soil, is also quite stable. The shear strength of municipal solid waste was studied by Singh and Murphy (1990) and their results were presented in Figure 6.7. The data is felt to be quite conservative and, even so, represents a material with considerable shear strength (i.e., a relatively high c- and/or ϕ-value). In spite of the high shear strength of waste materials, waste materials can (and have) failed because of related situations, the most common of which are the following:

(i) Weak subsoils beneath the waste mass [recall Figures 13.1(c) and (d)], which initially fail within the foundation and then continue progressively up through the waste mass.

(ii) High leachate levels in the waste create hydrostatic pressure, which decreases the effective stresses at the base. This, in turn, decreases the shear strength of an otherwise stable material.

(iii) Leachate injection into landfills, (see Chapter 15) has the same effect of decreasing effective stresses and the shear strength of the waste materials.

Thus, the shear strength of solid waste is an important issue, and the reader is referred to the appropriate sections in Chapter 6.

13.3.3 In-Situ Soil Slope and Subsoil Properties

A thorough geotechnical investigation in parallel with a hydrogeological study to reliably identify soil and site stratigraphy is necessary. During the investigation, information such as Standard Penetration Test (SPT) blow counts, visual description of soil samples, depth to groundwater, etc., should be recorded. Undisturbed samples are to be collected for laboratory analysis. Laboratory or in-situ tests are then conducted to evaluate the engineering properties of the soils. These properties primarily include the short- and long-term shear parameters, unit weight, and moisture content (Hassini, 1992).

"Short term" usually refers to the elapsed time up to the end of construction (i.e., the excavation time). Short-term failure is generally due to excessively steep slopes and may occur within a relatively short time after excavation. For saturated clays, excavation causes a rapid change of stress condition within the slope. In the zone of potential failure (i.e., high induced shear stress and/or low shear strength), a slight positive excess pore water pressure may initially result, which will cause a reduction in the effective shear strength, and hence increase the likelihood of failure.

"Long term" generally refers to long duration, post-construction times. As deformation along a potential failure zone increases beyond a critical limit, significant negative excess pore pressure begins to develop, which temporarily increases the strength in the potential failure zone (Humphrey and Leonards, 1986). The dissipation of this negative excess pore pressure can trigger failure of inadequately designed slopes. The rate of dissipation of negative excess pore pressure depends primarily on the coefficient of consolidation of the clay and the average depth to the potential failure zone. Drained shear strength conditions are attained at approximately the same time the excess pressure dissipates. Using Terzaghi's theory of consolidation (Terzaghi, 1943), it is possible to estimate the time for negative pore pressure release. For instance, let the characteristics of the slope be as follows:

Consolidation coefficient, c_v = 0.11 ft²/day (0.01 m²/day);
Depth to potential failure zone, H = 16.5 ft (5.0 m); and
Time factor for 90% consolidation, T_v = 0.95.

The time for pore water pressure release by substituting these values into the time-for-consolidation formulation of Equation 13.2 is about 6 years. The relevant equation is

$$t = \frac{T_v \cdot H^2}{C_v} \tag{13.3}$$

The critical factor of safety for slope stability analysis usually extends to the end of dissipation of the excess negative pore pressure. It is necessary to perform stability analysis for the following two conditions (Hassini, 1992):

(i) Immediately after excavation: The undrained shear parameters are used with a correction taking into account the rapid, but temporary, slight increase in pore pressure.

(ii) After a period corresponding to dissipation of negative excess pore pressure: The drained shear parameters are used. The effect of the reduction in confining stress (causing swelling or expansion) on shear parameters should be included.

For landfill covers, long-term conditions are most likely to be critical. Effective stress analysis is to be used. Drained shear parameters are determined from consolidated-drained tests or consolidated-undrained tests with pore pressure measurements. The pore pressure should be determined in the field from flow-net or seepage analysis. For the stability analysis, a factor of safety of 1.5 is recommended (Hassini, 1992).

Representative soil samples should be tested in the laboratory to determine soil parameters necessary for the appropriate calculations. The testing should include natural moisture contents, in-situ dry densities, grain size distributions, Atterberg Limits, unconfined compressive strengths, triaxial tests, and one-dimensional consolidation tests.

Consolidated-undrained (CU) triaxial tests with pore water pressure measurements and unconsolidated-undrained (UU) triaxial tests should be performed on selected Shelby tube samples. In general, the triaxial test consists of two stages. The first stage consists of the application of an all-around cell confining pressure. The second stage consists of the application of a principal stress to the top of the sample to induce shear stresses and eventual failure.

In the CU triaxial test, the sample is allowed to fully consolidate (i.e., all excess pore water pressures are dissipated) during the first stage. During the second stage, the sample fails slowly, at a strain rate of 0.002 in/min (0.05 mm/min), generating excess negative or positive pore water pressures. During the CU test, the pore water pressure is constantly measured during the second phase to allow a determination of the parameters necessary for an evaluation of the shear strength of the soils using an effective stress analysis. The effective stress analysis treats the soil as an essentially frictional material and is normally used for the evaluation of the long-term shear strength of cohesive soils. A three-point consolidated-undrained (CU) triaxial test is run using selected Shelby tube samples. Three consolidation pressures are selected to approximate the different confining pressures of the in-situ samples.

An unconsolidated-undrained (UU) triaxial test is performed to verify the short-term strength of the soil. In the UU triaxial test, the sample is not allowed to consolidate during the application of the cell pressure. The results of the UU test simulate the behavior of cohesive soils that fail in rapid shear over a short period of time (i.e., at a strain rate of approximately 1.4 mm/min). The results are therefore considered applicable to short-term slope stability analyses.

In order to determine average soil parameters for the slope stability analyses, the following parameters are to be plotted with respect to the site-specific stratigraphy: natural moisture content, in-place dry density, "N" values from the Standard Penetration Test, and unconfined compressive strength. The plots of these parameters versus depth are used in conjunction with the results from the triaxial tests to develop design parameters for the different soil strata defined in each critical cross section. The typical liner cross section and total unit weights and shear-strength values for these soil strata are then presented on individual cross sections to be used for the slope stability analyses.

13.4 VENEER SLOPE STABILITY ANALYSES

This section treats the standard veneer slope stability problem [as shown in Figure 13.1(a) and (b)] and then superimposes upon it a number of situations, all which tend to destabilize slopes. Included are gravitational, construction equipment, seepage and seismic forces, respectively. Each will be illustrated by a design graph and a numeric example.

13.4.1 Cover Soil (Gravitational) Forces

Figure 13.3 illustrates the common situation of a finite-length, uniformly-thick cover soil placed over a liner material at a slope angle β. It includes a passive wedge at the toe and has a tension crack on the crest. The analysis that follows is from Koerner and Soong (1998), but it is similar to Koerner and Hwu (1991). Comparable analyses are also available from Giroud and Beech (1989), McKelvey and Deutsch (1991), and others.

The symbols used in Figure 13.3 are defined a follows:

W_A = total weight of the active wedge
W_P = total weight of the passive wedge
N_A = effective force normal to the failure plane of the active wedge
N_P = effective force normal to the failure plane of the passive wedge
γ = unit weight of the cover soil
h = thickness of the cover soil
L = length of slope measured along the geomembrane
β = soil slope angle beneath the geomembrane

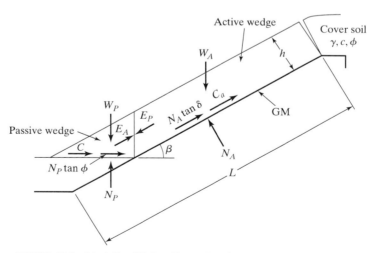

FIGURE 13.3 Limit Equilibrium Forces Involved in a Finite Length Slope Analysis for a Uniformly Thick Cover Soil

ϕ = friction angle of the cover soil
δ = interface friction angle between cover soil and geomembrane
C_a = adhesive force between cover soil of the active wedge and the geomembrane
c_a = adhesion between cover soil of the active wedge and the geomembrane
C = cohesive force along the failure plane of the passive wedge
c = cohesion of the cover soil
E_A = interwedge force acting on the active wedge from the passive wedge
E_p = interwedge force acting on the passive wedge from the active wedge
FS = factor of safety against cover soil sliding on the geomembrane.

The expression for determining the factor of safety can be derived as follows:

Considering the active wedge, the forces acting on it are

$$W_A = \gamma \cdot h^2 \cdot (L/h - 1/\sin\beta - \tan\beta/2) \tag{13.4}$$

$$N_A = W_A \cdot \cos\beta \tag{13.5}$$

$$C_a = c_a \cdot (L - h/\sin\beta) \tag{13.6}$$

By balancing the forces in the vertical direction, the following formulation results:

$$E_A \cdot \sin\beta = W_A - N_A \cdot \cos\beta - \frac{N_A \cdot \tan\delta + C_a}{FS}$$

Hence, the interwedge force acting on the active wedge is

$$E_A = \frac{(FS)(W_A - N_A \cdot \cos\beta) - (N_A \cdot \tan\delta + C_a) \cdot \sin\beta}{\sin\beta \cdot (FS)} \tag{13.7}$$

The passive wedge can be considered in a similar manner:

$$W_P = \frac{\gamma \cdot h^2}{\sin 2\beta}$$

$$N_P = W_P + E_P \cdot \sin\beta$$

$$C = \frac{c \cdot h}{\sin\beta} \tag{13.8}$$

By balancing the forces in the horizontal direction, the following formulation results:

$$E_P \cdot \cos\beta = \frac{C + N_P \cdot \tan\phi}{FS}$$

Hence, the interwedge force acting on the passive wedge is

$$E_P = \frac{C + W_P \cdot \tan\phi}{\cos\beta \cdot (FS) - \sin\beta \cdot \tan\phi}$$

By setting $E_A = E_P$, the resulting equation can be arranged in the form of the quadratic equation $ax^2 + bx + c = 0$, which in this case, using FS-values, results in

$$a \cdot FS^2 + b \cdot FS + c = 0$$

The resulting FS-value is then obtained from the conventional solution of the quadratic equation, which gives

$$FS = \frac{-b \pm (b^2 - 4 \cdot a \cdot c)^{0.5}}{2 \cdot a} \qquad (13.9)$$

where $a = (W_A - N_A \cdot \cos\beta) \cdot \cos\beta$
$b = -[(W_A - N_A \cdot \cos\beta) \cdot \sin\beta \cdot \tan\phi + (N_A \cdot \tan\delta + C_a) \cdot \sin\beta \cdot \cos\beta + (C + W_P \cdot \tan\phi) \cdot \sin\beta]$
$c = (N_A \cdot \tan\delta + C_a) \cdot \sin^2\beta \cdot \tan\phi$

When the calculated FS-value falls below 1.0, sliding of the cover soil on the geomembrane is to be anticipated. Thus, a value of greater than 1.0 must be targeted as being the minimum factor of safety. How much greater than 1.0 the FS-value should be, is a design and/or regulatory issue. Recommendations for minimum allowable FS-values under different conditions are available in Koerner and Soong (1998). In order to better illustrate the implications of Equations 13.9, typical design curves for various FS-values as a function of slope angle and interface friction angle are given in Figure 13.4. Note that the curves are developed specifically for the variables stated in the legend of the figure. Example 13.1 illustrates the use of the analytic development and the

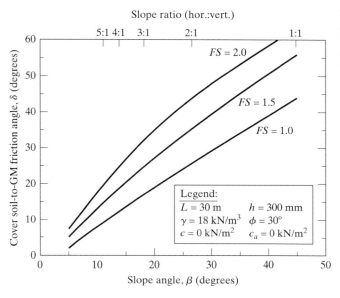

FIGURE 13.4 Design Curves for Stability of Uniform-Thickness Cohesionless Cover Soils on Linear Failure Planes for Various Global Factors of Safety

resulting design curves in what will be the standard example to which other examples will be considered as compared.

EXAMPLE 13.1

The following are given: a 30-m slope with a uniformly thick 300-mm-deep cover soil at a unit weight of 18 kN/m³. The soil has a friction angle of 30° and zero cohesion (i.e., it is a sand). The cover soil is placed directly on a geomembrane as shown in Figure 13.3. Direct shear testing has resulted in an interface friction angle between the cover soil and geomembrane of 22° with zero adhesion. What is the *FS*-value at a slope angle of 3(H)-to-1(V) (i.e., 18.4°)?

Solution Using Equation 13.9 to solve for the *FS*-value results in a value of 1.25, which is seen to be in agreement with the curves of Figure 13.4:

$$a = 14.7 \text{ kN/m}$$
$$b = -21.3 \text{ kN/m}$$
$$c = 3.5 \text{ kN/m}$$

Thus, $FS = 1.25$

This value can be confirmed using Figure 13.4.

Comment In general, this is too low of a value for a final cover soil factor-of-safety and a redesign is necessary. There are many possible options to increase the value (e.g., changing the geometry of the situation, the use of toe berms, tapered cover soil thickness, and veneer reinforcement, see Koerner and Soong, 1998). Nevertheless, this general problem will be used throughout this section for comparison with other cover soil slope stability situations.

13.4.2 Tracked Construction Equipment Forces

The placement of cover soil on a slope with a relatively low shear strength interface (like a geomembrane) should always start at the toe and move upward to the crest. Figure 13.5(a) shows the recommended method. In doing so, the gravitational forces of the cover soil and live load of the construction equipment are compacting previously placed soil and working with an ever-present passive wedge and a stable lower portion beneath the active wedge. While it is necessary to specify low ground pressure equipment to place the soil, the reduction in the *FS*-value for this situation of equipment working up the slope will be seen to be relatively small.

For soil placement down the slope, however, a stability analysis cannot rely on toe buttressing and also a dynamic stress should be included in the calculation. These conditions decrease the *FS*-value—in some cases, to a great extent. Figure 13.5(b) shows this procedure. Unless absolutely necessary, it is not recommended that cover soil be placed on a slope in this manner. If it is necessary, the design must consider the unsupported soil mass and the possible dynamic force of the specific type of construction equipment and its manner of operation.

For the *first case* of a bulldozer pushing cover soil up from the toe of the slope to the crest, the analysis uses the free body diagram of Figure 13.6(a). The analysis uses a

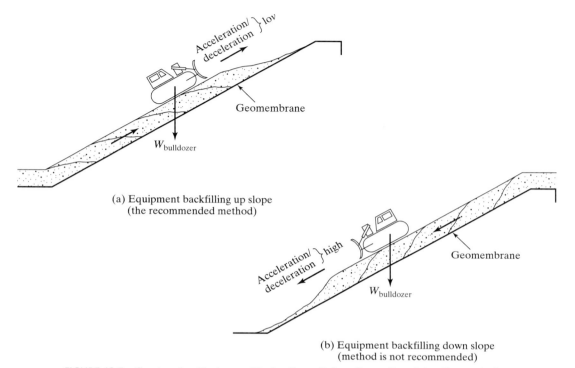

FIGURE 13.5 Construction Equipment Placing Cover Soil on Slopes Containing Geosynthetics

known type of construction equipment (such as a bulldozer characterized by its ground contact pressure) and dissipates this force or stress through the cover soil thickness to the surface of the geomembrane. A Boussinesq analysis is used (see Poulos and Davis, 1974). This results in an equipment force per unit width of

$$W_e = q \cdot w \cdot I \qquad (13.10)$$

where W_e = equivalent equipment force per unit width at the geomembrane interface;
$q = W_b/(2 \cdot w \cdot b)$;
W_b = actual weight of equipment (e.g., a bulldozer);
w = length of equipment track;
b = width of equipment track;
I = influence factor at the geomembrane interface (see Figure 13.7).

Upon determining the additional equipment force at the cover soil-to-geomembrane interface, the analysis proceeds as described in Section 13.3.1 for gravitational forces only. In essence, the equipment moving up the slope adds an additional term (W_e) to the W_A-force in Equation 13.4. Note, however, that this involves the generation of a resisting force as well. Thus, the net effect of increasing the driving force as well as the resisting force is somewhat neutralized insofar as the resulting *FS*-value is

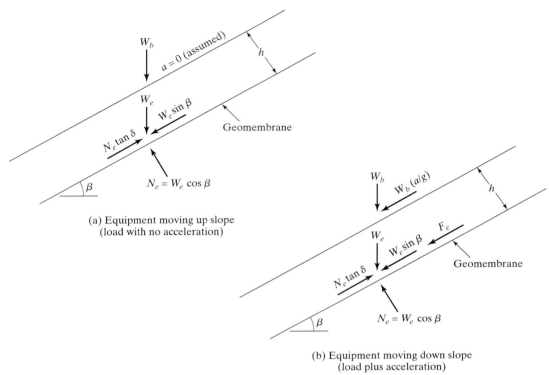

FIGURE 13.6 Additional (to Gravitational Forces) Limit Equilibrium Forces due to Construction Equipment Moving on Cover Soil (see Figure 13.3 for the gravitational soil force to which the above forces are added).

concerned. It should also be noted that no acceleration/deceleration forces are included in this analysis, which is somewhat idealistic. Using these concepts (the same equations used in Section 13.3.1 are used here), typical design curves for various *FS*-values as a function of equivalent ground contact equipment pressures and cover soil thicknesses are given in Figure 13.8. Note that the curves are developed specifically for the variables stated in the legend. Example 13.2 illustrates the use of the formulation.

EXAMPLE 13.2

The following are given: a 30-m-long slope with uniform cover soil of 300 mm thickness at a unit weight of 18 kN/m³. The soil has a friction angle of 30° and zero cohesion (i.e., it is a sand). It is placed on the slope using a bulldozer moving from the toe of the slope up to the crest. The bulldozer has a ground pressure of 30 kN/m² and tracks that are 3.0 m long and 0.6 m wide. The cover soil to geomembrane friction angle is 22° with zero adhesion. What is the *FS*-value at a slope angle 3(H)-to-1(V)(i.e., 18.4°)?

Solution This problem follows Example 13.1 exactly except for the addition of the bulldozer moving up the slope. Using the additional equipment load, Equation 13.10 substituted into Equation 13.9 results in the following:

$$a = 73.1 \text{ kN/m}$$
$$b = -104.3 \text{ kN/m}$$
$$c = 17.0 \text{ kN/m}$$

Thus, $FS = 1.24$

This value can be confirmed using Figure 13.8.

Comment While the resulting *FS*-value is still low, the result is important to assess by comparing it with Example 13.1 (i.e., the same problem except without the bulldozer). It is seen that the *FS*-value has only decreased from 1.25 to 1.24. Thus, in general, a low ground contact pressure bulldozer placing cover soil up the slope with negligible acceleration/deceleration forces does not significantly decrease the factor-of-safety.

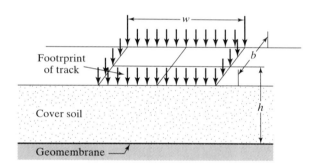

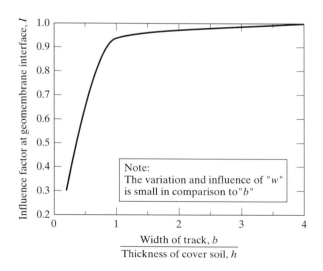

FIGURE 13.7 Values of Influence Factor, "I", for Use in Equation 13.10 to Dissipate Surface Force through the Cover Soil to the Geomembrane Interface (after Soong and Koerner, 1996)

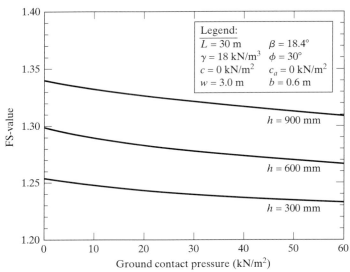

FIGURE 13.8 Design Curves for Stability of Different Thickness of Cover Soil for Various Construction Equipment Ground Contact Pressure

For the *second case* of a bulldozer pushing cover soil down from the crest of the slope to the toe as shown in Figure 13.5b, the analysis uses the force diagram of Figure 13.6(b). While the weight of the equipment is treated as just described, the lack of a passive wedge along with an additional force due to acceleration (or deceleration) of the equipment significantly decreases the resulting *FS*-values. This analysis again uses a specific piece of construction equipment operated in a specific manner. It produces a force parallel to the slope equivalent to $W_b \cdot (a/g)$, where W_b = the weight of the bulldozer, a = acceleration of the bulldozer, and g = acceleration due to gravity. Its magnitude is equipment operator dependent and related to both the equipment speed and time to reach such a speed (see Figure 13.9).

The acceleration of the bulldozer, coupled with an influence factor I from Figure 13.7, results in the dynamic force per unit width at the cover soil to geomembrane interface F_e. The relationship is given by

$$F_e = W_e \cdot (a/g) \qquad (13.11)$$

where F_e = dynamic force per unit width parallel to the slope at the geomembrane interface;
  W_e = equivalent equipment (e.g., bulldozer) force per unit width at geomembrane interface, recall Equation 13.10;
  β = soil slope angle beneath geomembrane;
  a = acceleration of the construction equipment;
  g = acceleration due to gravity.

Using these concepts, the new force parallel to the cover soil surface is dissipated through the thickness of the cover soil to the interface of the geomembrane. Again, a

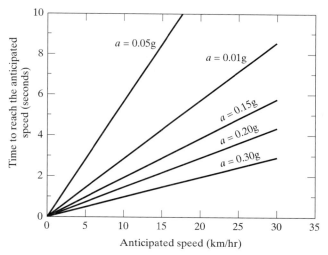

FIGURE 13.9 Graphic Relationship of Construction Equipment Speed and Rise Time to Obtain Equipment Acceleration.

Boussinesq analysis is used (see Poulos and Davis, 1974). The expression for determining the FS-value is derived next.

Considering the active wedge and balancing the forces in the direction parallel to the slope, the resulting formulation is

$$E_A + \frac{(N_e + N_A) \cdot \tan\delta + C_a}{FS} = (W_A + W_e) \cdot \sin\beta + F_e$$

where

N_e = effective equipment force normal to the failure plane of the active wedge.

$$N_e = W_E \cdot \cos\beta \quad (13.12)$$

Note that all the other symbols have been previously defined.

The interwedge force acting on the active wedge can now be expressed as

$$E_A = \frac{(FS)[(W_A + W_e) \cdot \sin\beta + F_e]}{FS} - \frac{[(N_A + N_e) \cdot \tan\delta + C_a]}{FS}$$

The passive wedge can be treated in a similar manner. The following formulation of the interwedge force acting on the passive wedge results:

$$E_P = \frac{C + W_P \cdot \tan\phi}{\cos\beta \cdot (FS) - \sin\beta \cdot \tan\phi}$$

By setting $E_A = E_P$, the resulting equation can be arranged in the form of the quadratic equation $ax^2 + bx + c = 0$ which in this case, using FS-values, is

$$a \cdot FS^2 + b \cdot FS + c = 0$$

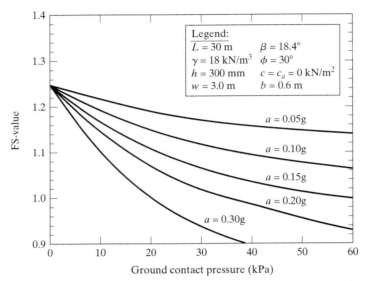

FIGURE 13.10 Design Curves for Stability of Different Construction Equipment Ground Contact Pressure for Various Equipment Accelerations

The resulting *FS*-value is then obtained from the conventional solution of the quadratic equation

$$FS = \frac{-b \pm (b^2 - 4 \cdot a \cdot c)^{0.5}}{2 \cdot a} \tag{13.13}$$

where $a = [(W_A + W_e) \cdot \sin\beta + F_e] \cdot \cos\beta$
$b = -\{[(N_A + N_e) \cdot \tan\delta + C_a] \cdot \cos\beta$
$\quad + [(W_A + W_e) \cdot \sin\beta + F_e] \cdot \sin\beta \cdot \tan\phi + (C + W_P \cdot \tan\phi)\}$
$c = [(N_A + N_e) \cdot \tan\delta + C_a] \cdot \sin\beta \cdot \tan\phi$

Using these concepts, typical design curves for various *FS*-values as a function of equipment ground contact pressure and equipment acceleration can be developed (see Figure 13.10). Note that the curves are developed specifically for the variables stated in the legend. Example 13.3 illustrates the use of the formulation.

EXAMPLE 13.3

The following are given: a 30-m-long slope with uniform cover soil of 300-mm thickness at a unit weight of 18 kN/m³. The soil has a friction angle of 30° and zero cohesion (i.e., it is a sand). It is placed on the slope using a bulldozer moving from the crest of the slope down to the toe. The bulldozer has a ground contact pressure of 30 kN/m² and tracks that are 3.0 m long and 0.6 m wide. The estimated equipment speed is 20 km/hr, and the time to reach this speed is 3.0 seconds. The cover soil to geomembrane friction angle is 22 degrees with zero adhesion. What is the *FS*-value at a slope angle of 3(H)-to-1(V) (i.e., 18.4°)?

Solution Using the design curves of Figure 13.10 along with Equation 13.13, the solution can be obtained.

- From Figure 13.9, at 20 km/hr and 3.0 seconds, the bulldozer's acceleration is 0.19g.
- From Equation 13.13,

$$a = 88.8 \text{ kN/m}$$
$$b = -107.3 \text{ kN/m}$$
$$c = 17.0 \text{ kN/m}$$

Thus, $FS = 1.03$

This value can be confirmed using Figure 13.10.

Comment This problem solution can now be compared with those of the previous two examples:

Example 13.1.	Cover soil along with no bulldozer loading:	$FS = 1.25$
Example 13.2.	Cover soil plus bulldozer moving up slope:	$FS = 1.24$
Example 13.3.	Cover soil plus bulldozer moving down slope:	$FS = 1.03$

The inherent danger of a bulldozer moving down the slope is readily apparent. Note, that the same result comes about by the bulldozer decelerating instead of accelerating. The sharp breaking action of the bulldozer is arguably the more severe condition, due to the extremely short times involved when stopping forward motion. Clearly, only in unavoidable situations should the cover soil placement equipment be allowed to work down the slope. If it is unavoidable, an analysis should be made of the specific stability situation and the construction specifications should reflect the precise conditions made in the design. The maximum weight and ground contact pressure of the equipment should be stated along with suggested operator movement of the cover soil placement operations. Truck traffic on the slopes can also give stresses as high or even higher than illustrated here and should be avoided in all circumstances.

13.4.3 Inclusion of Seepage Forces

The previous sections presented the general problem of slope stability analysis of cover soils placed on slopes under different conditions. The tacit assumption throughout was that either permeable soil or a drainage layer was placed above the barrier layer with adequate flow capacity to efficiently and safely remove permeating water away from the cross section. The amount of water to be removed is obviously a site-specific situation. Note that, in extremely arid areas, or with very low permeability cover soils, drainage may not be required, although this is generally the exception.

Unfortunately, adequate drainage of final covers has sometimes not been available and seepage-induced slope stability problems have occurred. Figure 13.11 shows a final cover slope failure during a heavy raining. The following situations have resulted in seepage-induced slides:

- Drainage soils with hydraulic conductivity (permeability) too low for site-specific conditions.
- Inadequate drainage capacity at the toe of long slopes, where seepage quantities accumulate and are at their maximum.
- Fines from quarried drainage stone either clogging the drainage layer or accumulating at the toe of the slope, thereby decreasing the as-constructed permeability over time.

498 Chapter 13 Landfill Stability Analysis

FIGURE 13.11 Final Cover Slope Failure during a Heavy Raining

- Fine, cohesionless, cover soil particles migrating through the filter (if one is present) either clogging the drainage layer, or accumulating at the toe of the slope, thereby decreasing the as-constructed outlet permeability over time.
- Freezing of the outlet drainage at the toe of the slope, while the top of the slope thaws, thereby mobilizing seepage forces against the ice wedge at the toe.

If seepage forces of the types described occur, a variation in slope stability design methodology is required. Such an analysis is the focus of this subsection. (See Koerner and Soong, 1998; and Qian, 1997; also, Thiel and Stewart, 1993; and Soong and Koerner, 1996.)

Consider a cover soil of uniform thickness placed directly above a geomembrane at a slope angle of β, as shown in Figure 13.12. What is different from previous examples, however, is that within the cover soil there can exist a saturated soil zone for part or all of the thickness. The saturated boundary is shown as two possibly different phreatic surface orientations. This is because seepage can be built up in the cover soil in two different ways: a horizontal buildup from the toe upward, or a parallel-to-slope buildup outward. These two hypotheses are defined and quantified as a horizontal submergence ratio (HSR) and a parallel submergence ratio (PSR). The dimensional definitions of both ratios are given in Figure 13.12.

When analyzing the stability of slopes using the limit equilibrium method, free-body diagrams of the passive and active wedges are taken with the appropriate forces

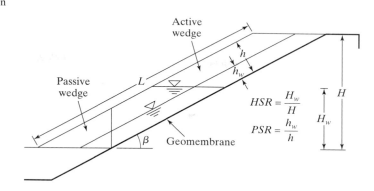

FIGURE 13.12 Cross Section of a Uniform Thickness Cover Soil on a Geomembrane Illustrating Different Submergence Assumptions and Related Definitions (Soong and Koerner, 1996)

being applied (now including pore water pressures). The formulation for the resulting factor of safety for horizontal seepage buildup and also for parallel-to-slope seepage buildup is described next.

13.4.3.1 The Case of the Horizontal Seepage Buildup. Figure 13.13 shows the free-body diagram of both the active and passive wedge assuming horizontal seepage building. Horizontal seepage buildup can occur when toe blockage occurs due to inadequate outlet capacity, contamination or physical blocking of outlets, or freezing conditions at the outlets.

All symbols used in Figure 13.13 were previously defined except the following:

γ_{sat} = saturated unit weight of the cover soil
γ_t = dry unit weight of the cover soil
γ_w = unit weight of water
H = vertical height of the slope measured from the toe
H_w = vertical height of the free water surface measured from the toe
U_h = resultant of the pore pressures acting on the interwedge surfaces
U_n = resultant of the pore pressures acting perpendicular to the slope
U_v = resultant of the vertical pore pressures acting on the passive wedge

The expression for determining the factor of safety can be derived as follows:

Considering the active wedge,

$$W_A = \frac{\gamma_{sat} \cdot h \cdot (2 \cdot H_w \cdot \cos\beta - h)}{\sin 2\beta} + \frac{\gamma_{dry} \cdot h \cdot (H - H_w)}{\sin\beta} \qquad (13.14)$$

$$U_n = \frac{\gamma_w \cdot h \cdot \cos\beta \cdot (2 \cdot H_w \cdot \cos\beta - h)}{\sin 2\beta} \qquad (13.15)$$

$$U_h = 0.5 \cdot \gamma_w \cdot h^2 \qquad (13.16)$$

$$N_A = W_A \cdot \cos\beta + U_h \cdot \sin\beta - U_n \qquad (13.17)$$

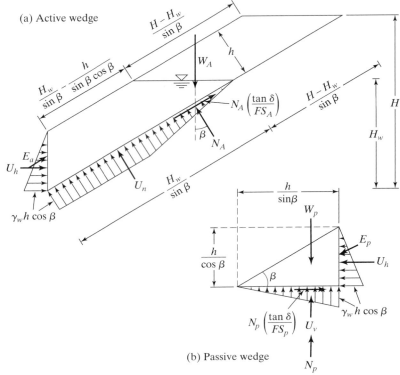

FIGURE 13.13 Limit Equilibrium Forces Involved in a Finite Length Slope of Uniform Cover Soil with Horizontal Seepage Buildup

The interwedge force acting on the active wedge can then be expressed as

$$E_A = W_A \cdot \sin\beta + U_h \cdot \cos\beta - \frac{N_A \cdot \tan\delta}{FS}$$

The passive wedge can be considered in a similar manner and the following expressions result:

$$W_P = \frac{\gamma_{sat} \cdot h^2}{\sin 2\beta} \qquad (13.18)$$

$$U_v = U_h \cdot \cot\beta \qquad (13.19)$$

The interwedge force acting on the passive wedge can then be expressed as

$$E_P = \frac{U_h \cdot (FS) - (W_P - U_v) \cdot \tan\phi}{\sin\beta \cdot \tan\phi - \cos\beta \cdot (FS)}$$

By setting $E_A = E_P$, the following equation can be arranged in the form of $ax^2 + bx + c = 0$, which in this case is

$$a \cdot FS^2 + b \cdot FS + c = 0$$

The resulting *FS*-value is then obtained from the conventional solution of the quadratic equation as

$$FS = \frac{-b \pm (b^2 - 4 \cdot a \cdot c)^{0.5}}{2 \cdot a} \qquad (13.20)$$

where $a = W_A \cdot \sin\beta \cdot \cos\beta - U_h \cdot \cos^2\beta + U_h$
$b = -W_A \cdot \sin^2\beta \cdot \tan\phi + U_h \cdot \sin\beta \cdot \cos\beta \cdot \tan\phi - N_A \cdot \cos\beta \cdot \tan\delta$
$\quad - (W_P - U_v) \cdot \tan\phi$
$c = N_A \cdot \sin\beta \cdot \tan\delta \cdot \tan\phi$

13.4.3.2 The Case of Parallel-to-Slope Seepage Buildup. Figure 13.14 shows the free body diagrams of both the active and passive wedges with seepage buildup in the direction parallel to the slope. Parallel seepage buildup can occur when soils placed above a geomembrane are initially too low in their hydraulic conductivity, or become too low due to long-term clogging from overlying soils that are not filtered. The individual forces, friction angles, and slope angles involved in Figure 13.14 are listed as follows:

W_A = weight of the active wedge (area times unit weight), lb/ft or kN/m;
W_P = weight of the passive wedge (area times unit weight), lb/ft or kN/m;
β = angle of the slope, degree;
H = height of the cover soil slope from the toe of the cover soil to the top of the slope (see Figure 13.14), ft or m;
h = thickness of the soil layer (perpendicular to the slope), ft or m;
h_w = depth of seepage water in the soil layer (perpendicular to the slope), ft or m;
γ = moisture unit weight of the soil layer, lb/ft^3 or kN/m^3;
γ_{sat} = saturated unit weight of the soil layer, lb/ft^3 or kN/m^3;
γ_w = unit weight of water, 62.4 lb/ft^3 or 9.81 kN/m^3;
ϕ = friction angle of the cover soil, degree;
δ = interface friction angle between the soil layer and geomembrane, degree;
N_A = normal force acting on bottom of the active wedge, lb/ft or kN/m;
F_A = frictional force acting on bottom of the active wedge, lb/ft;
U_{AN} = resultant of the pore water pressures acting on bottom of the active wedge (perpendicular to the slope), lb/ft or kN/m;
U_{AH} = resultant of the pore water pressures acting on lower lateral side of the active wedge (perpendicular to the interface between the active and passive wedges), lb/ft or kN/m;
E_A = force from passive wedge acting on active wedge (unknown in magnitude but assumed direction parallel to the slope), lb/ft or kN/m;
N_P = normal force acting on the bottom of passive wedge, lb/ft or kN/m;
F_P = frictional force acting on the bottom of passive wedge, lb/ft or kN/m;
U_{PH} = resultant of the pore water pressures acting on lateral side of the passive wedge (perpendicular to the lateral side), lb/ft or kN/m;

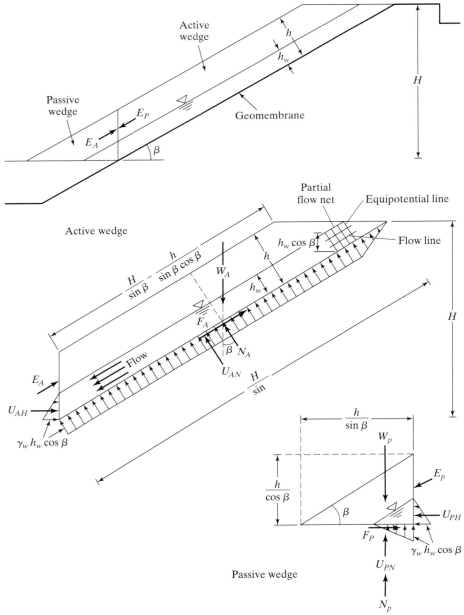

FIGURE 13.14 Cross Section of Sand Layer over Geomembrane on Side Slope with Seepage Parallel to Slope.

U_H = resultant of the pore water pressures acting on lateral side of the active wedge or passive wedge (perpendicular to the lateral side), lb/ft or kN/m, $U_H = U_{AH} = U_{PH}$;

U_{PN} = resultant of the pore water pressures acting on bottom of the passive wedge (perpendicular to bottom of the passive wedge), lb/ft or kN/m;

E_P = force from active wedge acting on passive wedge (unknown in magnitude but assumed direction parallel to the slope), lb/ft or kN/m, $E_A = E_P$;

FS = factor of safety for stability of the cover soil mass.

Considering the force equilibrium of the active wedge (Figure 13.14), we obtain

$\Sigma F_Y = 0$:
$$N_A + U_{AN} = W_A \cdot \cos\beta + U_{AH} \cdot \sin\beta$$
$$N_A = W_A \cdot \cos\beta - U_{AN} + U_{AH} \cdot \sin\beta \quad (13.21)$$

$\Sigma F_X = 0$:
$$F_A + E_A + U_{AH} \cdot \cos\beta = W_A \cdot \sin\beta$$
$$E_A = W_A \cdot \sin\beta - U_{AH} \cdot \cos\beta - F_A \quad (13.22)$$
$$F_A = N_A \cdot \tan\delta / FS \quad (13.23)$$

Substituting Equation 13.21 into Equation 13.23 gives
$$F_A = (W_A \cdot \cos\beta - U_A + U_{AH} \cdot \sin\beta) \cdot \tan\delta / FS \quad (13.24)$$

Substituting Equation 13.24 into Equation 13.22 gives
$$E_A = W_A \cdot \sin\beta - U_{AH} \cdot \cos\beta - (W_A \cdot \cos\beta - U_A + U_{AH} \cdot \sin\beta) \cdot \tan\delta / FS \quad (13.25)$$

Considering the force equilibrium of the passive wedge (Figure 13.14) yields
$$E_P = E_A \quad (13.26)$$

$\Sigma F_Y = 0$:
$$N_P + U_{PN} = W_P + E_P \cdot \sin\beta \quad (13.27)$$

Substituting Equation 13.26 into Equation 13.27 gives
$$N_P = W_P + E_A \cdot \sin\beta - U_{PN} \quad (13.28)$$

Substituting Equation 13.25 into Equation 13.28 gives
$$N_P = W_P - U_{PN} + [W_A \cdot \sin\beta - U_{AH} \cdot \cos\beta - (W_A \cdot \cos\beta - U_A + U_{AH} \cdot \sin\beta) \cdot \tan\delta / FS] \cdot \sin\beta$$
$$N_P = W_P - U_{PN} + W_A \cdot \sin^2\beta - U_{AH} \cdot \sin\beta \cdot \cos\beta - (W_A \cdot \cos\beta - U_A + U_{AH} \cdot \sin\beta) \cdot \sin\beta \cdot \tan\delta / FS \quad (13.29)$$

$\Sigma F_X = 0$:
$$F_P = U_{PH} + E_P \cdot \cos\beta \quad (13.30)$$

Substituting Equation 13.26 into Equation 13.30 gives
$$F_P = U_{PH} + E_A \cdot \cos\beta \quad (13.31)$$

Substituting Equation 13.25 into Equation 13.31 gives
$$F_P = U_{PH} + W_A \cdot \sin\beta \cdot \cos\beta - U_{AH} \cdot \cos^2\beta - (W_A \cdot \cos\beta - U_{AN} + U_{AH} \cdot \sin\beta) \cdot \cos\beta \quad (13.32)$$

$$FS = \frac{N_P \cdot \tan\phi}{F_P} \quad (13.33)$$

Substituting Equations 13.29 and 13.32 into Equation 13.33 gives

$$FS = \frac{(W_P - U_{PN} + W_A \cdot \sin^2\beta - U_{AH} \cdot \sin\beta \cdot \cos\beta) \cdot \tan\phi - (W_A \cdot \cos\beta - U_A + U_{AH} \cdot \sin\beta) \cdot \sin\beta \cdot \tan\delta \cdot \tan\phi/FS}{U_{PH} + W_A \cdot \sin\beta \cdot \cos\beta - U_{AH} \cdot \cos^2\beta - (W_A \cdot \cos\beta - U_{AN} + U_{AH} \cdot \sin\beta) \cdot \cos\beta \cdot \tan\delta/FS}$$

$$(U_{PH} + W_A \cdot \sin\beta \cdot \cos\beta - U_{AH} \cdot \cos^2\beta) \cdot FS - (W_A \cdot \cos\beta - U_{AN} + U_{AH} \cdot \sin\beta) \cdot \cos\beta \cdot \tan\delta = (W_P - U_{PN} + W_A \cdot \sin^2\beta - U_{AH} \cdot \sin\beta \cdot \cos\beta) \cdot \tan\phi - (W_A \cdot \cos\beta - U_A + U_{AH} \cdot \sin\beta) \cdot \sin\beta \cdot \tan\delta \cdot \tan\phi/FS$$

$$(W_A \cdot \sin\beta \cdot \cos\beta + U_{PH} - U_{AH} \cdot \cos^2\beta) \cdot FS^2 - (W_A \cdot \cos\beta - U_{AN} + U_{AH} \cdot \sin\beta) \cdot \cos\beta \cdot \tan\delta \cdot FS = (W_P - U_{PN} + W_A \cdot \sin^2\beta - U_{AH} \cdot \sin\beta \cdot \cos\beta) \cdot \tan\phi \cdot FS - (W_A \cdot \cos\beta - U_A + U_{AH} \cdot \sin\beta) \cdot \sin\beta \cdot \tan\delta \cdot \tan\phi$$

$$(W_A \cdot \sin\beta \cdot \cos\beta + U_{PH} - U_{AH} \cdot \cos^2\beta) \cdot FS^2 - [W_P \cdot \tan\phi + W_A \cdot (\sin^2\beta \cdot \tan\phi + \cos^2\beta \cdot \tan\delta) - U_{AN} \cdot \cos\beta \cdot \tan\delta - U_{PN} \cdot \tan\phi + U_{AH} \cdot \sin\beta \cdot \cos\beta \cdot (\tan\phi - \tan\delta)] \cdot FS + (W_A \cdot \cos\beta - U_A + U_{AH} \cdot \sin\beta) \cdot \sin\beta \cdot \tan\delta \cdot \tan\phi = 0 \quad (13.34)$$

Because $U_H = U_{PH} = U_{AH}$,

$$[W_A \cdot \sin\beta \cdot \cos\beta + U_H \cdot (1 - \cos^2\beta)] \cdot FS^2 - [W_P \cdot \tan\phi + W_A \cdot (\sin^2\beta \cdot \tan\phi + \cos^2\beta \cdot \tan\delta) - U_{AN} \cdot \cos\beta \cdot \tan\delta - U_{PN} \cdot \tan\phi + U_H \cdot \sin\beta \cdot \cos\beta \cdot (\tan\phi - \tan\delta)] \cdot FS + (W_A \cdot \cos\beta - U_{AN} + U_H \cdot \sin\beta) \cdot \sin\beta \cdot \tan\delta \cdot \tan\phi = 0 \quad (13.35)$$

Using $a \cdot x^2 + b \cdot x + c = 0$

The resulting *FS* can be expressed as

$$FS = \frac{-b \pm (b^2 - 4 \cdot a \cdot c)^{0.5}}{2 \cdot a} \quad (13.36)$$

where

$$a = W_A \cdot \sin\beta \cdot \cos\beta + U_H \cdot (1 - \cos^2\beta)$$
$$b = -[W_P \cdot \tan\phi + W_A \cdot (\sin^2\beta \cdot \tan\phi + \cos^2\beta \cdot \tan\delta) - U_{AN} \cdot \cos\beta \cdot \tan\delta - U_{PN} \cdot \tan\phi + U_H \cdot \sin\beta \cdot \cos\beta \cdot (\tan\phi - \tan\delta)]$$
$$c = (W_A \cdot \cos\beta - U_{AN} + U_H \cdot \sin\beta) \cdot \sin\beta \cdot \tan\delta \cdot \tan\phi$$

$$U_{AN} = \gamma_w \cdot h_w \cdot (H - 0.5\, h_w \cdot \cos\beta)/\tan\beta \quad (13.37)$$
$$U_H = 0.5 \cdot \gamma_w \cdot h_w^2 \quad (13.38)$$
$$U_{PN} = 0.5 \cdot \gamma_w \cdot h_w^2/\tan\beta \quad (13.39)$$
$$W_A = 0.5 \cdot [\gamma \cdot (h - h_w)(2 \cdot H \cdot \cos\beta - h - h_w) + \gamma_{sat} \cdot h_w \cdot (2 \cdot H \cdot \cos\beta - h_w)]/(\sin\beta \cdot \cos\beta) \quad (13.40)$$
$$W_P = 0.5 \cdot [\gamma \cdot (h^2 - h_w^2) + \gamma_{sat} \cdot h_w^2]/(\sin\beta \cdot \cos\beta) \quad (13.41)$$

EXAMPLE 13.4

A 44-ft (13.2-m) high and 3(H):1(V) slope has cover sand with a uniform thickness of 2 ft (0.6 m) at a unit weight of 110 lb/ft³ (17.3 kN/m³). The cover sand has a friction angle of 32 degrees and zero cohesion. Seepage occurs parallel to the slope and the seepage water head in the sand layer is 6 inches (0.15 m). The saturated unit weight of sand is 115 lb/ft³ (18 kN/m³). The interface friction angle between sand drainage layer and geomembrane is 22 degrees and zero adhesion. What is the factor of safety at a slope of 3(H)-to-1(V)?

Solution The side slope angle is at 18.4° for a 3(H):1(V) slope. Hence,

$\sin\beta = \sin(18.4°) = 0.316$, $\cos\beta = \cos(18.4°) = 0.949$, $\tan\beta = \tan(18.4°) = 0.333$.

$H = 44$ ft (13.2 m), $h = 2$ ft (0.6 m), $h_w = 0.5$ ft (0.15 m), $\gamma = 110$ lb/ft³ (17.3 kN/m³),

$\gamma_{sat} = 115$ lb/ft³ (18 kN/m³), $\gamma_w = 62.4$ lb/ft³ (9.81 kN/m³), $\phi = 32°$, $\delta = 22°$.

$\tan\phi = \tan(32°) = 0.625$, $\tan\delta = \tan(22°) = 0.404$.

$$U_{AN} = \gamma_w \cdot h_w \cdot (H - 0.5\, h_w \cdot \cos\beta)/\tan\beta \qquad (13.37)$$
$$= (62.4)(0.5)[44 - (0.5)(0.5)(0.949)]/(0.333) = 4{,}100.3 \text{ lb/ft } (58.02 \text{ kN/m})$$

$$U_H = 0.5 \cdot \gamma_w \cdot h_w^2 \qquad (13.38)$$
$$= (0.5)(62.4)(0.5)^2 = 7.8 \text{ lb/ft } (0.11 \text{ kN/m})$$

$$U_{PN} = 0.5 \cdot \gamma_w \cdot h_w^2/\tan\beta \qquad (13.39)$$
$$= (0.5)(62.4)(0.5)^2/(0.333) = 23.4 \text{ lb/ft } (0.33 \text{ kN/m})$$

$$W_A = 0.5 \cdot [\gamma \cdot (h - h_w)(2 \cdot H \cdot \cos\beta - h - h_w)$$
$$+ \gamma_{sat} \cdot h_w \cdot (2 \cdot H \cdot \cos\beta - h_w)]/(\sin\beta \cdot \cos\beta) \qquad (13.40)$$
$$= (0.5)\{(110)(2 - 0.5)[(2)(44)(0.949) - 2 - 0.5]$$
$$+ (115)(0.5)[(2)(44)(0.949) - 0.5]\}/[(0.316)(0.949)]$$
$$= (0.5)(13{,}366.98 + 4{,}773.19)/[(0.316)(0.949)] = 30{,}245.3 \text{ lb/ft } (427.6 \text{ kN/m})$$

$$W_P = 0.5 \cdot [\gamma \cdot (h^2 - h_w^2) + \gamma_{sat} \cdot h_w^2]/(\sin\beta \cdot \cos\beta) \qquad (13.41)$$
$$= (0.5)\{(110)[(2)^2 - (0.5)^2] + (115)(0.5)^2\}/[(0.316)(0.949)] = 735.7 \text{ lb/ft } (10.4 \text{ kN/m})$$

Using Equation 13.36,

$a = W_A \cdot \sin\beta \cdot \cos\beta + U_H \cdot (1 - \cos^2\beta)$
$= (30{,}245.3)(0.316)(0.949) + (7.8)[1 - (0.949)^2] = 9{,}071$ (128 for SI units)

$b = -[W_P \cdot \tan\phi + W_A \cdot (\sin^2\beta \cdot \tan\phi + \cos^2\beta \cdot \tan\delta) - U_{AN} \cdot \cos\beta \cdot \tan\delta - U_{PN} \cdot \tan\phi$
$\quad + U_H \cdot \sin\beta \cdot \cos\beta \cdot (\tan\phi - \tan\delta)]$
$= -\{(735.7)(0.625) + (30{,}245.3)[(0.316)^2(0.625) + (0.949)^2(0.104)] - (4{,}100.3)(0.949)(0.404)$
$\quad - (23.4)(0.625) + (7.8)(0.316)(0.949)(0.625 - 0.404)\}$
$= -(459.8 + 12{,}892.1 - 1{,}572.0 - 14.6 + 0.5) = -11{,}766$ (-166 for SI units)

$c = (W_A \cdot \cos\beta - U_{AN} + U_H \cdot \sin\beta) \cdot \sin\beta \cdot \tan\delta \cdot \tan\phi$
$= [(30{,}245.3)(0.949) - 4{,}100.3 + (7.8)(0.316)](0.316)(0.625)(0.404) = 1{,}963$ (28 for SI units)

$$FS = \frac{-b \pm (b^2 - 4 \cdot a \cdot c)^{0.5}}{2 \cdot a} \qquad (13.36)$$

$$= \frac{11{,}766 + [(-11{,}766)^2 - (4)(9{,}071)(1{,}963)]^{0.5}}{(2)(9{,}071)}$$

FIGURE 13.15 Sand Layer Failure along Sideslope Caused by Seepage Force

$$= \frac{11{,}766 + 8{,}198}{(2)(9{,}071)}$$
$$= 1.10$$

Comment The seriousness of seepage forces in a slope of this type is immediately obvious. Had the saturation been 100% of the drainage layer thickness, the FS-value would have been still lower. Furthermore, the result using a horizontal assumption of saturated cover soil with the same saturation ratio will give essentially identical low *FS*-values. Clearly, the teaching of this example problem is that adequate long-term drainage above the barrier layer in cover soil slopes must be provided to avoid seepage forces from occurring. Figure 13.15 shows a sand layer sliding failure along sideslope caused by seepage force.

An incremental placement method should be implemented for sideslopes higher than the maximum height that can be built in a single lift with a minimum required factor of safety, such as the previous example. Based on the incremental placement method, the first step is to place the sand drainage layer on the sideslope to the maximum unsupported height. As waste is filled against the sideslope to approximately 2 feet (0.6 m) below the protective layer, the next lift of the layer can proceed. This procedure that is illustrated in Figure 13.16 should be continued until the protective layer reaches the top of the sideslope. The heights of the following lifts of the sand drainage layer should not be higher than the calculated maximum unsupported height minus 2

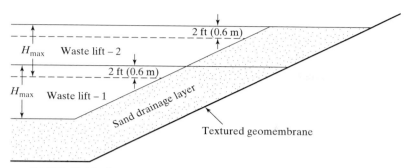

FIGURE 13.16 Incremental Placement of Soil Drainage Layer on Sideslope

feet (0.6 m). The height of the first lift of sand placement can be calculated as shown in the equations that follow (Qian, 1997):

In U.S. units,

$$H = (H_{total} - 2)/n + 2 \qquad (13.42)$$

In SI units,

$$H = (H_{total} - 0.6)/n + 0.6 \qquad (13.43)$$

where H = height of the first step of sand placement on the sideslope (see Figure 13.14), ft or m;
   H_{total} = total height of the cover sand slope from the toe of the cover sand to the top of the slope (see Figure 13.16), ft or m;
   n = number of the placement steps.

EXAMPLE 13.5

Continue the calculations of Example 13.4 and use the incremental method to achieve a factor of safety no less than 1.2 for the cover sand resting on the sideslope?

Solution   Use the incremental method to place drainage sand on the side slope to achieve a minimum factor of safety of 1.2. Try three steps of sand placement ($n = 3$) on the sideslope.

$$H = (H_{total} - 2)/n + 2 \qquad (13.42)$$
$$= (44 - 2)/3 + 2 = 14 + 2 = 16 \text{ ft (4.8 m)}$$

So,

$H = 16$ ft (4.8 m), $h = 2$ ft (0.6 m), $h_w = 0.5$ ft (0.15 m), $\gamma = 110$ lb/ft³ (17.3 kN/m³),
$\gamma_{sat} = 115$ lb/ft³ (18 kN/m³), $\gamma_w = 62.4$ lb/ft³ (9.81 kN/m³), $\phi = 32°$, $\delta = 22°$.
$\tan\phi = \tan(32°) = 0.625$, $\tan\delta = \tan(22°) = 0.404$,
$\sin\beta = \sin(18.4°) = 0.316$, $\cos\beta = \cos(18.4°) = 0.949$, $\tan\beta = \tan(18.4°) = 0.333$.

$$U_{AN} = \gamma_w \cdot h_w \cdot (H - 0.5\ h_w \cdot \cos\beta)/\tan\beta \qquad (13.37)$$
$$= (62.4)(0.5)[16 - (0.5)(0.5)(0.949)]/(0.333) = 1{,}476.9\ \text{lb/ft}\ (20.90\ \text{kN/m})$$
$$U_H = 0.5 \cdot \gamma_w \cdot h_w^2 \qquad (13.38)$$
$$= (0.5)(62.4)(0.5)^2 = 7.8\ \text{lb/ft}\ (0.11\ \text{kN/m})$$
$$U_{PN} = 0.5 \cdot \gamma_w \cdot h_w^2/\tan\beta \qquad (13.39)$$
$$= (0.5)(62.4)(0.5)^2/(0.333) = 23.4\ \text{lb/ft}\ (0.33\ \text{kN/m})$$
$$W_A = 0.5 \cdot [\gamma \cdot (h - h_w)(2 \cdot H \cdot \cos\beta - h - h_w)$$
$$+ \gamma_{sat} \cdot h_w \cdot (2 \cdot H \cdot \cos\beta - h_w)]/(\sin\beta \cdot \cos\beta) \qquad (13.40)$$
$$= (0.5)\{(110)(2 - 0.5)[(2)(16)(0.949) - 2 - 0.5]$$
$$+ (115)(0.5)[(2)(16)(0.949) - 0.5]\}/[(0.316)(0.949)]$$
$$= (0.5)(4{,}598.22 + 1{,}717.41)/[(0.316)(0.949)] = 10{,}530.1\ \text{lb/ft}\ (148.9\ \text{kN/m})$$
$$W_P = 0.5 \cdot [\gamma \cdot (h^2 - h_w^2) + \gamma_{sat} \cdot h_w^2]/(\sin\beta \cdot \cos\beta) \qquad (13.41)$$
$$= (0.5)\{(110)[(2)^2 - (0.5)^2] + (115)(0.5)^2\}/[(0.316)(0.949)] = 735.7\ \text{lb/ft}\ (10.4\ \text{kN/m})$$

Equation 13.36 yields
$$a = W_A \cdot \sin\beta \cdot \cos\beta + U_H \cdot (1 - \cos^2\beta)$$
$$= (10{,}530.1)(0.316)(0.949) + (7.8)[1 - (0.949)^2] = 3{,}159\ (45\ \text{for SI units})$$
$$b = -[W_P \cdot \tan\phi + W_A \cdot (\sin^2\beta \cdot \tan\phi + \cos^2\beta \cdot \tan\delta) - U_{AN} \cdot \cos\beta \cdot \tan\delta$$
$$- U_{PN} \cdot \tan\phi + U_H \cdot \sin\beta \cdot \cos\beta \cdot (\tan\phi - \tan\delta)]$$
$$= -\{(735.7)(0.625) + (10{,}530.1)[(0.316)^2(0.625) + (0.949)^2(0.404)] - (1{,}476.9)(0.949)(0.404)$$
$$- (23.4)(0.625) + (7.8)(0.316)(0.949)(0.625 - 0.404)\}$$
$$= -(459.8 + 4{,}488.5 - 566.2 - 14.6 + 0.5) = -4{,}368\ (-62\ \text{for SI units})$$
$$c = (W_A \cdot \cos\beta - U_{AN} + U_H \cdot \sin\beta) \cdot \sin\beta \cdot \tan\delta \cdot \tan\phi$$
$$= [(10{,}530.1)(0.949) - 1{,}476.9 + (7.8)(0.316)](0.316)(0.625)(0.404) = 680\ (10\ \text{for SI units})$$

$$FS = \frac{-b \pm (b^2 - 4 \cdot a \cdot c)^{0.5}}{2 \cdot a} \qquad (13.36)$$
$$= \frac{4{,}368 + [(-4{,}368)^2 - (4)(3{,}159)(680)]^{0.5}}{(2)(3{,}159)}$$
$$= \frac{4{,}368 + 3{,}238}{(2)(3{,}159)}$$
$$= 1.20$$

Thus, based on the above calculation, the first step is to place the drainage sand on the sideslope to a height of 16 feet (4.8 m). As waste is filled against the sideslope to approximately 2 feet (0.6 m) below the protective layer, the next lift of 14 feet (4.2 m) can be placed. This procedure should be continued until the protective layer reaches the top of the sideslope.

13.4.4 Inclusion of Seismic Forces

In areas of anticipated earthquake activity, the slope stability analysis of a final cover soil over an engineered landfill, abandoned dump, or remediated site must consider seismic forces. In the United States, the Environmental Protection Agency (EPA) regulations require such an analysis for sites that have a probability of ≥ 10% of experiencing a 0.10-g peak horizontal acceleration within 250 years. For the continental United States, this includes not only the western states, but major sections of the Midwest and northeast states, as well. If practiced worldwide, such a criterion would have huge implications.

The seismic analysis of cover soils of the type under consideration in this section is a two-part process:

(i) The calculation of a *FS*-value using a pseudostatic analysis via the addition of a horizontal force acting at the centroid of the cover soil cross section.

(ii) If the *FS*-value in the above calculation is less than 1.0, a permanent deformation analysis is required. The calculated deformation is then assessed in light of the potential damage to the cover soil section and is either accepted, or the slope requires an appropriate redesign. The redesign is then analyzed until the situation becomes acceptable.

The *first part* of the analysis is a pseudostatic approach that follows the previous examples except for the addition of a horizontal force at the centroid of the cover soil in proportion to the anticipated seismic activity. It is first necessary to obtain an average seismic coefficient (C_S) from a representative seismic zone map (e.g., as in Algermissen, 1969). Such maps are available on a worldwide basis. The value of C_S is nondimensional and is a ratio of the bedrock acceleration to gravitational acceleration. This value of C_S is modified using available computer codes such as "SHAKE" (see Schnabel et al., 1972) for propagation to the site and then to the landfill cover as shown in Figure 13.17. The computational process within such programs is quite intricate. For detailed discussion, see Seed and Idriss (1982) and Idriss (1990). The analysis is nonetheless similar to those previously presented.

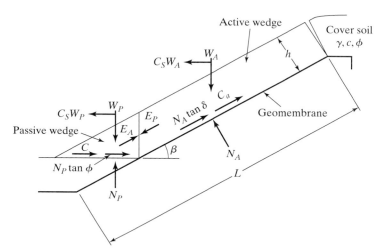

FIGURE 13.17 Limit Equilibrium Forces Involved in Pseudostatic Analysis Using a Average Seismic Coefficient.

Using Figure 13.17, the additional seismic force is $C_S \cdot W_A$ acting horizontally on the active wedge. All additional symbols used in Figure 13.17 have been previously defined, and the expression for finding the *FS*-value is derived next.

Considering the active wedge, by balancing the forces in the horizontal direction, the resulting formulation is

$$E_A \cdot \cos\beta + \frac{(N_A \cdot \tan\delta + C_a) \cdot \cos\beta}{FS} = C_S \cdot W_A + N_A \cdot \sin\beta$$

Hence, the interwedge force acting on the active wedge is

$$E_A = \frac{(FS)(C_S \cdot W_A + N_A \cdot \sin\beta) - (N_A \cdot \tan\delta + C_a) \cdot \cos\beta}{(FS) \cdot \cos\beta}$$

The passive wedge can be considered in a similar manner, and the following formulation results:

$$E_P \cdot \cos\beta + C_S \cdot W_P = \frac{C + N_P \cdot \tan\delta}{FS}$$

Hence, the interwedge force acting on the passive wedge is

$$E_P = \frac{C + W_P \cdot \tan\phi - C_S \cdot W_P \cdot (FS)}{(FS) \cdot \cos\beta - \sin\beta \cdot \tan\phi}$$

Again, by setting $E_A = E_P$, the equation can be arranged in the form of $ax^2 + bx + c = 0$, which in this case is

$$a \cdot FS^2 + b \cdot FS + c = 0$$

The resulting *FS* can be expressed as

$$FS = \frac{-b \pm (b^2 - 4 \cdot a \cdot c)^{0.5}}{2 \cdot a} \tag{13.44}$$

where $a = (C_S \cdot W_A + N_A \cdot \sin\beta) \cdot \cos\beta + C_S \cdot W_P \cdot \cos\beta$
$b = -[(C_S \cdot W_A + N_A \cdot \sin\beta) \cdot \sin\beta \cdot \tan\phi + N_A \cdot \tan\delta + C_a) \cdot \cos^2\beta$
$\quad + (C + W_P \cdot \tan\phi) \cdot \cos\beta]$
$c = (N_A \cdot \tan\delta + C_a) \cdot \sin\beta \cdot \cos\beta \cdot \tan\phi$

Using these concepts, a design curve for the general problem under consideration as a function of seismic coefficient can be developed. (See Figure 13.18.) Note that the curve is developed specifically for the variables stated in the legend. Example 13.6 illustrates the use of the equations and of the curve.

EXAMPLE 13.6

The following are given: a-30-m-long slope with uniform thickness cover soil of 300 mm at a unit weight of 18 kN/m³. The soil has a friction angle of 30° and zero cohesion (i.e., it is a sand). The cover soil is on a geomembrane as shown in Figure 13.17. Direct shear testing has resulted in an interface friction angle of 22° with zero adhesion. The slope angle is 3(H)-to-1(V)

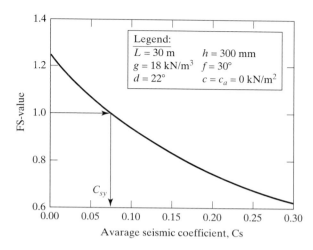

FIGURE 13.18 Design Curve for an Uniformly Thick Cover Soil Pseudostatic Seismic Analysis with Varying Average Seismic Coefficients.

(i.e., 18.4°). A design earthquake appropriately transferred to the site's cover soil results in an average seismic coefficient of 0.10. What is the *FS*-value?

Solution Solving Equation 13.44 for the values

$$a = 59.6 \text{ kN/m}$$
$$b = -66.9 \text{ kN/m}$$
$$c = 10.4 \text{ kN/m}$$

results in *FS* = 0.94

Note that the value of *FS* = 0.94 agrees with the design curve of Figure 13.18 at a seismic coefficient of 0.10.

Comment Had the above *FS*-value been greater than 1.0, the analysis would have been complete. The assumption is that cover soil stability can withstand the short-term excitation of an earthquake and still not slide. However, since the value in this example is less than 1.0, a second part of the analysis is required.

The *second part* of the analysis is directed toward calculating the estimated deformation of the lowest shear-strength interface in the cross section under consideration. The deformation is then assessed in light of the potential damage that may be imposed on the system.

To begin the permanent deformation analysis, a yield acceleration C_{sy} is obtained from a pseudostatic analysis under an assumed *FS* = 1.0. Figure 13.18 illustrates this procedure for the assumptions stated in the legend. It results in a value of C_{sy} = 0.075. Coupling this value with the time history response obtained for the actual site location and cross section results in a comparison as shown in Figure 13.19(a). If the earthquake time history response never exceeds the value of C_{sy}, there is no anticipated permanent deformation. However, whenever any part of the time history exceeds the value of C_{sy},

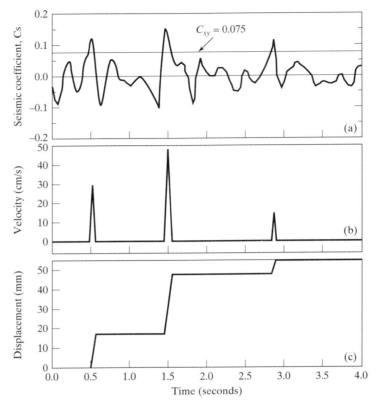

FIGURE 13.19 Design Curves to Obtain Permanent Deformation Utilizing
(a) Acceleration,
(b) Velocity, and
(c) Displacement curves.

permanent deformation is expected. By double integration of the acceleration time history curve to velocity [Figure 13.19(b)], and then to displacement [Figure 13.19(c)], the anticipated value of deformation can be obtained. This value is considered to be permanent deformation and is then assessed based on the site-specific implications of damage to the final cover system. Empirical charts (e.g., Makdisi and Seed, 1978) also can be used to estimate the permanent deformation. Example 13.7 continues the previous pseudostatic analysis into the deformation calculation.

EXAMPLE 13.7

Continue Example 13.6 and determine the anticipated permanent deformation of the weakest interface in the cover soil system. The site-specific seismic time-history diagram is given in Figure 13.19(a).

Solution The interface of concern is the cover soil-to-geomembrane ratio for this particular example. With a yield acceleration of 0.075 from Figure 13.18, and the site-specific design time history shown in Figures 13.19(a), integration produces Figure 13.19(b) and then 13.19(c). The three peaks exceeding the yield acceleration value of 0.075 produce a cumulative deformation of approximately 54 mm (2.1 inches). This value is now viewed in light of the deformation capability of the cover soil above the particular interface used at the site. Note that some references limit the deformation to either 100 or 300 mm (4 to 12 inches), depending on site-specific situations. (See Richardson et al., 1995.)

Comments An assessment of the implications of deformation [in this example it is 54 mm (2.1 inches)] is very subjective. For example, this problem could easily have been framed to produce much higher permanent deformation. Such deformation can readily be envisioned in high seismic-prone areas. (See Anderson and Kavazajian, 1995, and Matasovic et al., 1995.) Discussion is ongoing in this regard. In addition to an assessment of cover soil stability, the concerns for appurtenances and ancillary piping must also be addressed.

13.4.5 General Remarks

As seen from the previous analytic development and examples in this section, veneer slope failures are fully capable of being evaluated and, as such, of being avoided. Unfortunately, many failures of this type exist and the literature is abundantly clear in this regard (e.g., Thiel and Steward, 1993; McKelvey, 1994; Giroud et al., 1995a and 1995b).

Whatever analysis is selected by the designer (the foregoing followed that of Koerner and Soong, 1998; and Qian, 1997), the following items are overriding considerations:

(i) Proper assessment of interface shear strengths is critically necessary.
(ii) The designer must make a reasonable estimate of the equipment and the contractor's practice used to construct the system.
(iii) The designer must make a proper assessment of the design storm event.
(iv) A proper assessment of the design seismic event is necessary assuming that the site is in a potentially seismically active area.

It should also be noted that the designer has a number of alternatives with which to make a given slope stable. A higher factor of safety can result by modifying the geometry, adding downslope soil berms, or including veneer reinforcement. See Koerner and Soong (1998) for an extension of the methodology of this section into these various stabilizing considerations.

13.5 SUBSOIL FOUNDATION FAILURES

Designers regularly perform calculations to verify the safety of natural slopes, excavated slopes, and constructed embankments. These procedures are directly applicable to the design and analysis of solid waste materials that are placed in landfills [recall Figures 13.1(c), (d) and (e)]. Such calculations serve as a basis for choosing either waste slope angles and waste slope lengths with specified factor-of-safety (*FS*) values before waste placement, or for the redesign after obtaining an unacceptable value or analyzing a failure. The procedure involves determining the shear stresses developed along the most critical failure surface and comparing them with the shearing resistances of the materials through which the surface passes. The entire procedure is called a slope stability analysis and it is well developed in the geotechnical engineering literature (e.g., see Sherard et al., 1963; Hirschfield and Poulos, 1973; and others).

13.5.1 Method of Analysis

By far, the majority of slope stability procedures that are performed are based on two-dimensional (2-D) cross sections and analyses. Using such procedures, there are many analysis methods, but all assume that the critical cross section resulting in the lowest *FS*-value can be identified. Since numerous iterations are invariably required, computer codes are commonplace in order to identify the critical cross section. A cross section of a circular arc failure is shown in Figure 13.20(a). In the conventional manner, it is subdivided into n-slices where the ith slice is shown in Figure 13.20(b). From this point, a number of different calculation methods can be followed.

For the analysis of the case histories to follow which failed along a circular arc, the simplified Bishop method was used, see Koerner and Soong (2000). The derivation is readily available and leads to the following equation for the FS-value:

$$FS = \frac{\sum_{i=1}^{n}[c \cdot \Delta b_i + (W_i - u_i \cdot \Delta b_i) \cdot \tan\phi]/m_i}{\sum_{i=1}^{n} W_i \cdot \sin\theta_i} \quad (13.45)$$

Here, $\quad m_i = \cos\theta_i \cdot (1 + \tan\phi \cdot \tan\theta_i/FS) \quad (13.46)$

See Figure 13.20 for definitions of the terms.

13.5.2 Case Histories

To illustrate the importance of assessing the subgrade soil upon which the waste mass is to be placed, three case histories of failures are presented next. All were circular-arc-type failures initiating in the soft foundation soils beneath the solid waste and then propagating up through the waste mass itself.

Case History R-1

Case history R-1 is a municipal solid waste landfill that failed in 1984. The failure was rotational and involved approximately 110,000 m³ of solid waste. Divinoff and Munion (1986) and Erdogan et al., (1986) have reported on this case history.

Background. The municipal solid waste (MSW) landfill site in this case history covered 22 ha. It was bounded on three sides by a tidal marsh. The landfill was in operation for approximately 15 years and rose 44 m above the marsh at its highest location. The side

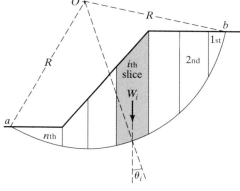

(a) Cross section

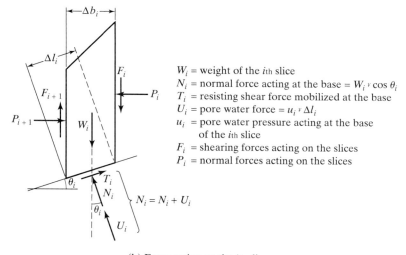

W_i = weight of the ith slice
N_i = normal force acting at the base = $W_i \cdot \cos\theta_i$
T_i = resisting shear force mobilized at the base
U_i = pore water force = $u_i \cdot \Delta l_i$
u_i = pore water pressure acting at the base of the ith slice
F_i = shearing forces acting on the slices
P_i = normal forces acting on the slices

$N_i = N_i + U_i$

(b) Force acting on the ith slice

FIGURE 13.20 Procedure Showing Circular Arc Subdivided into Slices and Analysis of the ith Slice

slopes were generally 4(H)-to-1(V) and a small toe berm was constructed around the edge of the slope. The development of a particular portion of the site was such that waste placement was temporarily postponed because a small stream entered the landfill at this location. Eventually, this area was filled and waste was placed quite rapidly (in 4 or 5 months) just before the failure occurred; it was the eventual location of the failure.

Description of Failure. Approximately 70 mm of rain fell for three days prior to the failure, which caused the water level of the adjacent marsh to rise approximately 3.2 m. A near-vertical crack opened at the top of the waste, eventually measuring 12-m deep, 18-m wide, and 180-m long(see Figure 13.21). The opening was crescent-shaped in plan view conforming closely to the ground contours before waste was placed at the

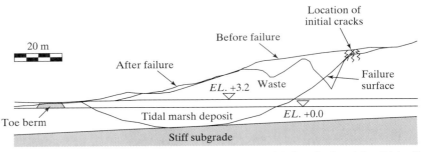

FIGURE 13.21 The Critical Cross Section of the Rotational Failure Surface (after Divinoff and Munion, 1986)

location. Subsidiary cracks opened down-slope in the moving waste mass as failure occurred. The toe berm was lifted several meters and was severely distorted as it moved in front of the failing waste mass, which extended for approximately 60 m. The critical cross section is shown in Figure 13.21. Note the difference in liquid level within the waste mass as just described.

Foundation Soil Conditions. The thickness of the tidal marsh beneath the waste was variable, but it can be described as having a 3-m surface crust (called "meadow mat") consisting of fine sand, silt, brick fragments, roots and decomposed solid waste. Some 7 to 12 m of silty clay/clayey silt with water content of 88% to 117% and undrained shear strength of 4.3 to 12.5 kPa was beneath the meadow mat. Stiff, dense silty fine sand was beneath the tidal marsh and its upper surface eventually formed the tangent point for the circular arc failure in the overlying tidal marsh and waste mass.

Waste Condition. The waste was placed directly on the meadow mat. There was neither a liner nor a leachate collection system in use at this landfill. Compaction of the waste was only nominal and a relatively low unit weight of 10.2 kN/m^3 was estimated. The shear strength of the waste was unknown, although Divinoff and Munion (1986) found by back analysis that an undrained strength of 37 kPa resulted in an $FS = 1.0$. The waste mass during its time of placement and before the failure was acting as a surcharge on the tidal marsh, which resulted in some (unknown) amount of consolidation.

Case History R-2

Case history R-2 is a municipal waste landfill that failed in 1989. The failure was multirotational in nature and involved approximately 500,000 m^3 of solid waste. Reynolds (1991) and Richardson and Reynolds (1990) have reported on the case history.

Background. The operation of the landfill site in this case history began in the 1970's. It was started as a relatively small, commercially-operated, municipal solid waste disposal site. Over the years, the facility was expanded and incorporated additional waste piles of ash/sludge and asbestos. The landfill was an above-grade landfill and was underlain with a thick deposit of clayey soil.

To the west of the site, work was starting on an expansion to the existing landfill. The expanded area was to be lined with a composite liner system consisting of a geomembrane overlying a 0.6-m-thick layer of recompacted clay. In order to facilitate the liner system construction and to increase the capacity of the future landfill, the plans called for the removal of up to 3 m of the upper stiff clay in the expansion area. Such excavation activity eventually became the triggering mechanism of the landfill failure.

Description of Failure. On August 14, 1989, after approximately 120 mm of rain fell for 10 days prior to the incident, a multirotational slope failure occurred toward and into the proposed landfill expansion area west of the existing landfill. During the movement, six large crevasses (one after the other) opened up in the waste mass. Figure 13.22(a) shows the critical cross section of the failed landfill. The entire movement lasted about 15 seconds. Analysis of the failure indicated that a single rotational failure first occurred under the original landfill slope as in Figure 13.22(a). This initial failure left a steep, unsupported slope within the remaining waste pile and the underlying clay. This directly led to five sequential slides of similar types progressing eastward into the existing landfill mass. Blocks of waste and clay followed the direction of the initial movement. As seen in Figure 13.22(b), the blocks formed progressively from the west to the east and moved horizontally as much as 50 m. Some of the crevasses were approximately 15-m wide and up to 10-m deep. Traces of remolded clay were discovered up to 120 m beyond the original toe of the landfill. Due to the remolding, the clay lost 80 to 90% of its original undrained shear strength.

Foundation Soil Conditions. The landfill in this case history is underlain by 15 to 21 m of clayey soil. The top 3 m of the thick clay deposit is weathered and fissured and is designated as a stiff clay. Below the stiff clay, there is the soft clay zone with thickness

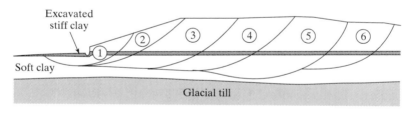

(a) Block formation

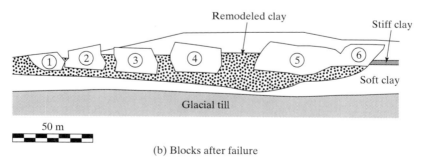

(b) Blocks after failure

FIGURE 13.22 Critical Cross Section of Landfill R-2 Showing the "Block" Formation (after Reynold, 1991)

varying from 12 to 18 meters. The vertical permeability of this zone is in the range of 10^{-7} to 10^{-8} cm/sec. Laboratory tests conducted prior to the failure indicated undrained shear strength (cohesion) of approximately 20 kPa for the underlying soft clayey soil. (See Reynolds, 1991; and Richardson and Reynolds, 1990.)

Waste Condition and Placement. Based on an estimated unit weight of waste of 5.8 kN/m^3, a height limitation of 16.8 m was originally set on the existing MSW landfill in mid-1986. By mid-1987, the landfill reached a height of approximately 12.2 m. Data obtained at that point indicated a waste unit weight of 12.2 kN/m^3. This is probably due to a higher compactive effort and the heavy (sand and gravel) daily cover material. While the increased unit weight of the waste worked against the slope stability, the strength gain due to the consolidation of the clayey soil permitted a gradual increase in the height of the landfill. As a result, by early 1989, the MSW landfill actually exceeded the height limitation to a value of 18.3 m. After the occurrence of the failure, several large pits of undisturbed waste were excavated and the removed material was weighed. The pit dimensions were also measured to determine the actual volumes for calculating the actual unit weight of the waste. The average of ten such tests produced a value of 14.7 kN/m^3. As to the shear strength parameters of the municipal solid waste, values of 17 kPa for cohesion and 20° for friction angle were determined, Reynolds (1991).

Case History R-3

Case History R-3 is a contaminated soil landfill expansion that failed in 1995 and involved approximately 110,000 m^3. De Santayana and Pinto (1998) have reported on the case history.

Background. The site is adjacent to a major river and was first used for soil disposal beginning in 1970. Disposal was directly on top of the in-situ estuarine and alluvial soft clay deposit. Landfilling commenced in 1985 and then ceased in 1990. The area was 17 ha in area and the height was approximately 15 m. A lateral expansion was then necessary consisting of an additional 7.5 ha. This was the eventual area that failed.

The foundation soil profile at the expansion site consisted of (from top to bottom); (i) a 4- to 5-m-thick, silty-clay fill layer, (ii) a 20- to 25-m-thick estuarine and alluvial soft clay deposit, (iii) a thin, irregular, sometimes absent, basal sand and gravel deposit, and (iv) bedrock, consisting of alternating layers mainly of sandy limestone and calcareous sandstone.

The Lining System. Due to the nature of the contaminated soils being placed in the landfill, a lining system with leachate collection capability was designed and constructed. A perimeter berm surrounding the three-cell area was first constructed and then the lining system was constructed. The lining system consisted of (from bottom to top) (i) a granular drainage layer for collection of consolidation water, (ii) a geotextile filter, (iii) a 0.5-m-thick compacted clay liner, (iv) a geosynthetic clay liner, (v) a 1.5-mm-thick, high-density polyethylene liner, and (vi) a leachate collection layer composed of a geonet drainage layer, a geotextile filter, and 0.3 m of a granular protective layer.

Description of the Failure. After two minor indications of instability were observed (as evidenced by tension cracks on top of the waste), a major failure occurred on June 25,

1995. About 110,000 m³ failed with a front length of about 270 m, affecting the area loaded with contaminated soils. (See Figure 13.23.) The area within the tension cracks suddenly sank several meters and moved towards the river. At the top of the slide mass, two main scarps appeared several meters apart. The failure surface was clearly visible at the scarps and was almost vertical. It resulted in a settlement of about 4 m and a horizontal movement of several meters toward the river. The geosynthetic materials were visibly torn at the scarps. There were numerous minor cracks, particularly at both sides of the area affected by the slide, and they were mostly parallel arranged, with opening widths of about a few centimeters to several decimeters.

The slide extended well into the river, propagating before it an undulating shape of mud waves and depressions. The east berm of the cells, situated approximately in the middle zone of the slide mass, underwent little vertical movement, but significant horizontal displacement towards the river occured. The berm was clearly displaced at the south side of the slide, and consequently, the pipes installed under the berm (underdrain pipes) and in the berm (leachate force main) were also broken.

Analysis of the Failure. According to de Santayna and Pinto (1998), the strength of the soft clay was overestimated in the original design. The soft clay layer was considered to

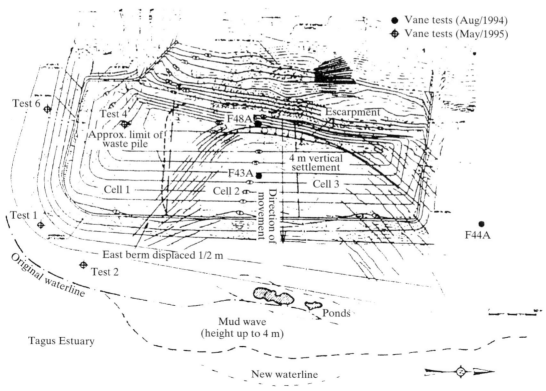

FIGURE 13.23 Plan View of Case History R-3 (after de Santayana and Pinto, 1998)

be normally consolidated under the surcharge of about 4 m of fill. The soft clay layer, however, was underconsolidated below the fill layer. The excess pore pressures caused by the placement of the fill in the 1970s and 1980s had experienced very little dissipation—particularly between elevations of -10 and -20 m—at the time waste placement started. In the middle zone of the soft clay layer, the difference between the actual undrained strength and the one used in the stability analyses was of the order of 10 kN/m^2. The original short-term stability analysis did not consider the possibility of failure surfaces extending to the river (like the one that actually happened), where there was no fill layer over the soft clay, and, hence, the soft clay did not have the undrained strength assumed in the stability calculations.

As noted, this case history had a geosynthetic lining system that failed along with the rotational movement. However, the lining system could not (and was not) a contributing issue to the failure. The little reinforcement benefit that may have been provided by the geosynthetic layer is negligible in the context of this large of a waste mass. This, as with the previous two case histories, was completely a geotechnical-related failure of the classical rotational failure mode except now a portion of the failure surface passes through waste materials.

13.5.3 General Remarks

It should be obvious from these three case histories that proper site characterization during the design stage and well before waste placement is critical. Irrespective of the high shear strength of waste materials, if the soil foundation fails, it will eventually propagate through the waste mass and cause the entire system to fail. Once a crack is observed on the surface of the waste mass, the entire failure surface beneath it has been mobilized. Failure of the mass is then imminent.

The situation is obviously important when dealing with soft, fine-grained soils. Typically, but certainly not always, such soils are near rivers, harbors, and estuaries. Best available geotechnical practice must be followed (recall Section 13.3.3). Even beyond site investigation, laboratory testing, and design which lead to site-specific plans and specifications, one should consider field instrumentation. Piezometers placed in the subsoil and inclinometers placed at the toe of the waste slope (and beyond) could be most valuable in providing an instantaneous assessment of the landfill as waste is being placed. Unfortunately, such instrumentation is rarely provided, even for sensitive site situations.

13.6 WASTE MASS FAILURES

The relatively low interface shear strengths of components within liner systems can lead to translational failures of the type shown in Figure 13.1(f). However, failure can only occur if the toe of the waste mass is unsupported by an opposing slope or large soil berm. Unfortunately, unsupported toe conditions are often the case. Canyon landfills are very common in areas of mountainous or rolling topography. Even when an excavation is dug for a landfill, the waste mass during filling is generally left unsupported at its toe. This section deals with the instability of such situations.

13.6.1 Translational Failure Analysis

While the approach to translational failures is generally similar to that described in Section 13.5.1, the failure surface is not circular, but usually piecewise linear. Thus, the simplified Bishop method is not applicable. A translational (or two-wedge) failure analysis is used to calculate the factor of safety for the landfill against possible mass movement of the type of "translational (or wedge) failure along liner" [Figure 13.1(f)] in the interim filling condition.

The waste mass shown in Figure 13.24(a) can be divided into two discrete parts, one active wedge lying on the side slope and tending to cause failure, and another passive wedge lying on the cell bottom floor and tending to resist failure. The forces acting on the active and passive wedges are shown in Figure 13.24(a). The individual forces, friction angles, and slope angles involved in the analysis are listed as follows:

W_P = weight of the passive wedge;

N_P = normal force acting on the bottom of the passive wedge;

F_P = frictional force acting on the bottom of the passive wedge (parallel to the bottom of the passive wedge);

E_{HP} = normal force from the active wedge acting on the passive wedge (unknown in magnitude, but with the direction perpendicular to the interface of the active and passive wedges);

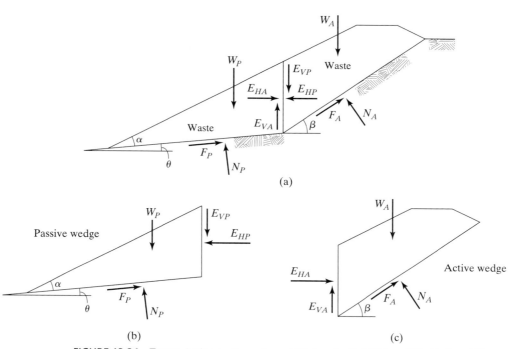

FIGURE 13.24 Forces Acting on Two adjacent Wedges for Solid Waste Filled in Landfill

E_{VP} = frictional force acting on the side of the passive wedge (unknown in magnitude, but with the direction parallel to the interface of the active and passive wedges);

FS_P = factor of safety for the passive wedge;

δ_P = minimum interface friction angle of multi-layer liner components beneath the passive wedge;

ϕ_s = friction angle of the solid waste;

α = angle of the solid waste slope, measured from horizontal, degrees;

θ = angle of the landfill cell subgrade, measured from horizontal, degrees;

W_A = weight of the active wedge;

W_T = total weight of the active and passive wedges;

N_A = normal force acting on the bottom of the active wedge;

F_A = frictional force acting on the bottom of the active wedge (parallel to the bottom of the active wedge);

E_{HA} = normal force from passive wedge acting on the active wedge (unknown in magnitude, but with the direction perpendicular to the interface of the active and passive wedges), $E_{HA} = E_{HP}$;

E_{VA} = frictional force acting on the side of the active wedge (unknown in magnitude, but with the direction parallel to the interface of the active and passive wedges), $E_{VA} = E_{VP}$;

FS_A = factor of safety for the active wedge;

δ_A = minimum interface friction angle of multi-layer liner components beneath the active wedge;

β = angle of the side slope, measured from horizontal, degrees;

FS = factor of safety for the entire solid waste mass.

Considering the force equilibrium of the passive wedge [Figure 13.24(b)], the forces acting on it are

$\Sigma F_Y = 0$:

$$W_P + E_{VP} = N_P \cdot \cos\theta + F_P \cdot \sin\theta \qquad (13.47)$$

$$F_P = N_P \cdot \tan\delta_P / FS_P \qquad (13.48)$$

$$E_{VP} = E_{HP} \cdot \tan\phi_s / FS_P \qquad (13.49)$$

Substituting Equations 13.48 and 13.49 into Equation 13.47 gives

$$W_P + E_{HP} \cdot \tan\phi_s / FS_P = N_P \cdot (\cos\theta + \sin\theta \cdot \tan\delta_P / FS_P), \text{ and} \qquad (13.50)$$

when $\Sigma F_X = 0$,

$$F_P \cdot \cos\theta = E_{HP} + N_P \cdot \sin\theta \qquad (13.51)$$

Substituting Equation (13.48) into Equation (13.51) gives

$$N_P \cdot \cos\theta \cdot \tan\delta_P / FS_P = E_{HP} + N_P \cdot \sin\theta$$

$$N_P \cdot (\cos\theta \cdot \tan\delta_P / FS_P - \sin\theta) = E_{HP}$$

$$N_P = \frac{E_{HP}}{\cos\theta \cdot \tan\delta_P/FS_P - \sin\theta} \tag{13.52}$$

Substituting Equation 13.52 into Equation 13.50 gives

$$W_P + E_{HP} \cdot \tan\phi_s/FS_P = \frac{E_{HP} \cdot (\cos\theta + \sin\theta \cdot \tan\delta_P/FS_P)}{\cos\theta \cdot \tan\delta_P/FS_P - \sin\theta}$$

$$E_{HP} \cdot (\cos\theta + \sin\theta \cdot \tan\delta_P/FS_P) = W_P \cdot (\cos\theta \cdot \tan\delta_P/FS_P - \sin\theta)$$
$$+ E_{HP} \cdot (\cos\theta \cdot \tan\delta_P/FS_P - \sin\theta) \cdot \tan\phi_s/FS_P$$

$$E_{HP} \cdot (\cos\theta + \sin\theta \cdot \tan\delta_P/FS_P - \cos\theta \cdot \tan\delta_P \cdot \tan\phi_s/FS_P^2 + \sin\theta \cdot \tan\phi_s/FS_P)$$
$$= W_P \cdot (\cos\theta \cdot \tan\delta_P/FS_P - \sin\theta)$$

$$E_{HP} = \frac{W_P \cdot (\cos\theta \cdot \tan\delta_P/FS_P - \sin\theta)}{\cos\theta + (\tan\delta_P + \tan\phi_s) \cdot \sin\theta/FS_P - \cos\theta \cdot \tan\delta_P \cdot \tan\phi_s/FS_P^2} \tag{13.53}$$

Considering the force equilibrium of the active wedge [Figure 13.12(c)] yields

$\Sigma F_Y = 0$:

$$W_A = F_A \cdot \sin\beta + N_A \cdot \cos\beta + E_{VA} \tag{13.54}$$
$$F_A = N_A \cdot \tan\delta_A/FS_A \tag{13.55}$$
$$E_{VA} = E_{HA} \cdot \tan\phi_s/FS_A \tag{13.56}$$

Substituting Equations 13.55 and 13.56 into Equation 13.54 gives

$$W_A = N_A \cdot (\cos\beta + \sin\beta \cdot \tan\delta_A/FS_A) + E_{HA} \cdot \tan\phi_s/FS_A \tag{13.57}$$

$\Sigma F_X = 0$:

$$F_A \cdot \cos\beta + E_{HA} = N_A \cdot \sin\beta \tag{13.58}$$

Substituting Equation 13.55 into Equation 13.58 gives

$$E_{HA} = N_A \cdot (\sin\beta - \cos\beta \cdot \tan\delta_A/FS_A)$$
$$N_A = \frac{E_{HA}}{\sin\beta - \cos\beta \cdot \tan\delta_A/FS_A} \tag{13.59}$$

Substituting Equation 13.59 into Equation 13.57 gives

$$W_A = E_{HA} \cdot \frac{\cos\beta + \sin\beta \cdot \tan\delta_A/FS_A}{\sin\beta - \cos\beta \cdot \tan\delta_A/FS_A} + E_{HA} \cdot \tan\phi_s/FS_A$$

$$E_{HA} \cdot \frac{\cos\beta + \sin\beta \cdot \tan\delta_A/FS_A + \sin\beta \cdot \tan\phi_s/FS_A - \cos\beta \cdot \tan\delta_A \cdot \tan\phi_s/FS_A^2}{\sin\beta - \cos\beta \cdot \tan\delta_A/FS_A} = W_A$$

$$E_{HA} = \frac{W_A \cdot (\sin\beta - \cos\beta \cdot \tan\delta_A/FS_A)}{\cos\beta + (\tan\delta_A + \tan\phi_s) \cdot \sin\beta/FS_A - \cos\beta \cdot \tan\delta_A \cdot \tan\phi_s/FS_A^2} \tag{13.60}$$

Because $E_{HA} = E_{HP}$ and $FS_A = FS_P = FS$, Equation 13.60 must equal Equation 13.53, giving

$$\frac{W_A \cdot (\sin\beta - \cos\beta \cdot \tan\delta_A/FS)}{\cos\beta + (\tan\delta_A + \tan\phi_s) \cdot \sin\beta/FS - \cos\beta \cdot \tan\delta_A \cdot \tan\phi_s/FS^2}$$
$$= \frac{W_P \cdot (\cos\theta \cdot \tan\delta_P/FS - \sin\theta)}{\cos\theta + (\tan\delta_P + \tan\phi_s) \cdot \sin\theta/FS - \cos\theta \cdot \tan\delta_P \cdot \tan\phi_s/FS^2}$$

$W_A \cdot (\sin\beta - \cos\beta \cdot \tan\delta_A/FS)[\cos\theta + (\tan\delta_P + \tan\phi_s) \cdot \sin\theta/FS - \cos\theta \cdot \tan\delta_P \cdot \tan\phi_s/FS^2]$
$= W_P \cdot (\cos\theta \cdot \tan\delta_P/FS - \sin\theta)[\cos\beta + (\tan\delta_A + \tan\phi_s) \cdot \sin\beta/FS - \cos\beta \cdot \tan\delta_A \cdot \tan\phi_s/FS^2]$

$(W_A \cdot \sin\beta - W_A \cdot \cos\beta \cdot \tan\delta_A/FS)[\cos\theta + (\tan\delta_P + \tan\phi_s) \cdot \sin\theta/FS - \cos\theta \cdot \tan\delta_P \cdot \tan\phi_s/FS^2]$
$= (W_P \cdot \cos\theta \cdot \tan\delta_P/FS - W_P \cdot \sin\theta)[\cos\beta + (\tan\delta_A + \tan\phi_s) \cdot \sin\beta/FS - \cos\beta \cdot \tan\delta_A \cdot \tan\phi_s/FS^2]$

$W_A \cdot \sin\beta \cdot \cos\theta + W_A \cdot (\tan\delta_P + \tan\phi_s) \cdot \sin\beta \cdot \sin\theta/FS - W_A \cdot \sin\beta \cdot \cos\theta \cdot \tan\delta_P \cdot \tan\phi_s/FS^2$
$- W_A \cdot \cos\beta \cdot \cos\theta \cdot \tan\delta_A/FS - W_A \cdot (\tan\delta_P + \tan\phi_s) \cdot \cos\beta \cdot \sin\theta \cdot \tan\delta_A/FS^2$
$+ W_A \cdot \cos\beta \cdot \cos\theta \cdot \tan\delta_A \cdot \tan\delta_P \cdot \tan\phi_s/FS^3 = W_P \cdot \cos\beta \cdot \cos\theta \cdot \tan\delta_P/FS$
$+ W_P \cdot (\tan\delta_A + \tan\phi_s) \cdot \sin\beta \cdot \cos\theta \cdot \tan\delta_P/FS^2 - W_P \cdot \cos\beta \cdot \cos\theta \cdot \tan\delta_A \cdot \tan\delta_P \cdot \tan\phi_s/FS^3$
$- W_P \cdot \cos\beta \cdot \sin\theta - W_P \cdot (\tan\delta_A + \tan\phi_s) \cdot \sin\beta \cdot \sin\theta/FS + W_P \cdot \cos\beta \cdot \sin\theta \cdot \tan\delta_A \cdot \tan\phi_s/FS^2$

$(W_A \cdot \sin\beta \cdot \cos\theta + W_P \cdot \cos\beta \cdot \sin\theta) \cdot FS^3 + [W_A \cdot (\tan\delta_P + \tan\phi_s) \cdot \sin\beta \cdot \sin\theta$
$+ W_P \cdot (\tan\delta_P + \tan\phi_s) \cdot \sin\beta \cdot \sin\theta - W_A \cdot \cos\beta \cdot \cos\theta \cdot \tan\delta_A - W_P \cdot \cos\beta \cdot \cos\theta \cdot \tan\delta_P] \cdot FS^2$
$- [W_A \cdot (\tan\delta_P + \tan\phi_s) \cdot \cos\beta \cdot \sin\theta \cdot \tan\delta_A + W_P \cdot (\tan\delta_A + \tan\phi_s) \cdot \sin\beta \cdot \cos\theta \cdot \tan\delta_P$
$+ W_A \cdot \sin\beta \cdot \cos\theta \cdot \tan\delta_P \cdot \tan\phi_S + W_P \cdot \cos\beta \cdot \sin\theta \cdot \tan\delta_A \cdot \tan\phi_s] \cdot FS$
$+ (W_A \cdot \cos\beta \cdot \cos\theta \cdot \tan\delta_A \cdot \tan\delta_P \cdot \tan\phi_s + W_P \cdot \cos\beta \cdot \cos\theta \cdot \tan\delta_A \cdot \tan\delta_P \cdot \tan\phi_s) = 0$

$(W_A \cdot \sin\beta \cdot \cos\theta + W_P \cdot \cos\beta \cdot \sin\theta) \cdot FS^3 + [(W_A \cdot \tan\delta_P + W_P \cdot \tan\delta_A + W_T \cdot \tan\phi_s) \cdot \sin\beta \cdot \sin\theta$
$- (W_A \cdot \tan\delta_A + W_P \cdot \tan\delta_P) \cdot \cos\beta \cdot \cos\theta] \cdot FS^2 - [W_T \cdot \tan\phi_s \cdot (\sin\beta \cdot \cos\theta \cdot \tan\delta_P$
$+ \cos\beta \cdot \sin\theta \cdot \tan\delta_A) + (W_A \cdot \cos\beta \cdot \sin\theta + W_P \cdot \sin\beta \cdot \cos\theta) \cdot \tan\delta_A \cdot \tan\delta_P] \cdot FS$
$+ W_T \cdot \cos\beta \cdot \cos\theta \cdot \tan\delta_A \cdot \tan\delta_P \cdot \tan\phi_s = 0 \qquad (13.61)$

Equation 13.61 is now solved as follows:

$$a \cdot FS^3 + b \cdot FS^2 + c \cdot FS + d = 0 \qquad (13.62)$$

$a = W_A \cdot \sin\beta \cdot \cos\theta + W_P \cdot \cos\beta \cdot \sin\theta$

$b = (W_A \cdot \tan\delta_P + W_P \cdot \tan\delta_A + W_T \cdot \tan\phi_s) \cdot \sin\beta \cdot \sin\theta$
$\qquad - (W_A \cdot \tan\delta_A + W_P \cdot \tan\delta_P) \cdot \cos\beta \cdot \cos\theta$

$c = -[W_T \cdot \tan\phi_s \cdot (\sin\beta \cdot \cos\theta \cdot \tan\delta_P + \cos\beta \cdot \sin\theta \cdot \tan\delta_A)$
$\qquad + (W_A \cdot \cos\beta \cdot \sin\theta + W_P \cdot \sin\beta \cdot \cos\theta) \cdot \tan\delta_A \cdot \tan\delta_P]$

$d = W_T \cdot \cos\beta \cdot \cos\theta \cdot \tan\delta_A \cdot \tan\delta_P \cdot \tan\phi_s$

When the cell subgrade is very small (i.e., $\theta \approx 0$), $\sin\theta \approx 0$, and $\cos\theta \approx 1$, Equation 13.62 then becomes

$$a \cdot FS^3 + b \cdot FS^2 + c \cdot FS + d = 0 \qquad (13.63)$$

where $a = W_A \cdot \sin\beta$
$\qquad b = -(W_A \cdot \tan\delta_A + W_P \cdot \tan\delta_P) \cdot \cos\beta$

$$c = -(W_T \cdot \tan\phi_s + W_P \cdot \tan\delta_A) \cdot \sin\beta \cdot \tan\delta_P$$
$$d = W_T \cdot \cos\beta \cdot \tan\delta_A \cdot \tan\delta_P \cdot \tan\phi_s$$

In the conventional translational (or two-wedge) failure analysis method, the direction of the resultant force E_P of E_{HP} and E_{VP} (or the resultant force E_A of E_{HA} and E_{VA}), which acts on the interface between the passive wedge and active wedge, is usually assumed to be parallel to waste filling slope. The effect of the waste property of the interface between the active and passive wedges (i.e., shear strength of the waste) on the stability is not considered for this assumption. Actually, the real direction of the resultant force E_A of E_{HA} and E_{VA} (or the direction of the interwedge force) should be calculated as

$$\begin{aligned}
\tan\omega &= E_{VP}/E_{HP} \\
&= (E_{HP} \cdot \tan\phi_s/FS)/E_{HP} \\
&= \tan\phi_s/FS \\
\omega &= \tan^{-1}(\tan\phi_s/FS)
\end{aligned} \qquad (13.64)$$

where ω = inclination angle of the interwedge force (i.e., the resultant force of E_{HP} and E_{VP}), measured from horizontal, degrees;
  ϕ_s = friction angle of solid waste;
  FS = factor of safety for the entire solid waste mass.

Municipal solid waste usually settles a considerable amount during the filling operation. Review of field settlements from several landfills indicates that municipal solid waste landfills usually settle approximately 15 to 30% of the initial height because of placement and decomposition. The large settlement of the waste fill induces shear stresses in the liner system on the side slope, all of which tends to displace the liner downslope. The large settlement of the waste fill also causes the large deformation of the landfill cover to induce shear stresses in the final cover system. These shear stresses induce shear displacements along specific interfaces in the liner and cover systems that may lead to the mobilization of a residual interface strength. In addition, thermal expansion and contraction of the side slope liner and cover systems during construction and filling may also contribute to the accumulation of shear displacements and the mobilization of a residual interface shear strength in the liner system (Qian, 1994; Stark and Poeppel, 1994).

Earthquake loading can provide permanent displacements along landfill liner interfaces, resulting in a permanent reduction in their available shear resistance following the completion of the dynamic loading. Post-earthquake static stability must therefore be evaluated using shear strengths that are compatible with the shear displacements predicted to be experienced during the earthquake. In areas of high seismicity, this probably implies that the static stability of the final configuration of the landfill should be assured assuming the mobilization of full residual strength conditions (Byrne, 1994).

526 Chapter 13 Landfill Stability Analysis

Landfill stability should be considered not only during construction and operation periods, but also for the duration of the closure period. Land development of closed landfills should be also considered in the future. Thus, the shear strengths (e.g., δ_P, δ_A, and ϕ_s) used in stability analysis must be carefully selected based on actual site-specific conditions.

EXAMPLE 13.8

Calculate the factor of safety for a landfill filling shown in Figure 13.25. Use a translational failure analysis and the following information:

Minimum interface friction angle of bottom liner system, $\delta_P = 20°$;

Minimum interface residual friction angle of side slope liner system, $\delta_A = 14°$;

Friction angle of solid waste, $\phi_s = 33°$;

Waste unit weight $= 10.2$ kN/m^3;

Landfill subgrade is 2% [50(H):1(V)];

Waste filling slope is 25% [4(H):1(V)];

Side slope angle, $\beta = 18.4°$;

Height of side slope is 30 m;

Distance between the top edge of waste and the top edge of side slope is 20 m.

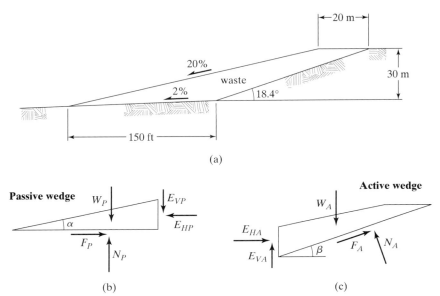

FIGURE 13.25 Cross Section of a Solid Waste Landfill during Filling Condition

Section 13.6 Waste Mass Failures

Solution The forces acting on the solid waste mass are shown in Figure 13.25.
The side slope angle is at 18.4° and the slope angle of cell subgrade is 1.15° according to a 2% slope; hence,

$$\sin\beta = \sin(18.4°) = 0.3162, \cos\beta = \cos(18.4°) = 0.9487,$$
$$\sin\theta = \sin(1.15°) = 0.0200, \cos\theta = \cos(1.15°) = 0.9998$$
$$\tan\delta_A = \tan(14°) = 0.2493, \tan\delta_P = \tan(20°) = 0.3640,$$
$$\tan\phi_s = \tan(33°) = 0.6494.$$

The total weight of solid waste mass is

$$W_T = 10{,}987 \text{ kN/m}$$

The weight of the passive wedge is

$$W_P = 3{,}465 \text{ kN/m}$$

The weight of the active wedge is

$$W_A = W_T - W_P = 10{,}987 - 3{,}465 = 7{,}522 \text{ kN/m}$$

Use Equation 13.62 to calculate *FS*.

Calculate the coefficients of *a*, *b*, *c*, and *d* in Equation 13.62:

$$a = W_A \cdot \sin\beta \cdot \cos\theta + W_P \cdot \cos\beta \cdot \sin\theta$$
$$= 7{,}522 \times 0.3162 \times 0.9998 + 3{,}465 \times 0.9487 \times 0.0200$$
$$= 2{,}444 \text{ kN/m}$$

$$b = (W_A \cdot \tan\delta_P + W_P \cdot \tan\delta_A + W_T \cdot \tan\phi_s) \cdot \sin\phi \cdot \sin\theta - (W_A \cdot \tan\delta_A + W_P \cdot \tan\delta_P) \cdot \cos\beta \cdot \cos\theta$$
$$= (7{,}522 \times 0.3640 + 3{,}465 \times 0.2493 + 10{,}987 \times 0.6494) \times 0.3162 \times 0.0200 -$$
$$(7{,}522 \times 0.2493 + 3{,}465 \times 0.3640 \times 0.9487 \times 0.9998$$
$$= -2{,}907 \text{ kN/m}$$

$$c = -[W_T \cdot \tan\phi_s \cdot (\sin\beta \cdot \cos\theta \cdot \tan\delta_P + \cos\beta \cdot \sin\theta \cdot \tan\delta_A) +$$
$$(W_A \cdot \cos\beta \cdot \sin\theta \cdot W_P \cdot \sin\beta \cdot \cos\theta) \cdot \tan\delta_A \cdot \tan\delta_P]$$
$$= -[10{,}987 \times 0.6494 \times (0.3162 \times 0.9998 \times 0.3640 + 0.9487 \times 0.0200 \times 0.2493) +$$
$$(7{,}522 \times 0.9487 \times 0.0200 + 3{,}465 \times 0.3162 \times 0.9998) \times 0.2493 \times 0.3640]$$
$$= -967 \text{ kN/m}$$

$$d = W_T \cdot \cos\beta \cdot \cos\theta \cdot \tan\delta_A \cdot \tan\delta_P \cdot \tan\phi_s$$
$$= 10{,}987 \times 0.9487 \times 0.9998 \times 0.2493 \times 0.3640 \times 0.6494$$
$$= 614 \text{ kN/m}$$

$$a \cdot FS^3 + b \cdot FS^2 + c \cdot FS + d = 0 \qquad (13.62)$$
$$2{,}444 \cdot FS^3 - 2{,}907 \cdot FS^2 - 967 \cdot FS + 614 = 0$$
$$FS^3 - 1.189 \cdot FS^2 - 0.396 \cdot FS + 0.251 = 0$$
$$FS^3 + 0.251 = 1.189 \cdot FS^2 + 0.396 \cdot FS$$

which is solved by trial and error as in the following table:

Assumed FS	$FS^3 + 0.251$	$1.189 \cdot FS^2 + 0.396 \cdot FS$	Closure
(1)	(2)	(3)	(2) − (3)
1.5	3.626	3.269	0.357
1.4	2.995	2.885	0.110
1.3	2.448	2.524	−0.076
1.35	2.711	2.702	0.009
1.34	2.657	2.666	−0.009
1.345	2.684	2.684	0

Thus, $FS = 1.345$.

The direction of the resultant force of E_{HP} and E_{VP} (i.e., direction of the interwedge force) can be calculated from Equation 13.34 as

$$\tan \omega = \tan \phi_s / FS \tag{13.64}$$
$$= \tan(33°)/1.345$$
$$= 0.649/1.345$$
$$= 0.483$$
$$\omega = 25.8°$$

Recall that the inclination of waste filling slope is 20%, which is only 11.3°. Thus, the direction of the resultant force of E_{HP} and E_{VP} is definitely not parallel to the waste filling slope as is often assumed in these types of calculations (Corps of Engineers, 1960).

13.6.2 Case Histories

Alternatively, for the analysis of the case histories that follow, which failed in a translational manner, the simplified Janbu method was used. (See Koerner and Soong, 2000.) This derivation is also readily available in the literature and leads to a similar equation for the FS-value, but it is now modified with an f_o-value. The resulting equation is

$$FS = (f_o) \cdot \frac{\sum_{i=1}^{n} [c \cdot \Delta b_i + (W_i - u_i \cdot \Delta b_i) \cdot \tan \phi]/m_i}{\sum_{i=1}^{n} W_i \cdot \sin \theta_i} \tag{13.65}$$

where m_i is defined in Equation 13.31, and f_o is a function of the curvature ratio of the failure surface and the type of soil. Since these surfaces are linear, however, the depth-to-length ratio is zero and the value of $f_o = 1.0$. The analysis becomes quite straightforward. (See Schuster and Krizek, 1978.)

To illustrate the seriousness of translational failures (they have represented the largest waste mass failures to date), three case histories are presented next.

Case History T-1

Case history T-1 is a municipal solid waste landfill that experienced a major failure in 1993. The failure was translational in nature and involved approximately 470,000 m^3 of solid waste. The sliding waste mass buried numerous homes in its path and resulted in the loss of 27 lives. The only reference to our knowledge is the consultants report to the municipality owner/operator (Anonymous, 1994).

Background. The operation of the landfill in this case history began in 1972. There was neither a liner nor leachate collection system present. The waste was apparently placed directly on the native ground surface. In the subsequent years, the facility was receiving approximately 1,500 metric tons of municipal solid waste daily. The landfill site is situated within the upper portion of a southward sloping valley that discharges runoff into a local stream. As shown in Figure 13.26(a), a village is located on the south side of the stream directly opposite the valley at an elevation significantly below the landfill. Figure 13.26(b) gives the critical 2-D cross section of the landfill prior to the failure. Of necessity, the scales are significantly distorted. The landfill had been built up into a relatively level plateau at a surface elevation of approximately 150 m. The southern side slope of the landfill, which faced the stream, was very steep, approximately 35 to 40°, which averages to 1(V)-to-1.3(H).

Waste Placement. The sequence of the municipal solid waste placement began at a location approximately 250 m from the landfill entrance [i.e., at the southern end of the landfill, as shown in Figure 13.26(a)]. The landfilling activities took place at this location for approximately 11 years until 1983. At that point, the waste mass had been built up into a level plateau at an average elevation of 130 m. The waste mass was not covered with soil when the landfilling operation was terminated at this particular location. Waste in this area rapidly decomposed over the years and turned black in color. A subsequent landfilling operation was started in the northern and eastern part of the site at a location near the entrance of the landfill. It was continued until 1990 when the waste mass reached an elevation of 140. Again, no cover soil was placed above the waste mass. The final phase of the landfilling activity was initiated in 1990 and continued until the time of failure. This phase initially took place over the top of the second phase waste mass (near the landfill entrance) and continued over the top of the original waste mass at the south portion of the landfill. By early 1993, the existing decomposed waste was entirely covered by an additional 20 to 25 m of waste that brought the surface of the landfill to an elevation of approximately 150 m. Note that several portions of the waste mass partially blocked the drainage route of storm water that ran into the tributary valley. This blockage created water ponds on both the eastern and western sites of the landfill [see Figure 13.26(a)]. The average elevation of these water ponds was approximately 130 m.

Description of Failure. On April 28, 1993, a sudden and massive waste failure occurred in this 50-m high landfill. The decomposed waste mass moved down the valley at high speed for approximately 500 m into the stream and continued up-gradient into the northern portion of a nearby village [see Figure 13.26(a)]. Slide debris, approximately 15- to 20-m in thickness, buried a number of homes and resulted in the loss of 27 lives.

530 Chapter 13 Landfill Stability Analysis

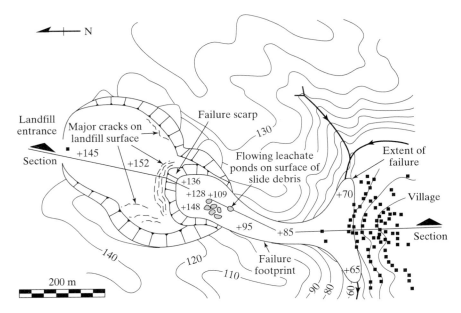

(a) Plan view of the site after failure

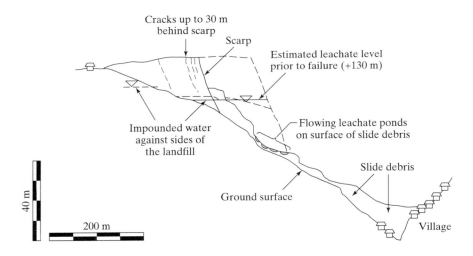

(b) Critical cross section after the failure

FIGURE 13.26 Plan View and Critical Cross Section of Case History T-1 after the Failure

The failure involved approximately 470,000 m³ of solid waste. The investigation concluded that the failure surfaces passed through the waste mass at about 110-m behind the crest of the waste slope and to the base of the decomposed waste adjacent to the natural ground surface [see Figure 13.26(b)]. Although many factors could have contributed to the failure (i.e., over-built steep slopes or even gas uplift pressures), the triggering mechanism of the failure was likely an excessive leachate level within the old, decomposed waste caused by continued water infiltrating from the adjacent surface water ponds. The flowing leachate ponds observed on surface of the slide debris tend to support this conclusion.

Case History T-2

Case history T-2 is a municipal solid waste landfill in which a major section failed in 1997. The failure was translational in nature and involved approximately 1,100,000 m³ of solid waste. Kenter et al. (1996), Stark et al. (1997), Stark and Evans (1997), Schmucker and Hendron (1997), Schmucker (1998), and Stark et al. (2000b) have reported on this case history.

Background. The landfill began as an outgrowth of farm waste storage in 1945 and transitioned into one that accepted mixed wastes shortly thereafter. Over the intervening years, the facility grew in both area and height as it continued to accept residential solid waste, commercial solid waste, industrial waste, and asbestos. The majority of the overall site was developed prior to the existence of requirements for engineered environmental control. The older areas of the landfill have neither liners nor leachate collection systems. These areas were, however, constructed over low permeability colluvium soils that overlie interbedded shales and limestones, all of which tend to act as a natural liner. Beginning in 1988, a clay liner was required and this was further upgraded to a composite liner in 1994. Thus, the initial phase consisted of a clay liner for its initial 5.7 ha and a composite liner was intended for the remaining 1.5 ha. As the northern portion of this phase (i.e., the 1.5 ha area) was being prepared for composite liner placement, excavation of the next phase was ongoing. It should be noted that both of these areas were at a significantly lower elevation than the existing landfill. Rock (shale) blasting was used to reach the lower elevation and was ongoing at the time of the failure.

Description of Failure. At about noon on March 9, 1996, five days after the first crack appeared on the top of the existing landfill, the landfill began to move as a large mass northward into the open excavation areas. A waste mass of approximately 1,100,000-m³ translated some 50 to 60 m in less than five minutes. Figure 13.27(a) shows an aerial photograph taken after the failure. Based on field observations and the results of a subsequent subsurface investigation, the failure surface passed through the solid waste at a very steep inclination down to the underlying colluvium soil [see Figure 13.27(b)]. From this point, the failure plane extended within the colluvium soil until it exited at the toe of the slope. The leachate head within the waste mass prior to failure was estimated at the level shown in Figure 13.27(b), with a maximum depth of approximately 13 m, Schmucker and Hendron (1997). It is concluded by Schmucker and Hendron (1997) and Schmucker (1998) that the likely triggering mechanism for the failure was the additional buildup of leachate head in the landfill due to ice forma-

532 Chapter 13 Landfill Stability Analysis

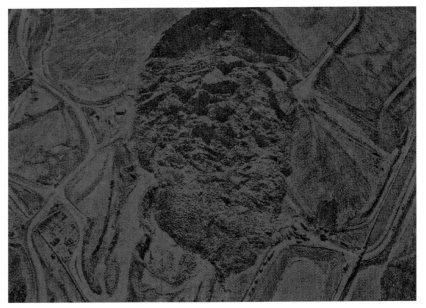

(a) Areial Photograph of Failure

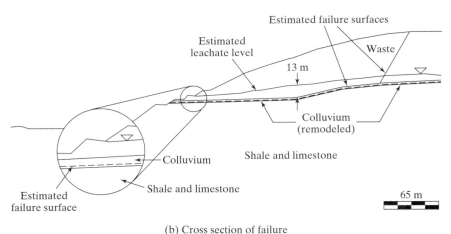

(b) Cross section of failure

FIGURE 13.27 Details of Case History T-2

tion at the exposed waste face near the toe of the slope. However, this hypothesis is contested by Stark et al. (2000b), who claim that the excessive height was the main contributing factor.

Foundation Soil Conditions. The existing landfill in this case history was placed immediately above the in-situ colluvium soils. Such soils are commonly found in this geographic area with thickness up to 10 m, although at the site the layer was approximately 3 to 4 m thick. It was postulated by Stark et al. (2000a) that the colluvium deposit was marginally stable because it was in a residual shear strength state due to the constant down-slope deformation caused by waste placement and its gravitational stresses. While this is uncertain, it is possible that dynamic stresses caused by blasting at the toe of the slope may have been an additional destabilizing factor.

Case History T-3

Case History T-3 is a hazardous waste landfill that failed in 1988. It was translational in that the entire waste body of approximately 490,000 m^3 moved as a mass within a period of seven hours. The failure resulted in 10.7 m of lateral displacement and 4.3 m of vertical settlement at the top of the solid waste as it moved down-gradient. Seed et al. (1990), Mitchell et al. (1990), Byrne et al. (1992), and Stark and Poepple (1994) have reported on this case history.

Background. The landfill site in this case history occupied 15 ha. The configuration consisted of a large oval-shaped bowl excavated into the ground to a depth of about 30 m. The bowl had a nearly horizontal base and side slopes of either 2 (H)-to-1 (V) or 3 (H)-to-1 (V) inclination, but was open-sided on its southeast side. Figure 13.28 shows the general configuration. Both the base and side slopes of the excavation were lined with a very complex multilayered geomembrane, compacted clay liner, and leachate drainage system.

The northern portion of the facility was completed first. The placement of hazardous solid waste was initiated in this section of the facility in early 1987, while the liner system for the future phase was being constructed southward. The waste mass placed in the southerly cell eventually was to provide a buttress against the previously placed waste.

Liner System. The liner system at the base of the landfill consisted of the following layers, from the top to the bottom:

- protective soil layer;
- leachate collection and removal system (consisting of a geotextile filter/separator, a 0.3-m-thick, granular soil layer, another geotextile filter/separator, and a geonet);
- primary composite liner (consisting of a 1.5-mm-thick, smooth HDPE geomembrane and a 0.5-m-thick compacted clay liner);
- leak detection, collection, and removal system (consisting of a geotextile filter/separator, a 0.3-m-thick, granular soil layer, and another geotextile filter/separator);
- secondary composite liner (consisting of a 1.5-mm-thick, smooth HDPE geomembrane and a 1.1-m-thick compacted clay liner);
- vadose zone de-watering system (consisting of a geotextile filter/separator, a 0.3-m-thick, granular soil layer and another geotextile filter/separator);
- 2.0-mm-thick, smooth HDPE geomembrane; and
- in-situ compacted soil subgrade.

534 Chapter 13 Landfill Stability Analysis

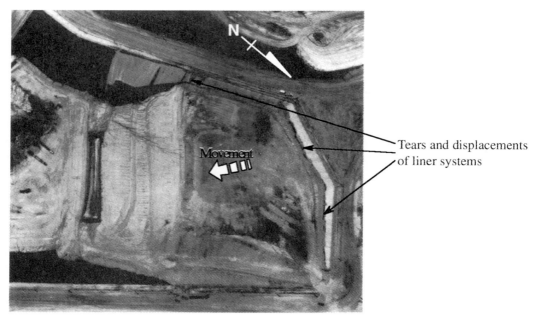

(a) Aerial Photograph Taken after the Failure of Landfill T-3

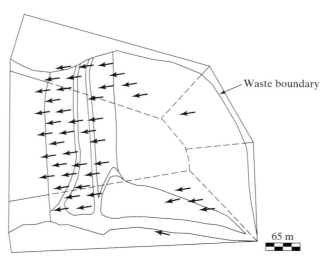

(b) Waste boundary after failure showing displacement vectors

FIGURE 13.28 Landfill T-3 after Failure (after Seed et al., 1990) Used with permission of ASCE.

The liner system on the side slopes of the landfill consisted of the following layers, from the top to the bottom:

- protective soil layer;
- leachate collection and removal system (consisting of a geotextile filter/separator and a geonet);
- primary 1.5-mm-thick, smooth HDPE geomembrane liner;
- leak detection, collection, and removal system (consisting of a geotextile filter/separator and a geonet);
- secondary composite liner (consisting of a 1.5-mm-thick, smooth HDPE geomembrane and a 1.1-m-thick compacted clay liner); and
- in-situ compacted soil subgrade.

Waste Properties. The waste material was estimated to have a unit weight of 17.3 kN/m^3. This relatively high value is indicative of a large amount of soil surrounding the containers of hazardous waste and the nature of the solid waste itself. The shear strength of the waste was not relevant to the analysis since the failure surfaces did not pass through any of the waste material.

Description of the Failure. On March 19, 1988, a slope-stability failure occurred at this particular landfill. The first sign of failure was a 12-mm-wide crack across a truck ramp at the northeast corner of the landfill. It was observed at about 6:30 AM. Since it was Saturday, few personnel were at the site, and its seriousness was neither suspected nor communicated to others. At approximately 9:30 AM, 75- to 100-mm-wide cracks with 150- to 200-mm vertical offsets were observed along the crests of the 2(H)-to-1(V) slopes. The main failure, which was quite abrupt, was reported to have occurred at 1:30 PM. Approximately 490,000 m^3 of waste was involved in the abrupt translational slide that occurred subsequently. Surface cracking was clearly visible, as were major tears and displacements on the exposed portions of the liner system along all three of the side slopes. The direction of the waste movement is shown in Figure 13.28(b) using displacement vectors. The maximum fill height was approximately 27 m at the time of the failure.

Failure Surfaces. As described by Mitchell et al. (1990) and Seed et al. (1990), the actual failure surfaces were the geomembrane-to-CCL interface of the secondary composite liner along the base, and the primary geomembrane-to-underlying geotextile interface on the side slopes of the landfill. As noted, all of the geomembranes were smooth, since textured liners were not available at the time of the failure. The triggering mechanism that led to the waste failure, however, was likely to have been the excessive wetness of the geomembrane-to-compacted clay liner interface at the base of the landfill. Rainfall during construction and waste placement, as well as the consolidation water expelled from the CCL, was felt to have caused an excessively wetted clay interface with the overlying geomembrane.

13.6.3 General Remarks

Translational failures of the type presented in these three case histories represent the largest landfill failures that have occurred insofar as waste volumes are concerned. As was seen, such failures can be life-threatening as well. As was the situation for rotational failures, the analysis is straightforward and many computer codes are available. The critical issue is the proper assessment of interface shear strengths. Both product-specific materials and site-specific conditions must be properly determined (recall Section 13.3.1).

As pointed out by Koerner and Soong (2000), excessive moisture was invariably the triggering mechanism in the failures that they evaluated. In their study of 10 landfill failures (5 of which are included in this section), all were due, at least in part, to excessive liquids. There were three locations of the excess liquids:

(i) Leachate buildup within the waste mass, resulting in the failure surfaces to be above the liner system that was at the base of the landfill.

(ii) Extremely wet compacted clay liners beneath the geomembrane in composite liner systems, resulting in failure surfaces immediately beneath the geomembrane and above, or nominally within, the excessively wet clay.

(iii) Wet foundation or soft backfill soils resulting in failure surfaces within the subgrade soil beneath the waste mass.

Additional information on these three situations, and the particular triggering mechanism involved in each of the failures is given in Table 13.1.

Clearly, excessive liquids above, below, or within the failure surfaces were involved in the failures in all 10 case histories presented and analyzed by Koerner and Soong (2000).

TABLE 13.1 Summary of Triggering Mechanisms Involved in the Case Histories Presented by Koerner and Soong (2000)

Case History	Reason for low initial FS-value	Triggering mechanism
U-3 U-4 L-4 L-5	Leachate buildup within waste mass	Excessive buildup of leachate level due to ponding Excessive buildup of leachate level due to ice formation Excessive buildup of leachate level due to liquid waste Excessive buildup of leachate level due to leachate injection
L-1 L-2 L-3	Wet clay beneath geomembrane (i.e., GM/CCL composite)	Excessive wetness of the GM/CCL interface Excessive wetness of the GM/CCL interface Excessive wetness of the bentonite in an unreinforced GCL
U-1 U-2 U-5	Wet foundation or soft backfill soil	Rapid rise in leachate level within the waste mass Foundation soil excavation exposing soft clay Excessive buildup of perched leachate level on clay liner

U = unlined (or clay lined) sites
L = geomembrane or composite (GM/CCL) lined sites
Used with permission of ASCE.

13.7 CONCLUDING REMARKS

The general concept of a waste containment system is to prevent leakage of leachate (and escape of gases) from a landfill. As such, most past efforts have rightly been focused on these issues. What particular liner system is best for a given site and given waste from a leakage perspective is obviously important and has been the focus of much of this text.

However, when one compares leakage from a waste containment system (even relatively large leakage) with the massive failures given in this chapter, the message is obvious. Waste failures simply are not acceptable. Nevertheless, some have occurred. Table 13.2 offers a glimpse of the magnitude and seriousness that such situations can engender.

Of major importance is that all of the failures in Table 13.2 were straightforwardly analyzed in light of current geotechnical and geosynthetics practice. That being said, it is necessary that we not only do the forensic analysis after the failures, but also that we do the proper design initially to avoid such failures in the first place. This requires that a knowledgeable design consultant be utilized throughout the process for each of the following stages:

(i) the liner system beneath the waste,
(ii) the waste placement at all critical stages, and
(iii) the cover system above the waste.

While liner and cover systems have generally been designed and constructed with carefully controlled CQC and CQA, waste placement has rarely been designed, and even more rarely has it been controlled. This latter issue is invariably left to the landfill operator. Since many of the failures in Table 13.1 were during waste filling, the operations of waste placement must be carefully planned and then executed accordingly.

TABLE 13.2 Summary of Waste Failures Analyzed by Koerner and Soong (2000)

Identification	Year	Location	Type of Failure	Quantity of Waste Involved in Failure (m^3)
Unlined Sites				
U-1	1984	N. America	Single Rotational	110,000
U-2	1989	N. America	Multiple Rotational	500,000
U-3	1993	Europe	Translational	470,000
U-4	1997	N. America	Translational	1,100,000
U-5	1997	N. America	Single Rotational	100,000
Lined Sites				
L-1	1988	N. America	Translational	490,000
L-2	1994	Europe	Translational	100,000
L-3	1997	N. America	Translational	300,000
L-4	1997	Africa	Translational	300,000
L-5	1997	S. America	Translational	1,200,000

Used with permission of ASCE.

This necessitates construction quality assurance (CQA) personnel being on the site during landfill placement operations on challenging sites where a failure might occur. Obviously, not all sites require such scrutiny, but certainly some do.

In general, the stability of a landfill should be evaluated by performing stability analyses for conditions that will exist at different stages of excavation, construction, operation, and closure of the facility. The analyses should address the following five conditions:

(i) Side slope stability during excavation;
(ii) Liner system stability during construction;
(iii) Waste mass stability during filling stage;
(iv) Final cover system stability; and
(v) Landfill postclosure stability.

With proper design and construction, it is hoped that the failures illustrated in this chapter will cease to exist.

PROBLEMS

13.1 Rank the seriousness of the six potential landfill instability situations described in Section 13.1 and illustrated in Figure 13.1.

13.2 Describe the general remedies for the six potential landfill instability situations described in Section 13.1 and illustrated in Figure 13.1.

13.3 The three materials that are possibly involved in landfill failures are geosynthetics, natural soils, and solid waste. (Recall Section 13.3.) What other types of materials might be involved under less common situations?

13.4 In the laboratory, determination of shear strength of geosynthetics, natural soils, and solid waste (recall Section 3.3), (a) What are the appropriate ASTM test methods in current use? and (b) How does the size of the test device possibly influence the results?

13.5 In the direct shear testing of geosynthetics, a strength intercept (i.e., an adhesion value) at zero normal stress is sometimes observed. For what types of materials can this value be justifiably used in stability analyses?

13.6 How does the shear strength of a compacted clay liner (CCL) vary with moisture content? Illustrate your answer on a graph of moisture content versus dry density.

13.7 What is the difference between peak shear strength, high-deformation shear strength, and residual shear strength? Illustrate your answer on a shear deformation versus shear strength graph.

13.8 Regarding Example 13.1, recalculate the *FS*-values for the following variations and plot your response curves. (Other variables than listed below remain the same as in the example).
   (a) Slope lengths from 10 to 100 m.
   (b) Cover soil thickness from 200 to 1000 mm.
   (c) Slope angles from 2(H)-to-1(V) to 5(H)-to-1(V).
   (d) Cover soil friction angles from 15 to 40°.
   (e) Interface friction angles from 10 to 35°.

13.9 Regarding Example 13.2, recalculate the *FS*-values for the following variations and plot your response curves. (Variables not given in the list that follows remain the same as in the example).
  (a) Bulldozer ground pressure from 20 to 150 kN/m².
  (b) Bulldozer tracks from 2.0 to 5.0 m long.
  (c) Bulldozer tracks from 0.4 to 1.0 m wide.

13.10 Regarding Example 13.3, recalculate the *FS*-values for the bulldozer time to reach 20 km/hour from 1 to 10 seconds and plot your response curves. (Other variables than those listed remain the same as in the example).

13.11 Two seepage force scenarios were presented in Section 13.3.3: horizontal and parallel. Give situations where each could occur by filling the the following:

Seepage Scenario	Leachate Collection System	Final Cover System
Horizontal		
Parallel		

13.12 A 36-ft- (10.8-m)-high and 3(H):1(V) slope has cover sand with a uniform thickness of 2 ft (0.6 m) and a unit weight of 115 lb/ft³ (18 kN/m³). The cover sand has a friction angle of 30° and zero cohesion. Seepage occurs parallel to the slope and the seepage water head in the sand layer is 5.2 inches (0.13 m). The saturated unit weight of the sand is 120 lb/ft³ (19 kN/m³). The interface friction angle between the sand drainage layer and geomembrane is 20° with zero adhesion. Calculate the factor of safety for the cover soil on the side slope. If the factor of safety is less than 1.2, use the incremental method to achieve a factor of safety no less than 1.2 for the cover sand resting on the sideslope.

13.13 Regarding Example 13.6, recalculate the *FS*-values for the seismic coefficient from 0.0 to 0.30 and plot your response curves. (Other variables than those listed remain the same as in the example).

13.14 For the above problem, what are the implications (i.e., regarding further analysis) if the *FS*-value is above or below 1.0?

13.15 In assessing the results of a permanent deformation analysis, as in Example 13.7, what factors influence the establishment of an allowable deformation value?

13.16 Three case histories of subsoil foundation failures were presented in Section 13.5. Assemble them in table form (e.g., height, slope, area, soils involved, waste involved, failure mass, and failure triggering action) and provide commentary. Also, include what possible preventative measures could have been taken to avoid the failures.

13.17 Calculate the factor of safety for a filling landfill like that shown in Figure 13.25 against possible mass movement. Use a two-wedge analysis and the following information:
  Interface friction angle of bottom liner, $\delta_P = 18°$;
  Interface residual friction angle of side slope liner, $\delta_A = 10°$;
  Friction angle of solid waste, $\phi_s = 33°$;
  Waste unit weight = 70 lb/ft³;
  Landfill subgrade is 2% [50 (H):1(V)], $\theta = 1.15°$;
  Side slope angle, $\beta = 18.4°$ [3(H):1(V)];
  Waste filling slope is 25% [4(H):1(V)];
  Height of side slope is 50 feet;
  Distance between the top edge of waste and the top edge of side slope is 50 feet.
  If the waste filling slope is changed to 33% [3(H):1(V)], recalculate the factor of safety.

13.18 Calculate the factor of safety for a filling landfill like the one shown in Figure 13.25 against possible mass movement. Use a two-wedge analysis and the following information:

Interface friction angle of bottom liner, $\delta_P = 18°$;
Interface residual friction angle of side slope liner, $\delta_A = 10°$;
Friction angle of solid waste, $\phi_s = 33°$;
Waste unit weight $= 70$ lb/ft^3;
Landfill subgrade is 15%, $\theta = 8.5°$;
Side slope angle, $\beta = 18.4°$ [3(H):1(V)];
Waste filling slope is 25% [4(H):1(V)];
Height of side slope is 50 feet;
Distance between the top edge of waste and the top edge of side slope is 50 feet.

13.19 Three case histories of waste mass failure were presented in Section 13.6. Assemble them in table form (e.g., height, slope, area, geosynthetics involved, waste involved, failure mass, and failure triggering mechanism) and provide commentary. Also, include what possible preventative measures could have been taken to avoid the failures.

13.20 The six landfill failures of Section 13.5 and 13.6 were all serious and involved many organizations and individuals. Interestingly, none of them had any instrumentation. What types of instrumentation could have been used in such situations and what information could have been generated to foreworn of such incidents?

REFERENCES

Algermissen, S. T., (1969) "Seismic Risk Studies in the United States," *Proc. 4th World Conference on Earthquake Engineering,* Vol. 1, Santiago, Chile, pp. A1-14–27.

Anderson, D. G. and Kavazajian, E. Jr., (1995) "Performance of Landfills Under Seismic Loading," *Proceedings of Third International Conference on Recent Advances in Geotechnical Earthquake Engineering and Soil Dynamics,* University of Missouri, Rolla, MO, April, Vol. 3, pp. 2–7.

Anonymous (1994), "Emergency Consulting Engineering and Design Services to Stabilize the Ümbaniye Dump Site and Evaluation of Potential Safety Problem at Other Solid Wate Dumps in Istanbul," Prepared for the Municipality of Greater Istanbul, Turkey, CH2M Hill International, Ltd.

Byrne, R. J., Kendall, J., and Brown, S., (1992) "Cause and Mechanism of Failure of Kettleman Hills Landfill," *Proceedings of ASCE Specialty Conference on Stability and Performance of Slope and Embankments—II,* Berkeley, CA, June 28–July 1, pp. 1–23.

Byrne, R. J., (1994) "Design Issues with Strain-Softening Interfaces in Landfill Liners," *Waste Tech '94,* Landfill Technology, Technical Proceedings, Charleston, SC, January 13–14.

Cancelli, A. and Rimoldi, P., (1989) "Design Criteria for Geosynthetic Drainage Systems in Waste Disposal," *Proc. of Sardinia '89,* 2nd Intl. Landfill Symposium, Porto Conte, Sassari, Italy.

Concoran, G. T. and McKelvy, J. A., (1995) "Stability of Soil Layers on Compound Geosynthetic Slopes," *Proc. Waste Tech '95,* New Orleans, LA, Environ Industry Assoc., pp. 301–304.

Corps of Engineers, (1960) "Manual EM 1110—1902," U.S. Army, Washington DC, 195 pages.

Daniel, D. E., Shan, H.-Y., and Anderson, J. D., (1993) "Effects of Partial Wetting on the Performance of the Bentonite Component of a Geosynthetic Clay Liner," *Proc. Geosynthetics '93,* IFAI, St. Paul, MN, pp. 1483–1496.

de Santayana, F. P. and Pinto, A. A. V., (1998) "The Bierolos Landfill Eastern Expansion Landslide," *Proc. Environmental Geotechnics,* S. e. Pinto (Ed.), Balkema Publ. Co., Rotterdam, pp. 905–910.

Divinoff, A. H. and Munion, D. W., (1986) "Stability Failure of a Sanitary Landfill," Proc. Intl. Symp. on Environmental Geotechnology, H.-Y. Fang, Ed., Envo Publ. Co., Inc. pp. 25–35.

Erdogan, H., Sadat, M. M., and Hsieh, N. N., (1986) "Stability Analysis of Slope Failures in Landfills: A Case Study," *Proc. of 9th Annual Madison Waste Conference,* Univ. of Wisconson-Madison, pp. 168–177.

Giroud, J. P. and Beech, J. F., (1989) "Stability of Soil Layers on Geosynthetic Lining Systems," *Geosynthetics '89 Proceedings,* Vol. 1, pp. 35–46.

Giroud, J. P., Williams, N. D., Pelte, T., and Beech, J. F., (1995a) "Stability of Geosynthetic-Soil Layered Systems on Slopes," *Geosynthetic International,* Vol. 2, No. 6, pp. 1115–1148.

Giroud, J. P., Bachus, R. C., and Bonaparte, R., (1995b) "Influence of Water Flow on the Stability of Geosynthetic-Soil Layered Systems on Slopes," *Geosynthetic International,* Vol. 2, No. 6, pp. 1149–1180.

Hassini, S., (1992) "Some Aspects of Landfill Design," *Environmental Geotechnology,* Proceedings of the Mediterranean Conference on Environmental Geotechnology, A. A. Balkema Publishers, pp. 137–144.

Hirshfield, R. C. and Poulos, S. J., Eds., (1973) *Embankment—Dam Engineering,* John Wiley and Sons, New York, NY 454 pgs.

Humphrey, D. N. and Leonards, G. A., (1986) "Slide Upstream Slope of Lake Shelbyville Dam," *Journal of Geotechnical Engineering,* ASCE, Volume 112, No. 5, pp. 564–577.

Idriss, I. M., (1990) "Response of Soft Soil Sites During Earthquake," *Proc. Symposium to Honor Professor H. B. Seed,* Berkeley, California.

Kenter, R. J., Schmucker, B. O., and Miller, K. R., (1996) "The Day the Earth Didn't Stand Still: The Rumpke Landslide," *Waste Age,* Atlanta, GA, March, pp. 36–41.

Koerner, R. M., (1994) *Designing with Geosynthetics,* 3rd Ed., Prentice Hall Book Co., Englewood Cliffs, NJ, 783 pgs.

Koerner, R. M. and Hwu, B. -L., (1991) "Stability and Tension Considerations Regarding Cover Soils on Geomembrane Lined Slopes," *Journal. of Geotextiles and Geomembranes,* Vol. 10, No. 4, pp. 335–355.

Koerner, R. M. and Soong, T. -Y., (1998) "Analysis and Design of Veneer Cover Soils," *Proc. 6th IGS,* IFAI, Roseville, MN, pp. 1–26.

Koerner, R. M. and Soong, T. -Y., (2000) "Stability Assessment of Ten Large Landfill Failures," *Advances in Transportation and Geoenvironmental Systems Using Geosynthetics, Proceedings of Sessions of GeoDenver 2000,* ASCE Geotechnical Special Publication No. 103, pp. 1–38.

Ling, H. I. and Leschinsky, D., (1997) "Seismic Stability and Permanent Displacement of Landfill Cover Systems," *Journal of Geotechnical and Geoenvironmental Engineering,* ASCE, Vol. 123, No. 2, pp. 113–122.

Makdisi, F. I. and Seed, H. B., (1978) "Simplified Procedure for Estimating Dam and Embankment Earthquake-Induced Deformations," *Journal of Geotechnical Engineering,* ASCE, Vol. 104, No. GT7, pp. 849–867.

Matasovic, N., Kavazanjian, E., Jr., Augello, A. J., Bray, J. D., and Seed, R. B., (1995) "Solid Waste Landfill Damage Caused by January 17, 1994 Northridge Earthquake," In: Woods, M.C. and Seiple, R. W., Eds., *The Northridge, California, Earthquake of 17 January 1994,* California Department of Conservation, Division of Mines, and Geology Special Publication 116, Sacramento, California, pp. 43–101.

Matasovic, N., Kavazanjian E. Jr., and Yan, L., (1997) "Newmark Deformation Analysis with Degrading Yield Acceleration," Proc. Geosynthetics '97, IFAI, St. Paul, MN, pp. 989–1000.

McKelvey, J. A. and Deutsch, W. L., (1991) "The Effect of Equipment Loading and Tapered Cover Soil Layers on Geosynthetic Lined Landfill Slopes, " *Proceedings of the 14th Annual Madison Avenyue Conference,* Madison, WI, University of Wisconsin, pp. 395–411.

McKelvey, J. A., (1994) "Consideration of Equipment Loadings in Geosynthetic Lined Slope Design," *Proc. 8th Intl. Conf. of the Intl. Assoc. for Computer Methods and Advancement in Geomechanics,* Morgantown, WV, Blakema, pp. 1371–1377.

Mitchell, J. K., Seed, R. B., and Seed, H. B., (1990) "Kettleman Hills Waste Landfill Slope Failure I: Liner System Properties," *J. of Geotech. Engrg.,* ASCE, Vol. 116, No. 4, pp. 647–668.

Mitchell, R. A. and Mitchell, J. K., (1992) "Stability Evaluation of Waste Landfills," *Proceedings of ASCE Specialty Conference on Stability and Performance of Slope and Embankments—II,* Berkeley, CA, June 28–July 1, pp. 1152–1187.

Mitchell, J. K., Seed, R. B., and Seed, H. B., (1990) "Kettleman Hills Waste Landfill Slope Failure I: Liner System Properties," *Journal of Geotechnical Engineering,* ASCE, Volume 116, No. 4, pp. 647–668.

Qian, Xuede, (1994) "Stability Analyses for Vertical and Lateral Expansions of Landfill," Michigan Department of Environmental Quality, Waste Management Division, Lansing, MI.

Qian, Xuede, (1997) "Stability Analysis of Cover Soil over Geosynthetic Layered Slope with Seepage Force," Michigan Department of Environmental Quality, Waste Management Division, Lansing, MI.

Reynolds, R. T., (1991) "Geotechnical Field Techniques Used in Monitoring Slope Stability at a Landfill," *Field Measurements in Geotechnics,* Sorum (ed.), Balkema, Rotterdam, ISBN 90 5410 0257.

Richardson, G. N., Kavazanjian, E. Jr., and Matasovic, N., (1995) "RCRA Subtitle D (258) Seismic Design Guidance for Municipal Solid Waste Landfill Facilities," EPA/600/R-95/051, U. S. Environmental Protection Agency, Cincinnati, OH, 143 pgs.

Richardson, G. N. and Reynolds, R. T., (1990) "Landslide at a Naturally Lined Landfill," ASCE Materials Conf., Denver, CO, 25 pgs.

Schmucker, B. O., (1998) "The Rumpke Waste Failure," *Waste Age,* March, pp. 30–38.

Schmucker, B. O. and Hendron, D. M., (1997) "Forensic Analysis of 9 March 1996 Landslide at the Rumpke Sanitary Landfill, Hamilton County, Ohio," *Proc. Seminar on Slope Stability in Waste Systems,* ASCE Central Ohio, Cincinnati and Toledo Sections.

Schnabel, P. B., Lysmer, J., and Seed, H. B., (1972) "SHAKE: A Computer Program for Earthquake Response Analysis of Horizontally Layered sites," Report No EERC 72–12, Earthquake Engineering Research Center, University of California, Berkeley, CA.

Schuster, R. L. and Krezik, R. J., (1978) *Landslides: Analyses and Control,* Natl. Academy of Sciences, Washington, DC, 427 pages.

Seed, H. B. and Idriss, I. M., (1982) *"Ground Motions and Soil Liquefaction During Earthquakes,"* Monographs No. 5, Earthquake Engineering Research Center, University of California, Berkeley, CA, 134 pgs.

Seed, R. B., Mitchell, J. K., and Seed, H. B., (1990) "Kettleman Hills Waste Landfill Slope Failure II: Stability Analyses," *Journal of Geotechnical Engineering,* ASCE, Volume 116, No. 4, pp. 669–690.

Sherard, J. L., Woodward, R. J., Gizienski, S. F., and Clevenger, W. A., (1963) *Earth and Earth-Rock Dams,* John Wiley & Sons, New York, 727 pgs.

Singh, S. and Murphy, P., (1990) "Evaluation of the Stability of Sanitary Landfills, ASTM STP 1070, ASTM, W. Conshohocken, PA, pp. 240–258.

Soong, T. -Y. and Koerner, R. M., (1996) *Cover Soil Slope Stability Involving Geosynthetic Interfaces,* Geosynthetic Research Institute Report 18, Philadelphia, PA, 87 pages.

Soong, T. -Y., Hungr, O., and Koerner, R. M., (1998) "Analysis of Selected Landfill Failures by 2-D and 3-D Stability Methods," *Proc. 12 GRI Conference,* GII Publications, Folsom, PA (in print).

Soong, T. -Y. and Koerner, R. M., (1999) "Stability Analysis (2-D and 3-D) and Assessment of Ten Lined Landfills," GRI Report #21, 250 pgs.

Soong, T. -Y. and Koerner, R. M., (1996) "Seepage Induced Slope Instability," *Journal of Geotextiles and Geomembranes,* Vol. 14, No. 7/8, pp. 425–445.

Stark, T. D., Eid, H. T., Evans, W. D., and Sherry, P. E., (2000a) "Municipal Solid Waste Slope Failure: Part I – Waste and Foundation Soil Properties," *Journal Geotechnical and Geoenvironmental Engineering,* ASCE, Vol. 126, No. 5, pp. 397–407.

Stark, T. D., Eid, H. T., Evans, W. D., and Sherry, P. E., (2000b) "Municipal Solid Waste Slope Failure: Part II – Stability Analyses," *Journal of Geotechnical and Geoenvironmental Engineering,* ASCE, Vol. 126, No. 5, pp. 408–419.

Stark, T. D. and Evans, W. D. (1997), "Balancing Act," Civil Engineering, ASCE, August, pp. 8A–11A.

Stark, T. D. and Poeppel, A. R., (1994) "Landfill Liner Interface Strengths from Torsional Ring Stress Tests," *Journal of Geotechnical Engineering,* ASCE, Vol. 120, No. 3, pp. 597–617.

Terzaghi, K., (1943) *Theoretical Soil Mechanics,* John Wiley & Sons, New York.

Thiel, R. S. and Stewart, M. G., (1993) "Geosynthetic Landfill Cover Design Methodology and Construction Experience in the Pacific Northwest," Proc. Geosynthetics '93, IFAI, St. Paul, MN, pp. 1131–1144.

CHAPTER 14

Vertical Landfill Expansions

14.1 CONSIDERATIONS INVOLVED IN VERTICAL EXPANSIONS
14.2 LINER SYSTEMS FOR VERTICAL EXPANSION
14.3 SETTLEMENT OF EXISTING LANDFILL
14.4 ESTIMATION OF DIFFERENTIAL SETTLEMENT DUE TO WASTE HETEROGENEITY
     14.4.1 CURRENT METHODS FOR ESTIMATING LOCALIZED SUBSIDENCE
     14.4.2 ELASTIC SOLUTION METHOD APPLIED TO A VERTICAL EXPANSION
14.5 VERTICAL EXPANSION OVER UNLINED LANDFILLS
14.6 DESIGN CONSIDERATIONS FOR LANDFILL STRUCTURES
14.7 GEOSYNTHETIC REINFORCEMENT DESIGN FOR VERTICAL EXPANSIONS
     14.7.1 THEORETICAL BACKGROUND FOR GEOSYNTHETIC REINFORCEMENT
     14.7.2 ASSUMPTIONS FOR GEOSYNTHETIC REINFORCEMENT DESIGN
     14.7.3 SELECTION OF MATERIAL PROPERTIES
     14.7.4 DETERMINATION OF GEOMETRIC AND LOADING PARAMETERS
     14.7.5 DESIGN CRITERIA
     14.7.6 SELECTION OF ALLOWABLE REINFORCEMENT STRAIN
     14.7.7 SELECTION OF LONG TERM ALLOWABLE DESIGN TENSILE LOAD
     14.7.8 DESIGN STEPS
     14.7.9 SPECIAL DESIGN CASES
     14.7.10 DESIGN EXAMPLE
14.8 STABILITY ANALYSIS FOR VERTICAL EXPANSIONS
     PROBLEMS
     REFERENCES

The acquisition and permitting of new landfill sites poses several difficulties. An attractive alternative to landfill owners is to consider expansions to existing landfills. This option may entail the design and permitting of a vertical expansion over old landfill areas. The advantages of vertical landfill expansion include (1) optimal use of landfill area, (2) high waste volume filled per unit area, (3) low construction cost, (4) less public opposition, and (5) easier permitting. Expansion can occur by vertical and/or lateral expansion in which the old landfill is encapsulated by the new (vertical and lateral expansion), or by placement of new landfill atop the old (piggyback expansion). Figure 14.1 shows a cross-section of a vertical and lateral expansion landfill; Figure 14.2 shows a cross section of a piggyback vertical expansion landfill.

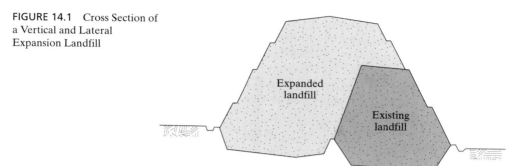

FIGURE 14.1 Cross Section of a Vertical and Lateral Expansion Landfill

14.1 CONSIDERATIONS INVOLVED IN VERTICAL EXPANSIONS

The additional waste fill from a vertical landfill expansion will cause settlement of the existing landfill and result in liner system and slope stability problems for both existing and expanded landfills. A gas collection system in the existing landfill may also be of concern due to the large deformation of the solid waste surrounding gas collection pipes. A liner and leachate collection system constructed on an existing landfill may experience large differential settlements. The long-term performance of these systems is thus a major design consideration.

Large differential settlements within existing refuse may occur because of the collapse or degradation of large objects, which have been deposited in old landfills. These settlements could result in tensile strains in a liner system placed on the top of an old landfill. If the tensile strain within the liner exceeds the tensile capacity of the material, whether it is a soil or a geosynthetic, tension cracks or tensile failure will develop. The tension cracks will reduce the effectiveness of the liner as a hydraulic barrier by providing a direct flow path through the liner system (Jang and Montero, 1993). Under extreme conditions, large differential settlements could result in the reversal of leachate flow gradients and directions. If grade reversal takes place at the surface of a liner and leachate collection system, leachate will pond on the liner, and increase the potential for infiltration of the leachate into the old landfill.

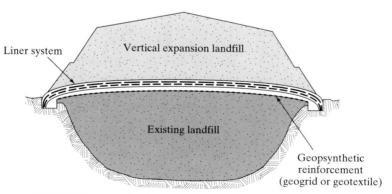

FIGURE 14.2 Cross Section of a Piggyback Vertical Expansion Landfill

The additional waste fill due to the vertical expansion may also affect the bottom topography or subgrade elevations beneath the existing landfill and cause ponding problems on the bottom liner. Almost certainly, this extra load will increase the deflection and wall stress of the leachate and gas collection pipes buried in the existing landfill (if any are present) and may also cause pipe failure or pipe wall stability problems.

Major design steps and considerations for vertical landfill expansion include (Qian, 1996)

(i) Selecting a suitable composite liner system for placement over the existing landfill.
(ii) Estimating the overall total settlement and differential settlement of the existing landfill caused by new waste fill.
(iii) Estimating the differential settlement due to the degradation of large objects in the old landfill, or reinforcing the liner system to minimize this differential settlement.
(iv) Calculating subgrade elevation changes beneath the existing landfill caused by the differential settlements due to both existing and extra waste filling.
(v) Evaluating the deformation and stability conditions of the leachate and gas collection pipes in the existing landfill due to the extra waste fill.
(vi) Evaluating the stability of the soil mass, liner system, waste mass, and final cover system in various conditions (e.g., excavation, construction, operation, and closure conditions).

14.2 LINER SYSTEMS FOR VERTICAL EXPANSION

The existing solid waste mass, which is relatively compressible, must provide the foundation of the liner system for the vertical expansion landfill. Tensile strains and stresses can develop within the various bottom liner components as a result of differential settlements due to the compression of the underlying solid waste landfill. These tensile strains and stresses can adversely affect the integrity of the liner components. If a compacted clay liner is proposed, it must be recognized that it possesses very little tensile strength (allowable tensile strain is less than 1.0 percent) and is susceptible to cracking as a result of differential settlement. Thus, it is likely that the effectiveness of a compacted clay liner as a hydraulic barrier would be seriously compromised in a vertical expansion landfill. As such, compacted clay liners are generally not recommended for vertical or lateral expansions. A geosynthetic clay liner (GCL) can be used as an alternative to a compacted clay liner. Geosynthetic clay liners are considerably more effective as impervious barriers. They can withstand relatively high in-plane tensile strains and stresses induced by differential settlement (recall Section 5.3). The allowable tensile strain of geosynthetic clay liners range from 6 to 20 percent, contrasted to less than 1 percent for a compacted clay liner.

With respect to the geomembrane components of a composite liner system placed over an existing landfill, several different geomembranes can be selected. These include linear low density polyethylene (LLDPE), flexible polypropylene (fPP), and polyvinyl chloride (PVC) geomembranes. It should be noted that

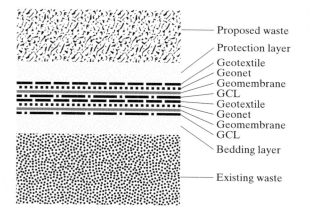

FIGURE 14.3 Double Composite Liner System over Existing Waste

high density polyethylene (HDPE) can also be considered if the tensile strain is mobilized slowly. The reason HDPE is often not used in these situations is that the test method used to simulate differential subsidence (ASTM D5617) applies load very fast in comparison to actual conditions in a landfill. The default pressure rate is 1.0 lb/in^2/min; thus, stress relaxation does not occur and the HDPE fails at relatively low strains of approximately 25%. The other geomembranes cited fail at strains from 75 to 100%. A textured geomembrane should generally be selected to provide a relatively greater interface strength between geomembrane and geosynthetic clay liner or geosynthetic composite drainage layer. Because of the magnitude of the settlements that the liner system will experience and the possibility of "local" liner deformations due to localized subsidence effects, it is important to select a geomembrane with superior extension properties. For a number of reasons (differential settlement, substandard liner under existing waste, etc.) a double liner system is desirable under a vertical expansion.

Cross sections of typical double-composite liner systems used in vertical expansions of landfills are shown in Figures 14.3 and 14.4. A geogrid or high strength geotextile is placed beneath the bottom of the liner system to reinforce the liner system in

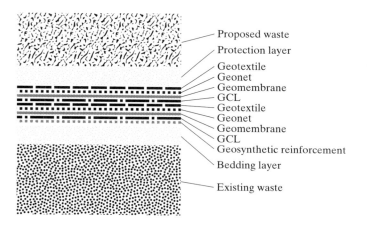

FIGURE 14.4 Double Composite Liner System Reinforced with Geosynthetic Reinforcement over Existing Waste

FIGURE 14.5 A Liner System Placed over an Existing Landfill for a Vertical Expansion Project

Figure 14.4. The geosynthetic reinforcement can prevent excessive tensile strain in the liner system over the existing landfill. Figure 14.5 shows a double composite liner system placed over an existing landfill for a vertical expansion project.

14.3 SETTLEMENT OF EXISTING LANDFILL

The area-wide (also the "total") settlement of waste when subjected to an increase in overburden pressure due to the vertical expansion is characterized by two components, a rapid primary settlement and a long-term, time-dependent secondary settlement. The primary and secondary settlements of the existing landfill due to the vertical expansion can be calculated by using the following equations from Chapter 12:

Primary Settlement of Existing Landfill

$$\Delta Z_c = C'_c \cdot H_o \cdot \log \frac{\sigma_o + \Delta \sigma}{\sigma_o} \tag{14.1}$$

where ΔZ_c = primary settlement of existing landfill;
H_o = initial thickness of the waste layer of the existing landfill;
C'_c = modified primary compression index. $C'_c = 0.17 \sim 0.36$;
σ_o = existing overburden pressure acting at the mid level of the waste layer;
$\Delta \sigma$ = increment of overburden pressure due to vertical expansion.

Secondary Settlement of Existing Landfill

$$\Delta Z_\alpha = C'_\alpha \cdot H_o \cdot \log \frac{t_2}{t_1} \qquad (14.2)$$

where ΔZ_c = secondary settlement of existing landfill;
H_o = initial thickness of the waste layer before starting secondary settlement;
C'_α = modified secondary compression index. $C'_\alpha = 0.03 \sim 0.1$;
t_1 = starting time of the secondary settlement. It is assumed to be equal to the age of existing landfill for vertical expansion project;
t_2 = ending time of the secondary settlement.

Total Settlement of Existing Landfill

$$\Delta Z = \Delta Z_c + \Delta Z_\alpha \qquad (14.3)$$

where ΔZ = total settlement of existing landfill;
ΔZ_c = primary settlement of existing landfill;
ΔZ_α = long-term secondary settlement of existing landfill.

The calculations should be performed at discrete points along several selected settlement lines over the existing landfill. At each point, the thickness of the existing waste in the existing landfill and the thickness of the proposed waste to be placed in the vertical expansion (i.e., the overburden pressure) can be estimated. As an example, see the cross-section, shown in Figure 14.6. The value of the total settlement at each point depends on both the thickness of the existing waste and the load due to the proposed waste fill.

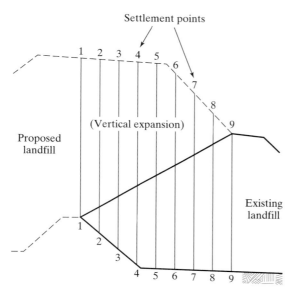

FIGURE 14.6 Cross Section of Existing and Proposed Landfills along Settlement Lines

The differential settlements resulting in tensile strains of liner system materials and leachate collection pipes over the existing landfill, and final grades between adjacent settlement points after settlement can be evaluated from the calculated values of the total settlements at various settlement points along each settlement line on the landfill subgrade.

The differential settlement between adjacent points can be calculated using the equation

$$\Delta Z_{i,i+1} = Z_{i+1} - Z_i \tag{14.4}$$

where $\Delta Z_{i,i+1}$ = differential settlement between points i and $i+1$,
Z_i = total settlement of point i,
Z_{i+1} = total settlement of point $i+1$.

The final slope angle between adjacent points after settlement can be calculated using the equation

$$\tan \beta_{Fnl} = \frac{X_{i,i+1} \cdot \tan \beta_{Int} - \Delta Z_{i,i+1}}{X_{i,i+1}} \tag{14.5}$$

where $X_{i,i+1}$ = horizontal distance between points i and $i+1$,
$\Delta Z_{i,i+1}$ = differential settlement between points i and $i+1$,
β_{Int} = initial slope angle between points i and $i+1$,
β_{Fnl} = final slope angle between points i and $i+1$ after settlement.

The landfill subgrade changes along each settlement line due to different settlements can be calculated from the preceding equation.

The tensile strains of the liner system and leachate collection system resulting from the settlements can be estimated using the equation

$$\varepsilon_{i,i+1} = \frac{(L_{i,i+1})_{Fnl} - (L_{i,i+1})_{Int}}{(L_{i,i+1})_{Int}} \times 100\% \tag{14.6}$$

where $\varepsilon_{i,i+1}$ = tensile strain in liner system between points i and $i+1$,
$(L_{i,i+1})_{Int}$ = distance between points i and $i+1$ in their initial positions,
$(L_{i,i+1})_{Fnl}$ = distance between points i and $i+1$ in their post-settlement positions.

The distance between points i and $i+1$ in their initial positions can be calculated using the equation

$$(L_{i,i+1})_{Int} = [(X_{i,i+1})^2 + (X_{i,i+1} \cdot \tan \beta_{Int})^2]^{1/2} \tag{14.7}$$

The distance between points i and $i+1$ in their post-settlement positions can be calculated using the equation

$$(L_{i,i+1})_{Fnl} = [(X_{i,i+1})^2 + (X_{i,i+1} \cdot \tan \beta_{Int} - \Delta Z_{i,i+1})^2]^{1/2} \tag{14.8}$$

The maximum acceptable tensile strains (i.e., the elongations at yield) of various liner system and leachate collection system components can be obtained from the product specific laboratory testing.

14.4 ESTIMATION OF DIFFERENTIAL SETTLEMENT DUE TO WASTE HETEROGENEITY

An approach for analyzing the differential settlement at the surface of a liner system caused by the collapse of a void within an existing landfill is outlined in this section. On the basis of this analysis, a new design method is presented and described. Currently available methods to estimate "void-induced" differential settlements are first summarized briefly. One of these methods, which is based on an elastic solution, is further discussed and illustrated by an example.

14.4.1 Current Methods for Estimating Localized Subsidence

To the authors' knowledge there is no specific methodology used to quantify the settlement resulting from the presence of a void within a landfill. However, similar situations are often encountered in mining and other geotechnical applications (e.g., collapse of buried sinkholes). Several methods to analyze "void-induced" settlement have been developed and are documented in mining and geotechnical journals and conference proceedings. The following four methods of analysis are briefly summarized herein based on a review by Jang and Montero (1993):

Mining Subsidence Empirical Methods. A mechanism similar to the collapse of a large object within an existing landfill and the resulting differential settlement at the surface of a liner system exists in mining operations (Brauner, 1973; BNCB, 1975). Mining subsidence occurs in a bowl-shaped pattern. Empirical methods have been developed to analyze the subsidence of "long-wall" mines, an underground mining technique mainly used in rock formations. The method is applied to a specific geographical region and is uniquely based on the geological characteristics of the region. Because there is no generic solution available for all geographical regions, this method is not suitable for the analysis of differential settlement for a landfill in a vertical expansion configuration.

Numerical Methods. Finite element analyses of a void within a soil layer have been conducted in several research fields other than landfill engineering. Wang and Badie (1985) used a finite element analysis and a physical model to analyze the bearing capacity of a shallow footing above a void embedded in clay. The results of Wang's finite element analysis were confirmed by experimental model tests. Drumm et al. (1987) also used finite element analysis to evaluate the deformation of highly plastic soils in contact with cavitose bedrock. The calculated settlements were presented as a function of cavity size.

Displacement Method. "Closed-form" solutions for the strain field in an initially isotropic and homogeneous incompressible soil due to near-surface ground loss were presented by Sagaseta (1987). The differential settlement of a point on a plane is calculated in this method as a function of the displacement of other points. The applications of the closed-form solutions to some typical problems indicate that the calculated movements agree quite well with experimental observations and compare favorably with other commonly used numerical methods.

Elastic Solution. An analytical elastic method to evaluate settlements caused by voids at depth was presented by Tsur-Lavie et al. (1980, 1988). This method can be used to calculate the surface settlement as a function of the dimension of the void, thickness of medium (soil/rock) over the void, and Poisson's ratios. The method presented is based on a solution developed by Golecki (1978, 1979) for stresses and displacements in an infinite homogeneous elastic half space, with discontinuous step-like uniform boundary displacement representing the collapsing of a void. The displacement in the surrounding medium and the resulting differential settlement at the medium surface is then calculated by an elastic method. The results obtained from the analytical elastic method were compared with British National Coal Board (BNCB) mining subsidence field measurement data by Tsur-Lavie et al. and are in close agreement with one another.

Any of the methods discussed previously can be used to calculate differential settlements resulting from the existence of a void at depth. The numerical methods discussed (i.e., the finite element and finite difference methods), are suitable for the analysis of problems with nonhomogeneous, anisotropical materials. On the other hand, the displacement method and the elastic solution method require little or no material properties for the analysis, and therefore they can be applied quite readily to a vertical expansion design. Of these two methods, the elastic solution has several advantages. First, it has been calibrated by field measurements; second, it is amenable to sensitivity analyses based on different soil or waste characteristics. The elastic solution method is much easier to apply than the finite element analysis, and since it neglects arching in the waste, it is conservative.

The section that follows describes how Jang and Montero (1993) employed the elastic solution method developed by Tsur-Lavie et al. to analyze the effect of "void-induced" differential settlement on a liner system constructed underneath a proposed vertical expansion to an existing landfill.

14.4.2 Elastic Solution Method Applied to a Vertical Expansion

14.4.2.1 Differential Settlement Mechanism.

Differential settlement in a vertical expansion liner system takes place when the surrounding waste in an existing landfill moves into the voids created by the degradation of a large object such as a large household appliance or unfilled box/container. A model was developed to analyze this mechanism and is shown in Figure 14.7 (Jang and Montero, 1993). In this model, soil and waste are represented by a half-space medium. Surface differential settlements in the model are influenced by the following factors:

(i) Engineering properties of the medium;
(ii) The thickness of the medium over the void, T;
(iii) The containment liner system grade, α;
(iv) The void size, length L_x × width L_y × depth D.

To quantify the influence of these factors on surface deformation, an influence function (Jang and Montero, 1993) is first developed. The differential settlement can then be calculated and evaluated with the influence function.

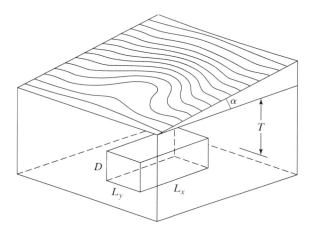

FIGURE 14.7 Differential Settlement Model—Three Dimensional Elastic Solution Model (Jang and Montero, 1993)

14.4.2.2 Approach. A computer program was developed to solve the three-dimensional equations presented in the analytical elasticity solution (Tsur-Lavie et al., 1988). This computer program evaluates the differential settlement caused by voids at depth. The input variables in the program consist of the following:

(i) The void size, defined by $L_x \times L_y \times D$;
(ii) The soil thickness T over the void; and
(iii) Poisson's ratio of the subsiding material.

The computer program was verified by comparing its output with the characteristic curves presented in the paper by Tsur-Lavie et al. (1988).

14.4.2.3 Assumptions. Three basic assumptions were made to model the deformation of a surface due to the presence of a void at depth. These assumptions are as follows:

(i) The liner surface is stress free;
(ii) Differential settlement on a horizontal plane is projected to a graded surface;
(iii) A constant Poisson's ratio.

For further discussion of each of these assumptions and their significance, see Jang and Montero (1993).

14.4.2.4 Example Calculation for the Design of a Sloped Liner System. As noted previously, evaluation of the surface deformation of a sloped liner surface requires an approximation that is made by projecting the differential settlements calculated on a horizontal surface to a sloped surface. This procedure results in a conservative approximation because the distance between the points along the sloping surface and the void are always greater than or equal to those in the horizontal case. Therefore, the differential settlements calculated in the sloped surface will be slightly higher than those in the horizontal case.

Design Problem. A landfill-buried void 3 ft (0.9 m) long by 3 ft (0.9 m) wide by 6 ft (1.8 m) deep is used in the analysis. These dimensions represent a typical household refrigerator—a large object likely to be disposed of at a sanitary landfill.

The analysis of a liner system sloping at a 7% grade is presented as an example (modified from Jang and Montero, 1993). The 7% sloping surface and soil/waste thicknesses of 6, 7.5, 9, and 12 ft (1.8, 2.25, 2.7, and 3.6 m) were analyzed over an area 21 ft (6.4 m) long by 21 ft (6.4 m) wide. The void is assumed to be located in the center of this area.

Presentation and Discussion of Results. In a contour plot, a contour line represents the same elevation along a line. A closed contour therefore indicates either a depression or a mound. The presence of closed contours representing a deep depression is an unacceptable condition in the design of a liner system.

The deformed surface elevations in each grid point are calculated by the computer program. The deformed surface contours on the 7% sloped surface are then plotted for each case, as shown in Figure 14.8 (Jang and Montero, 1993). Where the thickness of the soil/waste layer is 6 ft (1.8 m), as shown in Figure 14.8(a), the presence of the void will create a depression 0.3 ft (90 mm) deep. This depression could trap and accumulate liquids, hindering the free flow of leachate towards a collection point. When the thickness of the soil/waste layer above the void is increased to 7.5 ft (2.25 m), the depression is reduced to less than 0.1 ft (30 mm), as shown in Figure 14.8(b). With a 9-ft- (2.7-m)-thick soil/waste layer over the void, the depression disappears. The disturbance in the liner surface decreases as the depth to the void increases, as seen when the thickness of the soil/waste layer is increased to 12 ft (3.6 m) [Figure 14.8(c) and Figure 14.8(d)].

To evaluate the potential for grade reversal, a characteristic curve of required soil waste thicknesses, T, over a 3-ft- (0.9-m)-deep by 6-ft- (1.8-m)-long void versus liner surface grades, α, is illustrated in Figure 14.9.

For each analysis, the tensile strains were calculated from the deformed spacing between adjacent grid points in the surface against the initial grid spacing. A characteristic curve of the maximum tensile strains versus ratio of void width L_y to controlled fill thickness T was developed from the analyses' results. This curve is presented in Figure 14.10.

The maximum tensile strain generated by the differential settlements was examined to evaluate the integrity of a composite liner, which included a geomembrane and a clay layer (see Figure 14.10). The geomembrane can sustain tensile strains higher than those that the clay component of the composite liner can sustain before tensile failure or tension cracks develop within the clay component. Therefore, the tensile strain limit of the clay component can be used as an acceptable design criterion to evaluate the integrity of composite liners.

From the maximum tensile strains versus soil/waste thickness (T) curve, a soil layer thickness T equal to 9 ft (2.7 m), the maximum tensile strain caused by the 3-ft-(0.9-m)-wide by 3-ft- (0.9-m)-deep by 6-ft- (1.8-m)-long void is 0.2%. This strain is within acceptable limits for a clay layer as shown in Figure 14.11 (after Gilbert and Murphy, 1987).

Section 14.4 Estimation of Differential Settlement Due to Waste Heterogeneity

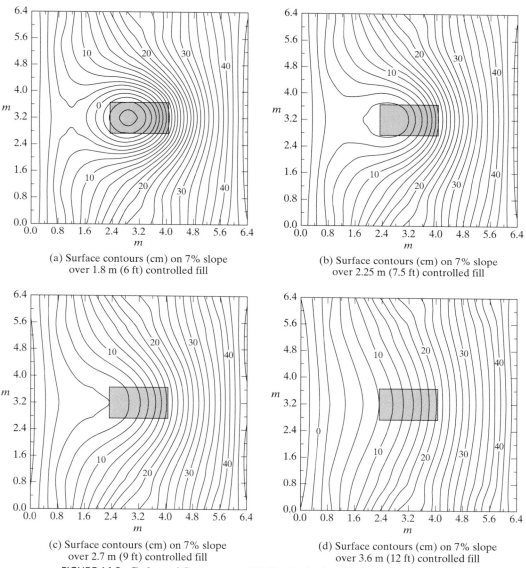

(a) Surface contours (cm) on 7% slope over 1.8 m (6 ft) controlled fill

(b) Surface contours (cm) on 7% slope over 2.25 m (7.5 ft) controlled fill

(c) Surface contours (cm) on 7% slope over 2.7 m (9 ft) controlled fill

(d) Surface contours (cm) on 7% slope over 3.6 m (12 ft) controlled fill

FIGURE 14.8 Deformed Contours on a 7% Sloping Surface (Jang and Montero, 1993)

On the basis of the above example analysis, a 2.7-m- (9-ft)-thick soil/waste layer can serve as a strain transition zone to prevent grade reversal, excessive tensile strains and stresses from developing in a liner system. Therefore, a 9-ft- (2.7-m)-thick layer of soil or "selected" waste should be placed, in this example, before constructing the vertical landfill containment liner.

556 Chapter 14 Vertical Landfill Expansion

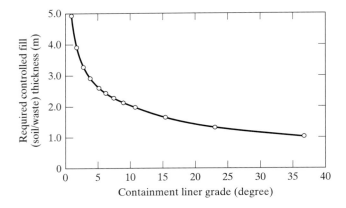

FIGURE 14.9 Required Controlled Fill Thickness to prevent Grade Reversal, T, versus Containment Liner Grade, α, on 0.9 m × 0.9 m × 1.8 m (3 ft × 3 ft × 6 ft) Void (Jang and Montero, 1993)

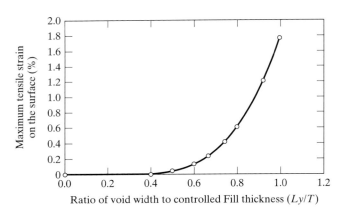

FIGURE 14.10 Maximum Tensile Strain versus Ratio of Void Width to Controlled Fill Thickness on 0.9 m × 0.9 m × 1.8 m (3 ft × 3 ft × 6 ft) Void (Jang and Montero, 1993)

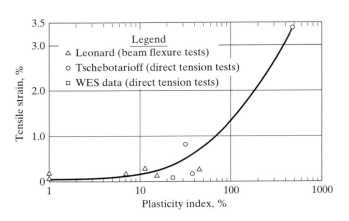

FIGURE 14.11 Tensile Strain versus Plasticity Index (Gilbert and Murphy, 1987)

For slope angles other than the 7% used in the example analysis, the tensile strains and potential for grade reversal on a liner surface can be evaluated (Jang and Montero, 1993) according to the thickness versus maximum tensile strain and thickness versus liner grade characteristic curves, settlement contours and containment liner design criteria. From this evaluation, the required backfill thicknesses in different liner grades and liner systems can be determined and designed to prevent grade reversal and excessive tensile strains on a vertical expansion liner and leachate collection system.

14.5 VERTICAL EXPANSION OVER UNLINED LANDFILLS

The moisture content of solid waste placed in landfills located in semiarid or arid areas are generally below field capacity. In this case leachate will not be released from the waste to impact underlying groundwater even though some of these old landfills lack liners. On the other hand, if the waste is compressed sufficiently, its available moisture-holding capacity will decrease and its moisture content may eventually reach field capacity. In this case, additional compression beyond this point will squeeze leachate from the waste (Zornberg et al., 2000). If a vertical expansion is planned over an unlined landfill located in a semiarid or arid area, a key consideration is to determine the minimum allowable compressed waste thickness beneath the vertical expansion portion of the landfill. This minimum thickness corresponds to a waste saturation or water content at which leachate stored within old waste in an existing unlined landfill can be released upon further compression.

Zornberg et al. (2000) developed a method to estimate the minimum allowable waste thickness without releasing leachate stored within the waste for a 60-m unlined landfill located in southern California. The field capacity, in-situ moisture content, and unit weight profiles of the waste in this selected landfill were all determined by lab and field testing techniques. These experimental data were used to evaluate the ability of the landfill to continue retaining moisture after additional waste placement. Analysis of their data indicated that if the final waste filling depth is kept below the calculated minimum allowable waste thickness, which ranged from 97 m to 109 m, the moisture content of the waste will not reach its field capacity. Therefore, the leachate should still remain within the waste mass and not impact the groundwater after vertical expansion. Details of the analysis procedures can be found in Zornberg et al. (2000).

14.6 DESIGN CONSIDERATIONS FOR LANDFILL STRUCTURES

If the existing landfill has leachate and gas collection systems, it is necessary to evaluate the effects of the vertical expansion on the operational condition of the leachate and gas collection systems of the existing landfill. Large settlements and differential settlements in the existing landfill may cause large deformation or failure of the gas exaction wells and gas collection header pipes of the gas collection system. The pipe deflection and pipe wall bucking of both leachate and gas collection pipes in the existing landfill must be recalculated by adding the extra load caused by the vertical expansion.

The total settlement of the existing landfill subgrade including elastic, primary and secondary settlements must be estimated again by considering the effect of the vertical expansion. The vertical effective load used to calculate the elastic and primary settlement should be equal to the sum of the existing and extra waste filling. Equations 12.19, 12.21, 12.22, and 12.23 can be used to calculate the elastic, primary, secondary, and total settlements, respectively. The final differential settlement, subgrade changes, and tensile strains of liner system and leachate collection system resulting from the settlements should be carefully estimated using Equations 14.4, 14.5, and 14.6, respectively. Grade reversal and ponding of the bottom liner of the existing landfill is not allowable after vertical expansion.

The landfill structures that may be affected by vertical expansion include the following:

(i) Existing and new liner systems,
(ii) Existing and new leachate collection and detection systems,
(iii) Existing gas collection system,
(iv) Existing waste mass,
(v) Foundation of existing waste mass,
(vi) Existing and new final cover systems, and
(vii) Underdrain system.

The structural considerations that affect the design of a vertical expansion landfill are summarized in Table 14.1:

14.7 GEOSYNTHETIC REINFORCEMENT DESIGN FOR VERTICAL EXPANSIONS

Presently, two methods are being applied to liner systems to minimize the deformation of a liner and leachate collection system constructed between an existing landfill and a proposed vertical expansion. These methods consist of either reinforcing the liner system with a geogrid or high strength geotextile or evaluating the potential differential settlement caused by a void within an existing landfill and allowing the liner system to deform. In the latter method, a required backfill thickness adjustment is calculated to prevent grade reversal and excessive tensile strains on a vertical expansion liner and leachate collection system. The design procedure for reinforcing the liner system with a geosynthetic will be described in this section. The next section describes how to use an elastic solution model to evaluate the potential differential settlement caused by a void within an existing landfill and how to determine the required backfill thickness adjustment to prevent grade reversal.

Liners and collection systems overlying existing landfill areas may require some type of reinforcement. The purpose of the reinforcement is to minimize the tensile strains in the overlying liner and leachate collection systems assuming that a void occurs. Voids are created by progressive degradation and collapse of large objects buried in the old landfill areas.

TABLE 14.1 Structural Considerations for Vertical Expansions

Structure	Design Considerations
Liner [geomembrane, compacted clay liner (NR), and geosynthetic clay liner]	• Tensile strain of new liners over the existing waste, • Stability of new liner system over the existing waste, • Slope changes of the existing liner system.
Pipe (leachate, riser, gas, and underdrain pipes)	• Strength and stability (bucking, crushing, and deflection), • Slope changes.
Geosynthetic Drainage Layer (geocomposite and geonet used in the existing leachate collection and detection system and underdrain system)	• Drainage capacity of geonet and geocomposite will be reduced due to extra waste fill.
Vertical Structures in the Existing Landfill (manholes, riser pipes and gas extraction pipes)	• Negative skin friction force due to waste settlement, • Bearing capacity and stability of the vertical manhole and riser pipe foundations due to negative skin friction force and extra waste fill.
Final Cover [geomembrane, compacted clay liner (NR), and geosynthetic clay liner]	• Tensile strain for the elements of the existing landfill cover caused by the extra settlement of the existing waste due to the extra waste fill, • Stability of new final cover.
Landfill Subgrade	• Subgrade changes of the existing landfill caused by foundation soil settlement due to extra waste fill, • Subgrade changes of the new landfill cause by the settlement of the existing waste.
Landfill and Foundation Stability	• Stability of the existing waste during the new waste filling, • Stability of the soil foundation due to extra loading, • Stability of combination of the existing and new landfills in various conditions.

NR = not recommended for liner systems of vertical expansions

Geogrid reinforcement is often used in vertical landfill expansion on the top of existing landfills (Figure 14.4), although high-strength geotextiles can function in this application as well. The design of the geosynthetic reinforcement is based on a worst-case scenario assumption that a void is located immediately underneath the liner. The liner is then treated as a plate bridging over the void and carrying the load from the proposed overlying waste. Geosynthetics are placed to support and protect the integrity of the liner system. The design methodology used in geosynthetic reinforcement is based on the tensioned membrane theory. When differential settlement takes place, the geosynthetic deflects into the depression as tensile stresses develop in the reinforced material.

The use of uniaxially oriented, high polyethylene (HDPE) geogrids to support a landfill lining for a vertical expansion and cover system over either a circular or a long narrow depression is described in a stepwise fashion as follows (modified from TENSAR, 1989). A similar approach is used with high-strength geotextiles as well as with other types of geogrids.

14.7.1 Theoretical Background for Geosynthetic Reinforcement

When a depression forms below a layer of geosynthetic reinforcement supporting a landfill lining or cover system, the reinforcement deflects into the depression. This deflection has two effects: bending of the fill materials overlying the reinforcement; and tensioning of the reinforcement (Figure 14.12). The bending of the fill materials generates arching inside the material, which transfers part of the applied load away from the depression. As a result, the vertical stress, σ_{v1}, acting on the reinforcement over the depression is smaller than the vertical stress, σ_{v2}, due to the applied load, which is equal to the weight of the overlying fill materials plus any applied surcharge, q, as shown in Figure 14.12(b). The tensioning of the reinforcement mobilizes a portion of the materials' tensile strength [Figure 14.12(a)]. As a result, the reinforcement acts as a "tensioned membrane" and normally can carry a load applied to the surface. The reinforcement will deflect until the resistance generated by arching of the fill materials and tensioning of the reinforcement balances the applied load.

The method described herein for the design of a geosynthetic to support a landfill lining system was developed by combining arching theory for the fill materials overlying the reinforcement with tensioned membrane theory for the reinforcement. This method has been successively developed by Giroud (1982), Bonaparte and Berg (1987a), and Giroud et al. (1988).

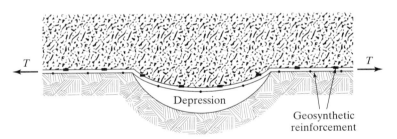

(a) Arching of fill material and tensioning of geosynthetic reinforcement

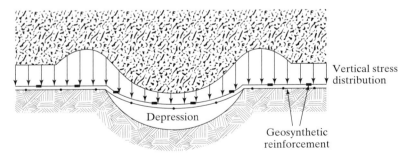

(b) Effect of fill material arching on vertical stress distribution

FIGURE 14.12   Load Carrying Mechanism (after TENSAR, 1989)

Arching of the materials overlying a layer of geosynthetic is dependent on the effective stress friction angles of the materials, the thickness of the materials, and the size of the depression. Materials with effective stress friction angles less than 25° may not develop a significant arch; therefore, it is assumed that the minimum effective stress friction angle of the materials overlying the geosynthetic is at least 25°. Arching will not be fully developed unless the total thickness of the materials overlying the geosynthetic is approximately six times greater than the radius of a circular depression or the half-width of a long, narrow depression. This approximation is for soils, whereas it is felt that the value is lower for municipal solid waste. The exact value awaits further investigation. Because of this, arching is usually developed in the materials above the reinforcement supporting a landfill lining system, but it is usually not developed above the reinforcement supporting a cover system if the size of the depression is greater than one-sixth of the thickness of the cover system.

14.7.2 Assumptions for Geosynthetic Reinforcement Design

The step-by-step design method described herein is directly applicable to geogrid or high strength geotextile support of lining systems under the following assumptions:

(i) The cross-section of the depression is circular. (Long and narrow depressions are addressed in Section 14.7.9.)

(ii) The depression is spanned by two layers of uniaxially oriented geogrid placed in perpendicular directions or by a geogrid or geotextile with equal tensile properties in its machine and cross directions wherein seam strength could well be the limiting parameter.

(iii) The depth of the depression is equal to or greater than the deflection of the reinforcement (i.e., none of the pressure on the reinforcement over the depression is transferred to the bottom of the depression).

(iv) Surcharge loads act uniformly on top of the waste placed above the lining system.

(v) Uniform material properties exist within each distinct zone (soil components of the lining or cover system, and waste placed above the lining system).

(vi) The soil components of the lining or cover system and the waste placed above the lining system have a minimum effective stress friction angle of 25° and effective stress cohesion of zero.

(vii) The bedding soil, if placed above the reinforcement is a granular soil with a minimum effective stress friction angle of 25° and effective stress cohesion of zero.

(viii) There are no hydrostatic (static or excess pore water) pressures within the lining system.

(ix) The long-term (e.g., 120 years) stress-strain characteristics of the reinforcement are accurately defined.

(x) The geogrids interlock with the bedding soil or the friction of the geotextile is adequate to mobilize the stresses that are generated.

(xi) Arching of the fill materials directly above the reinforcement is not affected by the synthetic components of the lining or cover system (i.e., geomembrane, geonet, and geotextile).

14.7.3 Selection of Material Properties

This design method (modified from TENSAR, 1989) assumes that the minimum effective stress friction angle of waste (for design of lining system support), soil components of lining and cover systems, and bedding soil is 25°. As shown in Table 14.2, this assumption is met by most soils. The equations given subsequently in this section were derived by Giroud et al. (1988) and incorporate a minimum effective stress friction angle of 20°. Accordingly, to be conservative, a minimum value of 25° has been recommended.

Bedding Soil. The bedding soil placed above or below the reinforcement is typically a well-compacted sandy soil. The moist unit weight of the bedding soil, γ_b (lb/ft^3 or kN/m^3), and the effective stress friction angle, ϕ'_b (degrees), should be measured or estimated using correlations with published values. For preliminary design purposes, assume that the moist unit weight of the bedding soil is 120 lb/ft^3 (19 kN/m^3) and the effective stress friction angle is at least 25° (see Table 14.2). However, these assumed values should be verified prior to final design.

Municipal Solid Waste. The moist unit weight, γ_{sw} (lb/ft^3 or kN/m^3), and the effective stress friction angle, ϕ'_{sw} (degrees), of the material contained above the lining system should be measured or estimated. The discussion below assumes that this material is municipal solid waste (MSW). For preliminary design purpose, assume that the average moist unit weight of the municipal solid waste (including daily cover) is 60 lb/ft^3 (9.4 kN/m^3) and the effective stress friction angle of the municipal solid waste is at least 25°. However, it is important that a site-specific evaluation of these properties be made. The project engineer should verify the values of ϕ'_{sw} and γ_{sw} used in the final design.

Lining Components. The lining components can consist of geomembranes, geosynthetic clay layers, geonet drainage layers, granular drainage layers, geotextiles, and protective soil cover layers. The material properties of the lining system that must be determined are the moist unit weights of soil components (e.g., granular drainage layers), γ_l (lb/ft^3 or kN/m^3), the effective stress friction angles of the soil components of the lining system, ϕ'_l (degrees), and the minimum yield strain, ε_y, of the lining system components.

TABLE 14.2 Representative Range of Effective Stress Friction Angle Values (Bowles, 1982)

Soil	Effective Stress Friction Angle
Gravel	
Medium Size	40° to 50°
Sandy	35° to 50°
Sand	
Loose Dry	28° to 35°
Dense Dry	35° to 46°
Silt or Silt sand	
Loose	27° to 30°
Dense	30° to 35°
Clay	20° to 30°

Section 14.7 Geosynthetic Reinforcement Design for Vertical Expansions

The moist unit weight of the soil components must be determined so that overburden stress on the reinforcement due to the lining system can be calculated. For preliminary design purposes, the moist unit weight of the soil component can be taken as 120 lb/ft^3 (19 kN/m^3), and the effective stress friction angle of the soil components of the lining system can be assumed to be at least 25°. (See Table 14.2.) However, the designer should verify these values prior to final design.

The minimum yield strain, ε_y, of the lining system components must be determined so that the allowable strain of the lining system, ε_l, can be calculated. The minimum yield strain of a lining system will be the yield strain of the compacted clay component, if one is present. Otherwise, the yield strain of the geomembrane or GCL is likely to be critical. In any event, the yield strains of all components should be considered. For HDPE geomembranes, an acceptable yield strain for preliminary design is 10%. This criterion is also applicable to polyester geogrids and geotextiles, and to polypropylene geotextiles. A factor of safety is applied to this value to calculate the allowable strain. (See discussion in Section 14.7.5.) The actual yield strain of the geosynthetic should be verified prior to the final design.

14.7.4 Determination of Geometric and Loading Parameters

Geometric Parameters. Simplified geometrical models for a geosynthetic supporting a lining system spanning a depression are shown in Figure 14.13. All the geometrical parameters shown must be defined to determine the required long-term reinforcement

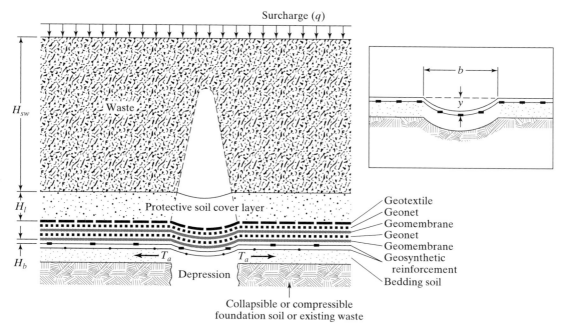

FIGURE 14.13 Simplified Model for Analysis of Stresses in Geosynthetic Reinforcement Supporting a Landfill Lining System Spanning a Depression (TENSAR, 1989)
(Note: Only deflected portion of fill materials exerts pressure on the reinforcement)

tensile load. Selection of the radius of the design depression, r (ft or m), and the allowable deflection of the geogrids, y (ft or m), are discussed in Section 14.7.5.

Loading Parameters. A uniformly distributed surcharge, q (lb/ft^2 or kN/m^2), may be incorporated into the design. This surcharge load is assumed to act on top of the waste contained above the lining system or on top of the cover system.

14.7.5 Design Criteria

Criteria to be used in designing the lining system must be specified. These include such considerations as the size of the depression, allowable deflection into the depression, and factor of safety. These are described next.

Size of Depression. The lining system underlying the reinforcement is assumed to rest initially on a firm subgrade. At some point in time, a circular depression of radius, r (ft or m), is assumed to develop directly below the support. (Long, narrow depressions are addressed in Section 14.7.9.) Subsidence of the underlying subgrade is assumed to be caused by localized collapse of the subgrade soil into depression below the support. Conditions causing subsidence include collapse of a karstic subgrade beneath a lining system or collapse of waste with large voids, such as barrels, household appliances, or furniture, beneath a lining or cover system. The "rusted refrigerator" assumption is commonly used to analyze lining and cover systems placed over municipal solid waste. For this case, the design depression is usually assumed to have a radius of 3 ft (0.9 m), which is the approximate radius of the depression created by a refrigerator that rusts and collapses in the underlying subgrade.

Allowable Deflection. The allowable deflection into a depression, y (ft or m), of the reinforcement is governed by the following criteria:

(i) Allowable strain of the lining system components (discussed in Section 14.7.3) and allowable strain of the reinforcement (discussed in Section 14.7.6).

(ii) Allowable depth of ponding of liquid in the lining system.

Strain Criteria: The reinforcement must be designed to support the lining system at a strain less than or equal to the allowable strain of the system, ε_l. (The allowable strain of a lining or cover system is equal to the minimum yield strain of the system, ε_y, divided by a factor of safety against rupture, F_R, as described in this section and given by Equation 14.11.) The reinforcement must also be designed to function at a strain less than its allowable strain, ε_g (Section 14.7.6).

Drainage Criteria: The deflection into a depression of the drainage layer of a lining system could impede drainage. The change in slope of a lining system drainage layer is dependent not only on the vertical deflection and radius of the depression, but also on the initial slope of the drainage layer. Furthermore, the effect of a depression on the hydraulic performance of a lining system is dependent on the allowable design head for that system. The designer should consider the drainage criterion once specific information regarding drainage layer slopes and allowable hydraulic heads are available.

Factor of Safety. A factor of safety F_R against rupture is incorporated into the design and applied both to the minimum long-term yield strain ε_y of the lining or cover system and to the rupture strain ε_r of the reinforcement. A factor of safety against yield or rupture of 1.5 is recommended.

14.7.6 Selection of Allowable Reinforcement Strain

The material properties of the reinforcement that should be determined are the allowable strain of the material, ε_g, and the long-term allowable design tensile load of the material, T_a. The selection of the allowable strain is discussed in this subsection; the long-term allowable design tensile load is discussed next in Section 14.7.7.

The allowable strain of geosynthetic reinforcement is equal to the rupture strain of the material divided by a factor of safety against rupture, F_R. Wrigley (1987) reported rupture strain values, ε_r, for HDPE geogrids of at least 15%. Since a factor of safety against rupture of 1.5 is recommended here (Section 14.7.5), the calculated allowable strain, ε_g, of HDPE geogrids is at least 10%, because of the following relationship:

$$\varepsilon_g = \varepsilon_r / F_R \tag{14.9}$$

For this method, the allowable strain of HDPE geogrids is conservatively set at $\varepsilon_g = 10\%$. This agrees very well with polyester geogrids and polyester geotextiles. Thus, a 10% strain criterion can be used for all types of currently available geosynthetic reinforcement materials.

14.7.7 Selection of Long Term Allowable Design Tensile Load

Methodology. The short-term, wide-width tensile strength is readily available for all types of geosynthetic reinforcement. It is performed using the ASTM D4595 testing procedure. The maximum stress is referred to as the ultimate strength, or "T_{ult}". In order to obtain the long-term allowable tensile strength, "T_{allow}", the ultimate value must be modified using reduction factors. Typically, these reduction factors are established for creep, installation damage, and chemical/biological degradation. The formulation is given by Equation 14.10 as.

$$T_{allow} = \frac{T_{ult}}{RF_{CR} \times RF_{ID} \times RF_{CBD}} \tag{14.10}$$

where T_{allow} = long-term allowable design tensile load (lb/ft or kN/m);
T_{ult} = short-term ultimate load from tensile tests (lb/ft or kN/m);
RF_{CR} = reduction factor which accounts for creep;
RF_{ID} = reduction factor which accounts for installation damage.
RF_{CBD} = reduction factor which accounts for chemical/biological degradation;

Other reduction factors (e.g., to account for seams or holes) can be included on a site-specific and product-specific basis. Some discussion on the various reduction factors follows. For additional commentary, see Koerner (1998).

Long Term Deformation (Creep). Since the wide width test is conducted quickly in comparison to field situations, a reduction factor for long-term creep is necessary. The customary procedure is covered in ASTM D5262. The tests usually take 10,000 hours for each load utilized. The resulting values are dependent upon the type of resin and the production process. Approximate values follow:

- oriented, homogeneous HDPE geogrids ≈ 2.5
- oriented, homogeneous PP geogrids ≈ 2.0
- flexible, woven PET geogrids ≈ 2.0
- woven PET geogrids ≈ 2.0
- woven PP geotextiles ≈ 2.5

Since the tests are very time consuming, accelerated methods are being currently pursued. Time-temperature superposition using the stepped isothermal method (SIM) is one such method that is showing promising results, (See Thorton et al., 1999.)

Installation Damage. It should be recognized that harsh installation stresses can cause damage to geosynthetics. In some cases, such stresses can exceed the design stresses for the particular application. If these stresses cause damage, the load carrying capability of the material is compromised. Many studies have been conducted (e.g., Bonaparte and Berg, 1987b; Koerner and Koerner, 1990). Typical values are given in Table 14.3.

Chemical/Biological Degradation. The major issue with long-term chemical and biological degradation for vertical expressions of landfills is the agressiveness of the leachate and gases that are generated. This is clearly a site-specific situation; however, it is probably not serious since polyolefins (PE and PP) are very stable, as is the case with high tenacity PET. A nominal value of 1.1 or 1.2 can be used.

Other Reduction Factors. If the site-specific conditions warrants, reduction factor for seams and/or holes purposely put in the reinforcement for piping can be included in Equation 14.10.

14.7.8 Design Steps

The following is a step-by-step procedure for the design of geogrid reinforcement to support lining and cover systems adapted from Giroud et al. (1988).

(i) Verify design assumptions (see Section 14.7.2).
(ii) Define material properties (see Section 14.7.3).
(iii) Define geometrical and loading parameters (see Section 14.7.4).
(iv) Define design criteria (see Section 14.7.5).

TABLE 14.3 Reasonable Installation Damage Reduction Factors for Geosynthetics

Type of Soil Subgrade and Cover	Placement and Backfilling Equipment		
	Light	Moderate	Heavy
SMOOTH—no stones	1.1	1.2	1.3
Intermediate	1.2	1.3	1.4
ROUGH—with stones	1.3	1.4	1.5

Section 14.7 Geosynthetic Reinforcement Design for Vertical Expansions

(v) Determine the allowable strain of the lining or cover system.

The allowable strain, is given by

$$\varepsilon_l = \varepsilon_y / F_R \tag{14.11}$$

where ε_y = minimum yield strain of the lining or cover system components;
F_R = factor of safety against rupture, generally ≥ 1.5.

The above calculation assumes the allowable deflection of the lining or cover system is governed by the strain presented in Section 14.7.5 and not by drainage criteria.

(vi) Determine the allowable design strain of the geosynthetic reinforcement. The allowable design strain ε_d is the lesser of the allowable strain of the lining system, ε_l, and the allowable strain of the reinforcement, ε_g, (equal to 10%, from Section 14.7.6). This allowable design strain will be used in steps (ix) and (xi).

(vii) Calculate the weighted average wet unit weight of the materials located from the top of the reinforcement layer to a distance of six times the depression radius ($6r$) above the top of the layer. The weighted average wet unit weight of the materials located up to a distance of $6r$ from the reinforcement layer beneath a lining system can be obtained using the following equations.

For $H_1 + H_b < 6r \leq H_1 + H_b + H_{sw}$,

$$\gamma_{avg} = \frac{\gamma_1 \cdot H_1 + \gamma_b \cdot H_b + \gamma_{sw} \cdot (6r - H_1 - H_b)}{6r} \tag{14.12}$$

For $H_1 + H_b \geq 6r$,

$$\gamma_{avg} = \frac{\gamma_1 \cdot (6r - H_b) + \gamma_b \cdot H_b}{6r} \tag{14.13}$$

For $H_1 + H_b + H_{sw} < 6r$ and

$$\gamma_{avg} = \frac{\gamma_1 \cdot H_1 + \gamma_b \cdot H_b + \gamma_{sw} \cdot H_{sw}}{H_1 + H_b + H_{sw}} \tag{14.14}$$

where γ_{avg} = weighted average wet unit weight of the materials located from the top of the layer to six times the depression radius above the reinforcement layer (lb/ft^3 or kN/m^3);
γ_1 = wet unit weight of the soil components of the lining or cover system (lb/ft^3 or kN/m^3);
γ_{sw} = wet unit weight of the solid waste contained above the lining system (lb/ft^3 or kN/m^3);
γ_b = moist unit weight of the bedding soil (lb/ft^3 or kN/m^3);
H_b = thickness of bedding soil layer above the reinforcement layer (ft or m). If the reinforcement is placed above the bedding soil layer, $H_b = 0$;

H_l = thickness of soil layer components of lining or cover system (ft or m);
H_{sw} = thickness of the solid waste contained above the lining system (ft or m);
r = radius of depression (ft or m).

(viii) Calculate the normal pressure over the depression, p, exerted on the reinforcement due to the overlying material:

$$p = 2 \cdot \gamma_{avg} \cdot r \cdot (1 - e^{-0.5(H_l + H_b + H_{sw})/r}) + q \cdot e^{-0.5(H_l + H_b + H_{sw})/r} \quad (14.15)$$

In this equation,

p = normal pressure exerted on the reinforcement over the depression (lb/ft² or kN/m²);

q = uniformly distributed surcharge acting on the waste contained above the lining system.

(ix) Determine the dimensionless factor Ω from Table 14.4, using the allowable design strain, ε_d, from (vi). The factor Ω is defined as:

$$\Omega = 0.25 \cdot (y/r + r/y) \quad (14.16)$$

where y = allowable deflection of the reinforcement into a depression; and ε_d is related to Ω by

$$1 + \varepsilon_d = 2 \cdot \Omega \cdot \sin^{-1}[1/(2 \cdot \Omega)] \quad \text{if } y/r < 0.5 \quad (14.17)$$

(x) Calculate the required uniaxial tensile load, T_r, in reinforcement (Assumption 2, Section 14.7.2) due to the normal pressure; that is,

$$T_r = p \cdot r \cdot \Omega \quad (14.18)$$

where T_r = required long-term uniaxial design tensile load (lb/ft or kN/m);
Ω = dimensionless factor, from Table 14.4.

(xi) Select the specific reinforcement product to provide the design tensile load.

TABLE 14.4 Dimensionless Factor Ω Used for Tension Membrane Analysis (Giroud et al., 1990)

$y/(2r)$ or y/b	ε (%)	Ω
0.138	5.00	0.97
0.151	6.00	0.90
0.160	6.69	0.86
0.165	7.00	0.84
0.175	8.00	0.80
0.186	9.00	0.76
0.197	10.00	0.73

14.7.9 Special Design Cases

Three design cases are addressed here: (i) design to span a circular depression, (ii) design to bridge a long narrow depression, and (iii) design of a single layer of biaxial geogrid or geotextile.

Single-Layer Reinforcement Spanning a Circular Depression. If a single layer of uniaxial geogrid is used to span a circular depression, the required long-term uniaxial geogrid design tensile load of the reinforcement [calculated in (x) of Section 14.7.8] should be increased by a factor of 2 (i.e., the required tensile load become $2T_r$). This technique is conservative because it assumes the tensile load in the geogrids are approximately that resulting from a geogrid layer spanning an infinitely long void (of width $b = 2r$). However, in the actual case the void is not infinitely long.

Single Layer of Uniaxial Geogrid Reinforcement Spanning a Long Narrow Depression. The steps presented in Section 14.7.8 for the design of geogrids to span a circular depression can be modified as follows for a single layer of geogrids spanning a long narrow depression (Giroud et al., 1988):

(i) to (iv):	No change.
(v) and (vi):	These steps should be carried out with the allowable uniaxial strain of the lining or cover system.
(vii)	Substitute the width of the long narrow depression, b, for the radius of the depression, r, in Equations 14.12, 14.13, and 14.14.
(viii)	Substitute b for r in Equation 14.15.
(ix)	Substitute b for $2r$ when using Table 14.4 or Equation 14.16 to determine the dimensionless factor Ω.
(x)	Substitute b for r in Equation 14.18.
(xi)	No change.

Single Layer of Biaxial Geogrid or Geotextile: Biaxial geogrids and geotextiles can be made equally strong in their machine and cross directions. As far as orientation is concerned, the procedure used for uniaxial geogrids over a long narrow depression can be used as noted above.

14.7.10 Design Example

Objective

Design a uniaxial geogrid reinforcement to support a municipal solid waste landfill lining system constructed over an existing municipal solid waste landfill that has a circular depression of dimensions 3 ft (0.9 m) diameter.

Problem Statement

- The reinforcement will be designed to span a circular depression that may develop in the existing municipal solid waste beneath the lining system.

- The design life of the landfill lining system is 50 years.
- The components of the lining system are as follows, from top to bottom:
  - 2 ft (0.6 m) thick granular drainage layer;
  - Geomembrane;
  - Geonet;
  - Geomembrane.
- The maximum slope angle of the lining system is 9.5° (6H:1V).
- The maximum height of the municipal solid waste placed on the lining system will be 40 ft (12 m).
- The relevant material properties for design are as follows:
  - Waste:

    $$\gamma_{sw} = 70 \text{ lb/ft}^3 \text{ (11 kN/m}^3\text{)}$$
    $$\phi'_{sw} = 25°$$

  - Granular drainage layer:

    $$\gamma_l = 120 \text{ lb/ft}^3 \text{ (19 kN/m}^3\text{)}$$
    $$\phi'_l = 35°$$

- The cohesion of all materials is assumed to be equal to zero.
- The tensile strengths of the geomembranes and geonet are considered to be negligible. The yield strain of the geomembrane is assumed to be $\varepsilon_y \gg 10\%$; thus, the reinforcement strain controls.

Design Method:

The following step-by-step design method adapted from TENSAR (1989) follows the procedures outlined in Section 14.7.8:

(i) Verify design assumptions (Section 14.7.2):
The design assumptions listed in Section 14.7.2 are met. Therefore, the design guideline outlined in Section 14.7.8 is applicable.

(ii) Define material properties (Section 14.7.3):
- Waste and granular drainage layer:
  Properties are as previously defined
- Yield strain of Geomembrane:
  $\varepsilon_y \gg 10\%$

(iii) Define geometrical and loading parameters (Section 14.7.4):
- Geometrical parameters:
  $H_l = 2$ ft (0.6 m); $H_{sw} = 40$ ft (12 m)
  r is determined in Step (iv).
- Loading parameter:
  $q = 0$

(iv) Define design criteria (Section 14.7.5):
- The radius r of the depression is assumed to be 3 ft (0.9 m) ("rusted refrigerator assumption").

- The allowable deflection on the reinforcement is governed by the strain criterion. The allowable uniaxial strain of the reinforcement is determined in Step (v).

(v) Select allowable uniaxial design strain (Section 14.7.8):
  - The allowable uniaxial strain of reinforcement is $\varepsilon_g = 10\%$ (Section 14.7.6).
  - Calculate the allowable strain ε_l of the geomembrane:

$$\varepsilon_l = \varepsilon_y/F_R$$
$$= (10\%)/1.5$$
$$= 6.7\%$$

  - Select the allowable design strain ε_d:

  Because ε_d is the lesser of ε_g and ε_l, we have $\varepsilon_d = 6.7\%$

(vi) Calculate the required design tensile load of the reinforcement (Section 14.7.8):
  - Calculate the weighted average wet unit weight, γ_{avg}, of the materials located from the top of the reinforcement to a distance of $6r$ above:
  $6r = 6 \times 3 = 18$ ft
  $H_1 = 2$ ft $< 6r$; therefore, use Equation 12.10 to get

$$\gamma_{avg} = \frac{\gamma_1 \cdot H_1 + \gamma_{sw} \cdot (6r - H_1)}{6r}$$

$$= \frac{(120)(2) + (70)[(6)(3) - 2]}{(6)(3)}$$

$$= 75.6 \text{ lb/ft}^3 \ (11.9 \text{ kN/m}^3)$$

  - Calculate the normal pressure p exerted on the geogrids over the depression:

$$p = 2 \cdot \gamma_{avg} \cdot r \cdot (1 - e^{-0.5(H_1 + H_b + H_{sw}/r)}) + q \cdot e^{-0.5(H_1 + H_b + H_{sw})/r} \quad (14.15)$$
$$= (2)(75.6)(3)[1 - e^{-0.5(2 + 40)/3}]$$
$$= 453 \text{ lb/ft}^2 \ (21.7 \text{ kN/m}^2)$$

  - Determine the dimensionless factor, Ω, using Table 14.4:
  For $\varepsilon_d = 6.7\%$, $\Omega = 0.86$
  - Calculate the required design tensile load of reinforcement:

$$T_r = p \cdot r \cdot \Omega \quad (14.18)$$
$$= (453)(3)(0.86)$$
$$= 1170 \text{ lb/ft} \ (17.1 \text{ kN/m})$$

(vii) Select the specific reinforcement (geogrid or geotextile) product to provide the required design tensile load: (Section 14.7.8)

The specific reinforcement must provide the calculated required design tensile load of 1,170 lb/ft (17.1 kN/m) at an allowable design strain of 6.7% over the 50-year design life of the reinforced soil structure. The allowable long-term (120 years) tensile loads of various candidate reinforcements at a strain of 6.7% are given in Table 14.5:

Of the products listed in Table 14.5, the GG-2 material with $T_{allow} = 1,700$ lb/ft (25 kN/m) would be selected. However, this is considerably stronger than the required value of 1,170 lb/ft (17.1 kN/m). Thus, the designer should look for other geogrids or geotextiles that more nearly fit the required strength value. In addition to the technical design presented, cost also plays a significant role. This falls under a product-specific situation.

TABLE 14.5 Assumed Properties of Candidate Reinforcement Products

Product	T_{ult}		Reduction Factors			T_{allow}	
	lb/ft	kN/m	RF_{CR}	RF_{CR}	RF_{ID}	lb/ft	kN/m
GG1	3,300	48	2.5	1.1	1.2	1,000	15
GG2	4,860	71	2.0	1.1	1.3	1,700	25
GG3	8,460	123	2.5	1.1	1.3	2,350	34

14.8 STABILITY ANALYSIS FOR VERTICAL EXPANSIONS

The stability of the whole landfill including both the existing and new filling portions should be evaluated by performing slope stability analyses for conditions in effect at different stages of construction and operation of the facility. The analyses should address the following conditions:

(i) Stability of the soil mass of the side slopes during excavation (during excavation condition).

(ii) Stability of the liner system of the vertical expansion portion on the side slopes prior to waste placement (post construction condition).

(iii) Stability of the proposed new portion during waste placement (interim configuration or during filling condition).

(iv) Stability of the proposed new portion after filling when the landfill is in its final configuration. The evaluation of final configuration stability considering potential slip surface through the waste mass, along liner system interface, and through the landfill foundation (final configuration or post filling condition).

(v) Stability of the final cover system (after closure condition).

The slope stability analysis methods for these conditions have been described in detail in Chapter 13. Values of the factor of safety of at least 1.5 are generally accepted by regulatory agencies as representing a long-term stable condition.

Experience has shown that potential slip surfaces in landfills having the lowest factors of safety are often along soil-geosynthetic or geosynthetic-geosynthetic interfaces within the liner system and cover system. Therefore, shear strength parameters for interfaces within the following systems are required for slope stability analyses:

(i) Liner system over natural ground;

(ii) Liner system on landfill side slope;

(iii) Liner system in the portions of the landfill that have previously been constructed;

(iv) Final cover system.

Each system contains several interfaces and materials along which (or in which) shear failure could potentially occur. The shear strength of each system is characterized by the shear parameters of the weakest material or interface in the system. These parameters are called the *critical shear strength* parameters. In addition, evaluation of the shear strength parameters for the waste itself are required to perform slope stability analyses for final and interim landfill configurations.

Municipal solid waste usually settles a considerable amount during the filling operation (recall Figure 12.1). Here it can be seen that municipal solid waste landfills settle approximately 10 to 30% of their initial height. The large settlement of the waste fill induces shear stresses in the liner system on the side slope, which tends to displace the liner downslope. The large settlement of the waste fill and the large deformation of the landfill cover tend to induce shear stresses in the final cover system. These shear stresses induce large shear displacements along specific interfaces in the liner and cover systems that may lead to the mobilization of a reduced or residual interface strength. In addition, thermal expansion and contraction of the side slope liner and cover systems during construction and filling may also contribute to the accumulation of shear displacements and the mobilization of a residual interface strength (Qian, 1994; Stark and Poeppel, 1994; Qian, 1996).

A landfill should be designed for long-term performance. Accordingly, landfill safety should be considered not only during the relatively brief construction and operation periods, but also during a closure period lasting potentially hundreds of years. The potential for development of other uses for closed landfills should be also considered. Therefore, residual interface strengths should be assessed and considered for all side slopes for design purposes to make a landfill stable and safe even after being subjected to large settlements.

PROBLEMS

14.1 What are advantages of a vertical landfill expansion?

14.2 Explain why a vertical landfill expansion project is more complicated than a normal landfill project.

14.3 What are main considerations for design of a vertical landfill expansion? What are the principal differences compared to the design of a normal landfill?

14.4 What type of liner system would you select over an existing landfill for a vertical expansion project? Explain the reasons for your selection.

14.5 What two types of differential settlement should be considered during design of a vertical landfill expansion? What causes these two types of differential settlement?

14.6 Explain how to minimize differential settlement caused by waste heterogeneity without having to reinforce the liner.

14.7 List the internal landfill structures that may be impacted by vertical landfill expansion. Explain what causes these impacts.

14.8 How can a liner system be designed to minimize potential damage to a liner and leachate collection system constructed over an existing landfill?

14.9 What assumptions are used for designing geogrids or high strength geotextiles to reinforce a liner system over an existing landfill?

14.10 Explain what are the design criteria for using geogrids or high strength geotextiles to reinforce the liner system over an existing landfill?

14.11 How many different conditions should be considered for stability analysis of proposed vertical expansion designs?

REFERENCES

BNCB, (1975) "Subsidence Engineers' Handbook," British National Coal Board, Mining Department, London, U. K.

Bonaparte, R. and Berg, R. R., (1987a) "The use of Polymer Geosynthetics to Support Roadways over Sinkhole Prone Areas," *Proceedings, Second Multidisciplinary Conference on Sinkholes and Environmental Impacts of Karsts,* Orlando, FL, February, pp. 437–445.

Bonaparte, R. and Berg, R. R., (1987b) "Long-Term-Allowable Tensile Load for Geosynthetic Reinforcement," *Proceedings of Geosynthetics '87,* New Orleans, February, pp. 181–192.

Bowles, J. E., (1982) *Foundation Analysis and Design,* McGraw-Hill, Inc., New York, NY, 816 pages.

Brauner, G., (1973) "Subsidence due to Underground Mining," U. S. Department of the Interior, Bureau of Mines, Denver Mining Research Center, Denver, CO.

Drumm, E. C., Kettelle, R. H., and Manrod, W. E., (1987) "Analysis of Plastic Soil in Contact with Cavitose Bedrock," *Proceedings of Geotechnical Practice for Waste Disposal '87,* Ann Arbor, Michigan, June, pp. 418–431.

Gilbert, P. A. and Murphy, W. L., (1987) "Prediction/Mitigation of Subsidence Damage to Hazardous Waste Landfill Covers," EPA/600/2-87/025, U. S. Environmental Protection Agency, Cincinnati, OH, March.

Giroud, J. P., (1982) "Design of Geotextiles Associated with Geomembranes," *Proceedings of the International Conference on Geotextiles,* Vol. 4, Las Vegas, NV, August.

Giroud, J. P., Bonaparte, R., Beech, J. F., and Gross, B. A., (1988) "Load-Carrying Capacity of a Soil Layer Supported by a Geosynthetic," *Proceedings of International Geotechnical Symposium on Theory and Practice of Earth Reinforcement,* Fukuoka, Japan, October, pp. 185–190.

Giroud, J. P., Bonaparte, R., Beech, J. F., and Gross, R. A. (1990) "Design of Soil Layer-Geosynthetic Systems Overlying Voids," *Journal of Geotextiles and Geomembranes,* Vol. 9, No. 1, pp. 11–50.

Golecki, J. J., (1978) "Boussinesq-Trefftz Solution and Related Representations in Compressible and Incompressible Elastostatics," *Solid Mechs. Archives,* Vol. 3, No 2, May, pp. 109–129.

Golecki, J. J., (1979) "Displacement Potentials and Corresponding Stress Function in Two-Dimensional Problems of Elasticity," *Solid Mechanics Archives,* Vol. 4, Issue 3, August, pp. 183–205.

Jang, D. J. and Montero, C., (1993) "Design of Liner Systems under Vertical Expansions: An Alternative to Geogrids," *Proceedings of Geosynthetics, '93,* Vol. 3, Vancouver, B.C. Canada, March, pp. 1497–1510.

Koerner, G. R. and Koerner, R. M., (1990) "Installation Survivability of Geotextiles and Geogrids." *Procedings 4th IGS Conference on Geotextiles, Geomembranes and Related Products,* Rotterdam: A.A. Balkema, pp. 597–602.

McGown, A., Paine, N., and DuBois, D. D., (1984) "Use of Geogrid Properties in Limit Equilibrium Analysis," *Proceedings of the Symposium on Polymer Grid Reinforcement in Civil Engineering,* The Institution of Civil Engineers, London, March, pp. 31–35.

Netlon, Ltd., (1986) "TENSAR SR55-SR80-SR110 Geogrids," Product Brochure, Blackburn, England (available from the TENSAR Corporation, Morrow, GA).

Qian, Xuede, (1994) "Stability Analyses for Vertical and Lateral Expansions of Landfill," Michigan Department of Environmental Quality, Waste Management Division, Lansing, MI, July.

Qian, Xuede, (1996) "Design of Vertical Landfill Expansions," Michigan Department of Environmental Quality, Waste Management Division, Lansing, MI, September.

Sagaseta, C., (1987) "Analysis of Undrained Soil Deformation due to Ground Loss," *Geotechnique,* Vol. 37, No. 3, September, pp. 301–320.

Stark, T. D. and Poeppel, A. R., (1994) "Landfill Liner Interface Strengths from Torsional-Ring-Shear Tests," *Journal of Geotechnical Engineering,* ASCE, Vol. 120, No. 3, pp. 597–615.

TENSAR, (1989) "Design of Tensar Geogrid Reinforcement to Support Landfill Lining Cover System," *Structural Synthetic Geogrid for Waste Facility Applications,* TTN: WM3, Tensar Environmental Systems, Inc., Morrow, GA.

Thorton, T. S., Sprague, C. J., Klompmaker, J., and Waddy, D. B., (1999) "The Relationship of Creep Curves to Rapid Loading Stress-Strain Curves for Polyester Geogrids," *Proceedings Geosynthetics, '99,* IFAI, pp. 735–744.

Tsur-Lavie, Y. and Denekamp, S., (1980) "A Boundary Element Method for the Analysis of Subsidence Associated with Longwall Mining," *Proceedings of 2nd International Conference on Ground Movements and Structures,* Cardiff, U. K., April, pp. 67–74.

Tsur-Lavie, Y., Denekamp, S., and Fainstein, G., (1988) "Surface Subsidence Associated with Longwall Mining: Two and Three Dimensional Boundary Element Model," *Engineering Geology of Underground Movements,* pp. 225–231.

Wang, M. C. and Badie, A., (1985) "Effect of Underground Void on Foundation Stability," *Journal of Geotechnical Engineering,* ASCE, Vol. 111, No. 8, pp. 1008–1019.

Wrigley, N. E., (1987) "Durability and Long-Term Performance of TENSAR Polymer Grids for Soil Reinforcement," *Materials Science and Technology,* Volume 3, March, pp. 161–170.

Zornberg, J. G., Jernigan, B. L., Sanglerat, T. R., and Cooley, B. H., (1999) "Retention of Free Liquids in Landfills Undergoing Vertical Expansion," *Journal of Geotechnical and Geoenvironmental Engineering,* ASCE, Vol. 125, No. 7, pp. 583–594.

CHAPTER 15

Bioreactor Landfills

15.1 INTRODUCTION
15.2 LIQUIDS MANAGEMENT STRATEGIES
    15.2.1 NATURAL ATTENUATION
    15.2.2 LONG TERM CONTAINMENT
    15.2.3 LEACHATE RECYCLING
    15.2.4 COMPARISON OF STRATEGIES
15.3 CONCEPTS OF WASTE DEGRADATION
    15.3.1 PHASES OF DEGRADATION
    15.3.2 FIELD CAPACITY MOISTURE CONTENT
    15.3.3 RELATED ASPECTS
15.4 LEACHATE RECYCLING METHODS
    15.4.1 SURFACE SPRAYING
    15.4.2 SURFACE PONDING
    15.4.3 LEACH FIELDS
    15.4.4 SHALLOW WELLS
    15.4.5 DEEP WELLS
    15.4.5 COMPARISON OF METHODS
15.5 BIOREACTOR LANDFILL ISSUES AND CONCERNS
    15.5.1 LINER SYSTEM INTEGRITY
    15.5.2 LEACHATE COLLECTION SYSTEM
    15.5.3 LEACHATE REMOVAL SYSTEM
    15.5.4 FILTER AND/OR OPERATIONS LAYER
    15.5.5 DAILY COVER MATERIAL
    15.5.6 FINAL COVER ISSUES
    15.5.7 WASTE STABILITY CONCERNS
15.6 PERFORMANCE-TO-DATE
15.7 SUMMARY COMMENTS
    PROBLEMS
    REFERENCES

The main approach to landfill design and construction to date has been to contain and isolate the contents. An engineering design and constructed envelope that incorporates both drainage and barrier layers is used to encapsulate and drain the waste and to prevent, or at least minimize, the escape of landfill leachate into the environment. Most of the preceding chapters of the book are devoted to the design of this envelope system and its appurtenant features. This approach requires no active intervention nor management of the waste itself. Critics of this approach (Lee and Jones, 1991a, 1992) have described landfills designed and constructed in this manner as "dry tombs". An alternative approach is to actively manage and treat the contents to decrease toxicity and to reduce the time required to reach a stable end condition. It is important to note that this alternative approach still requires the design and implementation of containment measures, but in addition, it entails an active waste treatment and management program as well.

Several considerations support an active intervention and management approach. First and foremost, a municipal waste landfill is a very large, de facto biochemical reactor. A variety of biological and biochemical processes occur within the landfill envelope, which over time modify the properties of the contained waste and also produce liquid leachate and gas. These processes, while admittedly difficult to predict and precisely quantify, are nevertheless amenable to treatment and modification. Suitable intervention can result in faster rates of decomposition, reduced toxicity, and shorter time to reach stable equilibrium. By supplying sufficient liquid to the waste and holding this status under proper environmental conditions (i.e., proper pH and nutrients), waste degradation and the stripping of contaminants can occur within a 5- to 20-year time frame. A variant of this approach, referred to as a *wet cell* landfill (Lee and Jones, 1991b), purposely introduces water into the landfill via a controlled, reverse groundwater gradient liner system; this introduced water promotes fermentation and leaching of the waste itself. A faster rate of decomposition and shorter "reactor" period eliminate one of the main criticisms of the present approach, which is that an "unmanaged" or passive landfill consisting of a liner encapsulated waste mass can exist largely in its as-placed state until a catastrophic event occurs, or until the lifetime of the liner system is exceeded.

15.1 INTRODUCTION

To recycle leachate back into, or onto, a landfill is certainly not a new idea. Owners and operators have long used this liquids strategy, not only to temporarily avoid leachate treatment but also to enhance waste degradation and generate gas (e.g., see Cheremisinoff and Morresi, 1976). What is relatively new is to consider the landfill as a massive anaerobic bioreactor operated under controlled conditions. Several workshops convened to explore this approach are worth noting in this regard. The first was an EPA Workshop convened in 1981 where it was established that there are two main operating conditions that must be maintained:

 (i) the moisture content must be held at field capacity, and
 (ii) the pH must be controlled near to neutrality.

The second workshop was convened by the Swiss Federal Institute for Water Resources and Pollution in 1988. A number of contributed papers were included in the proceedings of the Workshop (Baccini, 1988) describing and characterizing biological and chemical processes that occur in municipal waste landfills. In addition, participants at the Workshop identified specific measures to accelerate the decomposition processes of organic materials, and by this means shorten the time of the "reactor" period of a landfill. Major methods of acceleration consist of

(i) shredding or "homogenizing" the waste.
(ii) controlled moisture additions *(Note: In this case, "controlled" refers to monitoring pH and not adding water in amounts to make the pH acidic nor to exceed field capacity);*
(iii) controlled sewage sludge additions; and
(iv) the use of aerobically or partially decomposed waste to reduce the acid phase *(Note: this will reduce the methane generation)*

Field capacity implies not only utilizing the site-specific leachate, but (almost always) it also implies adding significantly more liquid to the waste mass. Candidate liquids in this regard are site precipitation water or snow melt, gray water from industrial processes, sewage sludge, and possibly storm/sanitary water outflows. A controlled inward hydraulic gradient landfill design is yet another method of introducing water (groundwater) into a landfill (albeit in relatively nonquantifiable amounts). The advantage of an inward gradient system is that the inward, advective gradient opposes outward leakage of contaminant solutes by diffusion. Furthermore, a controlled inward gradient would largely eliminate the problem of advective leakage through any defects (holes, cracks, fissures, etc.) in the bottom liner.

Whatever the method of adding liquid, a liquids management strategy leading to a bioreactor landfill necessitates changes in both landfill operations and design. This is a distinct contrast to the conventional strategy of leachate collection, removal, treatment, and disposal. Design, and operations issues discussed in this chapter include the following:

(i) base liner systems;
(ii) leachate collection systems;
(iii) leachate removal systems;
(iv) filter and/or operations layers;
(v) daily cover placement materials;
(vi) final cover placement materials;
(vii) final cover issues;
(viii) waste stability concerns;

Recent experiences utilizing the concept of bioreactor landfills will be presented, as well as concluding remarks in the form of recommendations.

15.2 LIQUIDS MANAGEMENT STRATEGIES

Of the many possible liquids management strategies for the handling of landfill leachate, most are subsets of either natural attenuation, remove/treat/discharge, or leachate recycling.

15.2.1 Natural Attenuation

Prior to 1982, when the first regulations were promulgated requiring geomembrane liners, natural attenuation was the de facto liquids management strategy. Even with low-permeability clay soil liners, absorption coupled with attenuation was generally felt to be adequate. Of course, such a strategy must be assessed on the basis of the type of waste, site location, environmental sensitivity, etc., and generally such a strategy anticipates that no significant groundwater impacts will occur. Also, for most landfill situations, natural attenuation has become an outmoded concept.

Recently, however, natural attenuation has made somewhat of a resurgence for arid and semiarid landfill sites (see Dwyer, 1995). Arid and semiarid climates were defined previously in Section 11.2.1. Site selection, however, remains contentious, and as a result this liquids management strategy has attracted considerable regulatory and owner concern.

15.2.2 Remove, Treat, and Discharge

The current liquids management strategy of the vast majority of landfills is to contain the waste within liner and cover systems of extremely low permeability. This has been the tacit approach taken throughout this book. The strategy has been called, in a somewhat derogatory manner, the *dry tomb* concept. This term actually infers long-term containment, where *long-term* is admittedly a subjective term. Essentially all states require containment times of over 30 years, and often one hears of required containment times of 100 years. Still longer containment times are often discussed for hazardous wastes and landfills located at particularly environmentally sensitive sites.

The following are some considerations that long-term containment implies (or requires):

(i) provide at least the local regulatory composite required liner system (and oftentimes double liner systems);
(ii) remove the leachate at the base of the landfill on demand;
(iii) properly treat and dispose of the collected leachate;
(iv) limit the hydraulic head on the liner to 300 mm so as to minimize leakage;
(v) place the final cover within 1-year after reaching the permitted height; and
(vi) provide for a minimum of 30-year postclosure maintenance period.

Thus, with long-term containment as the liquids management strategy, liner system durability and functioning is a key issue necessitating long lifetimes and adequate performance of the associated materials. This applies to both natural soils and geosynthetics.

15.2.3 Leachate Recycling

Leachate recycling, or recirculation, is currently allowed for municipal solid waste landfills (see EPA, 1993). The concept is to remove the leachate from the base of the landfill and then to reintroduce it onto, or into, the waste mass by any one of a number

TABLE 15.1 Leachate Storage Capacity (in Gallons) of Typical MSW Landfills per Acre of Surface Area

Height of Landfill	Initial Saturation of Waste (w)		
	10%	30%	50%
50 ft	3.6×10^6	2.8×10^6	2.1×10^6
100 ft	7.2×10^6	5.6×10^6	4.2×10^6
200 ft	14.4×10^6	11.2×10^6	8.4×10^6

Notes: $w = W_w/W_s$, whereas W_w = weight of liquid and W_s = weight of solid; also calculations assume an average specific gravity, $G_s = 1.5$; and average unit weight, $\gamma = 70$ lb/ft^3.

of methods. The various methods will be described later. The leachate can be treated, but more generally it is used directly as retrieved from the base of the landfill.

Considering the void space that is available in a typical landfill, MSW can accept huge quantities of liquid. (See Table 15.1.) This is a very attractive feature of leachate recycling from the economic perspective of an owner and/or operator.

There are several important aspects of leachate recycling that are of interest to regulators, but the most important one is to achieve "sustainability". The goals of sustainability are the following:

(i) for the operator to completely manage the outputs (liquids and gases);
(ii) to provide for environmentally acceptable residues;
(iii) to avoid long postclosure care periods; and
(iv) to potentially use the closed sites for beneficial purposes.

This last item is particularly provocative since regulatory guides are being formulated to use closed landfill sites for production and community uses. Clearly, such uses are possible on stable, non-gas-producing landfills (e.g., ash monofills and demolition waste sites). MSW landfills, however, pose a significant challenge insofar as large settlement is concerned (Spikula, 1996), as well as ongoing methane gas generation. If, however, waste degradation is essentially complete in a relatively short period and a geomembrane is used in the final cover, MSW landfill sites can be considered as possible locations for similar beneficial uses as mentioned previously.

15.2.4 Comparison of Strategies

The liquids management strategies just discussed are compared and contrasted in Table 15.2. Here it is seen that the low long-term cost of leachate recycling, along with the potential to use the site for productive purposes (sport fields, golf courses, paths and walkways, parking lots, etc.), are compelling reasons for investigating this strategy.

According to Reinhart et al. (2000), there were 12 states in the U.S. allowing for leachate recycling in 1993, and by 1997, there were 130 such landfills involved. There also are landfills in Germany, Sweden, Italy, and the United Kingdom that practice the

TABLE 15.2 Comparison of Different Liquids Management Strategies

Strategy	Groundwater Pollution Potential	Short Term Cost	Long Term Cost	Regulatory Concern	Current Status
Natural Attenuation	High	Low	Unknown	Great	Considered at Arid Sites
Remove, Treat and Discharge	Low	High	Unknown	Accepted	Ongoing Strategy
Leachate Recycling	Unknown	High	Low	Considerable	Attractive Strategy

technique. An EPA conference in 1995 was significant in framing the concept of leachate recycling (the topic actually was bioreactor landfills), design issues, operating parameters and presentation of several case studies. All of these aspects (and others) will be expanded upon in the text that follows.

15.3 CONCEPTS OF WASTE DEGRADATION

The ultimate goal of leachate recycling is to force the landfill to be a massive anaerobic reactor. As such, the waste is reduced to methane (which is captured) and relatively inert biomass with major losses of cellulose, the principal biodegradable constituent of MSW. On the other hand, the leachate will contain the inorganic contaminants (principally heavy metals, such as, lead, zinc, arsenic, mercury, etc.) that have been stripped from the solids. They will have to be removed from the leachate after the last pass-through at the end of the process. The organic contaminants will be converted through biodegradation, conversion to daughter products, oxidation/reduction processes, and/or physical stripping from the solids. Among others, Barlaz et al. (1990) describes the processes that are involved.

15.3.1 The Phases of Degradation

The generally accepted phases that leachate undergoes in a bioreactor landfill are presented in Figure 15.1, after Pohland and Harper (1986). These phases are as follows:

Phase I Initial adjustment phase, where moisture is added and the waste supports an active microbial community.

Phase II Oxygen depletion phase, where the waste transitions from aerobic to anaerobic. Both chemical oxygen demand (COD) and total volatile acids (TVA) appear in the leachate.

Phase III Acid is formed with a solubilization of the waste mass along with conversion of the biodegradable organic content. Both COD and TVA in the leachate are maximized.

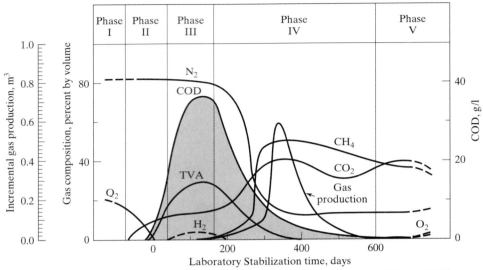

FIGURE 15.1 Five Phases of Landfill Stabilization (adapted from Pohland and Harper, 1986)

Phase IV   Acids are consumed and converted into methane and carbon dioxide in this phase. Heavy metals are removed from the leachate by complexation and precipitated onto the remaining solids.

Phase V   The final, or maturation phase, has the biological activity diminishing along with diminished gas production. Leachate strength is at much lower concentrations than in all previous phases.

The entire process can occur within 3 to 5 years under the optimal laboratory conditions for which the information was developed. Note that the degradation of the waste, per se, lags somewhat behind this time frame (e.g., 5 to 10 years). In this regard, it is important to realize that these studies were essentially laboratory (or bench) scale, using well-sorted and small waste masses, with carefully controlled and uniformly distributed moisture, often under controlled temperature and humidity conditions. Nevertheless, the processes and just-described phases have completely substantiated that degradation and stabilization of the waste can occur in the manner described. See Reinhart and Townsend (1998) for greater detail in this regard. The Solid Waste Association of North America (SWANA) has developed the following definition of a bioreactor landfill (Pacey et al., 1998):

> A bioreactor landfill is a sanitary landfill operated for the purpose of transforming and stabilizing the readily and moderately decomposable organic waste constituents within five to ten years following closure by purposeful control to enhance microbiological processes. The bioreactor landfill significantly increases the extent of waste decomposition, conversion rates and process effectiveness over what would otherwise occur within the landfill.

15.3.2 Field Capacity Moisture Content

It is critical that careful control of the moisture content of the waste mass be maintained in order to produce the various phases and time frames described in Figure 15.1. The target moisture content is called *field capacity*. This value of waste moisture content must eventually be met and held at every location within the landfill in order to complete the degradation process.

Field capacity is defined as that moisture content of the waste at which the maximum amount of liquid is held in the voids against gravity drainage. The mechanisms involved are absorption and capillary forces. At moisture contents greater than field capacity, gravitational drainage occurs. The excess moisture either provides field capacity conditions at lower elevations in the landfill, or enters the leachate collection and removal system, where it is captured and recycled.

Field capacity for MSW can be calculated on the basis of wet weight or dry weight. (See Table 15.3.) In either case, it is a very high value, particularly considering the low unit weight of MSW. It should be noted that the unit weight data of Table 15.3 are from relatively old (probably poorly compacted) landfills. Current waste compaction is such that considerably higher densities usually result.

Clearly, higher unit weights result in higher values of field capacity (Fungaroli and Steiner, 1979). Also, the finer the particle size of the waste, the greater is its field capacity. Thus, shredding the waste has the effect of increasing the field capacity (Hentrich et al., 1979).

The intent of this discussion on field capacity values is to recognize that a considerable amount of moisture must be added to MSW to reach its targeted moisture content. By way of illustration, Figure 15.2 presents the quantity of liquid required on the basis of incoming waste volume. The quantities are considerably higher than the typical leachate generation of most landfills. Thus, landfills will generally require additional liquid to be added to the quantity which is recycled from the base of the

TABLE 15.3 Field Capacity Moisture Values (modified from Reinhart and Townsend, 1998)

Unit Weight of Waste			Gravimetric Field Capacity	
kg/m^3	lb/yd^3	lb/ft^3	% wet wt.	% dry wt.
500–800	843–1350	27–50	54	117
500–800*	843–1350*	27–50*	43–50*	75–100
690–950*	1160–1600*	43–59*	53*	113
710	1200	44	47	89
688	1160	43	20–35	25–54
310	520	19	37	59
287	485	18	29	41
503	850	31	31–48	45–92
440	735	27	48	92
474	800	30	35	54

*shredded waste

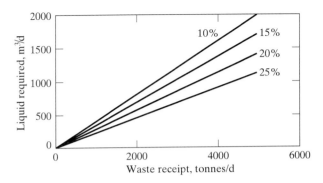

FIGURE 15.2 Liquid Addition Requirements to Meet Field Capacity (assume $w_{fc} = 50\%$, wet basis as a function of incoming waste mass) (after Reinhart et al., 2000)

particular site in question. Even further, Pacey (2000) recommends that the moisture content be maintained at a level somewhat higher than field capacity!

15.3.3 Related Aspects

There are many related aspects of the process (in addition to field capacity considerations) that are important to achieve a successful bioreactor landfill.

The comingling of old degraded waste with new incoming waste appears to be helpful in the degradation process (Barlaz et al., 1990). Apparently, old waste helps keep the pH from getting too low (a process called "souring"), which inhibits degradation. In this regard, other nutrient additions such as ammonia, glucose, or peptones can be added to the leachate being recycled (Filip and Kuster, 1979).

The addition of anaerobic sewage sludge is recognized by many as a practice that both augments the moisture content and tends to inhibit souring. In the latter situation, calcium carbonate can also be added (Barlaz et al., 1990).

Lastly, there are attempts at introducing air into the landfill and thereby promoting aerobic degradation of the waste. A few attempts are ongoing, but the potential for fire must be carefully considered.

As far as full-scale bioreactor landfills, Reinhart et al. (2000) report that 5 states in the U.S. are actively involved (NY, FL, CA, DE, IA). Conversely, 10 states have prohibited bioreactor landfills (AZ, IL, KA, KY, MA, MD, NE, NH, NH, OH). Thus, full-scale bioreactor landfills are quite rare, but leachate recirculation is not. As a result of this distinction, we may have entered a (temporary?) period of working with "liquid-starved" bioreactor landfills. Future research should investigate the degradation of waste at levels lower than field capacity.

15.4 LEACHATE RECYCLING METHODS

Next, we present five different leachate recycling methods and compare them with one another. While they are presented as separate methods, it also is possible to use them in combination. For example, Reinhart et al. (2000) suggest that "the most efficient approach to field capacity is to increase moisture content through wetting the waste at the working face and then [to] uniformly reach field capacity through leachate surface application or injection."

15.4.1 Surface Spraying

As the name implies, surface spraying utilizes tank trucks with an attached spray bar applying leachate to the surface of the waste mass. (See Figure 15.3(a).) Good coverage results from spraying on each lift. The method is low cost from both delivery and operations perspectives. Odors, vectors, and litter may be concerns depending on site-specific considerations.

15.4.2 Surface Ponding

Utilizing berms of waste (perhaps covered with a thin geomembrane), a temporary pond of leachate can be created. This method is well adapted to delivering large quantities of leachate and results in excellent coverage. Clearly, field capacity can be reached with the excess leachate draining to lower elevations. Operations essentially cease during ponding if the method is used on intermediate lifts. If ponding is undertaken after full height is reached, the method can be most effective. Odors, vectors, and litter may be concerns, depending on site-specific considerations.

15.4.3 Leach Fields

Leach fields are a variation of surface ponding wherein the leachate is delivered through well-defined drainage paths. (See Figure 15.3(b).) These leach fields are placed on a square or rectangular pattern beneath a temporary or final cover. Thus, odors, vectors, and litter are controlled. The injection rate, however, is limited and the initial implementation cost is high.

15.4.4 Shallow Wells

As the name implies, shallow wells penetrate into the waste and leachate is pumped as shown in Figure 15.3(c). A temporary or permanent cover is included and the wells must penetrate this cover system. The wells are perforated and placed at spacings varying from 10 to 30 meters. Although coverage is poor in the upper lifts, it probably is good at the lower elevations. The injection rate is limited and injection costs are relatively high. Odors, vectors and litter are completely controlled. If gas removal is practiced, short circuiting via leachate in gas wells has occurred; thus, careful planning and operations are required.

15.4.5 Deep Wells

As with shallow wells, deep wells can be installed throughout the waste mass. A temporary or permanent cover is used as shown in Figure 15.3(c). The wells are perforated

586 Chapter 15 Bioreactor Landfills

FIGURE 15.3 Injection Methods for Leachate Recirculation Treatment of Municipal Solid Waste

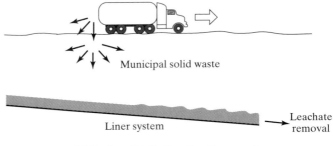

(a) Surface distribution directly on waste

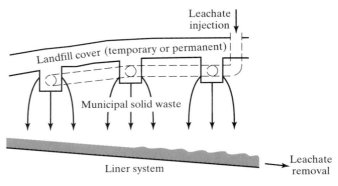

(b) Distribution beneath cover using manifold system

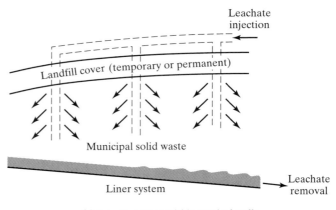

(c) Injection from within vertical wells

TABLE 15.4 Leachate Reintroduction Methods

Method	Odor/Vector/Liter Control	Injection Rate	Coverage	Injection Cost	Impact on Operations
Surface Spraying	Poor	Fast	Good	Low	Moderate
Surface Ponding	Poor	Fast	Good	Low	Low-to-High
Leach Fields	Good	Moderate	Moderate	Moderate	Low
Shallow Wells	Good	Slow	Poor	High	Low
Deep Wells	Good	Moderate	Moderate	High	Low

along their length allowing for isolation packers to inject the leachate at specific elevations. Coverage depends on the well spacings that are usually in the 20- to 50-meter range. The injection rate is limited and short circuiting with respect to gas extraction wells is a concern. Obviously, care must be exercised to avoiding penetration of the bottom liner (see Parsons, 2000). For this reason, and the fact that excess leachate will drain downward with gravity, deep wells are not as favored as shallow wells for leachate injection.

15.4.6 Comparison of Methods

Clearly, there is no universally accepted method of leachate recycling. As noted in the introduction to this section, the preferred strategy may be a combination of methods. Still, a comparison of methods is of interest. Table 15.4 presents the major issues with respect to leachate injection methods. The case histories to be presented later will give further insight into the methods that are most commonly used.

15.5 BIOREACTOR LANDFILL ISSUES AND/OR CONCERNS

With the goal being a bioreactor landfill operating through the various phases described in Figure 15.1, there are several issues and/or concerns that have been expressed by various parties involved. In general, they are either operations oriented or design oriented. This section discusses a number of these issues, all of which have to do with the required liquids bringing the waste mass to its field capacity moisture content.

15.5.1 Liner System Integrity

Considering the potential of the hydraulic head on the liner being well in excess of the common regulatory maximum of 300 mm, the integrity of the liner system must be assured. This, in turn, requires a geomembrane/compacted clay liner (GM/CCL) or a composite liner, as a minimum. Regulations generally call for the geomembrane to be at least 0.75 mm (30 mils) thick, with HDPE being at least 1.5 mm (60 mils) thick. The CCL must be 600 mm (2.0 ft.) thick with a permeability of 1.0×10^{-7} cm/sec, or lower.

However, this is not the only composite liner that should be considered. Recently available leakage data show that a GM/CCL is not the most leak-free system and that a geomembrane/geosynthetic clay liner (GM/GCL) is superior in this regard. Table 15.5 presents data from 91 double-lined landfills with leak detection systems. The data set included 287 individual cells. The table is subdivided according to GM, GM/CCL, and GM/GCL primary liners, with both sand and geonets (GN) as leak detection materials. The data were accumulated by GeoSyntec Consultants, Inc. and are reported in Bonaparte et al. (2001).

The average flow data from Table 15.5 was plotted in Figure 5.27 and clearly indicated that the GM/GCL outperforms both the GM by itself and the GM/CCL composite.

Recognizing that a GM/GCL composite liner is relatively thin in its cross section, a compromise consisting of three components is also possible (i.e., GM/GCL/CCL). In this case, however, the CCL component need not be at a permeability of 1.0×10^{-7} cm/sec, and permeabilities 10 to 100 times higher are adequate. This three-component composite liner system will often be less expensive than a traditional GM/CCL, especially in many areas where low-permeability native clay is not available.

15.5.2 Leachate Collection System

With the leachate moving through the waste mass in the quantities necessary for field capacity to be achieved, the leachate collection system must be relatively open insofar as its drainage capability is concerned (i.e., it must have a high permeability). Unfortunately, state regulations require relatively low minimum values of leachate

TABLE 15.5 Leakage Rates from Leak Detection Systems of Double Lined Landfills (after Bonaparte et al., 2001) [All Flow Rates in liter/hectare/day (lphad)]

Liner/LDS Type	Type I (GM-Sand)			Type II (GM-GN)			Type III (GM/CCL-Sand)		
Life Cycle Stage	1	2	3	1	2	3	1	2	3
Average Flow	380	170	64	90	100	ND	210	140	64
Minimum Flow	7.6	0.0	0.2	4.8	1.4	ND	1.2	22	0.0
Maximum Flow	2140	1480	240	370	360	ND	1180	660	270
No. of "points"	30	32	8	7	11	ND	31	41	15
No. of landfills	11	11	4	4	6	ND	11	11	4

Liner/LDS Type	Type IV (GM/CCL-GN)			Type V (GM/GCL-Sand)			Type VI (GM/GCL-GN)		
Life Cycle Stage	1	2	3	1	2	3	1	2	3
Average Flow	170	83	65	130	22	0.3	6.5	2.6	ND
Minimum Flow	0.0	0.0	0.0	0.0	0.0	0.0	0.0	0.0	ND
Maximum Flow	690	500	130	970	280	0.9	34	9.0	ND
No. of "points"	21	27	12	19	19	4	6	4	ND
No. of landfills	6	9	3	3	3	1	2	2	ND

Life Cycle Stages:
Stage 1–Initial Life
Stage 2–Active Life
Stage 3–Post Closure

"points" = Number of measuring points (i.e., outlets of single or multiple cells)
LDS = leak detection system
ND = No Data

TABLE 15.6 Results of 1998 Survey of State Regulation Requirements for the Permeability of Leachate Collection Soil

State Required Permeability (k) Value	Number of States	
	Subtitle D "MSW"	Subtitle C "Hazardous"
$k \geq 1.0 \times 10^{-2}$ cm/sec	11	20
$k \geq 1.0 \times 10^{-3}$ cm/sec	15	1
$k \geq 1.0 \times 10^{-4}$ cm/sec	1	0
Issue not addressed	20	1
States not responding	3	3
No such facilities in state	n/a	25
Total	50	50

collection soil and generally these values are the short-term values. Table 15.6 presents data from a recent survey of state regulations (Koerner et al., 1998).

It has been suggested that these values are entirely too low for the bioreactor landfill concept to be realized. Koerner et al. (1994) have exhumed four leachate collection systems that were clogged, and some were even excessively clogged. The main clogging mechanisms were fine particulates and microorganism growth. Key indicators of progressive clogging were total suspended solids (TSS) and biochemical oxygen demand (BOD). When these values are in excess of approximately 2500 mg/l, excessive clogging can be anticipated. While not known to be the case, it is expected that leachate which is being recycled in the earlier phases shown in Figure 15.1 can easily achieve such values of TSS and BOD. As such, leachate collection soil should have a minimum initial permeability of 1.0 cm/sec. This is typical of a coarse-gravel soil (e.g., AASHTO #57 gravel). Alternatively, a sand with a drainage geocomposite could result in a comparable situation.

If gravel is used as the leachate collection system, it cannot be placed directly on a geomembrane. However, a needle-punched, nonwoven geotextile can provide an excellent protection (cushion) layer. A design for the minimum mass per unit area of such a geotextile is available from Koerner (1998). The method uses the conventional factor-of-safety equation, written as

$$FS = \frac{p_{\text{allow}}}{p_{\text{act}}} \quad (15.1)$$

where FS = factor of safety (against geomembrane puncture),
p_{act} = actual pressure due to the solid waste mass, and
p_{allow} = allowable pressure using different types of geotextiles and site-specific conditions.

Based on a large number of ASTM 5514 puncture experiments, the following empirical relationship for p_{allow} has been obtained:

$$p_{\text{allow}} = \left(50 + 0.00045 \frac{M}{H^2}\right)\left[\frac{1}{\text{MF}_S \times \text{MF}_{\text{PD}} \times \text{MF}_A}\right]\left[\frac{1}{\text{RF}_{\text{CR}} \times \text{RF}_{\text{CBD}}}\right] \quad (15.2)$$

Here,

p_{allow} = allowable pressure (kPa),
M = geotextile mass per unit area (g/m²)
H = protrusion height (m),
MF_S = modification factor for protrusion shape,
MF_{PD} = modification factor for packing density,
MF_A = modification factor for arching in solids,
RF_{CR} = reduction factor for long-term creep, and
RF_{CBD} = reduction factor for long-term chemical/biological degradation.

Note that in this formula, all MF-values ≤ 1.0 and all RF-values ≥ 1.0. Note also that the formula requires the set of modification factors and reduction factors given in Table 15.7.

The situation can be approached from a given mass per unit area geotextile to determine the unknown factor-of-safety value, or from an unknown mass per unit area geotextile and a given factor-of-safety value. The following example uses the latter approach.

TABLE 15.7 Modification Factors and Reduction Factors for Geomembrane Protection Design Using Nonwoven Needle-Punched Geotextiles

Modification Factors

MF_S		MF_{PD}		MF_A	
Angular	1.0	Isolated	1.0	Hydrostatic	1.0
Subrounded	0.5	Dense, 38 mm	0.83	Geostatic, shallow	0.75
Rounded	0.25	Dense, 25 mm	0.67	Geostatic, mod.	0.50
		Dense, 12 mm	0.50	Geostatic, deep	0.25

Reduction Factors

			RF_{CR}		
RF_{CBD}		Mass per unit area	Protrusion (mm)		
		(g/m²)	38	25	12
Mild leachate	1.1	Geomembrane alone	N/R	N/R	N/R
Moderate leachate	1.3	270	N/R	N/R	>1.5
Harsh leachate	1.5	550	N/R	1.5	1.3
		1100	1.3	1.2	1.1
		>1100	$\cong$1.2	$\cong$1.1	$\cong$1.0

N/R = Not recommended

EXAMPLE 15.1

Given a coarse gravel (subrounded with $d_{50} = 38$ mm) leachate collection layer with a permeability of approximately 1.0 cm/sec to be placed on a 1.5 mm thick HDPE geomembrane under a 50 m high landfill. What geotextile mass per unit area is necessary for a *FS*-value of 3.0? Assume that the solid waste weighs 12 kN/m³.

Solution: Use H = 25 mm = 0.025 m (1 inch), which is an estimate since the gravel particles are not isolated, but are adjacent to one another, $MF_S = 0.5$ for shape, $MF_{PD} = 0.83$ for packing density, $MF_A = 0.25$ for arching, $RF_{CR} = 1.5$ for creep, and $RF_{CBD} = 1.3$ for long-term degradation. Now calculate the value of p_{allow} using Equation 15.1:

$$FS = \frac{p_{allow}}{p_{act}}$$

$$3.0 = \frac{p_{allow}}{(50)(12)}$$

$$p_{allow} = 1800 \text{ kN/m}^2$$

Then, calculate the required mass per unit area of the geotextile using Equation 15.2:

$$p_{allow} = \left(50 + 0.00045 \frac{M}{H^2}\right)\left[\frac{1}{MF_S \times MF_{PD} \times MF_A}\right]\left[\frac{1}{RF_{CR} \times RF_{CBD}}\right]$$

$$1800 = \left[50 + 0.00045 \frac{M}{(0.025)^2}\right]\left[\frac{1}{0.5 \times 0.83 \times 0.25}\right]\left[\frac{1}{1.5 \times 1.3}\right]$$

$$M = 436 \text{ g/m}^2, \quad \text{hence use a } 500 \text{g/m}^2 \text{ geotextile}$$

15.5.3 Leachate Removal System

Embedded within the leachate collection material is an interconnected perforated pipe network called the leachate removal system. In recent years this pipe system has favored the use of corrugated high density polyethylene in sizes from 100 to 300 mm diameter. The pipe is slotted within the valleys of the corrugations and is joined by means of couplings mated to the particular pipe's outer surface configuration.

Some states (approximately seven) require cleanouts to verify that the pipe is free from defects and obstructions. If 100% cleanout is required, the inlet/outlet design will dominate the layout of the entire facility.

For relatively high landfills (sometimes called *megafills*), pipe deformations can be significant. Concern over end connections and pipe intersections have also been raised. The current thrust is to analyze the situation using finite element analysis, but it appears that additional investigations are necessary, particularly for landfills over 50 m in height.

15.5.4 Filter and/or Operations Layer

Above the leachate collection and removal system is often placed a filter layer (150 mm of sand or a geotextile) and/or an operations layer (300 to 450 mm of locally available soil). Operations layers are also sometimes called *protection layers*. For landfills that practice leachate recycling leading to a bioreactor landfill, the omission of both the filter and the operations layer is suggested. In the case of the filter, bioclogging is a distinct possibility (see Koerner et al., 1994). In the case of operations layers, they appear destined to become de facto liners as they attempt to transmit leachate which is high in suspended solids and microorganisms.

Instead, the scheme for an open drainage system at the base of a landfill of the type considered herein should be a lift of *select waste* placed directly on the leachate collection and removal system. This select waste should be carefully sorted so that large bulky and protruding objects are not present. Neither should it have excessively fine particles (e.g., fly ash) nor be liquid sludges, which could cause excessive clogging of the leachate collection system. The first lift (approx. 3 m in height) of select waste should not be compacted other than trafficking by the haulage and spreading equipment. Subsequent lifts of wastes can be placed and compacted as with normal operations. Figure 15.4 shows photographs of select waste being placed directly on the stone of a leachate collection system.

15.5.5 Daily Cover Material

Daily cover material refers to the 150 mm of soil placed over each lift of waste for the prevention of vectors, odors, and fires, and for control of water and gas. Unfortunately, this soil layer can also provide a de facto barrier to the downward flow of leachate. Thus, for landfills to function as bioreactors, an alternative daily cover material (ADCM) is recommended. In this regard, there are many candidate materials available. (See Table 15.8 after Pohland and Graven, 1993.) All of these materials essentially allow for free flow of leachate from lift to lift within the waste mass.

As far as the technical equivalency of a particular ADCM to the regulated 150 mm of soil is concerned, the author suggests a rating matrix as shown in Table 15.9. Here the soil is rated on an item-by-item basis for the various site-specific concerns. The total is forced to equal a numeric value of 100. Next, the product-specific ADCM is rated for each of the same concerns. For the ADCM to be equivalent, the total must be 100 or more. Furthermore, each concern must be equal or greater. If (as in the example) several concerns are not met, a decision must be reached on how critical they are, or the operations permit must be modified to account for the deficiency. For example, if fire was a critical concern, the permit could be amended to require hot wastes to be offloaded and watered to bring the temperature to ambient before placement in the landfill.

15.5.6 Final Cover Issues

A final cover for the long-term closure of a landfill consists of a foundation layer, gas collection layer, barrier layer, surface water drainage layer, cover soil layer, and topsoil with vegetative growth (Koerner and Daniel, 1997). While all components are important, the gas collection layer is particularly important. The gas generation is very concentrated over a relatively short period of time. Thiel (1998) presents a detailed discussion on how to design a gas collection system. Typical regulations call for the final cover construction to be within one year after placement of the final lift of waste.

FIGURE 15.4 Photographs of "Select" Waste Being Placed Directly on Leachate Collection Layer without Filter or Operation Layers

TABLE 15.8 Alternative Daily Cover Materials (after Pohland and Graven, 1993)

Polymer Foams
- Rusmar
- Sanifoam
- Terrafoam
- Topcoat

Slurry Sprays
- Con Cover (paper)
- Land-Cover (clay/polymer)
- Posishell (paper)

Sludges and Indigenous Materials
- Naturite/Naturfill
- N-Viro Soil
- Chemfix
- Green waste/compost
- Ash-based
- Auto fluff
- Foundry Sand
- Shredded Tires

Reusable Geosynthetics
- Air Space Saver
- Griffolyn
- Cormier
- FabriSoil
- Covertech
- Sanicover
- Aqua-Shed
- Polyfelt
- Tarpmatic
- Typar

For landfills practicing leachate recycling with the goal of a bioreactor landfill, *the final cover should not be placed until the majority of settlement occurs.* This is probably within a 5- to 20-year time frame (recall Figure 12.9). Instead, a temporary cover should be placed.

If odors are a concern, a geomembrane or GCL must be provided within the temporary cover. If odors are not a concern, a soil cover is acceptable unless gas is to

TABLE 15.9 Suggested Method to Assess Technical Equivalency of ADCMs*

Regulatory Concerns over Daily Cover	150 mm Soil Rating	ADCM Product "×" Rating	Relative Equivalency
Control of vectors (birds, animals, flies)	25	20	Slightly less**
Air-borne controls (blowing litter, odors)	25	20	Slightly less**
Control of fires (waste at surface and at depth)	20	15	Slightly less**
Control of water infiltration	15	25	Better
Control of gas movement	15	25	Better
Total Score	100	105	Better

*Table considers regulatory issues only; not the value added to the owner/operator.
** Issue must be judged as being noncritical or permit adjusted accordingly in order to approve the ADCM.

be collected. In this latter case, a geomembrane should be provided. The ideal situation is a geosynthetic material that allows water to enter the waste mass, but that prevents the gas from escaping (see Hullings and Swyka, 1999).

As provocative as it may seem, it is conceivable to grade the cover in a concave (rather than convex) shape. Thus, the cover would provide a pool or reservoir to capture rain and snow, allowing it to percolate into the waste mass and thereby augment the recycling of leachate. Clearly, this concept represents a paradigm shift in our thinking, but it is technically feasible and should therefore be considered.

15.5.7 Waste Stability Concerns

The worst-case scenario in practicing leachate recycling with the goal of a bioreactor landfill is to add liquid to the point where the entire waste mass becomes unstable and fails. Unfortunately, this has occurred on at least two occasions (see Koerner and Soong, 1999).

The first case history (called "L-4") was a codisposed municipal and hazardous waste landfill that failed in 1997. The failure involved approximately 300,000 m^3 of waste. This case history has not appeared in the published literature, and the following discussion is based on a preliminary report to the site owner and personal communications with various parties involved.

Beneath the recently constructed section of a large landfill, a CCL was placed over a geotextile/geomembrane/geotextile liner system (a very atypical liner system). Active leachate recycling was being practiced at the site. The failure occurred after a period of 48 hours of rainfall. The failure was very abrupt, and the waste actually liquefied. The failure surface was the sloping old-to-recently placed waste interface. This consisted of liquid waste saturating a series of edge-control wood bark berms. From here, the failure surface transitioned to the liner system beneath the waste with the actual slip surface being the upper geotextile-to-geomembrane interface. See Figure 15.5 for an aerial photograph of the site and the profile of the critical 2-D cross section. The triggering mechanism was felt to be excessive leachate placement into the already-saturated wood bark between the old and the recent sections of the landfill.

The second case history (called "L-5") was a municipal solid waste landfill that failed in 1997. It involved approximately 1,200,000 m^3 of waste. The leachate management practice at the site involved pressure injection of leachate into the waste using perforated deep wells at several locations. This case history also has not appeared in the published literature, and the following discussion is based on personal communications with various parties involved.

The 200-ha landfill was developed in the late 1980s. Waste placement operations on a geomembrane/CCL-lined section began in late 1995. This section was the area where the failure eventually occurred. The mobilized slide lasted for approximately 20 minutes, and the failure mass moved downslope approximately 1,500 m in distance. Based on field observations, the failure surface was estimated to have passed through the waste mass at a very steep inclination down to the top of the composite liner. From this point, the failure surface extended along the sand-to-geomembrane interface until it exited at the toe of the slope. See Figure 15.6 for the plan view of the site and the

596 Chapter 15 Bioreactor Landfills

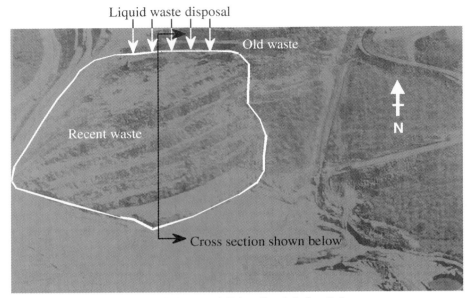

(a) Aerial Photograph Taken Shortly before Failure

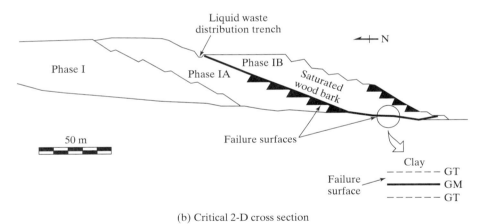

(b) Critical 2-D cross section

FIGURE 15.5 Aerial Photograph and Critical Cross Section of Landfill in Case History L-4

profile of the critical 2-D cross section. The triggering mechanism was thought to be the increase in leachate head within the waste mass due to the aggressive leachate injection operations.

15.6 PERFORMANCE-TO-DATE

A modern landfill represents a truly massive structure. To experiment at the full-scale level represents a challenge of similar proportions. As a result, there are very few

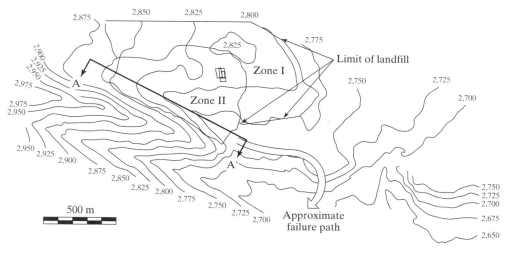

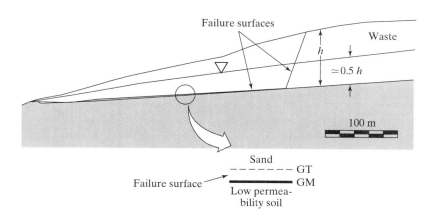

(b) Critical 2-D cross section (A-A') showing estimated leachate level prior to failure

FIGURE 15.6 Plan View and Profile of the Critical 2-D Cross Section of Case History L-5

leachate recycling/bioreactor landfills that are being monitored and reported in the open literature. Even further, most of the projects are being sponsored or funded by the owner and/or operator. Those reported in the literature are general leachate recycled cases, although a few have added sewage sludge and additional water. Reinhart and Townsend (1998) report on 11 on-going "bioreactor landfill" studies. Of the group, 10 practice leachate recycling, 5 use or add sewage sludge, 1 adds water and 1 injects air. Monitoring is usually via leachate and gas generation. Settlements are sometimes measured, as is microbial activity and the amount of degradation. In the latter cases (i.e., measurement of microbes and degradation), the method is unknown. The authors feel that 11 bioreactor landfills represents a rather sparse amount of data, particularly since the concept of bioreactor landfills has been known for over 25 years.

As mentioned in Section 15.2.4, there were 130 landfills practicing leachate recycling in 1997. It is felt that at least qualitative data should be more available than the 11 cases mentioned earlier. Regarding quantitative data, the situation is very poor. To the author's knowledge, modern instrumentation—such as piezometers, observation wells, in-situ gas monitors, in-situ temperature monitors, etc—has either not been used or has not been reported in the literature.

15.7 SUMMARY COMMENTS

After 25 years of experimentation with waste degradation by many researchers, it is abundantly clear that municipal solid waste held at or above field capacity and near neutral pH conditions will rapidly decompose. This decomposition will result in large amounts of methane and carbon dioxide, as well as the ultimate leachate being significantly less threatening to the environment. The time frame can be as few as 5-years, although 10 to 20 years might be required to decompose all zones within a large landfill. The concept is certainly appealing to everyone involved in MSW landfilling technology. The quest for zero long-term maintenance clearly is a goal worth pursuing.

This chapter focused mainly on the liquids management method of leachate recirculation. It has attempted to bring out the realization that the optimum waste degradation moisture content, called *field capacity*, invariably requires the addition of moisture beyond that available from the site-specific cell or landfill. In some cases, it requires significant additions of liquids. This most logically comes from adding microbial-containing liquids, such as sewage sludge and previously decomposed waste (e.g., from landfill mining of old cells). Other additions of liquids can come from industrial gray water, locally captured rainwater and snow melt, or from local streams and water courses. However, both leachate recycling and the addition of other liquids pose operations and design challenges.

On an *operations* basis, the distribution of the liquid onto, or into, the waste is not a trivial matter. All of the waste mass must eventually be at or above field capacity and have the proper environment for the microbes to act. Various methods have been presented in this chapter. Shredded waste has been suggested, but the costs often are prohibitive. Also, the possibility of using two or more liquid introduction methods was discussed. Whatever method is used, preferential flow paths must be avoided so all of the waste has adequate liquid available throughout the degradation process.

As for design challenges, there are also concerns, but they can be handled with proper planning and implementation. This chapter includes sections addressing the following topics, along with the related recommendations in parentheses:

- *Liner system integrity* (GM/GCL composite liners have performed admirably and should be considered in addition to GM/GCL/CCL composites with the CCL component being on the order of 1.0×10^{-5} cm/sec).

- *Leachate collection soil* (must be of high permeability—for example, greater than 1.0 cm/sec such as a gravel—and can use a superposition of sand placed over a geosynthetic drainage material).
- *Leachate removal system* (concern is over pipe connections and deformations under very high landfills).
- *Filter and operations layers* (it is recommended that both of these layers be eliminated).
- *Daily cover materials* (it is recommended that alternate materials be used to provide free flow between lifts of waste).
- *Final cover issues* (it is recommended that placement be postponed until after primary settlement is completed—that is, 5 years or longer).
- Waste stability concerns (concern is over high hydraulic heads leading to instability).

At this point, it is felt that more data from existing landfills are needed to assess the general performance and particular nuisances of bioreactor landfills. Further, these landfills must be monitored using currently available instrumentation. While quantities of gas and leachate are good indications that the intended processes are working, much more is necessary to control and assess the situation. In this regard, liquid monitoring piezometers, gas monitoring probes, temperature probes, and settlement monitoring at discrete elevations throughout the waste mass are recommended. Realizing this is beyond the financial capability of most owners and operators, however, it seems as though Federal agencies, state agencies, and large waste authorities should be showing leadership in this regard. Obviously, a mixture of public and private funds also is a possibility.

Leachate recycling leading to bioreactor landfills is an exciting and worthwhile pursuit that can revolutionize current practice. What is needed is the will and financial resources to justify the practice and corroborate the relatively sparse field information that has been developed to date.

PROBLEMS

15.1 Other than scale, how does a bioreactor landfill differ from the degradation and decomposition of compost as practiced in the backyards of many homes?

15.2 In the case of a bioreactor landfill, describe the nature of the final (end-point) products insofar as the following:
  (a) the residual solids left in the landfill;
  (b) the leachate at the end of the process; and
  (c) the remaining gases left in the landfill.

15.3 What factors could negatively influence the time for "complete" degradation as optimally proposed in Figure 15.1?

15.4 Wellpont extraction of landfill gas along with leachate injection using shallow or deep wells can be somewhat problematic. What is the main concern and how can it be resolved?

15.5 Replot the data of Table 15.5 for minimum and maximum flows in a manner similar to that presented in Figure 5.27 for the average data. Show the "points" as a bar with maximum, average, and minimum values indicated for each type of situation.

15.6 Recalculate Example 15.1 for different protrusion heights varying from 6 mm to 100 mm and plot the corresponding response curve.

15.7 In Germany, the minimum mass per unit area geotextile for puncture protection is 2,000 g/m^2. Using the same data as given in Example 15.1 backcalculate to determine the *FS*-value using this type of geotextile. Then, redo the problem again using a 3,000 g/m^2 geotextile, as that is sometimes used.

15.8 It has been suggested that leachate can exist in landfills in five separate ways: (a) discontinuous leachate, (b) perched leachate, (c) leachate head on liner, (d) leachate head above gas or liner, and (e) leachate under excess pore pressure. Sketch what these different scenarios look like. [Hint: See reference by Koerner and Soong (2000)]

15.9 For the above different leachate scenarios, describe how each of them influences the stability of the waste mass.

15.10 Regarding the instrumentation of bioreactor landfills (or for that matter standard types of landfills), describe what the following types do and how they function.

  (a) piezometers;

  (b) slope indicators;

  (c) settlement anchor points.

15.11 For each of the landfill sketches shown in Figure 13.1(d), (e) and (f), sketch and describe an instrumentation monitoring program if instability is a concern.

REFERENCES

Baccini, P., (1998) "The Landfill: Reactor and Final Storage," Lecture Notes in the Earth Sciences No. 20 Proceedings, Workshop on Land Disposal of Solid Wastes, Gersenzee, Switzerland, March 14–17, 1988, Springer-Verlag, New York.

Barlaz, M. A., Ham, R. K., and Schaefer, D. M., (1990) "Methane Production from Municipal Refuse: A Review of Enhancement Techniques and Microbial Dynamics," Critical Reviews in Environmental Control, Vol. 19, No. 6, CRC Press Inc., pp. 557–584.

Bonaparte, R., Daniel, D. E., and Koerner, R. M., (2001) "Assessment and Recommendations for Optimal Performance of Waste Containment Systems," Final Report, Grant No. CR-821448, U. S. EPA, Cincinnati, OH (in press).

Cheremisinoff, P. N. and Morresi, A. C., (1976) *Energy from Solid Wastes,* Marcel Dekker Inc., New York.

Dwyer, S. F., (1995) "Alternative Landfill Cover Demonstration," *Landfill Closures Environmental Protection and Land Recovery,* Geotechnical Special Publication No. 53, R. J. Dunn and U. P Singh, Eds., ASCE, pp. 19–34.

Filip, A. and Kuster, E., (1979) "Microbial Activity and the Turnover of Organic Matter in a Municipal Refuse Disposed of in a Landfill," *European Jour. Appl. Microbiol. Biotechnol.,* Vol. 7, pp. 371–379.

Fungaroli, A. A. and Steiner, R. L., (1979) *Investigation of Sanitary Landfill Behavior,* EPA/600/2-79, 312 pgs.

Heintrich, R. L., Swartzbaugh, J. T., and Thomas, J. A., (1979) "Influence of MSW Processing on Gas and Leachate Production," *Proceedings of the Fifth Annual Research Symposium on Municipal Solid Waste Land Disposal,* EPA-600/9-79-023a.

Hullings, D. E. and Swyka, M. A., (1999) "Geosynthetics in Bioreactor Designs," *Proc. GRI-13 Conference,* GSI, Folsom, PA, pp. 254–262.

Koerner, G. R., Koerner, R. M., and Martin, J. P., (1994) "Geotextile Filters Used for Leachate Collection Systems: Testing, Design, and Field Behavior," *Journal of Geotechnical Engineering,* ASCE, Vol. 120, No. 10, Oct. 1994, pp. 1792–1803.

Koerner, J. R., Soong. T.-Y., and Koerner, R. M., (1998) "A Survey of Solid Waste Landfill Liner and Cover Regulations: Part I—USA Status," GRI Report #21, Folsom, PA, 211 pgs.

Koerner, R. M., (1998) *Designing with Geosynthetics,* 4th Edition, Prentice Hall Publ. Co., Englewood Cliffs, NJ, 761 pgs.

Koerner, R. M. and Daniel, D. E., (1997) "Final Covers for Solid Waste Landfills and Abandoned Dumps," ASCE Press, Reston, VA, 256 pgs.

Koerner, R. M. and Soong, T.-Y., (1999) "Assessment of Ten Landfill Failures Using 2-D and 3-D Stability Analysis Procedures," *Proc. 2nd Austrian Geotechnical Conference,* H. Brandl, Ed., Vienna, Austria, pp. 1–41.

Koerner, R. M. and Soong, T.-Y., (2000) "Leachate in Landfills: The Stability Issues," *Jour. Geotextiles and Geomembranes,* Vol. 18, No. 5, pp. 293–309.

Lee, G. F. and Jones, R. A., (1991a) "Landfills and Ground-Water Quality" Guest editorial, *Journ. of Ground Water,* Vol. 29, pp. 482–486.

Lee, G. F. and Jones, R. A., (1991b) "Managed Fermentation and Leaching: An Alternative to MSW Landfills," Biocycle, Vol. 31, No. 5, pp. 78–83.

Lee, G. F. and Jones, R. A., (1992) "Municipal Solid Waste Management in Lined, 'Dry-Tomb' Landfills: A Technologically Flawed Approach for Protection of Groundwater Quality," Report published by G. Fred Lee & Associates, El Macero, CA, 66 pp.

Pacey, J. P., Yazdani, R., Reinhart, D., Morck, R., and Augenstein, D., (1998) The Bioreactor Landfill: An Innovation in Solid Waste Management, SWANA Position Paper.

Pacey, J. P., (2000) personal communication.

Parsons, J., (2000) "Innovative Pour Applied Polyurea Liner Repair," *Proc. Great Lakes Region Solid Waste Management Conference,* Engr. Soc. Detroit, MI, March 15, 2000, 16 pgs.

Pohland, F. and Graven, J. P., (1993) "The Use of Alternative Materials for Daily Cover at Municipal Solid Waste Landfills," US EPA Report EPA/600/R-93/172, US Environmental Protection Agency, Washington, DC.

Pohland, F. G. and Harper, S. R., (1986) "Critical Review and Summary of Leachate and Gas Production from Landfills," EPA/600/2-86/073, US Environmental Protection Agency, Cincinnati, OH, 212 pgs.

Reinhart, D. R., McCreanor, P. T., and Townsend, T., (2000) "The Bioreactor Landfill: Research and Development Needs," *Proceedings of GRI-14 Conference on Hot Topics in Geosynthetics-I,* Geosynthetic Information Institute, Folsom PA, pp. 1–8.

Reinhart, D. R. and Townsend, T. G., (1998) *Landfill Bioreactor Design and Operation,* Lewis Publishers, CRC Press LLC, Boca Raton, 189 pgs.

Spikula, D. R., (1996) "Subsidence Performance of Landfills," *Proc. GRI-10 Conference on Field Performance of Geosynthetics and Geosynthetic Related Systems,* GSI, Folsom, PA, pp. 237–244.

Thiel, R., (1998) "Design Methodology for a Gas Pressure Relief Layer Below a Geomembrane Landfill Cover to Improve Slope Stability," Geosynthetics International, Vol. 5, No. 6, pp. 589–617.

US Environmental Protection Agency, (1995) "Landfill Bioreactor Design and Operation," EPA/600/R-95/146, Seminar Publication, Risk Reduction Engineering Laboratory, Cincinnati, OH, 230 pgs.

US Environmental Protection Agency, (1993) "Leachate Recirculation for Remedial Action at MSW Landfills," Final Report, Risk Reduction Engineering Laboratory, Cincinnati, OH, 230 pgs.

US Environmental Protection Agency, (1981) "Landfill Methane Utilization Technology Workbook," Contract 31-109-38-5686, J. L. Baron and R. C. Eberhart Eds., CPE-8101, 122 pgs.

CHAPTER 16

Construction of Compacted Clay Liners

16.1 SUBGRADE PREPARATION
16.2 SOIL MATERIALS FOR COMPACTED CLAY LINERS
16.3 COMPACTION OBJECTIVES AND CHOICES
    16.3.1 DESTRUCTION OF SOIL CLODS
    16.3.2 MOLDING WATER CONTENT
    16.3.3 DRY UNIT WEIGHT
    16.3.4 TYPE OF COMPACTION
    16.3.5 COMPACTIVE ENERGY
    16.3.6 LIFT INTERFACES
16.4 INITIAL SATURATION SPECIFICATIONS
16.5 CLAY LINER COMPACTION CONSIDERATIONS
16.6 COMPACTION SPECIFICATIONS
16.7 LEACHATE COLLECTION TRENCH CONSTRUCTION
16.8 PROTECTION OF COMPACTED SOIL
    16.8.1 PROTECTION AGAINST DESICCATION
    16.8.2 PROTECTION AGAINST FREEZING
    16.8.3 EXCESS SURFACE WATER
16.9 FIELD MEASUREMENT OF WATER CONTENT AND DRY UNIT WEIGHT
16.10 CONSTRUCTION QUALITY ASSURANCE AND QUALITY CONTROL
    16.10.1 CRITICAL QUALITY ASSURANCE AND QUALITY CONTROL ISSUES
    16.10.2 QUALITY ASSURANCE AND QUALITY CONTROL FOR COMPACTED CLAY LINER CONSTRUCTION
    16.10.3 DOCUMENTATION REPORT
    PROBLEMS
    REFERENCES

This chapter describes the proper construction and placement of a landfill compacted clay liner (CCL). The success of a CCL depends on the correct selection of soil materials, careful evaluation of engineering properties of selected soil materials, and adequate quality assurance and quality control during construction. It also depends on the protection of the CCL until an adequate cover of the intended overlying materials are properly placed.

    A well-constructed landfill liner protects groundwater from leachate contamination in perpetuity. Clay is the most important constituent of a soil liner because of its low hydraulic conductivity and sorptive properties. On the other hand, clay is a difficult engineering material to work with because of its highly moisture-dependent physical properties. A properly constructed CCL consists of a minimum 2-foot- (0.6-m)-thick layer of compacted clay with a hydraulic conductivity of less than 1.0×10^{-7} cm/sec. To meet these requirements, the following steps should be taken during the construction of a compacted clay liner:

(i) Select suitable soil materials.
(ii) Meet moisture-density criteria.
(iii) Break up soil clods.
(iv) Conduct proper compaction.
(v) Eliminate lift interfaces.
(vi) Avoid desiccation.

The factors that most influence the design and construction of a compacted soil liner to meet a permeability performance standard of 1.0×10^{-7} cm/sec are listed in Table 16.1. Regulations worldwide call for the CCL to have a hydraulic conductivity of 1.0×10^{-7} cm/sec (or lower). This value is often contested, but it appears to have technical merit. Above this threshhold value the predominant flow is advective,

TABLE 16.1 Key Factors that Influence Permeability of Compacted Clay Liners (Elsbury, et al., 1990)

Principal Group	Key Factors
Design Stage	
Soil Type	Workability
	Gradation
	Swell Potential
Other Considerations	Overburden Stress
	Liner Thickness
	Foundation Stability
Construction Stage	
Basic Compaction Objectives	Destruction of Clods
	Interlift Bonding
Essential Choices (to Achieving the Basic Compaction Objectives)	Lift Thickness
	Water Content of Soil
	Type and Weight of Roller
	Number of Passes and Coverages
	Size of Clods
Supporting Elements (That are included in or subsidiary to essential choices)	Dry Density
	Degree of Saturation
Other Considerations	Soil Preparation
	Construction Quality Assurance
Postconstruction Stage	
Environmental Influences	Desiccation
	Freezing

Used with permission of ASCE.

while below such a value the flow is primarily diffusive. Thus, the criteria of 1.0×10^{-7} cm/sec is a good choice for the maximum allowable hydraulic conductivity for a CCL.

16.1 SUBGRADE PREPARATION

The subgrade on which a CCL is placed should be properly prepared. The subgrade must be able to provide adequate support for the maximum static loading of the landfill and various dynamic loads of construction and placement equipment. A compacted clay liner may be placed on natural soil, free draining granular material, geosynthetic material, or waste material, depending on the particular design and individual components in the liner or cover system.

If a CCL is the lowest component of the liner system, native soil generally forms the subgrade. In such a case, the subgrade should be compacted to eliminate any soft spots. Water should be added or removed as necessary to produce a suitably firm subgrade per specification requirements. If the natural soil subgrade is not compacted, it may be difficult to compact the first one or two lifts of the soil liner to 90 or 95% modified Proctor density. The natural soil subgrade should be compacted to the same degree as the CCL itself. If the subgrade material is sandy then a different compaction procedure should be followed. The Proctor density versus moisture relationship is not directly applicable for coarsely graded materials (ASTM D698 and ASTM D1557). Instead a relative density should be specified and the standard relative density test (ASTM D4253) should be used. A sandy subgraded should be compacted to 85 to 90% relative density.

The density should be checked at 100-feet (30-m) grid points. A smaller or larger grid may be chosen (Bagchi, 1994), depending on the waste type, soil properties, and reliability of the contractor. A vibratory roller may be used for compacting sandy soil. However, if the sand is already in a dense condition, application of vibration will loosen the sand. Therefore, the in situ density of the sand should be checked prior to choosing the compaction equipment. A nuclear device (ASTM D3017) or other standard method [e.g., sand cone (ASTM D1556) or rubber balloon (ASTM D2167)] can be used to determine the in-place density.

In other instances, a CCL may be placed on top of geosynthetic components of the liner system (e.g., a geotextile). In this case, the main concern is the smoothness of the geosynthetic material on which soil is placed and conformity of the geosynthetic material to the underlying material (e.g., no bridging should occur over ruts left by vehicle traffic).

Consolidation and rebound characteristics of the subbase should be evaluated during design of the landfill site. If the landfill subbase is predominantly sandy then checking volume stability is less important than it is for a clayey subbase. The maximum possible overburden pressure for a modern municipal solid waste landfill can be very high. A good example of a huge modern landfill is the Eagle Mountain Landfill near Palm Springs, California. If permits are finalized and construction proceeds as planned, it will be approximately 800 ft (240 m) high. If the unit weight is assumed to be 70 lb/ft^3 (11 kN/m^3), the overburden pressure on the subbase of this landfill will be 56,000 lb/ft^2 (8100 kPa). While this landfill will be on rock subgrade, large landfills on soil should have settlements calculated. Also, rebound of the subbase due to excavation of overburden should be checked if the subbase is either clayey or, a thick clayey stratum is identified a short distance below the subbase grade. Rebound of the subbase may cause uneven heave of the liner resulting in failure of the leachate collection system. When estimating rebound, the weight of the liner may be considered as a weight-restricting rebound if quick construction of the liner is planned. On the other hand, the total weight of the waste should not be taken into account because too much time may elapse between liner construction and waste placement up to the final design height for total rebound to take place (Bagchi, 1994). This issue should be investigated during landfill design whenever necessary.

Rock subgrades beneath the CCL present an entirely different set of circumstances. If relatively flat, rock surfaces make the ideal subgrade. If undulating or jagged, the voids must be carefully backfilled with soils compacted to a relative density of 95% of standard Proctor. The most troublesome rock type is limestone; particularly limestones which are karstic by nature. Potential sinkholes are most troublesome from a design perspective. To be sure, geosynthetic reinforcement (as discussed in Section 14.4) could be employed, but the estimated void opening is a troublesome issue and one that requires active research and investigation.

16.2 SOIL MATERIALS FOR COMPACTED CLAY LINERS

The primary requirement for a compacted clay liner material is that it be compacted to produce a very low hydraulic conductivity. The following selection criteria apply:

Fines. The soil should contain at least 20 percent of silt or clay sized material (i.e., "fines"). Fines herein are defined as the percentage, on a dry-weight basis, of material passing the No.200 sieve that has openings of 0.075 mm.

Plasticity Index. The soil should have a minimum plasticity index of at least 10%, although some soils with a slightly lower plasticity index may be suitable. Soils with plasticity indices less than about 10 percent have very little clay and usually will not produce the necessary low hydraulic conductivity. Soils with plasticity index greater than 30 to 40% are difficult to work with, as they form hard chunks when dry and sticky clods when wet, which make them difficult to work with in the field. Such soils also tend to a have high shrink/swell potential and poor volume stability. Soils with plasticity indices between approximately 10 and 35% are generally ideal.

Percentage of Gravel. The percentage of gravel (defined as material retained on the No.4 sieve, which has openings of 4.76 mm) should not be excessive. A maximum amount of 10% gravel by weight is reasonable. For many soils, however, larger amounts may not necessarily be deleterious if the gravel is uniformly distributed in the soil and does not interfere with compaction by footed rollers. For example, Shakoor and Cook (1990) and Shelley and Daniel (1993) found that the hydraulic conductivity of a compacted, clayey soil was insensitive to the amount of gravel present, as long as the gravel content did not exceed 50 to 60%. Gravel is only deleterious if the pores between gravel particles are not filled with clayey soil, and the gravel forms a continuous pathway through the liner. The key problem to be avoided is segregation of gravel in pockets that contain little or no fine-grained soil.

Stones and Rocks. No stones or rocks larger than 1 to 2 inches (25 to 50 mm) in diameter should be present in the soil liner material.

Soil preparation includes blending and homogenizing the soil, removing gravel or other off-specification materials, and adding or removing water to achieve the desired water content. An important purpose for blending is to achieve reasonably uniform moisture content. Blending the soil on site with a pulverizing mixer is helpful in reducing clod size and producing a more uniform moisture content.

16.3 COMPACTION OBJECTIVES AND CHOICES

Destruction of soil clods and good bonding between lifts are basic compaction goals. Roller type and weight, compaction energy, lift thickness, lift interface, and water content are key considerations in achieving these compaction goals.

16.3.1 Destruction of Soil Clods

Two theories (the particle-orientation theory and the clod theory) have been advanced (Elsbury et al., 1990) to explain the flow of water through compacted clayey soils.

The particle-orientation theory proposed by Lambe (1958a) relates the hydraulic conductivity of compacted soil to the orientation of the soil particles. Lambe suggested that the soil particles are oriented in a so called "flocculated" pattern when soil is compacted dry of optimum water content and in a "dispersed" pattern when soil is compacted wet of optimum water content. The relationship between particle orientation and water content is depicted in Figure 16.1. Lambe (1958a) concluded that the hydraulic conductivity of soil compacted wet of optimum is less than the hydraulic

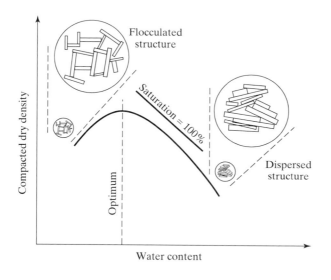

FIGURE 16.1 Effects of Water Content on Soil Structure (Herrmann and Elsbury, 1987). Used with permission of ASCE.

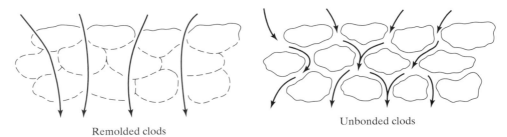

FIGURE 16.2 Influence of Remolding of Soil Clods on Permeability of Lift of Compacted Clay (Herrmann and Elsbury, 1987). Used with permission of ASCE.

conductivity of soil compacted dry of optimum because soil with a flocculated particle structure has larger average pore sizes than a soil with a dispersed structure. Thus, according to the particle-orientation theory, the arrangement of individual particles, which is influenced by molding water content, controls the hydraulic conductivity.

Studies of laboratory-compacted soils have shown that the presence of clods and interclod pores have a major influence on permeability. The clod theory proposes that most of the water flow that takes place in compacted clay occurs in the relatively large pore spaces located between clods of clay, rather than through the microscopic pore spaces among soil particles within the clods themselves, as shown in Figure 16.2. Daniel (1990) described the influence of the fate and size of clods on the permeability of liners in the field. He noted that "to achieve low hydraulic conductivity, it is necessary to destroy the clods and to eliminate large interclod pores ... either by wetting the soil to a high water content or using a large compactive effort."

16.3.2 Molding Water Content

The degree of saturation of a clay soil liner material at the time of compaction is perhaps the single most important variable that controls the engineering properties of the compacted material. The influence of molding water content and compactive effort on hydraulic conductivity of compacted clay soils is shown in Figure 16.3. Soils compacted at water contents less than optimum (dry of optimum) tend to have a relatively high hydraulic conductivity; soils compacted at water contents greater than optimum (wet of optimum) tend to have a low hydraulic conductivity and low shear strength. For some soils, the water content relative to the plastic limit may indicate the degree to which the soil can be compacted to yield low hydraulic conductivity. In general, if the water content is greater than the plastic limit, the soil is in a plastic state and should be amenable to remolding to a low-hydraulic-conductivity material (USEPA, 1993). Soils with water contents dry of the plastic limit will exhibit very little "plasticity" and thus may be difficult to compact into a low-hydraulic-conductivity mass without delivering enormous compactive energy to the soil.

It is usually preferable to compact the soil wet of optimum to minimize hydraulic conductivity. However, the soil must not be placed at a water content that is too high, or other properties may be adversely affected. For example, the shear strength may be too low, soil placement and handling may be more difficult, the compaction plant may

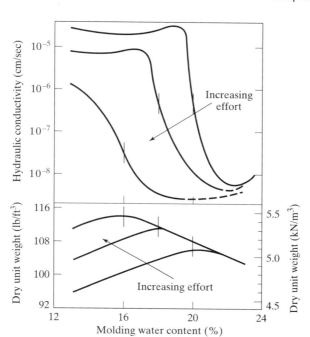

FIGURE 16.3 Effects of Compactive Effort on Maximum Dry Unit Weight and Hydraulic Conductivity of Compacted Soils (USEPA, 1989)

malfunction, desiccation cracks can appear if the soil dries, and ruts may form when construction vehicles pass over the liner. Figure 16.4 shows that the relationship between the optimum water content and maximum dry unit weight can be represented with a simple equation nearly identical to the relationship reported by Blotz et al. (1998). This equation has the following forms:

In U.S. units,

$$\gamma_{dmax} = 7 \cdot w_o^{-0.34} \tag{16.1}$$

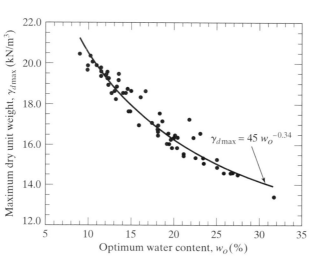

FIGURE 16.4 Maximum Dry Unit Weight versus Optimum Water Content for Soils (Benson et al., 1999). Used with permission of ASCE.

$\gamma_{dmax} = 45 \, w_o^{-0.34}$

where γ_{dmax} = maximum dry unit weight of soil, lb/ft³, and
w_o = optimum water content of soil, %.

In SI Units,

$$\gamma_{dmax} = 45 \cdot w_o^{-0.34} \qquad (16.2)$$

where γ_{dmax} = maximum dry unit weight of soil, kN/m³, and
w_o = optimum water content of soil, %.

16.3.3 Dry Unit Weight

The dry unit weight of a compacted soil is widely perceived as an important parameter in the construction of compacted clay liners. Construction specifications commonly require (Elsbury et al., 1990) that the dry unit weight of the liner exceed either 95% of the maximum dry unit weight measured with standard Proctor compaction or 90% of the maximum dry unit weight measured with modified Proctor compaction. An important note of caution is in order here. Although a target dry density is frequently specified, a high dry density, per se, is not the main purpose of compaction. Instead the main purpose of compaction is to achieve low hydraulic conductivity, preserve a reasonable shear strength, and minimize volume changes.

16.3.4 Type of Compaction

Experience-based laboratory results have shown (USEPA, 1993) that the type of compaction can affect hydraulic conductivity (e.g., as shown in Figure 16.5). Kneading the soil helps to break down clods and remold the soil into a homogenous mass that is free of voids or large pores. A kneading type compaction is particularly beneficial for highly plastic soils. For certain bentonite-soil blends that do not form clods, kneading

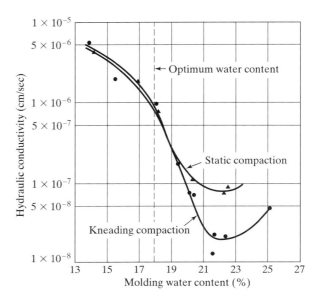

FIGURE 16.5 Effect of Type of Compaction on Hydraulic Conductivity (Mitchell et al., 1965). Used with permission of ASCE.

FIGURE 16.6 A "Footed" Compaction Roller

is not necessary. Most soil liners are constructed with "footed" rollers. Figure 16.6 shows a "footed" or cleated compaction roller. The "feet" on the roller penetrate into a loose lift and knead the underlying soil with repeated passages of the roller. The dimensions of the feet on rollers varies considerably. Footed rollers with short feet ($\approx$ 3 inches or 75 mm) are called "pad foot" rollers; the feet are said to be "partly penetrating" because the foot is too short to penetrate fully a typical loose lift of soil. Footed rollers with long feet ($\approx$ 8 inches or 200 mm) are often called "sheepsfoot" rollers; the feet fully penetrate a typical loose lift. Figure 16.7 contrasts rollers with partly and fully penetrating feet.

16.3.5 Compactive Energy

Another important variable controlling the engineering properties of soil liner material is the compactive energy. As shown in Figure 16.3, increasing the compactive energy increases the dry unit weight of the soil, decreases the optimum water content, and reduces hydraulic conductivity. The hydraulic conductivity of a soil that is compacted wet of optimum could be lowered by one to two orders of magnitude by increasing the compactive energy, even though the dry unit weight of the soil is not increased measurably. More compactive energy helps to remold clods of soil, realign soil particles, reduce the size or degree of connection of the largest pores in the soil, and lower hydraulic conductivity. With respect to type of compaction, Mitchell et al.

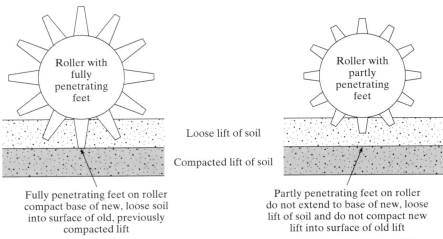

FIGURE 16.7 Footed Rollers with Partly and Fully Penetrating Feet (USEPA, 1993)

(1965) found that kneading compaction produced lower permeability than other compactive methods. This suggests that footed rollers are preferred because the feet remold the clods better.

The specific compactive energy (e.g., lb-ft/ft^3 or kN-m/m^3) delivered to soil depends on the weight of the roller, the number of passes of the roller over a given area, and the thickness of the soil lift being compacted. Increasing the weight and number of passes can increase the compactive effort. The best combination of these factors to use when compacting low hydraulic conductivity soil liners depends on the water content of the soil and the firmness of the subbase.

Heavy rollers cannot be used if the soil is very wet or if the foundation is weak and compressible [e.g., if municipal solid waste is located just 1 to 2 ft (300 to 600 mm) below the layer to be compacted for a landfill cover]. Rollers with static weights of at least 30,000 to 40,000 lb (13,600 to 18,100 kg) are recommended for compacting low hydraulic conductivity layers in a cover system. Rollers that weigh up to 70,000 lb (32,000 kg) are available and may be desirable for compacting bottom liners of landfills, but such rollers are too heavy for many cover systems because of the presence of compressible waste material a short distance below the cover. The roller must make a sufficient number of passes over a given area to ensure adequate compaction. The minimum number of passes will vary, but at least 5 to 10 passes are usually required to deliver sufficient compactive energy and to provide adequate coverage (USEPA, 1991).

16.3.6 Lift Interfaces

Clay liners are constructed by compacting the clay in horizontal layers commonly called lifts. During construction of a compacted clay liner, if a new lift is applied directly to the unscarified surface of a previous lift, a zone of higher permeability and

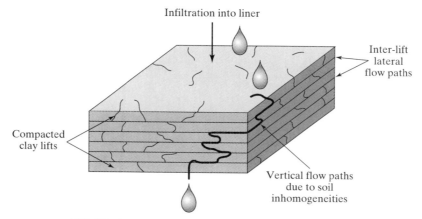

FIGURE 16.8 Flow Pathway Created by Poorly Bonded Lifts

low strength forms at the interface. Moisture moving through the liner could then spread quickly across this interface to a lower lift (Figure 16.8). To avoid formation of interface flow paths, the surface of the previous lift should be scarified before another compacted clay lift is added. When soil is scarified it is usually roughened to a depth of about 1 inch (25 mm). If the soil is scarified, the scarified zone becomes part of the loose lift of soil and should be counted in measuring the loose lift thickness.

At some point, there is a maximum lift thickness beyond which a given roller cannot remold and bond the clods or achieve good interlift bonding for a given water content. It is common practice to use a loose lift thickness of 8 to 9 inches (200 to 300 mm) and a compacted lift thickness of 6 inches (150 mm) (Herrmann and Elsbury, 1987). A fully penetrating sheepsfoot roller can be used to compact a new lift and bond it intimately to the previous lift. "Fully penetrating" means that the height of the feet on the compaction wheels are greater than the thickness of the loose soil placed to form the new lift, so the feet can fully penetrate the loose lift (including loose soil at the top of the old lift) and fully compact the lift interface (Figure 16.7).

16.4 INITIAL SATURATION SPECIFICATIONS

A minimum initial degree of saturation can be used to define acceptable compaction conditions because the initial degree of saturation depends upon both compaction water content and dry unit weight, and contours of constant initial degree of saturation are approximately parallel to the line of optimums (e.g., Boutwell and Hedges, 1989; Benson and Boutwell, 1992; and Benson et al., 1994b). Thus, lower hydraulic conductivity should be obtained at a higher initial degree of saturation.

The data shown in Figure 16.9 exhibit an approximate trend of decreasing field hydraulic conductivity with increasing initial degree of saturation. Because the specific gravity of solids was unknown for most field testing sites, it was back-calculated

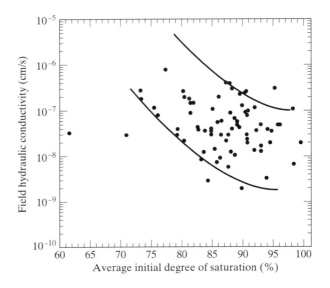

FIGURE 16.9 Field Hydraulic Conductivity versus Initial Degree of Saturation (Benson et al., 1999). Used with permission of ASCE.

(Benson et al., 1999) from the optimum water content and maximum dry unit weight assuming the initial degree of saturation at optimum was 85%. Consequently, the initial degrees of saturation shown in Figure 16.9 are estimates. There is significant scatter in the field hydraulic conductivity versus average initial degree of saturation because the true values of specific gravity were not available, and small changes in specific gravity cause larger changes in initial degree of saturation.

16.5 CLAY LINER COMPACTION CONSIDERATIONS

A laboratory study has shown that increasing clod size from 1/16 to 3/9 inch (1.6 to 9.5 mm) increased the permeability of the clay by a factor of 30 (Daniel, 1984). Thus, it is important to reduce clod size. Allowable clod sizes suggested by design engineers vary between 1 inch (25 mm) and the thickness of the lift (Goldman et al., 1986). The clod size becomes even more important if the preconstruction moisture content of the soil is less than the optimum moisture at which compaction is proposed. Wetting of clay takes time. If the clod size used in the field for permeability testing is larger than that used in the laboratory, the entire clay mass may not attain the same moisture content within the short time period that commonly occurs in the field.

Adequate time should be allowed to ensure uniform distribution of moisture when water is added to a soil in the field. Nonuniform moisture distribution in a clay liner can be caused by large clod sizes, uneven water distribution by sprinklers, and insufficient "curing time" allowed for moisture penetration (Ghassemi et al., 1983). The moisture content of compacted portions of the liner should be maintained or preserved during inactive periods to prevent drying (which may lead to desiccation cracks) or overwetting (which requires long drying periods). Moisture content can be maintained by "proof rolling" the liner with a rubber tire, smooth steel drum vehicle, or covering the liner with a plastic sheet. Once started, a clay liner project should be finished completely. If construction is halted due to onset of winter (in cold regions) then the top foot of the liner should be scarified before commencing again (Bagchi, 1994).

Three main types of rollers have been used for field compaction: sheepsfoot roller (self-propelled or towed), vibratory (smooth drum or sheepsfoot), and rubber tire. A nonvibratory sheepsfoot is generally recommended for constructing a clay liner because it provides a kneading action and has the ability to break down clods. A vibratory type roller or rubber tire roller may not provide the same degree of kneading action and particle orientation desired for constructing a low permeability liner. Vibration may temporarily decrease pore pressure within clay clods, which increases their shear strength; as a result more pressure will be needed to break down clods and construct a uniform, homogeneous well-compacted layer. Static compaction produces a flocculated structure with large pores (Mitchell et al., 1965). Comparisons of the effect of kneading versus static compaction at various water content on soil permeability show that the lowest permeability can be achieved by using kneading compaction at wet of optimum moisture (Mitchell et al., 1965). The use of a heavy weight sheepsfoot roller is generally preferred for compacting clay (Hilf, 1975) because of its kneading action, ability to break up clods, and deep penetration into a lift of soil. The size of the landfill should also be considered in choosing appropriate compaction plants. Because large compaction equipment needs a large turning radius, suitable smaller equipment should be selected for small landfills. However, small equipment may be incapable of providing the pressure required to achieve a low permeability liner. Thus, design of smaller clay-lined sites poses a construction-related problem, which should be considered during the planning stages (Bagchi, 1994).

Typically, the maximum allowable final (after-compaction) lift thickness is 6 inches (150 mm). Final elevation surveys should be used to establish thicknesses of completed earthwork segments. The specified maximum lift thickness is a nominal value. The actual value may be determined by surveys on the surface of each completed lift, but an acceptable practice is to survey the liner after construction and calculate the average thickness of each lift by dividing the total thickness by the number of lifts.

Sometimes it is necessary to "tie in" a new section of soil liner to an old one (e.g., when a landfill is being expanded laterally). A lateral excavation should be made about 10 to 20 ft (3 to 6 m) into the existing soil liner, and the existing liner stair-stepped as shown Figure 16.10 to tie the new liner into the old one. The surface of each of the steps in the old liner should be scarified (USEPA, 1993) to maximize bonding between the new and old sections. Figure 16.11 shows compaction of a primary clay liner over a secondary leachate drainage layer.

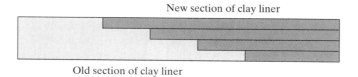

FIGURE 16.10 Stair-Step Tie-In of New Clay Liner to Existing Clay Liner

FIGURE 16.11 Compaction of a Primary Clay Liner over Secondary Leachate Drainage Layer

16.6 COMPACTION SPECIFICATIONS

One of the most contentious issues in quality assurance testing (Benson et al., 1999) is the degree to which hydraulic conductivity obtained from laboratory tests (k_L) conducted on undisturbed, 3-inch (75-mm) diameter specimens provide an accurate indication of field hydraulic conductivity (k_F). Several investigations have shown that for poorly built liners, k_L can be several orders of magnitude lower than k_F, whereas the two values are often similar for well-built liners (Daniel, 1984; Day and Daniel, 1985; Reades et al., 1990; Benson and Boutwell, 1992; Benson et al., 1994a).

Investigators (e.g., Benson and Boutwell, 1992; Benson et al., 1994a) have cited many reasons to account for differences between k_F and k_L. The concensus is that k_F and k_L are similar under the following conditions: (i) when evaluated at similar effective stress, and (ii) when the clay liner in the field is devoid of macroscopic features (e.g., macropores, cracks, fissures, and heterogeneous materials) or is at effective stresses high enough to close macroscopic features (Boynton and Daniel, 1985). Liners compacted wet of the line of optimums are usually devoid of macroscopic features and thus have low k_F. Under such conditions, only microscale pores are present to conduct flow, and they are readily represented in small laboratory-scale specimens. In contrast, macroscopic defects, such as interclod voids, macropores, cracks, fissures, etc., are not accurately represented nor simulated in small laboratory-scale specimens. Another possibility is that these void-volume defects are closed during sampling or testing.

Macroscopic defects are more common when construction practice is poor, and their presence results in high k_F, and as a result, k_F will be much greater than k_L (Benson et al., 1999).

Numerous laboratory studies over the past 40 years have shown that the hydraulic conductivity of compacted clay can vary several orders of magnitude from the dry to the wet side of the line of optimums (Lambe, 1954; Mitchell et al., 1965; Garcia-Bengochea et al., 1979; Acar and Oliveri, 1990; Benson and Daniel, 1990). In contrast, the composition (e.g., index properties) usually has a much smaller effect on hydraulic conductivity (Daniel, 1987; Benson et al., 1994b). The compaction water content is so influential that it masks the importance of the other variables (Benson and Trast, 1995).

To ensure that compacted clay liners will have low hydraulic conductivity, construction specifications typically require that the water content fall within a specified range (often 0–4% wet of optimum) and that the percent compaction equal or exceed a specified minimum (typically 90% of the maximum dry unit weight from modified Proctor compaction, or 95% of the maximum dry unit weight from standard Proctor compaction). The objective of this "percent compaction" type of specification is to achieve compaction wet of optimum water content. Unfortunately, this type of specification may unwittingly result in compaction dry of the line of optimums even if compaction is wet of a particular optimum water content, because the optimum water content also varies with compactive effort (Daniel, 1984; Daniel and Benson, 1990; Benson and Boutwell, 1992). Nevertheless, nearly all specifications in the past have been written in this manner (Benson et al., 1999). This failure to recognize that conventional compaction specifications, based on a minimum percent compaction and minimum water content, often lead to difficulties because this type of specification does not ensure that compaction will be wet of the line of optimums. Despite widespread publication of procedures that will avoid this problem (e.g., Daniel and Benson, 1990; Daniel and Koerner, 1995), many designers and specification writers continue to repeat the mistake. This type of inappropriate specification, still in common use, which is NOT recommended, is shown in Figure 16.12. The more appropriate and recommended approach, on the other hand, is shown in Figure 16.13 (Benson

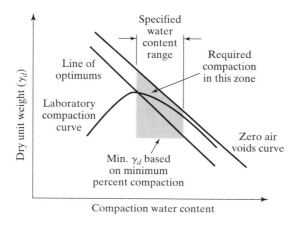

FIGURE 16.12 Schematic of Conventional Percent Compaction Specification (Benson et al., 1999). Used with permission of ASCE.

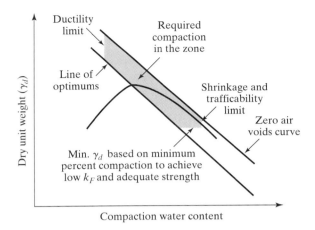

FIGURE 16.13 Schematic of Recommended Compaction Specification (Benson et al., 1999). Used with permission of ASCE.

et al., 1999). In order to avoid the conventional percent compaction specification (minimum dry unit weight and wet of a single optimum water content), it is necessary to craft a specification that ensures compaction on or above the line of optimums as shown in Figure 16.13. In addition to meeting the requirement of low hydraulic conductivity, a specification writer must also ensure that the clay liner has adequate shear strength also and is constructable for the range of water contents and dry unit weights that are specified (Benson et al., 1999). The issue of adequate shear strength has also proved to be important, given the fact that several very large landfill failures have occurred either (i) in the clay liner or (ii) at the interface between a compacted clay liner and an overlying geomembrane (recall Chapter 13). Too little attention has been given to this issue considering that wet-of-optimum moisture contents in clay soils tend to be associated with very low undrained shear strengths.

16.7 LEACHATE COLLECTION TRENCH CONSTRUCTION

The spacing and slope of leachate collection pipes are critical items for controlling leachate head over the liner and minimizing leakage through the liner. The location of the leachate collection line, as shown in the design drawings, should be strictly followed. When the collection pipe is to be installed in a trench, care must be taken to ensure that the trench has the design slope (minimum of 1.0%) toward the sump (Bagchi, 1994). A detailed construction sequence for a leachate collection trench for a double composite liner system includes the following steps:

(i) Grade and compact subgrade and secondary clay liner [Figure 16.14(a)].
(ii) Cut secondary leachate collection trench [Figure 16.14(b)].
(iii) Install secondary geomembrane liner [Figure 16.14(c)].
(iv) Install geotextile cushion and perforated pipe [Figure 16.14(d)].
(v) Place aggregate envelope [Figure 16.14(e)].
(vi) Place geotextile/geonet drain [Figure 16.14(f)].

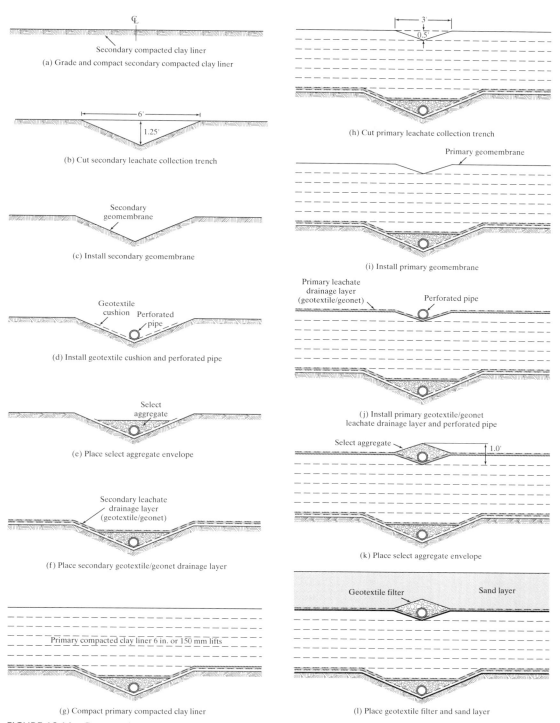

FIGURE 16.14 Construction Procedure of Leachate Collection Trench for Double Composite Liner System

620 Chapter 16   Construction of Compacted Clay Liners

- **(vii)** Compact primary clay liner [Figure 16.14(g)].
- **(viii)** Cut primary leachate collection trench [Figure 16.14(h)].
- **(ix)** Install primary geomembrane liner [Figure 16.14(i)].
- **(x)** Install geotextile/geonet drain and perforated pipe [Figure16.14(j)].
- **(xi)** Place aggregate envelope [Figure 16.14(k)].
- **(xii)** Place geotextile filter and granular blanket. To facilitate construction a granular blanket may be placed in non-trench areas prior to aggregate placement to allow equipment access to trenches [Figure 16.14(l)].

Seaming of the geomembrane liner should be avoided within three feet (0.9 m) of the leachate collection trench. The gravel used in the trench and the leachate collection pipes should be brought in carefully. Place the collection pipes near the trench site so they do not have to be dragged on the liner because dragging may tear the geomembrane. Bagchi (1994) recommends spreading a granular drainage or protective blanket over the entire geomembrane first, excluding 6 inches (150 mm) away from both sides of the collection trench. This procedure will allow movement of light vehicles on the liner for placement of gravel and collection pipes in the trench and avoid damaging the geomembrane.

16.8   PROTECTION OF COMPACTED SOIL

Thermal desiccation or drying from atmospheric exposure can damage and compromise the effectiveness of a liner (Daniel, 1984; Daniel and Wu, 1993). Many laboratory and field studies have shown that freeze-thaw can increase the hydraulic conductivity of compacted fine-grained soils by one to two orders of magnitude (Zimmie et al., 1992; Benson and Othman, 1993). These latter studies have shown that ice lenses formed during freeze-thaw result in a network of cracks. As a result, the hydraulic conductivity increases. Therefore, a compacted clay liner should be protected against both desiccation and freezing.

16.8.1   Protection against Desiccation

Figure 16.15 shows a desiccation crack in a compacted clay liner. There are several ways to prevent compacted soil liner materials from drying and desiccating. For example, a soil may be smoothly rolled with a steel drum roller to produce a thin, dense skin of soil on the surface. This thin, dense skin of soil helps to minimize transfer of water into or out of the underlying material. However, the smooth-rolled surface should be scarified prior to placement of a new lift of soil.

A better preventive measure is to water the clay liner periodically. Care must be taken to deliver water uniformly to the soil and not to create zones of excessively wet soil. Adding water by hand is not recommended because water is not delivered uniformly to the soil.

An alternative preventive measure is to cover the soil temporarily with a geomembrane, moist geotextile, or moist soil. A geomembrane acts as a thin vapor barrier that can be sacrificed after the project is completed. If a geomembrane or geotextile is used, it should be weighted down with sandbags or other materials to prevent

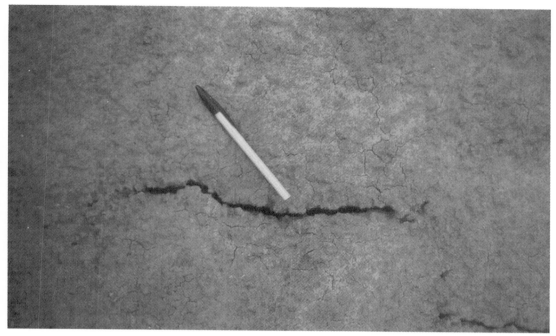

FIGURE 16.15 Desiccation Cracks in a Compacted Clay Liner

transfer of air between the geosynthetic cover and soil. Care should be taken in the case of a geomembrane cover to ensure that the underlying soil does not become heated and desiccate. A light-colored geomembrane works better in this regard. If moist soil is placed over the soil liner, the moist soil is removed later using grading equipment.

If a soil desiccates to a depth less than or equal to the thickness of a single lift, the desiccated lift may be disked, moistened, and recompacted. However, disking may produce large hard clods of clay that will require pulverization. It is also important to recognize that if a soil is wetted, time must be allowed for water to be absorbed into the clods of clay and hydration to take place uniformly. For this reason, it may be necessary to remove the desiccated soil from the construction area, to process the lift in a separate processing area, and to replace the soil accordingly (USEPA, 1993).

16.8.2 Protection against Freezing

Frozen soil should never be used to construct soil liners. Frozen soils form hard pieces that cannot be properly remolded and compacted. Inspectors should be on the lookout for frozen chunks of soil when construction takes place in freezing temperatures.

Freezing of soil liner materials can produce significant increases in hydraulic conductivity. Accordingly, soil liners must be protected from freezing before and after construction. Several methods can be used to insulate CCLs against freezing including the use of lightweight geofoams.

If superficial freezing takes place on the surface of a lift of soil, the surface should be scarified and recompacted. If an entire lift has been frozen, the entire lift should be disked, pulverized, and recompacted. If the soil is frozen to a depth greater than one lift, it may be necessary to strip away and replace the frozen material (USEPA, 1997).

16.8.3 Excess Surface Water

In some cases exposed lifts of liner material, or the completed liner, are subjected to heavy rains that soften the soil. Surface water creates a problem if the surface is uneven (e.g., if a footed roller has been used and the surface has not been smooth rolled with a smooth, steel-wheeled roller): numerous small puddles of water will develop in the depressions or low areas. Puddles of water should be removed before further lifts of material, or other components of the liner or cover system, are constructed. The material should also be disked repeatedly to allow the soil to dry, and only when the soil is at the proper water content is compaction advisable. Alternatively, the wet soil may be removed and replaced.

Even if puddles have not formed, the soils may be too soft to permit construction equipment to operate on the soil without creating ruts. To deal with this problem, the soil can be allowed to dry slightly by natural processes (but care must be taken to ensure that it does not dry too much and does not crack excessively during the drying process). Alternatively, the soil may be disked, allowed to dry while it is periodically disked, and than compacted.

If soil is reworked and recompacted, construction quality assurance and quality control tests should be performed at the same frequency as for the rest of the project (USEPA, 1993). However, if the required reworking is very small (e.g., in a sump, tests should be performed in the confined area to confirm proper compaction, even if this requires sampling at a greater frequency).

16.9 FIELD MEASUREMENT OF WATER CONTENT AND DRY UNIT WEIGHT

The most widely used method of measuring the water content and dry unit weight of compacted soil in the field is the nuclear method (ASTM D3017 and D2922). Measurement of water content with a nuclear device involves the modulation or thermalization of neutrons provided by a source of fast neutrons. The radioactive source of fast neutrons is embedded in the interior part of a nuclear water content/density device (Figure 16.16). The number of thermal neutrons that are encountered over a given period of time is a function of the number of fast neutrons that are emitted from the source and the density of hydrogen atoms in the soil located immediately below the nuclear device. Through appropriate calibration, and with the assumption that the only source of hydrogen in the soil is water, the nuclear device provides a measure of the water content of the soil over an average depth of about 8 inches (200 mm) (USEPA, 1993).

There are a number of potential sources of error with a nuclear water content measuring device. The most important potential source of error is extraneous hydrogen atoms not associated with water. Possible sources of hydrogen other than water

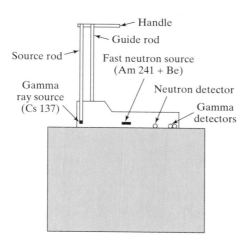

FIGURE 16.16 Schematic Diagram of Nuclear Water Content/Density Device (USEPA, 1993)

include hydrocarbons, methane gas, hydrous minerals (e.g., gypsum), hydrogen-bearing minerals (e.g., kaolinite, illite, and montmorillonite), and organic matter in the soil. Under extremely unfavorable conditions, the nuclear device can yield water content measurements that are as much as ten percentage points in error (almost always on the high side). Under favorable conditions, measurement error is less than 1.0 %. The nuclear device should be calibrated for site specific soils and changing conditions within a given site. For additional information about potential sources of error, see USEPA (1993).

Construction quality control and quality assurance personnel should be well versed in the proper use of nuclear water content measurement devices. There are many opportunities for error if personnel are not properly trained or do not correctly use the equipment. A nuclear device should be checked occasionally against other types of moisture measuring equipment to ensure that site-specific variables are not influencing test results. Nuclear equipment can also be checked against other nuclear devices (particularly new devices or recently calibrated devices) to minimize the potential for errors (USEPA, 1993).

Unit weight can be measured with a nuclear device operated in one of two ways, as shown in Figure 16.17. The most common usage is called *direct transmission,* in which a source of gamma radiation is lowered down a hole made in the soil to be tested [Figure 16.17(a)]. Detectors located in the nuclear density device sense the intensity of gamma radiation at the ground surface. The intensity of gamma radiation detected at the surface is a function of the intensity of gamma radiation at the source and the total unit weight of the soil material. The second mode of operation of the nuclear density device is called backscattering. With this technique, the source of gamma radiation is located at the ground surface [Figure 16.17(b)]. The intensity of gamma radiation detected at the surface is a function of the density of the soil as well as the radioactivity of the source. With the backscattering technique, the measurement is heavily dependent upon the density of the soil within the upper 1 to 2 inches (25 to 50 mm) of soil. The direct transmission method is recommended for soil liners (USEPA, 1993)

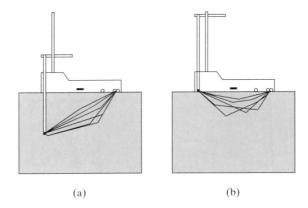

FIGURE 16.17 Measurement with Nuclear Device by (a) Direct Transmission and (b) Backscattering (USEPA, 1993)

because direct transmission provides a measurement averaged over a greater depth than backscattering. See USEPA (1993) for detailed procedures and precautions when measuring unit weight using a nuclear device in the direct transmission mode.

After total unit weight has been determined, the measured water content is used to compute dry unit weight. The potential sources of error with the nuclear device are fewer and less significant in the density-measuring mode compared to the water content measuring mode. The most serious potential source of error is improper use of the nuclear density device by the operator. One common gross error is to drive the source rod into the soil rather than inserting the source rod into a hole that has been preformed with the drive rod. Other potential sources of error include improper separation of the source from the base of the hole, inadequate period of counting, inadequate warm-up, spurious sources of gamma radiation, and inadequate calibration (USEPA, 1993).

Figure 16.18 shows a field technician conducting field measurement of water content and unit weight with a nuclear water content/density device.

16.10 CONSTRUCTION QUALITY ASSURANCE AND QUALITY CONTROL

Proper construction quality assurance and quality control (QA/QC) directly impacts the hydraulic conductivity of a compacted clay liner. There are several phases or aspects of proper quality assurance and quality control for compacted clay liner construction. They are: (i) quality assurance and quality control before and during clay liner construction, (ii) presence of qualified quality assurance and quality control personnel, and (iii) preparation and filing of a documentation report. Each item is important in providing proper quality assurance and quality control for successful construction and functioning of a landfill. Construction of a landfill involves a substantial amount of money. In addition, failure of a landfill could result in contamination of the groundwater and soil, and cleaning up this contamination could cost millions of dollars. Therefore, sufficient attention should be given to quality assurance and quality control during construction of such structures. Expenditures of 10 to 20% of the cost of the project for overall quality assurance and quality control purposes are customary for all civil engineering projects (Bagchi, 1994). Landfills are no exception.

FIGURE 16.18 Conducting Field Measurement of Water Content and Unit Weight with a Nuclear Water Content/Density Device

16.10.1 Critical Quality Assurance and Quality Control Issues

The construction quality assurance and quality control (QA/QC) processes for soil liners are intended to accomplish three objectives (USEPA, 1993):

(i) Ensure that soil liner materials are suitable.
(ii) Ensure that soil liner materials are properly placed and compacted.
(iii) Ensure that the completed liner is properly protected.

Some of these issues, such as protection of the liner from desiccation after completion, simply require application of common sense procedures. Other issues, such as preprocessing of materials, are potentially much more complicated because, depending on the material, many construction steps may be involved. Furthermore, tests alone will not adequately address many of the critical quality assurance and quality control issues—visual observations by qualified personnel, supplemented by intelligently selected tests, provide the best approach to ensuring quality of the constructed soil liner.

The objective of construction quality assurance is to ensure that the final product meets site specific plans, specifications, and QA documents. A detailed program of tests and observations is necessary to accomplish this objective. The objective of construction quality control is to control the manufacturing or construction process to meet project specifications. With geosynthetics, the distinction between quality control and quality assurance is obvious—the geosynthetics manufacturer or installer performs manufacturing and construction quality control respectively while an independent organization conducts manufacturing and construction quality assurance. In contrast, construction quality control and quality assurance activities for soils are more closely linked than in geosynthetics installation. For example, on many earthwork projects the construction quality assurance inspector will typically determine the water content of the soil and report the value to the contractor; in effect, the quality assurance inspector is also providing quality control input to the contractor. On some projects, the contractor is required to perform extensive tests as part of the construction quality control process, and the construction quality assurance inspector performs tests to check or confirm the results of construction quality control tests (USEPA, 1993).

16.10.2 Quality Assurance and Quality Control for Compacted Clay Liner Construction

The steps involved in choosing a borrow source for a soil liner, discussed in Section 2.4, should be followed. The composition of the soil, required compactive effort and moisture content at compaction are thus known before liner construction is undertaken. Proper testing is required to implement the proposed compaction during construction. The items that need attention are lift thickness, number of construction plant passes (e.g., with a roller), frequency of testing compaction density and moisture content, and permeability of the compacted liner.

Soils are normally placed in loose lifts not exceeding 9 inches (225 mm) in thickness and with the compacted lift not exceeding 6 inches (150 mm). Extreme care must be used in compacting the first lift so as not to damage underlying components nor to mix dissimilar soils into the recompacted clay. The soil should be compacted to achieve a permeability not greater than 1.0×10^{-7} cm/sec, based on laboratory compaction density/permeability testing, and it shall be compacted to not less than 90% of the maximum density as determined by the Modified Proctor Test, ASTM D1557. Each lift is integrated into the previous lift by techniques such as scarifying each lift and by using compaction equipment capable of penetrating the thickness of each compacted lift, except that such a compactor shall not be used in the first two lifts immediately above a synthetic liner, leak detection system, or other sensitive liner system component (MDEQ, 1999). An improperly tied lift may exhibit higher permeability in the horizontal direction along the lift boundary. A sheepfoot roller tends to eliminate the sharp discontinuity between successive lifts and thus is a better choice of equipment for compaction.

The compactive effort delivered by rollers, which is expressed as ft-lb/ft^3 (kN-m/m^3), is a function of the number of passes of the roller over a given area of soil. A pass may be defined as one pass of the construction equipment or one pass of a drum over a given point in the soil liner. It does not matter whether a pass is defined as a pass of the equipment or a pass of a drum, but the construction specifications and construction quality assurance plan should define what is meant by a pass (USEPA,

1993). Normally, one pass of the vehicle constitutes a pass for self-propelled rollers, and one pass of a drum constitutes a pass for towed rollers.

Some construction documents require a minimum coverage, which (after USEPA, 1993) is defined as

$$C = (A_f/A_d) \times N \times 100\% \qquad (16.3)$$

where C = coverage,
   A_f = sum of the area of the feet on the drums of the roller,
   A_d = area of the drum itself,
   N = number of passes.

Construction specifications sometimes require 150 to 200% coverage of the roller. For a given roller and minimum percent coverage, the minimum number of passes (N) may be computed.

The number of passes of a compactor over the soil can have an important influence on the overall hydraulic conductivity of the soil liner. Periodic observations should be made of the number of passes of the roller over a given point. Approximately three observations per hectare per lift (one observation per acre per lift) is the recommended frequency of measurement (USEPA, 1993). The minimum number of passes that is reasonable depends upon many factors and cannot be stated in general terms. However, experience has shown that at least 5 to 15 passes of a compactor over a given point is usually necessary to remold and compact clay liner materials thoroughly. As an example, the Michigan Department of Environmental Quality specifies 4 to 6 passes of self-propelled soil compaction equipment for each lift with foot contact pressure of between 200 and 400 psi (1,380 and 2,760 kN/m^2). (Note: Foot contact pressure should be regulated so the soil is not sheared on the third or fourth pass.) The specified foot contact area varies between 5 and 14 in^2 (3200 and 9000 mm^2).

A common field approach is to measure density more frequently at the beginning of a compaction project. The number of passes required for a particular piece of equipment to achieve the specified density at wet of optimum is standardized based on the initial test runs on a trial section. Bagchi (1994) recommends the use of the DM-7 [DM-7, 1971] guidelines for foot contact pressure. Usually the foot shears the soil initially, but after a few passes the foot rides on top of the clay (i.e., it no longer shears it). The expression "walking out" is sometimes used to describe this process. If the foot contact pressure is excessively high and the layer continues to be sheared even after seven or eight passes, then foot contact pressure should be reduced by reducing the weight of the drum. On the other hand, if the foot does not shear the clay to at least two-thirds its length from the beginning, then the weight of the drum should be increased.

The number of density, moisture content, and permeability tests to be done for quality assurance and quality control purposes should be specified. The frequency of the quality assurance and quality control tests specified by design engineers varies widely. A large number of density and moisture content tests tend to be specified during construction to ensure uniformity. These tests are relatively fast and less complicated to run. Fewer permeability tests (laboratory or field) are usually specified because of their higher cost, complexity, and time requirements. A detailed study of compacted density, moisture content, and permeability of a clay liner has indicated some variability of these properties within the liner (Daniel and Benson, 1990). However, these variations fell within reasonable limits. The testing program for construction quality control proposed by the Michigan Department of Environmental Quality is listed in Table 16.2, and it is presented here as an example.

The construction quality control and quality assurance for compacted clay liners essentially consist of using suitable materials, placing and compacting the materials properly, and protecting the completed liner. The steps required to fulfill these requirements are summarized as follows:

(i) The subgrade on which the soil liner will be placed should be properly prepared.

(ii) The materials employed in constructing the soil liner should be suitable and should conform to the plans and specifications for the project.

(iii) The soil liner material should be preprocessed, if necessary, to adjust the water content, to remove oversized particles, to break down clods of soil, or to add amendments such as bentonite.

(iv) The soil should be placed in lifts of appropriate thickness and then be properly remolded and compacted.

TABLE 16.2 Testing Program for Construction of Compacted Clay Liners (modified from Sherman, 1992)

Facility Component	Factor	Frequency	Test Method
Clay Borrow Source	Unified Soil Classification	5,000 yd^3 (3,800 m^3) and all changes in material	ASTM D2487
	Moisture-Density Curve	5,000 yd^3 (3,800 m^3) and all changes in material	ASTM D1557
	Moisture/Density/ Permeability Relationship	Initially and all changes in material	ASTM D1557 ASTM D5084
Compacted Subbase	Moisture/Density (Nuclear)	5 tests/acre/lift (12 tests/hectare/lift) or 1 test/100 feet grid/lift (3 test/100 m grid/lift), 3 tests/day minimum	ASTM D2922
Clay Liners	Moisture/Density (Nuclear)	5 tests/acre/lift (12 tests/hectare/lift) or 1 test/100 feet grid/lift (3 test/100 m grid/lift), 3 tests/day minimum	ASTM D2922
	In-Situ Permeability Testing	10,000 yd^3 (7,600 m^3) changes and all in material, 3 tests minimum per liner/cover	ASTM D5084
	Thickness	5 tests/acre/lift (12 tests/hectare/lift) or 1 test/100 feet grid/lift (3 test/100 m grid/lift)	Surveying or direct measure
	Slope		Surveying

Note: 1 acre × 3 feet = 4840 yard^3
1 acre × 2 feet = 3230 yard^3
1 acre × 6 inches = 807 yard^3 (1 test/161 yard^3 @ 5 tests/acre)
6 inches on 100 feet grid = 185 yard^3
1 hectare = 2.5 acres
1 m = 3.28 ft
1 m^3 = 1.3 yard^3

FIGURE 16.19 A Completed Compacted Clay Liner

(v) The compacted clay liner should be protected from damage caused by desiccation or freezing temperatures.

(vi) The final surface of the compacted clay liner should be properly prepared to support the next layer (typically a geomembrane) that will be placed on top of the soil liner.

Figure 16.19 shows a completed compacted clay liner that is ready for placement of the geomembrane.

16.10.3 Documentation Report

Good documentation of all the construction items is essential. Clear and concise documentation provides construction details, notes any departure from the original proposal, and discusses reasons for such departures; it is also helpful in case of disputes. This means that considerable importance should be given to the documentation reports. The location of any test must be clearly indicated. Since compaction, moisture content, hydraulic conductivity, and other tests are done on different lifts, it is advisable to draw each lift separately and then show the appropriate test locations on the corresponding lift. The drawings and narrative should be clear and concise. A daily log of construction activities and quality assurance and quality control tests should be maintained at the site by the field QA/QC technicians. Such field logs are important documents, especially in case of a dispute. Writing over and erasing must be avoided in the log book. In case of an error, the mistake should simply be crossed out. Bagchi

(1994) recommends that the report be reviewed in detail by all the relevant parties (i.e., the owner, contractor, and quality assurance team) before submitting it to the regulatory agency. Finally, a detailed construction certification report should be prepared by both the design and quality assurance firms for submission to a regulatory agency for its review and approval.

PROBLEMS

16.1 In order to meet the design criteria of compacted clay liners, what precautionary steps should be taken during construction?

16.2 What are the criteria for selecting soil materials for a compacted soil liner?

16.3 What are the basic compaction goals for construction of a soil liner? What are key considerations in achieving these compaction goals?

16.4 Explain why it is usually preferable to compact a clay soil used in a liner at a water content greater than optimum (i.e., wet of optimum); explain why soil cannot be placed at a water content that is too high.

16.5 What are the benefits that can be achieved when increasing compactive energy?

16.6 Why will poor interlift bonding increase hydraulic conductivity of soil liners?

16.7 Explain how to achieve good interlift bonding during construction of a soil liner?

16.8 What type of compaction roller is recommended for constructing a soil liner? Why?

16.9 Explain how to make a good lateral connection between a new section of soil liner and an old one?

16.10 Explain what the traditional "percent compaction" specification is.

16.11 Describe the problems associated with the traditional "percent compaction" specification. Describe how these problems can be corrected using a different approach.

16.12 What measures can be used to prevent compacted soil liner materials from drying and desiccating?

16.13 If an uncompleted compacted clay liner was left exposed to the elements over a winter season, what measures should be taken when starting construction again the following spring?

16.14 What objectives are construction quality assurance and quality control (QA/QC) processes for soil liners intended to accomplish? Explain how these objectives can be achieved?

REFERENCES

Acar, Y. B. and Oliveri, I., (1990) "Pore Fluid Effects on the Fabric and Hydraulic Conductivity of Laboratory-Compacted Clay," *Transp. Res. Rec. 1219,* Transportation Research Board, Washington, D.C., pp. 144–159.

Bagchi, A., (1994) *Design, Construction, and Monitoring of Sanitary Landfill,* 2nd Edition, John Wiley & Sons, Inc., New York, NY.

Benson, C. H., and Boutwell, G. P., (1992) "Compaction Control and Scale-Dependent Hydraulic Conductivity of Clay Liners," *Proceedings of 15th Annual Madison Waste Conference,* University of Wisconsin, Madison, Wisconsin, pp. 62–83.

Benson, C. H. and Daniel, D. E., (1990) "Influence of Clods on Hydraulic Conductivity of Compacted Clay," *Journal of Geotechnical Engineering,* ASCE, Vol. 116, No. 8, pp. 1231–1248.

Benson, C. H., Daniel, D. E., and Boutwell, G. P., (1999) "Field Performance of Compacted Clay Liners," *Journal of Geotechnical and Geoenvironmental Engineering,* ASCE, Vol. 125, No. 5, pp. 390–403.

Benson, C. H., Hardianto, F. S., and Motan, E. S., (1994a) "Representative Specimen Size for Hydraulic Conductivity Assessment of Compacted Soil Liners," *Hydraulic Conductivity and Waste Contaminant Transport in Soil,* ASTM STP 1142, Edited by D. E. Daniel and S. J. Trautwein, American Society for Testing and Materials, Philadelphia, pp. 3–29.

Benson, C. H. and Othman, M. A., (1993) "Hydraulic Conductivity of Compacted Clay Frozen and Thawed In Situ," *Journal of Geotechnical Engineering,* ASCE, Vol. 119, No. 2, pp. 276–294.

Benson, C. H. and Trast, J., (1995) "Hydraulic Conductivity of Thirteen Compacted Clays," *Clays and Clay Minerals,* Vol. 43, No. 6, pp. 669–681.

Benson, C. H., Zhai, H., and Wang, X., (1994b) "Estimating the Hydraulic Conductivity of Compacted Clay Liners," *Journal of Geotechnical Engineering,* ASCE, Vol. 120, No. 2, pp. 367–387.

Blotz, L., Benson, C. H., and Boutwell, G. P., (1998) "Estimating Optimum Water Content and Maximum Dry Unit Weight for Compacted Clays," *Journal of Geotechnical and Geoenvironmental Engineering,* ASCE, Vol. 124, No. 9, pp. 907–912.

Boutwell, G. P. and Hedges, S., (1989) "Evaluation of Water-Retention Liners by Multivariate Statistics," *Proceedings of 12th ICSMFE,* Balkema, Rotterdam, The Netherlands, pp. 815–818.

Boynton, S. S. and Daniel, D. E., (1985) "Hydraulic Conductivity Test on Compacted Clay," *Journal of Geotechnical Engineering,* ASCE, Volume 111, No. 4, pp. 465–478.

Daniel, D. E., (1984) "Predicting Hydraulic Conductivity of Clay Liners," *Journal of Geotechnical Engineering,* ASCE, Volume 110, No. 4, pp. 285–300.

Daniel, D. E., (1987) "Earth Liners for Waste Disposal Facilities," *Geotechnical Practice for Waste Disposal '87,* R. D. Woods, ed., ASCE, Ann Arbor, Michigan, pp. 21–39.

Daniel, D. E., (1990) "Summary Review of Construction Quality Control for Compacted Soil Liners," *Waste Containment System: Construction, Regulation, and Performance,* ASCE, R. Bonaparte, ed., New York, NY, pp. 175–189.

Daniel, D. E. and Benson, C. H., (1990) "Water Content-Density Criteria for Compacted Soil Liners," *Journal of Geotechnical Engineering,* ASCE, Volume 116, No. 12, pp. 1811–1830.

Daniel, D. E. and Koerner, R. M., (1995) "Waste Containment Facilities: Guidance for Construction, Quality Assurance and Quality Control of Liner and Cover Systems," ASCE Press New York, NY.

Daniel, D. E. and Wu, Y. -K., (1993) "Compacted Clay Liners and Covers for Arid Sites," *Journal of Geotechnical Engineering,* ASCE, Volume 119, No. 2, pp. 223–237.

Day, S. R. and Daniel, D. E., (1985) "Hydraulic Conductivity of Two Prototype Clay Liners," *Journal of Geotechnical Engineering,* ASCE, Volume 111, No. 8, pp. 957–970.

DM-7, (1971) "Design Manual, Soil Mechanics, Foundations and Earth Structures," NAVFAC DM-7, U. S. Department of the Navy, Washington, DC.

Elsbury, B. R., Daniel, D. E., Sraders, G. A., and Anderson, D. C., (1990) "Lessons Learned from Compacted Clay Liner," *Journal of Geotechnical Engineering,* ASCE, Volume 116, No. 11, pp. 1641–1660.

Garcia-Bengochea, I., Lovell, C., and Altschaeffl, A., (1979) "Pore Distribution and Permeability of Silty Clays," *Journal of Geotechnical Engineering,* ASCE, Volume 105, No. 7, pp. 839–856.

Ghassemi, M., Haro, M., Metzgar, J., et al., (1983) "Assessment of Technology for Constructing Cover and Bottom Liner Systems for Hazardous Waste Facilities," EPA/68-02/3174. U. S. Environmental Protection Agency, Cincinnati, OH.

Goldman, L. J., Truesdale, R. W., Kingsbury, G. L., Northeim, C. M., and Damle, A. S., (1986) "Design Construction and Evaluation of Clay Liners for Waste Management Facilities," EPA/530-SW-86/007. US Environmental Protection Agency, Cincinnati, OH.

Herrmann, J. G. and Elsbury, B. R., (1987) "Influential Factors in Soil Liner Construction for Waste Disposal Facilities," *Proceedings of Geotechnical Practice for Waste Disposal '87,* Ann Arbor, MI, ASCE, pp. 522–536.

Hilf, J. W., (1975) "Compacted Fill," *Foundation Engineering Handbook,* Edited by H. F. Winterkorn and H. Y. Fang, Van Nostrand-Reinhold, New York, pp. 244–341.

Lambe, T. W., (1954) "The Permeability of Compacted Fine-Grained Soils," *ASTM STP 163,* ASTM, West Conshohocken, PA, pp. 56–67.

Lambe, T. W., (1958a) "The Structure of Compacted Clay," *Journal of Soil Mechanics and Foundation Engineering,* ASCE, Volume 84, No. 2, Paper 1654, pp. 1–34.

Lambe, T. W., (1958b) "The Engineering Behavior of Compacted Clay," *Journal of Soil Mechanics and Foundation Engineering,* ASCE, Volume 84, No. 2, Paper 1655, pp. 1–35.

MDEQ, (1999) "Solid Waste Management Act Administrative Rules Promulgated Pursuant to Part 115 of the Natural Resources and Environmental Protection Act, 1994 PA 451, as amended (Effective April 12, 1999)," Michigan Department of Environmental Quality, Waste Management Division, Lansing, MI.

Mitchell, J. K., Hopper, D. R., and Campanella, R. G., (1965) "Permeability of Compacted Clay," *Journal of Soil Mechanics and Foundation Engineering,* ASCE, Vol. 91, No. 4, pp. 41–65.

Olsen, H. W., (1962) "Hydraulic Flow through Saturated Clays," *Clays and Clay Minerals,* Vol. 9, No. 2, pp. 131–161.

Reades, D., Lahti, L., Quigley, R., and Bacopoulos, A., (1990) "Detailed Case History of Clay Liner Performance," Proceedings of Waste Containment System, R. Bonaparte, ed., ASCE, Reston, VA, pp. 156–174.

Seed, H. B. and Chan, C. K., (1959) "Structure and Strength Characteristics of Compacted Clays," *Journal of Soil Mechanics and Foundation Engineering,* ASCE, Vol. 85, No. 5, pp. 87–128.

Shakoor, A. and Cook, B. D., (1990) "The Effect of Stone Content, Size, and Shape on the Engineering Properties of a Compacted Silty Clay," Bulletin of Association of Engineering, Geologists, XXVII(2), pp. 245–253.

Shelley, T. L. and Daniel, D. E., (1993) "Effect of Gravel on Hydraulic Conductivity of Compacted Soil Liners," *Journal of Geotechnical Engineering,* ASCE, Volume 119, No.1, pp. 54–68.

Sherman, V. W., (1992) "Construction Quality Control and Construction Quality Assurance for Low Permeability Clay Barrier Soils," Michigan Department of Environmental Quality, Lansing, MI.

USEPA, (1989) "Requirements for Hazardous Waste Landfill Design, Construction, and Closure," EPA/625/4-89/022, Center for Environmental Research Information, Office of Research and Development, US Environmental Protection Agency, Cincinnati, OH, August.

USEPA, (1991) "Design and Construction of RCRA/CERCLA Final Covers," EPA/625/4-91/025, Office of Research and Development, US Environmental Protection Agency, Washington, DC, May.

USEPA, (1993) "Quality Assurance and Quality Control for Waste Containment Facilities," Technical Guidance Document, EPA/600/R-93/182, US Environmental Protection Agency, Office of Research and Development, Washington, DC, September.

Zimmie, T. F., LaPlante, C. M., and Bronson, D., (1992) "The Effects of Freezing and Thawing on the Permeability of Compacted Clay Landfill Covers and Liners," *Environmental Geotechnology,* Proceedings of the Mediterranean Conference on Environmental Geotechnology, A. A. Balkema Publishers, pp. 213–218.

CHAPTER 17

Installation of Geosynthetic Materials

17.1 MATERIAL DELIVERY AND CONFORMANCE TESTS
17.2 INSTALLATION OF GEOMEMBRANES
    17.2.1 GEOMEMBRANE PLACEMENT
    17.2.2 GEOMEMBRANE SEAMING
    17.2.3 GEOMEMBRANE SEAM TESTS
    17.2.4 GEOMEMBRANE DEFECTS AND REPAIRS
    17.2.5 GEOMEMBRANE PROTECTION AND BACKFILLING
17.3 INSTALLATION OF GEONETS
    17.3.1 GEONET PLACEMENT
    17.3.2 GEONET JOINING
    17.3.3 GEONET REPAIRS
17.4 INSTALLATION OF GEOTEXTILES
    17.4.1 GEOTEXTILE PLACEMENT
    17.4.2 GEOTEXTILE OVERLAPPING AND SEAMING
    17.4.3 GEOTEXTILE DEFECTS AND REPAIRS
    17.4.4 GEOTEXTILE BACKFILLING AND COVERING
17.5 INSTALLATION OF GEOCOMPOSITES
    17.5.1 GEOCOMPOSITE PLACEMENT
    17.5.2 GEOCOMPOSITE JOINING AND REPAIRS
    17.5.3 GEOTEXTILE COVERING
17.6 INSTALLATION OF GEOSYNTHETIC CLAY LINERS
    17.6.1 GEOSYNTHETIC CLAY LINER PLACEMENT
    17.6.2 GEOSYNTHETIC CLAY LINER JOINING
    17.6.3 GEOSYNTHETIC CLAY LINER REPAIRS
    17.6.4 GEOSYNTHETIC CLAY LINER BACKFILLING OR COVERING
    PROBLEMS
    REFERENCES

Geosynthetic material installation includes the placement and attachment of the following components: geomembranes, geonets, geotextiles, geocomposites, geosynthetic clay liners, and, occasionally, geogrids. Proper construction quality assurance and quality control must be provided during geosynthetic material installation. Quality assurance refers to means and actions employed to assure conformity of the geosynthetic system production and installation with the project-specific quality assurance plan, drawings, specifications, and contractual and regulatory requirements. Quality assurance is provided by an organization separate from production and installation. Quality control refers only to those actions taken to ensure that materials and workmanship meet the requirements of the plans and specifications. Quality control is provided by the manufacturers and installers of the various components of the geosynthetic system.

In this regard, geosynthetic materials have four levels of quality management associated with them:

   (i) Manufacturing quality control (MQC),
   (ii) Manufacturing quality assurance (MQA),
   (iii) Construction quality control (CQC),
   (iv) Construction quality assurance (CQA).

Quite often, MQA and CQA are performed by the same organization. Conversely, MQC and CQC are often performed by different organizations—the manufacturer and the installer. Of course, many of the larger manufacturers have their own installation crews.

17.1 MATERIAL DELIVERY AND CONFORMANCE TESTS

A construction quality assurance (CQA) representative should be present, whenever possible, to observe material delivery and unloading on site. The CQA representative must note any damaged materials received in addition to any material damaged during unloading. Stored materials must be protected adequately against dirt, theft, vandalism, and passage of vehicles.

Upon delivery of the rolls of geosynthetic materials, the construction quality assurance (CQA) consultant should ensure that conformance test samples are obtained. These samples should then be forwarded to the geosynthetic quality assurance laboratory for testing to ensure conformance with the site-specific specifications.

The rolls to be sampled should be selected by the CQA personnel. Samples should be taken across the entire width of the roll, but must not include the first three feet. Unless otherwise specified, samples should be three feet across the roll width. Sampling rate is site-specific, but a possible guide is that samples should be taken at a rate of at least one per 100,000 ft^2 (10,000 m^2) for a geomembrane, geonet, geotextile, and geosynthetic clay liner, respectively, and 50,000 ft^2 (5,000 m^2) for a geocomposite. It is reasonable that the following conformance tests be conducted:

Geomembrane

   (i) Thickness (ASTM D5199),
   (ii) Tensile strength and elongation (ASTM D638 for HDPE and LLDPE, ASTM D882 for PVC, and ASTM D751 for CSPE-R),
   (iii) Puncture (D4833 for HDPE and LLDPE),
   (iv) Tear resistance (ASTM D1004, Die C for HDPE, LLDPE, and PVC),
   (v) Ply adhesion (ASTM D413) for CSPE-R,
   (vi) Density (optional) ASTM D1505 or ASTM D792

Geonet

(i) Density (ASTM D1505 or ASTM D792),
(ii) Mass per unit area (ASTM D5261),
(iii) Thickness (ASTM D5199),
(iv) Compression (ASTM D1621) (optional),
(v) Transmissivity (ASTM D4716) (optional).

Geotextile

(i) Mass per unit area (ASTM D5261),
(ii) Grab tensile strength (ASTM D4632),
(iii) Trapezoidal tear strength (ASTM D4533),
(iv) Burst strength (ASTM D3786),
(v) Puncture strength (ASTM D4833),
(vi) Possibly apparent opening size (ASTM D4751),
(vii) Possibly permittivity (ASTM D4491).

Geocomposite

(i) Thickness (ASTM D5199),
(ii) Ply Adhesion (ASTM D413),
(iii) Compression (ASTM D1621) (optional),
(iv) Transmissivity (ASTM D4716) (optional).

Note that the geotextile should not be peeled from the geocomposite and tested separately. Damage invariably occurs during stripping.

Geosynthetic Clay Liner (GCL)

(i) Mass per unit area (ASTM D5261), sampling frequency based on ASTM D4354,
(ii) Free swell of the clay component (ASTM D5890), sampling frequency based on ASTM D4354,
(iii) Hydraulic conductivity [ASTM D5084 (mod.) or GRI-GCL2],
(iv) Direct shear testing (ASTM D6243),
(v) Peel testing for needle punched or stitch bonded GCL [ASTM D413 (mod.)], one test per 20,000 ft^2 or 2,000 m^2.

The construction quality assurance (CQA) engineer should review all conformance test results from laboratory conformance testing and report any nonconformance to the project manager prior to the deployment of the geosynthetic materials. The CQA personnel are responsible for checking that all test results meet or exceed the values listed in the project specifications.

17.2 INSTALLATION OF GEOMEMBRANES

Installation of geomembranes is critical and relatively complicated, compared with installations of other geosynthetic materials.

17.2.1 Geomembrane Placement

When the subgrade or subbase (either soil or some other geosynthetic materials) is approved as acceptable, the rolls or pallets of the temporarily stored geomembranes are brought to their intended location, unrolled or unfolded, and accurately spotted for field seaming.

17.2.1.1 Earthwork. The general and/or earthwork contractor should be responsible for preparing and maintaining the subgrade in a condition suitable for installation of the geomembrane. The following general minimum guidelines for earthwork quality control should be followed:

(i) The surface to be lined should be smooth and free of debris, roots, and angular or sharp rocks. All fill should consist of well-graded materials, free of organic, trash, clayballs, or other deleterious material that may cause damage to the geomembrane. The upper six inches of the finished subgrade should not contain stones or debris larger than one-half inch in diameter. The subgrade should be compacted in accordance with design specifications, but in no event less than what is required to provide a firm, unyielding foundation sufficient to permit the movement of vehicles and welding equipment over the subgrade without causing rutting or other deleterious effects. The subgrade should have no sudden sharp or abrupt changes in grade.

(ii) The earthwork contractor should protect the subgrade from desiccation, flooding, and freezing. If required, protection may consist of a thin plastic protective cover (or other material as approved by the engineer) installed over the completed subgrade until the placement of the geomembrane liner begins. A subgrade with desiccation cracks greater than one-half inch in width or depth, or one exhibiting swelling, heaving, or other similar conditions, must be replaced or reworked by the general and/or earthwork contractor to remove these defects.

(iii) Upon request, the installer's site supervisor will provide the owner's and/or contractor's representative with a written acceptance of the surface to be lined. This acceptance will be limited to the amount of area that the installer is capable of lining during a particular work shift. Direction and control of any subsequent repairs to the subgrade, including the subgrade surface, remain the responsibility of the earthwork contractor.

The anchor trench should be excavated by the general and/or the earthwork contractor to the lines, grades, and widths shown on the design drawings prior to geomembrane placement. Anchor trenches excavated in clay soils susceptible to desiccation cracks should be excavated only the distance required for that days liner placement to minimize the potential of desiccation cracking of the clay soils. Corners in the anchor trench should be rounded where the geomembrane enters the trench to minimize sharp bends in a geomembrane. Figure 17.1 shows a subgrade that has been prepared for placement of a geomembrane.

FIGURE 17.1   A Soil Subgrade Ready for Geomembrane Placement

17.2.1.2 Field Panel Placement. Prior to commencement of liner deployment, layout drawings should be produced to indicate the panel configuration and general location of field seams for the project. Each panel used for the installation should be given a numeric or alpha-numeric identifier. This panel identification number should be associated with a manufacturing roll number on the panel placement form that identifies the resin type, batch number, and date of manufacture.

Geomembranes should not be deployed during any precipitation, in the presence of excessive moisture (i.e., fog, dew), in high or low ambient temperature, in an area of standing water, or during high winds (greater than approximately 20 miles/hour or 32 km/hour).

Install field panels as indicated on the layout drawing. If a panel is deployed in a location other than that indicated on the layout drawings, the revised location must be noted on an "as-built" drawing. The as-built drawing must be modified at the completion of the project to reflect actual panel locations. As-built drawings will be maintained and submitted by the lining company and/or the third party construction quality assurance company representative as required by project specifications and contract documents. The following steps in the deployment method should be followed:

(i) The method and equipment used to deploy the panels must not damage the geomembrane or the supporting subgrade surface.

(ii) Personnel working on the geomembrane must not wear shoes that can damage the geomembrane or engage in actions that could result in damage to the geomembrane.

FIGURE 17.2 Geomembrane Placement

(iii) Adequate temporary loading and/or anchoring (i.e., sandbags, tires) must be done to prevent uplift of the geomembrane by wind.

(iv) The geomembrane must be deployed in a manner to minimize or avoid wrinkles. The QA plan must be specific in this regard.

(v) Any area of a panel seriously damaged (torn, twisted, or crimped) must be marked and suitably repaired.

Information relating to geomembrane panel placement, including date, time, panel number, and panel dimensions, will be maintained on the panel placement form. If a portion of a roll is set aside to be used at another time, the roll number must be written on the remainder of the roll in several places. Figure 17.2 shows geomembranes being deployed panel by panel in a landfill cell.

17.2.2 Geomembrane Seaming

The field seaming of the deployed geomembrane rolls or panels is a critical aspect of their successful functioning as a barrier to liquid flow. This section describes the various seaming methods in current use and describes the concept and importance of test strips (or trial seams).

17.2.2.1 Seam Requirements and Preparation

Layout

In general, seams should be oriented parallel to the slope (i.e., oriented along, not across the slope). Whenever possible, horizontal seams should be located on the base of the cell, at least five feet from the toe of the slope. Each seam made in the field should be numbered and indicated on the record drawings. Seaming information including seam number, welder ID, machine number, temperature setting, and weather conditions must be included on the panel seaming form.

Personnel Qualification

The construction quality assurance (CQA) inspector (also called the monitor) must ensure that personnel performing field seaming have the following qualifications:

(i) The master seamer should have a minimum of 1,000,000 feet^2 (100,000 m^2) of acceptable geomembrane seaming experience, using the type of seaming apparatus proposed for use on the current project. A master seamer must be present on site during all seaming operations.

(ii) Other seamers should have acceptably seamed a minimum of 100,000 feet^2 (10,000 m^2) of geomembrane. The master seamer provides direct supervision over the other seamers.

Geomembrane Panel Seaming Methods

All geomembranes, regardless of type, should be seamed by thermal methods. Geomembrane panel seaming can be performed by two methods: fusion (hot wedge) welding and extrusion welding. Fusion (hot wedge) welding is generally used for long seams. Extrusion welding is used for capping and patch repairs, and for joining panels at locations where fusion welding is not practical due to joint configuration. Hot air welding is also possible, but it is very sensitive to workmanship and is only used when other methods are not applicable.

Fusion (Hot Wedge) Welding: Fusion welding consists of placing a heated wedge, mounted on a self propelled vehicular unit, between two overlapped sheets such that the surface of both sheets are heated above the polyethylene's melting point. Figure 17.3 shows a dual track fusion weld, the most common type. A single track fusion weld is also possible, but is without the air channel for nondestructive testing. After being heated by the wedge, the overlapped panels pass through a set of preset pressure wheels that compress the two panels together to form the weld. The fusion welder is equipped with a device that continuously monitors the temperature of the wedge.

Other specifications for fusion welding include the following:

(i) The panels of geomembrane must be overlapped approximately four to six inches prior to welding. The panels must be adjusted so that the seams are aligned with the least possible number of wrinkles and "fishmouths".

FIGURE 17.3 Cross Section of a Fusion Hot Wedge Welded Seam

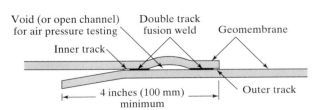

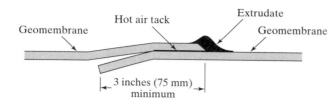

FIGURE 17.4 Cross Section of an Extrusion Fillet Welded Seam

(ii) The seam area must be prepared prior to seaming to assure the area is clean and free of moisture, dust, dirt, or debris of any kind. Grinding is not required for fusion welding.

(iii) At the discretion of the installer's project superintendent a movable protective layer may be used directly below the overlap of geomembrane to be seamed to prevent build-up of dirt or moisture between the panels.

Extrusion Fillet Welding: Extrusion fillet welding consists of introducing a ribbon of molten resin along the edge of the overlap of the two geomembrane sheets to be welded (Figure 17.4). Grinding of the geomembrane surface is necessary for proper bonding of the molten extrudate. A hot-air preheat and the addition of molten polymer (extrudate) causes some of the material of each sheet to be liquefied resulting in a homogeneous bond between the molten weld bead and the surfaces of the overlapped sheets. The extrusion welder is equipped with gauges giving the temperature in the apparatus and a numerical setting for the pre-heating unit.

Other specifications for extrusion fillet welding include the following:

(i) For geomembranes thicker than 40 mils (1.0 mm), extrusion welded seams are prebeveled prior to heat tacking in place.

(ii) The geomembrane panel must be overlapped a minimum of 3 inches (75 mm). The seam areas must be prepared prior to seaming to assure the area is clean and free of moisture, dust, dirt, and debris of any kind.

(iii) A hot-air device must be used to temporarily bond the geomembrane panels, taking care not to damage the geomembrane.

(iv) The seam overlap must be ground within one hour of the welding operation. Grinding must be done in a manner that does not damage the geomembrane. Grind marks on both sheets should be covered with extrudate.

(v) The extruder must be purged prior to beginning of each seam to remove all heat-degraded extrudate from the barrel. The purged extrudate will be placed on scrap material to prevent contact with installed geomembrane.

(vi) The welding rod must be of the same density as the geomembrane being seamed and kept clean and dry.

17.2.2.2 Trial Welds. Trial welds should be conducted by welding technicians at the following minimum frequency: at all start-ups, immediately prior to planned shut-downs, whenever the equipment requires start-up after a breakdown, at a minimum of four-hour intervals, as weather conditions dictate, or as directed by the CQA monitor if welding problems are suspected.

All trial welds should be conducted under the same conditions as actual seaming. Once qualified by a passing trial weld, welding technicians should not change welding parameters (temperature, speed, etc.) without performing another trial weld. Figure 17.5 shows a welding technician conducting a trial weld.

Trial Weld Length

The trial weld should be made by joining two pieces of excess geomembrane. Trial welds for fusion welds must be at least 15 feet (4.5 m) long. Extrusion weld trial seams must be at least 4 feet (1.2 m) long.

Suggested Field Testing Procedure:

(i) The CQA monitor should visually inspect the seam for squeeze out, indentation, thinning, and general appearance.

(ii) Three one-inch (25 mm) wide strips should be cut from the trial seam. One of them should be cut from near the middle of the seam and others shall be cut from near each end of the test seam. Two of the samples should be tested in peel, and one in shear using a digital readout field tensiometer.

(iii) A sample is considered to pass when the following results are achieved (for double-wedge welding) both welds (i.e., inner and outer tracks) should be tested in shear. The EPA (1993) recommends that shear test results be 90% of yield or break strength, while peel test results be 62% of yield or break strength. What must be clearly defined in applying such percentages is the base strength values.

FIGURE 17.5 A Trial (or Test) Weld Being Conducted by a Welding Technician

For HDPE the GRI GM13 specification values are recommended, while for LLDPE the GRI GM17 specification values should be used.

(iv) An example guide for an acceptable trial weld sample is when the specimens pass peel and shear tests according to the criteria in Table 17.1.

(v) The trial weld must be repeated in its entirety if any of the trial weld samples fail in either peel or shear.

(vi) When second trial welds fail, the seaming apparatus and seamer should not be used for production welding until the deficiencies or conditions are corrected and two consecutive successful trial welds are achieved (USEPA, 1993).

Trial Weld Documentation

The CQA monitor must be present during peel and shear testing and must record the date, time, operator, machine number, ambient and operating temperatures, speed setting, peel values, and whether the trial seam passes or fails. All trial weld records should be maintained on a trial weld form. The CQA monitor will give final approval to proceed with welding after observing trial welds.

17.2.2.3 General Seaming Procedures

(i) The CQC monitor must inspect the seam area to assure it is clean and free of moisture, dust, dirt, and debris of any kind. The rolls of geomembrane should be overlapped by a minimum of four inches for fusion welding and three inches for extrusion welding.

TABLE 17.1 Suggested Guide for Trial Seam Strengths (see also GRI GM20)

Type of Material	Thickness		Fusion Seam				Extrusion Seam			
			Shear		Peel		Shear		Peel	
	mils	mm	lb/in	kN/m	lb/in	kN/m	lb/in	kN/m	lb/in	KN/m
HDPE	30	0.75	63	11	49	8.6	63	11	35	6.1
	40	1	86	15	67	12	86	15	48	8.4
	60	1.5	126	22	98	17	126	22	70	12
	80	2	171	30	133	23	171	30	95	17
	100	2.5	216	38	168	29	216	38	115	20
Textured HDPE	40	1	76	13	60	11	76	13	42	7.4
	60	1.5	113	20	88	15	113	20	63	11
	80	2	151	26	118	21	151	26	84	15
LLDPE	20	0.5	20	3.5	17	3.0	20	3.5	17	3.0
	30	0.75	30	5.3	23	4.0	30	5.3	23	4.0
	40	1	40	7.0	35	6.1	40	7.0	35	6.1
	60	1.5	50	8.8	52	9.1	50	8.8	52	9.1
	80	2	60	11	70	12	60	11	70	12
	100	2.5	100	18	88	15	100	18	88	15
Textured LLDPE	40	1	40	7.0	30	5.3	40	7.0	35	6.1
	60	1.5	60	11	50	8.8	60	11	52	5.1

644 Chapter 17 Installation of Geosynthetic Materials

- **(ii)** The welding technicians must periodically check the machine operating temperature and speed, and mark this information on the geomembrane. For fusion welding, a movable protective layer of plastic may be required to be placed directly below the overlapped membranes being seamed. This is to prevent any moisture buildup between the sheets to be welded and/or to provide consistent rate of speed for the wedge welding device.
- **(iii)** "Fishmouths" or wrinkles at seam overlaps should cut along the ridge of the wrinkle in order to achieve a flat overlap. The cut "fishmouth" or wrinkles should be extrusion seamed and any portion where the overlap is inadequate should then be patched with an oval or round patch of the same geomembrane, extending a minimum of six inches beyond the cut in all directions.
- **(iv)** Seaming should extend to the outside edge of the panel to be placed in the anchor trench. All cross seams are to be extrusion welded where they intersect. The top flap of geomembrane is removed in the area to be extrusion welded and the weld area is ground parallel to the seam prior to welding.
- **(v)** Whenever possible, welding technicians should cut a specimen at the end of every seam. Prior to welding the next seam, the specimen will be tested using the field tensiometer. The CQA monitor may request additional trial welds, based on the testing result.
- **(vi)** In the event that there are noncomplying seam test strips, the welding machine will be taken out of service until a passing trial weld is obtained. Additional specimens must be taken to localize the flaw.
- **(vii)** The CQA monitor may take destructive samples from any seam, if defects are suspected. Results of field seam test strips will be maintained in the destructive test column on the panel seaming form.

Figure 17.6 shows a welding technician conducting seaming using a hot wedge machine.

17.2.2.4 Seaming Weather Conditions. A very important factor affecting seaming quality is the weather condition during seaming. The following are some general guidelines; however, the site-specific QA document will always take precedence over these suggestions.

Normal Weather Conditions
Normal weather conditions for seaming are as follows:

- **(i)** Temperature between 40°F and 104°F (4.5°C and 40°C).
- **(ii)** Dry conditions (i.e., no precipitation or other excessive moisture, such as fog or dew).
- **(iii)** Winds less than of 20 miles/hour (32 km/hour).

The Construction Quality Assurance (CQA) monitor should verify that these weather conditions are fulfilled and notify the Project Manager in writing if they are not. Ambient temperature shall be measured by the CQA Monitor in the area where the panels are to be placed. The Project Manager will then decide if the installation is to

FIGURE 17.6 Hot Wedge Fusion Seaming

be stopped, or if special procedures are to be used. Perhaps more important than ambient conditions is the temperature of the sheets to be seamed. Light-colored sheets can be seamed at significantly higher ambient temperatures than those listed previously.

Cold Weather Conditions

To ensure a quality installation, if seaming is conducted when the ambient temperature is below 40°F (4.5°C), the following conditions must be met (Blayer, 1993; GLSI, 1993):

(i) Geomembrane surface temperatures should be determined by the CQA monitor at intervals of at least once per 100 feet (30 m) of seam length to determine if preheating is required. For extrusion welding, preheating is required if the surface temperature of the geomembrane is below 40°F (4.5°C).

(ii) Preheating may be waived by the project manager based on a recommendation from the CQA engineer, if the Installer demonstrates to the CQA engineer's satisfaction that welds of equivalent quality may be obtained without preheating at the expected temperature of installation.

(iii) If preheating is required, the CQA monitor shall inspect all areas of geomembrane that have been preheated by a hot air device prior to seaming to ensure that they have not been overheated. All preheating devices should be approved prior to use by the project manager.

(iv) Care should be taken to confirm that the surface temperature is not lower than the minimum surface temperature specified for welding, due to winds or other adverse conditions. It may be necessary to provide wind protection for the seam area.

(v) Additional destructive tests should be taken at an interval between 500 feet (150 m) and 250 feet (75 m) of seam length, at the discretion of the CQA monitor.

(vi) Sheet grinding may be performed before preheating, if applicable.

(vii) Trial seaming should be conducted under the same ambient temperature and preheating conditions as the actual seams. Under cold weather conditions, new trial seams should be conducted if the ambient temperature changes by more than 5°F degrees (or 3°C degrees) from the initial trial seam test conditions.

Warm Weather Conditions

At high ambient (or sheet) temperatures, considerable concern over seaming of the geomembrane should be expressed unless the installer can demonstrate, to the satisfaction of the project manager, that geomembrane seam quality is not compromised. Trial seaming should be conducted under the same ambient temperature conditions as the actual seams. At the option of the CQA monitor, additional destructive tests may be required for any suspect areas.

17.2.2.5 Seaming Documentation. All seaming operations must be documented by CQA monitor or a designated assistant. At the start of each seam, the welding technicians must record the following information on the liner using permanent markers: date, time, welding technician ID, machine number, machine operating temperature, and speed. The CQA monitor or assistant records date, time, seam number, technician ID, machine ID, set temperature, speed, and weather conditions on the panel seaming form.

The welding technicians periodically check the operating temperature and speed, then mark the information on the seam. The CQA monitor must make periodic checks of the welding operations to verify overlap, cleanliness, etc.

17.2.3 Geomembrane Seam Tests

The welded seam created by the dual track fusion welding process is composed of an outer (primary) seam and an inner (secondary) track that creates an unwelded channel. The presence of an unwelded channel permits the fusion seams to be tested by inflating the sealed channel with air to a predetermined pressure and observing the stability of the pressurized channel over time.

17.2.3.1 Air Pressure Test. An air pressure test is a nondestructive test used to check the quality of fusion-welded seams.

Equipment for Air Testing

(i) An air pump capable of generating and sustaining a pressure from 20 to 60 psi (138 to 414 kN/m^2).

(ii) A rubber hose with fittings and connections.

FIGURE 17.7 Air Pressure Test for Dual Channel Fusion Hot Wedge Seams

(iii) A sharp hollow needle, or other approved pressure feed device with a pressure gauge capable of reading and sustaining a pressure between 0 and 60 psi (0 to 414 kN/m^2).

Procedure for Air Testing

(i) Seal both ends of the seam to be tested. Insert the needle or other approved pressure feed device into the sealed channel created by the fusion weld (Figure 17.7).

(ii) According to the procedures of ASTM D5820 and the specific values for different geomembranes per GRI-GM6 the inflation pressures and maximum pressure drops are given in Tables 17.2 to 17.7.

TABLE 17.2 Air Pressure Inflation Schedule for HDPE and Coextruded Polyethylene Geomembranes (both Smooth and Textured Types)

Geomembrane Thickness		Minimum Pressure		Maximum Pressure	
mil	mm	lb/in^2	kPa	lb/in^2	kPa
40	1.0	24	165	30	205
60	1.5	27	185	30	205
80	2.0	30	205	35	240
100	2.5	30	205	35	240

TABLE 17.3 Air Pressure Inflation Schedule for VLDPE and LLDPE (Smooth and Textured) and PVC, fPP Geomembranes and Other Nonreinforced Flexible Geomembranes

Geomembrane Thickness		Minimum Pressure		Maximum Pressure	
mil	mm	lb/in^2	kPa	lb/in^2	kPa
20	0.5	10	70	20	140
30	0.75	15	105	25	170
40	1.0	20	140	30	205
50	1.25	25	170	35	240
60	1.5	25	170	35	240

(iii) At the conclusion of all pressure tests, the end of the air-channel opposite the pressure gauge is cut. A decrease in gauge pressure must be observed, or the air channel is considered "blocked" and the test will have to be repeated from the point of blockage. If the point of blockage cannot be found, cut the air channel in the middle of the seam and treat each half as a separate test.

(v) Remove the pressure feed needle and seal the resulting hole by extrusion welding.

Checking Procedure for a Noncomplying Air Test

(i) Check seam end seals and retest seams. If a seam will not maintain the specified pressure, the seam should be visually inspected to localize the flaw. If this method is unsuccessful, cut a 1 inch sample from each end of the seam.

(ii) Perform destructive peel tests on the samples using the field tensiometer.

(iii) If all samples pass destructive testing, remove the overlap left by the wedge welder and vacuum test the entire length of the seam.
  (a) If a leak is located by the vacuum test, repair by extrusion fillet welding. Test the repair by vacuum testing.
  (b) If no leak is discovered by vacuum testing, the seam will be considered to have passed non-destructive testing.

(iv) If one or more peel specimens are not in compliance, additional samples must be taken.

TABLE 17.4 Air Pressure Inflation Schedule for CSPE-R and fPP-R Geomembranes (and Other Scrim Reinforced Flexible Geomembranes)

Geomembrane Thickness		Minimum Pressure		Maximum Pressure	
mil	mm	lb/in^2	kPa	lb/in^2	kPa
36	0.90	20	140	30	205
45	1.15	25	170	35	240

TABLE 17.5 Maximum Pressure Drop Schedule for HDPE and Coextruded Polyethylene Geomembranes (both Smooth and Textured Types)

Geomembrane Thickness		Maximum Pressure over 5 minutes	
mil	mm	lb/in^2	kPa
40	1.0	4.0	27
60	1.5	3.0	21
80	2.0	2.0	14
100	2.5	2.0	14

(a) When two passing samples are located, the length of seam bounded by the two passing test locations will be considered non-complying. The overlap left by the wedge welder will be heat tacked in place along the entire length of seam and the non-complying portion of seam will be extrusion fillet welded.

(b) Test the entire length of the repair seam by vacuum testing.

Air Pressure Testing Documentation

All information regarding air-pressure testing, (date, initial time and pressure, final time and pressure, pass/fail designation, and technicians initials) must be written at both ends of the seam, or portion of seam tested. All of the above information must also be recorded on the non-destructive testing form by the CQA monitor.

17.2.3.2 Vacuum Test. A vacuum test is a nondestructive test, which is used to check the quality of extrusion welded seams and locations where repairs (e.g., destructive tests) have occurred. Vacuum tests also can be used on fusion welded seams when the geometry of a fusion weld makes air pressure testing impossible or impractical, and when attempting to locate the precise location of a defect believed to exist after air pressure testing.

TABLE 17.6 Maximum Pressure Drop Schedule for VLDPE and LLDPE(Smooth and Textured) and PVC, fPP Geomembranes and Other Nonreinforced Flexible Geomembranes

Geomembrane Thickness		Maximum Pressure over 2 minutes	
mil	mm	lb/in^2	kPa
20	0.5	5.0	35
30	0.75	5.0	35
40	1.0	4.0	27
50	1.25	4.0	27
60	1.5	3.0	20

TABLE 17.7 Maximum Pressure Drop Schedule for CSPE-R and fPP-R Geomembranes (and Other Scrim Reinforced Flexible Geomembranes)

Geomembrane Thickness		Maximum Pressure over 3 minutes	
mil	mm	lb/in^2	kPa
36	0.90	4.0	27
45	1.15	3.0	20

Equipment for Vacuum Testing

  (i) Vacuum box assembly consisting of a rigid housing with a soft neoprene gasket attached to the bottom, a transparent viewing window, port hole or valve assembly, and a vacuum gauge.
  (ii) Vacuum pump assembly equipped with a pressure controller and pipe connection.
  (iii) A rubber pressure/vacuum hose with fittings and connections.
  (iv) A bucket and means to apply a soapy solution.
  (v) A detergent soap and water solution.

Procedure for Vacuum Testing

  (i) Trim excess overlap (if any) from the seam.
  (ii) Turn on the vacuum pump to reduce the vacuum box to approximately 10 inches (250 mm) of mercury [i.e., 5 psi (34.5 kN/m^2) gauge].
  (iii) Apply a generous amount of a strong solution of liquid detergent and water to the area to be tested.
  (iv) Place the vacuum box over the area to be tested and apply sufficient downward pressure to "seat" the seal strip against the liner (see Figure 17.8).
  (v) Close the bleed valve and open the vacuum valve.
  (vi) Apply approximately 5 psi (34.5 kN/m^2) vacuum to the area as indicated by the gauge on the vacuum box.
  (vii) Ensure that a leak tight seal is created.
  (viii) For a period of approximately 10 to 15 seconds, examine the geomembrane through the viewing window for the presence of soap bubbles.
  (ix) If no bubbles appear after 15 seconds, close the vacuum valve, open the bleed valve, and move the box over to the next adjoining area. Repeat the process with a minimum three-inch overlap of the previous test.
  (x) Special attention shall be exercised when vacuum testing "T" seams or patch intersections with seams.
  (xi) Mark and repair all areas where the soap bubbles. Retest repaired areas.

FIGURE 17.8 Vacuum Test for Extrusion Fillet Seams

Vacuum Testing Documentation

The vacuum testing crew must use permanent markers to write on the liner indicating the operators initials, date, and pass/fail designation on all areas tested. Records of vacuum testing must be documented by the GQA monitor on the non-destructive testing form.

17.2.3.3 Destructive Tests. The purpose of destructive testing is to determine and evaluate seam strength. These tests require direct sampling and subsequent patching. Therefore, destructive testing should be held to a minimum to reduce the amount of repairs to the geomembrane.

Sampling Procedure for Destructive Testing

(i) Destructive test samples should be marked and cut out randomly according to the site-specific CQA document (see Figure 17.9). This is often a minimum of one test per 500 ft (150 m) of seam length. A strict requirement such as this can have the effect of penalizing good seamers and rewarding poor seamers. Alternatively, one should consider the methods of control charts or the method of attributes (Richardson, 1992). The method of attributes has been formalized as the GRI-GM14 test method. These alternative procedures have the effect of opening up the testing interval as destructive tests consistently pass. Conversely, they close the interval as more-and-more destructive tests fail.

652 Chapter 17 Installation of Geosynthetic Materials

FIGURE 17.9 Cutting a Destructive Test Sample

- **(ii)** The location of destructive samples should be selected by the CQA monitor, with samples cut by the installer. Additional samples for destructive tests will be taken at the direction of the CQA monitor in areas of contamination, visual flaws or other potential areas of faulty welds.
- **(iii)** Destructive samples should be taken and tested as soon as possible after the seams are welded (the same day), in order to receive test results in a timely manner.
- **(iv)** The CQA monitor observes all destructive testing and records the date, time, seam number, location, and test results on the destructive testing form.
- **(v)** All holes in the geomembrane resulting from obtaining the seam samples should be repaired immediately. All patches should be vacuum tested.
- **(vi)** Sample locations must be located on the layout drawing.
- **(vii)** Suggested size of samples:
  - **(a)** The sample should be 12 inches (0.3 m) wide by 36 inches (0.9 m) long with the seam centered lengthwise. The sample may be increased in size to account for independent laboratory testing by the owner, the owner's request, or because of specific project specifications. The sample is cut in thirds, with two pieces given to the CQA monitor and the other given to the installer.

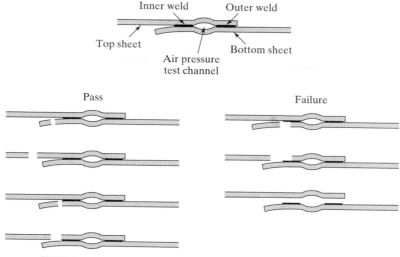

FIGURE 17.10 Destructive Sample Test Codes for Hot Wedge Welds

(b) The CQA monitor's laboratory should cut out ten one-inch wide replicate specimens from the destructive sample for testing. Five specimens should be tested for shear strength and the other five specimens, tested for peel strength. To be acceptable, all of the test specimens must pass, in accordance with the criteria in Table 17.1 or GRI GM20 and Figures 17.10 and 17.11. The figures describe the difference between a film tear bond (FTB) in which the sheet fails and is acceptable and a nonfilm tear bond (non-FTB) which is an unacceptable delamination of the seam.
(c) In cases of dispute, the installer uses his piece to conduct the same test.
(d) The third piece of sample is archived in the project files.

Laboratory Testing of Destructive Seam Samples

Seam destructive samples are packaged by the CQA monitor and sent to an independent laboratory for analysis. Destructive samples are tested for "Shear Strength" and "Peel Adhesion" although these terms are antiquated in the context of today's practices. Five specimens should be tested for each test method. Some CQA plans call for all five specimens of each type (shear and peel) to pass. Others call for 4 out of 5 passing, with the fifth being at least 80% (for example) of the specified value.

Procedure for Non-Complying Destructive Tests

The following procedures can apply whenever a sample fails a field or laboratory destructive test:

(i) The Installer can retrace the welding path to an intermediate location (a minimum of 10 feet from each side of the failed test) at the CQA monitor's discretion, and take a small sample for an additional field test. If this test passes, then laboratory samples can be cut and forwarded to the laboratory for full testing. If the test fails, the process is repeated.

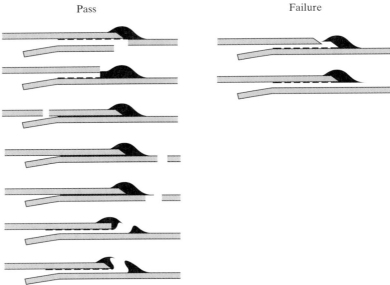

FIGURE 17.11   Destructive Sample Test Codes for Fillet Extrusion Welds

(ii) If the laboratory samples pass, then the seam is cap-stripped between the two passing sample locations.

(iii) If either of the samples are still not in compliance, then additional samples are taken in accordance with the above procedure until two passing samples are found to establish the zone in which the seam should be reconstructed.

(iv) All passing seams must be bounded by two locations from which samples passing laboratory destructive tests have been taken.

(v) In cases of reconstructed seams exceeding 150 feet (45 m), a sample from within the zone which the seam has been reconstructed should be taken and pass destructive testing.

(vi) All destructive seam samples should be numbered and recorded on a destructive test form.

17.2.4 Geomembrane Defects and Repairs

As each panel is deployed, the geomembrane surface should be observed by the CQA monitor for defects, holes, blisters, undispersed raw materials, and any sign of contamination by foreign matter. Because light reflected by the geomembrane helps detect defects, the surface of the geomembrane should be clean at the time of observation. Reflecting light will cause imperfection of the surface of the geomembrane to appear white or light in color. The geomembrane surface should be brushed, blown, or washed by the installer if the amount of dust, dirt or mud inhibits observation.

Each location suspected of having defects must be nondestructively tested as appropriate in the presence of the CQA monitor. Each location that fails nondestructive testing should be marked by CQA monitor, and repaired accordingly. A final "walk through" will be preformed to accept all portions of the geomembrane deployed; once this has occurred, the area should be partitioned off and no personnel should be permitted in this area.

Repair Procedures

Any portion of the geomembrane exhibiting a flaw or failing a destructive or non-destructive test must be repaired. Some suggestions in this regard are as follows:

(i) Small holes should be repaired by extrusion welding. If the hole is larger than 1/4 inch, it must be patched.
(ii) Failed seams must be cap-stripped.
(iii) Tears should be repaired by patching. If the tear is on a slope and has a sharp end, it should be rounded by cutting prior to patching.
(iv) Blisters, large holes, undispersed raw materials, and contamination by foreign matter should be repaired by large patches.
(v) Surfaces of the geomembrane that are to be patched should be abraded and cleaned no more than 15 minutes prior to the repair.
(vi) A folded geomembrane should be carefully inspected and possibly replaced. The QA document should be specific in how a wind-blown geomembrane is to be inspected, repaired, and possibly replaced. Patching should be permitted at the approval of the CQA monitor provided seams are parallel, and not transverse to the slope.

Patches should be round or oval in shape, made of the same geomembrane, and extend a minimum of 6 inches (150 mm) beyond all edges of the defect. All patches should be of the same compound and thickness as the geomembrane specified. All patches should have their top edge beveled with a grinder prior to placement on the geomembrane (see Figure 17.12). Patches must be applied using approved methods only. The extrudate rod used to make the repairs should be of the same geomembrane material as the sheet itself.

All surfaces must be clean and dry at the time of repairs. All seaming equipment used in repairs should be approved by the CQA monitor and installer. All repair procedures, materials, and techniques should be approved in advance of the specific repairs by the CQA monitor and installer.

Verification of Repairs

Each repair should be nondestructively tested, except when the CQA monitor requires a destructive seam sample obtained from a repaired seam. Repairs that pass the nondestructive test will generally be considered adequate. Failed tests indicate that the repair should be repeated and retested until passing test results are achieved.

FIGURE 17.12 Patching an Area where a Destructive Sample has been taken

Daily documentation of all nondestructive and destructive tests should be recorded by the CQA monitor. This documentation should identify all seams that initially failed the test, and evidence that these seams were repaired and successfully retested. Panel and seam location are recorded. The type of repair should be documented (i.e., patch, cap, etc). Identification of any cap strips that are repaired for failing a destructive test should be recorded. The repair locations should be identified on the record drawing. Figure 17.13 shows a landfill cell that is covered by a geomembrane. Figure 17.14 shows a geomembrane panel layout drawing with all defect and repair locations.

17.2.5 Geomembrane Protection and Backfilling

A field-deployed and seamed geomembrane must be backfilled with soil or covered with a subsequent layer of geosynthetics in a timely manner after its acceptance by the CQA personnel. If the covering layer is soil, it generally will be a drainage material, such as sand or gravel, depending upon the required permeability of the overlying layer. Depending upon the particle size, hardness, and angularity of this soil, a geotextile or other type of protection layer may be necessary. If the covering layer is a geosynthetic material, it will generally be a geonet or geocomposite drain, which is usually placed directly upon the geomembrane. This is obviously a critical step because geomembranes are relatively thin materials with limited puncture and tear strengths. Specifications should be clear and unequivocal regarding this final step in the installation survivability of geomembranes (USEPA, 1993).

FIGURE 17.13 A Landfill Cell Base Covered by a Geomembrane

17.2.5.1 Soil Backfilling of Geomembranes. There are at least three important considerations (USEPA, 1993) that affect soil backfilling of geomembranes: (i) type of soil backfill material, (ii) type of placement equipment, and (iii) considerations of slack in the geomembrane.

With regard to the type of soil backfill material used, its particle-size characteristics, hardness, and angularity are important. In general, the maximum soil particle size is the most important consideration. Poor gradation, increased angularity and increased hardness are of the greatest concern. Past research on puncture resistance of geomembranes have shown that HDPE and CSPE-R geomembranes are more sensitive to puncture than LLDPE and PVC geomembranes for conventional thicknesses of the respective types of geomembranes (Hullings and Koerner, 1991). The maximum backfill particle size for use with HDPE and CSPE-R geomembranes should not exceed 0.5 to 1.0 inch (12 to 25 mm). LLDPE and PVC geomembranes appear to be able to accommodate larger soil backfill particle sizes. If the soil particle size must exceed the approximate limits given (e.g., to provide high permeability in a drainage layer), then a protection material must be placed on top of the geomembrane and beneath the soil. Needle-punched nonwoven geotextiles, as well as other protection materials, have been used in this regard. New materials (e.g., recycled fiber geotextiles and rubber matting) are being evaluated as well.

With regard to the method and type of placement equipment, the initial lift height of the backfill soil is a primary consideration. (Note that construction equipment should *never* be allowed to move directly on any deployed geomembrane. This

658 Chapter 17 Installation of Geosynthetic Materials

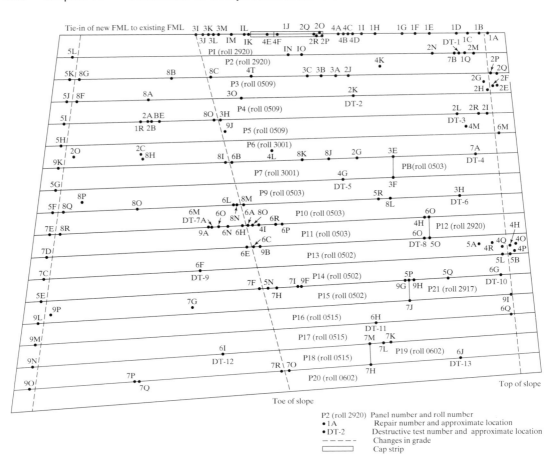

FIGURE 17.14 Geomembrane Panel Layout Drawing with Defect and Repair Locations

includes rubber-tired vehicles such as automobiles and pickup trucks, but does not include lightweight equipment like all-terrain vehicles (ATV's). The minimum initial lift height should be determined for the type of placement equipment and soil under consideration, however, 6 inches (150 mm) is usually considered to be a minimum. Between this lift height and approximately 12 inches (300 mm), low ground pressure placement equipment should be specified. Ground contact pressure equipment of less than 5.0 lb/in^2 (34.5 kPa) is recommended. For lift heights greater than 12 inches (300 mm), proportionately heavier placement equipment can be used.

Placement of soil backfilling should proceed from a stable working area adjacent to the deployed geomembrane and gradually progress outward. Soil is never to be dropped from dump trucks or front end loaders directly onto the geomembrane. The soil should be pushed forward in an upward tumbling action so as not to impact directly on the geomembrane. It should be placed by a bulldozer or front end loader, never by a motor grader which would necessarily have its front wheels riding directly on the geomembrane. Sometimes "fingers" of backfill are pushed out over the

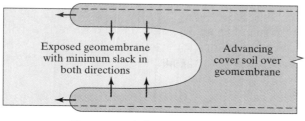

FIGURE 17.15 Advancing Primary Leachate Drainage Gravel in "Fingers" over the Deployed Geomembrane (USEPA, 1993)

geomembrane with controlled amounts of slack between them. Figure 17.15 shows a sketch of this type of soil covering placement. The backfill is then widened so as to connect the "fingers", with a minimum slack being induced into the geomembrane. This procedure is at the discretion of the design engineer and depends on site specific materials and conditions.

The issue of backfilling geomembranes that have waves (or "wrinkles") in them are particularly contentious (Soong et al., 1998). Most regulations call for "intimate contact" between the geomembrane and the underlying CCL or GCL. If the QA document calls for a literal interpretation, and if the CQA monitor insists on the geomembrane being perfectly flat, there will be major implications as far as installation costs are concerned. This issue must be discussed at the prebid construction meeting so that everyone involved understands this intent and the associated implications.

17.2.5.2 Geosynthetic Covering of Geomembranes. Various geosynthetic materials may be selected to cover the deployed and seamed geomembrane. Often, a geotextile or a geonet will be selected as the covering material. Sometimes, however, it will be a geogrid, as in the case of cover soil reinforcement on slopes or subbase reinforcement of the liner system placed over an existing landfill for a vertical landfill expansion project. A geogrid could even be added to as a drainage geocomposite on slopes to avoid instability of natural drainage soils. As with the previous discussion on soil covering, no construction vehicles of any type should be allowed to move directly on the geomembrane (or any other geosynthetic material for that matter). Generators, low tire inflation all-terrain vehicles (ATV's), hydraulic lifting devices, and other seaming-related equipment are allowed, as long as they do not damage the geomembrane. Proper planning and sequencing of the operations is important for logistical control. The geosynthetic materials are laid directly on the geomembrane with out allowing any type of bonding to the geomembrane. For example, thermally fusing a geonet to a geomembrane should not be permitted.

The geosynthetics placed above the geomembrane will either be overlapped (as with some geotextiles), sewn (as with other geotextiles), connected with plastic ties (as with geonets), mechanically joined with plastic rods or bars (as with geogrids), or male/female joined (as with some drainage geocomposites). These self-tying or attachment methods should not damage the underlying geomembrane.

17.3 INSTALLATION OF GEONETS

Geonets are always covered with either a geomembrane or a geotextile, i.e., they are never directly soil covered since the soil particles would fill the apertures of the geonet rendering it useless.

17.3.1 Geonet Placement

Geonet cleanliness is essential to its performance. Therefore, the geonet rolls should be protected against dust and dirt during shipment and storage. The CQA monitor should verify that the geonet is free of dirt and dust prior to installation. The CQA monitor should report the outcome of this verification to the project manager, and if the geonet is dirty or dusty, it should be washed by the installer prior to installation. Washing operations should be observed by the CQA monitor and improper washing operations should be reported to the project manager.

The installer should handle a geonet in a manner that ensures that it is not damaged in any way. The following guidelines should be considered:

(i) On slopes, the geonet should be secured in the anchor trench and then rolled down the slope in such a manner as to continually keep the geonet sheet in tension. If necessary, the geonet can be positioned by hand after being unrolled to minimize waves or wrinkles. Geonets can be placed in the horizontal direction (across the slope) in some special locations (e.g., at the toe of a slope, or where an extra layer of geonet is required). Such locations should be identified by the design engineer in the project drawings.

(ii) It is critical that triplanar geonets be oriented with their major flow direction aligned with the maximum gradient of the slope or base of the landfill.

(iii) Geonets should not be welded to geomembranes with extrusion welders. A geonet should be cut using approved cutters (i.e., hook blade, scissors, etc). Care should be taken to prevent damage to underlying layers.

(iv) Care must be taken not to entrap dirt in the geonet that could cause clogging of the drainage system, and/or stones that could damage the adjacent geomembrane.

Figure 17.16 shows two roles of geonet placed on the top of a slope ready for deployment.

17.3.2 Geonet Joining

When several layers of geonet are installed, care should be taken to prevent the strands of one layer from penetrating the channels of the next layer, thereby significantly reducing the transmissivity. Layered geonets must be placed in the same direction and never laid perpendicular to the underlying geonet (USEPA, 1993). Some suggestions about the joining of geonets are as follows:

(i) Adjacent roll edges of geonets should be overlapped by at least 3 to 4 inches (75 to 100 mm). The roll end of geonets should be overlapped 6 to 8 inches (150 to 200 mm). Since flow in the machine direction for triplanar geonets, they must be aligned properly (i.e., in the direction of maximum gradient).

FIGURE 17.16 Rolls of Geonets and Covering Geotextile Ready for Placement

- (ii) All overlaps should be joined by ties with plastic fasteners or polymeric braids (see Figure 17.17). Metallic ties or fasteners should not be used. The tying devices should be white or yellow, to contrast with the black geonet (for easy inspection).
- (iii) Tying should be done every five feet (1.5 m) along the edges, every six inches (150 mm) along the ends and in anchor trenches.
- (iv) Horizontal seams should not be allowed on side slopes. This requires that the length of the geonet should be at least as long as the sideslope, anchor trench, and a minimum run out at the bottom of the facility. If horizontal seams are allowed, they should be staggered from one roll to the adjacent roll.
- (v) In the corners of the sideslopes (where overlaps between perpendicular geonet strips are required), an extra layer of geonet should be unrolled along the slope, on top of the previously installed geonets, from top to bottom of the slope.
- (vi) If double geonets are used, they should be layered on top of one another so that interlocking does not occur, and the roll edges and ends should be staggered so that the joints do not lie above one another.

17.3.3 Geonet Repairs

Any holes or tears in geonets should be repaired by placing a geonet patch extending a minimum of 12 inches (300 mm) beyond the edges of the hole or tear. The patch is secured to the original geonet by tying to the underlying geonet at 6 inches (150 mm)

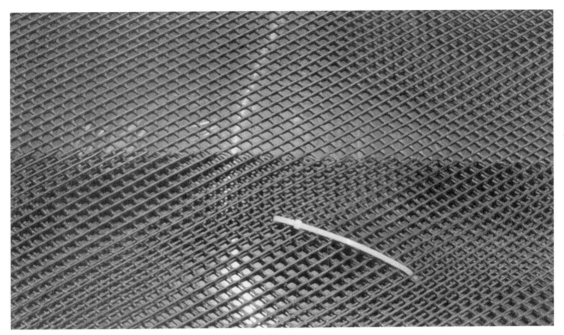

FIGURE 17.17 Geonet Tying

spacings. If the hole or tear width across the roll is more than 50% of the width of the roll, the damaged area should be cut out and the two potions of the geonet shall be joined.

17.4 INSTALLATION OF GEOTEXTILES

Geotextiles are usually placed over a geomembrane, geonet, or drainage soil. The overlying material can be a geomembrane, geosynthetic clay liner, compacted clay liner, geonet or drainage soil.

17.4.1 Geotextile Placement

During shipment and storage, the geotextile should be protected from ultraviolet light exposure, precipitation or other inundation, mud, dirt, dust, puncture, cutting, or any other damaging or deleterious conditions. Geotextile rolls are shipped and stored in relatively opaque and watertight wrappings for protection against ultraviolet exposure. The protective wrappings from the geotextile rolls should be removed and deployed only after the substrate layer, soil, or other geosynthetic material, has been approved by the CQA personnel. Some suggested guidelines (USEPA, 1993) that apply to geotextile placement are as follows:

(i) The installer should take the necessary precautions to protect the underlying layers upon which the geotextile will be placed. If the substrate is soil, construction equipment can be used provided that excess rutting is not created. In some cases one inch (25 mm) is the maximum rut depth allowed. If the ground freezes, the depth of ruts should be reduced to a lesser specified value. If the substrate is a geosynthetic material, deployment must be by hand, by use of small pneumatic jack lifts on pneumatic tires having low ground contact pressure, or by use of all-terrain vehicles (ATVs), having low ground contact pressure.

(ii) During placement, care must be taken not to entrap (either within or beneath the geotextile) stones, excessive dust or moisture that could damage a geomembrane, cause clogging of drains or filters, or hamper subsequent seaming.

(iii) On side slopes, the geotextiles should be anchored at the top and then unrolled down the slope so as to keep the geotextile free of folds and wrinkles.

(iv) Geotextiles should be cut using an approved cutter. If the material is being cut in place, special care must be taken to protect underlying geosynthetic materials from damage.

(v) Nonwoven geotextiles placed on textured geomembranes can be troublesome due to sticking and are difficult to align or even separate after they are placed on one another. A thin sheet of plastic on the geomembrane during deployment of the geotextile can be helpful in this regard. The sheet is removed after correct positioning of the geotextile.

(vi) All geotextiles should be weighted with sandbags, or the equivalent, to provide resistance against wind uplift. Such sandbags are installed during deployment and should remain until replaced with cover material.

(vii) After deployment a visual examination of the geotextile should be carried out over the entire surface to ensure that no potentially harmful foreign objects are present, such as stones, sharp objects, small tools, sandbags, etc.

Figure 17.18 shows several rolls of geotextile set on the top of the sideslope and positioned for placement down the slope.

17.4.2 Geotextile Overlapping and Seaming

Seaming of geotextiles, by sewing (see Figure 17.19), is common for geotextiles placed in waste facilities. Sewn seams are normally required for geotextiles used in filtration, but may be waived for geotextiles used in separation (e.g., as protective layers for geomembranes), or drainage (e.g., as gas collection layers above the waste) as per plans and specifications. In such cases, heat bonding is also an acceptable alternate method of joining separation geotextiles.

There are three types of sewn geotextile seams: *flat* or *prayer* seams, *J* seams, and *butterfly* seams. While each can be made by a single thread or by a two-thread chain stitch (as illustrated), the latter stitch is recommended. Furthermore, single, double, or even triple rows of stitches can be made. The following guidelines are suggested (USEPA, 1993):

(i) The type of seam, type of stitch, stitch count or number of stitches per inch and number of rows should be specified based on the tendency of the fabric to fray, strength need and toughness of the fabric. For filtration and separation

664 Chapter 17　Installation of Geosynthetic Materials

FIGURE 17.18　Geotextiles Positioned for Placement

FIGURE 17.19　Seaming of Geotextiles by Sewing

geotextiles a flat seam using a two-thread chain stitch and one row is usually specified. For reinforcement geotextiles, stronger and more complex seams are utilized. Alternatively, a minimum seam strength, per ASTM D4884, can be specified.

(ii) The seams should be continuous (i.e., spot sewing is generally not allowed).

(iii) On slopes greater than approximately 5(H):1(V), geotextiles are normally sewn continuously along the entire length of the seam. Geotextiles should be overlapped a minimum of 4 inches prior to sewing. In general, no horizontal seams should be allowed on side slopes (i.e. seams should be along, not across, the slope), except as small patches and repairs.

(iv) The thread must be polymeric with chemical and ultraviolet light resistant properties equal or greater than that of the geotextile itself.

(v) The color of the sewing thread should contrast that of the color of the geotextile for ease in visual inspection. This may not be possible due to polymer composition in some cases.

(vi) Heat seaming of geotextiles may be permitted for certain seams. A number of methods are available such as hot plate, hot knife and ultrasonic devices. If thermally bonded, the geotextile should be overlapped a minimum of eight inches prior to seaming.

(vii) Overlapping (without seaming) of geotextiles may be permitted for certain situations. The overlap distance depends on the specific site conditions and should be noted in the specifications or QA document.

17.4.3 Geotextile Defects and Repairs

Any holes or tears in geotextiles made during placement or any time before backfilling should be repaired. Some suggestions are as follows:

(i) The patch material used for repair of a hole or tear should be the same type of polymeric material as the damaged geotextile, or as approved by the CQA engineer.

(ii) The patch should extend at least 12 inches (300 mm) beyond any portion of the damaged geotextile.

(iii) The patch should be sewn in place by hand or machine so as not to accidentally shift out of position or be moved during backfilling or covering operations.

(iv) The machine direction of the patch should be aligned with the machine direction of the geotextile being repaired.

(v) The thread should be of contrasting color to the geotextile and of chemical and ultraviolet light resistance properties equal or greater than that of the geotextile itself.

Figure 17.20 shows a landfill cell that has been covered by geotextile panels that have been sewn together.

FIGURE 17.20 A Landfill Cell Covered by Geotextile Panels that have been Seamed Together

17.4.4 Geotextile Backfilling and Covering

The layer of material placed above the deployed geotextile will be either soil, waste or another geosynthetic material. Soils will vary from compacted clay layers to coarse aggregate drainage layers. Waste placed above a geotextile should be only what is referred to as "select" waste, (i.e., carefully separated from large puncturing objects and placed so as not to cause damage). Geosynthetics will vary from geomembranes to geosynthetic clay liners. The following suggestions apply to geotextile backfilling and covering (USEPA, 1993):

(i) If soil is used to cover the geotextile it should be done in such a way that the geotextile is not shifted from its intended position and underlying materials are not exposed or damaged.

(ii) If a geosynthetic material covers the geotextile, both the underlying geotextile and the newly deployed material should not be damaged during the process.

(iii) If solid waste covers the geotextile, the type of waste should be carefully specified and observed by CQA personnel.

(iv) The overlying material should not be deployed such that excess tensile stress is mobilized in the geotextile. On side slopes, this requires soil backfilling to proceed from the bottom of the slope upward.

(v) Soil backfilling or covering by another geosynthetic, should be done within the time frame stipulated for the particular type of geotextile. Typical time frames for geotextiles are within 14 days for polypropylene and 28 days for polyester geotextiles.

17.5 INSTALLATION OF GEOCOMPOSITES

Drainage geocomposites are usually placed on the sideslopes for drainage so as to reduce seepage pressure and provide stability. This stability is further provided by thermally bonding the geotextile(s) to the geonet in the factory.

17.5.1 Geocomposite Placement

The placement of drainage geocomposite in the field is similar to that described for geonet and geotextile. (Refer to Sections 17.3.1 and 17.4.1 for details.)

17.5.2 Geocomposite Joining and Repairs

Drainage geocomposites are usually joined together by folding back the geotextile from the geonet cores, overlapping the geonet cores a minimum distance (typically 3 to 4 inches or 75 to 100 mm along the edges, and 6 to 8 inches or 150 to 200 mm along the ends), and using plastic fasteners or polymer braid to tie adjacent ribs together at minimum intervals (typically every 5 feet or 1.5 m along the edges and every 6 inches or 150 mm along the ends) (see Figure 17.21). The geotextile must be refolded over the connection area (top and bottom) assuring a complete covering of the core's surface.

The following suggestions apply to the joining of drainage geocomposite materials:

(i) Adjacent edges of drainage cores should be overlapped for at least three ribs (or two apertures), or 3 to 4 inches (75 to 100 mm) for geonet core.

(ii) The ends of drainage cores (in the direction of flow) should be overlapped for at least five ribs or four apertures or 6 to 8 inches (150 to 200 mm) for geonet core.

(iii) The geotextiles covering the joined cores must provide a complete seal against backfill soil entering into the core (see Figure 17.22).

(iv) Horizontal seams should not be allowed on side slopes. This requires that the drainage geocomposite be provided in rolls that are at least as long as the side slope.

(v) Holes or tears in drainage cores are repaired by placing a patch of the same type of material over the damaged area. The patch should extend at least five ribs or four apertures or 6 to 8 inches (150 to 200 mm) beyond the edges of the hole or tear.

(vi) Holes or tears of more than 50% of the width of the drainage core on side slopes should require the entire length of the drainage core to be removed and replaced.

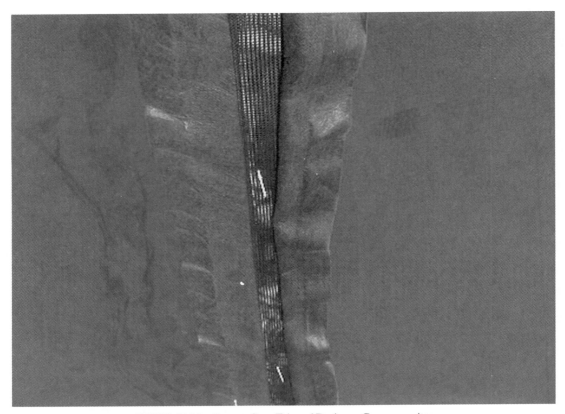

FIGURE 17.21   Geonet Core Tying of Drainage Geocomposites

(vii) Holes or tears in the geotextile covering the drainage core should be repaired as described in Section 16.4.3.

17.5.3 Geocomposite Covering

Drainage geocomposites with a geotextile are covered with either soil, waste, or in some cases, a geomembrane. The following suggestions should be considered (USEPA, 1993):

  (i) The core of the drainage geocomposite should be free of soil, dust and accumulated debris before backfilling or covering with a geomembrane. In extreme cases this may require washing of the core to accumulate the particulate material to the low-end (sump) area for removal.
 (ii) Placement of the backfilling soil, waste or geomembrane should not shift the position of the drainage geocomposite nor damage the underlying drainage geocomposite, geotextile or core.
(iii) When using soil or waste as backfill on a side slope, work begins at the toe of the slope and extends upward.

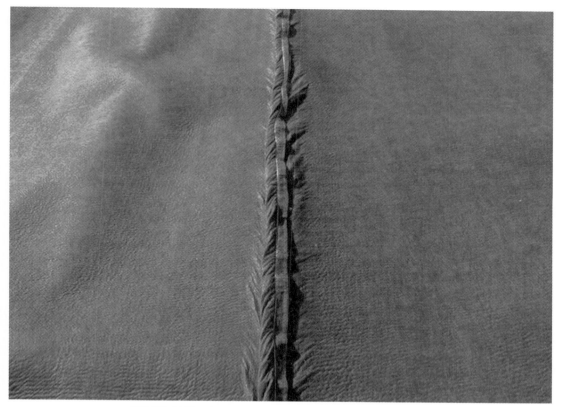

FIGURE 17.22 Geotextile Seaming of Drainage Geocomposites

17.6 INSTALLATION OF GEOSYNTHETIC CLAY LINERS

This section provides procedures for the placement, joining, repairing and covering of geosynthetic clay liners. The performance of the geosynthetic clay liner is wholly dependent on the quality of its installation (USEPA, 1993; Koerner et al, 1995; Well, 1997).

17.6.1 Geosynthetic Clay Liner Placement

The installation contractor should remove the protective wrapping from the rolls to be deployed only after the substrate layer (soil or other geosynthetic material) in the field has been approved by CQA personnel. The wrapping is necessary to protect the geotextiles from ultraviolet exposure and the bentonite from water absorption. Figure 17.23 shows a roll of geosynthetic clay liner that has been shipped to a construction site and removed from its protective wrapping. The specification and CQA documents should be written in such a manner as to ensure that the GCLs are not damaged in any way. A CQA monitor should be present at all times during the handling, placement and covering of GCLs. The following suggestions also should be considered (USEPA, 1993):

670 Chapter 17 Installation of Geosynthetic Materials

FIGURE 17.23 A Roll of Geosynthetic Clay Liner Removed from its Protective Wrapping

(i) The installer should take the necessary precautions to protect materials underlying the GCL. If the substrate is soil, construction equipment can be used to deploy the GCL providing excessive rutting is not created. Excessive rutting should be clearly defined and qualified. In some cases one inch (25 mm) is the maximum rut depth allowed. If the ground freezes, the depth of ruts should be reduced to a lesser value. If the substrate is a geosynthetic material, GCL deployment should be by hand, or by use of pneumatic jack lift or lightweight equipment on tires having low ground contact pressure.

(ii) There are four different deployment methods that should be considered (manual unroll, gravity roll release, stationary roll pull, and moving roll pull), see Tranger and Tewes (1995) for details.

(iii) The minimum overlap distance that is specified should be verified. This is typically 6 to 12 inches (150 to 300 mm) depending upon the particular product and site conditions.

(iv) Additional bentonite is introduced into the overlap region with certain types of GCLs. There are typically those with needle punched nonwoven geotextiles on their opposing surfaces. The clay is usually added by using a lime spreader or line chalker with the bentonite clay in a dry state. Alternatively, a bentonite clay paste, in the mixture range of 4 to 6 parts water to 1 part of clay, can be extruded in the overlap region. The manufacturer's recommendations on type and quantity of clay to be added should be followed.

(v) During placement, care must be taken not to entrap in or beneath the GCL, fugitive clay, stones, or sand that could damage a geomembrane, cause clogging of drains or filters, or hamper subsequent seaming of materials either beneath or above the GCL.

(vi) On side slopes, the GCL should be anchored at the top and then unrolled to keep the material free of wrinkles and folds.

(vii) Trimming of the GCL should be done with great care so that fugitive clay particles do not come in contact with drainage materials, such as geonets, geocomposites, or natural drainage materials.

(viii) The deployed GCL should be visually inspected to ensure that no potentially harmful objects are present, e.g., stones, cutting blades, small tools, sandbags, etc.

(ix) The deployed GCL should be inspected with a metal detector to assure that no broken needles from needle punching or stitch bonding operations are present.

Figure 17.24 shows geosynthetic clay liners being deployed panel by panel in a landfill cell.

17.6.2 Geosynthetic Clay Liner Joining

Joining of GCLs is generally accomplished by overlapping without sewing or other mechanical connections. The overlap distance requirements should be clearly stated. For all GCLs, the required overlap distance should be marked on the underlying layer

FIGURE 17.24 Geosynthetic Clay Liner Placement

by a pair of continuous guidelines that are usually printed on the surface of the GCLs. The overlap distance is typically 6 to 12 inches (150 to 300 mm). For those GCLs with needle-punched nonwoven geotextiles on their surfaces, dry bentonite is generally placed in the overlapped region. In this case, extreme care should be taken to avoid fugitive bentonite particles from coming into contact with leachate collection systems. Fugitive bentonite is also troublesome for seaming of geomembranes, particularly texture types. A preferred alternative is to extrude a moistened tube of bentonite into the overlapped region.

17.6.3 Geosynthetic Clay Liner Repairs

For the geotextile-related GCLs, holes, tears or rips in the covering geotextiles made during transportation, handling, placement, or anytime before backfilling, should be repaired by patching using a geotextile. If the bentonite component of the GCL is disturbed—either by loss of material or by shifting—it should be covered using a full GCL patch of the same type of product.

The following suggestions apply (USEPA, 1993):

 (i) Any patch used for repair of a tear or rip in the geotextile should be done using the same type as the damaged geotextile or other approved geotextile by the CQA engineer.
 (ii) The size of the geotextile patch must extend at least 12 inches (30 cm) beyond any portion of the damaged geotextile and be adhesive or heat bonded to the product to avoid shifting during backfilling with soil or covering with another geosynthetic material.
 (iii) If bentonite particles are lost from within the GCL or if the clay has shifted, the patch should consist of the full GCL product. It should extend at least 12 inches (30 cm) beyond the extent of the damage at all locations. For those GCLs requiring additional bentonite clay in overlap seaming, a similar procedure should be use for patching.
 (iv) Particular care should be exercised in using a GCL patch since fugitive clay can be lost, which then can find its way into drainage materials or onto geomembranes in areas that eventually are to be seamed together.

17.6.4 Geosynthetic Clay Liner Backfilling or Covering

The layer of material placed above the deployed GCL will be either soil or another geosynthetic material. Soil types will vary from compacted clay liners to coarse aggregate drainage layers. Geosynthetics will generally be geomembranes, although other geosynthetics may also be encountered, depending on the specific site design. The GCL should generally be covered before a rainfall or snow event occurs. The reason for covering with the adhesive-bonded GCLs is that hydration before covering can cause changes in thickness as a result of uneven swelling or whenever compressive or shear loads are encountered. Hydration before covering may be less of a concern for the needle-punched types of GCLs, but migration of the fully hydrated clay in these products might also be possible under sustained compressive or shear loading.

If soil covers the GCL, it should be done in such a way that the GCL or underlying materials are not damaged. The direction of backfilling should proceed in the direction of downgradient shingling of the GCL overlaps.

The overlying material should not be deployed such that excess tensile stress is mobilized in the GCL. On side slopes, this requires soil backfill to proceed from the bottom of the slope upward.

PROBLEMS

17.1 What is the difference between construction quality assurance (CQA) and construction quality control (CQC)?

17.2 Describe the purpose of conformance tests?

17.3 What are the sampling frequencies of conformance tests for various geosynthetic materials?

17.4 What requirements should a subgrade meet before deployment of a geomembrane?

17.5 Describe two different seaming methods for geomembrane panels.

17.6 What is the purpose of trial welds? How often should they be conducted?

17.7 List the factors influencing seeming quality.

17.8 Under what type of weather conditions should field seaming be conducted?

17.9 Describe the basic principle of the vacuum test that is used to check the seam quality.

17.10 How would you find the total length of the seam that must be reconstructed and repaired if a sample fails a destructive laboratory test?

17.11 Explain how to repair various geomembrane defects and defective seams and how to make sure the repaired spots meet quality standards.

17.12 Describe how to protect a geomembrane after placement.

17.13 Why should the geonet be washed prior to installation if it is judged to be dirty or dusty?

17.14 Explain how to join two geonet panels together.

17.15 What are the requirements for effective geotextile seaming?

17.16 Explain how to prevent damaging a geosynthetic clay liner during placement.

17.17 How is a geosynthetic clay liner joined together?

REFERENCES

Blayer, S. R., (1993) "Brent Run Landfill Cold Weather Seaming," Michigan Department of Environmental Quality, Waste Management Division, Morrice, MI.

Diaz, V. A. and Myles, B., (1990) "The Field Seaming of Geosynthetics," Industrial Fabrics Association International, St. Paul, MN.

Hullings, D. E. and Koerner, R. M., (1991) "Puncture Resistance of Geomembranes Using a Truncated Cone Test," *Proceedings,* Geosynthetics '91 Conf., Vol. 1, pp. 273–285.

Koerner, R. M., Gartung, E., and Zanzinger, H., (1995) *Geosynthetic Clay Liners,* A. A. Balkema, Rotterdam, 237 pp.

Richardson, G. N., (1992) "Construction Quality Management for Remedial Action and Remedial Design Waste Containment Systems," EPA/540/R-92/07/073, EPA, Washington, DC.

Soong, T-. Y. and Koerner, R. M., (1998) "Laboratory Study of HDPE Geomembrane Waves," Proceedings, 6ICG, Atlanta, GA, IFAI, pp. 301–306.

Tranger, R. and Tewes, K., (1995) "Design and Distillation of a State-of-the-Art Landfill Liner System," In *Geosynthetic Clay Liners,* Koerner R. M. et al. Eds., A. A. Balkema, Rotterdam, pp. 175–182.

Well, L. W. Ed., (1997) "Testing and Acceptance Criteria for Geosynthetic Clay Liners," STP 1308, ASTM, West Conshohockten, PA, 208 pp.

USEPA, (1993) "Quality Assurance and Quality Control for Waste Containment Facilities," Technical Guidance Document, EPA/600/R-93/182, U.S. Environmental Protection Agency, Office of Research and Development, Washington, DC., September.

CHAPTER 18

Postclosure Uses of MSW Landfills

18.1 ATHLETIC AND RECREATIONAL FACILITIES
    18.1.1 GOLF COURSES/DRIVING RANGES
    18.1.2 SPORT FIELDS
    18.1.3 PATHS, TRAILS, AND NATURE WALKS
    18.1.4 WILDLIFE AND CONSERVATION AREAS
    18.1.5 MULTIPLE USE FACILITIES
18.2 INDUSTRIAL DEVELOPMENT
    18.2.1 PARKING LOTS
    18.2.2 EQUIPMENT/MATERIAL STORAGE
    18.2.3 LIGHT INDUSTRIAL BUILDINGS
18.3 AESTHETICS
18.4 CONCLUDING REMARKS
    PROBLEMS
    REFERENCES

Closed landfills represent some of the largest human-made structures that exist. Furthermore, they often are located near populated areas where visibility by the public is high and potential utilization of the closed site would be advantageous. There are certain potential risks or hazards associated with landfill reuse that may affect a particular use. These must be evaluated on a case-by-case basis. These risks include methane gas generation, differential settlement, and possible conflicts with gas wellheads and leachate collection lines. In regard to gas generation, it is important to recognize that with a geomembrane in the final cover the unpleasant odor of methane is essentially eliminated. A closed landfill represents a quasi-stable system, with ongoing settlement; nevertheless, beneficial or compatible uses still can be found. In this regard, it is important to assess the remaining settlement (recall Chapter 12 for the calculation procedure), since this value will be an essential input parameter into many of the suggested uses that will be described in this concluding chapter of the book.

18.1 ATHLETIC AND RECREATIONAL FACILITIES

The most feasible and widely practiced use of closed MSW landfills are for various types of athletic activities. Some reasons for this are the following:

    (i) relatively short term periodic use of the area

(ii) no overnight use of the area

(iii) general tolerance to limited settlement

(iv) no intersecting streets or vehicular activity

(v) often conveniently located to populated areas

18.1.1 Golf Courses/Driving Ranges

Golf courses and driving ranges constitute a major use of closed MSW landfills. All of the reasons cited previously play into such uses. The oldest ongoing golf course of this type is located in Industry Hills, California, which is about 20 miles east of Los Angeles, California. Here, two 18-hole courses were built in the 1970s and have functioned ever since. The landfill itself was placed from 1940 to 1960, and thus settlements and gas release are both in Phase V of the degradation process (as shown in Figures 10.1 and 15.1). Grading is an ongoing maintenance item because of continuing settlement; on the other hand, the location's convenience as a golf course is appealing enough that the costs are readily justified.

Converting closed landfills to golf courses currently is so prevalent that the National Golf Foundation is in its second edition of a compendium of articles on the subject (National Golf Foundation, 1999). This collection of 61 articles presents some of the advantages and disadvantages of developing golf facilities over landfills, stripmines, quarries, volcanic areas, and Superfund cleanup sites. Often, an article describes the case history of how one of these sites was converted into a golf course. Some examples are a Delaware facility built using coal ash from a nearby factory as fill, a new facility on the shore of Lake Michigan built over an old cement factory and gravel quarry, a New Jersey course built over the site of an old landfill, an award-winning course built on a former coal mine, and many others. Several articles address various problems associated with these types of alternative sites for building golf courses and driving ranges.

Conrad (2000) reports on the Britannia Hills Golf Course located on 80 acres (32 ha) of the 200 acres (81 ha) making up the Britannia Sanitary Landfill Site in the region of Peel, Canada. The location is adjacent to the city of Toronto, Ontario. Three factors make the project innovative:

(i) The co-existence of two facilities on one property: a recycling depot and household hazardous waste area, along with the golf course.

(ii) Ongoing design and construction of the course while waste filling of adjacent areas of the landfill were still active.

(iii) A close partnership between the city, region and environmental agency.

A composite liner covered the site with a relatively thick layer of cover soil. The cap could not be altered to create the necessary golf course contouring. Thus, all contours were made out of fill added to the top of the existing site. A computerized irrigation system that provides the proper amount of water for grass growth was designed and installed. Gas is an ongoing concern, and the site has numerous gas monitors along with a small collection system around the clubhouse that was built at the edge of the site. When the remaining active portion of the landfill is closed, the golf course will expand from 18 to 27 holes. By this process, an already active golf course will be transformed into a championship golf course with revenue of more than $1 M per year.

18.1.2 Sport Fields

Somewhat more sensitive to postclosure settlement, yet still in the context of the attractiveness of athletic facilities to closed landfill sites, are various types of sport fields.

Mackey (1996) reports on the development of the 33-acre (13-ha) Sanlando Landfill facility located 10 miles (16 km) north of Orlando, Florida into five softball fields (see Figure 18.1). The landfill was operated from 1950 to 1970. Thus, by the time of site development the waste was 25 to 45 years old. With heavy proof rolling most of the settlement and gas generation was concluded. The final cover barrier layer was a compacted clay liner. There were several areas, however, which required the use of a geomembrane—for example, under several small buildings to eliminate gas migration, under stormwater detention areas, and under the entrance area fountain. Particularly difficult was the geomembrane beneath the central administrative/concession building since it was supported by 14 piles penetrating the clay cover and waste into the firm foundation soils. Note that most sites, particularly those with geomembranes, could not have such penetrations. The piles were rectangular and necessitated a round cylinder of concrete to be placed to allow for the geomembrane in the cover to be attached. Water and sewer pipes for the same building were spaced too close together for easy geomembrane boot placement. At least 24-inch (600-mm) spacing should generally be allowed. Individual electrical lines should not be used; instead, they should be bundled together and placed in a pipe conduit so that a geomembrane boot connection can be made. Regardless of the difficulties encountered, the site is currently completely functional and a good revenue generator for the county.

18.1.3 Paths, Trails, and Nature Walks

Depending on the location of the closed landfill, particularly its climate and hydrology, vegetation can be encouraged to grow, and a number of foot-traffic uses are possible. In general, shallow-rooted bushes, shrubs and vegetated plants are acceptable. Deep rooted trees are of concern because of the possibility of penetration into the underlying waste. This concern is greatest for soil barrier layers (i.e., CCLs and GCLs) and least for geomembranes. However, even in the case of compacted clay liners, root penetration is limited by depth and low porosity because of a root's need for air and oxygen. Furthermore, roots avoid not only unfavorable surroundings (viz., sunlight, lack of oxygen, lack of moisture, presence of methane) but they avoid mechanical barriers or obstacles as well. This bioadaptive characteristic of plant roots is termed *edaphoecotropism* (Vanicek, 1973), or more simply, *stress avoidance.* If a geomembrane is used, it should be relatively thick with high mechanical properties, such as puncture, tear and tensile resistance. Unfortunately, there is no research available to our knowledge regarding root penetration of intact geomembranes.

Mackey (1996) reports on the Dyer Boulevard Landfill in Palm Beach County, Florida that is a relatively high mounded closure. An unpaved mountain bike course

678 Chapter 18 Postclosure Uses of MSW Landfills

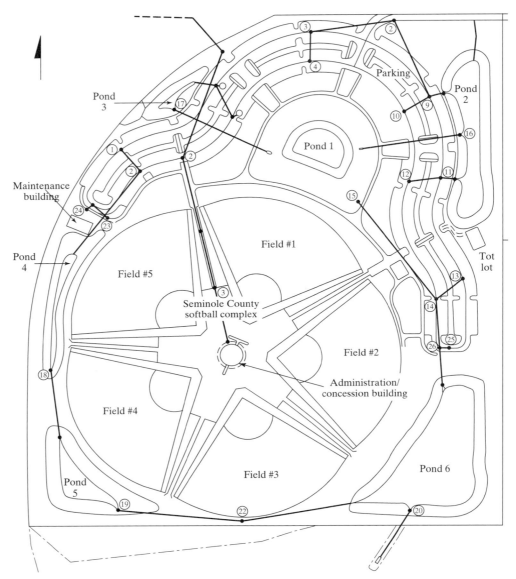

FIGURE 18.1 Sanlando Softball Complex (after Mackey, 1996)

was constructed atop the landfill cover. Some rutting was anticipated and a geotextile was placed under all of the trails. The functions of the geotextile were as (i) a reinforcement layer, (ii) a separation layer, and (iii) as an indicator of excessive rutting. Whenever the geotextile becomes exposed the trail must be regraded and filled accordingly.

18.1.4 Wildlife and Conservation Areas

There have been several successful attempts to reclaim closed landfills as wildlife areas and to provide facilities for environmental education and passive recreation. A good case in point is the Kingsland Overlook, a 6-acre reclaimed section of a 200-acre (81-ha) former MSW landfill in Lyndhurst, New Jersey. The site development was undertaken by the Hackensack Meadowlands Development Commission (HMDC) in 1990. This was the first time this agency had developed a major installation of diverse native plant communities atop a capped landfill. The lack of a natural soil profile and the shallow soil depth limited plant types to those with fibrous, horizontal rooting patterns.

The project was intended to serve as a wildlife management model for future closed landfills in the district. An evaluation period of five years following installation showed that the park or wildlife area was functioning as expected. Plant communities established on the site have generally thrived, the diversity of animal species has increased, and a greater number and varieties of birds were observed as well. The park also attracted more visitors to the site and facilitated an expansion of the HMDC Environment Center's education program.

18.1.5 Multiple Use Facilities

Some MSW landfills have been converted to multipurpose recreational facilities that include active sport facilities (e.g., baseball, football, and soccer fields) in combination with more passive recreational facilities (e.g., playgrounds, picnic sites, and bicycle/walking paths). A good illustration of this approach is the Lyon Township Community Park in Wayne County, Michigan. The park is perched atop a former landfill overlooking Interstate Highway 96.

The Park was constructed over a period of years and developed in several phases. The first phase was completed in 1992, and included construction of three softball fields and a peripheral bicycle path. The second phase was finished in 1994, and included a little league baseball field, a soccer field, a football field, and several tennis courts. A restroom/concession building and paved parking lots were also included in this second phase. The final phase included development of the playgrounds, picnic facilities with pavilions, basketball and volleyball courts, and an ice-skating rink and warming house. A diagram or map showing the location of all the different facilities atop the landfill is shown in Figure 18.2. An aerial photo of the site taken in mid-phase construction (see Figure 18.3) shows views of the baseball fields, parking lots, tennis courts, and peripheral bike path. The Park comprises approximately 170 acres (69 ha) and extends from I-96 south to Grand River Avenue.

The Lyon Community Township Park has attracted much attention as an example of a successful reuse of a site that would otherwise have been considered wasted land. The facility is owned by a private company (Browning-Ferris Industries (BFI)), which, as part of a joint closure plan, shaped the landfill cap, planted vegetation, constructed roads and playgrounds, and seeded picnic areas. The Park was developed in a partnership or cooperative arrangement between the private sector (BFI), a utility company (Detroit Edison), the Michigan Department of Natural Resources (MDNR), and Lyon Township. The MDNR allocated some $525,000 toward the project. In return, the Township and BFI committed over $1.7 million to the park. The Lyon

680 Chapter 18 Postclosure Uses of MSW Landfills

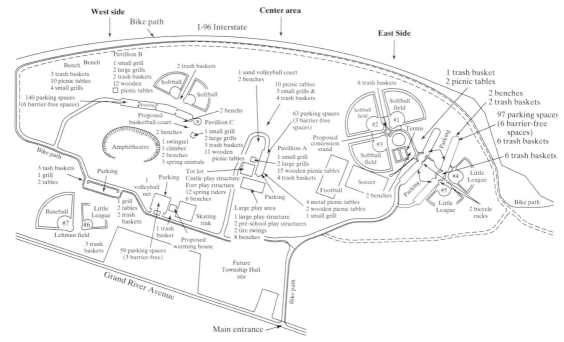

FIGURE 18.2 Location Diagram of Sports and Recreational Facilities, South Lyon Township Park

Township Community Park serves as a good model of successful landfill redevelopment resulting from a concerted and cooperative effort on the part of both private and public organizations.

18.2 INDUSTRIAL DEVELOPMENT

The use of closed landfills for some types of industrial development is possible on a site-specific basis. While this development is significantly less feasible than for athletic facilities, a few situations and case histories are available.

18.2.1 Parking Lots

From a minimum exposure perspective, closed MSW landfills are reasonably suited for parking lots. The surcharge effect on settlement is the major issue, and design must be carefully considered in this regard. The general steps in the design process are as follows:

(i) An assessment of site suitability is conducted, addressing area, grades, and ingress and egress.

(ii) As far as settlement and stability of the underlying waste mass is concerned (recall Chapters 12 and 13), the maximum loads and duration must be determined.

FIGURE 18.3 Airphoto Showing Views of Baseball Fields, Tennis Courts, and Bike Path, South Lyon Township Park

- (iii) Pavement design is another major consideration and must reflect the anticipated amount of maintenance to be undertaken on an ongoing basis.
- (iv) Subsurface drainage and storm water management both require careful planning and design.
- (v) Gas management must be considered in light of the type and age of the underlying waste versus the number of hours spent by personnel working at the site.
- (vi) Many ancillary items also need careful attention to detail: light poles, guard rails, car stops, electrical conduits, signage, and landscaping (and particularly the need for watering associated with it).
- (vii) Light buildings or enclosures for tollbooths, utility meters, maintenance equipment storage, etc., need proper placement and design. These small structures should be placed along the perimeter of the site (if possible) where the waste thickness is minimal or off the edge of the waste completely.

Overriding issues in the above list are site suitability and pavement design. Regarding *site suitability* the type, age, and thickness of the waste should be known. From a settlement perspective, the site should be heavily proof rolled prior to construction. Rolling, however, is depth limited and if insufficient, ground modification techniques should be considered. Hausmann (1990) provides good insight into the various methods. Of the many that are available, deep dynamic compaction or *pounding* has been used most commonly. Galenti, et al. (1992) used the technique to increase density and obtain additional air space with excellent results. Obviously, the bearing capacity was increased, and the settlement potential decreased. The depth of influence of the pounding depends on the mass of the falling weight and its drop height. A widely used empirical formula (after Dobson and Slocomb, 1982) is

$$0.4 \cdot (W \cdot H)^{0.5} < D < (W \cdot H)^{0.5} \tag{18.1}$$

where D = effective depth, m;
        W = weight being dropped, metric ton;
        H = height of drop, m.

The preceding formula, which has been sustained for depths up to 50 ft (15 m), was developed for soils. It might very well be less effective in MSWs than in soil deposits. The relationship for MSWs remains for further investigation.

Regarding *pavement design,* asphalt is generally preferred over Portland cement and other inflexible materials as a surface pavement for final covers, particularly when subsidence and differential settlement is anticipated (Keech, 1995). The options are bituminous chip/seal, binder-only asphalt pavement and binder/top asphalt pavement. The flexibility of asphaltic concrete offers some protection against settlement, and can serve as an auxiliary hydraulic barrier provided that it is properly designed and placed. The case histories that follow serve to illustrate the cross sections that might be considered.

The City of Tacoma (Washington) Sanitary Landfill was designed to withstand the stresses of fully loaded trucks riding on its surface while maintaining a permeability of less than 1.0×10^{-7} cm/sec over the entire landfill cover (Schlert, 1991). From the ground surface down to the underlying waste, the cover consisted of the following layers:

(i) hot mix asphalt surface course: 3 in. (75 mm)
(ii) crushed stone: 8 in. (200 mm)
(iii) sand drainage layer: 12 in. (300 mm)
(iv) geomembrane
(v) sand gas transmission layer: 12 in. (300 mm)
(vi) compacted waste mass

At another location at the same site where police officers are being trained in high-speed driving techniques, the profile is somewhat more conservative. From the ground surface down to the waste mass, it is as follows:

(i) hot mix asphalt surface course: 2 in. (50 mm)
(ii) geotextile paving fabric
(iii) permeable asphalt layer: 2 in. (50 mm)
(iv) asphalt treated base course: 2–4 in. (50–100 mm)
(v) crushed stone base: varies in thickness
(vi) compacted waste mass

The geotextile is included as both an interface crack-retarding layer and a waterproofing layer, since it is bitumen impregnated as part of the lay-down process.

At a Superfund site in East Palo Alto, California, Audibert and Lew (1991) report on a three-layer asphalt cover over soil that had been deep mixed with Portland cement and silicates. From the ground surface down to the underlying waste, the layers were as follows:

- **(i)** dense graded asphalt: 3 in. (75 mm)
- **(ii)** open graded asphalt drainage layer: 3 in. (75 mm)
- **(iii)** hydraulic asphalt: 3 in. (75 mm)
- **(iv)** deep mixed contaminated soil/cement/silicate: 26 ft (7.9 m)

For this particular site, there was no conventional geomembrane or soil barrier layer. Note, however, that neither settlement nor gas was an issue, and that is very unlikely at a municipal solid waste site.

18.2.2 Equipment/Material Storage

Loads exceeding those typical of parking lots are equipment/material storage areas. Since access is required, such sites necessarily involve both roadways and parking areas. The major difference from parking lots is that storage areas have long-term loads that can be very great. As such, MSW landfill sites must be carefully selected as to their suitability. In general, one would not consider such a site unless it was mixed with significant soil and/or other nondegradable material (e.g., fly ash, tailings, slag, etc.).

An example of a site that has been successfully used as an equipment storage area is located in Harmens, Maryland. However, this example is not a closed MSW site, rather it is an arsenic-contaminated soil site. The soils were heavily proof rolled, and then the following cross section was placed (from the ground surface down to the contaminated soil):

- **(i)** hot mix asphalt surface course: 1.5 in. (37 mm)
- **(ii)** asphalt base course: 2.0 in. (50 mm)
- **(iii)** crushed stone aggregate layer: 6.0 in. (150 mm)
- **(iv)** select contaminated soil layer: 6.0 in. (150 mm)
- **(v)** bulk of contaminated soil

To use an MSW landfill site for an equipment/material storage site, the waste should be fully degraded with little remaining settlement, or the owner should be willing to provide ongoing maintenance.

18.2.3 Light Industrial Buildings

It is within the state of the practice of geotechnical engineering to design and construct light industrial buildings on closed landfills providing postconstruction settlements are carefully considered. Invariably, such buildings will be on a mat foundation covering the entire building footprint. The pressure that is exerted (depending on the building

684 Chapter 18   Postclosure Uses of MSW Landfills

mass and its contents) must be compared with the bearing capacity of the underlying waste material. If excessive, the concept of a *compensated* (also called *floating*) foundation can be considered. The geotechnical literature is abundant on these topics. For many industrial waste materials, such design strategies are possible. For MSW materials, however, relatively large postconstruction settlements can generally be anticipated. The amount and time for such settlement to occur must be communicated to the building owner and other parties that may be involved. If such settlements can be tolerated, for example, by an ongoing maintenance plan, the structure can be designed accordingly. What remains for careful consideration and construction are utility connections and access for people and equipment. Regarding utilities, flexible connections should be considered, which will be a major design issue, particularly for water and gas services. Regarding access, modular paving units should be considered. They can be used repeatedly as is necessary.

The issue of landfill gas blocking and/or rerouting is also of paramount importance. A geomembrane vapor barrier will be necessary with positive fixity to all footings and edge walls. Such fixity should be by thermal welding of the geomembrane to polymer inserts that are cast in the concrete during construction, not merely with bitumen or an adhesive. There must also be a drainage layer beneath the concrete foundation made from gravel or geocomposite sheet drains. Grading of this drainage layer is necessary, as is proper venting beyond the limits of the building.

To be sure, proper design and construction of light industrial buildings in MSWs is possible, in the short term. Keech (1996) provides a general overview of the design process and its inherent issues and potential solutions. The role of the geotechnical and civil engineer in addressing these issues is described as well. The long-term situation is another matter, however. At the minimum, the designer should provide ongoing monitoring. This is particularly the case for landfill gas monitoring and also for natural gas utilities, if used. Settlement monitoring should also be considered—not only absolute settlements, but also relative settlements to the adjacent surface. Such monitoring can be periodic or continuous. If continuous, limits might be set beyond which a signal or alarm is sounded. While there is an obvious expense to such attention to detail, at least the technology and required equipment are available.

18.3 AESTHETICS

Whatever the final postclosure situation of a landfill, in general, landfills are difficult to hide from the view of the public. Trees around the site and carefully considered vegetation are certainly helpful, but many of the larger landfills tower over the landscape as a large hill, or even a small mountain. At numerous times different groups have given the situation some thought (both positive and negative). Perhaps the most unique result of positive thinking in this regard is the following conceptual design of the surface of a closed MSW landfill.

The New Jersey Meadowlands Commission has a number of closed landfills in the northern part of New Jersey. This is a highly populated area with highways, railroads, airfields, and harbors all adding to the complex infrastructure that exists. One triangular 65-acre- (26-ha)-site is essentially isolated by a major interstate roadway, a

Section 18.3 Aesthetics **685**

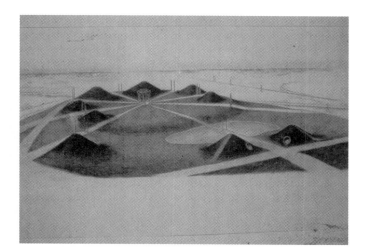
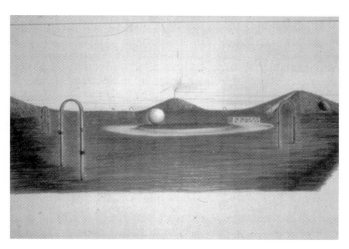

FIGURE 18.4 Proposed Visual Artwork on Closed Landfill (after Pinyan, 1987)

busy three-lane highway, and a railroad. As reported by Pinyan (1987) a group of graphic artists were commissioned to provide the closed landfill surface with a visual artwork. The result is quite provocative. Figure 18.4 shows three of the sketches. Note the purposeful hilly terrain, the landfill gas loops and flares, and the astrological motif of the entire site.

While it goes without saying that not every closed landfill realistically can become artwork, creative thinking nonetheless should be part of the closure process. Waste containment is a dynamic process with changes accepted wherever they can be technically justified.

18.4 CONCLUDING REMARKS

This concluding chapter has presented some relatively modern concepts in the utilization of closed MSW landfill sites. Such uses, however, are almost always contentious, and they involve many groups that must reach consensus (i.e., owners, regulators, local community, politicians and others). The situation is one in which a long-term strategy should be considered as early as possible. Bonaparte (1995) lists the following seven possible strategies:

(i) Standard landfill with perpetual postclosure period.
(ii) Standard landfill with limited postclosure period.
(iii) Standard landfill with clean closure.
(iv) Recirculation landfill with perpetual postclosure period.
(v) Recirculation landfill with limited postclosure period.
(vi) Recirculation landfill with clean closure.
(vii) Inward-gradient landfill.

Strategies 4, 5 and 6 are part of the concept of bioreactor landfills, which was the topic of Chapter 15. Furthermore, the "clean closure" strategies fit nicely into the postclosure scenarios described in this chapter.

If an MSW landfill is within or close to this category, its utilization can and should be considered. Interestingly, all but one of the case histories presented in this chapter were publicly owned facilities. Thus, it appears to the authors that public agencies (cities, municipalities, townships) can and have made reasonable strides toward utilization of their landfills.

However, the privately owned landfills are often the most appealing sites. With few exceptions, privately owned landfills are enormous in size and are often perfectly located for development. Yet, a paradox exists. To our knowledge, no municipality has been willing to relieve ownership of a privately owned landfill for public use as described herein. This "logjam" is unfortunate. There is no fundamental difference between publicly owned MSW sites and privately owned MSW sites. They should be considered in the same light and the political/societal obstacles of relieving ownership of private landfills for beneficial uses should begin. We sincerely hope that all types of closed MSW landfills can be viewed as potentially positive influences on the local environment and can be used by the widest possible sector of the population.

PROBLEMS

18.1 In utilizing a closed landfill as a golf course, the inclusion of water ponds and obstacles is attractive.
   (a) What type of cross section would you consider?
   (b) Periodic drainage would be necessary. What type of concerns would arise?
   (c) Sand traps are also necessary. What cross section should be considered? [Hint: sand traps need drainage so that play can resume right after a rainstorm]

18.2 Differential settlement poses a major concern for sport fields built on closed MSW landfills. How could such abrupt settlements be minimized or avoided?

18.3 Sport fields on closed MSW landfills (as with all others) need lighting for evening games. This requires lights on tall poles. What types of foundations should be considered? Illustrate your answer with some sketches.

18.4 A designer has a number of choices for pavements of parking lots placed on closed MSW landfills. Fill out the following table in this regard:

Pavement Type	Advantages	Disadvantages
1. Crushed stone with dust palliative		
2. Stone chips/bitumen sealed		
3. Asphalt—binder only		
4. Asphalt—binder and top		
5. Portland cement concrete		

18.5 What does a flexible connection (recall Section 18.2.3) look like for the following utilities servicing a light industrial building on a closed MSW landfill?
   (a) Water service.
   (b) Sewer outlet.
   (c) Natural gas service.
   (d) Electrical utility conduit.

18.6 What other types of visual artworks might be considered for closed MSW landfills? [Recall the sketches of Figure 18.4]

REFERENCES

Audibert, J. M. E. and Lew, L. R., (1991) "Asphalt Used for Environmental Caps in Texas," *Asphalt,* Vol. 5, No. 3, Winter, pp. 112–115.

Bonaparte, R., (1995) "Long-Term Performance of Landfills," *Proceedings of ASCE Specialty Conference 2000,* ASCE STP 46, Y. B. Acar and D. E. Daniel, Eds., pp. 415–553.

Conrad, L. G., (2000) "A Canadian Perspective of Using the Britannia Sanitary Landfill Site as a Golf Course," *Proceedings of 5th Landfill Symposium of the Solid Waste Association of North America,* Austin, Texas, June 27–30, pp. 73–79.

Dobson, T. and Slocombe, B., (1982) "Deep Densification of Granular Fills," *Proceedings of Second Geotechnical Conference on Design and Construction,* Las Vegas, NV, April 26–28, 1982, 21 pp.

Galenti, V., Eith, A. E., Leonard, M. S. M., and Fenn, P. S., (1992) "An Assessment of Deep Dynamic Compaction as a Means to Increase Refuse Density for an Operating Municipal Waste Landfill," *Proc. on the Planning and Engineering of Landfills,* Midland Geotechnical Society, United Kingdom, July, pp. 183–193.

Hausmann, M. R., (1990) *Engineering Principles of Ground Modification,* McGraw-Hill Inc., New York, 632 pages.

Keech, M. A., (1995) "Design of Civil Infrastructure Over Landfills," in *Landfill Closures—Environmental Protection and Land Recovery,* ASCE, Reston, VA, pp. 160–183.

Keech, M .A., (1996) "Construction of Retail Buildings on Closed Landfill Sites," *Public Works,* Vol. 127, No. 3, pp. 40–43.

Mackey, R. E., (1996) "Three End-Uses for Closed Landfills and Their Impacts to the Geosynthetic Design," *Proceedings of GRI-9 Geosynthetics in Infrastructure Enhancement and Remediation,* Geosynthetic Information Institute, Folsom, PA, pp. 226–236.

National Golf Foundation, (1999) *Developing Golf Courses on Landfills, Strip-Mines and Other Unusual Locations,* 2nd Ed., Jupiter, FL, 105 pages.

Pinyan, C., (1987) "Sky Mound to Raise from Dump," *ENR,* June 11, pp. 28–29.

Schlect, E. D., (1991) "Tacoma Asphalt Cap is Tough and Impermeable," *Asphalt,* Vol. 5, No. 1, Summer, pp. 37–42.

Vanicek, V., (1973) "The Soil Protective Role of Specially Shaped Plant Roots," *Biol. Conservation,* Vol. 5, No. 3, pp. 175–180.

APPENDIX I

HELP Model Input and Output—Active Condition

INPUT DATA

Liner System Input Information for Active Condition

Layer Number	Layer Description	Type of Layer	Thickness (inch)	k (cm/sec)	Soil Type and Texture Number
1	Daily Cover	Percolation Layer	6	1.0×10^{-3}	Silty Sand, #5
2	Waste	Percolation Layer	120	2.0×10^{-4}	Waste, #43
3	Sand Protective Layer	Percolation Layer	24	1.0×10^{-3}	Sand, #44
4	Geocomposite	Drainage Layer	0.20	33	Drainage Net, #34
5	60-mil HDPE Geomembrane	Geomembrane Liner	0.06	2.0×10^{-13}	Geomembrane, #35
6	Geosynthetic Clay Liner	Barrier Soil Liner	0.23	3.0×10^{-9}	Bentonite Mat, #17
7	Geocomposite	Drainage Layer	0.20	33	Drainage Net, #34
8	60-mil HDPE Geomembrane Natural Clay	Geomembrane Liner	0.06	2.0×10^{-13}	Geomembrane, #35

Other Input Data for Active Condition

Information	Active Condition
Landfill Location	Detroit, Michigan
Total Area (acre)	1
Area Allowing Runoff (%)	0
Surface Slope (%)	2
Surface Slope Length (ft)	100
Surface Condition	Bare Ground
Maximum Leaf Area Index	0
Evaporative Zone Depth	6
Drainage Length in Liner	100
Drainage Slope Liner	2

OUTPUT—CALCULATION RESULTS

```
PRECIPITATION DATA FILE:      C:\HELP3\ACTIVE.D4
TEMPERATURE DATA FILE:        C:\HELP3\ACTIVE.D7
SOLAR RADIATION DATA FILE:    C:\HELP3\ACTIVE.D13
EVAPOTRANSPIRATION DATA:      C:\HELP3\ACTIVE.D11
SOIL AND DESIGN DATA FILE:    C:\HELP3\ACTIVE.D10
OUTPUT DATA FILE:             C:\HELP3\ACTIVE.OUT
```

****

LANDFILL PROJECT I - ACTIVE CONDITION

****

NOTE: INITIAL MOISTURE CONTENT OF THE LAYERS AND SNOW WATER WERE
        COMPUTED AS NEARLY STEADY-STATE VALUES BY THE PROGRAM.

LAYER 1

TYPE 1 - VERTICAL PERCOLATION LAYER
MATERIAL TEXTURE NUMBER   5

```
THICKNESS                  =           6.00   INCHES
POROSITY                   =           0.4570 VOL/VOL
FIELD CAPACITY             =           0.1310 VOL/VOL
WILTING POINT              =           0.0580 VOL/VOL
INITIAL SOIL WATER CONTENT =           0.3741 VOL/VOL
EFFECTIVE SAT. HYD. COND.  =    0.100000005000E-02 CM/SEC
```

NOTE: SATURATED HYDRAULIC CONDUCTIVITY IS MULTIPLIED BY 1.80
        FOR ROOT CHANNELS IN TOP HALF OF EVAPORATIVE ZONE.

LAYER 2

TYPE 1 - VERTICAL PERCOLATION LAYER
MATERIAL TEXTURE NUMBER   0

```
THICKNESS                  =         120.00   INCHES
POROSITY                   =           0.5200 VOL/VOL
FIELD CAPACITY             =           0.2950 VOL/VOL
WILTING POINT              =           0.1400 VOL/VOL
INITIAL SOIL WATER CONTENT =           0.3465 VOL/VOL
EFFECTIVE SAT. HYD. COND.  =    0.199999995000E-03 CM/SEC
```

LAYER 3

TYPE 1 - VERTICAL PERCOLATION LAYER
MATERIAL TEXTURE NUMBER   0

```
THICKNESS                   =         24.00   INCHES
POROSITY                    =          0.3570 VOL/VOL
FIELD CAPACITY              =          0.1130 VOL/VOL
WILTING POINT               =          0.0580 VOL/VOL
INITIAL SOIL WATER CONTENT  =          0.2049 VOL/VOL
EFFECTIVE SAT. HYD. COND.   =    0.100000005000E-02 CM/SEC
```

LAYER 4

TYPE 2 - LATERAL DRAINAGE LAYER
MATERIAL TEXTURE NUMBER   34

```
THICKNESS                   =          0.20   INCHES
POROSITY                    =          0.8500 VOL/VOL
FIELD CAPACITY              =          0.0100 VOL/VOL
WILTING POINT               =          0.0050 VOL/VOL
INITIAL SOIL WATER CONTENT  =          0.0110 VOL/VOL
EFFECTIVE SAT. HYD. COND.   =         33.0000000000   CM/SEC
SLOPE                       =          2.00   PERCENT
DRAINAGE LENGTH             =        100.0    FEET
```

LAYER 5

TYPE 4 - FLEXIBLE MEMBRANE LINER
MATERIAL TEXTURE NUMBER   35

```
THICKNESS                   =          0.06   INCHES
POROSITY                    =          0.0000 VOL/VOL
FIELD CAPACITY              =          0.0000 VOL/VOL
WILTING POINT               =          0.0000 VOL/VOL
INITIAL SOIL WATER CONTENT  =          0.0000 VOL/VOL
EFFECTIVE SAT. HYD. COND.   =    0.199999996000E-12 CM/SEC
FML PINHOLE DENSITY         =          1.00   HOLES/ACRE
FML INSTALLATION DEFECTS    =          1.00   HOLES/ACRE
FML PLACEMENT QUALITY       =       3 - GOOD
```

LAYER 6

TYPE 3 - BARRIER SOIL LINER
MATERIAL TEXTURE NUMBER 17

THICKNESS	=	0.23 INCHES
POROSITY	=	0.7500 VOL/VOL
FIELD CAPACITY	=	0.7470 VOL/VOL
WILTING POINT	=	0.4000 VOL/VOL
INITIAL SOIL WATER CONTENT	=	0.7500 VOL/VOL
EFFECTIVE SAT. HYD. COND.	=	0.300000003000E-08 CM/SEC

LAYER 7

TYPE 2 - LATERAL DRAINAGE LAYER
MATERIAL TEXTURE NUMBER 34

THICKNESS	=	0.20 INCHES
POROSITY	=	0.8500 VOL/VOL
FIELD CAPACITY	=	0.0100 VOL/VOL
WILTING POINT	=	0.0050 VOL/VOL
INITIAL SOIL WATER CONTENT	=	0.0100 VOL/VOL
EFFECTIVE SAT. HYD. COND.	=	33.0000000000 CM/SEC
SLOPE	=	2.00 PERCENT
DRAINAGE LENGTH	=	100.0 FEET

LAYER 8

TYPE 4 - FLEXIBLE MEMBRANE LINER
MATERIAL TEXTURE NUMBER 35

THICKNESS	=	0.06 INCHES
POROSITY	=	0.0000 VOL/VOL
FIELD CAPACITY	=	0.0000 VOL/VOL
WILTING POINT	=	0.0000 VOL/VOL
INITIAL SOIL WATER CONTENT	=	0.0000 VOL/VOL
EFFECTIVE SAT. HYD. COND.	=	0.199999996000E-12 CM/SEC
FML PINHOLE DENSITY	=	1.00 HOLES/ACRE
FML INSTALLATION DEFECTS	=	1.00 HOLES/ACRE
FML PLACEMENT QUALITY	=	3 - GOOD

GENERAL DESIGN AND EVAPORATIVE ZONE DATA

NOTE: SCS RUNOFF CURVE NUMBER WAS COMPUTED FROM DEFAULT SOIL DATA BASE USING SOIL TEXTURE #5 WITH BARE GROUND CONDITIONS, A SURFACE SLOPE OF 2.% AND A SLOPE LENGTH OF 100. FEET.

```
SCS RUNOFF CURVE NUMBER              =   84.40
FRACTION OF AREA ALLOWING RUNOFF     =    0.0   PERCENT
AREA PROJECTED ON HORIZONTAL PLANE   =    1.000 ACRES
EVAPORATIVE ZONE DEPTH               =    6.0   INCHES
INITIAL WATER IN EVAPORATIVE ZONE    =    2.245 INCHES
UPPER LIMIT OF EVAPORATIVE STORAGE   =    2.742 INCHES
LOWER LIMIT OF EVAPORATIVE STORAGE   =    0.348 INCHES
INITIAL SNOW WATER                   =    0.000 INCHES
INITIAL WATER IN LAYER MATERIALS     =   48.922 INCHES
TOTAL INITIAL WATER                  =   48.922 INCHES
TOTAL SUBSURFACE INFLOW              =    0.00  INCHES/YEAR
```

EVAPOTRANSPIRATION AND WEATHER DATA

NOTE: EVAPOTRANSPIRATION DATA WAS OBTAINED FROM DETROIT MICHIGAN

```
MAXIMUM LEAF AREA INDEX                 =    1.00
START OF GROWING SEASON (JULIAN DATE)   =    121
END OF GROWING SEASON (JULIAN DATE)     =    286
AVERAGE ANNUAL WIND SPEED               =   10.20 MPH
AVERAGE 1ST QUARTER RELATIVE HUMIDITY   =   73.00 %
AVERAGE 2ND QUARTER RELATIVE HUMIDITY   =   67.00 %
AVERAGE 3RD QUARTER RELATIVE HUMIDITY   =   71.00 %
AVERAGE 4TH QUARTER RELATIVE HUMIDITY   =   75.00 %
```

NOTE: PRECIPITATION DATA WAS SYNTHETICALLY GENERATED USING COEFFICIENTS FOR DETROIT, MICHIGAN

NORMAL MEAN MONTHLY PRECIPITATION (INCHES)

JAN/JUL	FEB/AUG	MAR/SEP	APR/OCT	MAY/NOV	JUN/DEC
1.86	1.69	2.54	3.15	2.77	3.43
3.10	3.21	2.25	2.12	2.33	2.52

NOTE: TEMPERATURE DATA WAS SYNTHETICALLY GENERATED USING COEFFICIENTS FOR DETROIT, MICHIGAN

694 Appendix I HELP Model Input and Output—Active Condition

NORMAL MEAN MONTHLY TEMPERATURE (DEGREES FAHRENHEIT)

JAN/JUL	FEB/AUG	MAR/SEP	APR/OCT	MAY/NOV	JUN/DEC
23.40	25.80	35.00	47.40	58.10	67.70
71.90	70.50	63.30	51.90	39.50	28.50

NOTE: SOLAR RADIATION DATA WAS SYNTHETICALLY GENERATED USING COEFFICIENTS FOR DETROIT, MICHIGAN

STATION LATITUDE = 42.40 DEGREES

****

AVERAGE MONTHLY VALUES IN INCHES FOR YEARS   1 THROUGH   30

****

	JAN/JUL	FEB/AUG	MAR/SEP	APR/OCT	MAY/NOV	JUN/DEC
PRECIPITATION						
TOTALS	1.82	1.75	2.38	3.33	2.97	3.36
	2.93	2.98	2.31	1.66	2.36	2.63
STD. DEVIATIONS	0.65	0.80	1.07	1.30	1.13	1.41
	1.26	1.64	1.37	1.06	1.00	1.05
RUNOFF						
TOTALS	0.000	0.000	0.000	0.000	0.000	0.000
	0.000	0.000	0.000	0.000	0.000	0.000
STD. DEVIATIONS	0.000	0.000	0.000	0.000	0.000	0.000
	0.000	0.000	0.000	0.000	0.000	0.000
EVAPOTRANSPIRATION						
TOTALS	0.701	0.783	1.816	2.302	1.485	1.670
	1.443	1.321	1.049	0.573	0.691	0.588
STD. DEVIATIONS	0.163	0.228	0.387	0.692	0.778	0.870
	0.932	0.735	0.723	0.402	0.338	0.155
LATERAL DRAINAGE COLLECTED FROM LAYER 4						
TOTALS	1.6330	1.1699	0.9742	0.8310	0.9431	1.4285
	1.6783	1.7222	1.6809	1.4847	1.3780	1.2074
STD. DEVIATIONS	0.4614	0.2938	0.3854	0.4456	0.5853	0.5298
	0.5853	0.4103	0.5841	0.3834	0.4356	0.5077

Appendix I   HELP Model Input and Output—Active Condition

PERCOLATION/LEAKAGE THROUGH LAYER 6

TOTALS	0.0000	0.0000	0.0000	0.0000	0.0000	0.0000
	0.0000	0.0000	0.0000	0.0000	0.0000	0.0000
STD. DEVIATIONS	0.0000	0.0000	0.0000	0.0000	0.0000	0.0000
	0.0000	0.0000	0.0000	0.0000	0.0000	0.0000

LATERAL DRAINAGE COLLECTED FROM LAYER 7

TOTALS	0.0000	0.0000	0.0000	0.0000	0.0000	0.0000
	0.0000	0.0000	0.0000	0.0000	0.0000	0.0000
STD. DEVIATIONS	0.0000	0.0000	0.0000	0.0000	0.0000	0.0000
	0.0000	0.0000	0.0000	0.0000	0.0000	0.0000

PERCOLATION/LEAKAGE THROUGH LAYER 8

TOTALS	0.0000	0.0000	0.0000	0.0000	0.0000	0.0000
	0.0000	0.0000	0.0000	0.0000	0.0000	0.0000
STD. DEVIATIONS	0.0000	0.0000	0.0000	0.0000	0.0000	0.0000
	0.0000	0.0000	0.0000	0.0000	0.0000	0.0000

AVERAGES OF MONTHLY AVERAGED DAILY HEADS (INCHES)

DAILY AVERAGE HEAD ACROSS LAYER 6

AVERAGES	0.0014	0.0011	0.0008	0.0007	0.0008	0.0013
	0.0014	0.0015	0.0015	0.0013	0.0012	0.0010
STD. DEVIATIONS	0.0004	0.0003	0.0003	0.0004	0.0005	0.0005
	0.0005	0.0004	0.0005	0.0003	0.0004	0.0004

DAILY AVERAGE HEAD ACROSS LAYER 8

AVERAGES	0.0000	0.0000	0.0000	0.0000	0.0000	0.0000
	0.0000	0.0000	0.0000	0.0000	0.0000	0.0000
STD. DEVIATIONS	0.0000	0.0000	0.0000	0.0000	0.0000	0.0000
	0.0000	0.0000	0.0000	0.0000	0.0000	0.0000

****

Appendix I HELP Model Input and Output—Active Condition

```
******************************************************************************
```

AVERAGE ANNUAL TOTALS & (STD. DEVIATIONS) FOR YEARS 1 THROUGH 30

	INCHES		CU. FEET	PERCENT
PRECIPITATION	30.46	(3.646)	110581.9	100.00
RUNOFF	0.000	(0.0000)	0.00	0.000
EVAPOTRANSPIRATION	14.423	(1.8046)	52354.00	47.344
LATERAL DRAINAGE COLLECTED FROM LAYER 4	16.13115	(2.30515)	58556.078	52.95268
PERCOLATION/LEAKAGE THROUGH FROM LAYER 6	0.00000	(0.00000)	0.009	0.00001
AVERAGE HEAD ACROSS TOP OF LAYER 6	0.001	(0.000)		
LATERAL DRAINAGE COLLECTED FROM LAYER 7	0.00000	(0.00000)	0.000	0.00000
PERCOLATION/LEAKAGE THROUGH FROM LAYER 8	0.00000	(0.00000)	0.009	0.00001
AVERAGE HEAD ACROSS TOP OF LAYER 8	0.000	(0.000)		
CHANGE IN WATER STORAGE	−0.090	(1.6092)	−328.19	−0.297

```
******************************************************************************


******************************************************************************
```

PEAK DAILY VALUES FOR YEARS 1 THROUGH 30

	(INCHES)	(CU. FT.)
PRECIPITATION	2.92	10599.601
RUNOFF	0.000	0.0000
DRAINAGE COLLECTED FROM LAYER 4	0.18551	673.40155
PERCOLATION/LEAKAGE THROUGH LAYER 6	0.000000	0.00003

AVERAGE HEAD ACROSS LAYER 6	0.005	
DRAINAGE COLLECTED FROM LAYER 7	0.00000	0.00000
PERCOLATION/LEAKAGE THROUGH LAYER 8	0.000000	0.00003
AVERAGE HEAD ACROSS LAYER 8	0.000	
SNOW WATER	4.01	14573.7139
MAXIMUM VEG. SOIL WATER (VOL/VOL)	0.4570	
MINIMUM VEG. SOIL WATER (VOL/VOL)	0.0191	

****

****

FINAL WATER STORAGE AT END OF YEAR 30

LAYER	(INCHES)	(VOL/VOL)
1	0.3481	0.0580
2	40.7863	0.3399
3	4.5837	0.1910
4	0.0029	0.0143
5	0.0000	0.0000
6	0.1725	0.7500
7	0.0020	0.0100
8	0.0000	0.0000
SNOW WATER	0.000	

****
****

APPENDIX II

HELP Model Input and Output— Postclosure Condition

INPUT DATA

Liner System and Final Cover Input Information for Postclosure Condition

Layer Number	Layer Description	Type of Layer	Thickness (inch)	k (cm/sec)	Soil Type and Texture Number
1	Vegetative/Protective Layer	Percolation Layer	30	1.9×10^{-4}	Silt, #9
2	Geocomposite	Drainage Layer	0.20	33	Drainage Net, #34
3	40-mil HDPE Geomembrane	Geomembrane Liner	0.04	2.0×10^{-13}	Geomembrane, #35
4	Compacted Soil Liner	Barrier Soil Liner	18	1.0×10^{-5}	Clay, #46
5	Daily Cover	Percolation Layer	6	1.0×10^{-3}	Silty Sand, #5
6	Waste	Percolation Layer	1920	2.0×10^{-4}	Waste, #43
7	Sand Protective Layer	Percolation Layer	24	1.0×10^{-3}	Sand, #44
8	Geocomposite	Drainage Layer	0.20	33	Drainage Net, #34
9	60-mil HDPE Geomembrane	Geomembrane Liner	0.06	2.0×10^{-13}	Geomembrane, #35
10	Geosynthetic Clay Liner	Barrier Soil Liner	0.23	3.0×10^{-9}	Bentonite Mat, #17
11	Geocomposite	Drainage Layer	0.20	33	Drainage Net, #34
12	60-mil HDPE Geomembrane Natural Clay	Geomembrane Liner	0.06	2.0×10^{-13}	Geomembrane, #35

Other Input Data for Postclosure Condition

Information	Postclosure Condition
Landfill Location	Detroit, Michigan
Total Area (acre)	1
Area Allowing Runoff (%)	100
Surface Slope (%)	25
Surface Slope Length (ft)	1000
Surface Condition	Good Grass
Maximum Leaf Area Index	1
Evaporative Zone Depth	6
Drainage Length in Liner	100
Drainage Slope Liner	2

OUTPUT—CALCULATION RESULTS

```
PRECIPITATION DATA FILE:      C:\HELP3\CLOSE.D4
TEMPERATURE DATA FILE:        C:\HELP3\CLOSE.D7
SOLAR RADIATION DATA FILE:    C:\HELP3\CLOSE.D13
EVAPOTRANSPIRATION DATA:      C:\HELP3\CLOSE.D11
SOIL AND DESIGN DATA FILE:    C:\HELP3\CLOSE.D10
OUTPUT DATA FILE:             C:\HELP3\CLOSE.OUT
```

****

LANDFILL PROJECT II - POSTCLOSURE CONDITION

****

NOTE: INITIAL MOISTURE CONTENT OF THE LAYERS AND SNOW WATER WERE COMPUTED AS NEARLY STEADY-STATE VALUES BY THE PROGRAM.

LAYER 1

TYPE 1 - VERTICAL PERCOLATION LAYER
MATERIAL TEXTURE NUMBER   9

```
THICKNESS                  =        30.00      INCHES
POROSITY                   =         0.5010    VOL/VOL
FIELD CAPACITY             =         0.2840    VOL/VOL
WILTING POINT              =         0.1350    VOL/VOL
INITIAL SOIL WATER CONTENT =         0.3260    VOL/VOL
EFFECTIVE SAT. HYD. COND.  =     0.190000006000E-03 CM/SEC
```

NOTE: SATURATED HYDRAULIC CONDUCTIVITY IS MULTIPLIED BY 1.80 FOR ROOT CHANNELS IN TOP HALF OF EVAPORATIVE ZONE.

LAYER 2

TYPE 2 - LATERAL DRAINAGE LAYER
MATERIAL TEXTURE NUMBER   34

```
THICKNESS                  =         0.20      INCHES
POROSITY                   =         0.8500    VOL/VOL
FIELD CAPACITY             =         0.0100    VOL/VOL
WILTING POINT              =         0.0050    VOL/VOL
INITIAL SOIL WATER CONTENT =         0.0144    VOL/VOL
EFFECTIVE SAT. HYD. COND.  =        33.0000000000 CM/SEC
SLOPE                      =        25.00      PERCENT
DRAINAGE LENGTH            =      1000.0       FEET
```

LAYER 3

TYPE 4 - FLEXIBLE MEMBRANE LINER
MATERIAL TEXTURE NUMBER 35

THICKNESS	=	0.04 INCHES
POROSITY	=	0.0000 VOL/VOL
FIELD CAPACITY	=	0.0000 VOL/VOL
WILTING POINT	=	0.0000 VOL/VOL
INITIAL SOIL WATER CONTENT	=	0.0000 VOL/VOL
EFFECTIVE SAT. HYD. COND.	=	0.199999996000E-12 CM/SEC
FML PINHOLE DENSITY	=	1.00 HOLES/ACRE
FML INSTALLATION DEFECTS	=	1.00 HOLES/ACRE
FML PLACEMENT QUALITY	=	3 - GOOD

LAYER 4

TYPE 3 - BARRIER SOIL LINER
MATERIAL TEXTURE NUMBER 0

THICKNESS	=	18.00 INCHES
POROSITY	=	0.4270 VOL/VOL
FIELD CAPACITY	=	0.4180 VOL/VOL
WILTING POINT	=	0.3670 VOL/VOL
INITIAL SOIL WATER CONTENT	=	0.4270 VOL/VOL
EFFECTIVE SAT. HYD. COND.	=	0.999999975000E-05 CM/SEC

LAYER 5

TYPE 1 - VERTICAL PERCOLATION LAYER
MATERIAL TEXTURE NUMBER 5

THICKNESS	=	6.00 INCHES
POROSITY	=	0.4570 VOL/VOL
FIELD CAPACITY	=	0.1310 VOL/VOL
WILTING POINT	=	0.0580 VOL/VOL
INITIAL SOIL WATER CONTENT	=	0.1310 VOL/VOL
EFFECTIVE SAT. HYD. COND.	=	0.100000005000E-02 CM/SEC

LAYER 6

TYPE 1 - VERTICAL PERCOLATION LAYER
MATERIAL TEXTURE NUMBER   0

THICKNESS	=	1920.00 INCHES
POROSITY	=	0.5200 VOL/VOL
FIELD CAPACITY	=	0.2950 VOL/VOL
WILTING POINT	=	0.1400 VOL/VOL
INITIAL SOIL WATER CONTENT	=	0.2950 VOL/VOL
EFFECTIVE SAT. HYD. COND.	=	0.199999005000E-03 CM/SEC

LAYER 7

TYPE 1 - VERTICAL PERCOLATION LAYER
MATERIAL TEXTURE NUMBER   0

THICKNESS	=	24.00 INCHES
POROSITY	=	0.3570 VOL/VOL
FIELD CAPACITY	=	0.1130 VOL/VOL
WILTING POINT	=	0.0580 VOL/VOL
INITIAL SOIL WATER CONTENT	=	0.1130 VOL/VOL
EFFECTIVE SAT. HYD. COND.	=	0.100000005000E-02 CM/SEC

LAYER 8

TYPE 2 - LATERAL DRAINAGE LAYER
MATERIAL TEXTURE NUMBER   34

THICKNESS	=	0.20 INCHES
POROSITY	=	0.8500 VOL/VOL
FIELD CAPACITY	=	0.0100 VOL/VOL
WILTING POINT	=	0.0050 VOL/VOL
INITIAL SOIL WATER CONTENT	=	0.0100 VOL/VOL
EFFECTIVE SAT. HYD. COND.	=	33.0000000000 CM/SEC
SLOPE	=	2.00 PERCENT
DRAINAGE LENGTH	=	100.0 FEET

LAYER 9

TYPE 4 - FLEXIBLE MEMBRANE LINER
MATERIAL TEXTURE NUMBER 35

THICKNESS	=	0.06 INCHES
POROSITY	=	0.0000 VOL/VOL
FIELD CAPACITY	=	0.0000 VOL/VOL
WILTING POINT	=	0.0000 VOL/VOL
INITIAL SOIL WATER CONTENT	=	0.0000 VOL/VOL
EFFECTIVE SAT. HYD. COND.	=	0.199999996000E-12 CM/SEC
FML PINHOLE DENSITY	=	1.00 HOLES/ACRE
FML INSTALLATION DEFECTS	=	1.00 HOLES/ACRE
FML PLACEMENT QUALITY	=	3 - GOOD

LAYER 10

TYPE 3 - BARRIER SOIL LINER
MATERIAL TEXTURE NUMBER 17

THICKNESS	=	0.23 INCHES
POROSITY	=	0.7500 VOL/VOL
FIELD CAPACITY	=	0.7470 VOL/VOL
WILTING POINT	=	0.4000 VOL/VOL
INITIAL SOIL WATER CONTENT	=	0.7500 VOL/VOL
EFFECTIVE SAT. HYD. COND.	=	0.300000003000E-08 CM/SEC

LAYER 11

TYPE 2 - LATERAL DRAINAGE LAYER
MATERIAL TEXTURE NUMBER 34

THICKNESS	=	0.20 INCHES
POROSITY	=	0.8500 VOL/VOL
FIELD CAPACITY	=	0.0100 VOL/VOL
WILTING POINT	=	0.0050 VOL/VOL
INITIAL SOIL WATER CONTENT	=	0.0100 VOL/VOL
EFFECTIVE SAT. HYD. COND.	=	33.0000000000 CM/SEC
SLOPE	=	2.00 PERCENT
DRAINAGE LENGTH	=	100.0 FEET

LAYER 12

TYPE 4 - FLEXIBLE MEMBRANE LINER
MATERIAL TEXTURE NUMBER 35

THICKNESS	=	0.06 INCHES
POROSITY	=	0.0000 VOL/VOL
FIELD CAPACITY	=	0.0000 VOL/VOL
WILTING POINT	=	0.0000 VOL/VOL
INITIAL SOIL WATER CONTENT	=	0.0000 VOL/VOL
EFFECTIVE SAT. HYD. COND.	=	0.199999996000E-12 CM/SEC
FML PINHOLE DENSITY	=	1.00 HOLES/ACRE
FML INSTALLATION DEFECTS	=	1.00 HOLES/ACRE
FML PLACEMENT QUALITY	=	3 - GOOD

GENERAL DESIGN AND EVAPORATIVE ZONE DATA

NOTE: SCS RUNOFF CURVE NUMBER WAS COMPUTED FROM DEFAULT
SOIL DATA BASE USING SOIL TEXTURE #9 WITH A
GOOD STAND OF GRAS, A SURFACE SLOPE OF 25.%
AND A SLOPE LENGTH OF 1000. FEET.

SCS RUNOFF CURVE NUMBER	=	75.30
FRACTION OF AREA ALLOWING RUNOFF	=	100.0 PERCENT
AREA PROJECTED ON HORIZONTAL PLANE	=	1.000 ACRES
EVAPORATIVE ZONE DEPTH	=	6.0 INCHES
INITIAL WATER IN EVAPORATIVE ZONE	=	2.333 INCHES
UPPER LIMIT OF EVAPORATIVE STORAGE	=	3.006 INCHES
LOWER LIMIT OF EVAPORATIVE STORAGE	=	0.810 INCHES
INITIAL SNOW WATER	=	0.000 INCHES
INITIAL WATER IN LAYER MATERIALS	=	587.544 INCHES
TOTAL INITIAL WATER	=	587.544 INCHES
TOTAL SUBSURFACE INFLOW	=	0.00 INCHES/YEAR

Appendix II HELP Model Input and Output—Postclosure Condition

```
                    EVAPOTRANSPIRATION AND WEATHER DATA
                    ------------------------------------------------------------


    NOTE:    EVAPOTRANSPIRATION DATA WAS OBTAINED FROM
                 DETROIT                 MICHIGAN

             MAXIMUM LEAF AREA INDEX                 =     1.00
             START OF GROWING SEASON (JULIAN DATE)   =      121
             END OF GROWING SEASON (JULIAN DATE)     =      286
             AVERAGE ANNUAL WIND SPEED               =    10.20 MPH
             AVERAGE 1ST QUARTER RELATIVE HUMIDITY   =    73.00 %
             AVERAGE 2ND QUARTER RELATIVE HUMIDITY   =    67.00 %
             AVERAGE 3RD QUARTER RELATIVE HUMIDITY   =    71.00 %
             AVERAGE 4TH QUARTER RELATIVE HUMIDITY   =    75.00 %


    NOTE:    PRECIPITATION DATA WAS SYNTHETICALLY GENERATED USING
                 COEFFICIENTS FOR DETROIT MICHIGAN




                    NORMAL MEAN MONTHLY PRECIPITATION (INCHES)

      JAN/JUL      FEB/AUG      MAR/SEP      APR/OCT      MAY/NOV      JUN/DEC
      -------      -------      -------      -------      -------      -------
        1.86         1.69         2.54         3.15         2.77         3.43
        3.10         3.21         2.25         2.12         2.33         2.52

    NOTE:    TEMPERATURE DATA WAS SYNTHETICALLY GENERATED USING
                 COEFFICIENTS FOR       DETROIT       MICHIGAN




                  NORMAL MEAN MONTHLY TEMPERATURE (DEGREES FAHRENHEIT)

      JAN/JUL      FEB/AUG      MAR/SEP      APR/OCT      MAY/NOV      JUN/DEC
      -------      -------      -------      -------      -------      -------
       23.40        25.80        35.00        47.40        58.10        67.70
       71.90        70.50        63.30        51.90        39.50        28.50

    NOTE:    SOLAR RADIATION DATA WAS SYNTHETICALLY GENERATED USING
                 COEFFICIENTS FOR       DETROIT       MICHIGAN

                    STATION LATITUDE  =   42.40 DEGREES


****************************************************************************************
```

Appendix II HELP Model Input and Output—Postclosure Condition

```
AVERAGE MONTHLY VALUES IN INCHES FOR YEARS   1 THROUGH   30
*************************************************************************
```

	JAN/JUL	FEB/AUG	MAR/SEP	APR/OCT	MAY/NOV	JUN/DEC
PRECIPITATION						
TOTALS	1.82	1.75	2.38	3.33	2.97	3.36
	2.93	2.98	2.31	1.66	2.36	2.63
STD. DEVIATIONS	0.65	0.80	1.07	1.30	1.13	1.41
	1.26	1.64	1.37	1.06	1.00	1.05
RUNOFF						
TOTALS	0.774	1.125	0.976	0.453	0.007	0.005
	0.007	0.044	0.001	0.001	0.001	0.315
STD. DEVIATIONS	0.794	0.969	0.900	0.919	0.027	0.014
	0.022	0.103	0.006	0.006	0.004	0.535
EVAPOTRANSPIRATION						
TOTALS	0.695	0.778	2.002	2.888	2.502	2.988
	2.632	2.041	1.890	1.180	1.012	0.703
STD. DEVIATIONS	0.146	0.228	0.382	0.711	0.985	0.892
	1.079	0.893	0.907	0.532	0.259	0.146
LATERAL DRAINAGE COLLECTED FROM LAYER 2						
TOTALS	0.3599	0.0050	0.0035	0.3476	0.6147	0.5312
	0.4462	0.5986	0.5186	0.3752	0.5153	1.1363
STD. DEVIATIONS	0.3230	0.0234	0.0095	0.4967	0.4231	0.3885
	0.5013	0.7486	0.4714	0.4070	0.6277	0.8691
PERCOLATION/LEAKAGE THROUGH LAYER 4						
TOTALS	0.0000	0.0000	0.0000	0.0000	0.0000	0.0000
	0.0000	0.0000	0.0000	0.0000	0.0000	0.0000
STD. DEVIATIONS	0.0000	0.0000	0.0000	0.0000	0.0000	0.0000
	0.0000	0.0000	0.0000	0.0000	0.0000	0.0000

706 Appendix II HELP Model Input and Output—Postclosure Condition

LATERAL DRAINAGE COLLECTED FROM LAYER 8

TOTALS	0.0000	0.0000	0.0000	0.0000	0.0000	0.0000
	0.0000	0.0000	0.0000	0.0000	0.0000	0.0000
STD. DEVIATIONS	0.0000	0.0000	0.0000	0.0000	0.0000	0.0000
	0.0000	0.0000	0.0000	0.0000	0.0000	0.0000

PERCOLATION/LEAKAGE THROUGH LAYER 10

TOTALS	0.0000	0.0000	0.0000	0.0000	0.0000	0.0000
	0.0000	0.0000	0.0000	0.0000	0.0000	0.0000
STD. DEVIATIONS	0.0000	0.0000	0.0000	0.0000	0.0000	0.0000
	0.0000	0.0000	0.0000	0.0000	0.0000	0.0000

LATERAL DRAINAGE COLLECTED FROM LAYER 11

TOTALS	0.0000	0.0000	0.0000	0.0000	0.0000	0.0000
	0.0000	0.0000	0.0000	0.0000	0.0000	0.0000
STD. DEVIATIONS	0.0000	0.0000	0.0000	0.0000	0.0000	0.0000
	0.0000	0.0000	0.0000	0.0000	0.0000	0.0000

PERCOLATION/LEAKAGE THROUGH LAYER 12

TOTALS	0.0000	0.0000	0.0000	0.0000	0.0000	0.0000
	0.0000	0.0000	0.0000	0.0000	0.0000	0.0000
STD. DEVIATIONS	0.0000	0.0000	0.0000	0.0000	0.0000	0.0000
	0.0000	0.0000	0.0000	0.0000	0.0000	0.0000

AVERAGES OF MONTHLY AVERAGED DAILY HEADS (INCHES)

DAILY AVERAGE HEAD ACROSS LAYER 4

AVERAGES	0.0003	0.0000	0.0000	0.0003	0.0005	0.0004
	0.0003	0.0004	0.0004	0.0003	0.0004	0.0008
STD. DEVIATIONS	0.0002	0.0000	0.0000	0.0004	0.0003	0.0003
	0.0004	0.0005	0.0004	0.0003	0.0005	0.0006

DAILY AVERAGE HEAD ACROSS LAYER 10

    AVERAGES   0.0000   0.0000   0.0000   0.0000   0.0000   0.0000
                          0.0000   0.0000   0.0000   0.0000   0.0000   0.0000

    STD. DEVIATIONS   0.0000   0.0000   0.0000   0.0000   0.0000   0.0000
                          0.0000   0.0000   0.0000   0.0000   0.0000   0.0000

DAILY AVERAGE HEAD ACROSS LAYER 12

    AVERAGES   0.0000   0.0000   0.0000   0.0000   0.0000   0.0000
                          0.0000   0.0000   0.0000   0.0000   0.0000   0.0000

    STD. DEVIATIONS   0.0000   0.0000   0.0000   0.0000   0.0000   0.0000
                          0.0000   0.0000   0.0000   0.0000   0.0000   0.0000

****

****

             AVERAGE ANNUAL TOTALS & (STD. DEVIATIONS) FOR YEARS   1 THROUGH   30

                                        INCHES   CU. FEET   PERCENT
                                     ----------------   ------------   ------------
PRECIPITATION   30.46   (3.646)   110581.9   100.00

RUNOFF   3.710   (1.6749)   13468.42   12.180

EVAPOTRANSPIRATION   21.311   (2.3186)   77360.32   69.957

LATERAL DRAINAGE COLLECTED   5.45222 (2.22275)   19791.564   17.89766
  FROM LAYER 2

PERCOLATION/LEAKAGE THROUGH   0.00001 (0.00000)   0.030   0.00003
  FROM LAYER 4

AVERAGE HEAD ACROSS TOP   0.000   (0.000)
  OF LAYER 4

LATERAL DRAINAGE COLLECTED   0.00001 (0.00000)   0.029   0.00003
  FROM LAYER 8

PERCOLATION/LEAKAGE THROUGH   0.00000 (0.00000)   0.000   0.00000
  FROM LAYER 10

AVERAGE HEAD ACROSS TOP   0.000   (0.000)
  OF LAYER 10

LATERAL DRAINAGE COLLECTED FROM LAYER 11	0.00000	(0.00000)	0.000	0.00000
PERCOLATION/LEAKAGE THROUGH FROM LAYER 12	0.00000	(0.00000)	0.000	0.00000
AVERAGE HEAD ACROSS TOP OF LAYER 12	0.000	(0.000)		
CHANGE IN WATER STORAGE	-0.011	(0.8877)	-38.44	-0.035

****

****

PEAK DAILY VALUES FOR YEARS 1 THROUGH 30

	(INCHES)	(CU. FT.)
PRECIPITATION	2.92	10599.601
RUNOFF	3.024	10975.5254
DRAINAGE COLLECTED FROM LAYER 2	0.59150	2147.12842
PERCOLATION/LEAKAGE THROUGH LAYER 4	0.000001	0.00211
AVERAGE HEAD ACROSS LAYER 4	0.013	
DRAINAGE COLLECTED FROM LAYER 8	0.00000	0.00211
PERCOLATION/LEAKAGE THROUGH LAYER 10	0.000000	0.00000
AVERAGE HEAD ACROSS LAYER 10	0.000	
DRAINAGE COLLECTED FROM LAYER 11	0.00000	0.00000
PERCOLATION/LEAKAGE THROUGH LAYER 12	0.000000	0.00000
AVERAGE HEAD ACROSS LAYER 12	0.000	
SNOW WATER	4.01	14573.7139
MAXIMUM VEG. SOIL WATER (VOL/VOL)		0.4891
MINIMUM VEG. SOIL WATER (VOL/VOL)		0.0819

****

Appendix II HELP Model Input and Output—Postclosure Condition

```
******************************************************************
              FINAL WATER STORAGE AT END OF YEAR    30
------------------------------------------------------------------
              LAYER            (INCHES)           (VOL/VOL)
              -----            --------           ---------
                1               9.1494             0.3050

                2               0.0029             0.0144

                3               0.0000             0.0000

                4               7.6860             0.4270

                5               0.7860             0.1310

                6             566.3999             0.2950

                7               2.7120             0.1130

                8               0.0020             0.0100

                9               0.0000             0.0000

               10               0.1725             0.7500

               11               0.0020             0.0100

               12               0.0000             0.0000

          SNOW WATER            0.000

******************************************************************
******************************************************************
```

INDEX

A

Absorptive capacity, 214
Acceleration, 492, 497
    Bedrock motion, 509
    Gravitational, 509
    Peak horizontal, 509
    Yield acceleration, 511
Active Condition, 218
Active gas collection system. *See* Gas collection systems
Active period, 354, 355, 361
Acceptable zone, 52, 63, 65, 68
Accumulated contaminant mass, 158, 159, 166
Adsorption, 23
Advection, 14, 15, 154
Advective mass flux ratio, 155
Age of waste mass, 355, 361. *See also* Waste characteristics
Alternative daily cover material (ADCM), 592
Alternative landfill cover, 412
Ambient temperature, 213, 225, 638, 645
Anaerobic bioreactor, 577
Anaerobic reactor, 581
Anchor trench, 104, 637
    Rectangular, 106
    V-shaped, 114
Apparent cohesion, 194

B

Bacteria, 212, 334
Bacterial content, 387
Backfill (Backfilling), 558, 656, 666, 672
Barrier
    Anisotropic, 417
    Capillary, 413
    Monolayer, 416
Bedding soil, 562
Bending stress, 432
Bends, on collection pipes, 324
Bentonite, 131, 144, 156, 481
Bernoulii equation, 121, 300
Biochemical oxygen demand (BOD), 212, 589
Biodegradation, 442
Biological clogging (bioclogging), 262, 268, 592
Bioreactor landfill, 338, 578, 584, 592
    Biochemical reactor, 577
    Massive anaerobic reactor, 581
    Waste degradation, 580
    Phases of degradation, 581
    See also Leachate recycling methods; liquid management strategy
Borings, 38. *See also* Landfill siting
Borrow source, 39
Blower, 343, 347, 350
Breakthrough, 154. *See also* Combined advection-diffusion
Breakthrough time, 157, 159, 161, 164
Buckling, 311

C

Carbon dioxide, 212, 333, 335
Case history, 514, 528

Chemical clogging, 268
Chemical oxygen demand (COD), 212, 581
Climate, 214, 419
    Arid, 413
    Data, 233
    Semiarid, 413
Clods, 69, 604, 607
Clogging, 252, 261, 592
Closed landfill site, 677
Coarse-textured soil. *See* Soil texture classification
Coefficient
    Flow, 376, 377
    Runoff, 226
    Transition loss, 323, 376, 377
    See also Hydraulic conductivity; compressibility; effective diffusion; seismic
Coextrusion, 89
Collapse, of large object, 545
Combined advection-diffusion, 17, 158
Compacted lift, 626
Compacted soil layer, 407
Compaction: 610
    Curve, 57
    Test, 56
Compactive energy, 611
Compatibility, 261
Complimentary error function, 17
Composite action, 12
Comprehensive Environmental Response, Compensation and Liability Act (CERCLA), 231
Compressibility, 199
Compression index:
    Modified primary, 200, 449
    Modified secondary, 201, 203, 450
    Primary, 200, 449
    Secondary, 201, 450
Concentration gradient, 119, 156
Concentration ratio, of dissolved solids, 158, 161, 170
See also Solute concentration
Condensate: 346, 369
    Collection and pump station, 343, 346
    Drainage, 366
    Storage tank, 343, 347
Confining stress, 135
Conformance Tests, 635
Constrained modulus, 310, 469

Construction, 603, 618
    Quality assurance (CQA), 624, 626
    Quality control (CQC), 624, 626
Construction quality assurance (CQA)
    Consultant, 635
    Monitor, 641, 644, 652
    Personnel, 635, 662
    Representative, 635
Contact conditions, of geomembrane with underlying soil
    Good, 122, 123
    Poor, 122, 123
Contaminant mass, 158, 164. *See also* Leakage mechanisms
Contaminant transport, 154
Cover soil, 497
Coverage, of the roller, 627
Cracking, of clay liners and cover, 432
Creep, 565
Critical buckling pressure, 311
Critical cross section, 514, 516
Critical failure surface, 514
Cropping (vegetative) management factor, 425
Cycles
    Freeze-thaw, 78, 139, 406
    Wet-dry, 136, 406

D

Daily cover, 442
Darcy friction factor, 372
Darcy's Law (formula), 120, 229
Darcy-Weisbach equation, 372
Decomposition, 334, 338, 339, 458, 577
    Aerobic, 212, 333
    Anaerobic, 212, 334
    Biochemical, 441, 459
    Waste, 212
Defects, 232, 237, 578, 654, 665
Deflection ratio, 308
Degradation, 545, 546, 581
    Aerobic, 584
    Biological, 565, 566
    Chemical, 565, 566
Deployment method, 638
Depression, 559, 564, 569
Desiccation, 65, 406, 408, 604, 620, 637
Destructive test samples, 651, 653
Diffusion, 14, 87, 119, 339, 578
    Steady, 16, 155
    Unsteady, 17

Diffusive mass flux ratio, 156
Distortion, 141, 406, 431
Downward drag force, 315
Drainage, 254
    Cross-plane, 12
    In-plane, 12
    Material, 252, 253
Drawing
    As-built, 638
    Layout, 638

E
Earth work, 637
Effective diffusion coefficient, 17, 156, 160, 161
Effective stress, 486
Elbows, on collection pipes, 324
Energy recovery, 351
Erosion, 399, 418
    Control, 417
    Control materials, 430
    Control practice factor, 427
    Control principles, 429
    Protection, 418
    See also Soil erodibility; soil loss prediction
Evaporation, 214, 221, 228, 412
Evapotranspiration, 214, 228, 229, 232, 240, 412, 413, 416
Equipment force, effect on stability, 491
Expansion, of landfill,
    Lateral, 544
    Piggyback vertical, 544
    Vertical, 173, 442, 544, 552, 557

F
Failure, 478, 537
    Foundation, 478, 513
    Landfill, 478
    Rotational, 480
    Sliding, 478, 479
    Stability, 478
    Translational, 480, 521
    Waste mass, 520
Fault, geological, 33
Field capacity, 190, 192, 217, 221, 229, 235, 557, 583
Field measurement, 622
Fill
    Above and below ground, 5-6
    Area, 6
    Canyon, 6
    Trench, 6
Filter, 253, 404, 591
    Geotextile, 259, 296, 404
    Materials, 253, 301
    Soil, 259, 296
    Specifications, 253
Filtration, 254
Final cover, 9, 217, 223, 399, 592
Fine-textured soil. *See* Soil texture classification
Fittings, on collection pipes, 323, 376
Flare system, 350
Flexible membrane liner, 86
Flexible polypropylene (fPP), 87, 91
Floodplain, 32
Flow
    Cross-plane, 254
    In-plane, 254
Flow rate
    Allowable, 266
    Required, 269
    Ultimate, 266
Fraction of liquid gas condensate, 367
Free drainage condition, 277, 289
Freezing and thawing, 75, 139
Frozen soil, 621
Friction angle, 194, 561
Friction head loss, 322
Frost penetration, 401

G
Gas
    Collection and control system, 9, 332, 352
    Collection head pipe, 343, 345, 557
    Condensate, 230, 346, 366
    Extraction trench, 343
    Extraction well, 342, 343, 359, 370
    Extraction wellhead, 343, 345
    Flaring, 350
    Flow rate, 370, 372
    Generation, 336, 338
    Generation rate, 338, 354, 360, 372
    Migration, 339, 340, 406
    Migration probe, 343, 347
    Pressure, 339, 367
    Temperature, 367
    Vent layer, 409
Gas-to-electricity power plant, 351

Index 713

Geocomposite, 251, 264, 341, 636, 667
Geogrid, 542, 558, 564, 566
Geomembrane, 86, 235, 406, 546, 635, 637
    Defects, 654
    Liner, 121
    Maximum strain, 93
    Maximum stress, 93
    Modulus, 93
    Placement, 637
    Repair, 654
    Seam tests, 646
    Seaming, 639
    Ultimate strain, 93
    Ultimate stress, 93
Geonet, 251, 263, 636, 660
    Biplanar, 263
    Triplanar, 263
Geosynthetic reinforcement, 558, 560, 561
Geotextile, 251, 254, 636, 662
    High strength, 547, 558
    Nonwoven needle punched, 254
Geosynthetic clay liner (GCL), 131, 172, 408, 546, 636, 669
    Geomembrane-supported, adhesive-bonded, 134
    Geotextile-encased, adhesive-bonded, 132
    Geotextile-encased, needle-punched, 133
    Geotextile-encased, stitch-bonded, 133
    Reinforced, 133, 143, 144, 149, 151
    Unreinforced (Non-reinforced), 132, 143, 149, 151
Grade reversal, 558
Gravel content, 73
Grinding, 641
Groundwater, 1, 211, 213, 294

H
Hazen's formula, 252
Heterogeneous, 458
Hydration, 135, 483
Hydraulic barriers, 119
Hydraulic conductivity, 52, 58, 61, 134, 189, 230, 233, 252, 404, 604
Hydraulic contact, 12, 125
Hydraulic Evaluation of Landfill Performance (HELP), 191, 226, 230, 405
Hydraulic gradient, 120

Hydraulic radius, 299
Hydrodynamic dispersion coefficient, 19

I
In-place density, 605
In-plane, 12, 264, 409
Incremental placement, 506
Industrial development, 18-6
Infiltration, 213, 238, 399, 545
    Snowmelt, 225
    Surface water, 213
Inflow rate, 257, 275
Influence area, 361
Influence radius, 362
Installation, 634, 637, 660, 662, 667, 669
Interwedge force, 488, 495, 510
Intimate contact, 12, 125
Intrusion, 268
Inward hydraulic gradient, 578

J
Joining
    Geocomposite, 667
    Geonet, 660
    Geosynthetic clay liner, 671

K
Kneading, 610

L
Laboratory Tests
    Compressibility, 48
    Hydraulic conductivity (permeability), 47, 58
    Moisture-density relationship, 46
    Particle size distribution, 46
    Shear strength, 48
    Water content, 46
Land disposal, 1
Landfill, 2, 4
    Capacity, 454
    Components, 6
    Envelope, 1-6
    Gas, 212, 333
    Lifetime, 161, 170
    Settlement, 440, 449, 458
    Siting, 29, 35
    Stability, 480
    Structures, 557
    See also Bioreactor landfill; postclosure uses

Lateral drainage, 229
Layer
　Coarser-grained, 413
　Drainage, 247, 252, 404, 497
　Erosion control, 401
　Filter, 253
　Finer-grained, 413
　Foundation, 409
　Gas vent, 409
　Hydraulic barrier, 405
　Infiltration, 399
　Lateral drainage, 237
　Leachate collection and leak detection, 247
　Leachate drainage, 247, 283
　Primary leachate drainage, 248
　Protection, 248, 401
　Secondary leachate drainage, 248
　Vertical percolation, 237
Leachate, 8, 211, 213, 221, 294, 557
　Characterization, 212
　Collection and removal system, 8, 294
　Collection trench, 295, 618
　Drainage, 264
　Generation (production) rate, 217, 218, 223
　Head, 247, 275, 531
　Injection, 484
　Level, 442, 484
　Primary, 211
　Quality, 213
　Quantity, 213, 217
　Recirculation (Recycling), 213, 219, 235, 442, 579, 580
　Recycling methods, 584
　Secondary, 211
Leak detection, 12, 13, 236, 588
Leakage, 119, 122, 232, 237, 294, 578, 587
Leakage mechanism, 154
Lifetime, 154, 577
Lift: 343, 506, 605, 621
　Interface, 604, 612
　Thickness, 219, 615
Liner
　Compacted clay (CCL), 52, 61, 120, 153, 154, 546, 603, 628
　Composite, 9, 12, 122, 235, 248, 250
　Double composite, 9, 11, 13, 248
　Geomembrane, 9, 121, 237
　Geosynthetic clay (GCL), 131, 153, 154, 172, 408, 546
　Soil, 237
Liner system, 7, 247, 481, 587
Liquid blockage, 366
Liquid management strategy, 578, 580
Live load, 490
Location restrictions, 29, 31
Looped piping system, 371
Loose lift, 626
Low pressure Mueller equation, 373

M

Manhole, 315
Manning's equation, 298
Mass flux, 154
　Advective, 1-11, 154, 163,
　Diffusive, 1-14, 154, 156, 163
　See also Combined advection-diffusion
Maximum dry density (unit weight), 57
Maximum leachate head, 274, 283, 288
Maximum liquid head, 274, 288
Maximum saturated depth, 274, 288
Mechanical compression, 441
Megafills, 591
Methane, 212, 333, 334, 675
　Gas transmission, 87
　Generation potential, 353, 355
　Generation rate constant, 352, 355
Minimum number of passes, 612
Minor head loss, 323
Modification factor, 590
Modified Iowa formula, 305
Moisture (water) content: 185, 336, 557, 583
　Dry gravimetric, 185, 186, 187
　Initial, 221, 229
　Volumetric, 186
　Wet gravimetric, 185
Moisture extraction, 230
Moisture-holding capacity, 557
Moisture index, 413
Moisture management. *See* Liquid management strategy
Monitoring,
　Gas, 1
　Groundwater, 1
　Settlement, 684
Municipal solid waste (MSW), 180, 181, 182, 333, 440, 562

N

Natural attenuation, 579
Negative skin friction, 315
New Source Performance
    Standards/Emission Guidelines
    (NSPS/EG), 342, 352
NMOC emission rate, 352
Nonmethane organic components
    (NMOCs), 335, 342, 352
Normal Pressure, 270
Nuclear water content/density device, 622

O

Opening size
    Apparent (AOS), 259
    Equivalent (EOS), 259
Optimum water content, 57
Overburden pressure, 79, 442, 472
Overlap distance, 670, 671

P

Panel placement, 638
Pass, of compactor, 627
Passive gas collection system, 341
Patch, 655, 665, 672
Peak shear strength, 145, 151
Percent open area (POA), 261
Percolation, 211, 221, 225, 238, 241, 297, 412
Perforations, 297, 299, 319
Period
    Active, 355, 356, 361
    Closure, 355
Permanent deformation, 509
Permittivity, 257
Permeability (hydraulic conductivity) test, 42, 58
Permeameter
    Flexible-wall, 60
    Rigid-wall, 59
Physical-chemical change, 441, 459
Pipe
    HDPE, 297, 298, 345, 373
    Header, 345, 370
    Leachate collection, 274, 278, 295, 297, 304
    Perforated, 306, 341, 343
    PVC, 297, 298, 345, 373
Pipe
    Deflection, 304, 557
    Stiffness, 309
Piping gradient, 366
Piping system layout, 365
Placement, 658
    Geocomposite, 667
    Geomembrane, 637
    Geonet, 660
    Geosynthetic clay liner, 669
    Geotextile, 662
Plugging, 252, 253
Polyethylene
    Chlorosulfonated (CSPE), 87, 92
    High density (HDPE), 87, 88
    Linear low density (LLDPE), 87, 89
    Very low density (VLDPE), 87
Polymeric materials, 87
Polyvinyl chloride (PVC), 87, 91
Ponding, 295, 558
Pore (water) pressure, 485, 486, 499, 501
Porosity, 188, 235
Postclosure
    Condition, 231
    Period, 686
Precipitation, 211, 214, 218, 225, 240, 413
Pressure loss, 370, 376, 377
Proctor test
    Modified, 56
    Reduced, 57
    Standard, 56
Proof rolling, 614
Pseudostatic analysis, 509
Pump: 320
    Cycle time, 324
    On level, 319
    On time, 324
    Off level, 319
    Off time, 324

Q

Quality
    Assurance, 153, 626, 634
    Control, 626, 635

R

Rainfall energy factor, 421
Rainfall erosion, 418, 419
Raveling, 441
RCRA Subtitle D, 11, 29, 31
Reactions
    Biochemical, 213
    Biological, 212
    Chemical, 211, 212

Recreational facilities, 675, 679
Reduction factor, 256, 266, 565, 566, 590
Required leachate flow rate, 298
Repairs
    Geocomposite, 667
    Geomembrane, 654
    Geonet, 661
    Geosynthetic clay liner, 672
    Geotextile, 665
Residual shear strength, 146
Resource Conservation and Recovery Act (RCRA), 29, 231, 341, 399
Retardation, 261
Reynolds number, 372
Riser pipe, 314, 318, 341
Roller
    Footed, 611
    Self-propelled, 627
    Steel wheeled, 622
Root penetration, 229, 401
Runoff, 225, 225, 226, 412
Run-on, 218, 242
Runout, 104
Rupture strain, 565

S

Samples
    Disturbed, 38
    Undisturbed, 38
Scrim, 87, 92
Sealed double-ring infiltrometer (SDRI), 42
Seaming, 643, 663
    Method, 640
    Weather condition, 644
Seepage, 497
    Buildup, 499, 501
    Force, 252, 497, 506
    Velocity, 157, 161
Seismic
    Analysis, 509
    Coefficient, 509
    Forces, 508
    Impact zone, 33
Self-seal (Self sealing), 131, 138
Settlement, 199, 431, 440, 443, 451
    Differential, 141, 406, 409, 431, 472, 545, 546, 550, 551, 675
    Elastic, 469
    Foundation, 469
    Immediate, 200

Landfill, 440, 449, 458
Primary, 200, 449, 548
Rate, 444
Secondary, 450, 549
Solid waste, 440
Total, 431, 449, 469, 549
Shear strength, 64, 143, 193, 481
    Interface, 144, 148, 481
    Internal, 144
    Large displacement, 145
    Peak, 145, 151, 482
    Residual, 482
Shrinkage cracking, 65
Shrinkage potential, 65
Site suitability, 680
Slope-length factor, 424
Slope stability, 252, 487
Sludge, 217, 219
Soil
    Drainage, 252
    Erodibility, 419
    Erodibility factor, 421
    Erosion, 417, 429
    Loss prediction, 420
    Moisture storage, 413, 416
    Reaction modulus, 305, 307, 309
    Retention, 258
Soil Conservation Service (SCS) curve number method, 226
Soil texture classification, 421
    Course-textured, 415
    Fine- textured, 415
Soil-water characteristic curve, 414
Solar radiation, 232, 239, 413
Solid waste landfill, 1
Solute concentration ratio, 158
Specification, 53, 252, 613, 640, 641
    Compaction, 616
    Construction, 627
Stability
    Final cover system, 538
    Landfill postclosure, 538
    Liner system, 538
    Sideslope, 538
    Waste mass, 538
Standard dimension ratio (SDR), 299
Static head, 322
Steady advection, 15, 154
Steady diffusion, 16, 155
Steady flow state, 275

Strain softening, 97, 151
Strain rate, 486
Stress history, 442
Subbase, 605
Subgrade, 134, 144, 472, 472, 546, 558, 605, 628
Subgrade change, 472
Subsidence, 295, 407, 431
    Differential, 434
    Localized, 101, 551
Subsoil, 484
Sump, 314, 318
Surcharge effect, 680

T

Test
    Air pressure, 646
    Destructive, 651
    Nondestructive, 646, 649
    Vacuum, 649
Tensile failure, 545
Tensile strain, 142, 143, 432, 469, 472, 545, 550, 554
    Out-of-plane, 408, 435
Tension crack, 545
Textured geomembrane, 90, 150, 483
Theory, of soil structure,
    Particle-orientation, 607, 608
Toe drain, 405
Topsoil, 41, 217, 401
Total accumulated contaminant mass, 160, 166, 169
See also Combined advection-diffusion
Total discharge head, 322
Total suspended solids (TSS), 589
Total volatile acids (TVA), 581
Transit times, 22
Transmissivity, 230, 269
Transpiration, 228, 238, 412
Trial seam strengths, 643
Trial weld, 641, 643
Triaxial test
    Consolidated undrained (CU), 486
    Unconsolidated undrained (UU), 486
Two-stage borehole test (TSB), 45

U

Unbalance friction forces, 98
Underdrain system, 558
Unified Classification System (UCS), 73
Unit weight, 182
Universal Soil Loss Equation (USLE), 420, 429
Unlined landfill, 557
Unstable area, 34
Unsteady diffusion, 17
U.S. Army Corps of Engineers, 301
U.S. Department of Agriculture (USDA), 231
U.S. Environmental Protection Agency (EPA), 5, 11, 124, 230, 352

V

Vacuum source, 343, 347
Valve, 323, 376
Vapor
    Diffusion, 237
    Pressure, 367, 368
    Transmission, 87
Vegetation, 217, 401, 412, 425
Vertical expansion, 173, 442, 544, 552, 557
Volumetric shrinkage test, 67
Volumetric strain, 67

W

Waste characteristics
    Age, 337, 355
    Composition, 336
    Condition, 216, 516
    Heterogeneity, 551
Waste behavior
    Compaction, 338, 442
    Degradation, 557, 580
    Settlement, 454
    Stability, 595
Water balance, 412
Water vapor, 335, 368
Wedge
    Active, 488, 503, 521
    Passive, 488, 503, 521
Welding
    Extrusion fillet, 641
    Fusion (hot wedge), 640
Welding rod, 641
Wetting and drying, 137
Wicking layer, 414
Wilting point, 190, 192, 229, 235
Wrinkles, 644, 659